Me
the
Un
mc
fo:
Bo
an

DUE FOR

11 JA
DUE FOI

12 J/

D.

2

DUE FOR

Charge-Coupled Devices: Technology and Applications

OTHER IEEE PRESS BOOKS

Charge-Coupled Devices: Technology and Applications

Edited by

Roger Melen
Associate Director
Integrated Circuits Laboratory
Stanford University

Dennis Buss
Branch Manager
Central Research Laboratory
Texas Instruments

A volume in the IEEE PRESS Selected Reprint Series, prepared under the sponsorship of the IEEE Solid-State Circuits Council.

IEEE PRESS

The Institute of Electrical and Electronics Engineers, Inc. New York

Copyright © 1977 by
THE INSTITUTE OF ELECTRICAL AND ELECTRONICS ENGINEERS, INC.
345 East 47 Street, New York, NY 10017
All rights reserved.

PRINTED IN THE UNITED STATES OF AMERICA

Library of Congress Catalog Card Number 76-20887

IEEE International Standard Book Numbers: Clothbound 0-87942-083-9
Paperbound 0-87942-084-7

Sole Worldwide Distributor (Exclusive of the IEEE):

JOHN WILEY & SONS, INC.
605 Third Ave.
New York, NY 10016

Wiley Order Numbers: Clothbound 0-471-02570-4
Paperbound 0-471-02571-2

Contents

Part I
Introduction

It has been a scant six years since the invention of the charge-coupled device (CCD) was announced by Bell Laboratories in March of 1970, and already CCD products are making their debut in the marketplace. The CCD represents an entirely new concept in metal–oxide–semiconductor (MOS) circuitry which has application to optical imaging, digital memory, and analog signal processing. During the past six years, CCD's have been the subject of active research at laboratories throughout the world, as reflected by over 1000 technical papers, and it is the purpose of this reprint book to summarize this activity.

This reprint book is divided into five parts. Part I introduces the reader to CCD's with a brief historical perspective. Part II presents papers dealing with CCD physics and technology, and the remaining three parts deal with the three application areas of CCD's: optical imaging in Part III, digital memory in Part IV, and analog signal processing in Part V. This book is primarily applications-oriented and, apart from the Introduction, the papers are selected to represent the state of the art in such a way that the reader can quickly appreciate the current status of any aspect of CCD technology. Since this volume overviews all aspects of CCD's, it includes papers of a review nature emphasizing the current state of the art. The reader interested in pursuing a given aspect of CCD technology will find ample references to earlier work in the papers included here.

The charge-coupling principle was first presented and demonstrated in the classic papers by Kosonocky and Sauer and by Boyle and Smith (papers 1 and 2 in this part) which appeared in March 1970. The invention of the CCD reportedly resulted from an effort on the part of Boyle and Smith to find a semiconductor equivalent to the magnetic bubble [1], which at the time was an emerging technology offering competition to semiconductor memories. The CCD presented in the second paper in this part utilized a particularly simple single-level MOS structure. However, as subsequent developments demonstrated, practical CCD's require a more complex multilevel electrode structure.

Although the CCD represented a new MOS concept, it had a "distant cousin" called the bucket-brigade device (BBD), which is presented briefly in paper 3 of this part. The BBD is an IC realization of a discrete concept which is functionally equivalent to the CCD, although different in structure. For a number of years, CCD's and BBD's were developed in parallel under the generic name of charge-transfer devices (CTD's), and in some of the application papers reprinted in this volume, BBD's are discussed. However, it appears at present that CCD's have superior electrical performance and higher packing density than BBD's, and that CCD's will fill most CTD applications in the future.

The CCD is a monolithic array of closely spaced MOS capacitors which transfers an analog signal charge from one capacitor to the next. The mechanism by which this transfer occurs has no discrete counterpart, and represents a new type of MOS circuit element which can be combined in integrated circuit (IC) form with conventional MOS. It is the analog capability of CCD's which is responsible for their application to self-scanned optical imaging and to analog signal processing, and it is their high packing density which makes them attractive for digital memory applications. The basic CCD concepts are reviewed in paper 4 of this part.

It is the Editors' hope that this reprint volume will provide the applications engineer with the necessary background to understand CCD's and to generate new uses for this novel and versatile device. For further discussion of CCD's and BBD's, the reader is referred to the book by Séquin and Tompsett [2].

REFERENCES

[1] A. H. Bobeck, P. I. Bonyhard, and J. E. Geusic, "Magnetic bubbles—An emerging new memory technology," *Proc. IEEE*, vol. 63, pp. 1176–1195, Aug. 1975.
[2] C. H. Séquin and M. F. Tompsett, *Charge Transfer Devices*, suppl. 8 to *Advances in Electronics and Electron Physics*. New York: Academic, 1975.

The ABCs of CCDs. They're basically MOS analog registers that can be employed in either analog or digital applications. Here are pointers in evaluating them.

If you haven't designed with CCDs yet, it's probably only a question of time before you do. Charge-coupled devices are turning up in photo-sensor arrays, large-storage memories and such signal-processing components as variable delay lines, transversal filters and signal correlators.

In most applications you probably won't be involved in specifying the construction of a CCD. However, a basic knowledge of CCD operation will help you evaluate the devices from different manufacturers. And system design can be optimized around a specific device.

Actually a CCD is a simple device. In essence, it is a shift register formed by a string of closely spaced MOS capacitors. A CCD can store and transfer analog-charge signals—either electrons or holes—that may be introduced electrically or optically.

Charges stored and shifted

The storing and transferring of charge occurs between potential wells at or near a silicon-silicon dioxide interface. The MOS capacitors, pulsed by a multiphase clock voltage, form these wells. For a three-phase, n-channel CCD (Fig. 1), the charges transferred between potential wells are electrons.

The application of a positive step voltage to a gate electrode (like ϕ_1 at time t_1) forms a depletion region in the p-type silicon beneath a gate. The particular gate is the one that causes a minimum of electron energy—a potential well—to exist at the Si-SiO$_2$ interface (Fig. 1b). However, the potential well doesn't last indefinitely, and thermally generated electrons eventually fill the well completely. Thus the CCD is basically a dynamic device in which charge can be stored for much shorter times than the thermal relaxation time of the CCD's capacitors. Depending on device processing, this time may vary from one

second to several minutes at room temperature.

The introduction of minority-carrier signals reduces the depth of the well, much like the way a fluid fills up a container. Charges transfer from wells under the ϕ_1 electrodes to wells under ϕ_2 because of the surface potential changes due to clocking (Fig. 1c). A similar transfer moves charges from ϕ_2 to ϕ_3 and then from ϕ_3 to ϕ_1. After one complete clock cycle, the charge pattern has moved one stage (three gates) to the right. No significant amount of thermal charge accumulates in a particular well because the charges are continually being swept out.

Note that the three-phase structure is symmetrical and the direction of charge flow is determined by the clock-phase sequence. For example, by an interchange of the ϕ_1 and ϕ_3 clock lines, the charge could be made to transfer to the left. Operation with less than three-phase clocks requires an asymmetry in the CCD structure to determine the direction of signal flow.

The charge signals are laterally confined into a channel by means of channel stops (Fig. 1d). These can be heavily doped diffusions, thick-field oxides or another gate level (field shield) under the phase electrodes to which a dc bias is applied.

Charges transfer in 3 ways

Free charge moves from one well to another by three separate mechanisms: self-induced drift, thermal diffusion and fringing field drift. Self-induced drift, a charge-repulsion effect, is only important at relatively large signal-charge densities. It is the dominant mechanism in the transfer of the first 99% or so of charge signal.

Thermal diffusion results in an exponential decay of the remaining charge under the transferring electrode. The decay has a time constant that increases as the square of the center-to-center electrode spacing. Fringing field drift can help speed the charge-transfer process considerably. The fringing field is the electric field in the direction of charge flow, and it depends on process parameters and device geometry.

The fraction of charge transferred from one well to the next is referred to as the charge-trans-

Walter F. Kosonocky and **Donald J. Sauer**, Members of the Technical Staff, RCA Laboratories, Princeton, NJ, 08540.

Reprinted with permission from *Electron. Des.*, vol. 23, pp. 58–63, Apr. 12, 1975.

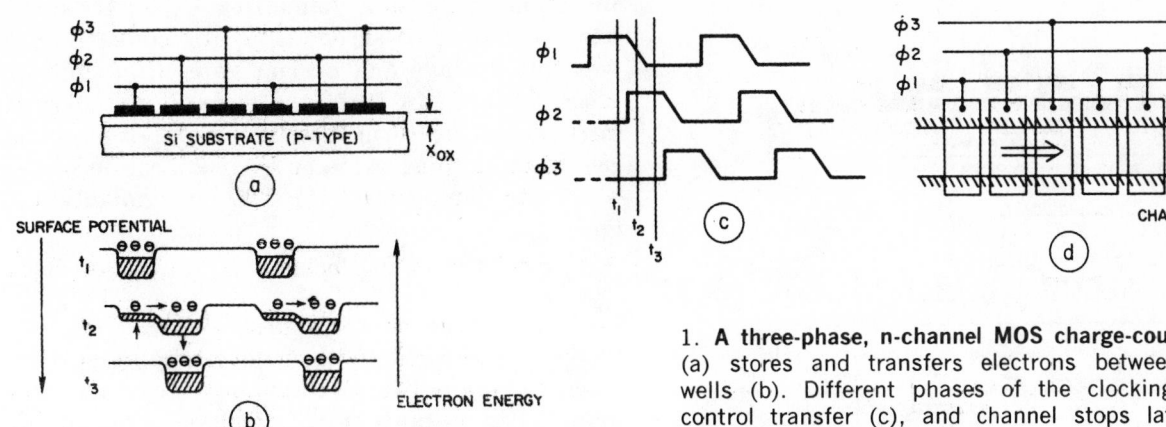

1. **A three-phase, n-channel MOS charge-coupled device** (a) stores and transfers electrons between potential wells (b). Different phases of the clocking waveform control transfer (c), and channel stops laterally confine the charge signals (d).

fer efficiency, η. The fraction left behind is the transfer loss, or transfer inefficiency, and it is denoted by ϵ, so that $\eta + \epsilon = 1$. Because η determines how many transfers can be made before the signal seriously distorts and becomes delayed, it is the most important performance parameter.

Boosting transfer efficiency

If a single charge pulse with an initial amplitude P_o transfers down a CCD register, after n transfers the amplitude, P_n, will be

$$P_n = P_o \eta^n \simeq P_o (1 - n\epsilon) \quad \text{(for small } \epsilon\text{).} \quad (1)$$

Clearly ϵ must be very small if many transfers are required. If you allow an $n\epsilon$ product of 0.1 and an over-all loss of 10%, a three-phase, 330-stage shift register requires $\epsilon < 10^{-4}$, or a transfer efficiency of 99.99%.

The maximum achievable value for η depends on two factors: how fast the free charge can transfer between adjacent gates and how much of the charge gets trapped at each gate location by stationary states. In surface-channel devices, charge trapping usually results from the fast states at the Si-SiO$_2$ interface. The trapping of

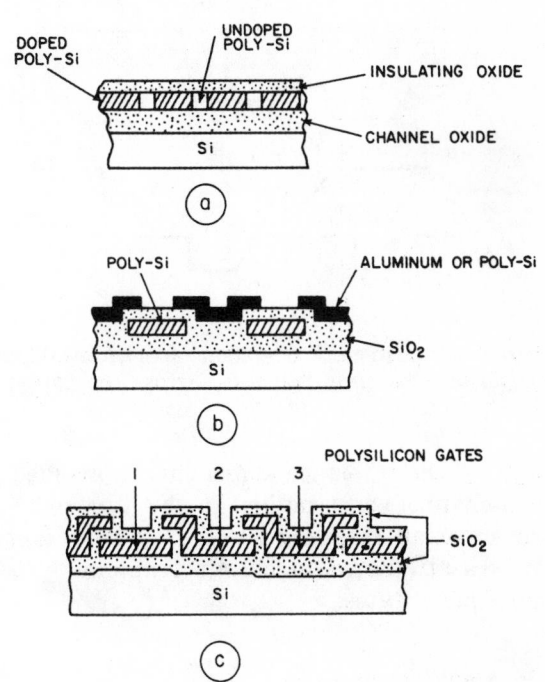

2. **CCD structures generally employ sealed-channel construction.** This can involve a single level of doped polysilicon gates (a), polysilicon-aluminum or two polysilicon levels (b), or three levels of polysilicon (c).

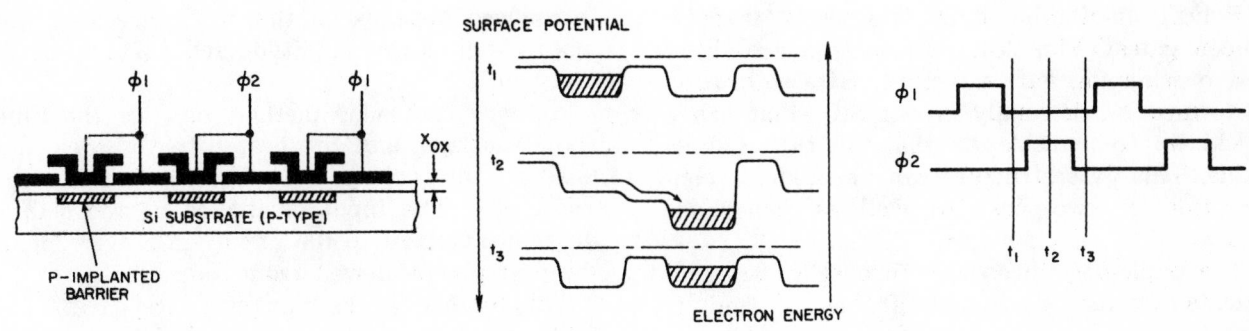

3. **Two-phase drop-clock operation** uses nonoverlapping clock pulses.

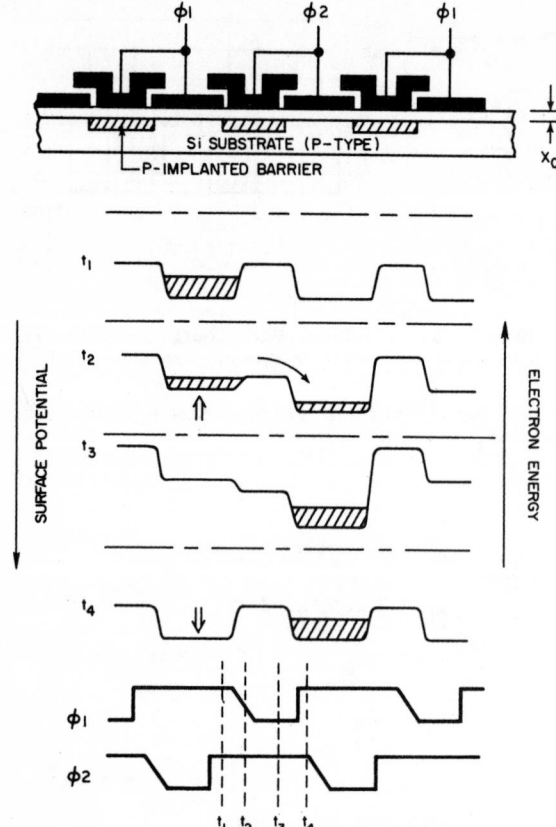

4. Two-phase push-clock operation employs overlapping clock pulses. Charge is "pushed" across the barrier.

charge by the interface states can be avoided by buried-channel construction. In this type of CCD, however, small trapping losses may be observed. They are attributed to charge trapping by stationary bulk states.

Up to 3 gate levels used

The most common CCD structures employ sealed-channel construction involving one, two or three levels of polysilicon (Fig. 2). The selectively doped single-layer structure (Fig. 2a) passivates the interlectrode spaces with high-resistivity polysilicon. A second-layer metallization (aluminum) forms interconnections.

Polysilicon-aluminum, or two levels of polysilicon gates (Fig. 2b), represents a self-aligning, overlapping gate structure. Gate separation is formed by thermally grown SiO_2 that has a thickness comparable to that of the channel oxide. This gate structure can be used for construction of two-phase as well as four-phase CCDs.

The triple-polysilicon structure (Fig. 2c) represents another alternative. It has the unique feature of a separate polysilicon level for each phase.

The number of clock phases can be reduced to two, or even one, if the potential wells are made

directional with an asymmetrical CCD-gate structure. This can be accomplished by connection of separate storage and barrier gates to a common clock voltage. In turn, the storage and barrier regions can be formed with two different thicknesses of channel oxide or by modification of the substrate doping level through ion implantation.

Push clocks vs drop clocks

Unlike three or four-phase CCDs, two-phase devices can operate with nonoverlapping positive clock-voltage pulses. Overlapping clock pulses are referred to as push clocks, while nonoverlapping clock pulses are called drop clocks.

In two-phase drop-clock operation (Fig. 3), the transfer barriers are formed by ion-implanted p-type regions under transfer gates. The shift of charge from the potential well under the ϕ_1 storage electrode to a well under ϕ_2 occurs during the positive ϕ_2 pulse. A similar process moves charge from ϕ_2 to ϕ_1 during the positive ϕ_1 pulse to complete the cycle.

In two-phase push-clock operation (Fig. 4), charge is "pushed" across the potential barrier during the fall time of the clock waveform.

The two-phase structure also may be operated from a single clock line by application of a dc bias to one of the phases. Then half of the transfers involve drop-clock operation, and the other half are push-clock.

Input stages accept electrical signals

Of course, for applications, means must be provided to introduce charge into the CCD register and then to detect signals at the output. An electrical input is usually introduced by one of the three ways shown in Fig. 5.

In the current-input method (Fig. 5a), the source diffusion, S, is dc-biased, and an input voltage pulse, V_{G1}, is applied to the first gate, G1. The combination forms an MOS current source that fills the first potential well under ϕ_1 for the duration, Δt, of the input pulse. This method is relatively critical, since the amount of charge introduced depends on the MOS threshold voltage as well as the amplitude and duration of the input pulse, V_{G1}.

A more controlled method samples the input signal voltage, and the first potential well then fills to the voltage of the source diffusion (Fig. 5b). The input is applied as the source-diffusion voltage, while input gate G1 isolates the first potential well from the source.

This method works best with a relatively slow fall time for the input-gate clock pulse. And though input charge isn't determined by the sample-pulse amplitude or duration, it does depend on the MOS threshold of gate ϕ_1.

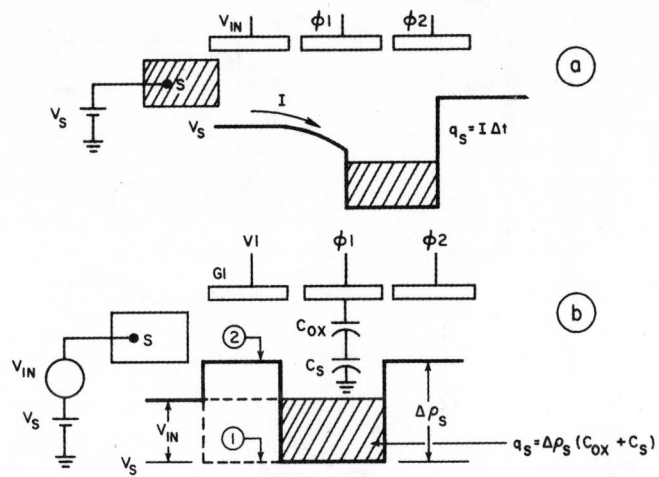

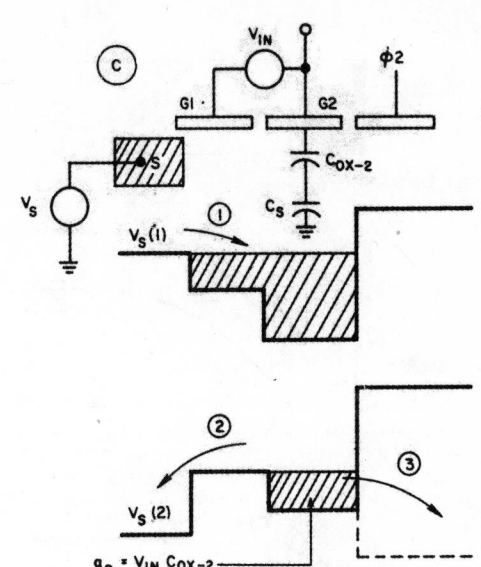

5. **Signals can be introduced in one of three ways.** The current-input method (a) has critical requirements, while a sampling approach (b) relaxes some of the conditions. A linear method (c) features low noise.

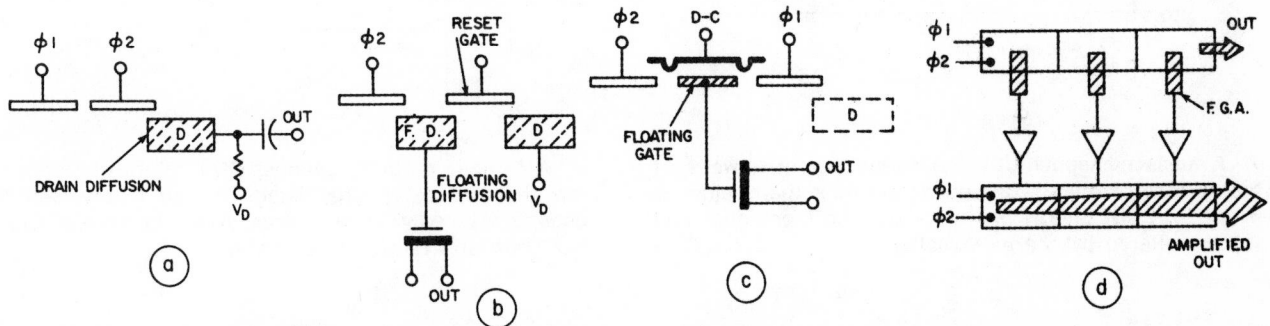

6. **Signals can be detected in one of several ways.** Current sensing (a) employs the drain diffusion of the output stage. Voltage or charge-sensing is obtained with an internal floating-diffusion amplifier (b). And nonde- structive sensing can be obtained with an internal floating-gate amplifier, in either single (c) or distributed form (d). With a floating-gate amplifier, only transfer noise is introduced into the signal.

A charge-presetting input method is linear, has the advantage of low noise, and it doesn't depend on threshold (Fig. 5c). The basic concept is to form a potential well at gates G1 and G2, with input gate G1 acting as a barrier between the source diffusion and the input well under G2. The input is applied as the relative voltage between gates G1 and G2.

The input well is first overfilled by raising the source potential above the G1 barrier. The excess input charge returns to the source diffusion when its potential is lowered. If the same channel oxide is used for both gates G1 and G2, the input charge signal, q_s, is

$$q_s = V_{in} C_{ox-2} \qquad (2)$$

where C_{ox-2} is the oxide capacitance of gate G2.

Three methods are also available for the detection of charge signals at the output (Fig. 6). Current sensing measures current flow in the drain of a CCD (Fig. 6a). The current results from charge signals coupled to the drain diffusion by the last gate electrode. And the output signal takes the form of a current spike at a relatively high capacitance terminal. In theory, the method provides a highly linear detection scheme.

Amplifier comes on the chip

The floating diffusion amplifier (Fig. 6b) is the most popular detection approach when an on-chip amplifier is used. The output circuit periodically resets the floating diffusion to a reference potential. The floating diffusion, in turn, is connected to the gate of an on-chip inverter or source-follower amplifier. The detected signal varies proportionally with the floating-diffusion voltage as a function of the charge signal. Hence the technique is referred to as a voltage, or charge-sensing, method.

With a floating-gate amplifier (Fig. 6c or 6d),

5

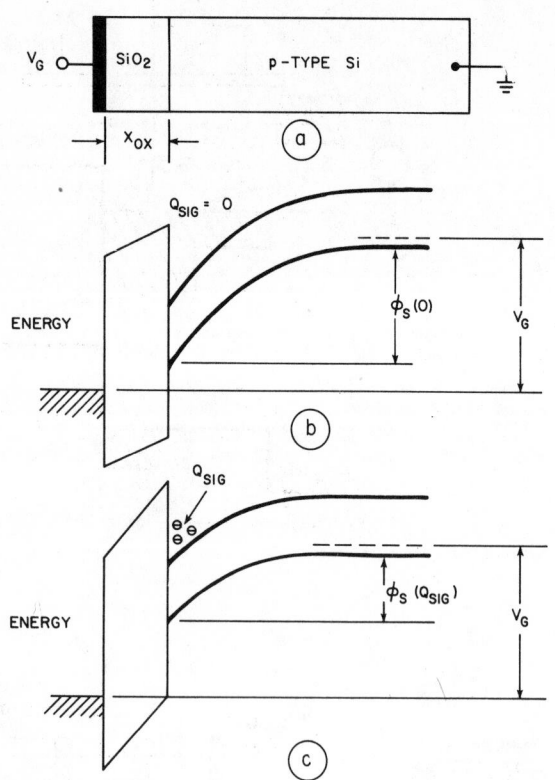

7. **A surface-channel CCD** (a) forms potential wells at the surface of the silicon substrate rather than below as in a bulk CCD. Energy bands for an empty potential well (b) are altered by charge signal (c).

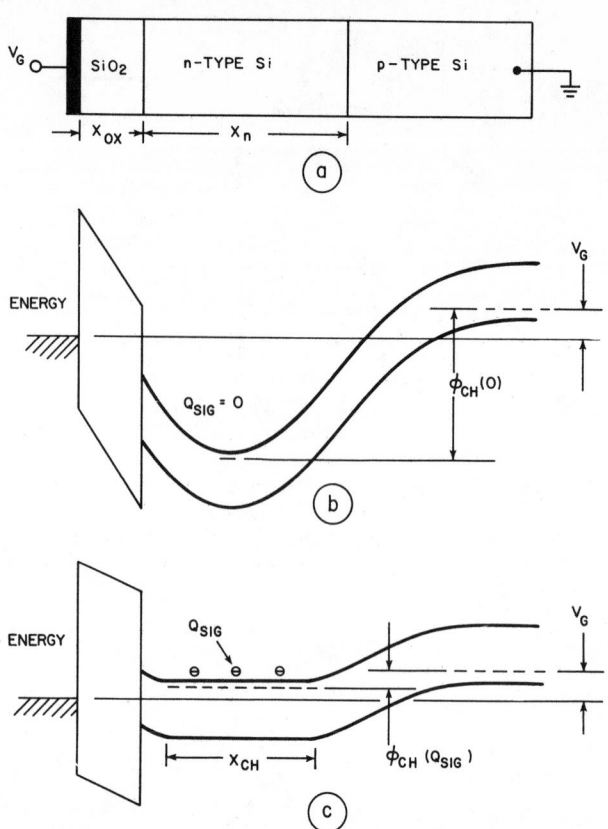

8. **A buried, or bulk, channel CCD** (a) forms wells below the surface. Energy bands for an empty well (b) change markedly when charge signal is present (c). A BCCD doesn't require bias charge.

the signal can be sensed nondestructively, so that only transfer noise is introduced into the charge signal. The floating gates are connected to an on-chip MOS amplifier.

Surface vs buried channel

The charge-coupled device can be constructed as a surface-channel device (SCCD) or as a buried-channel device (BCCD), as shown in Figs. 7 and 8, respectively.

The potential wells of an SCCD are formed at the Si-SiO₂ interface. In contrast, the BCCD forms wells below the silicon surface to avoid charge trapping by surface states. The silicon substrate has an additional thin layer whose conductivity type is the opposite of that of the substrate. During the operation of a BCCD and with no signal charge—$Q_{SIG} = 0$—the top layer is depleted of mobile charge. Hence the potential minimum forms below the surface of silicon.

Although the analysis and the design of BCCDs are somewhat more involved than that of SCCDs, the external operation of the two may differ only by the dc level of the clock voltage pulses. However, a BCCD doesn't require a fat zero—bias charge—for high efficiency, or $\eta > 0.9999$. Also,

it doesn't exhibit noise caused by the trapping of charge by the fast interface states, and it has a higher frequency response than an SCCD with the same dimensions.

In fact, a BCCD with a thick epitaxial top layer can achieve $\eta > 0.9999$ with clock frequencies in excess of 100 MHz. However, the maximum charge signal in a BCCD is up to three times smaller than that of SCCD. Also, the BCCD can have a higher dark current—a type of leakage—than the SCCD.

As with most other ICs, noise represents an important consideration in the evaluation of CCDs. Such applications as signal processing and imagers can require either very low signals or large dynamic range. Also, the theoretical minimum size of a CCD memory element depends on noise characteristics, since a memory's error rate is a function of the signal-to-noise ratio.

The general conclusions on noise in CCDs are as follows:

■ Transfer noise due to free-charge transfer is quite low. Usually it involves only the small amount of charge left behind.

■ Noise associated with trapping of charge by fast interface states and that caused by a fat-zero signal represent the major noise fluctuations

6

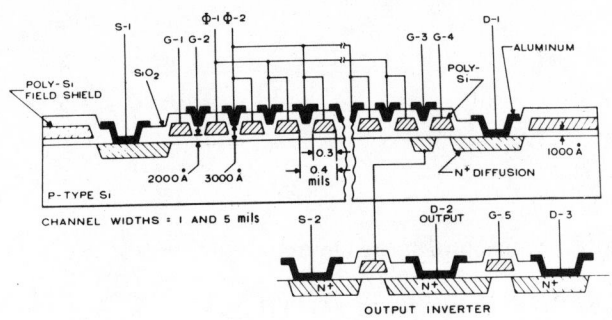

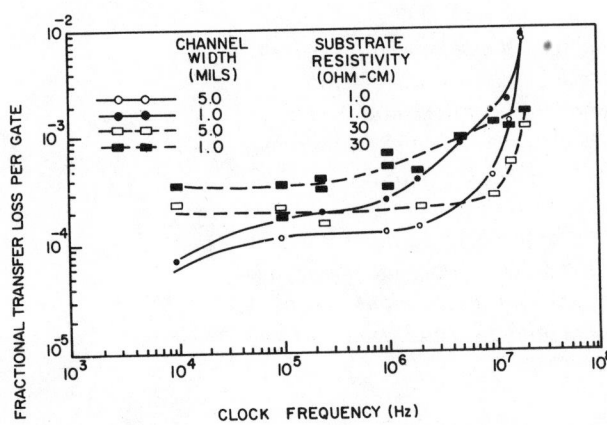

9. The fractional transfer loss of 500-stage, two-phase SCCDs with 30% bias charge is obtained for different channel widths and substrate resistivities.

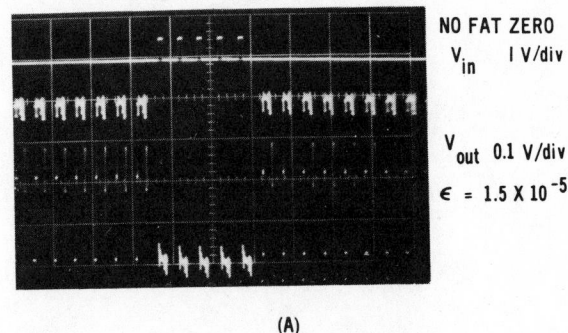

NO FAT ZERO
V_{in} 1 V/div

V_{out} 0.1 V/div

$\epsilon = 1.5 \times 10^{-5}$

(A)

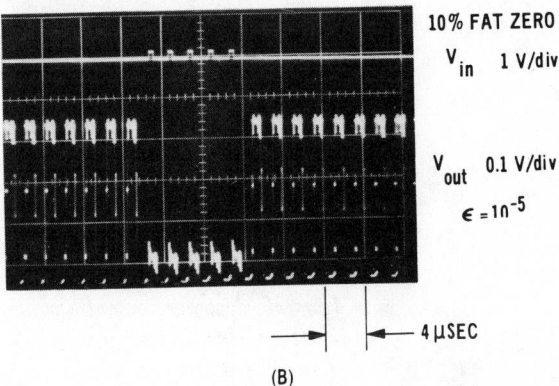

10% FAT ZERO
V_{in} 1 V/div

V_{out} 0.1 V/div

$\epsilon = 10^{-5}$

4 μSEC

(B)

10. Input/output waveforms for a 500-stage, two-phase BCCD show that bias charge, or fat zero, isn't necessary (a). But it can reduce transfer losses (b).

in surface-channel CCDs. A signal-to-noise ratio of 60 to 80 dB should still be possible with SCCDs.

■ Several low-noise input techniques are available. They have thermal-noise fluctuations that are comparable to those associated with charging of the capacitance of the input potential well.

■ Noise associated with the resetting of the output floating diffusion can be considerably reduced by a technique of synchronous double sampling.

■ Output circuits in the form of a floating-gate amplifier (FGA) or a distributed FGA are expected to lead to still smaller noise levels in the output circuit.

■ Analytical and experimental studies indicate that noise inherent in the charge-transfer action of CCDs won't impose any limitation on the size of memory elements in the foreseeable future.

What about CCD performance?

The two most important performance characteristics of CCDs are dark current and transfer inefficiency, or transfer losses, as a function of clock frequency. Charge-transfer efficiency, which was 0.99 for a three-phase SCCD in 1970, has

reached 0.99999 for two and three-phase SCCDs in 1974.

The dark-current characteristics of a CCD include the average thermally generated background charge as well as localized dark-current spikes that are very sensitive to the applied gate voltage. Dark current background levels as low as 5 to 10 nA/cm² have been reported. However, the control of dark current and its spikes remains one of the critical aspects of CCD manufacturing.

The typical performance of an SCCD is shown in Fig. 9. Transfer efficiency varies inversely with the density of fast-interface states. The lowest fractional transfer loss for an SCCD has been reported by Bell Laboratories: A 1600-stage 40 μm-wide triple-polysilicon-gate line imager had a loss of 1×10^{-5} at a clock frequency of 0.3 MHz.

For BCCDs, fractional transfer losses in the range of 10^{-4} to 10^{-5} have been reported by several companies. A fractional loss of 5×10^{-5} at a clock frequency of 135 MHz was reported by Philips.

The output waveform for a 500-stage, two-phase BCCD with and without fat zero are shown in Fig. 10. These waveforms illustrate the capability of BCCDs for high charge-transfer efficiency. ■■

Charge Coupled Semiconductor Devices

By W. S. BOYLE and G. E. SMITH

(Manuscript received January 29, 1970)

In this paper we describe a new semiconductor device concept. Basically, it consists of storing charge in potential wells created at the surface of a semiconductor and moving the charge (representing information) over the surface by moving the potential minima. We discuss schemes for creating, transferring, and detecting the presence or absence of the charge.

In particular, we consider minority carrier charge storage at the Si-SiO₂ interface of a MOS capacitor. This charge may be transferred to a closely adjacent capacitor on the same substrate by appropriate manipulation of electrode potentials. Examples of possible applications are as a shift register, as an imaging device, as a display device, and in performing logic.

A new semiconductor device concept has been devised which shows promise of having wide application. The essence of the scheme is to store minority carriers (or their absence) in a spatially defined depletion region (potential well) at the surface of a homogeneous semiconductor and to move this charge about the surface by moving the potential minimum. A variety of functions can then be performed by having a means of generating or injecting charge into the potential well, transferring this charge over the surface of a semiconductor, and detecting the magnitude of the charge at some location. One method of producing and moving the potential wells is to form an array of conductor-insulator-semiconductor capacitors and to create and move the potential minima by applying appropriate voltages to the conductors. The purpose of this paper is to describe the operation of this basic structure and some possible applications. We present calculations which show feasibility for a simple silicon-silicon dioxide MOS structure.

First consider a single MIS structure on an n-type semiconductor. A diagram of energy vs. distance is shown in Fig. 1 for an applied voltage difference V_1 in which the metal is negative with respect to the semiconductor and large enough to cause depletion. When the

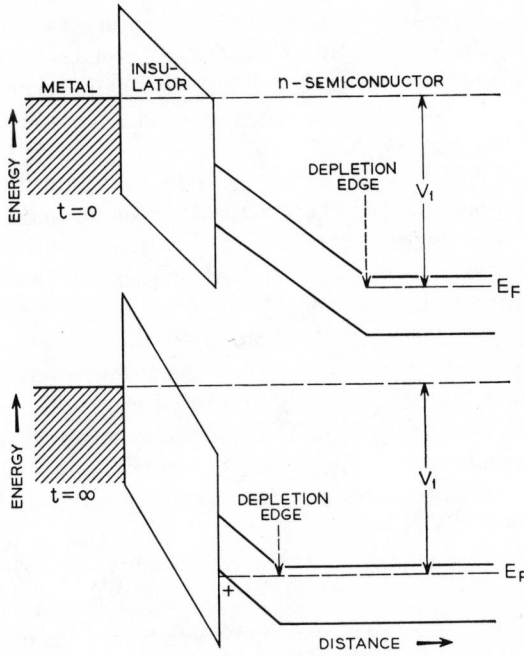

Fig. 1—A plot of electron energy vs distance through an MIS structure both with (at time $t = \infty$) and without (at time $t = 0$) charge stored at the surface.

voltage is first applied at $t = 0$, there are no holes at the semiconductor-insulator interface and the voltage is divided between the semiconductor and insulator as shown. If holes are introduced into the depletion region by some means, they will collect at the semiconductor interface causing the interface potential to become more positive. Eventually the situation shown in Fig. 1 for $t = \infty$ is reached. This is the steady state condition for the structure and it occurs when the valance band at the interface is approximately at the same energy as the Fermi level E_F in the bulk. Any further introduction of holes will cause the interface potential to become yet more positive and holes will be injected into the bulk until the steady state condition is again reached.

Now, consider the linear array of MIS structures on an n-type semiconductor as shown in Fig. 2 where every third electrode is connected to a common conductor. As an initial condition, a voltage $-V_2$ is applied to electrodes 1, 4, 7, and so on, and a voltage $-V_1$ $(V_2 > V_1)$ is applied to the other electrodes. The semiconductor is held at zero potential and the V_i's are taken as positive numbers. It is assumed that $V_1 > V_T$ where V_T is the threshold voltage for the production of inversion under steady state conditions. The edge of the depletion region is indicated by the dashed line. Also, as an example, positive charge is placed under electrodes 1 and 7 and none under electrode 4, as indicated in Fig. 2(a). Now a voltage $-V_3$ ($V_3 > V_2$) is applied to electrodes 2, 5, 8, and so on, as shown in Fig. 2(b) and the charge will transfer from electrode 1 to the potential minimum under electrode 2, and so on. The voltages are now changed to the condition of Fig. 2(c) and, as shown, charge has been shifted one spatial position and the sequence is ready to be continued.

It has been assumed in the foregoing that the voltages were applied and manipulated in a time shorter than the storage time τ where $\tau = Q/I_d$ is the time for the thermally generated current I_d to supply the equilibrium charge density Q. The thermal current I_d results from generation-recombination centers in the depletion region and at the semiconductor-insulator interface. Storage times of the order of seconds have been reported.[1-3]

It is of interest now to consider the capacitance and surface potential

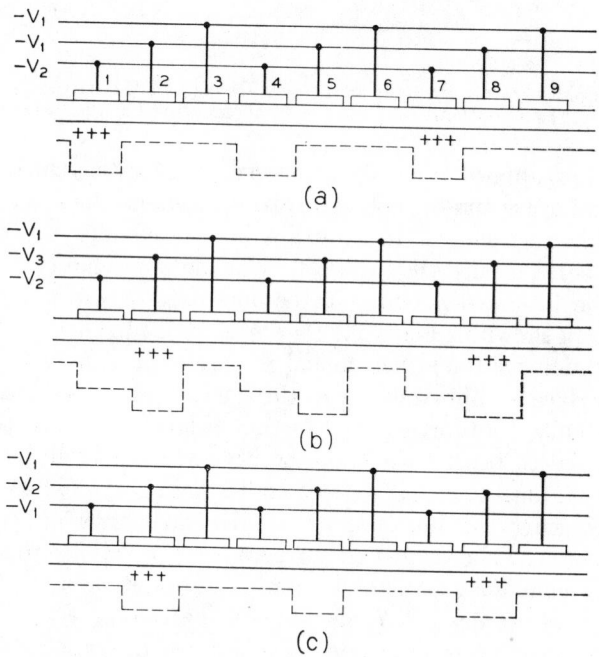

Fig. 2—Schematic of a three phase MIS charge coupled device.

of the structure as a function of stored charge. The capacitance can be used as a measure of the stored charge. Knowledge of the surface potential is necessary to design a structure that insures complete transfer of charge since, referring to Fig. 2, as charge flows from electrode 1 to electrode 2 the potential of 1 will fall and the potential of 2 will rise. Clearly the voltages V_3 and V_2 must be chosen such that the surface potential at electrode 2 is always lower. The steady state minority carrier density Q per unit area for a given gate voltage V_G is given by $Q = C_0(V_G - V_T)$ where $C_0 = K_0\epsilon_0/X_0$ is the oxide capacitance, K_0 is the oxide dielectric constant, and X_0 the thickness. For a charge density $Q' \leqq Q$, the potential φ_s at the semiconductor surface can be shown to be

$$\varphi_s = (V_G - V_{FB}) - \frac{Q'}{Q}(V_G - V_T)$$
$$- \frac{B}{C_0}\left(\left\{1 + \frac{2C_0(V_G - V_{FB})}{B}\left[1 - \frac{Q'}{Q}\left(\frac{V_G - V_T}{V_G - V_{FB}}\right)\right]\right\}^{\frac{1}{2}} - 1\right) \tag{1}$$

where $B = K_s q N_D X_0/K_0$, K_s is the silicon dielectric constant, V_{FB} is the flatband voltage and N_D the donor density. Similarly, the capacitance between gate and substrate can be shown to be

$$C = C_0\left\{1 + \frac{2C_0 V_G}{B}\left[1 - \frac{Q'}{Q}\left(\frac{V_G - V_T}{V_G - V_{FB}}\right)\right]\right\}^{-\frac{1}{2}} \tag{2}$$

These quantities are plotted as a function of Q'/Q in Fig. 3 for a representative structure with $V_T = 1.2V$. The depletion width $X_d = (2K_s\varphi_s\epsilon_0/qN_D)^{\frac{1}{2}}$ is also plotted. It is seen that these quantities are a reasonably strong function of Q'/Q for the parameters chosen. In a practical situation, the gate voltages chosen ($\sim 10V$) are readily attainable from silicon integrated circuits.

There are two interrelated quantities of interest in describing the transfer of charge from one electrode to the next. One is the time to transfer the charge and the other is the transfer efficiency which we define as the fraction of charge transferred from one electrode to the next. The time constant for transfer of charge from one electrode to another by diffusion will be of the order $\tau_0 = L^2/4D$ where L is the linear dimension of the electrode and D the diffusion constant. It is assumed that the spacings and the applied voltages are such that no potential barrier exists between the electrodes. For $L = 10^{-3}$ cm and $D = 10$ cm^2/sec, it is found that $\tau_0 = 2.5 \times 10^{-8}$ sec. The amount

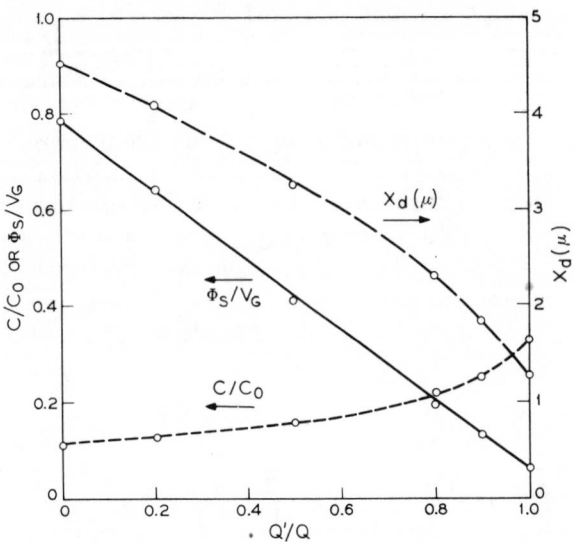

Fig. 3—A theoretical plot of depletion width (X_d), surface potential (φ_s) and capacitance (C) of an MIS structure as a function of charge at the interface (Q'). Edge effects are neglected. The values used in this calculation were:

$$V_G = 10v;$$
$$X_0 = 2 \times 10^{-5}cm;$$
$$N_D = 5 \times 10^{14}/cm^3;$$
$$C_0 = 1.7 \times 10^{-8}F/cm^2.$$

of charge remaining will decay in an almost exponential manner with this time constant if trapping in deep surface and bulk states is neglected. If trapping times are comparable to any transfer times of interest, such effects will also detract from the transfer efficiency and for the current Si-SiO₂ technology, it appears that surface states will be the limiting factor.

There will also be a field enhanced component resulting from the change of surface potential with charge density. A qualitative understanding of this is obtained by considering the situation in Fig. 2(b) immediately after $-V_3$ has been applied but before charge has transferred from electrode 1 to electrode 2. As charge from the right hand edge of electrode 1 flows into the potential well under electrode 2, the potential at that edge will become more negative by an amount given by equation (1). This effect results in a field parallel to the interface which adds to the diffusion component. This field will propagate back under the electrode and decrease in magnitude as charge flows but will always add to the diffusion component. The exact nature of this field has not been calculated but its effect is expected to be significant. For example, an average potential drop of 0.1 volts across the width of a 10^{-3} cm electrode will result in a transit time of 2.5×10^{-8} sec. assuming a mobility of 400 cm^2/v-sec. Further field enhancement may be obtained by making the electrode width plus interelectrode spacing comparable to the oxide thickness and using the fringing field of the neighboring electrode.

The structure in Fig. 2 may be used as a shift register with the addition of a charge generator at one end (input) and a detector at the other. The generation can be accomplished by a forward biased p-n junction, by surface avalanching in an MOS structure,[4] or by radiation induced pair creation. Detection may be accomplished by current detection with a reverse biased p-n junction or Schottky barrier or by utilizing the change of capacitance with charge [see equation (2)].

An estimate of the basic signal-to-noise limitations can be made by considering detection by a reversed biased diode put in place of the last electrode and connected to ground by a resistor. If Q is the average amount of charge stored in an element and f the transfer frequency, then the average signal current is simply $I_s = Qf$. For the example given in Fig. 2, $Q \approx 10^{-13}$ C for an electrode with an area of 10^{-6} cm^2. This results in a signal of 10^{-7} amperes at one megacycle. State of the art video amplifiers have an equivalent noise of about 10^{-9} amperes at one megacycle and would dominate the shot noise which for this example is 2×10^{-10} amperes.

The basic shift register concept may be used to construct a recirculating memory or used as a delay line for times up to the storage time. Clearly, charge transfer in two dimensions is possible as well as the ability to perform logic. An imaging device may be made by having a light image incident on the substrate side of the device creating electron-hole pairs. The holes will diffuse to the electrode side where they can be stored in the potential wells created by the electrodes. After an appropriate integration time, the information may be read out via shift register action. A display device[5] may be constructed by the inverse process of reading in the information (minority carriers) via shift register action and then forward biasing the MIS structure to force the minority carriers into the bulk where radiation recombination takes place.

Aside from problems of yield, the limit to the usefulness device will be determined largely by the speed of transfer, the fractional amount of charge not transferred, and the thermal discharge current. Preliminary experiments[6] show that for existing silicon technology, these parameters lie within the range of usefulness.

The authors wish to thank D. Kahng, C. N. Berglund and E. I. Gordon for stimulating discussions during the course of this work.

REFERENCES

1. Heiman, F. P., "On the Determination of Minority Carrier Lifetime from the Transient Response of an MOS Capacitor," IEEE Trans. on Electron Devices, *ED-14*, No. 11 (November 1967), pp. 781–784.
2. Hofstein, S. R., "Minority Carrier Lifetime Determination from Inversion Layer Transient Response," IEEE Trans. on Electron Devices, *ED-14*, No. 11 (November 1967), pp. 785–786.
3. Buck, T. M., Casey, H. C., Jr., Dalton, J. V., and Yamin, M., "Influence of Bulk and Surface Properties on Image Sensing Silicon Diode Arrays," B.S.T.J., *47*, No. 9 (November 1968), pp. 1827–1854.
4. Goetzberger, A., and Nicollian, E. H., Appl. Phys. Lett. *9*, No. 12 (December 1966), pp. 444–446.
5. Gordon, E. I., private communication.
6. Amelio, G. F., Tompsett, M. F., and Smith, G. E., "Experimental Verification of the Charge Coupled Device Concept," B.S.T.J., this issue, pp. 593–600.

Experimental Verification of the Charge Coupled Device Concept

By G. F. AMELIO, M. F. TOMPSETT and G. E. SMITH

(Manuscript received February 5, 1970)

Structures have been fabricated consisting of closely spaced MOS capacitors on an n-type silicon substrate. By forming a depletion region under one of the electrodes, minority carriers (holes) may be stored in the resulting potential well. This charge may then be transferred to an adjacent electrode by proper manipulation of electrode potentials. The assumption that this transfer will take place in reasonable times with a small fractional loss of charge is the basis of the charge coupled devices described in the preceding paper.[1] To test this assumption, devices were fabricated and measurements made. Charge transfer efficiencies greater than 98 percent for transfer times less than 100 nsec were observed.

The basic principles of the charge coupled device, as already described,[1] are very simple indeed, but it is not clear whether the properties of an MIS system are adequate to give viable devices. The purpose of this paper is to describe experiments which have been carried out using the silicon-silicon dioxide system to investigate these properties and their effect on device performance in terms of charge transfer speed and efficiency.

The requirements on the silicon-silicon dioxide interface and on the oxide itself are very demanding. One essential feature is a long storage time which is the time required for a pulsed MOS element to reach the steady state condition. The storage time is a function of the flat-band voltage, the pulse voltage and the number of generation-recombination centers at the interface and in the neighboring bulk. Ignoring bulk states and using the capacitance of a 1200 Å thick oxide, it is readily calculated that for zero threshold voltage, a pulse voltage of 20 V and the surface recombination current[2-4] of 3.7×10^{-8} A cm^{-2} appropriate to a fast state density of $2 \times 10^{+10}$ states/cm^2, the storage time is about 16 seconds.

The operational requirement of a charge coupled device is that it must be able to transfer charge with only minimal loss at high speeds. The object of our experiments has been to evaluate this. Estimates of the rate of charge transfer have been made[1] but estimates of transfer efficiency are much more speculative on account of ambiguity in the density of surface states in the energy region near the band edge for a particular oxide.

Several types of oxide on nominal 10 Ω-cm n-type (100) and (111) orientated silicon have been tried. Steam grown oxides with a fast surface state density as low as $N_{ST} = 2 \times 10^{10}$ states/cm^2 gave oxide storage times less than 100 msecs for a 20 V pulse. This unexpected result was attributed to generation-recombination centers caused by impurities which had diffused into the bulk. A silane deposited oxide had storage times greater than one second but was not stable with respect to migration of positive charge. The oxide which has given the best results so far is a dry oxide 1200 Å thick grown in oxygen at 1100°C for one hour and annealed in a nitrogen atmosphere for one hour at 400°C. The flatband potential for this oxide is typically −5 V.

The initial device configuration used in the experiments described below is a linear array of Cr-Au squares 0.1×0.1 mm and separated by 3 μm gaps. These squares were produced by conventional photolithography on the oxidized silicon slice. The slice was diced, each die mounted on a 10 pin header and each square gold-wire bonded to a pin.

The device as described above was designed principally for ease of fabrication and is in no way optimized in either material processing or geometry. Indeed, there are reasons for supposing that p-type material might be preferable since the minority carrier mobility will be greater than in n-type silicon.

Evaluation of charge coupled device performance is based on the measurement of several parameters including percentage of charge transferred (efficiency), the limiting speed of transfer and the storage time. For times greater than about one-half second, the latter property is easily measured by applying a negative step voltage and observing the change of capacitance with time using a capacitance bridge and an xy recorder. Storage times less than one-half second are normally associated with high interface state density oxides or a large number of bulk generation-recombination centers and are of no interest in the present application.

Observation of charge transfer including the determination of efficiency and speed requires a different approach. The experimental configuration chosen in the measurements is shown in Fig. (1). Capacitor P_0 in this figure is used to supply a source of holes to the other units $(P_1, P_2, \cdots)$ by surface avalanching. The avalanching pulse, is adjustable 0-200 volts with a full width half maximum of 60 nanoseconds. For the 10 Ω-cm substrate material and 1200 Å oxide thickness used, avalanching occurs in the vicinity of -165 volts. Unit pulse generators are attached to pads P_1, P_2, and P_3. In a many

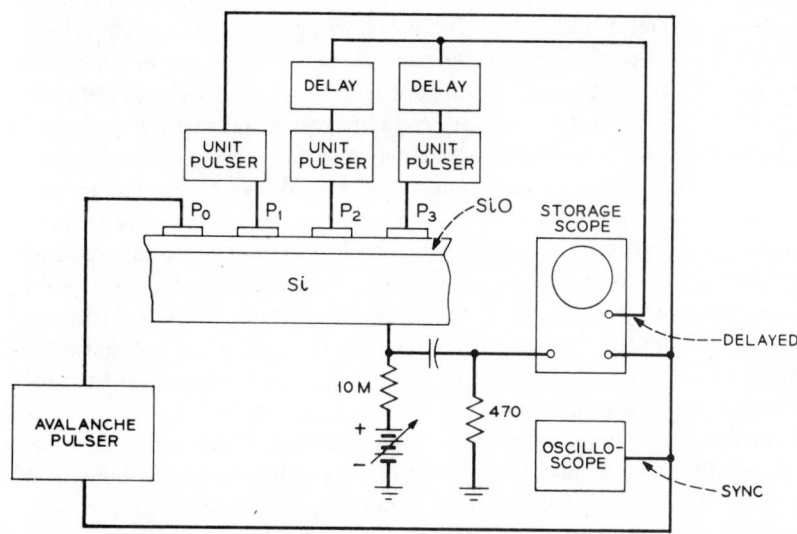

Fig. 1—Schematic of experimental configuration used to evaluate charge transfer.

transfer application, P_1 is attached to each P_{1+3n}, P_2 attached to each P_{2+3n} and P_3 attached to each P_{3+3n}.[1] The substrate is connected to a load resistor and a positive potential for the purpose of biasing the MOS elements beyond threshold voltage. The output signal is capacitively coupled to a relatively low impedance so that response times on the order of 100 nanoseconds are achieved. As each MOS capacitor is pulsed by the supply generators, a charging and discharging spike is observed at the oscilloscope which, for the circuit shown, is proportional to the current flow. The pulsing sequence for capacitors P_0, P_1 and P_2 is illustrated in Fig. (2) for the conditions when the P_1 and P_2 pulse voltages (V_{P_1} and V_{P_2}) do and do not overlap in time. Note the avalanche pulse occurs shortly after P_1 is turned on. Of interest is the charge transfer from P_1 to P_2, all others being ignored for the sake of simplicity. Below each pulse sequence is shown the expected (idealized) oscilloscope display with a positive going signal taken in the downward direction.

For the nonoverlap condition, there are two essentially separate events. Notice that the turn-off pulse of P_1 is larger than the turn-on pulse when avalanching of P_0 occurs during the on-time of P_1. This is easily understood. Each current pulse is given by the relation

$$i(t) = \frac{dQ}{dt} = \frac{dQ}{dV}\frac{dV}{dt} = c(V)\frac{dV}{dt} \qquad (1)$$

where $c(V)$ is the differential capacitance of the device, V is the voltage across the capacitor and Q is the charge flowing from ground. Assuming the turn-on and turn-off characteristics of each voltage pulser are made the same, the magnitude of the current is determined by the differential capacitance $c(V)$. At turn-on, the capacitance which must be charged to an additional V_p volts from the bias voltage V_b is represented by the oxide and depletion capacitance in series. When P_0 is avalanched, the holes generated diffuse to P_1 and invert the surface there. The depletion region under P_1 diminishes and the associated capacitance increases. Now at turn-off when the voltage across the device is returned to V_b, the relaxation pulse is of greater magnitude than the turn-on pulse by an amount related to the change in differential capacitance. When no holes are stored, as in the case of P_2, the turn-on and turn-off pulses are of equal amplitude.

Consider now the overlap case. There, instead of the turn-off pulse of P_1 resulting in a large hole injection into the bulk, the holes are transmitted to the adjacent MOS capacitor. The turn-off pulse amplitude of P_1 should therefore decrease. On the other hand, the turn-off

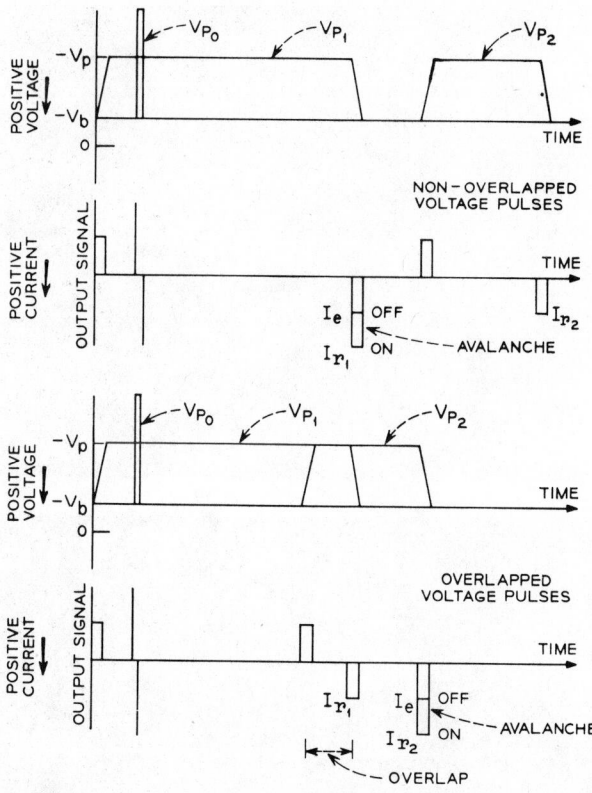

Fig. 2—The pulsing sequence for capacitors P_0, P_1 and P_2. Below each pulse sequence is shown the expected (idealized) oscilloscope display with positive going signal taken in the downward direction.

pulse of P_2 is expected to increase as a result of the charge transmitted to it from P_1.

The efficiency of a single charge transfer can be defined as

$$\eta = \frac{\text{charge arriving at } P_2}{\text{charge originally stored in } P_1} \qquad (2)$$

where the charge is given by the integral from V_b to $V_b + V_p$ of the difference in differential capacitance during turn-off and turn-on. This charge can be approximately related to the current pulses discussed above if equal voltage pulses and similar MOS capacitance properties for the two pads are assumed. Thus, for bias voltages significantly beyond threshold where the $c(V)$ curve is relatively flat, the efficiency

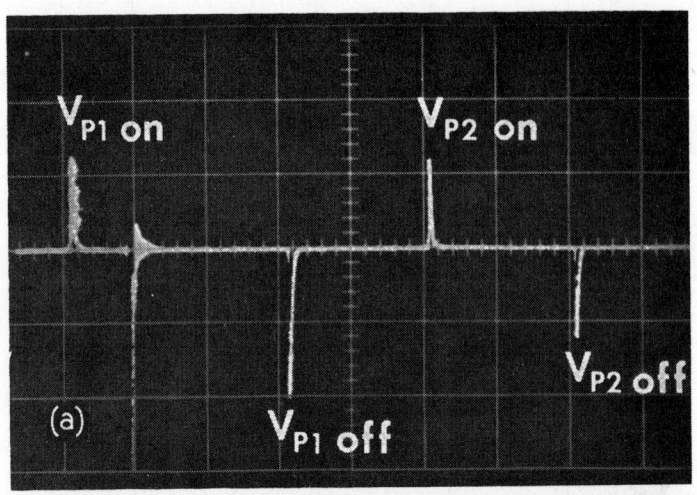

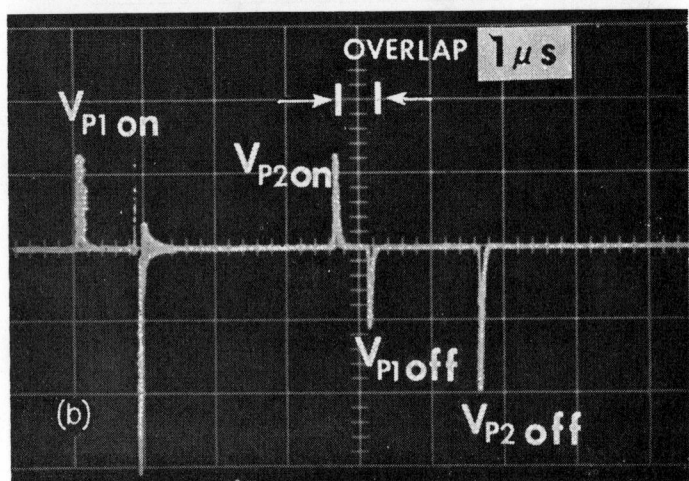

Fig. 3—Oscilloscope traces showing the (a) nonoverlap and (b) overlap charging pulses for an actual device with $V_b = -5V$ and $V_p = -20V$. The time base is 2 μs/cm.

can be reasonably approximated in terms of the peak amplitudes of the relaxation current pulses as

$$\eta \approx \frac{I_{r2} - I_e}{I_{r1} - I_e} \tag{3}$$

where I_{r2} is the turn-off pulse when P_2 contains charge, I_{r1} is the turn-off pulse when P_1 contains charge and I_e is the turn-off pulse when either is empty.

In Figs. 3(a) and (b), the nonoverlap and overlap charging pulses for an actual device with $V_b = -5$ V and $V_p = -20$ V are given. The threshold voltage for this unit is -1.5 volts. The evidence of charge transport is unmistakable. In Fig. 4 a similar, although somewhat more complicated, photograph shows the superposition of many events (using a storage oscilloscope) in which the pulse duration of V_{P_1} is increased until it overlaps V_{P_2} by one microsecond. Following this, V_{P2} is additionally delayed until there is once again no overlap. The rounding seen in the turn-off pulse of P_2 after it goes out of overlap is attributable to holes remaining in the vicinity of P_1 and P_2 after P_1 is turned off. The high transfer speed of the device is seen by the rapidity with which the relaxation pulse P_1 falls off as the pulses overlap. Although not evident in Fig. 2, the charge transfer efficiency is not a function of time for overlap times greater than 100 nanoseconds (the rise time of the pulses used in the experiment). Using enlarged photographs similar to that shown in Fig. 3 and equation (3), the measured efficiency in $\eta = 94 \pm 6$ percent. Measurements made in this manner have been performed on devices with wet, dry and deposited oxides on $\langle 111 \rangle$ and $\langle 100 \rangle$ oriented surfaces. To date, the best results have

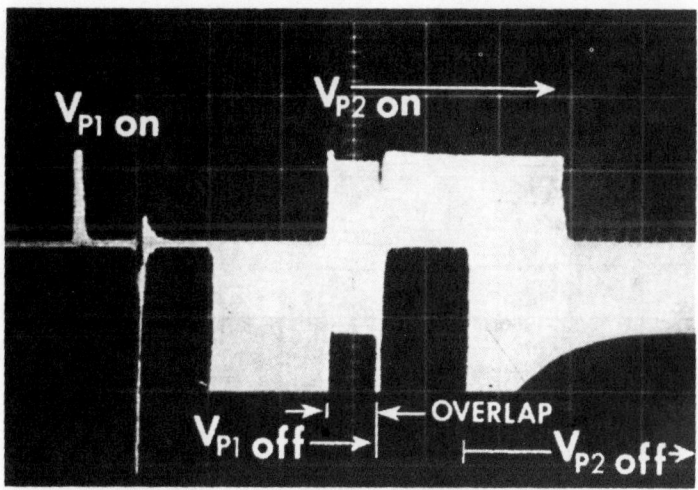

Fig. 4—Oscilloscope traces showing the superposition of many events (using a storage oscilloscope) in which the pulse duration of V_{P_1} is increased from 4 μs until it overlaps V_{P_2} by one microsecond. The turn-on of V_{P_2} is then additionally delayed until there is once again no overlap. The same device is used as for Fig. 3.

been obtained for a 1200 Å dry oxide on the $\langle 100 \rangle$ surface of silicon.

More recently, multiple transfer measurements have been conducted for which efficiencies greater than 90 percent have been demonstrated after five transfers, again with pulse widths of 3 μs and overlap times of 1 μs. This implies an η of over 98 percent. For these measurements, current integration has been employed for more accurate determination of the charge transferred.

The authors wish to acknowledge the help of R. A. Furnanage in fabricating the devices and the assistance of P. M. Ryan and E. J. Zimany, Jr. in making the measurements.

REFERENCES

1. Boyle, W. S., and Smith, G. E., "Charge Coupled Semiconductor Devices," B.S.T.J., this issue, pp. 587–593.
2. Heiman, F. P., "On the Determination of Minority Carrier Lifetime from the Transient Response of an MOS Capacitor," IEEE Trans. on Electron Devices, *ED-14*, No. 11 (November 1967), pp. 781–784.
3. Hofstein, S. R., "Minority Carrier Lifetime Determination from Inversion Layer Transient Response," IEEE Trans. on Electron Devices, *ED-14*, No. 11 (November 1967), pp. 785–786.
4. Buck, T. M., Casey, H. C., Jr., Dalton, J. V., and Yamin, M., "Influence of Bulk and Surface Properties on Image Sensing Silicon Diode Arrays," B.S.T.J., *47*, No. 9 (November 1968), pp. 1827–1854.

Integrated MOS and Bipolar Analog Delay Lines using Bucket-Brigade Capacitor Storage

F. L. J. Sangster

Philips Research Laboratories

Eindhoven-The Netherlands

A NEW ELECTRONIC variable delay line for analog data processing has been realized in integrated-circuit form.

Its basis is a chain of storage capacitors and charge-transfer circuits, acting as an analog shift register with externally variable shift rate. This configuration is known as a *bucket-brigade* circuit [1], because of the resemblance to a fire-brigade of old.

A feature of the new device is that information is stored in an array of capacitors not directly as charge level, but rather as a charge deficit. This concept leads to a simple circuit (Figure 1a) with only one transistor per storage capacitor[3], which is less complex than analog delay lines described elsewhere[4,5].

The delay line uses two complementary clock signals with a frequency equal to the sampling frequency applied to the input signal. Functionally, the device provides a delay line in which bandwidth and delay can be interchanged between wide limits. Signal delay can be accurately controlled or if needed, can be changed electronically.

As the storage capacitor is located between collector and base of the switching transistor, the delay line is simply a series connection of transistor with enlarged parasitic *Miller* capacitance, which is easily realized in integrated circuit form. Figure 1b shows a cross section of a bipolar integrated *bucket*. The enlarged junction between base and collector acts as the storage capacitor.

The performance of the circuit is, among other things, dependent on the interaction between successive signal samples proceeding along the capacitor chain. Although this interaction was only a small effect per stage in our first IC, it accumulated to an unacceptable level after hundreds of stages. This phenomenon, caused by the parasitic capacitance between collector and emitter, results in imperfect signal transients and reduces the bandwidth to less than the maximum value of half the clock frequency. Figure 3 shows a new integrated circuit in which parasitic coupling between the storage stages has been reduced to a sufficiently low level by locally narrowing the aluminum interconnection layer. This circuit contains 72 *buckets* with a capacitor value of 25 pF, a sampling circuit, an amplifier and an output buffer stage integrated together on a 2x2 mm chip. It was designed to obtain an experimental delay element for various applications. The amplifier included on the chip is necessary to compensate for signal attenuation in the *bucket* brigade. Charge deficits appearing at the emitter of a bipolar *bucket* are not fully transferred to the storage capacitor at the collector side, as a small amount of charge is supplied via the base contact. This causes some loss of information during each transfer, resulting in a considerable attenu-

[1] Janssen, J. M. L., "Discontinuous Low-Frequency Delay Line with Continuously Variable Delay," *Nature*, p. 148-149; Jan. 26, 1952.

[2] Hannan, W. J., et al, "Automatic Correction of Timing Errors in Magnetic Tape Recorders," *IEEE Transact. MIL*, p.246-254; July-Oct., 1965.

[3] Sangster, F. L. J., and Teer, K., "Bucket-Brigade Electronics-New Possibilities for Delay, Time Axis Conversion and Scanning," *IEEE J. Solid State Circuits*, p. 131-136; June, 1969.

[4] Krause, G., "Analogue Memory Chain: A New Circuit for Storing and Delaying Signals," *Electronics Letters*, p. 544-546; Dec., 1967.

[5] Mao, R., et al, "Integrated MOS Analog Delay Line," *ISSCC DIGEST OF TECHNICAL PAPERS*, p. 164-165, Feb., 1969.

[6] To be shown at the Physics Exhibition, Alexandra Palace, London, March 2-5, 1970.

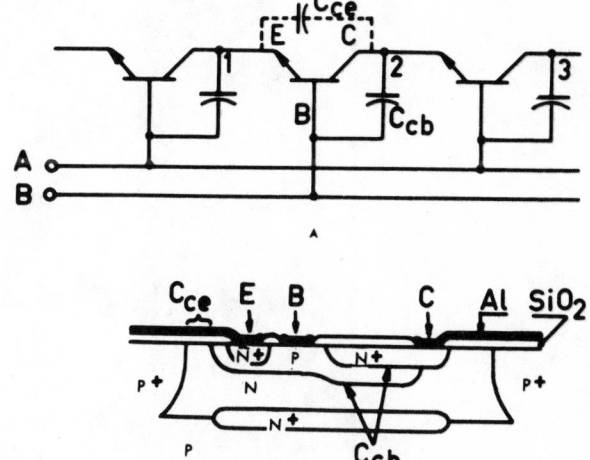

FIGURE 1—A bipolar *bucket-brigade* circuit is illustrated in (a). Sectional view of one monolithic integrated *bucket* is shown in *(b)*.

Reprinted from *1970 IEEE Int. Solid-State Circuits Conf., Dig. Tech. Papers*, vol. XIII, Feb. 18-20, 1970, pp. 74-75, 185.

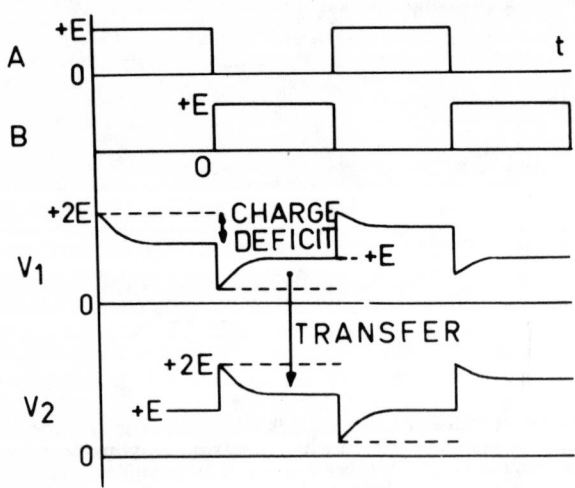

FIGURE 2—Principle of charge-deficit transfer.

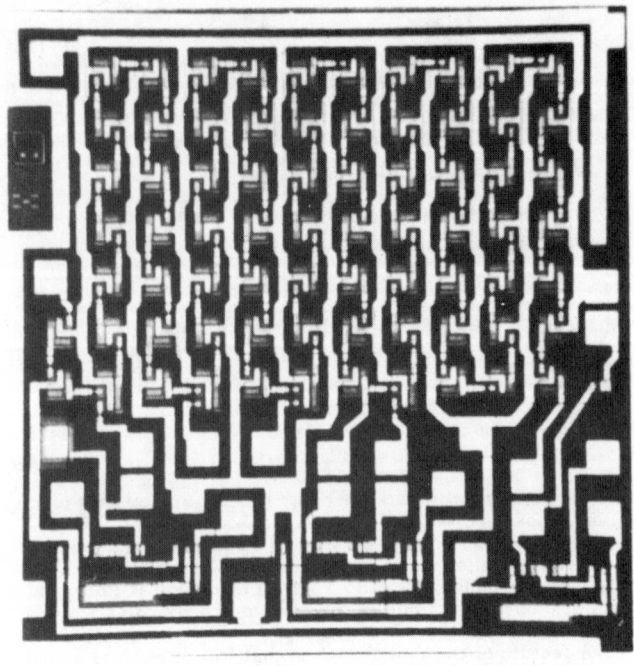

FIGURE 3—Bipolar integrated *bucket-brigade* delay line with 72 stages; chip size, 2 x 2 mm.

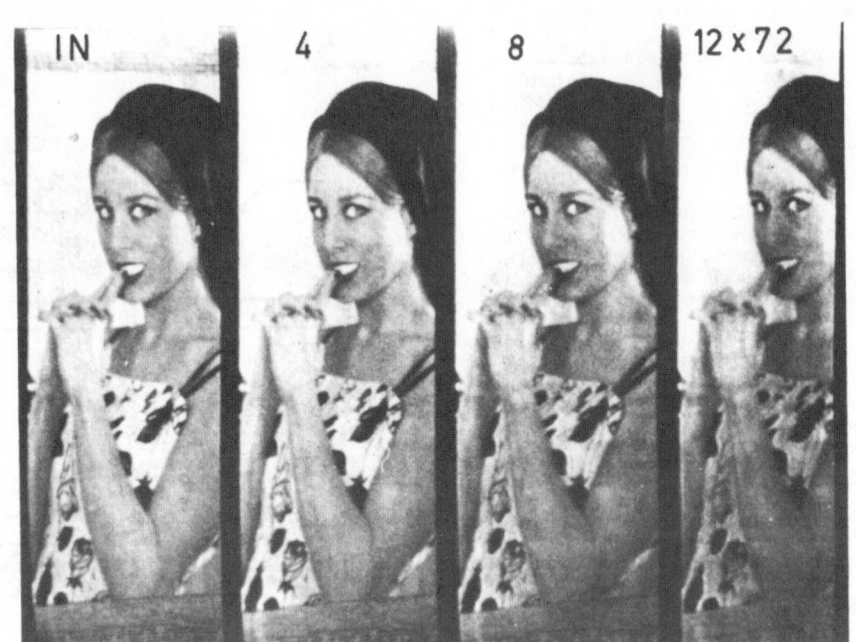

FIGURE 4—Delayed video pictures. left to right, original signal and output signals after 4, 8 and 12 *ICs* (each with 72 stages), respectively.

ation of the signal in large arrays; e.g., by a factor of 3 after 100 stages.

Some characteristics of this IC are: clock pulses 2 V, minimum and maximum clock rate 10 kHz and 30 MHz, signal-to noise ratio after 144 stages better than 60 dB; linearity after 144 stages within 1%. This circuit has been used successfully in time-fault correction systems for video recorders and in several other applications. To illustrate its performance a TV picture obtained with this bipolar circuit is shown in Figure 4. The four parts of the picture, from left to right represent the input signal and output signals after 288, 576 and 864 *buckets*, respectively. Signal degradation is small, even after several hundreds of stages[6].

MOS Integrated Circuit

In MOS technology, the *bucket-brigade* concept leads to an even simpler configuration than possible in bipolar technology. Input and output circuits with fewer components are possible. Due to the absence of dc gate currents, attenuation is negligible even after hundreds of stages and no amplifiers are necessary. Figure 5a represents a complete MOS delay line. A cross section of an MOS *bucket* is shown in Figure 5b. The storage capacitor is formed by the enlarged aluminum gate layer and the P area underneath. A striking feature of

the MOS *bucket* is that no interconnection pattern between adjacent stages is necessary as the drain of one stage also forms the source of the following one. Figure 6 shows a P-channel MOS integrated circuit comprising 72 stages with 8-pF storage capacitors on a 1.5 x 2.4 mm chip. This circuit uses two clock signals of 5-V amplitude. It has been tested between clock frequencies of 100 Hz and 3 MHz and it has been used experimentally for speech processing and audio delay[6].

The MOS circuit uses larger clock voltages and has more dissipation at a given frequency (roughly a factor of 10) than the bipolar circuit; this however is not a severe problem in audio applications. Its speed is not high enough for most video applications, except for color signal delay—part of signal handling in European PAL color-TV receivers; at present glass delay lines are in use for that purpose.

Acknowledgments

The author is very much indebted to the microcircuits group led by J. G. van Santen for fabricating the experimental samples, and gratefully acknowledges the valuable contributions of H. Heyns and C. Mulder in the layout design of the integrated circuits and of W. Ruiterkamp in the electronic experiments.

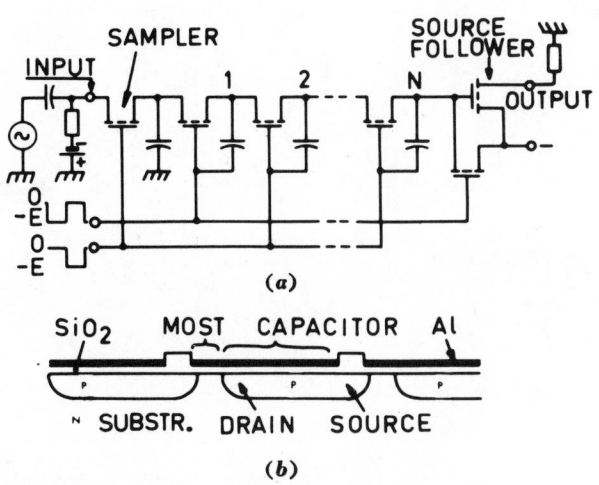

FIGURE 5—Circuit configuration of a MOS delay line is shown in (a); sectional view of the integrated circuit in (b).

FIGURE 6—MOS integrated *bucket brigade* delay line with 2 x 36 stages: chip size, 1.5 x 2.4 mm.

Part II
Device Physics and Technology

Charge-coupled device physics and technology progressed dramatically in the years 1969–1975. The transfer of charge between storage potential wells was analyzed by many investigators. Charge trapping at the semiconductor/oxide interface in a surface-channel CCD was deduced to be the limiting factor on transfer efficiency at moderate charge-transfer rates. Field-aided diffusion, beneath the electrode from which charge is transferred, was attributed to be the transfer-rate-limiting mechanism for most of the experimental devices built during this period.

New devices and technologies were invented. It was discovered that charge storage and transfer could be performed in the semiconductor bulk by selectively doping the semiconductor, resulting in the storage potential minimum being several thousand angstroms in the semiconductor from the semiconductor/oxide interface. These "buried" or "bulk" channel devices have reduced sensitivity to the trapping states at the semiconductor/oxide interface and greater transfer speed resulting from a greater influence of self-induced and fringing fields on the charge-transfer dynamics.

The following reprint papers in this part were selected to represent a sampling of the vast amount of journal literature available on CCD physics and technology. Paper 1 is an overview paper which should serve to introduce the reader to the concepts presented in the following papers. This is followed by papers by Carnes *et al.* (paper 2), Tompsett (paper 3, Brodersen *et al.* (paper 4), and Lee *et al.* (paper 5) in which the theoretical and experimental aspects of charge-transfer rate and charge trapping are analyzed. Paper 6 by Carnes and Kosonocky presents the sources of noise in CCD's and shows that dynamic range from 40 to 90 dB is typical.

Buried-channel CCD's are the topic of the last three papers in this part. The first two, papers 7 and 8, of these three present analyses of the buried charge storage, transfer, and trapping. Two major conclusions that can be drawn from these two papers are that charge transfer occurs more rapidly in buried-channel than in surface-channel CCD's and no background charge ("fat zero") is required for buried-channel CCD's to overcome the detrimental effects of the semiconductor surface traps. The avoidance of the effects of surface traps through use of buried channels is an important factor in low light level video CCD sensor applications which otherwise might require "bias" lighting or other dynamic-range-reducing background charge injection techniques. In paper 9, Esser demonstrates that in a deeply buried epitaxially grown channel, a buried-channel CCD can be simultaneously optimized for high charge transfer rates (135 MHz) and large charge-handling capability using what he refers to as the "peristaltic charge transfer" in which the majority of the charge is transferred along the surface and only the last portion of the transferring charge packet is transported by the buried-channel mechanism. These 135-MHz transfer rates are noteworthy because they are less than two orders of magnitude from those transfer rate limits set by a fundamental material properties of silicon (i.e., the saturation-limited velocity of carriers).

Again the Editors wish to emphasize that these selected papers do not represent a chronological account of the research effort in CCD physics and technology. Papers have been selected for their tutorial value, and the reader is referred to the references of each paper for more complete information.

CHARGE-COUPLED DEVICES - AN OVERVIEW

Walter F. Kosonocky
RCA Laboratories
Princeton, New Jersey 08540

SUMMARY

A comprehensive review of the present state-of-the-art in CCD's is presented. The results covered include the operation of four-, three- and two-phase CCD's, theory of charge transport in surface-channel devices, noise, device construction, and experimental performance data reported thus far. A comparison is made of the surface and bulk (buried) channel devices. Brief reviews are also presented of charge-coupled imagers, CCD memory devices, and analog signal processing applications of CCD's.

INTRODUCTION

The charge-coupled device (CCD) is a MOS integrated circuit device concept announced in the early part of 1970 by Bell Telephone Laboratories[1-3]. Basically it is a shift register for analog signals made in the form of a string of MOS capacitors[1-4]. Since CCD's can store and transfer (analog) charge signals, they can be used for various types of signal processing applications such as electronically variable delay lines, and a variety of transversal filters. CCD's can also be constructed to operate as very effective self-scanned photosensor arrays -- charge-coupled imagers. Finally, because of their small size CCD's are becoming new contenders for high density semiconductor memories. The progress in CCD's has been rapid. Initial devices operated with charge-transfer efficiencies of only 99%. Now, four years later, surface and bulk (buried) channel devices are operating with transfer efficiencies approaching 99.999%. While first devices made had only eight stages[3], 1600 stage registers were constructed as line imagers[5] and 512 x 320 element area imagers[6] represent probably the largest operational MOS arrays. The evolution of CCD's has entered the development stage. Devices such as 256 x 1, 500 x 1 and 100 x 100 element buried-channel CCI's are already commercially available from Fairchild. Among the companies reported to be actively engaged in the development of CCD's are Bell Telephone Laboratories, Northern Bell, RCA, T.I., G.E., Westinghouse, Hughes, Rockwell International, Honeywell, TRW, and Intel.

OPERATION OF CCD'S

The charge-coupled device operates by storing and transferring charge, representing information, between potential wells at or near the surface of silicon. The potential wells are formed by closely spaced MOS capacitors that are being pulsed into a deep-depletion mode by a multiphase clock voltage. Therefore, to understand CCD's it is important to have a thorough

understanding of MOS capacitors and of how the surface potential, ψ_S (the potential at the Si-SiO$_2$ interface relative to the potential of the bulk of silicon), depends upon the various parameters involved. The operation of an MOS capacitor (on a p-type substrate) in deep depletion (b) and in thermal equilibrium (c) is illustrated in Figure 1. After a step voltage is applied, the time required for the capacitor to thermally relax from the condition illustrated in Fig. 1 (b) to that of Fig. 1(c) is called the thermal relaxation time. Depending on the construction

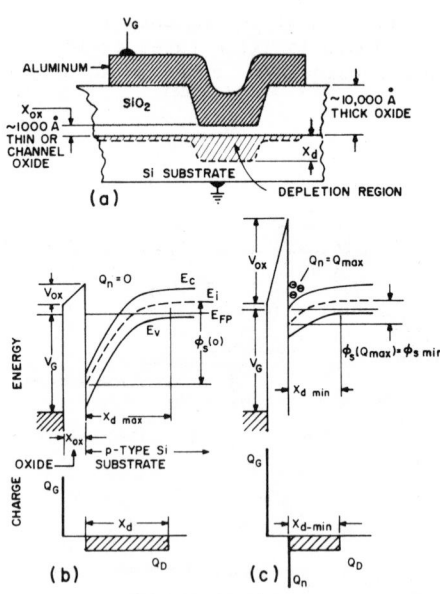

Figure 1 (a) Cross-sectional view of an MOS capacitor, an element for charge-coupled device; (b) energy diagram and charge distribution for MOS capacitor pulsed into deep depletion; (c) energy diagram and charge distribution for an MOS capacitor in thermal equilibrium.

and processing the thermal relaxation time of an MOS capacitor at room temperature has been measured from one second to several minutes. Since a useful potential well does not exist in thermal equilibrium, the CCD is basically a dynamic device in which charge can be stored only for times much shorter than the thermal relaxation time of its MOS capacitors. The solution of the one-dimensional Poisson's equation, subject to the appropriate boundary conditions and including two-dimensional sheets of charge at the Si-SiO$_2$ interface due to fixed oxide charge (Q_{ss})

Reprinted with permission from *1974 Western Electron. Show and Conv. Tech. Papers*, vol. 18, Sept. 10-13, 1974, pp. 2/1-2/20.

and signal charge represented by minority carriers (Q_{sig}), shows that the surface potential ψ_s is given by:

$$\psi_s = V_G' - B\left[\left(1 + \frac{2V_G'}{B}\right)^{1/2} - 1\right] \quad (1)$$

where
$V_G' = V_G + (X_{ox}/\varepsilon_{ox})(Q_{ss} + Q_{sig})$; V_G is the applied gate voltage; Q_{ss} is the fixed oxide charge per unit area; Q_{sig} is the signal charge of minority carriers (inversion layer charge); $B = qN_A\varepsilon_s X_{ox}^2/\varepsilon_{ox}^2 = 0.15\,(N_A/10^{15})\,(X_{ox}/1000\,\text{Å})^2$; q is the electronic charge in coulombs; N_A is the doping density in acceptors/cm^3; ε_{ox} is the dielectric constant of the oxide layer; and X_{ox} is the thickness of the oxide layer. Eq. 1 is the most important one in CCD design.

The formation of the potential well for one element of a CCD is illustrated again in Figure 2 for an MOS capacitor formed on an n-type substrate.

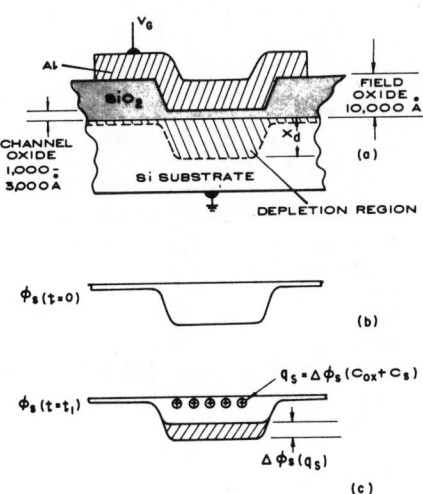

Figure 2(a) Cross-sectional view of an MOS capacitor representing element for charge-coupled circuit; (b) surface potential profile just after application of step voltage V_G; (c) surface potential profile with charge signal q_s in the potential well.

Since the charge signal can be approximately linearly related to a change of the surface potential, a fluid model is useful for describing the operation of a CCD. Thus, for times much shorter than the thermal relaxation time, the depth of the potential well is determined by the gate voltage; and when minority carriers are introduced as the signal into the potential well,

they reduce the depth of the well -- very much like a fluid fills up a container.

Basic Charge-Transfer Action

A Three-phase CCD is just a line of these MOS capacitors spaced very closely together with every third one connected to the same gate or clock voltage, as shown in Figure 3(a). If a higher negative voltage is applied to the ϕ_1 clock line than ϕ_2 and ϕ_3, the surface potential variation along the interface will be similar to Figure 3(b). If the device is illuminated by light, charge will accumulate in these wells. Charge can also be introduced electrically at one end of the line of capacitors from a source

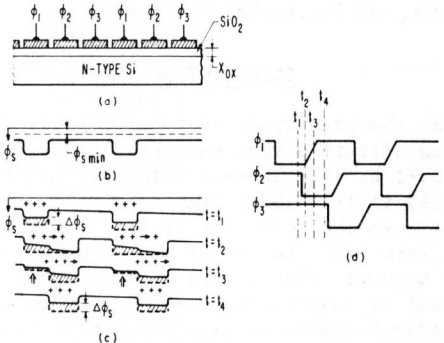

Figure 3: Operation of a 3-phase charge coupled shift register: (a) cross section of the structure along the channel oxide; (b) surface potential profile for $\phi_1 = -V$, $\phi_2 = 0$, $\phi_3 = 0$ forming a potential well under the phase-1 electrode; (c) transfer of charge from the potential wells under the phase-1 electrode to the potential wells under the phase-2 electrode illustrated by the profiles of surface potential at times shown in (d); (d) waveforms of the phase voltages.

diffusion controlled by an input gate. To transfer this charge to the right to a position under the ϕ_2 electrodes, a negative voltage is applied to the ϕ_2 line. The potential well there initially goes deeper than that under a ϕ_1 electrode, which is storing charge, and the charge tends to move-over under the ϕ_2 electrodes. Clearly, the capacitors have to be close together so that the depletion layers overlap strongly and the surface potential in the gap region is a smooth transition from one region to the next. Then the negative voltage on the ϕ_1 line is decreased to a small negative DC level, thus, increasing the surface potential under the ϕ_1 gates and spilling the charge into the ϕ_2 wells. The charge, at least most of it, now resides one-third of a stage to the right under the ϕ_2 gates. The charge is prevented from moving to the left by the barrier under the ϕ_3 gates. A similar process moves it from ϕ_2 to ϕ_3 and then from ϕ_3 to ϕ_1. After one complete cycle of a given clock voltage, the charge pattern moves one stage (three gates) to the right. No significant amount of thermal charge accumulates in a particular well because it is continually being swept out by the charge transfer action. The charge being transferred is eventually transferred into a reversed-biased drain diffusion and from there it is returned to the substrate. The charging current required once each cycle to maintain the drain diffusion at a fixed potential can be measured to determine the signal magnitude (current-sensing) or a resettable floating diffusion which controls the potential of a MOSFET gate can be employed (voltage-sensing). One can visualize a CCD shift register as a multi-gate MOSFET in which the charge signal is moved as charge packets from the source diffusion to the drain diffusion under the control of phase clock voltages applied to the gates.

A four-phase CCD can also be operated with symmetrical wells. Here, however, we have a choice of two types of clock waveforms[7] illustrated in Figure 4.

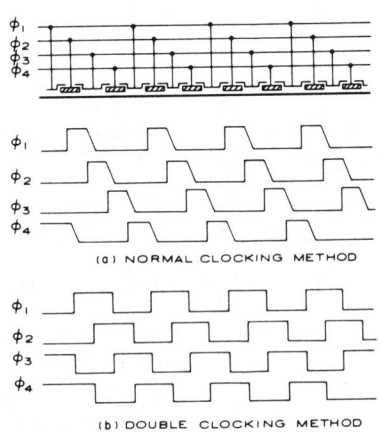

Figure 4: Two types of clock waveforms for operation of a four-phase CCD.

Two-Phase CCD's

For CCD's with symmetrical potential wells it takes three or more phase clocks to determine the directionality of the signal flow. However, if the potential wells are made directional as shown in Figure 5, a two-phase CCD can be constructed. Such two-phase CCD's can consist of separate storage and separate barrier gates connected to a common clock voltage. Alternately, barrier and gate storage regions may be formed under a single gate by modifying the channel-oxide (by means of two thicknesses of channel oxide, two different dielectrics or variations

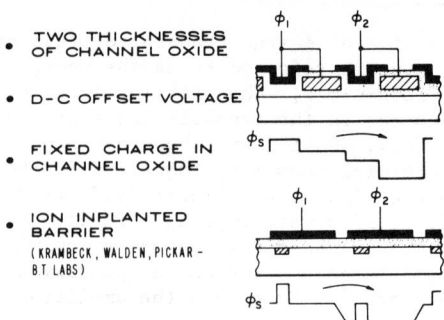

TWO-PHASE CCD's

- TWO THICKNESSES OF CHANNEL OXIDE
- D-C OFFSET VOLTAGE
- FIXED CHARGE IN CHANNEL OXIDE
- ION IMPLANTED BARRIER (KRAMBECK, WALDEN, PICKAR - BT LABS)

Figure 5: Two types of two-phase CCD's.

of fixed charges in the channel oxide) or by modifying the doping of the substrate by ion implantation. Double gate 2-phase CCD's have been made with stepped oxide barriers, d-c offset voltage, applied between the barrier and the storage gates, and implanted barriers.[8,9] Ion-implanted barriers have been used for construction of aluminum gate two-phase CCD's[10] and two-phase CCD's with single layer poly-Si[11] and self-aligned poly-Si aluminum[12] gates.

Unlike the three-or-four-phase CCD's, the two phase CCD's can be operated with non-overlapping clock-voltage pulses. The overlapping clock pulses are referred to as push-clock, and the non-overlapping clock pulses as drop-clock.[13]

An important consideration in the design and operation of two-phase CCD's is to assure the complete-charge-transfer mode (C-C mode) and avoid the bucket-brigade mode (B-B mode) also referred to as the bias-charge mode.[8,9] The difference between the two modes of operation is illustrated in Figure 6.

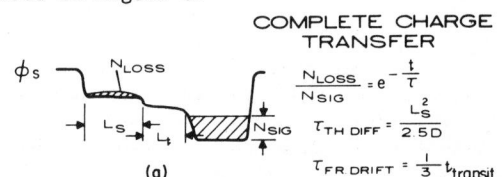

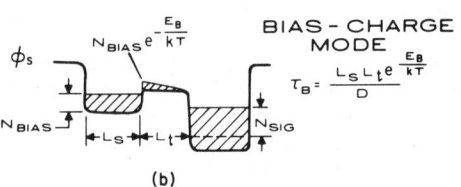

Figure 6: Comparison of the charge transfer characteristics (a) for the C-C mode and (b) for the B-B mode.

THEORY OF OPERATION

Charge-Transfer Efficiency

The fraction of charge transferred from one well to the next is referred to as the charge transfer efficiency, η. The fraction left behind per transfer is the transfer loss, or transfer inefficiency, denoted by ε, so that $\eta + \varepsilon = 1$. Because η determines how many transfers can be made before the signal is seriously distorted and delayed, it is a very important figure of merit for a CCD. If a single charge pulse with an initial amplitude P_O is transferred down a CCD register, after n transfers the amplitude P_n will be:

$$P_n = P_O \eta^n \cong P_O (1-n\varepsilon) \quad \text{(for small } \varepsilon) \quad (2)$$

Clearly, ε must be very small if a large number of transfers are required. If we allow an $n\varepsilon$ product of 0.1, an overall loss of 10 percent, then a 3ϕ, 330 stage shift register requires $\varepsilon < 10^{-4}$, or a transfer efficiency of 99.99 percent.

The maximum achievable value for η is limited by how fast the free charge can transfer between adjacent gates and how much of the charge gets trapped at each gate location by stationary trapping states. In surface-channel devices the charge trapping is normally attributed to the fast states at the $Si-SiO_2$ interface. The trapping of charge by the interface states can be avoided by means of bulk (buried) channel construction. In bulk-channel CCD's small trapping type transfer losses have also been observed and are being attributed to charge trapping by stationary bulk states.

Free-Charge Transfer Mechanisms

Three separate mechanisms cause the free charge to move from one well to another; self-induced drift, thermal diffusion, and fringing field drift.

Self-induced drift[14,15,16] is a charge-repulsion effect which is only important at large signal charge densities ($\geq 10^{10}$ charges/cm^2). The mechanism is important in transferring the first 99% or so of the charge signal and is responsible for improving the frequency response of certain devices operating with a large background charge, or "fat-zero".

Thermal diffusion[15,16,17] gives an exponential decay of charge under the transferring electrode with time constant

$$\tau_{th} = L^2/2.5D \quad (3)$$

where L is the center-to-center electrode spacing; and D is the diffusion constant.

By means of thermal diffusion alone, 99.99% of the charge is transferred each cycle at frequencies f_4(in Hz) given approximately by:

$$f_4 \text{(thermal diffusion)} = 5.6 \times 10^7/L^2 \quad (4)$$

assuming D = 6.75; L is center-to-center spacing measured in μm.

Fringing field drift[18,29] can help to speed up the charge transfer process considerably. The fringing field is the electric field along the direction of the charge propagation along the channel. This field will vary with distance along the gate with the minimum occurring at the center of the transferring gate. The magnitude of the fringing field increases with increasing oxide thickness and gate voltage and decreases with increasing gate length and doping density and in general is higher in bulk channel devices. The effect of the fringing field upon charge transfer is difficult to assess analytically. A computer simulation[18] of the transfer process under the influence of strong fringing fields has indicated that the charge remaining under the transferring electrode still decays exponentially with time.

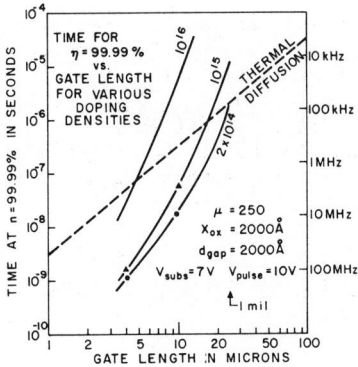

Figure 7: Time required to achieve η = 99.99% vs. gate length for various doping levels of n-type substrate. The thermal diffusion time is the maximum time required in any case.

Figure 7 shows the time required to reach η=99.99% as a function of gate length for various substrate doping densities. According to these calculations for a p-(surface) channel CCD, η=99.99% is possible at a clock frequency of 10 MHz with gate length L = 7 μm and substrate doping of 10^{15}cm^{-3}. This assumes that trapping effects are negligible.

Charge Trapping by Fast Interface States

Charges can be lost[13,19,20] from the signal into fast interface states because, while the filling rate of these states is proportional to the number of free carriers, the empty rate depends only upon the energy level of the interface stage. Thus, even though a roughly equal amount of time is available for filling and for emptying, many of the interface states can fill much faster than they can empty, and thus retain some of the signal charge and release it into trailing signal packets. This type of loss can be minimized by continually propagating a small zero-level charge or fat zero through the device.[19] This tends to keep the slower states filled so they do not have to be filled by the signal charge.

An analytical expression[9] for fractional loss into fast interface states ε_s is given by:

$$\varepsilon_s = \left(\frac{1/n_{s,o}}{n_s/n_{s,o} + 1} \right) kTN_{ss} \ \ln \left(1 + \frac{2f}{k_1 n_{s,o}} \right) \quad (5)$$

where $n_{s,o}$ is the fat zero carrier density in charges/cm^2; n_s is the signal density in chgs/cm^2; kT is in units of eV (0.026 at room temperature); N_{ss} is the fast state density in states/ (eV – cm^2); f is the clock frequency; and k_1 is a constant depending upon the trapping cross-section ($\sim 10^{-2} cm^2$/sec).

Eq. 5 implies that without fat zero, $\varepsilon_s \simeq 10^{-2}$ for $N_{ss} = 10^{11}$ $(cm^2-eV)^{-1}$, and $\varepsilon_s \simeq 10^{-3}$ for $N_{ss} = 10^{10}$ $(cm^2-eV)^{-1}$, at 1-MHz clock frequency. By introducing $n_{s,o}$ equal to approximately 10 to 25% of a full well, ($\sim 2 \times 10^{11} cm^{-2}$) interface state losses can be essentially eliminated. For example, suppose $n_{s,o} = 2 \times 10^{11}$, $n_s = 10^{12}$ and $f = 10^6$. Then:

$$\varepsilon_s = 2.2 \times 10^{-7} \text{ for } N_{ss} = 10^{10}$$
$$2.2 \times 10^{-6} \text{ for } N_{ss} = 10^{11}.$$

Edge Effect

Experimental measurement,[9] however, indicates that losses at intermediate frequencies, when operation with fat zero, are higher than that predicted by Eq. (5), and are consistent with the "edge-effect."[20] This is because the model used for Eq. (5) has assumed a potential well with steep walls so that the background charge, or fat zero, is present at every point where signal charge, when present, resides. However, actual potential wells have sloping sides and there is a region along the edges that does not benefit from the presence of fat zero. In this case, a simple model[9] predicts a fractional loss per transfer given by:

$$\varepsilon_{s, \text{ parallel edges}} = 3.9 \times 10^{-4} \left(\frac{1}{W_{mils}} \right) \left(\frac{N_{ss}}{10^{10}} \right) \left(\frac{10^{15}}{N_D} \right)^{1/2} \quad (6)$$

for clock frequency of 1.0 MHz, signal of 4V and oxide thickness of 1000Å; where W_{mils} is the channel width in mils and N_D is the substrate doping.

Bulk (Buried) Channel CCD's

The transfer loss due to interface state trapping and the subsequent requirement for fat zero can be essentially eliminated when the channel is moved away from the Si-SiO$_2$ interface. This is done in the bulk (buried) channel CCD[11,12,21-31] by including a thin layer of conductivity type opposite to that of the substrate as shown in Figure 8(a). When this layer is completely depleted of majority carriers by applying the appropriate potential to the drain diffusions, the depleted layer results in a parabolic potential well as seen in Figure 8(c). This well can store and transfer charge as described earlier.

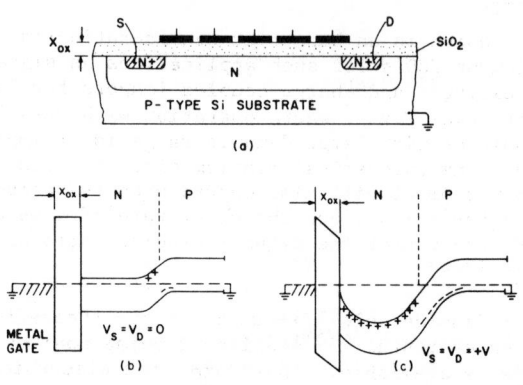

Figure 8: Bulk-Channel CCD: (a) Crossection of the device; (b) Energy-band diagram in thermal equilibrium; (c) Energy-band diagram where the n-type layer is depleted.

The bulk channel CCD, however, still is left with the bulk states which, although to a smaller degree than the surface states, also exhibit some trapping losses with ε in the range of 1 to 2×10^{-4}.

The buried-channel CCD was originally described by Smith at Bell Telephone Laboratories in 1971.[30] A 500-stage buried-channel line imager was announced by Fairchild in 1972. Esser[25-27] of Philips described in 1973 the performance of the peristaltic CCD, a bulk-channel CCD consisting of a 10-μm thick epitaxial layer and operating at clock frequency of 135 MHz with transfer efficiency of 0.9999. Since many reports[11,12,21-31] have been made on the operation of bulk channel CCD's, a comprehensive analysis of the operation of these devices is expected to become available in the near future.

The main advantage of bulk channel CCD's is that the signal is no longer trapped by fast interface states. Therefore, transfer efficiency in the range of 0.9999 can be obtained without fat zero at room temperature. In addition, the carrier mobility is higher since the transport is in the bulk rather than at the surface. Finally, these devices can be constructed to operate with large fringing fields by forming the bulk-channel further away from the electrodes thus allowing operation with very high clock frequency (in excess of 100 Mz).

On the other hand, bulk channel devices involve somewhat more complex processing and more critical design. They also have reduced signal handling capability because the well capacitance is smaller.

Noise in CCD's

Noise is an important consideration in evaluation of CCD's for such applications as signal processing[32] and charge-coupled imagers for low-light level T.V.[33] where operation with very low signals or with large dynamic range is of importance. The theoretical minimum size of a CCD memory element will also depend upon the noise characteristics since the error rate of a memory is a function of the signal-to-noise ratio of the device.[34,35]

Considerable literature on the theoretical[34,36-39] and experimental[32,40] studies of noise nources in CCD's is available. Therefore, the discussion here will be limited to the most important conclusions.

● The transfer noise due to free-charge transfer is quite low since normally it involves only a small amount of charge left behind.[37]

●The noise associated with trapping of charge by fast interface states and the noise due to a fat zero signal represent the major noise fluctuations in the surface channel CCD's. A signal-to-noise ratio of 60 to 80 db still should be possible with surface channel CCD's.[32,33,37]

●Several techniques for introduction of a low noise fat zero were proposed which lead to thermal noise fluctuations comparable to those associated with charging of the capacitance of the input potential well.[40-43]

●The noise associated with the resetting of the output floating diffusion can be considerably reduced by a technique of synchronous double sampling.[32,44]

●Output circuits in the form of a floating gate amplifier (FGA)[45] or a distributed FGA[45] are expected to lead to still smaller noise levels in the output circuit.[33,37]

●The analytical and experimental studies indicate that the noise inherent in the charge-transfer action of CCD's will not impose any limitation on the practical size CCD memory elements in the foreseeable future.[34,35]

●At room temperature the sensitivity of charge-coupled area imagers is mainly limited by local variations of dark currents rather than the random noise.

Input Circuits

As will be shown later, the optical input can be introduced to any point of a CCD structure. The three most important ways by which an electrical input can be introduced into a CCD are shown in Figure 9.

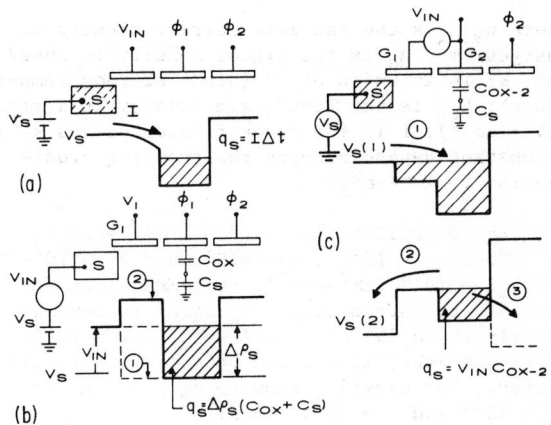

Figure 9: Introduction of electrical input into a CCD.

In Figure 9(a) the source diffusion, S, is d-c biased and the input G-1 voltage applied to the first gate (or gates). This forms essentially a current source that keeps filling the first potential well for a duration of the input voltage pulse, Δt.[3]

A somewhat less critical way of introducing the charge to the first potential well is shown in Figure 9(b). Here, the first potential well is filled to the voltage of the source diffusion.[44,47] The input is now applied as the source diffusion voltage, and the input gate G-1 is used to isolate the first potential well from the source diffusion. Equilibration of the source potential with the first potential well filled with the input charge signal requires a finite (slow) fall time of the input gate clock pulse.

A completely linear and least noisy way of introducing the input into the first potential well is shown in Figure 9(c). This was proposed independently by Tompsett[41,47] and Kosonocky[40]. The basic concept of this method is to form a potential well by gates G-1 and G-2 with the input gate G-1 acting as a barrier between the source diffusion and the input well under G-2. The input is applied as the relative voltage between G-1 and G-2 gates. The input well is first overfilled by raising the source diffusion potential or by dipping the input well below the source potential. After this operation [see (1)] the excess input charge is allowed to return to the source diffusion [See (2)]. If the same channel oxide is used for both gates, G-1 and G-2, the input charge signal, q_s, is

$$q_s = V_{in}C_{ox-2} \qquad (7)$$

where C_{ox-2} is oxide capacitance of the gate G-2. The rms noise fluctuations at room teperature for this input charge are estimated[40] as

$$\overline{N_n} = 400\sqrt{C_{ox-2}(pF)} \qquad (8)$$

where $C_{ox-2}(pF)$ is expressed in pF. For comparison, see also similar analysis of Thornber[38] and Boonstram and Sangster.[48]

Output Circuits

Current sensing of the output of a CCD consists of measuring the current, flowing in the drain of a CCD, as shown in Fig. 10(a). A signal

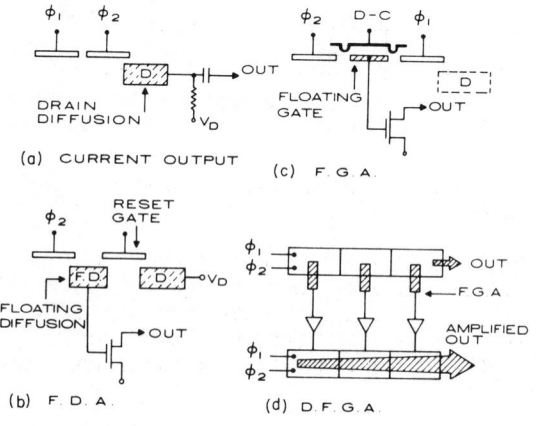

Figure 10: CCD Output Circuits.

in the form of a current spike at a relatively high capacitance output terminal is obtained. Potentially this can be a very linear method for detection of the output signal.[41,47]

Floating diffusion amplifier, FDA, shown in Figure 10(b) is the most popular approach for detection of output signal with an on-chip amplifier.[8,46] This output circuit operates by periodic resetting of the floating diffusion to a reference potential. The floating diffusion in turn is connected to the gate of an on-chip MOS amplifier (inverter or source-follower). The

detected signal is proportional to the variation of the floating diffusion voltage as a function of the charge signal, thus the method is referred to as a voltage or charge-sensing method.

The major noise associated with operation of this circuit is due to the periodic resetting of the floating diffusion.[37] This noise can be, in principle, eliminated by the technique of the double sychronous sampling.[32,44]

Floating Gate Amplifier (FGA)[45] illustrated in Figure 10(c) is a technique by which the signal can be sensed nondestructively so that only the transfer noise is introduced into the charge signal. The floating gates in this case are connected to the gate of an on-chip MOS amplifier.

Distributed floating gate amplifier, DFGA,[45] illustrated in Figure 10(d) has been proposed as a way of reducing the noise associated with the MOS output amplifier. In this case, if the signals detected by the FDA's can be efficiently injected into a parallel CCD, signal-to-noise ratio at the output will improve as $\sqrt{N}$, where N is the number of stages of DFGA. This is because the signals will add coherently while the noise should add incoherently.

DEVICE CONSTRUCTION-CCD TECHNOLOGIES

In order to fabricate a CCD one must decide upon:
- type of charge-coupling gate structure
- means for channel confinement
- surface vs. buried channel, and
- substrate doping and conductivity type.

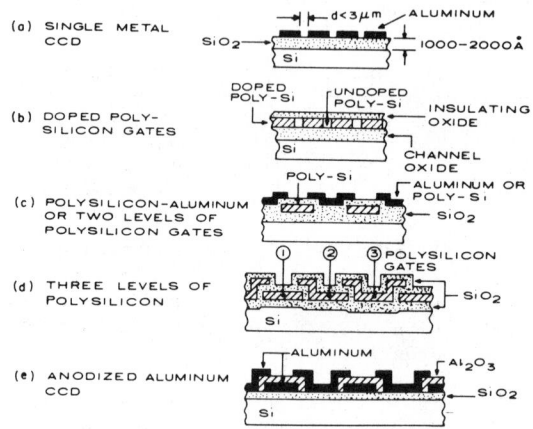

Figure 11: Charge-coupling structures.

Charge-Coupling Structures

The five most important charge-coupling gate structures are illustrated in Figure 11.

Single metal gates,[1-3] usually made with aluminum, where the gaps between the electrodes are made by etching represent the simplest CCD fabrication process. While the first CCD's were made in this form, the requirement of very small gaps between the gates, diffusion tunnels for interconnections and, most of all, the instabilities associated with the exposed channel oxide in the gaps lead to the development of more reliable sealed structures described below.[99]

Doped (Single Layer) polysilicon gate structure[49,50] passivates the interelectrode spaces by undoped high-resistivity polysilicon. The second layer metallization (aluminum) is used in this case for interconnections and light shielding.

Polysilicon-aluminum gates[8,9,46,51,52] represent a self-aligning, overlapping gate structure in which the separation between the gates is formed by thermally grown SiO_2 and corresponds to the thickness of the channel oxide. This gate structure can be used for construction of two-phase as well as four-phase CCD's. The fabrication of these devices is most compatible with the standard silicon-gate processing, thus this type of a device is sometimes referred to as "the standard two-phase CCD." Commonly used layout rules (0.2 mil spaces between and 0.1 mil gate-overlap) result in 1.2 mil center-to-center spaces between CCD stages. One of the limitations of this approach, that is common to all polysilicon gate structures, is the RC time delay in long gates due to the high resistivity of the polysilicon in comparison to aluminum.[8]

A special (newer) version of this structure employs polysilicon for both gate layers. The double-polysilicon gate structure employs aluminum metallization for interconnections.

Triple-polysilicon gate structures[53-54] are used for a sealed-channel three-phase CCD. Unlike in the use of the double polysilicon overlapping gates where size of the storage well is determined in the first level of polysilicon, the size of two of the wells in the triple polysilicon gate construction is subject to alignment tolerances. The main advantages of the triple-polysilicon construction are that for a given gate size, the spaces between the gates become maximized, the shorts between the gates on the same polysilicon level do not lead to catastrophic failures, and there is no need for small contact holes between the polysilicon gates and the aluminum interconnections as for most CCD structures.

Anodized aluminum[7,56] charge-coupling gate structure represents another form of sealed channel CCD. The main advantage of this construction is that both gate layers are made of low resistivity conductors, thus avoiding the delay time in the build-up of the clock voltage

on long polysilicon gates. Although in this gate structure the low-temperature anodized Al_2O_3 covers only very small regions of the channel oxide, the thermally grown SiO_2 is generally considered a preferable insulator. Other overlapping metal gate structures reported employ refractory metals such as molybdenum[4,14] or tungsten[57] and deposited oxide as an insulator.

Channel Confinement

The three different ways by which the width of the CCD channel can be defined are illustrated in Figure 12.

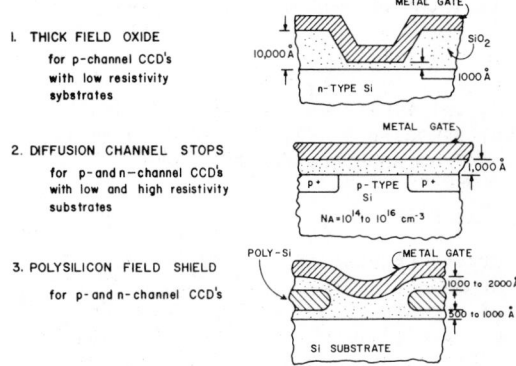

Figure 12: Methods of channel confinement

The thick field oxide approach is most compatible with p-channel devices made on low resistivity substrates.

The diffused channel stop method is most widely used for both surface and bulk channel devices. The diffused channel stops require additional area in the input and output circuits in order to maintain minimum spaces between the source and drain diffusions, and the heavily-doped channel stops.

The polysilicon field-shield used as a channel stop has the advantage of allowing a closer spacing in the peripheral circuits, also with a relatively planar surface topology, as well as some additional design flexibility as some polysilicon electrodes can be used as a channel stop or a transfer gate.

To what extent the choice of channel confinement affects the performance of a CCD has not been documented as yet. Experiments as well as calculations show that with proper choice of technology a channel stop width of 0.1 to 0.2 mil may be sufficient to separate two adjacent CCD channels.[8,52]

Surface vs. Bulk Channel

The bulk, or buried channel CCD's can be

constructed by the following methods:

(1) the different conductivity type top layer on the substrate can be formed by ion-implantation.[21,22,23] These devices had been named originally by G. Smith as buried-channel CCD's. Practical consideration normally limits the thickness of such top layers to less than 1.0 μm.

(2) A relatively thick (in the range of 10 μm) top layer is deposited as an epitaxial film. This type of device was named by Esser as the peristaltic CCD, or PCCD.[25,26]

(3) A profiled PCCD was described by Esser as a bulk channel CCD made in the form of low doped ($10^{15}cm^{-3}$ 10 μm thick epitaxial film with higher doped ($10^{16}cm^{-3}$) 0.5 μm-thick surface layer formed by ion-implantation. According to Esser, such profiled bulk channel CCD's combine advantages of the relatively large charge handling capability of (1) and the higher speed of operation of (2).[27]

The choice between surface and bulk channel depends on the application. The surface channel offers somewhat simpler processing, more conventional on-chip MOS circuits, and higher charge handling capability per unit area, but at the price of fat zero operation. The bulk channel devices do not need fat zero for achieving quite good transfer efficiency, although a little bit of fat zero -- "slim zero" helps.* Bulk channel devices can also be designed for very high frequency operation. But for many high speed devices the clock driver requirements and other distributed effects will be the limiting factors. The potential for very low noise operation, especially at lower than room temperature, is another advantage of bulk channel CCD's.[33] In addition to somewhat lower charge handling capability (about a factor of 3 or more) the basic limitation of bulk channel CCD's is inherently larger dark current because of larger thermal generation by the empty surface states[58] combined with inherently larger electric fields at the Si-SiO$_2$ interface and normally larger depletion layers.

Choice of Substrate

Most surface-channel CCD's have been made on a substrate with (100) orientation to minimize the interface-state trapping losses. The n-channel offers a higher mobility, i.e. corresponding to about a factor of two improvement in the frequency response. The choice of substrate doping or resistivity may also be related to the tradeoff between a frequency response and charge transfer efficiency. As is shown in Figure 15, devices made with low substrate doping have larger fringing fields resulting in better transfer efficiency at higher clock frequencies. On the other hand, at low clock frequencies, the higher doped substrates are better since edge effect trapping is reduced.

Another example where a proper choice of substrate doping is important is the two-phase CCD with stepped-oxide construction for achieving the signal directionality.[8,9]

PERFORMANCE

In addition to the usual parameters that characterize MOS devices, such as gate threshold, oxide charge and interface states, the two most important performance characteristics of CCD's are (1) dark current and (2) transfer inefficiency, or transfer losses, as a function of clock frequency and possibly also clock voltage.

Dark Current

The dark current or thermally generated background charge can be measured (1) either directly as the (average) drain current during the continuous operation of a CCD, (2) or the CCD clock can be periodically stopped for a fixed integration time while the thermally-generated charge is collected. The clocks are then cycled and the dark current profile is read-out and detected.[8] The second approach has the advantage of providing a complete profile of the dark current distribution including the magnitude of the localized dark current spikes that are very sensitive to the applied gate voltage.

*Another term "skinny zero" has been used in connection with digital operation of a CCD with zero background charge. Since certain amount of confusion arises in what is meant by the real zero, skinny zero, slim zero, and fat zero, the following suggestion is offered here:

1. Real zero should represent, in general, the operation with no introduced background charge;

2. Skinny or slim zero should be reserved to described operations in which intentionally (by electrical or optical inputs) or unintentionally (due to thermal generation) a small amount of background charge is introduced (on the order of 1% or less) into CCD's.

3. Fat zero should be used, in general, to describe a large amount of background charge or operation of CCD in which the amount of the background charge may range anywhere from 0 to more than 50% of a full well signal.

Dark current background levels as low as 5 to 10 nA/cm^2 have been reported. However, the control of dark current and dark current spikes is still one of the more critical aspects of CCD manufacturing.

Transfer Losses

Surface-Channel CCD's. Typical performance reported for surface-channel CCD's[8,9] is shown in Figures 14 and 15 for the two-phase CCD shown in crossection in Figure 13. The transfer efficiency obtainable with surface-channel CCD is generally recognized as directly related to the

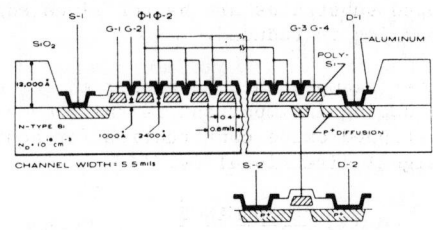

128 STAGE SHIFT REGISTER
(a)

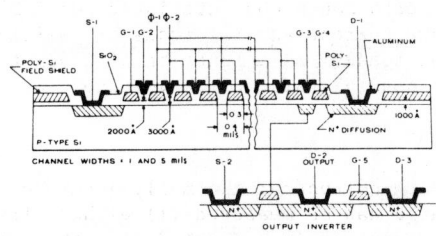

500 STAGE SHIFT REGISTER
(b)

Figure 13: Cross-sectional views of registers along the channel for 128-stage registers, and 500-stage registers.

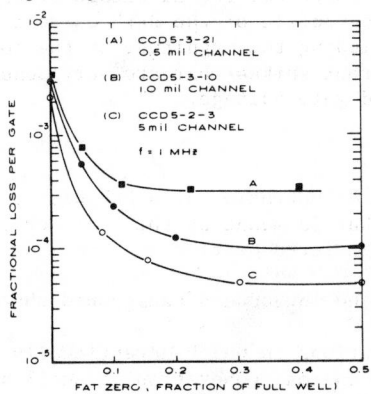

Figure 14: Fractional loss per transfer at 1 MHz vs amount of fat zero for 0.5-1.0 and 5.0-mil wide p-channel 128-stage registers made on (100) substrates are shown as Curves A, B and C, respectively.

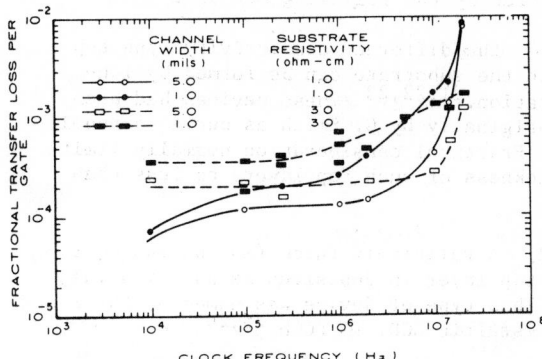

Figure 15: Transfer loss vs. clock frequency for 500-stage n-channel registers operating with 30% fat zero.

density of the fast interface states, N_{ss}. The measurements of the p-channel CCD's suggest an N_{ss} of 3×10^{10} $(cm^2\text{-}eV)^{-1}$ for devices on substrates with (100) orientation. The lowest fractional transfer loss for a surface channel CCD has been reported by the workers of Bell Telephone Laboratories[5]- the operation of a 1600 stage 40 μm-wide triple-polysilicon gate line imager, which was 1×10^{-5} at clock frequency of 0.3 MHz. According to the measurements of transfer noise in this device, N_{ss} is estimated to be in the low 10^9 $(cm^2\text{-}eV)^{-1}$ range.

Buried-Channel CCD

Fractional transfer losses in the range of 10^{-4} to 10^{-5} have been reported by a number of companies for the operation of bulk-channel CCD's. These devices were made in all the various forms of the sealed-channel gate structures shown in Figures 11 b,c,d and e.

The highest frequency performance thus far was reported by Esser.[27] The homogeneously-doped layer PCCD was operated at clock frequency of 135 MHz with a fractional transfer loss of 5×10^{-5} without fat zero at a charge density of 1.5×10^{11} electrons/cm^2. The profiled PCCD at the same clock frequency showed fractional transfer loss of 2×10^{-4} without fat zero and 10^{-4} with fat zero for signal charge density of 3×10^{11} electrons/cm^2. It might be noted that a surface channel CCD normally can operate with a signal charge density in excess of 10^{12} electrons/cm^2.

Data obtained at RCA Laboratories on operation of two-phase bulk channel CCD's made with an ion-implanted top layer and polysilicon-aluminum gates also gave transfer losses in the range of 10^{-4} for operation without fat zero. The introduction of a fat zero, however, still further improved the transfer efficiency.

Although it is not certain to what extent the announced transfer inefficiency numbers represent typical devices, it is clear that both surface- and bulk-channel CCD's are basically capable of operation with transfer losses in the range 10^{-4} to 10^{-5}.

APPLICATIONS

Charge-Coupled Imagers, CCI's

The CCD concept has introduced a revolutionary new approach to the design of self-scanned solid-stage image sensors.[5,6,11,28,44,50 54,55,57,59-71] One can think of the CCD as the semiconductor equivalent of an electron-beam tube in which the charge signal can be moved (transferred) and stored under the control of the clock voltage pulses free of pick-up and switching transients. The only limitations on the charge-coupling process comes about because the charge transfer is not 100% complete. The finite transfer loss (inefficiency) results in some distortion of the signal, and introduces transfer noise. The pick-up from the clock voltages, however, is limited to a single output stage and single frequency that is easily filtered out.[66-100]

Line Imagers shown in Figure 16 illustrate the two ways by which an optical signal can be detected.

integrated optical input are shifted down the CCI register and detected by a single output amplifier. To prevent smearing of the image, the optical integration time should be much longer than the total time required to transfer the detected image from the CCD line sensor. Since all charge elements are amplified by the same amplifier, non-uniformities -- usually a problem in optical arrays in which each sensor elements uses a separate amplifier -- are avoided. Since there is no direct coupling of the clock voltages to the charge signal in the CCD channel, the clock pick-up is limited only to a single output stage. In addition, since only the clock frequency, which is outside of the video bandpass, is used in CCD transfer, clock pick-up is not the problem as it is in x-y scanned arrays where one of the clocks occurs at the horizontal line frequency and cannot be simply removed by appropriate filtering.

A more effective CCD line sensor (or CCI line array) is shown schematically in Fig. 17a. Here the optical input can be continuously integrated by the linear array of photodiodes. During the operation, the detected line image is periodically transferred in parallel to the opaque CCD register from where it is read-out serially. It is essentially an analog parallel-to-series converter, with time-integrated optical input and electrical output. A dual CCD-channel line-sensor is shown in Fig. 17b.

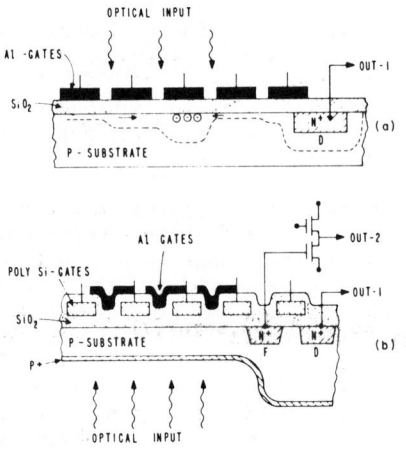

Figure 16: Crossectional view of (a) top (front) illuminated CCI; (b) back illuminated CCI.

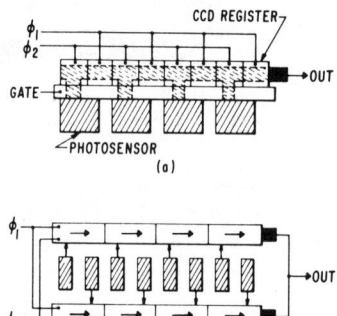

Figure 17: Charge-coupled line sensors: (a) with single CCD channel (b) with two parallel CCD channels.

Let us assume now that an optical input is applied to such a CCD register while the clock voltages are adjusted so that one potential well is created at each stage along the CCD channel. The photogenerated charge will collect in these wells during the optical integration time. At the end of the integration time, the accumulated charge packets representing the

500-stage 2-phase CCD's[9] and 500-stage 4-phase CCD's[56] have been operated as line sensors of the illuminated register type and were reported at RCA and T.I., respectively. 256 stage 2-phase and 500 stage 3-phase buried channel line CCI's with nonilluminated registers are commercially available from Fairchild.[28] The largest line sensors are the non-illuminated registers reported by Bell Telephone Laboratories. They include a 1500 element 4-phase CCI[57] with dual output registers and a 1600 element[55] triple poly-silicon line sensor with a single output register.

Area imagers. The two most popular area CCI's are the interline transfer system[11,28,65] and the frame transfer system,[61] shown in Figures 18 and 19, respectively. The interline

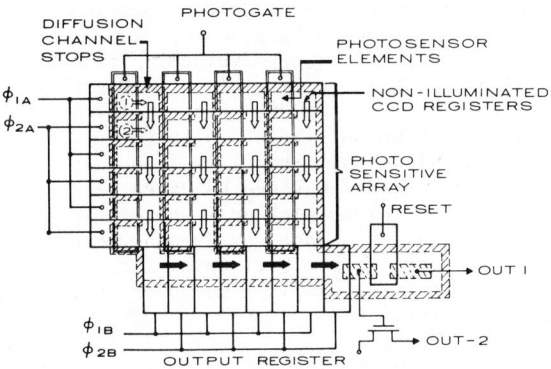

Figure 18: Interline transfer CCI.

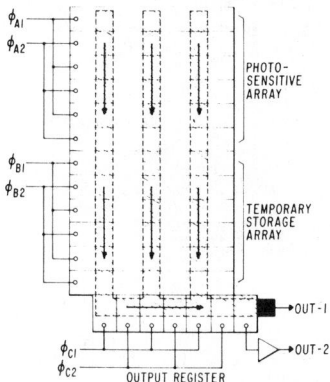

Figure 19: Frame-transfer CCI.

transfer system can be visualized as consisting of a parallel array of the line sensors with non-illuminated registers all leading in parallel into a single output register. The optical image is detected by vertical lines of photosensitive MOS capacitors formed by a transparent polysilicon photogate. The vertical line sensors are separated from each other by opaque vertical CCD registers. Since two photosensor elements can be read by one stage of the vertical register, the image is detected as two vertically interlaced fields. Once every frame time one field is transferred into the non-illuminated registers. Then, the entire detected image is shifted down in unison by clock A and transferred into the output register one (horizontal) line at a time. The horizontal

lines are then transferred out from the output register by the high frequency clock B before the next horizontal line is shifted in.

The frame transfer system that can be illuminated either from the top or from the back of the substrate is shown schematically in Figure 19. In this system the optical image is detected by a separate photosensitive area of CCD registers. Then, assuming a TV format with 1/60-s frame time, the detected image is transferred into the opaque temporary storage array by clocks A and B during the vertical blanking time (900 μs). From there, it is shifted down one horizontal line at a time into the output register and transferred out by the high speed clock C. The time available for parallel loading of the output register corresponds to the horizontal line retrace time of 10 μs, which leaves 50 μs for the read-out of the horizontal line from the output register.

Vertical interlacing[62] can be obtained in the frame transfer system simply by shifting the center of gravity of the detected charge by a half-stage of the A-registers from one field to the next.

Bell Telephone Laboratories announced the first area imager -- a 128 x 106 frame transfer CCI in the form of single metal (aluminum) gate structures[68] and a larger 256 x 220 device using three-phase, triple polysilicon gate structure.[54-55]

Fairchild has been the first to announce commercially available 100 x 100 buried-channel interline transfer CCI's.[11,28]

The largest area imager announced to date is 512 x 320 three-phase frame transfer device built by RCA.[6]

Large area devices represent the major goal in the development area CCI's. Other considerations, however, are control of blooming due to large local optical overloads,[69,70] electronic exposure control, and low light level operation.[28,32,33,44,66,67,71]

Digital Memories

Charge-coupled shift registers are basically analog devices. To store digital signals in these devices it is necessary to periodically refresh, regenerate, or in other words, to redigitize the charge signals. Floating diffusion or floating gate output can be used for construction of signal regeneration stages. The experimental data indicates that such signal refreshing stages can be constructed following procedures similar to those used in the construction of the refreshable MOS memories.[35,46,72,73]

Various system organizations can be considered for the construction of charge-coupled memories.[35,75-82] The choice of the system organization depends on the desired system performance. Two types of systems in the form of a single large-storage loop per chip are illustrated in Fig. 20.

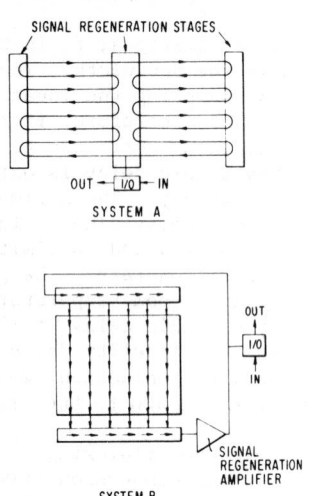

Figure 20: *Two types of memory systems using single storage loop.*

As is shown, the signal flow in system A follows a serpentine pattern and has signal refreshing stages at each corner. The system B on the other hand is in the form of two serial and one large parallel shift register (as well as a single signal regenerator) similar to the frame transfer CCI shown in Fig. 19. Two phase or multilayer charge-coupling gate structures are needed for an efficient design of system A. System B, on the other hand, could be constructed with either two-phase or three-phase structures. At this point, it should also be noted that the frame-transfer CCI with a separate store can be designed so that it can also be operated as a digital store, or as an electronic camera for digital imagers.

The system organizations for obtaining more parallel operation, smaller storage loops, and shorter access time are shown in Fig. 21. The addressing of the individual storage loops in system C is accomplished by a decoder. Another way of organizing randomly accessed CCD memory loops is illustrated by system D. In this case each CCD memory loop is operating in conjunction with a randomly accessed MOS memory cell.

Finally, the most parallel memory is system E, shown in Fig. 22. This system represents the charge-coupled version of one transistor-per-bit refreshable memory. The signal is stored in single charge-coupled elements that are periodically gated by the word lines into the digit-line diffusions. Each digit-line drives a

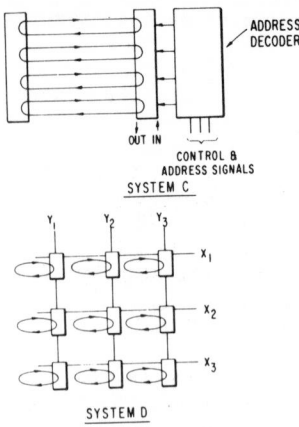

Figure 21: *Two memory system organizations for obtaining more parallel operation, smaller storage loops, and shorter access times.*

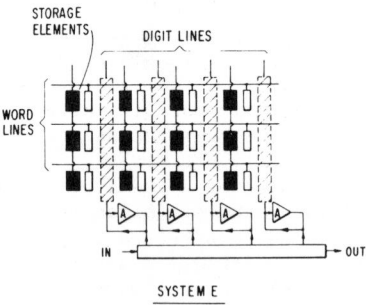

Figure 22: *Charge-coupled version of one transistor-per-bit refreshable random access memory.*

signal-regeneration or sense amplifier whose output regenerates the information in the selected bit location and also may be gated-out. The main difference is that, in systems A through D, the signal-regeneration amplifiers are driven by high impedance charge-coupled storage loops. Thus, high packing density can be accomplished since relatively high voltage (on the order of **several** volts) is available at the low capacitance input of the signal regeneration for small charge signals representing the information. However, system E has a large number of individual charge-storage elements connected in parallel to a single regeneration amplifier. Therefore, the input voltage available to the sense amplifier is attenuated by the capacitance divider corresponding to a relatively small storage capacitance and a considerably larger capacitance associated with digit-line amplifier input. It should be added here, that the use of charge-coupling principles for the design and operation of the signal regeneration amplifier of such a memory system was proposed.[81]

The performance of the CCD memory can be characterized by:

• data rate - mainly due to power requirements should not be above 5 MHz.

• CCD clock power -- that is proportional to the clock frequency. For standby (idle) operation the minimum clock frequency depends on the dark current of the device.

• Average access time determined by memory loop length and clock frequency.

The main interest in the CCD memories is their potential for low cost and, in some applications, low power requirements. Because of the serial nature of the CCD memory and because it represents a new technology, following the "Gordon Moore rule of 4", the CCD memory should be at least cheaper by a factor of 4 over the semiconductor refreshable RAM's. The low cost, in the range of 0.1 to 0.01 cent per bit, is expected because of the high packing density that is possible with CCD memories. But, in the final analysis, the low cost will have to be demonstrated by production yields as well.

For certain special systems with low power requirements, it can be estimated that a 10^8 bit CCD memory could be constructed for operation with megabit read-write rates and total power requirements of only several watts.

Finally, it may be noted that following the initial early proposals for CCD memories, recently several experimental memory arrays have been reported such as the work of Northern Bell[79,80] and RCA.[35] CCD memories are also scheduled to be described at 1974 WESCON by Fairchild and Intel.

ANALOG SIGNAL PROCESSING

Charge-coupled devices represent a new LSI technique for processing of analog information. The potential of lower cost, lower power requirements and in some cases a better performance are the expected advantages of CCD's in comparison to the more conventional digital techniques for processing of information. Assuming that CCD's can be processed for prices comparable with other LSI devices, CCD's are expected to be considerably cheaper for many applications than the digital filters because (1) CCD's eliminate the need for A to D and D to A conversion, and (2) a single CCD filter operating as a functional unit can replace a large amount of digital hardware.[87,91]

Fixed or electronically variable delay lines for video or audio signals represent one of the most direct and obvious applications of CCD's[47,75]. The time delay, τ_d, for a CCD delay line is

$$\tau_d = N(1/f_c) = N/(2\Delta B) \qquad (9)$$

where N is the number of stages, f_c is the clock frequency, and ΔB is the bandwidth. The CCD delay line operates by sampling the input signal once every clock cycle. Therefore, it is capable of signal bandwidth ΔB, of close to $f_c/2$. The electronically-variable delay is obtained by varying the clock frequency. The maximum time delay, $\tau_d$max, for a CCD delay line is independent of the number of stages and is limited by the dark current generation rate. The practical upper limit for $\tau_d$max at room temperature is 0.1 to 10 s. For example, audio delay of 10 to 20 ms with ΔB of 20 kHz (hi-fidelity delay for quadrasonic speaker synchronization) would require f_c = 40 kHz and N = 400 to 800 stages.

Multiplexing and demultiplexing,[83] time synchronization and time conversion (compression and expansion), frame storage and various types of frame format conversions are examples of signal processing functions that can be conveniently performed with CCD's.

Transversal filters with fixed and variable weights represent possibly the most interesting and most effective charge-coupled devices for processing of analog signals.[84-88]

The signals at each stage of CCD delay lines can be tapped, weighted, and summed by a technique illustrated in Figure 23, which is referred to as the electrode weighting and summing approach.

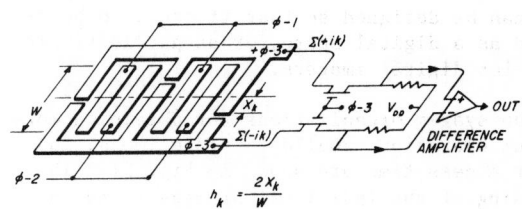

Figure 23: Three-phase transversal CCD filter employing the electrode weighting and summing approach.

The electrode weighting and summing approach is a very effective way for construction of transversal filters with fixed weights. Another approach for tapping, weighting and summing of signals, demonstrated by means of bucket-brigade devices, is the floating-diffusion signal tapping and source-follower weighting and summing.[88] A version of this approach employing a floating gate sensing is illustrated in Figure 24. As shown in this figure, external or on-chip control of the gate voltage to the source-follower load device offers a possibility of variable signal weighting.

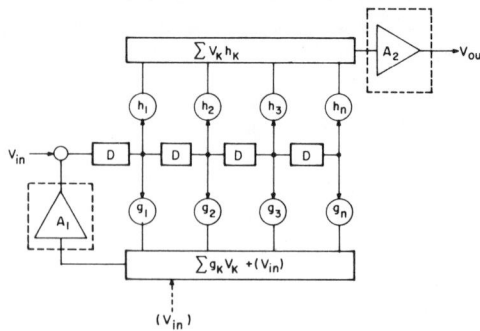

Figure 25: *Commercial form of recursive CCD filter. A_1 and A_2 are either on-chip or off-chip amplifiers.*

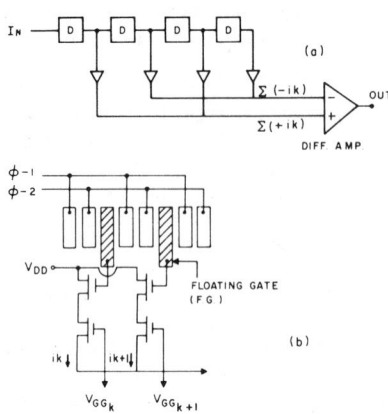

Figure 24: *Transversal CCD filter with floating gate signal tapping and source-follower weighting and summing: (a) block diagram; (b) two-and-half-phase CCD filter with externally controlled tap weights.*

Specific CCD implementations of transversal filters with fixed weights (using electrode weighting and summing) that have been already considered include match filters for various spread-spectrum communication applications,[84] and radar pulse compression band-pass filters, and chirp-z transform filters[87,90] for spectral analysis. The present accomplishments in this area are represented by 500-stage CCD band-pass filter[91] and a 500-stage chirp-z transform filter[92] reported by T.I.

Transversal filters with variable weights are more difficult to achieve. The source-follower weighting and summing provides a convenient way for obtaining binary weights (1 and 0 or +1 and -1). The approaches proposed for obtaining continuously variable weights include:

(1) Analog-binary serial CCD.[87]

(2) Analog-binary CCD also referred to as charge-sloshing and sequentially addressed memory (SAM).[93-95]

(3) MNOS variable conductance CCD.[96,97]

Recursive filters (see Figure 25) which include both feedback and feed forward have, up until now, only been implemented with bucket-brigate devices as the delay sections.[98] Unlike the transversal filters that can have only a finite impulse response the recursive filters, because of the presence of feedback, can have an infinite impulse response. Therefore, a very high quality bandpass recursive filter can be designed with only a small number of delay sections. The comparison of recursive and transversal filters, however, must include on-chip implementation of the feedback including a very precise control of amplitude. No device of this type has yet been reported.

CONCLUSIONS

The general status of the CCD technology is characterized by the following

• The basic operation of CCD's is well understood, design principles are evolving, and the CCD's have entered the developmental stage.

• Several ways for construction of CCD's are available. But the sealed-channel poly-silicon gate structure, with diffusion channel stops, is most prevalent.

• The choice between the surface and bulk (buried) channel CCD depends on the applications. The main advantage of the surface channel CCD is larger signal handling capability (by a factor of 3 to 4) and more conventional MOS on-chip devices. The bulk (buried) channel can operate with very low transfer losses without the need of fat zero and with lower noise. The dark current characteristics on the other hand are expected to be in favor of the surface-channel CCD.

• The development of both the CCD memories and the charge-coupled imagers (CCIs) is directed at large area devices (larger than the

conventional LSI chips). The CCD memories, however, must be price competitive with other types
of MOS RAMS. Although some CCI's will sell at
any price, a large volume production will also
require a high resolution (better than 100 x
100) as well as prices comparable with other
LSI products.

• CCD's present a new and very effective
approach for processing of analog signal information. These devices, at the present time,
are evolving as custom CCD integrated circuits.
The full potential of this new I.C. technology,
however, will be realized only as a result of
close cooperation between the CCD designer and
the signal processing system engineer.

• Some CCD's are already on the market. Thus,
one can anticipate that, as there will be more
available sources with large volume of CCD products, the prices of CCD components should follow
the path of other LSI products.

ACKNOWLEDGEMENT

The author wishes to express his appreciation to J. E. Carnes and K. H. Zaininger for
their contributions to this paper.

REFERENCES

1. W. S. Boyle and G. E. Smith, "Charge Coupled
Semiconductor Devices," Bell System Tech.
Journal, Briefs, 49, No. 4, p. 587, Apr. 1970.

2. G. F. Amelio, M. F. Tompsett and G. E. Smith,
"Experimental Verification of the Charge
Coupled Device Concept," Bell System Tech.
Journal, Briefs, 49, No. 4, p. 593, Apr. 1970.

3. M. F. Tompsett, G. F. Amelio and G. E. Smith,
"Charge Coupled 8-Bit Shift Register," Appl.
Physics Ltrs. 17, No. 3, p. 111, Aug. 1, 1970.

4. W. E. Engeler, J. J. Tiemann and R. D.
Baertsch, "Surface Charge Transport in Silicon," Applied Physics Ltrs. 17, No. 11,
p. 469, Dec. 1, 1970.

5. D. A. Sealer, C. H. Sequin and M. F.
Tompsett, "High Resolution Charge Coupled
Image Sensors," 1974 IEEE Intercon Technical
Papers, Session 2, New York, N.Y., Mar. 26-
29, 1974.

6. R. L. Rodgers, III, "Charge-Coupled Imager
for 525 Line Television, 1974 IEEE Intercon
Technical Papers, Session 2, New York, N.Y.
March 26-29, 1974.

7. D. R. Collins, W. C. Rhines, J. B. Barton,
S. R. Shortes, R. W. Brodersem, and A. F.
Tasch, Jr., "Electrical Characteristics of
500-Bit Al-Al$_2$O-Al CCD Shift Reigsters," IEEE
Proceedings, Vol. 62, No. 2, 282-284, Feb. 1974.

8. W. F. Kosonocky and J. E. Carnes, "Two-
Phase Charge-Coupled Devices with Over-
lapping Polysilicon and Aluminum Gates,"
RCA Review 34, No. 1, p. 164, Mar. 1973.

9. W. F. Kosonocky and J. E. Carnes, "Design
and Performance of Two-Phase Charge-Coupled
Devices with Overlapping Polysilicon and
Aluminum Gates," 1973 International Electron Devices Meeting Technical Digest,
p. 123, Washington, D.C., Dec. 3-5, 1973.

10. R. H. Krambeck, R. H. Walden and K. A.
Pikar, "A Doped Surface Two-Phase CCD,"
Bell System Tech. Journal 51, No. 8, p.
1849, October, 1972.

11. G. F. Amelio, "Physics and Applications of
Charge-Coupled Devices," 1973 IEEE Intercon
Technical Papers, Session 1, New York, N.Y.
March 26-30, 1973.

12. D. M. Erb, W. Kotyczka, S. C. Su, C. Wang,
and G. Clough, "An Overlapping Electrode
Buried Channel CCD", 1973 IEEE Int'l. Electron Devices Meeting Technical Digest, p. 24
Washington, D.C., Dec. 3-5, 1973.

13. A. M. Moshen, T. C. McGill, Y. Daimon and
C. A. Mead, "The Influence of Interface
States on Incomplete Charge Transfer in
Overlapping Gate Charge-Coupled Devices,"
IEEE Jour. of Solid-State Circuits, SC-8,
No. 2, p. 125, April 1973.

14. W. E. Engeler, J. J. Tiemann and R. D.
Baertsch, "The Surface-Charge Transistor,"
IEEE Trans. on Electron Dev. ED-18, No. 12,
p. 1125, Dec. 1971.

15. J. E. Carnes, W. F. Kosonocky and E. G.
Ramberg, "Free Charge Transfer in Charge-
Coupled Devices," IEEE Trans. on Electron
Dev. ED-19, No. 6, p. 798, June, 1972.

16. L. G. Heller and H. S. Lee, "Digital Signal
Transfer in Charge-Transfer Devices," IEEE
Jour. of Solid-State Circuits, SC-8, No. 2,
p. 116, April 1973.

17. C. K. Kim and M. Lenzlinger, "Charge transfer in charge-coupled devices," J. Appl.
Phys., vol. 42, pp. 3586-3594, Aug. 1971.

18. J. E. Carnes, W. F. Kosonocky, and E. G.
Ramberg, "Drift-aiding fringing fields in
charge-coupled devices," IEEE J. Solid-
State Circuits, vol. SC-6, pp 322-326,
Oct. 1971.

19. J. E. Carnes and W. F. Kosonocky, "Fast-
Interface-State Losses in Charge-Coupled
Devices," Appl. Phys. Lett. 20. No. 7,
p. 261, Apr. 1972.

20. M. F. Tompsett, "The Quantitative Effects of Interface States on the Performance of Charge-Coupled Devices," IEEE Trans. on Elec. Dev. ED-20, No. 1, p. 45, Jan. 1973.

21. R. H. Walden, R. H. Krambeck, R. J. Strain, J. McKenna, N. L. Schryer, and G. E. Smith, "The Buried Channel Charge Coupled Device," B.S.T.J. 51, No. 7, p. 1635, Sept. 1972.

22. C. K. Kim, J. M. Early, and G. F. Amelio, "Buried Channel Charge-Coupled Devices," 1972 NEREM, Boston, Mass., Nov. 1-3, 1972.

23. C. K. Kim, "Design and Operation of Buried Channel Charge-Coupled Devices," CCD Applications Conf. Proceedings, sponsored by the Naval Electronics Laboratory Center, San Diego, Ca., p. 7-11, Sept. 18-20, 1973.

24. K. C. Gunsagar, C. K. Kim, and J. D. Phillips, "Performance and Operation of Buried Channel Charge-Coupled Devices," 1973 IEEE Int'l Electron Devices Meeting Technical Digest, Washington, D.C., pp. 21-25, Dec. 3-5, 1973.

25. L. J. M. Esser, "The Peristaltic Charge-Coupled Device," CCD Applications Conf. Proceedings, pp. 269-277, San Diego, Ca., Sept. 18-20, 1973.

26. L. J. M. Esser, M. G. Collet, and J. G. van Santen," The Peristaltic Charge-Coupled Device," 1973 Int'l Electron Device Meeting, Technical Digest, pp. 17-20, Washington, D.C., Dec. 3-5, 1973.

27. J. M. Esser, "The Peristaltic Charge-Coupled Device for High Speed Charge Transfer," 1074 IEEE International Solid-State Circuits Conf Digest of TEchnical Papers, pp. 28-29, Philadelphia, Pa., Feb. 13-15, 1974.

28. A. L. Solomon, "Parallel-Transfer-Register Charge-Coupled Imaging Devices," 1974 IEEE Intercon Technical Papers, Session 2, New York, N.Y., March 26-28, 1974.

29. J. McKenna and N. L. Schryer, "The Potential in a Charge-Coupled Device with No Mobile Minority Carriers and Zero Plate Separation," B.S.T.J. Vol. 52, No. 5, p. 660, May-June 1973.

30. G. E. Smith, described at the Device Research Conference, Ann Arbor, Michigan, June 1971.

31. Y. Daimon, A. M. Mohsen and T. C. McGill, "Charge-Transfer in Buried Channel Charge-Coupled Devices," 1974 IEEE Int'l Solid-State Circuits Conf Digest of Technical Papers, Phila., Pa., p. 146-147, Feb. 13-15, 1974.

32. S. P. Emmons and D. D. Buss, "The Performance of CCD's in Signal Processing at Low Signal Levels," CCD Applications Conference Proceedings, pp 189-205, San Diego, Ca., Sept. 18-20, 1973.

33. J. E. Carnes and W. F. Kosonocky, "Sensitivity and Resolution of Charge-Coupled Imagers at Low Light Levels," RCA Review 33, p. 607, December 1972.

34. K. K. Thornber, "Operational Limitations of Charge Transfer Devies," Bell System Tech. J. 52, 9, p. 1453, Nov. 1973.

35. J. E. Carnes, W. F. Kosonocky, J. M. Chambers and D. J. Sauer, "Charge-Coupled Devices for Computer Memories," AFIPS Conference Proceedings, Vol. 43, 1974 National Computer Conference and Exposition, pp. 827-836, Chicago, Ill, May 6-10, 1974.

36. D. F. Barbe, "Noise and Distortion Consideration in CCD's." Electronic Letters, 8, 207, 1972.

37. J. E. Carnes and W. F. Kosonocky, "Noise Sources in Charge-Coupled Devices," RCA Review, 33, p. 327, June 1972.

38. K. K. Thornber, "Noise Suppression in Charge Transfer Devices," Proc. IEEE, 60, No. 9, p. 1113, Sept. 1972.

39. K. K. Thornber and M. F. Tompsett, "Spectral Density of Noise Generated in Charge Transfer Devices," IEEE Trans. on Electron Dev. ED-20, No. 4, p. 456, April, 1973.

40. J. E. Carnes, W. F. Kosonocky and P. A. Levine, "Measurements of Noise in Charge-Coupled Devices," RCA Review, 34, pp. 553-565, Dec. 1973.

41. M. F. Tompsett, "Using Charge-Coupled Devices for Analog Delay," CCD Applications Conference Proceedings, San Diego, Ca., pp 189-205, Sept. 18-20, 1973.

42. S. P. Emmons and D. D. Buss, "Techniques for Introducing a Low Noise Fat Zero in CCD's," presented at the Device Research Conference, Boulder, Colorado, June, 1973.

43. S. P. Simmons and D. D. Buss, "A Stable, Uniform, Low Noise Input for Charge-Coupled Devices," submitted to the Journal of Applied Physics.

44. M. H. White, D. R. Lampe, F. C. Alaha and I. A. Mack, "Characterization of Surface Channel CCD Image Arrays at Low Light Levels," IEEE J. Solid-State Circuits SC-9, pp. 1-12, Feb. 1974.

45. D. D. Wen and P. J. Salsbury, "Analysis and Design of a Single-Stage Floating Gate Amplifier," 1973 IEEE Int'l Solid-State Circuits Conference Digest of Technical Papers, Philadelphia, Pa. pp. 154-155, Feb. 14-16, 1973.

46. W. F. Kosonocky, "Charge-Coupled Digital Circuits," IEEE Jour. of Solid-State Circuits $\underline{SC-6}$, No. 5, p. 314, Oct. 1971.

47. M. F. Tompsett and E. J. Zimany, Jr., "Use of Charge-Coupled Devices for Delaying Analog Signals," IEEE Jour. of Solid-State Circuits $\underline{SC-8}$, No. 2, p. 151, Apr. 1973.

48. L. Boonstra and F. L. J. Sangster, "Progress on Bucket-Brigade Charge-Transfer Devices," 1972 IEEE Int'l Solid-State Circuit Conf. Digest of Technical Papers, pp. 140-141, Philadelphia, Pa., Feb. 1972.

49. C. K. Kim and E. H. Snow, "P-channel Charge-Coupled Device with Resistive Gate Structure," Appl. Phys. Ltrs., $\underline{20}$, 514, 1972.

50. L. Walsh and R. H. Dyke, "A New Charge-Coupled Area Imaging Device," CCD Applications Conference Proceedings, San Diego, Ca., pp. 21-22, Sept. 18-20, 1973.

51. W. F. Kosonocky and J. E. Carnes, "Two-Phase Charge-Coupled Shift Registers," 1972 IEEE Solid-State Circuit Conference Digest of Technical Papers, pp. 132-133, Philadelphia, Pa., Feb. 16-18, 1972.

52. N. A. Patrin, "Performance of Very High Density CCD," IBM Journal of Res. and Dev. Vol. 17, No. 3, pp. 241-248, May 1973.

53. W. J. Bertram, B. B. Kosicki, F. J. Morris, D. A. Sealer, C. H. Sequin, T. A. Shankoff, and M. F. Tompsett, "3-Phase Charge Coupled Devices Using 3 Levels of Polysilicon," presented as Late News Paper at the 1973 Int'l Electron Devices Meeting, Washington, D.C. Dec. 3-5, 1973.

54. C. H. Sequin, D. A. Sealer, F. J. Morris, R. R. Buckley, W. J. Bertram, and M. F. Tompsett, "Charge-Coupled Image-Sensing Devices Using Three Levels of Polysilicon," 1974 Int'l Solid-State Circuit Conference Digest of Technical Papers, pp.24,25,218, Philadelphia, Pa., Feb. 13-15, 1974.

55. D. A. Sealer, C. H. Sequin and M. F. Tompsett, "High Resolution Charge Coupled Image Sensors," 1974 IEEE Intercon Technical Papers, Session 2, New York, N.Y., Mar. 26-29, 1974.

56. D. R. Collins, W. C. Rhines, J. B. Barton, S. R. Shortes, R. W. Brodensen, and A. F. Tash, Jr., "Electrical Characteristics of Long CCD Shift Registers Fabricated Using Al-Al$_2$O$_3$-Al Double Level Metallization," 1973 Int'l Electron Devices Meeting, Technical Digest, p. 29, Washington, D.C., Dec. 3-5, 1973.

57. M. F. Tompsett, D. A. Sealer, C. H. Sequin and T. A. Shankoff, "Charge-Coupled Image Sensing: State-of-the-Art," 1973 IEEE Intercon Technical Papers, Session 1, Paper 1/4, New York, N.Y., March 26-30, 1973.

58. A. F. Tash, R. W. Brodensen, D. D. Buss, and R. T. Bate, "Dark Current and Charge Storage Considerations in Charge-Coupled Devices," CCD Applications Conference Proceedings, pp. 179-185, San Diego, Ca., Sept. 18-20, 1973.

59. G. F. Amelio, W. J.Bertram, Jr. and M. F. Tompsett, "Charge-Coupled Imaging Devices: Design Considerations," IEEE Trans. on Electron Dev., $\underline{ED-18}$, No. 11, p. 986, Nov., 1971.

60. M. F. Tompsett, G. F. Amelio, W. J. Bertram, Jr., M. F. Tompsett, R. R. Buckley, W. F. McNamara, J. C. Mikkelsen, Jr. and D. A. Sealer, "Charge-coupled Imaging Devices" Experimental Results," IEEE Trans. on Elec. Dev. $\underline{ED-18}$, No. 11, p. 992, Nov. 1971.

61. C. H. Sequin, D. A. Sealer , W. J. Bertram, Jr., M. F. Tompsett, R. R. Buckley, T. A. Shankoff and W. J. McNamara, "A Charge-Coupled Area Image Sensor and Frame Store," IEEE Trans. on Electron Dev. $\underline{ED-20}$, No. 3, p. 244, Mar. 1973.

62. C. H. Sequin, "Interlacing in Charge-Coupled Imaging Devices," IEEE Trans. on Electron Dev. $\underline{ED-20}$, No. 6, p. 535, June, 1973.

63. P. K. Weimer, W. S. Pike, M. G. Kovac and F. V. Shallcross, "Design and Operation of Charge-Coupled Image Sensors," 1973 Int'l Solid-State Circuit Conference, Digest of Technical Papers, pp. 132-133, Philadelphia, Pa., Feb. 14-16, 1973.

64. M. G. Kovac, F. V. Shallcross, W. S. Pike and and P. K. Weimer, "Design, Fabrication and Performance of a 128 x 160 Element Charge-Coupled Image Sensor," CCD Applications Conference, Proceedings, pp. 37-42, San Diego, Ca., Sept. 18-20, 1973.

65. D. F. Barbe and M. H. White, "A Tradeoff Analysis for CCD Area Imagers: Front-side Illumination Interline Transfer vs. Back-side Illuminated Frame Transfer," CCD App.

Conference, Proceedings, pp. 13-20, San Diego, Ca., Sept. 18-20, 1973.

66. S. B. Campana, "Charge-Coupled Devices of Low Light Level," CCD Applications Conference Proceedings, San Diego, Ca., pp. 235-246, Sept. 18-20, 1973.

67. M. H. White, D. R. Lampe, F. C. Blaha, and I. A. Mack, "Characterization of Surface Channel CCD Image Arrays at Low Light Levels," CCD Applications Conference Proceedings, pp. 23-25, San Diego, Ca., Sept. 18-20, 1973.

68. M. F. Tompsett, W. J. Bertram, D. A. Sealer and C. H. Sequin, "Charge-Coupling Improves its Image, Challenging Video Camera Tubes," Electronics, 46, No. 2, p. 162, Jan. 18, 1973.

69. W. F. Kosonocky, J. E. Carnes, M. G. Kovac, P. Levine, F. V. Shallcross and R. L. Rodgers, "Control of Blooming in Charge-Coupled Imagers," RCA Review, Vol. 35, No. 1, pp. 3-24, March, 1974.

70. C. H. Sequin, T. A. Shankoff and D. A. Sealer, "Measurements on a Charge-Coupled Area Image Sensor with Blooming Suppression" IEEE Trans. on Electron Devices, Vol. ED-21 pp. 331-341, June 1974.

71. W. F. Kosonocky and J. E. Carnes, "Charge Coupled Imagers for Low Light Television," Proceedings of Technical Program, Electro-Optical Systems Design Conference, New York, N.Y., pp. 212-222, Sept. 12-14, 1972.

72. M. F. Tompsett, "A Simple Charge Regenerator for Use with Charge-Transfer Devices and the Design of Functional Logic Arrays," IEEE Jour. of Solid State Circuits SC-7, No. 3, p. 237, June 1972.

73. W. E. Engeler, J. J. Tiemann and R. D. Baertsch, "A Memory System Based on Surface-Charge Transport" 1971 Int'l Solid-State Circuits Conference, Digest of Technical Papers, pp. 169-165, Philadelphia, Pa., Feb. 17-19, 1971

74. N. G. Vogl and T. Harroun, "Operational Memory System Using Charge-Coupled Devices" 1972 Int'l IEEE Solid-State Circuits Conf. Digest of Technical Papers, Philadelphia, Pa., pp. 20-21, Feb. 16-18, 1972.

75. M. F. Tompsett and E. J. Zimany, "Use of Charge-Coupled Devices for Analog Delay," 1972 IEEE Int'l Solid State Circuits Conf., Digest of Technical Papers, pp. 136-137, Philadelphia, Pa., Feb. 16-18, 1972.

76. D. R. Collins, J. B. Barton, D. D. Buss, A. R. Kmetz and J. E. Schroeder, "CCD Memory Options," 1973 Int'l IEEE Solid-State Circuits Conf., Digest of Technical Papers, p. 136, Philadelphia, Pa., Feb. 14-16, 1973.

77. H. A. R. Wegener, "Appraisal of Charge Transfer Technologies for Peripheral Memory Applications," CCD Applications Conf. Proceedings, pp. 43-54, San Diego, Ca., Sept. 18-20, 1973.

78. R. Agusta, and T. V. Harroun, "Conceptual Design of an Eight-Me abyte High Performance Charge-Coupled Storage Device," CCD Applications Conf. Proceedings, pp. 55-62, San Diego, Ca., Sept. 18-20, 1973.

79. A. Ibrahim and L. Sellars, "4096-Bit Charge Coupled Device Serial Memory Array," 1973 Int'l Electron Device Meeting, Technical Digest, pp. 141-143, Washington, D.C., Dec. 3-5, 1973.

80. S. D. Rosenbaum and J. T. Caves, "CCD Memory Array with Fast Access By On-Chip Decoding," 1974 IEEE Int'l Solid-State Circuits Conference, Digest of Technical Papers, pp. 210-211, Philadelphia, Pa. Feb. 13-15, 1974.

81. W. E.Engeler, J. J. Tiemann and R. D. Baertsch, "Surface-Charge RAM System," 1972 IEEE Int'l Solid State Circuits Conference, Digest of Technical Papers, pp. 18-19, Philadelphia, Pa., Feb. 16-18, 1972.

82. R. A. Belt, "CCD Digital Memory for Radar Applications," CCD Applications Conference Proceedings, pp. 63-68, San Diego, Ca., Sept. 18-20, 1973.

83. T. F. Cheek, J. Barton, S. P. Emmons, J. E. Schroeder and A. F. Tasch, "Design and Performance of Charge-Coupled Device Time-Division Analog Multiplexers," CCD Applications Conference Proceedings, pp. 127-139 San Diego, Ca., Sept. 18-20, 1973.

84. D. R. Collins, W. H. Bailey, W. M. Gosney and D. D. Buss, "Charge-Coupled Device Analog Matched Filters," Electron Lett., 8, p. 328, June 29, 1972.

85. D. R. Collins, W. H. Bailey, D. D. Buss, "Analog Matched Filter Using Charge-Coupled Devices, NEREM 72, Boston, Mass., November, 1972.

86. D. D. Buss, D. R. Collins, W. H. Bailey and C. R. Reeves, "Transversal Filtering Using Charge-Transfer Devices," IEEE Jour. of Solid-State Circuits, SC-8, No. 2, p. 138 April, 1973.

87. D. D. Buss and W. H. Bailey, "Applications of Charge Transfer Devices to Analog Signal Processing," 1974, IEEE Intercon Technical Papers, Session 9, paper 9/1; March 26-29, 1974, New York, N.Y.

88. D. D. Buss, W. H. Bailey and D. R. Collins, "Matched Filtering Using Tapped Bucket-Brigade Delay Lines," Electronic Letters $\underline{8}$ No. 4, Jan., 1972.

89. D. D. Buss, W. H. Bailey and D. R. Collins, "Spread Spectrum Communication Using Charge Transfer Devices," Proc. of 1973 Symposium on Spread Spectrum Communications, San Diego, Calif., March, 1973.

90. L. R. Rabiner, R. W. Schafer, and C. M. Rader, "The Chirp Z-Transform Algorithm," IEEE Trans. on Audio and Electroacoustic $\underline{AU-17}$, pp. 86-92, June 1969.

91. D. D. Buss, W. H. Bailey and A. F. Tasch, "Signal Processing Applications of Charge Coupled Devices," to be presented at the Int'l Conference on the Technology and Applications of Charge-Coupled Devices, Edinburg, September 25-27, 1974.

92. W. H. Bailey, R. W. Brodensen, W. L. Eorsole M. W. Whatley, L. R. Hite and D. D. Buss, "CCD's for Radar Pulse-Doppler Processing using the Chirp Z Transform," presented at the Government Microcircuits Applications Conference, Boulder, Colorado, June 25-27, 1974.

93. R. D. Baertsch, W. E. Engeler and J. J. Tiemann, "A New Surface Charge Analog Store," 1973 IEEE Int'l Electron Device Meeting, Technical Digest, pp. 134-137, Washington, D.C., Dec. 3-5, 1973.

94. J. J. Tiemann, R. D. Baertsch, and W. E. Engeler, "A Surface-Charge Correlator," 1974 IEEE Int'l Solid-State Circuits Conf. Digest of Technical Papers, pp. 154-155, Philadelphia, Pa., Feb. 13-15, 1974.

95. J. J. Tiemann, W. E. Engeler, R. D. Baertsch and D. M. Brown, "Intracell Charge-Transfer Structures for Signal Processing," IEEE Trans. Elec. Dev., $\underline{ED-21}$, pp. 300-308, May 1974.

96. M. H. White, D. R. Lampe and J. L. Fagan, "CCD MNOS Devices for Programmable Analog Signal Processing and Digital Monvilatile Memory," 1973 Int'l Electron Devices Meeting, Tech. Dig., pp. 130-133, Washington, D.C., Dec. 3-5, 1973.

97. D. R. Kampe, M. H. White, J. H. Mims, R. W. Web and G. A. Gilmour, "CCD's for Discrete Analog Signal Processing (DASP)," 1974 IEEE Intercon Technical Papers, Session 9, paper 9/2, New York, N.Y. March 26-29, 1974.

98. W. J. Butler, M. B. Barron and C. M. Puckette, "Practical Considerations for Analog Operation of Bucket-Brigade Circuits" IEEE J. of Solid State Circuits, $\underline{SC-8}$, 2, p. 157, April, 1973.

99. C. H. Sequin, "Experimental Investigation of a Linear 500 Element 3-phase CCD," Bell Telephone Journal, Vol. 53, pp. 581-610, April, 1974.

100. S. R. Shortes, W. W. Chan, W. C. Rhiner, J. B. Barton and D. R. Collins, "Characterization of Thinned Back-side Illuminated CCD Imager," Appl. Phys. Letters, Vol. 4, pp. 565-567, 1 June, 1974.

Free Charge Transfer in Charge-Coupled Devices

JAMES E. CARNES, MEMBER, IEEE, WALTER F. KOSONOCKY, SENIOR MEMBER, IEEE, AND
EDWARD G. RAMBERG, FELLOW, IEEE

Abstract—The free charge-transfer characteristics of charge-coupled devices (CCD's) are analyzed in terms of the charge motion due to *thermal diffusion*, *self-induced drift*, and *fringing field drift*. The charge-coupled structures considered have separations between the gates equal to the thickness of the channel oxide. The effect of each of the above mechanisms on charge transfer is first considered separately, and a new method is presented for the calculation of the self-induced field. Then the results of a computer simulation of the charge-transfer process that simultaneously considers all three charge-motion mechanisms is presented for three-phase CCD's with gate lengths of 4 and 10 μ. The analysis shows that while the majority of the charge is transferred by means of the self-induced drift that follows a hyperbolic time dependence, the last few percent of the charge decays exponentially under the influence of the fringing field drift or thermal diffusion, depending on the design of the structure. The analysis shows that in CCD's made on relatively high resistivity substrates, the transfer by fringing-field drift can be very fast, such that transfer efficiencies of 99.99 percent are expected at 5- to 10-MHz bit rates for 10-μ gate lengths and at up to 100 MHz for 4-μ gate lengths.

NOMENCLATURE

A Constant.

C_{ox} Oxide capacitance per unit area (F/cm²).

D Diffusion constant.

D_{eff} Effective diffusion constant associated with self-induced drift.

$$D_{\text{eff}} = \frac{\mu q}{C_{\text{ox}}} n(y, t).$$

$E_{F(y,t)}$ Fringing field at interface along direction of charge motion (V/cm).

$E_{S(y,t)}$ Self-induced field at interface along direction of charge motion (V/cm).

$h(y)$ Function describing the y dependence of $n(y, t)$ for all t.

$g(t)$ Function describing time dependence of $n(y, t)$ for all y.

$j_n(y, t)$ Electron particle current density.

k Boltzmann's constant.

$n(y, t)$ Concentration of signal carriers in charge-coupled device (CCD) inversion layer (cm⁻²).

n_{ave} Average value of signal carrier concentration, i.e.,

$$n_{\text{ave}} = \frac{1}{L} \int_0^L n(y, t)\, dy.$$

n_0 Uniform concentration of signal carriers (cm⁻²).

$N_{\text{tot}}(t)$ Total number of carriers remaining under electrode at time t, i.e.,

$$N_{\text{tot}}(t) = \int_0^L n(y, t)\, dy.$$

$N_{\text{tot}}(0)$ Total number of carriers under electrode at beginning of transfer process.

q Electronic charge.

t Time from beginning of transfer process.

t_0 Characteristic time associated with decay of charge via self-induced drift.

t_1 Time required for charge to decay to point where self-induced fields are equal to thermal fields.

t_4 Time required to reach 99.99 percent transfer efficiency.

$t_4{}^{\text{th}}$ Time required to reach 99.99 percent transfer efficiency in the absence of fringing fields, i.e., thermal diffusion dominates the transfer process.

T Absolute temperature (K).

V Clock voltage.

V_T Thermal voltage kT/q.

x Direction normal to Si-SiO₂ interface.

X_d Depletion layer thickness at center of transferring electrode.

X_0 Gate oxide thickness.

y Direction of charge motion.

ϵ_1 Dielectric permittivity of insulator.

ϵ_2 Dielectric permittivity of semiconductor.

ϵ ϵ_1/ϵ_2.

$\phi_s(y, t)$ Surface potential (V).

ϕ_{s0} Surface potential in the absence of signal carriers, i.e., when $n(y, t) = 0$.

λ Line charge density (C/cm).

η Charge transfer efficiency.

μ Field effect mobility.

$\sigma(y)$ Surface charge density.

τ_1 Time constant of exponential decay of final remaining charge, i.e., final decay constant.

τ_{th} Exponential decay constant for thermal diffusion.

τ_{tr} Single carrier transit time. The time required for a single carrier to drift the length of the electrode under influence of the fringing field.

Manuscript received September 13, 1971; revised January 24, 1972.

The authors are with the David Sarnoff Research Center, RCA Laboratories, Princeton, N. J. 08540.

Reprinted from *IEEE Trans. Electron Devices*, vol. ED-19, pp. 798–808, June 1972.

I. Introduction

BECAUSE charge-coupled devices (CCD's) [1], [2] are analog shift registers with no mechanism for gain, the attainable charge-transfer efficiency η (the percent of charge transferred from one gate to the next) is extremely important. In order to achieve practical devices of several hundred stages, extremely low loss per gate is required; e.g., a 100-stage three-phase device with an overall efficiency of 90 percent must have a single gate transfer efficiency of 99.97 percent.

There are two basic mechanisms that degrade transfer efficiency. The first is the incomplete transfer of free charge because insufficient time is allowed for transfer. The second is the trapping of charge in fast interface states. This paper is concerned only with the analysis of incomplete transfer of free charge.

In this study, the time required for free charge to transit from one potential well to the next has been determined by numerical solution of the continuity equation in which all the factors that cause charge motion have been simultaneously considered. This work was undertaken to determine the fundamental limitations on charge-transfer efficiency imposed by the free charge-transfer process and to determine the optimum device configuration for maximum transfer efficiency.

The three mechanisms that cause charge motion (thermal diffusion, self-induced drift, and fringing field drift) are discussed in Section II, including a new method for calculating the self-induced electric fields. The results of the numerical solutions of charge transfer along with the associated charge and field profiles are presented in Section III. The two methods for calculating the self-induced fields are compared in Section IV, and a discussion of the results and conclusions are presented in Section V.

II. Transfer Mechanisms

In the analysis of charge motion in a CCD it is convenient to identify two sources of electric field at the interface along the directions of charge propagation: 1) the fringing field $E_F(y, t)$ due to externally applied potentials on the gate electrodes and 2) the field change due to the presence of the signal charges themselves, the self-induced field $E_s(y, t)$. Thus, if it is assumed that the field effect mobility μ is independent of electric field, the electron particle current density $j_n(y, t)$ may be written as the sum of three terms:

$$j_n(y, t) = -n(y, t)\mu E_F(y, t) \quad \text{fringing field drift}$$
$$-n(y, t)\mu E_S(y, t) \quad \text{self-induced drift}$$
$$-n\frac{\partial n(y, t)}{\partial y} \quad \text{thermal diffusion} \quad (1)$$

where $n(y, t)$ is the carrier concentration of electrons (cm^{-2}) in the inversion layer, y is the direction of charge

propagation along the interface, and D is the diffusion constant.

The continuity equation

$$\frac{\partial n(y, t)}{\partial t} = -\frac{\partial j_n(y, t)}{\partial y} \quad (2)$$

describes the decay of charge, but an analytical solution including all of the terms in (1) is somewhat difficult. Consequently, in this section the three mechanisms will be considered separately in order to gain insight into their basic features. Numerical solutions considering all terms of (1) simultaneously will be presented in Section III.

Thermal Diffusion

Kim [3] has treated the transfer of free charge due to thermal diffusion. As derived in Appendix I, a Fourier analysis of the thermal diffusion process shows that for an initially uniform carrier concentration n_0, the carrier profile $n(y, t)$ remaining under the transferring electrode approaches the following expression asymptotically in time:

$$n(y, t) = \frac{4n_0}{\pi} \cos \frac{\pi y}{2L} \exp\left(-\frac{\pi^2 Dt}{4L^2}\right) \quad (3)$$

and the total number of carriers remaining at time t, $N_{\text{tot}}(t)$ by

$$N_{\text{tot}}(t) = \frac{8}{\pi^2} N_{\text{tot}}(0) \exp\left(-\frac{\pi^2 Dt}{4L^2}\right). \quad (4)$$

Aside from a small fraction of charge that decays very quickly, the decay of the total charge due to thermal diffusion is exponential with decay constant τ_{th} of $L^2/2.5D$.

Self-Induced Drift—Gradient Method

Self-induced drift occurs because of the electric fields induced by the signal charge itself. The magnitude of the self-induced fields directed along the Si-SiO$_2$ interface E_s was the first described by Engeler *et al.* [4] by taking the gradient along the interface of the surface potential ϕ_s. In this approximation ϕ_s is calculated for the one-dimensional MOS capacitor and is assumed to be proportional to the signal carrier concentration n_s. Thus, if the oxide capacitance C_{ox} is much greater than the depletion layer capacitance, one can write

$$\phi_s(y, t) = \phi_{s0} - \frac{q}{C_{\text{ox}}} n(y, t); \quad (5)$$

then

$$E_s(y, t) = \frac{q}{C_{\text{ox}}} \frac{\partial n}{\partial y}(y, t). \quad (6)$$

The particle current is then given by

$$j_n(y, t) = - \frac{\mu q n(y, t)}{C_{ox}} \frac{\partial n(y, t)}{\partial y} . \qquad (7)$$

This is equivalent to a diffusion current with a concentration-dependent diffusion coefficient. The continuity equation is given by

$$\frac{\partial n(y, t)}{\partial t} = \frac{\mu q}{C_{ox}} \frac{\partial}{\partial y} \left(n(y, t) \frac{\partial n(y, t)}{\partial y} \right) = \frac{\mu q}{2 C_{ox}} \frac{\partial^2}{\partial y^2} n^2(y, t). \qquad (8)$$

If the carrier concentration $n(y, t)$ is separable into a product function $h(y)g(t)$, then (as shown in Appendix II) the total charge remaining will decay according to the following asymptotic expression:

$$\frac{N_{tot}(t)}{N_{tot}(0)} = \frac{t_0}{t + t_0} \qquad (9)$$

where

$$t_0 = \frac{L^2 C_{ox}}{1.57 \mu q n_0} . \qquad (10)$$

Note that t_0 depends inversely upon the initial carrier concentration n_0.

When the charge level drops to the point where the average carrier concentration n_{ave} is such that

$$\frac{\mu q}{C_{ox}} n_{ave} = D \qquad (11)$$

the self-induced fields will be reduced to the level of thermal fields and further decay will be due to thermal diffusion. The time t_1 required to reach the condition (11) can be expressed as

$$t_1 = t_0 \left(\frac{n_0 \mu q}{C_{ox} D} - 1 \right) = 1.6 \tau_{th} \left(1 - \frac{kT}{q} \frac{C_{ox}}{q n_0} \right). \qquad (12)$$

Since in most cases $kT C_{ox}/q^2 n_0 \ll 1$, the transition from self-induced drift to thermal diffusion occurs, in the absence of any fringing field drift, after a time approximately equal to the thermal diffusion time constant $L^2/2.5D$.

Self-Induced Drift—Integral Method

Another more exact method of calculating the self-induced fields along the Si-SiO₂ interface will be referred to as the integral method. In this approach, the electric field along the Si-SiO₂ interface due to an infinite line charge, also at the interface, is calculated by the method of images for the case where the Si is an infinite dielectric with no space charge region. See Fig. 1(a) and Appendix III. The contributions of each line charge element that makes up any charge distribution (with spatial variations along the direction of the field and infinite extent perpendicular to it) are then summed by integration and the desired field is determined. Equa-

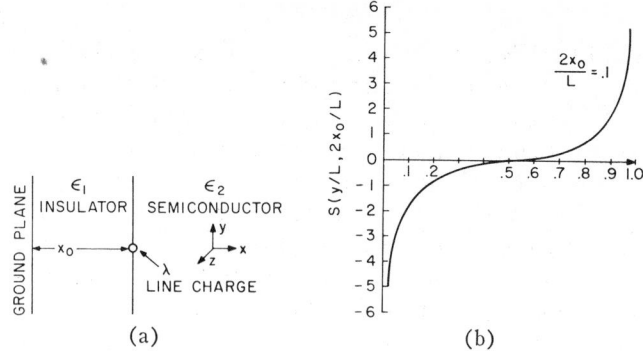

Fig. 1. (a) Cross-sectional view of a line charge at a dielectric interface and near a ground plane. The total y-directed field at the dielectric interface is determined by integrating over all the line charge elements that make up the actual charge distribution. (b) Shape function $S(y/L, 2X_0/L)$ for any uniform charge distribution of length L. The field magnitude for the Si-SiO₂ interface is $(n_0/10^{11}) 1.83 \times 10^3 S(y/L, 2X_0/L)$ V/cm.

tion (13) gives the expansion that for most cases must be evaluated numerically.

$$E_s(y) = \frac{1}{\pi \epsilon_2 (1 + \epsilon)} \left[\int_{y-L_1}^{L_2-y} \frac{\sigma(y) y \, dy}{y^2} - \frac{2\epsilon}{1 + \epsilon} \sum_1^\infty \left(\frac{1 - \epsilon}{1 + \epsilon} \right)^{n-1} \int_{y-L_1}^{L_2-y} \frac{\sigma(y) y \, dy}{(2n X_0)^2 + y^2} \right] \qquad (13)$$

where $\epsilon = \epsilon_1/\epsilon_2$.

The case of a uniform charge concentration of length L can be evaluated analytically as shown in Appendix III and the results provide useful insight into the question of self-induced field magnitudes. Equation (14) gives $E_s(y)$ for Si-SiO₂ for a uniform concentration n_0 of of length L.

$$E_s(y) = 1.8 \times 10^3 \left(\frac{n_0}{10^{11}} \right) S \left(\frac{y}{L}, \frac{2X_0}{L} \right). \qquad (14)$$

$E_s(y)$ is directly proportional to n_0, but the relative shape of $E_s(y)$ is independent of n_0. $S(y/L, 2X_0/L)$ is defined in Appendix III and is plotted in Fig. 1(b). Strictly speaking, for a perfect step function charge profile, $S(0, 2X_0/L) \to \infty$. However, for most of the cases encountered in our charge transfer studies where the charge profile is rounded, a maximum value of S was four or five for a $2X_0/L$ ratio of 0.1. Using $S = 4$ in (14) provides a very simple expression for the maximum self-induced field magnitude for approximately uniform charge profiles given by

$$E_{s\,max} \cong 7.2 \times 10^3 \left(\frac{n_0}{10^{11}} \right) \text{V/cm}. \qquad (15)$$

This expression provides a simple, approximate relationship between self-induced field magnitudes and the charge concentration.

The integral method also allows one to calculate how a charge will arrange itself within a potential well. For example, a charge placed in a square well with no ex-

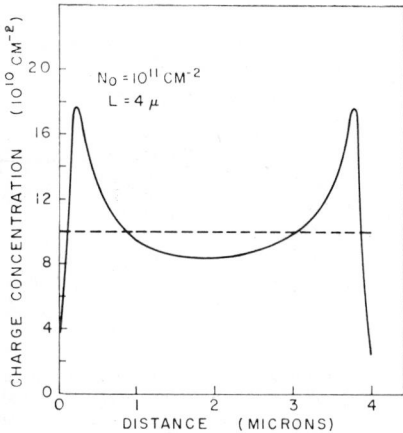

Fig. 2. Cusping of charge stored in a square potential well due to self-induced fields calculated by the integral method.

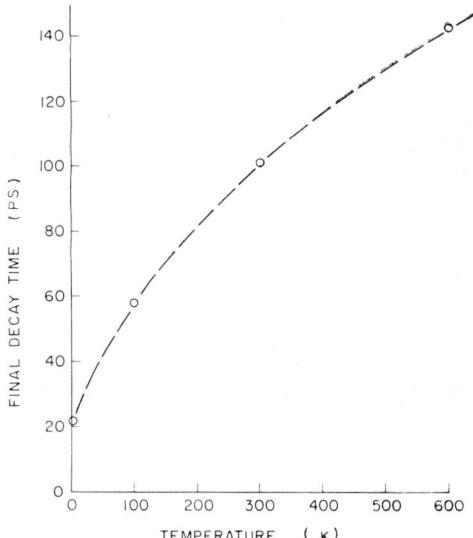

Fig. 3. Final decay time of total remaining charge τ_f versus temperature as determined by computer simulation for 4-μ gates and 10^{15} cm^{-3} doping.

ternally induced electric fields in the interior will tend to crowd at the edges of the well due to the mutually repulsive properties of like charges similar to the arrangement of charge on a conducting disk in free space [5]. The results of a computer solution of this effect using the integral method (13) to calculate $E_s(y)$ is shown in Fig. 2. The well is not exactly square—the walls have a finite slope which accounts for the low charge concentration near the edges. However, the bottom of the well is flat over the majority of the region and the cusping of charge is due to the self-induced fields. The gradient method cannot be used to solve this type of problem.

Fringing Field Drift

The drift-aiding fringing field along the Si-SiO₂ interface arises because the surface potential at any point is determined not only by the electrode directly above but also by adjacent electrodes. An analytical expression for three-phase CCD fringing fields [6] and computer solutions for both two- and three-phase fringing fields have previously been presented [6], [7]. These results indicate that in a certain range of gate lengths and substrate doping levels, the single carrier transit time (the time a single carrier would take to transit the length of the gate under influence of the drift-aiding fringing field) is small enough to indicate that fringing field drift may greatly aid the transfer process.

The numerical solutions to be discussed in Section III indicate that under the influence of fringing field drift and thermal diffusion, the relative charge profile becomes fixed or stationary and decays exponentially. When the spatial dependence of the charge profile becomes independent of time $n(y, t)$ can be expressed as a product solution, and for the exponentially decaying case can be written as

$$n(y, t) = h(y) \exp\left(-t/\tau_f\right) \quad (16)$$

where

$$N_{\text{tot}}(0) = \int_0^L h(y)\, dy. \quad (17)$$

τ_f is the final decay constant of the total remaining charge. Neglecting self-induced effects, the continuity equation becomes

$$V_T \frac{d^2h(y)}{dy^2} + E_F(y)\frac{dh(y)}{dy} + \left(\frac{dE_F(y)}{dy} + \frac{1}{\mu\tau_f}\right) h(y) = 0 \quad (18)$$

where $V_T = kT/q$.

Solution of this equation subject to the boundary conditions

$$\mu h(0) E_F(0) + D \left.\frac{dh(y)}{dy}\right|_0 = 0 \quad (19)$$

and

$$\mu h(L) E_F(L) + D \left.\frac{dh(y)}{dy}\right|_L = \frac{-1}{\tau_f} \int_0^L h(y)\, dy \quad (20)$$

allows one to determine $h(y)$ and τ_f. Only a numerical solution for a general $E_F(y)$ appears possible.

From (18) it is clear that when $V_T = 0$, at the peak of charge where

$$\frac{dh(y)}{dy} = 0$$

$$\tau_f = -\frac{1}{\mu \left.\dfrac{dE}{dy}\right|_{\text{peak}}}. \quad (21)$$

This has been substantiated by numerical solutions.

Numerical solutions also show that the charge will decay exponentially even at zero temperature, but increasing temperature causes τ_f to increase. Thus thermal diffusion acts to hinder fringing field drift. Fig. 3 shows τ_f versus T for one particular fringing field configuration studied numerically.

51

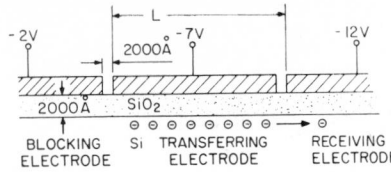

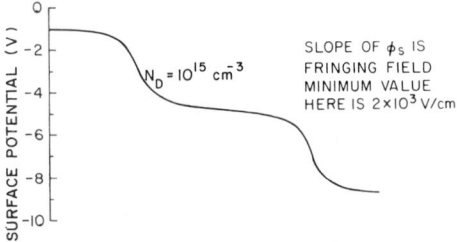

Fig. 4. Cross-sectional schematic of basic configuration studied by computer simulation along with the surface potential for the 10^{15} cm^{-3} doping case.

III. COMPUTER SIMULATION RESULTS

In order to better understand the interaction of the various transfer mechanisms in typical CCD structures, the complete continuity equation has been solved numerically using both the integral method and the gradient method for calculating self-induced fields. Both 4-μ and 10-μ gate lengths were considered as well as doping levels of 2×10^{14} and 10^{15} cm^{-3}. Oxide thickness and electrode separations were 2000Å in all cases [7]. (See Fig. 4.) Three-phase devices were assumed and the fringing field profile was determined numerically for the case where the transferring electrode potential was midway between the blocking and receiving electrode potentials. The voltages used were 2, 7, and 12 V in all cases. In actual practice, the transferring electrode potential will be changing from 12 to 2 V during the transfer process, but incorporating this into the solution unnecessarily complicates the problem. If anything, the electrode voltage configuration used probably underestimates the speed of transfer due to fringing field drift.

In the 4-μ gate studies, an initial uniform charge distribution of 4.6×10^{11} cm^{-2} was used. No cusping occurred because fringing fields balanced the self-induced fields so that a uniform concentration was maintained. In all cases the mobility was assumed to be constant and equal to 250 cm^2/V·s. All transfer times will scale inversely with mobility.

The configuration studied in the most detail was the 4-μ gate, 10^{15} doping case. The overall charge decay versus time results using the integral method is shown in Fig. 5. For times less than 500 ps the charge decay is dominated by self-induced drift. The dotted line shows how self-induced drift proceeds in the absence of any fringing field. This decay would eventually become exponential (after one thermal diffusion time of 10 ns) with decay constant of 10 ns. However, the fringing field drift causes the transfer to progress at a faster rate.

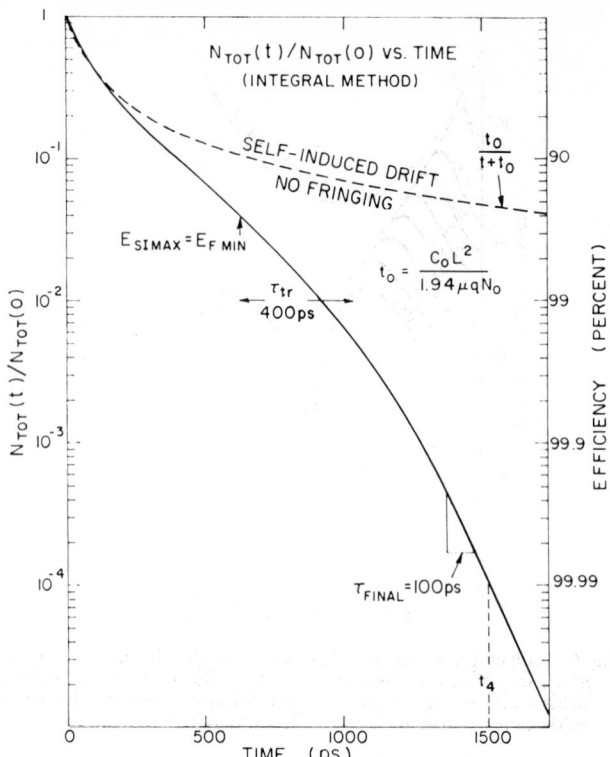

Fig. 5. Normalized total remaining charge versus time for 4-μ gate length and 10^{15} cm^{-3} doping. Dotted line indicates how the charge transfer would proceed in the absence of fringing fields.

The single carrier transit time τ_{tr} given by

$$\tau_{tr} = \frac{1}{\mu} \int_0^L \frac{dy}{E_F(y)} \quad (22)$$

is 400 ps for this configuration. Clearly, the charge does not transit the sample in 400 ps. One of the main reasons for this is the retarding effect of the self-induced fields on the left half of the transferring region. The self-induced field profiles for various times are shown in Fig. 6. Also shown is the drift-aiding fringing field. The self-induced fields remain as high as the minimum fringing field for up to 600–800 ps for this case and this tends to hold back the charge. Consequently, the charge remains well spread out over the entire gate region for times much longer than τ_{tr}.

The charge profiles are shown in Fig. 7. After approximately 800 ps the fringing fields dominate and self-induced effects are negligible. The charge profile then tends to drift to the right under the influence of the drift-aiding fringing field and one might expect it to continue to drift entirely out of the region into the receiving potential well. However, this does not happen; rather, the charge profile becomes stationary at approximately 1400 ps and decays exponentially thereafter.

This occurs basically because the fringing field is not uniform but has large negative gradients on the right side of the region. The final decay time for this case is 100 ps. As noted in Section II, the τ_f should be approximately

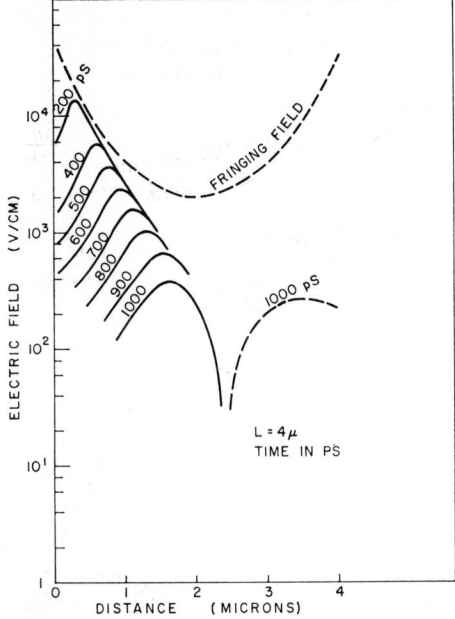

Fig. 6. Self-induced field profiles (calculated by the integral method) at various times for the same problem shown in Fig. 4. Solid line fields cause charge motion to left (hinder transfer), dotted line fields cause charge motion to the right (aid transfer).

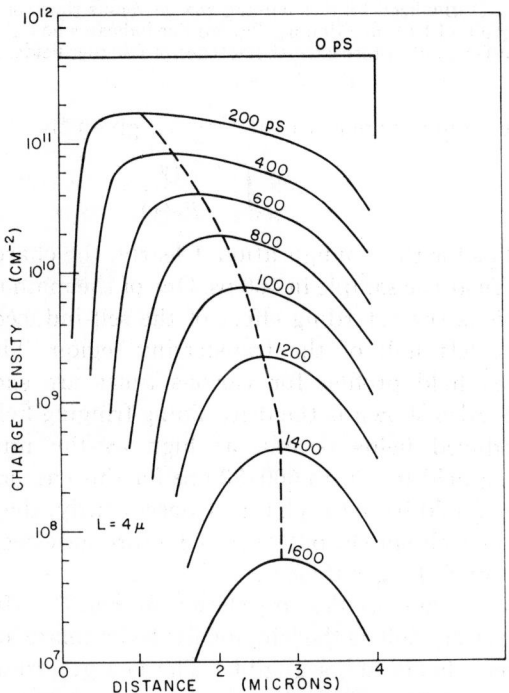

Fig. 7. Charge profiles for various times during the same problem shown in Fig. 4. Dotted line traces the charge peaks.

$$\tau_f = -\left[\mu \frac{dE_F(y)}{dy}\Big|_{\text{peak}}\right]^{-1}, \quad \text{for } V_T = 0. \quad (23)$$

In this case this quantity is 115 ps which compares favorably with the actual τ_f of 100 ps.

As expected, decreasing doping density with the resultant increase in fringing field magnitude results in faster transfer. Fig. 8 compares the charge decay curves

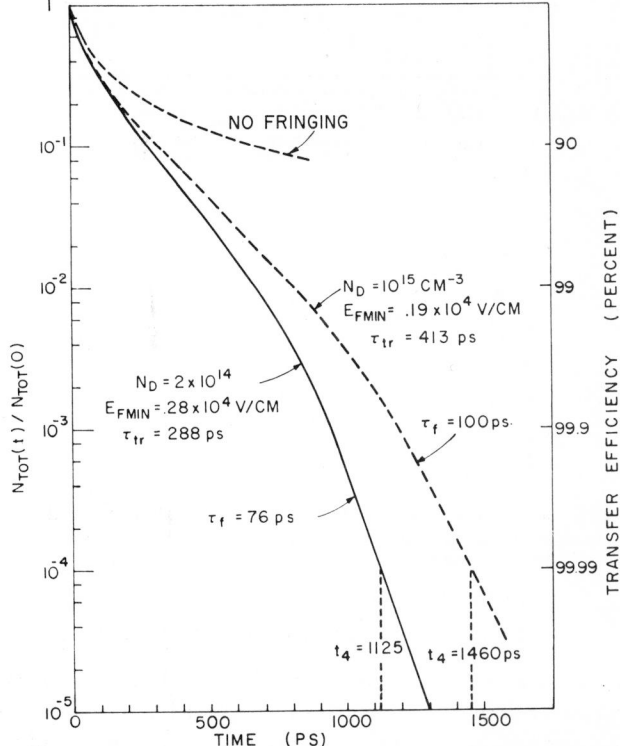

Fig. 8. Normalized total charge remaining versus time for the 10^{15} and 2×10^{14} cm^{-3} doping cases, $L=4$ μ. Lower doping results in higher minimum fringing field and faster charge transfer.

for 10^{15} and 2×10^{14} cm^{-3} doping. The higher fringing fields result in more complete transfer at all times and reduces the time t_4 required to achieve $\eta = 99.99$ percent from 1.46 ns to 1.125 ns. *In the absence of fringing fields, approximately 53 ns would be required to achieve $\eta = 99.99$ percent by means of self-induced drift and thermal diffusion alone.*

Fig. 9 shows charge decay results for the 10-μ gate case, both 2×10^{14} and 10^{15} doping. Here the initial charge level was small, approximately 10^{10} cm^{-2}, so that no self-induced effects are apparent. Here again the final decay is exponential and transfer is materially aided by the higher fringing fields of the lower doped substrate.

The results for the four cases studied ($n_0 = 10^{10}$ cm^{-2}) are summarized in Fig. 10, which shows the total charge remaining versus time normalized by the single-carrier transit time. All four cases with widely varying transit times generally follow the same curve with

$$\tau_f \cong 1/3\ \tau_{\text{tr}} \quad (24)$$

and the time t_4 to achieve $\eta = 99.99$ percent of

$$t_4 \cong 4\ \tau_{\text{tr}}. \quad (25)$$

These relationships apply only at room temperature where fringing field drift dominates the final transfer of charge. When τ_{tr} is greater than τ_{th}, thermal diffusion dominates and τ_f will be τ_{th}. The time required to reach $\eta = 99.99$ percent when thermal diffusion dominates t_4^{th} is given by

$$l_4{}^{\text{th}} = \tau_{\text{th}}\left[6.7 - \log \frac{qn_0}{C_{\text{ox}}}\right] \cong 5.5\,\tau_{\text{th}}. \qquad (26)$$

IV. GRADIENT METHOD VERSUS INTEGRAL METHOD FOR SELF-INDUCED FIELD CALCULATION

Since an alternative method of calculating the self-induced fields, the integral method (Section II) has been introduced, the natural question of comparison with the gradient method arises. The most meaningful comparison involves the electric fields and overall charge-transfer results for the two different methods. The degree of agreement between the gradient and integral methods for calculating self-induced field magnitudes depends upon the charge profile. Almost no agreement is obtained for the uniform charge concentration; the gradient method predicts zero field except at the edges, while the integral method prediction is given by (14) and Fig. 1. Fig. 11 shows a comparison for a charge profile that was picked at random during a computer simulation of charge transfer in a 4-μ gate case. To the left where the self-induced fields hinder charge transfer, the integral method predicts higher fields, while on the right where self-induced fields aid charge transfer, the integral method fields are lower. Thus the integral method should predict slower charge transfer than the gradient method and in fact, comparison of the charge transfer rates using the two methods tends to bear this out. Fig. 12 shows the overall charge decay versus time for the 4-μ gate, 10^{15} cm^{-3} doping case using the two methods. As expected, the integral method is slower with approximately twice the total charge remaining at any given time. However, because self-induced drift is important for only the first decade or so of charge transfer in what is essentially an exponentially decaying process, there are no significant time differences in the overall charge transfer characteristics for the two methods. In view of the fact that the integral method takes much more computer time, the gradient method is probably better suited for certain numerical solutions.

On the other hand, the integral method is required to accurately calculate charge profiles in storing potential wells and express the self-induced field magnitude as a function of the charge concentration rather than charge gradient.

V. DISCUSSION AND CONCLUSIONS

Equations (24) and (25) provide a rough estimate of the free charge-transfer performance at room temperature for three-phase charge-coupled structures which have strong drift-aiding fringing fields—at least for the closely spaced electrode structures studied. Larger gaps between electrodes will change the fringing field profile and may lead to relationships which differ from (24) and (25). Based upon (24) and (25) the expected performance of closely spaced, three-phase CCD's for various gate lengths and doping densities can be assessed based on knowledge of the fringing field profile alone, since

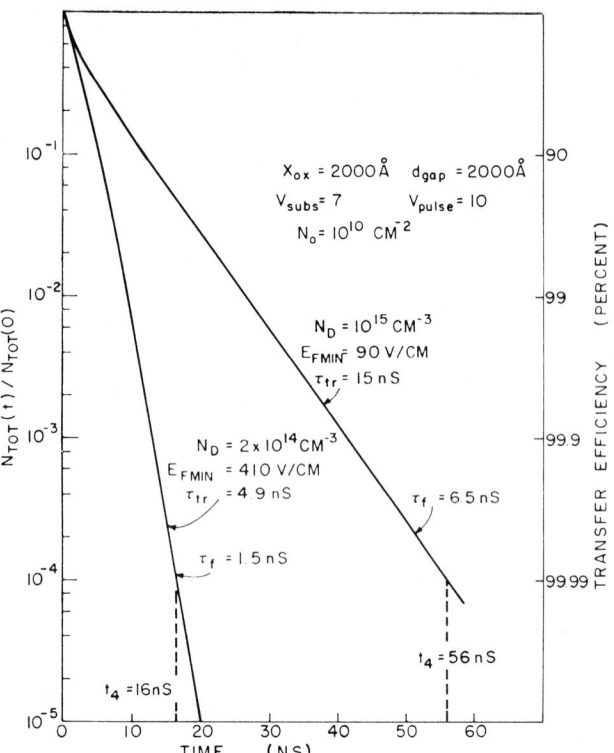

Fig. 9. Normalized total charge remaining versus time for 10-μ gate length, both 10^{15} and 2×10^{14} cm^{-3} doping. Initial charge concentration was low so that no self-induced drift effects are observed.

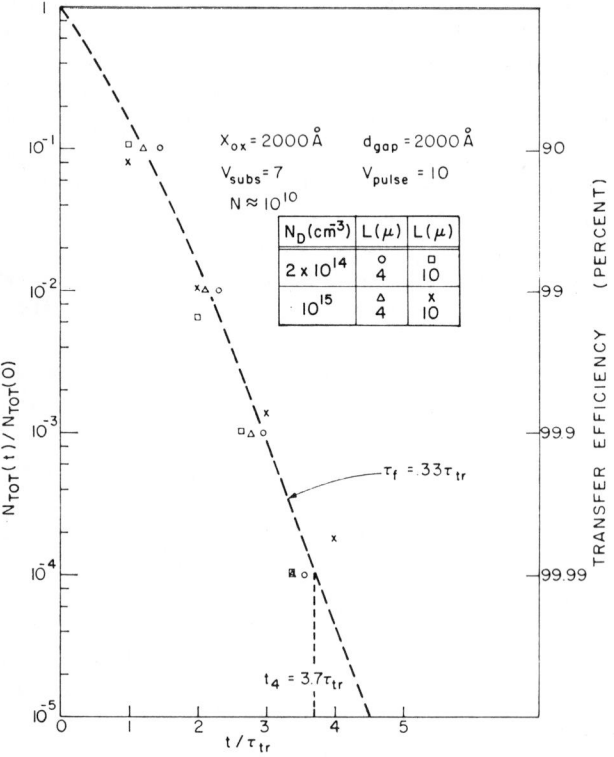

Fig. 10. Normalized total charge remaining versus time normalized by the single carrier transit time τ_{tr}.

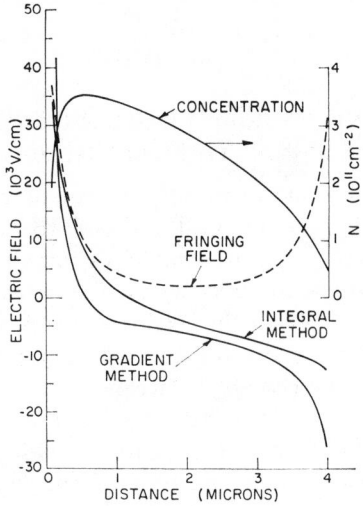

Fig. 11. Comparison of the self-induced fields calculated by the integral and gradient methods for charge profile picked at random during computer simulation of charge transfer in 4-μ gate case. Charge concentration profile is shown at top. Negative of fringing field is also shown for comparison purposes.

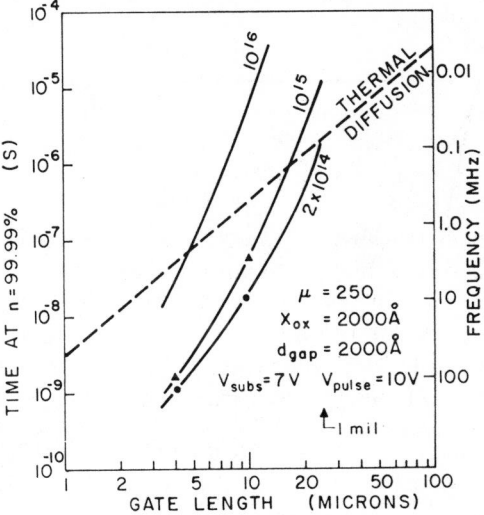

Fig. 13. Time required to achieve $\eta = 99.99$ percent versus gate length for various doping levels. Thermal diffusion line is the maximum time required in any case.

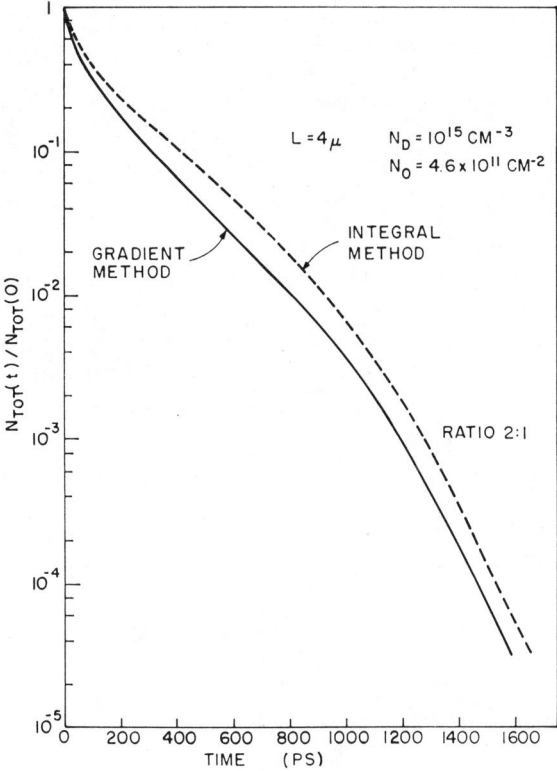

Fig. 12. Comparison of the normalized total charge remaining versus time for self-induced fields calculated by integral and gradient methods.

this determines τ_{tr}. Reference [6] describes an analytical solution for three-phase fringing field magnitudes and presents a relatively simple expression for the single carrier transit time (again probably valid only for closely spaced CCD structures) given by

$$\tau_{tr} = \frac{L^3}{6.5\mu X_{ox}V} \left[\frac{5X_d/L + 1}{5X_d/L}\right]^4 \qquad (27)$$

where V is the pulse voltage, X_d is the depletion width

at the center of the transferring electrode, and the other symbols have their usual meaning. The time to achieve $\eta = 99.99$ percent ($t_4 \cong 4\tau_{th}$) is a reasonable figure of merit since this is consistent with devices with several hundred gates. Based upon (25) and (27), Fig. 13 shows t_4 versus gate length for various doping levels. The points show the values predicted by computer simulation. Also shown by the dotted line is the t_4 predicted by thermal diffusion, (26), assuming $n_0 = 5 \times 10^{11}$ cm^{-2}. When the t_4 based upon fringing field considerations exceeds the thermal diffusion value, the latter time applies. In other words, the thermal diffusion t_4 is the longest time required for $\eta = 99.99$ percent for any configuration. Fig. 13 clearly shows the advantage to be gained in speed by using low doped substrates to maximize the fringing field—especially below 15-μ gate lengths. Noteworthy is the 10-MHz operation predicted for 10-μ gates and 100-MHz operation predicted for 4-μ gates. It must be kept in mind, however, that these predictions are based only upon the consideration of free charge transfer effects. No interface state trapping has been considered and this effect could possibly limit the transfer efficiency and frequency of operation to lower values than those predicted by Fig. 13.

However, it is clear from these results that fringing field drift can be the most important transfer mechanism in appropriately designed CCD structures. Ironically, fringing field drift is also the most difficult mechanism to analyze, requiring numerical solutions to determine the charge decay characteristics. The relationships of (24) and (25) were obtained for closely spaced electrodes and other configurations may give different results.

It is also clear from this study that in CCD structures that have strong drift-aiding fringing fields the transfer of free charge is sufficiently fast so that it should not be a factor in device design or operation. More specifically, incomplete transfer schemes such as "fat zero" operation

[8] do not appear necessary from a free charge transfer standpoint. However, such schemes may be useful in reducing surface state losses.

In conclusion, the three mechanisms causing free charge transfer in CCD's have been discussed. A new method for calculating self-induced fields as a function of charge concentration has been presented. A numerical solution of the charge transfer process shows that self-induced drift dominates during short times, but the majority of transfer is controlled by fringing field drift which is fast enough for 10-MHz operation with 10-μ gates and 100-MHz operation with 4-μ gates, provided interface state losses are negligible.

APPENDIX I

CHARGE TRANSFER BY THERMAL DIFFUSION

The charge transfer due to thermal diffusion alone can be determined by means of Fourier analysis of the charge profile [9]. The diffusion equation that describes the decay of charge is obtained by substituting (1) of the text, using only the thermal diffusion term, into (2) to obtain

$$\frac{\partial n(y, t)}{\partial t} = D \frac{\partial^2 n(y, t)}{\partial y^2}. \qquad \text{(I-1)}$$

For the case of zero carrier gradient at $y = 0$ and zero carrier concentration at $y = L$, a Fourier expansion yields the following general solution where the coefficients are determined by the initial carrier profile $n(y, 0)$:

$$n(y, t) = \sum_{k=0}^{\infty} \alpha_k \cos (\lambda_k y) \exp (-\lambda_k^2 D t) \qquad \text{(I-2)}$$

where

$$\lambda_k = \frac{\pi}{2L} (2k + 1)$$

and

$$\alpha_k = \frac{2}{L} \int_0^L n (y, 0) \cos \left(\frac{\pi(2k + 1)}{2L} y \right) dy.$$

If the initial concentration is uniform, i.e., $n(y, 0) = n_0$ for $0 \leq y \leq L$, then

$$\alpha_k = \frac{4n_0(-1)^k}{(2k + 1)\pi}, \qquad 0 < k < \infty \qquad \text{(I-3)}$$

and

$$n(y, t) = \sum_{0}^{\infty} \frac{4n_0(-1)^k}{(2k + 1)\pi} \cos \left(\frac{(2k + 1)\pi y}{2L} \right)$$
$$\cdot \exp \left(- \frac{(2k + 1)^2 \pi^2 D t}{4L^2} \right). \qquad \text{(I-4)}$$

Each of the terms in the sum decay exponentially with a decay time τ_k given by

$$\tau_k = \frac{4L^2}{(2k + 1)^2 \pi^2 D}. \qquad \text{(I-5)}$$

The higher order terms decay faster and the charge profile quickly decays to a cosine shape after which the charge profile is given by

$$n(y, t) = \frac{4n_0}{\pi} \cos \frac{\pi y}{2L} \exp \left(- \frac{\pi^2 D}{4L^2} t \right) \qquad \text{(I-6)}$$

with the total charge remaining after time t given by

$$N_{\text{tot}}(t) = \int_0^L n(y, t) \, dy = \frac{8}{\pi^2} N_{\text{tot}}(0) \exp \left(- \frac{\pi^2 D}{4L^2} t \right). \qquad \text{(I-7)}$$

Aside from a small fraction of charge which decays very quickly, the decay of the total charge due to thermal diffusion can be considered to be exponential with decay constant $L^2/2.5D$.

APPENDIX II

APPROXIMATE ANALYSIS OF SELF-INDUCED DRIFT

Equation (8) of the text can be solved numerically but the results do not in themselves provide useful insight into the self-induced drift process. The numerical results do point out, however, that after a very short time the decaying carrier profile maintains the same relative shape with all points having the same time dependence. This implies that the carrier concentration at the surface is separable into a product solution

$$n(y, t) = h(y)g(t). \qquad \text{(II-1)}$$

This removes the partial differentials from (8) and permits the separation of variables with each side of the equation necessarily equal to a constant according to

$$\frac{1}{g^2(t)} \frac{dg(t)}{dt} = -A = \frac{\mu q}{2C_{\text{ox}}} \frac{1}{h(y)} \frac{d^2 h^2(y)}{dy^2} \qquad \text{(II-2)}$$

where A is a constant depending upon the initial conditions. The time dependence is then given by

$$g(t) = \frac{g(0)}{1 + A g(0)t} \qquad \text{(II-3)}$$

and the decay of charge is given by the ratio

$$\frac{g(t)}{g(0)} = \frac{t_0}{t + t_0}, \qquad t_0 = \frac{1}{A g(0)}. \qquad \text{(II-4)}$$

In order to obtain an estimate for the constant t_0, we note that the continuity equation can be written formally as a diffusion equation

$$j_n(y, t) = -\mu n(y, t) \frac{q}{C_{\text{ox}}} \frac{\partial n(y, t)}{\partial y} = -D_{\text{eff}} \frac{\partial n(y, t)}{\partial y} \qquad \text{(II-5)}$$

with

$$D_{\text{eff}} = \frac{\mu q}{C_{\text{ox}}} n(y, t). \qquad \text{(II-6)}$$

Treating D_{eff} as a constant, we then write down a solution corresponding to (I-4), limiting ourselves, however, to the leading term of the Fourier series so as to satisfy the condition (II-1):

$$n(y, t) = h(y)g(t) = n_0 \cos \frac{\pi y}{2L} \exp\left(-\frac{\pi^2 D_{\text{eff}} t}{4L^2}\right). \quad \text{(II-7)}$$

We then evaluate the average value of $A g(0)$ over the electrode length L to find $1/t_0$:

$$\frac{1}{t_0} = A g(0) = -\frac{1}{L} \int_0^L \frac{1}{g^2(t)} \frac{dg(t)}{dt} g(0) \, dy$$

$$g(0) = 1$$

and using (II-2) and (II-7)

$$\frac{1}{g^2(t)} \frac{dg(t)}{dt} = -\frac{\pi^2 \mu q}{4L^2 C_{\text{ox}}} n_0 \cos \frac{\pi y}{2L}$$

$$\frac{1}{t_0} = \frac{\pi^2}{4L^3} \frac{\mu q n_0}{C_{\text{ox}}} \int_0^L \cos \frac{\pi y}{2L} \, dy = \frac{\mu q n_0 \pi}{2L^2 C_{\text{ox}}}. \quad \text{(II-8)}$$

Thus

$$t_0 = \frac{L^2 C_{\text{ox}}}{1.57 \mu q n_0}. \quad \text{(II-9)}$$

This is only an approximate solution but provides insight into the decay process and the parameters that affect t_0.

Appendix III

Integral Method for Calculating Self-Induced Fields

Consider an infinite line charge of λ C/cm along the z axis located at the interface between two dielectrics which form the $y-z$ plane [see Fig. 1(a)]. An infinite ground plane parallel to the dielectric is located a distance X_0 from the line charge. It is desired to calculate the electric field at the dielectric interface directed perpendicular to the line charge axis. This will be the self-induced field due to a line charge element which will affect charge motion in the direction of transfer.

The desired field can be calculated using the method of images, but since charge is reflected by both the ground plane and the dielectric interface, an infinite series of image line charges is required to satisfy all of the boundary conditions. The reflection of a line charge λ_n through the ground plane results in an image charge $-\lambda_n$ located an equal distance on the other side of the ground plane. When reflecting a line charge λ_n at $-X_n$ through the dielectric interface the following rules apply.

Rule 1: An image line charge located at $+X_n$

$$\lambda_n' = -\lambda_n \frac{1-\epsilon}{1+\epsilon} \qquad \epsilon = \epsilon_1/\epsilon_2 \quad \text{(III-1)}$$

is used along with λ_n to calculate the potential for $x < 0$.

This image charge must itself be reflected through the ground plane.

Rule 2: An image line charge located at $-X_n$

$$\lambda_n'' = \frac{2}{1+\epsilon} \lambda_n \quad \text{(III-2)}$$

is used alone to calculate the potential for $x > 0$. The original "real" line charge λ located at $X = 0$ is reflected along with its λ' image through the ground plane to form a $\lambda_1 = -\lambda 2\epsilon/(1+\epsilon)$ image at $x = 2X_0$. This charge contributes to the potential in both the ϵ_1 and ϵ_2 regions, and is then reflected by the dielectric interface to form the λ_1' image at $+2X_0$ which is used along with λ_1 to calculate the potential in the ϵ_1 region. The λ_1'' image at $-2X_0$ contributes to the potential in the ϵ_2 region. The λ_1' image is then reflected through the ground plane to form the $-\lambda_1'$ image at $-3X_0$ and the process repeats *ad infinitum* to form two infinite series of image charges.

Using this procedure, the potentials for the ϵ_1 region and ϵ_2 region are given by

$$\phi_1 = \frac{-\epsilon\lambda}{(1+\epsilon)\pi\epsilon_1} \sum_1^\infty \left(\frac{1-\epsilon}{1+\epsilon}\right)^{n-1}$$
$$\cdot \ln \frac{\sqrt{[x - 2(n-1)X_0]^2 + y^2}}{\sqrt{(x + 2nX_0)^2 + y^2}} \quad \text{(III-3)}$$

$$\phi_2 = -\frac{\lambda}{(1+\epsilon)\pi\epsilon_2}\left[\ln \sqrt{x^2+y^2} - \frac{2\epsilon}{1+\epsilon}\right.$$
$$\left.\cdot \sum_1^\infty \left(\frac{1-\epsilon}{1+\epsilon}\right)^{n-1} \ln \sqrt{(x+2nX_0)^2+y^2}\right]. \quad \text{(III-4)}$$

These potentials can be shown to satisfy the following boundary condition requirements.

1) Ground Plane:

$$\phi_1(-X_0, y) = 0.$$

2) Continuous Potential at Dielectric Interface:

$$\phi_1(0, y) = \phi_2(0, y).$$

3) Continuous Normal D at Dielectric Interface:

$$\epsilon_1 \left.\frac{\partial \phi_1}{\partial x}\right|_{x=0} = \epsilon_2 \left.\frac{\partial \phi_2}{\partial x}\right|_{x=0}.$$

4) Gauss' Law at Origin:

$$\lim_{r \to 0} \left\{\pi r \epsilon_1 \left(-\frac{\partial \phi_1}{\partial r}\right) + \pi r \epsilon_2 \left(-\frac{\partial \phi_2}{\partial r}\right)\right\} = \lambda.$$

The electric field in question is then given by

$$E_y(0, y) = -\frac{\partial \phi_2}{\partial y} = \frac{\lambda}{\pi\epsilon_2(1+\epsilon)}\left[\frac{1}{y} - \frac{2\epsilon}{1+\epsilon}\right.$$
$$\left.\cdot \sum_1^\infty \left(\frac{1-\epsilon}{1+\epsilon}\right)^{n-1} \frac{y}{(2nX_0)^2 + y^2}\right]. \quad \text{(III-5)}$$

This is just the field due to a line charge element. By integrating the contributions of all the line charge elements that make up a surface-charge distribution extending from L_1 to L_2 (with variations only along the y direction) the self-induced electric field is determined and is given by

$$E_s(y) = \frac{1}{\pi\epsilon_2(1+\epsilon)}\left[\int_{y-L_1}^{L_2-y}\frac{\sigma(y)y\,dy}{y^2} - \frac{2\epsilon}{1+\epsilon}\right.$$
$$\left.\cdot\sum_1^\infty\left(\frac{1-\epsilon}{1+\epsilon}\right)^{n-1}\int_{y-L_1}^{L_2-y}\frac{\sigma(y)y\,dy}{(2nX_0)^2+y^2}\right]. \quad \text{(III-6)}$$

For an arbitrary $\sigma(y)$ this calculation is best done by computer, but an analytical solution is possible for a uniform charge distribution σ_0 which extends from $y=0$ to $y=L$. If $\beta\equiv y/L$, then

$$E_s(y) = \frac{\sigma_0}{2\pi(\epsilon_1+\epsilon_2)}\sum_1^\infty\left[\frac{1-\epsilon}{1+\epsilon}\right]^{n-1}$$
$$\cdot\ln\left\{\frac{1+\left[\frac{2(n-1)X_0}{(1-\beta)L}\right]^2}{1+\left[\frac{2(n-1)X_0}{\beta L}\right]^2}\frac{1+\left[\frac{2nX_0}{\beta L}\right]^2}{1+\left[\frac{2nX_0}{(1-\beta)L}\right]^2}\right\}. \quad \text{(III-7)}$$

By defining the sum, which does not depend upon σ_0, as the shape function $S(y/L, 2X_0/L)$, we can write

$$E_s(y) = \frac{\sigma_0}{2\pi(\epsilon_1+\epsilon_2)}S\left(\frac{y}{L}, \frac{2X_0}{L}\right)$$
$$= 1.8\times10^3\left(\frac{n_0}{10^{11}}\right)S\left(\frac{y}{L}, \frac{2X_0}{L}\right)\text{V/cm},$$
$$\text{for Si-SiO}_2 \quad \text{(III-8)}$$

where $n_0=\sigma_0/q$.

Typical S plots are shown in Fig. 1(b).

REFERENCES

[1] W. S. Boyle and G. E. Smith, "Charge-coupled semiconductor devices," *Bell Syst. Tech. J.*, vol. 49, pp. 587–593, Apr. 1970.
[2] M. F. Tompsett *et al.*, "Charge-coupled 8-bit shift register," *Appl. Phys. Lett.*, vol. 17, pp. 111–115, Aug. 1970.
[3] C.-K. Kim and M. Lenzlinger, "Charge transfer in charge-coupled devices," *J. Appl. Phys.*, vol. 42, pp. 3586–3594, Aug. 1971.
[4] W. E. Engeler, J. J. Tiemann, and R. D. Baertsch, "Surface charge transport in silicon," *Appl. Phys. Lett.*, vol. 17, pp. 469–472, Dec. 1970.
[5] J. H. Jeans, *The Mathematical Theory of Electricity and Magnetism*, 5th ed. New York: Cambridge, 1963, p. 249.
[6] J. E. Carnes, W. F. Kosonocky, and E. G. Ramberg, "Drift-aiding fringing fields in charge-coupled devices," *IEEE J. Solid-State Circuits*, vol. SC-6, pp. 322–326, Oct. 1971.
[7] W. F. Kosonocky and J. E. Carnes, "Charge-coupled digital circuits," *IEEE J. Solid-State Circuits*, vol. SC-6, pp. 314–322, Oct. 1971.
[8] W. E. Engeler, J. J. Tiemann, and R. D. Baertsch, "A memory system based on surface-charge transport," in *Int. Solid-State Circuit Conf. Dig. Tech. Papers*, Feb. 1971, pp. 164–165.
[9] J. Crank, *The Mathematics of Diffusion*. Oxford: Clarendon, 1956.

The Quantitative Effects of Interface States on the Performance of Charge-Coupled Devices

MICHAEL F. TOMPSETT

Abstract—The several effects of interface states in limiting the performance of surface channel charge-coupled devices (CCD's) are described and evaluated. The limitations on transfer efficiency may be minimized by using a background charge in the device at all times. Experimental measurements of transfer inefficiency on three-phase devices and a two-phase device are presented and correlated with the predicted values, although measurements of the density and capture cross sections of interface states after device fabrication are required for accurate quantitative predictions of transfer inefficiencies. It is concluded that trapping effects are a limitation on the transfer efficiencies obtainable in surface channel charge-coupled devices, particularly, for example, at frequencies less than 1 MHz for devices having 10-μm-long transfer electrodes, but are not a direct limitation on the high-frequency performance.

The effect of interface states in adding transfer noise onto the charge packets is also described, and is shown to be small, although in some devices it may reduce the signal-to-noise ratios that might otherwise be possible.

LIST OF SYMBOLS

A Area directly under transfer electrode.

A_b, A_s Area of interface under transfer electrode covered by "static" background and signal charges, respectively.

A_t Area over which interface states are filled at each transfer.

C_o Capacitance per unit area of oxide under transfer electrode.

c_s Carrier density at interface.

E Magnitude of energy of interface state relative to band edge.

E_g Bandgap energy.

$E_{m(t)}$ Energy of interface states that are predominantly emptying at time t, after removal of free charge.

E_s Normal electric field at the interface.

$e(E)$ Fractional rate of emission of carriers from an interface state at energy E.

F^2 Mean square of the fluctuation in the number of charges introduced into the signal packet per transfer.

f Frequency component in the noise spectrum.

f_o Drive frequency.

k Boltzmann's constant.

K_b, K_s Factors by which the transfer time constant τ must be multiplied in the case of background and signal charge, respectively, to give the time following the initiation of charge transfer before there is a net emptying of interface states under the transfer electrodes.

L Length of transfer electrode in the direction of charge transfer.

m Number of phases.

N_c Density of states in either the conduction or valence bands.

N_e Net number of carriers per unit area retained by interface states from a signal charge packet.

Manuscript received May 12, 1972; revised August 17, 1972.

The author is with Bell Telephone Laboratories, Inc., Murray Hill, N. J. 07974.

Reprinted from *IEEE Trans. Electron Devices*, vol. ED-20, pp. 45–55, Jan. 1973.

N_f Total number of free carriers per unit area.

ΔN_f Number of free carriers per unit area in the gap or transfer region.

N_r Number of carriers per unit area retained by reason of variable trap occupation.

$N_s(E)$ Number of interface states at energy E per unit area per unit of energy in the bandgap.

N_{ss} Number of interface states per unit area per unit of energy in the bandgap and considered constant across the gap.

$N(t)$ Mean number of carriers per unit area emitted in time t from a set of traps initially full at time $t=0$.

N_t Number of carriers per unit area retained by reason of variable transfer time.

n Total number of transfers in a device.

$p, \bar{p}$ Probabilities of a trap being filled or not being filled, respectively, during charge transfer.

$\bar{p}_b, \bar{p}_s$ Probabilities of a trap not being filled during transfer of a background and a signal charge packet, respectively.

Q_b, Q_s Magnitudes of background and signal charge packets, respectively.

q Electronic charge.

v Number of FAT ZEROS between signal charge packets.

$S(f)$ Current noise spectrum in the output signal.

t Time.

Δt Time for a carrier to move across the transfer region.

t_w Waiting time between charge packets.

V_b, V_s Approximately the change in surface potential at the interface with and without background and signal charge, respectively.

V_N Total variance per unit area in the emptying of the interface states.

ΔV_N Variance per unit area in the emptying of interface states in an energy band ΔE.

$\bar{v}$ Mean thermal velocity of charge carriers.

v_f Field-aided drift velocity of carriers moving across transfer region.

α Fraction of traps empty at any given time.

β Attenuation of a charge packet in a CCD that has been empty for time t_w.

γ_e Ratio of extra area, over which signal charge spreads, giving rise to edge effect, to area A.

γ_g Ratio of area of the gap region that never sees "static" charge to area A.

ϵ Total transfer inefficiency per transfer.

ϵ_e Transfer inefficiency component arising from edge effect.

ϵ_e' Transfer inefficiency component arising from edge effect modified by charge trapped during transfer.

ϵ_g' Transfer inefficiency component arising from charges trapped during transfer.

ϵ_g Net transfer inefficiency given by the sum of ϵ_e' and ϵ_g'.

ϵ_i Individual components of transfer inefficiency.

ϵ_p Transfer inefficiency component arising from charge trapping under the transfer electrode in the absence of background charge.

ϵ_t Transfer inefficiency component arising from the effect of variable transfer time.

ϵ_v Transfer inefficiency component arising from variable trap occupation.

ρ Probability of carrier trapped at energy level E being emitted in time t.

$\sigma(E)$ Capture cross section of an interface state at energy level E.

τ Time constant for transfer of charge from under one electrode to the next.

τ_e Emission time constant of interface states.

τ_{mb} Emission time constant of midband interface states.

τ_t Lifetime of an empty trap.

I. Introduction

IN THE basic three-phase charge-coupled device (CCD) [1], [2], charge is transferred along the surface of a semiconductor by creating potential wells under closely spaced metal electrodes on an oxide grown over the semiconductor. Most of the charge is contained in the inversion layer at the surface and is free to move according to electrostatic and charge mobility considerations, but some can be involved in trapping processes occurring at the semiconductor–oxide interface. The objectives of this paper are to show the ways in which interface state processes limit the charge transfer efficiencies and the signal-to-noise ratios obtainable with CCD's, and to make quantitative predictions of the effects for different geometries and types of CCD's. Measurements on a 96-element 3-phase CCD [3], a 500-element 3-phase CCD [4], and a 250-element 2-phase CCD [5] will be described and correlated with the theoretical predictions.

In an operating CCD with low dark current and in the absence of charge input over a long period, the interface states become deeply depleted. If a charge packet is fed into one potential well and then transferred along the device, some charge will be trapped in the interface states under each electrode. These traps have a re-emission time constant τ_e that varies exponentially with the energy level of the trap, as described by the Shockley–Read–Hall equations to be discussed later. If the time constant for a given energy range is short compared to the time allowed for the transfer of the charge packet, then the majority of the trapped carriers in that energy range are re-emitted and transferred with the main packet. However, those carriers trapped in states having time constants on the order of the time allowed for transfer, or longer, will in part add to the main packet, while others will form a residual in the following packets. This will appear experimentally as transfer inefficiency, a term that will be defined more precisely later. This transfer inefficiency may be reduced by keeping background charge or FAT ZEROS [6] passing along the device at all times, and the use of such a background charge will be assumed throughout this paper. However, this is not a complete panacea, and the remaining limitations on transfer inefficiency arising from the effects of interface states will be presented herein. In addition, the statistical variations in the numbers of carriers re-emitted from traps gives rise to transfer noise in the signal packets of an amount that will be evaluated.

II. TRAPPING EFFECTS WITH BACKGROUND CHARGE

For the initial purposes of discussion and calculation, we consider that square pulses are applied to the transfer electrodes, and that the free charge transfers instantaneously from one electrode to the next at time intervals of $1/mf_o$, where f_o is the drive frequency and m is the number of phases. The sequence of events at one particular transfer electrode may be considered. After a charge has been transferred to the region under the electrode, it remains there for a time $1/mf_o$, during which period the interface states are filled. At the end of this period, the free charge packet is considered to transfer instantaneously under the next electrode, and for the next $1/mf_o$ time period, any carriers re-emitted from traps move forward to join the main packet. However, in the period following this, at least for a three-phase device, re-emitted carriers may diffuse either forward, or backward into the following element. Because of the asymmetrical potential distribution under the electrodes, more of these carriers will drift backward than forward, and in the analysis to follow, all carriers emitted at this time will be considered to move backward. This approximation becomes more accurate as the electrodes become shorter in a three-phase device, and it is true for two- and four-phase devices. If a charge packet is present in the next period, the traps will be refilled during the $1/mf_o$ period that this charge is present, and so on. These effects are seen either as a time sequence at one electrode or as a spatial sequence along the device, as shown in Fig. 1. Here the net trapping or release of carriers under each electrode is indicated.

A background current may be inserted at the input end of the device so that charge packets are passed along the device in the absence of signal charge, and if these packets are sufficiently large, the interface states directly under the transfer electrodes will be filled every cycle. The same fraction of this trapped charge will then be re-emitted into the succeeding packet each transfer cycle, so that each charge packet will lose as much net charge into the interface states as it gained from the previous packet. Since the interface states can be al-

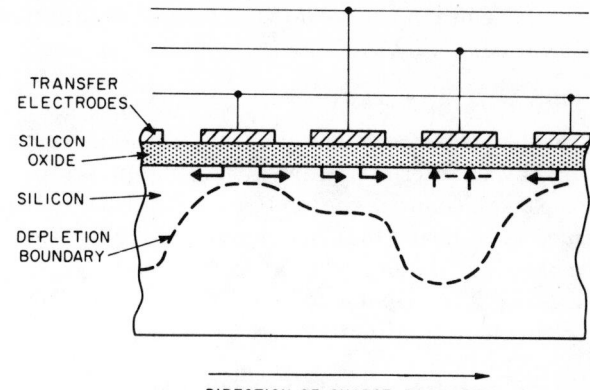

Fig. 1. Net trapping or release of carriers under each electrode as a charge packet transfers along a three-phase CCD.

most completely filled in the presence of the background charge, the same net trapping occurs even when a large-signal charge is present. Thus to a first order, the background charge is expected to neutralize the effect of the interface states. There are, however, several ways in which the transfer inefficiency associated with a signal charge packet passing along the device can arise, even when background charge is used in each element.

A. Edge Effect

One major source of transfer inefficiency arises in the contribution of the interface states under the edges of the transfer electrodes. This occurs because the area of interface over which the charge carriers of a large charge packet are stored is larger than that occupied by the background packet, and hence an increased fraction of the large packet is trapped.

B. Capture During Transfer

Charge capture during transfer, either under a transfer electrode or in the gap region between electrodes, also contributes to transfer inefficiency. However, this charge trapping reduces the edge effect referred to above.

C. Variable Transfer Time

Another source of transfer inefficiency arises when the transfer of free carriers is not instantaneous [7]. Specifically, the total time to transfer these carriers is much greater in the presence of a large charge packet than in the case of a small charge packet. While charge is still present under the electrode from which the charge is being transferred, the traps are being refilled, and a net emptying only occurs after all the free carriers have left. Hence the time available for the emission of trapped carriers is less in the case of the transfer of a large charge packet than in the case of a background packet, and more carriers will remain trapped at the end of the transfer period. These will be re-emitted into the following packets, and will be observed as transfer inefficiency. The effect of a finite rise time of the transfer pulse voltage, and the use of sinusoidal waveforms in the extreme case, will have a similar effect.

D. Variable Trap Occupation

Transfer inefficiency also arises from the higher mean occupancy of interface states during the presence of a large packet than in the presence of the background charge. This means that net charge will be left behind from the single large packet at each transfer.

In this paper we seek to assess the magnitude of all these effects in limiting the performance of CCD's.

III. TRANSFER INEFFICIENCY

In a CCD, a small fraction ϵ of a charge packet may be considered left behind at each transfer, and this is defined as the transfer inefficiency per transfer for a signal of a given size. Any uniform background charge is subtracted out in arriving at this definition. The charge that is left behind then emerges at later times than the charge transferred correctly. The transfer inefficiency ϵ may be written as the sum of transfer inefficiencies ϵ_i arising from different effects, and for a device with n transfers, an overall transfer inefficiency product $n\epsilon = n\sum_i \epsilon_i$ may be defined. The effect of this transfer inefficiency product on broadening and delaying single charge packets passed along the device, or on the frequency response of the device used with an analog input in the general case, has been discussed elsewhere [8]–[11]. All of these effects have been used to determine $n\epsilon$ in the experimental devices.

In the case where no background charge is used and a single charge packet is passed along a device that has not transferred charge for some appreciable period of time, there is an apparent attenuation of this packet as the charge is trapped into deep interface states. The fractional attenuation can be measured directly, as will be described in Section XIV, and is used as a measure of transfer inefficiency in this case.

IV. EMISSION OF CARRIERS FROM FILLED INTERFACE STATES

The fractional rate of emission $e(E)$ of carriers from a given interface state can be described using the Shockley–Read–Hall equations [12], [13] as

$$e(E) = \sigma(E)\bar{v}N_c \exp(-E/kT) \quad (1)$$

where E is the magnitude of the energy of that state relative to the conduction band for electrons or the valence band for holes, $\sigma(E)$ is the capture cross section of the interface state, $\bar{v}$ is the mean thermal velocity of the charge carriers, and N_c is the density of states in the conduction or valence bands, respectively. Hence the mean number of carriers emitted in time t per unit area from a set of traps initially full at time $t=0$ is given by

$$N(t) = \int_0^{E_g/2} N_s(E)$$
$$\cdot \{1 - \exp[-t\sigma(E)\bar{v}N_c \exp(-E/kT)]\}\, dE \quad (2)$$

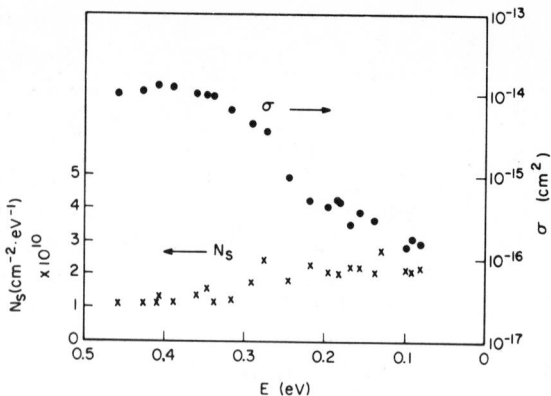

Fig. 2. Interface state density and electron capture cross section of interface states [15] as a function of energy below the conduction band for a double layer SiO₂–Al₂O₃ oxide similar to that used in the experimental CCD's, but grown on an n-type substrate.

where $N_s(E)$ is the interface state density per electron-volt of states having energy E. In the special case where $N_s(E)$ is independent of energy and equal to N_{ss}, and $\sigma(E)$ is a constant across the bandgap, it can be shown [14] that (2) becomes

$$N(t) = kTN_{ss} \ln(\sigma\bar{v}N_c t). \quad (3)$$

This expression is valid for $(\sigma\bar{v}N_c)^{-1} \ll t \ll \tau_{mb}$ where τ_{mb} is the emission time constant of the midband interface states. For electrons in silicon, $(\sigma\bar{v}N_c)^{-1} = 8 \times 10^{-12}$ s and $\tau_{mb} = 10^{-2}$ s. Therefore expression (3) could be expected to be valid in the time range 10^{-10}–10^{-3} s.

Recent work [15] indicates that σ is actually a very strong function of E, and $\sigma(E)$ measured for a double layer oxide, similar to that used in one of the experimental devices to be described, is shown plotted in Fig. 2. Measurements of the interface density $N_s(E)$ have been made [15] for this double layer oxide and are also shown in Fig. 2. A numerical integration of (2) was executed using values of $\sigma(E)$ taken from Fig. 2 and a constant $N_s(E) = 2 \times 10^{10}$ cm⁻²·eV⁻¹. The function $N(t)$ obtained is shown plotted in Fig. 3. Equation (3) is also shown plotted with $\sigma = 4.5 \times 10^{-16}$ cm² for comparison purposes. The effect of the higher capture cross sections near the center of the band is to increase the number of interface states that will be filled and emptied during each charge transfer cycle.

When filled interface states are allowed a time t in which to empty, aside from statistical variations, they empty to an energy level $E_m(t)$. The value of $E_m(t)$ for the two cases plotted in Fig. 3 is indicated on the right-hand axis of this figure, since in the case of $N_s(E)$ constant across the bandgap, E_m follows the same curve as $N(t)$. Curve (b) of Fig. 3 then shows that, for example, with a three-phase CCD operating at $f_o = 1$ MHz, the interface states in the first time interval of 0.33 μs empty to an energy level of $E_m \approx 0.35$ eV. At this level, Fig. 2 shows that $\sigma = 1 \times 10^{-14}$ cm². However, σ increases rapidly with increasing energy in the range 0.25–0.35 eV, and the use of (1) and considerations of the statistics of trap emptying show that almost as many of the states in this

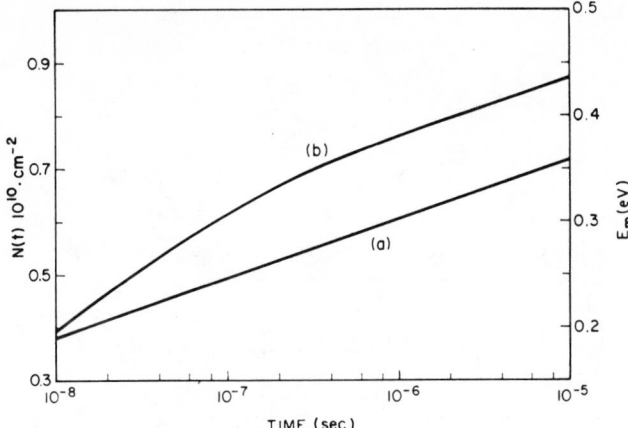

Fig. 3. Plot of $N(t)$ showing the number of interface states that empty in time t for (a) $\sigma = 4.5 \times 10^{-16}$ cm^{-2} and (b) σ varying as shown in Fig. 2. $N_{ss} = 2 \times 10^{10}$ cm$^{-2} \cdot$eV^{-1} in both cases. $E_m(t)$, the mean energy level to which the interface states have emptied in time t, is also represented by the same curves and is indicated by the second ordinate axis.

energy range are full as are states at energy levels of 0.35 eV. These are the states that are actively emptying and contributing to the transfer efficiency limiting processes at $f_o = 1$ MHz. Therefore, a mean $\sigma = 5 \times 10^{-15}$ cm^{-2} for these states will be assumed in later calculations.

V. EFFECT OF INTERFACE STATES IN THE ABSENCE OF BACKGROUND CHARGE

The results of the last section may be used to calculate the effect of interface states over the whole of the interface at which charge is stored under the transfer electrode in the absence of background charge. In the absence of any signal and background charge packets for time t_w, $N(t_w)$ interface states will empty. Upon the arrival of a signal charge packet, the charges extend over an area A_s of the interface and fill $A_s N(t_w)$ interface states. If the charge packet transfers instantaneously along the device to the next transfer electrode at time $t = 0$, then one time interval $1/mf_o$ occurs in which the carriers can move forward with the main charge packet rather than backward into the following location. Then the net number of carriers retained by the interface states from the signal charge packet is given by $A_s N_e$ where

$$N_e = \left[N(t_w) - N\left(\frac{1}{mf_o}\right) \right]. \qquad (4)$$

The best case, when the least number of carriers are retained, will be when t_w has its minimum value. In practice, this occurs when all the charge packets are full, but no useful information is then carried by the device. The limiting case occurs when every other charge packet passing along the device contains a full signal complement of carriers and those in between are empty. Then $t_w = (m+1)/mf_o$, and using (3) and (4), we obtain

$$N_e = kTN_{ss} \ln (m + 1). \qquad (5)$$

Equation (5) is subject to the assumption that $\sigma(E)$ is constant; otherwise N_e must be obtained by numerical integration of (3) as used to obtain curve (b) in Fig. 3. The transfer inefficiency ϵ_p arising from this effect may be written as

$$\epsilon_p = \frac{qA_s N_e}{\text{signal charge}}. \qquad (6)$$

Now the signal charge Q_s can be written as $Q_s = AC_o V_s$, where A is the area directly under the transfer electrode and is approximately equal to A_s, C_o is the oxide capacitance per unit area, and V_s is defined as the voltage difference required to satisfy the above expression for Q_s. V_s is approximately the change in surface potential at the interface with and without the signal charge. Therefore, (6) becomes

$$\epsilon_p = qN_e(C_o V_s)^{-1} = \frac{qkTN_{ss}}{C_o V_s} \ln (m + 1). \qquad (7)$$

It should be noted that in the first approximation, this quantity is independent of frequency.

VI. EDGE EFFECT

The results of the last section describe the effect of interface states directly under the transfer electrodes in the absence of a background charge but, as discussed earlier, the use of a background charge causes these interface states to be filled every cycle and their effect to be neutralized. However, the background charge will accumulate in the regions of highest surface potential and fill the interface states there, but a full charge packet spreads out further under the edges of the electrodes. The interface states over this additional area will therefore trap signal charge and re-emit it when the signal charge packet moves, and thus contribute an edge effect component of transfer inefficiency.

The situation is illustrated in Fig. 4, which shows the surface potentials [16] under the transfer electrodes in a three-phase CCD as a function of position along the device. Suitable voltages are shown applied to the electrodes. The curves are drawn for two geometries, one with electrodes that are 12 μm long with 3-μm gaps, and the other with electrodes that are 3 μm long with 3-μm gaps. For the latter case, a background and a signal charge packet, for which $V_b = 2$ V and $V_s = 4$ V, respectively, are shown "in the potential well." As can be seen, the signal charge packet spreads over an area A_s of the interface, which is larger than the area A_b covered by the background charge. If the area directly under the electrode is designated A and a parameter $\gamma_e = (A_s - A_b)/A$ is defined, then the additional area over which the signal charge spreads may be written $\gamma_e A$. The transfer inefficiency due to the edge effect is then obtained simply by inserting γ_e into (7). Therefore the transfer inefficiency arising from the edge effect may be denoted ϵ_e and is given by

$$\epsilon_e = q\gamma_e N_e(C_o V_s)^{-1}. \qquad (8)$$

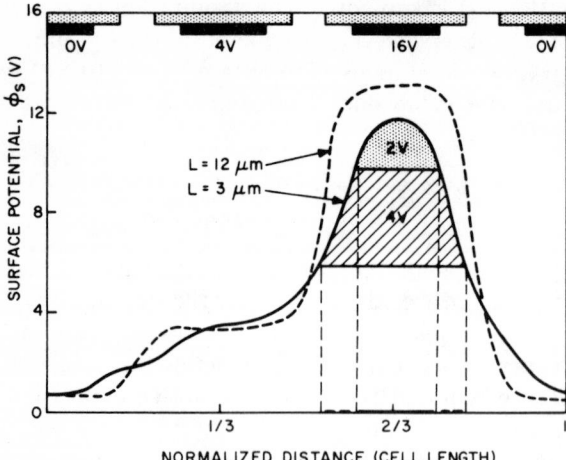

Fig. 4. Surface potential along a CCD computed [16] for transfer electrodes of different length L. The oxide thickness was 3000 Å, fixed charge at the interface 2×10^{11} cm^{-2}, gap width 3 μm, and electrode thickness 2000 Å. The potentials applied to the electrodes were 0, 4, and 16 V.

This basic edge effect will be modified, in ways to be discussed in the next section, by charges trapped during transfer. The values of γ_e can be obtained by examining the calculated surface potentials like those shown in Fig. 4, and by assuming that the sides of the well are not appreciably deformed by the presence of the charge. In order to minimize γ_e, the working range between the size of a background charge packet and that of a full signal charge must be chosen so as to work in the region where the walls of the potential well are steepest.

VII. CHARGE CAPTURE DURING TRANSFER

In order to calculate the quantitative effects of capture of charge carriers during transfer, the probability of an empty trap being filled by carriers moving along the interface must be calculated. In the static case, the lifetime τ_t of an empty trap before capture of a carrier in the presence of a carrier density c_s at the interface given by

$$\tau_t = \frac{1}{\sigma \bar{v} c} . \tag{9}$$

If a total nondegenerate free carrier density N_f in the inversion layer and a normal electric field E_s at the interface are assumed, then energy considerations normal to the interface lead to a value for c_s given by

$$c_s = \frac{q}{kT} E_s N_f \tag{10}$$

so that

$$\tau_t = \frac{kT}{q \bar{v} \sigma E_s N_f} . \tag{11}$$

As a first approximation to the transfer process, a uniform transfer of charge across the gap region may be considered. Each carrier is assumed to move with a velocity v_f across the gap region, and the total transfer

time of the charge packet is Δt. During this time, a mean-free carrier density of ΔN_f exists in the gap region and is given by

$$\Delta N_f = \frac{LC_o V}{q v_f \Delta t} \tag{12}$$

where L is the length of the electrode from under which the charge is being transferred, and $C_o V$ is defined in the same way as that following (6), except that V is now the change in surface potential for any given size of charge packet. Now, combining (11) and (12) with $N_f = \Delta N_f$ gives

$$\frac{\Delta t}{\tau_t} = \frac{\bar{v} \sigma L C_o E_s V}{kT v_f} . \tag{13}$$

The probabilities p and $\bar{p}$ of a trap in the gap region either being filled or not being filled, respectively, during charge transfer are given by

$$\bar{p} = 1 - p = \exp(-\Delta t / \tau_t). \tag{14}$$

The effect of this trapping on transfer inefficiency is twofold. First, the trapping of background charge over the area at the edge of the electrode reduces the edge effect by the fraction of states filled. Hence the modified transfer inefficiency ϵ_e' due to edge effect using (8) is given by

$$\epsilon_e' = q N_e \gamma_e \bar{p}_b / C_o V_s \tag{15}$$

where $\bar{p}_b$ is the probability of an interface state being left empty after the passage of a background charge packet.

Second, in that part of the gap region that never experiences a "static" charge, in the sense that charge temporarily held under a transfer electrode is static, a component of transfer inefficiency ϵ_g' is obtained from the difference in trapping of the signal charge packet and the background charge packet. This difference is proportional to $\bar{p}_b - \bar{p}_s$, where $\bar{p}_s$ is the fraction of states left empty following transfer of a signal charge packet. The trapped charge behaves as already described by (7), so that

$$\epsilon_g' = q N_e \gamma_g (C_o V_s)^{-1} (\bar{p}_b - \bar{p}_s) \tag{16}$$

where γ_g is the ratio of the area of the gap region that never sees "static" charge to the area of the electrode under which charge is stored. Equations (15) and (16) may be combined by adding the individual transfer inefficiencies and a transfer inefficiency ϵ_g for the overall effect of the gap obtained.

$$\epsilon_g = \epsilon_e' + \epsilon_g' = q N_e (C_o V_s)^{-1} [(\gamma_e + \gamma_g) \bar{p}_b - \gamma_g \bar{p}_s]. \tag{17}$$

Numerical evaluation of (13) and (17) will be carried out and discussed in Section XIII.

VIII. VARIABLE TRANSFER TIME

For a CCD operating at a frequency f_o, only the carriers emitted in the first $(1/m f_o)$ time period after transfer move into the main charge packet, while those emitted at later times, apart from some of those emitted in the second period as discussed in Section II, move into following packets. The difference in the charge emitted in the first $(1/mf_o)$ time period in the case of the transfer of a background charge packet and that of a single large-signal packet gives a measure of the extra charge retained in the latter case, and hence the transfer inefficiency. This difference arises because a large charge packet takes longer to transfer, and hence the interface states only start emptying at a later time. If square drive pulses are used and the transfer time constant [7] for transfer of charge from under one electrode to the next is τ, then a net emptying of interface states begins to occur when sufficient free charge has transferred, say in times $K_b\tau$ and $K_s\tau$ for the background charge and the signal charge packets, respectively, where K_b and K_s are factors to be obtained from considerations of charge motion during transfer as influenced by the geometry of the CCD and the waveforms used to drive it. The extra number N_t of carriers retained per unit area per transfer in the case of a signal charge packet is given by

$$N_t = N\left(\frac{1}{mf_o} - K_b\tau\right) - N\left(\frac{1}{mf_o} - K_s\tau\right). \quad (18)$$

Using (3) and (18), and assuming $K_b\tau$ and $K_s\tau \ll (mf_o)^{-1}$,

$$N_t = mkTN_{ss}(K_s - K_b)\tau f_o. \quad (19)$$

Transfer inefficiency ϵ_t arising from the effect of variable transfer time is defined in a similar way to that used in (6), so that

$$\epsilon_t = mqkTN_{ss}(K_s - K_b)\tau f_o(C_oV_s)^{-1}. \quad (20)$$

A similar result may be obtained if the transfer pulses have a rise time comparable to or in excess of the transfer time.

IX. VARIABLE TRAP OCCUPATION

Another possible cause of transfer inefficiency involving interface states arises from the variable trap occupation in the presence of small and large charge packets. In order to estimate the approximate magnitude of this effect, a measure of trap occupation as a function of charge packet size is required. For any given energy level, the fraction α of traps empty at any time will on the average be given by

$$\alpha = \frac{\tau_t}{\tau_e + \tau_t} \quad (21)$$

where τ_t is the lifetime of an empty trap and τ_e is the emission time of a trap at a given energy level. An expression for τ_t already has been presented in (11), so that combining this equation and (21) gives

$$\alpha = \left(1 + \frac{q\bar{v}\sigma E_s N_f \tau_e}{kT}\right)^{-1}. \quad (22)$$

Now, those interface states of importance in charge transfer are those for which $\tau_e \approx 1/mf_o$. Hence the fraction of interface states empty at any given moment in the energy range of interest is

$$a \approx \left(1 + \frac{q\bar{v}\sigma E_s N_f}{kTmf_o}\right)^{-1}. \quad (23)$$

An energy range kT centered about the energy of interface states for which $\tau_e = 1/mf_o$ is considered, and it is assumed that all the states in this range are filled in the presence of the signal charge packet, but the above fraction is empty on average in the presence of the background charge packet. Then in the case of the signal charge packet, if the extra filled states empty in the $1/mf_o$ time period, the number N_r of extra charge carriers per unit area retained is approximately given by

$$N_r \approx kTN_{ss}\left(1 + \frac{q\bar{v}\sigma E_s N_f}{kTmf_o}\right)^{-1} \quad (24)$$

where N_f is the free charge in the case of the background charge packet.

Transfer inefficiency ϵ_v arising from variable trap occupation is then defined in a similar way to that used in (6), and using (24) is given by

$$\epsilon_v = qkTN_{ss}(C_oV_s)^{-1}\left(1 + \frac{q\bar{v}\sigma E_s N_f}{kTmf_o}\right)^{-1}. \quad (25)$$

X. TRANSFER INEFFICIENCY FOR SEPARATED SIGNAL PACKETS

The calculations presented so far have all assumed the limiting case of signal charges in every other element. However, the case of charge packets separated by, say, r FAT ZEROS is an important one, and is readily solved by putting $t_w = (rm+1)/mf_o$ in (4) and using the new value of N_e obtained. The transfer inefficiency ϵ in this case is then given by

$$\epsilon_r = \epsilon\frac{\ln(rm+1)}{\ln(m+1)} \quad (26)$$

where ϵ is the value obtained in the previous calculations.

XI. ATTENUATION IN AN EMPTY CCD

An aspect of interface state trapping that should be briefly discussed is the large loss of charge per transfer that occurs when charge packets are first passed along a device that has been running empty for a time t_w. To avoid confusion, this is best referred to as an attenuation β per transfer. The first packet will be attenuated by the interface states under the transfer electrode that fill entirely and those in the gap regions that only fill partially. Hence β may be calculated from a variation of (17) using a value of N_e given by (4) and

$$\beta = q(C_oV_s)^{-1}[1 + \gamma_g p_s]\left[N(t_w) - N\left(\frac{1}{mf_o}\right)\right] \quad (27)$$

$$\beta = kTN_{ss}(1 + \gamma_g p_s)(C_oV_s)^{-1}\ln(mf_o t_w). \quad (28)$$

XII. INTRODUCTION OF SIGNAL NOISE BY INTERFACE STATES

The constant filling and emptying of interface states as a charge packet moves along a CCD is a source of fluctuations in the number of carriers in the packet. These fluctuations will then appear as noise in the output signal of the device. It is readily appreciated that neither the fastest states, which empty completely during the available transfer time, nor the slowest states, which do not empty at all in this time period, can be responsible for this noise, since the number of carriers emitted from these states is exactly determined. It is those states having re-emission time constants on the order of the transfer time that contribute most of the noise, for which an expression may be derived in the following way.

The fractional rate of emission $e(E)$ of trapped carriers from a given interface state of energy E is given by (1). All traps are considered to be full at time $t=0$ as before. The probability ρ of a carrier trapped at energy level E being emitted in time t is given by

$$\rho = 1 - \exp(-te(E)). \qquad (29)$$

In a small energy band ΔE of interface states at time t, the distribution of full and empty states will be binominal with a variance ΔV_N per unit area given by

$$\Delta V_N = \rho(1-\rho)N_s(E)\Delta E. \qquad (30)$$

The total variance V_N for all interface states is obtained by summing the individual variances over all possible energy levels, so that by integrating (30) and using (1) and (29)

$$
\begin{aligned}
V_N &= \int_0^{E_g/2} N_s(E)\rho(1-\rho)\,dE \\
&= \int_0^{E_g/2} N_s(E)\{[1 - \exp(-te(E))] \\
&\quad \cdot \exp(-te(E))\}\,dE.
\end{aligned} \qquad (31)
$$

Equation (31) may be put into two integrals of the same form as (2) and, assuming that $\sigma(t)$ and $N_s(E)$ are constant as a function of energy, may be integrated in the same way as (2) to give

$$V_N = kTN_{ss}\ln(2). \qquad (32)$$

This expression is independent of time, as expected, and is approximately equal to the shot noise in the carriers emitted from an energy band of interface states of width kT.

Now, as the charge packet moves along the device, an excess of carriers re-emitted into one charge packet leads to a deficit in the trailing packet because of the extra interface states that have to be filled. Hence a charge fluctuation in one packet must be accompanied by a fluctuation of the opposite sign in the trailing packet. Therefore, twice the variance of the number of carriers returned to the signal packet at each transfer equals the mean square of the noise or fluctuation F^2 in the number of charges introduced into the signal packet per transfer. F^2 is given by

$$F^2 = 2A_t V_N \qquad (33)$$

where A_t is the area over which the interface states are filled and is given by

$$A_t = A_s + A\gamma_g p_s \qquad (34)$$

where the fraction of interface states filled during charge transfer have been added onto the area A_s over which the static charge spreads. Hence the signal-to-noise ratio S/N in each charge packet after n transfers is given by

$$S/N = \frac{(Q_s/q)^2}{nF^2}. \qquad (35)$$

Using $Q_s = A_s C_o V_s$ and (32)–(34), (35) becomes

$$S/N \simeq \frac{A_s C_o^2 V_s^2}{2nqkTN_{ss}(1+\gamma_g p_s)\ln(2)}. \qquad (36)$$

There are, of course, other sources of noise present in the output signal, such as input noise [10] and thermal noise in the preamplifier [17], which will reduce the observed S/N ratio from the value given by (36). Equation (38) applies to the noise in each charge packet and is the appropriate expression to use for each charge packet in digital applications of CCD's. However, the noise between neighboring charge packets is correlated, since a charge excess in one packet appears as a deficit in the trailing packet, and in an analog application, the current noise spectrum $S(f)$ in the output signal will depend on the frequency component f and can be shown [18] to be given by

$$S(f) = 2nf_o F^2(1 - \cos 2\pi f/f_o). \qquad (37)$$

Using (32) and (33), (37) becomes

$$S(f) = \frac{4\ln(2)kT}{q}\,nf_o A_t N_{ss}(1 - \cos 2\pi f/f_o). \qquad (38)$$

This expression shows how the noise decreases with decreasing f, as would be expected from the correlation between the fluctuations in neighboring packets. The magnitude of the expected noise will be evaluated in the following.

XIII. NUMERICAL EVALUATION OF THE COMPONENTS OF TRANSFER INEFFICIENCY AND SIGNAL NOISE

Expressions for several components of the transfer inefficiency in surface channel CCD's have been derived, and it is now appropriate to evaluate their relative magnitudes. This will be done for an experimental three-phase CCD with 9.7-μm-long electrodes and a 3-μm gap to give a feel for the magnitudes involved. The results are shown in Table I, and several conclusions become apparent for this particular device. Without background charge, a poor transfer efficiency is obtained. On using background charge, the edge effect is the dominant com-

TABLE I

NUMERICAL COMPONENTS OF TRANSFER INEFFICIENCY FOR A
THREE-PHASE CCD WITH $L = 9.7$ µm AND A 3-µm GAP

Effect	Equation	Symbol	Value
Without background charge	(7)	ϵ_p	2.3×10^{-3}
With background charge:			
Edge	(8)	ϵ_e	4.8×10^{-4}
Modified edge	(15)	ϵ_e'	1.4×10^{-4}
Gap	(16)	ϵ_g'	6.8×10^{-5}
Overall Gap	(17)	ϵ_g	2.1×10^{-4}
Variable transfer time	(20)	ϵ_t	4.8×10^{-6}
Variable trap occupation	(25)	ϵ_v	4.0×10^{-7}

Values of parameters used in calculation

$C_o = 1.24 \times 10^{-8}$ F·cm^{-2} $N_{ss} = 2 \times 10^{10}$ cm^{-2}·eV^{-1}
$\Delta V_s = 4$ V, $\Delta V_b = 2$ V $\sigma = 5 \times 10^{-15}$ cm^2
$\gamma_e = 0.21, \gamma_g = 0.15$ $v_F = 5 \times 10^6$ cm·s^{-1}
$\tau = 1$ ns, $(K_s - K_b) = 1$ $E_s = 2 \times 10^4$ V·cm^{-1}
$f_o = 1$ MHz $\bar{p}_b = 0.29, \bar{p}_s = 0.09$

TABLE II

EVALUATION OF (13) AND (14) TO OBTAIN THE FRACTION OF INTERFACE STATES IN GAP REGION NOT FILLED DURING CHARGE TRANSFER

Device	E_s(V·cm^{-1})	C_o(F·cm^{-2})	v_f(cm·s^{-1})	$\Delta t / \tau \Delta V$	$\bar{p}_s$	$\bar{p}_b$
Three-phase CCD with $L = 9.7$ µm and a 3-µm gap	2×10^4	1.24×10^{-8}	5×10^6	0.62	0.09	0.29
Three-phase CCD with $L = 3$ µm and a 3-µm gap	2×10^4	1.8×10^{-8}	5×10^6	0.27	0.33	0.61
Two-phase CCD with $L = 11$ µm and a transfer region 7 µm long	1×10^4	1.8×10^{-8}	5×10^6	0.45	0.17	0.40

Assumed values: $\sigma = 5 \times 10^{-15}$ cm^2 $\Delta V_s = 4$ V $\Delta V_b = 2$ V

ponent, but this is significantly reduced by background charge trapped during transfer. The effects of variable trap occupation and variable transfer time at 1-MHz frequency of operation are both negligible. By comparing (17) and (20), the frequency at which the overall effect of the gap equals the effect of variable transfer time can be determined and is given by

$$f_o = \frac{[(\gamma_e + \gamma_g)\bar{p}_b - \gamma_g \bar{p}_s] \ln(m+1)}{m(K_s - K_b)\tau}. \quad (39)$$

For the device being considered, the drive frequency required to meet this condition would be given by $f_o = (20\tau)^{-1}$. Under this condition, the dynamics of charge motion sets a limit [7] on the performance, so that the effect of variable transfer time is not a primary limitation on the high-frequency performance of CCD's.

The above calculations indicate that the overall gap effect is the most important effect of interface states to be considered in surface channel CCD's. This effect will now be evaluated and compared for the three experimental devices to be described in the next section. Table II shows the values of some of the parameters assumed and the calculation of $\bar{p}_s$ and $\bar{p}_b$. Table III shows the estimated values of γ_e and γ_g and the calculated values of ϵ_g for the three experimental devices.

Equation (28) for the attenuation β of the first charge packet passed along a CCD operated without charge input for a time t_w may be evaluated for the three-phase CCD with $L = 9.7$ µm operated at $f_o = 1$ MHz. Then with $t_w = 2.5$ ms, for example, $\beta \approx 1 \times 10^{-3}$. This value indicates that a single charge packet would be completely attenuated in about 1000 transfers.

TABLE III

EVALUATION OF (17) TO OBTAIN THE
CCD TRANSFER INEFFICIENCY ϵ_g

Device	γ_e	γ_g	ϵ_g
Three-phase CCD with $L = 9.7$ µm and a 3-µm gap	0.21	0.15	2.1×10^{-4}
Three-phase CCD with $L = 3$ µm and a 3-µm gap	0.75	0.33	8.6×10^{-4}
Two-phase CCD with $L = 11$ µm and a transfer region 7 µm long	0.1	0.7	5×10^{-4}

The value of N_e used is that given by (5) with $N_{ss} = 2 \times 10^{10}$ cm^{-2}·eV^{-1}. $\Delta V_b = 2$ V; $\Delta V_s = 4$ V.

An expression for the noise introduced into the signal by statistical variations in the emptying of the interface states is given by (36). Using the same values of the parameters as used for the calculations above on the three-phase CCD with 9.7-µm electrodes, a signal-to-noise ratio $S/N = 6 \times 10^8/n$ is obtained. For the 288 transfers in this device, the signal-to-noise ratio becomes approximately 63 dB. For the 500-element device, the signal-to-noise ratio for each charge packet would be $6 \times 10^7/n$ or 46 dB for 1500 transfers. The same ratio of signal-to-total noise is obtained for the analog case, when (37) is integrated for a bandwidth of $f_o/2$.

In general, the magnitude of the noise arising from interface states with $N_{ss} = 2 \times 10^{10}$ cm^{-2}·eV^{-1} after 1000 transfers is approximately equal to the shot noise in a signal generated with an optical input. In some cases this may exceed the thermal noise generated in the preamplifier [17]. Thus the noise induced by the interface states in a device having many transfers, and particularly when using an electrical input [10], can be a limit-

TABLE IV
DESCRIPTION OF EXPERIMENTAL DEVICES

Number of Elements	Number of Phases	Length of Electrodes (μm)	Width of gaps or transfer region (μm)	Substrate conductivity ($\Omega \cdot$cm p-type)	C_o (F$\cdot$cm^{-2})	Width of Channel (μm)	Size of charge packet for $\Delta V = 1$ (pC)
96	3	9.7	3	20–40	1.24×10^{-8}	46	0.0625
500	3	3	3	20–40	1.8×10^{-8}	12	0.0073
250	2	11	7	1–5	1.8×10^{-8}	12	0.026
					0.73×10^{-8}		

ing factor on the obtainable signal-to-noise ratio. Fortunately, in image sensing applications of CCD's, where low-frequency noise is more objectionable than high-frequency noise, the magnitude of the noise decreases with decreasing frequency, as shown by (38).

XIV. EXPERIMENTAL RESULTS

Two n-channel, three-phase charge-coupled devices, one having 96 elements [3] and the second having 500 elements [4], have been used in measurements of transfer inefficiency. A 250-element n-channel undercut-isolated two-phase device has also been measured [5]. The essential features of these devices are summarized in Table IV. The 96-element device used a double oxide consisting of thermally grown silicon dioxide capped with 500 Å of alumina, while the 500-element device used a thermally grown silicon dioxide layer only. The two values of the oxide capacitance for the two-phase device correspond to values for the thin and thick regions of oxide under the transfer electrodes, capped with 500- and 1500-Å alumina, respectively.

Details of the measurements of transfer inefficiency on these devices are described elsewhere [4], [5], [19], but the results obtained on our best devices are summarized in Table V. Measurements of noise generated in CCD's have not been carried out at this time, but are planned.

XV. DISCUSSION

Those effects of interface states that determine the transfer inefficiencies of CCD's have been described. Transfer inefficiencies have been theoretically predicted and experimentally measured for devices with different electrode configurations. Reasonably good agreement is indicated between Tables III and V. In neither of the three-phase devices does transfer inefficiency increase with increasing frequency, as would be expected if the transfer inefficiency were determined by charge transport limitations, but remains essentially constant, as predicted for the effects of interface states. Indeed, for both of these devices, the transfer time constant is on the order of 1 ns or less, and no such limitation is expected. For the two-phase device fabricated on a more highly doped substrate with longer electrodes, the effects of charge transfer times are expected to be seen for frequencies in excess of 1 MHz.

There are many uncertainties in numerically evaluating the effects of interface states. The precise values of

TABLE V
MEASURED TRANSFER INEFFICIENCIES

Conditions of Measurement					
	V_b, V	2	2	0	0
	V_s, V	4	4	4	4
	f_o, Hz	10^5	10^6	10^6	10^6
	t_w, ms	0	0	0	2.5
Three-phase CCD with $L = 9.7$ μm and a 3-μm gap		7.4×10^{-4}	1.9×10^{-4}	4×10^{-4}	18.5×10^{-4}
Three-phase CCD with $L = 3$ μm and a 3-μm gap		40×10^{-4}	20×10^{-4}	30×10^{-4}	—
Two-phase CCD with $L = 11$ μm and a 7-μm transfer region		4×10^{-4}	4×10^{-4}	8×10^{-4}	26×10^{-4}

the interface state densities and the capture cross sections, as a function of energy, are very uncertain, and depend on the type of oxide used and the frequency of operation, although recent [15] work increases our knowledge of these parameters. However, the measurements have to be made on the opposite polarity of substrate from that on which the CCD is fabricated. Hence it is impossible to make a direct correlation of the experimental values of transfer inefficiency and the values of interface state density and capture cross section measured in the appropriate region of the bandgap. As has been shown, the improvement in performance arising from the charge trapped during transfer is appreciable, but it must be realized that the values of $\bar{p}_b$ and $\bar{p}_s$ used to calculate this improvement are only estimates. These values are derived from (13) by substitution of the appropriate values for σ, E_s, and v_f. The problem of assigning a value to σ has already been discussed, and E_s and v_f can only represent average values, both spatially and temporally, across the gap regions during transfer.

Table III shows the strong effect that gaps have on the transfer efficiency, particularly as the length of the transfer electrodes is reduced. This occurs because γ_e increases superlinearly with L^{-1} for a given gap size, and because there is less background charge available to fill the interface states in the gap region and hence neutralize their effect. In the two-phase device, the region under the thick oxide corresponds to the gap region in the three-phase device, but it is much longer. However, by using a higher substrate doping, the charge should transfer more slowly because of reduced fringing field, so that the interface states in this region are filled more completely by the background charge. Therefore, a similar transfer inefficiency to the 96-element device would be expected at frequencies below that at which charge motion becomes a limitation.

Some of the transfer inefficiencies quoted are very low indeed and permit the design of ambitious devices. Specific values are very dependent on device geometry, but certain design features should enable the effect of interface states in CCD's to be kept to a minimum. An adequate background charge is required to refill the interface states each cycle. Long transfer electrodes with narrow gaps minimize transfer inefficiencies. Use of electron-beam techniques to fabricate narrow gaps on the planar three-phase structure or narrow thick oxide regions on the undercut-isolated structure are appropriate ways of satisfying this requirement. The use of thinner oxides both confines the charge more closely, thus reducing γ_e, and also enables more charge to be stored under each electrode by increasing the oxide capacitance C_o. Similarly, a higher substrate doping reduces γ_e. Another method of maximizing charge trapping in the gap regions and reducing the edge effect, which is possible in a four-phase structure, is to store charge under two electrodes at any given time, so that the background charge covers all the interface including the gap regions.

A sophisiticated solution to the problem of avoiding trapping in interface states is to move the charge carriers in potential wells in the bulk rather than at the surface as in the buried channel CCD [20]. This device could be expected to have much higher transfer efficiencies, and would not have noise in the signal arising from the re-emission of trapped carriers.

High-frequency limitations of specific devices will not be discussed in detail. The requirements for minimizing transfer inefficiency arising from interface states and from charge transport are conflicting, particularly with respect to electrode length and substrate doping. As has been shown already, the interface state effect arising from variable transfer time is not a major effect. The other interface state effects should become worse at higher frequencies, since the capture cross section σ for the faster interface states decreases. Fig. 3 shows that the slope of curve (b) for $N(t)$ increases at shorter times when it is dominated by the faster states, and N_e therefore increases at higher operating frequencies. The smaller capture cross sections will also reduce capture during transfer in the relevant energy band and reduce the neutralizing effect of the background charge at higher operating frequencies.

The predictions and measurements of such small transfer inefficiencies confirms that high-performance CCD's can be designed and fabricated. Interface states do provide a limit to the performance of CCD's, but awareness of the effects enables designs to be chosen that minimize those effects that decrease transfer inefficiency, and should enable even the most ambitious charge-coupled arrays to be realized.

ACKNOWLEDGMENT

The author is grateful to W. J. Bertram, E. I. Gordon, G. E. Smith, and H. A. Watson for useful comments and discussions on the manuscript.

REFERENCES

[1] W. S. Boyle and G. E. Smith, "Charge coupled semiconductor devices," *Bell Syst. Tech. J.*, vol. 49, pp. 587–593, 1970.
[2] M. F. Tompsett, G. F. Amelio, and G. E. Smith, "Charge coupled 8-bit shift register," *Appl. Phys. Lett.*, vol. 17, pp. 111–115, 1970.
[3] M. F. Tompsett, G. F. Amelio, W. J. Bertram, Jr., R. R. Buckley, W. J. McNamara, J. C. Mikkelsen, Jr., and D. A. Sealer, "Charge-coupled imaging devices: Experimental results," *IEEE Trans. Electron Devices*, vol. ED-18, pp. 992–996, Nov. 1971.
[4] C. H. Séquin *et al.*, "A 500-element three-phase charge coupled device," unpublished.
[5] M. F. Tompsett, B. B. Kosicki, and D. Kahng, "Measurements of transfer inefficiency of 250-element undercut-isolated charge coupled devices," *Bell Syst. Tech. J.*, Jan. 1973.
[6] C. N. Berglund and H. J. Boll, "Performance limitations of the IGFET bucket-brigade shift register," *IEEE Trans. Electron Devices*, vol. ED-19, pp. 852–860, July 1972.
[7] R. J. Strain and N. L. Schryer, "A nonlinear analysis of charge coupled device transfer," *Bell Syst. Tech. J.*, vol. 50, pp. 1721–1740, 1971.
[8] C. N. Berglund, "Analog performance limitations of charge-transfer dynamic shift registers," *IEEE J. Solid-State Circuits*, vol. SC-6, pp. 391–394, Dec. 1971.
[9] W. B. Joyce and W. J. Bertram, "Linearized dispersion relation and Green's function for discrete charge transfer devices with incomplete transfer," *Bell Syst. Tech. J.*, vol. 50, pp. 1741–1759, 1971.
[10] M. F. Tompsett and E. J. Zimany, "Use of charge coupled devices for analog delay," in *IEEE Int. Solid-State Circuits Conf., Dig. Tech. Papers*, Feb. 1972, pp. 136–137; also submitted to *IEEE J. Solid-State Circuits*.
[11] M. F. Tompsett, "Charge transfer devices," *J. Vac. Sci. Technol.*, vol. 9, pp. 1166–1181, 1972.
[12] R. N. Hall, "Electron-hole recombination in germanium," *Phys. Rev.*, vol. 87, p. 387, 1952.
[13] W. Shockley and W. T. Read, "Statistics of recombination of holes and electrons," *Phys. Rev.*, vol. 87, pp. 835–842, 1952.
[14] R. J. Strain, "Properties of an idealized traveling-wave charge-coupled device," *IEEE Trans. Electron Devices*, vol. ED-19, pp. 1119–1130, Oct. 1972.
[15] H. Deuling, E. Klausmann, and A. Goetzberger, "Interface states in Si-SiO₂ interfaces," *Solid-State Electron.*, vol. 15, pp. 559–571, 1972.
[16] G. F. Amelio, "Computer modelling of charge-coupled device characteristics," *Bell Syst. Tech. J.*, vol. 51, pp. 705–730, 1972.
[17] G. F. Amelio, W. J. Bertram, Jr., and M. F. Tompsett, "Charge-coupled imaging devices: Design considerations," *IEEE Trans. Electron Devices*, vol. ED-18, pp. 986–992, Nov. 1971.
[18] K. K. Thornber and M. F. Tompsett, "Spectral density of noise generated in charge transfer devices," submitted to *IEEE Trans. Electron Devices*.
[19] M. F. Tompsett, "Measurements of transfer inefficiency on a 96-element three-phase charge coupled device," unpublished.
[20] R. H. Walden, R. H. Krambeck, R. J. Strain, J. McKenna, N. L. Schryer, and G. E. Smith, "A buried channel charge coupled device," *Bell Syst. Tech. J.*, vol. 51, pp. 1635–1640, 1972.

Experimental Characterization of Transfer Efficiency in Charge-Coupled Devices

ROBERT W. BRODERSEN, DENIS D. BUSS, AND AL F. TASCH, JR.

Abstract—The most important characteristic of a charge-coupled device is its charge transfer efficiency (CTE). There are three basic types of loss which degrade CTE: fixed loss, proportional loss, and nonlinear loss. Examples are given of each type of loss and techniques for measurement of all three types of loss are described. A method of determining the minimum fat zero which eliminates fixed loss is shown and an experiment is presented which confirms that fixed loss due to surface states can be completely eliminated by the use of a fat zero. The effect of interelectrode gaps on CTE is discussed in detail. A nonlinear loss model is used to describe the dispersion due to barriers in the gaps and the very detrimental effect of wells in the gap region is shown. The techniques presented in the analysis of these losses are very general and can be used whenever a detailed description of the transfer loss mechanism is required.

I. INTRODUCTION

A S THE sophistication of charge-coupled devices (CCD's) increases, so should the technique used to evaluate and characterize these devices. There has, however, been little progress in new methods of measuring and modeling the dispersion since Berglund [1] defined the inefficiency factor as

$$\epsilon = \frac{\Delta N}{N_{\text{sig}} - N_{\text{fz}}} \quad (1)$$

where ΔN is the net charge lost from a large charge packet; N_{sig} is the signal charge; and N_{fz} is the continuously introduced background charge of fat zero. In this paper, some new methods of measuring the transfer inefficiency will be described. These techniques will be used to show that fixed loss from surface states can be entirely eliminated from surface-channel devices by use of a fat zero and a method will be given to determine the minimum level of the fat zero required to eliminate this loss. Also, a generalization of the inefficiency parameter, $\epsilon(N)$, will be defined which is a function of the signal amplitude. In addition, a technique which can be used to evaluate $\epsilon(N)$ will be presented. An application of this technique will be made to a device which is operating with a dispersion which cannot be described by the usual ϵ. The usefulness of the generalized form, $\epsilon(N)$, is thereby demonstrated.

In order to show the application of these new techniques, two aspects of the charge-transfer mechanism will be discussed: 1) the effect of surface states and 2) the effect of the exposed interelectrode gap. For sim-

Manuscript received May 16, 1974; revised September 23, 1974. This work was supported in part by U. S. Navy NAVELEX under Contract N00039-73-C-0013.
The authors are with Texas Instruments Inc., Dallas, Tex. 75222.

plicity, an n-channel device was always used for measurements and analysis. A few general definitions and descriptions will now be given.

II. GENERAL CHARGE-TRANSFER LOSS CONSIDERATIONS

A. Definitions

The "loss" which is referred to as the charge-transfer loss is the redistribution of charge from the signal-charge packet into the trailing-charge packets. The total amount of charge in the signal- and trailing-charge packets is usually conserved during transfer along the device. To characterize the charge-transfer loss, it is useful to distinguish between three types of loss. A definition and some general characteristics of each type of loss will now be given.

1) Proportional Loss: Proportional loss results in a dispersion which can be described by an inefficiency parameter, ϵ. The amount of charge left behind, N_{Loss}, after a signal charge, N_{sig}, makes one transfer is given by

$$N_{\text{Loss}} = \epsilon N_{\text{sig}}. \quad (2)$$

The transfer inefficiency parameter, ϵ, for this type of loss is independent of signal amplitude, and, although it generally decreases with increasing fat-zero level, it usually cannot be eliminated.

2) Fixed Loss: Fixed loss is the loss of a fixed amount of charge, N_{Fixed}, during each transfer which is independent of the size of the signal charge.

If a surface-channel device is operated without a fat zero, the surface states give rise to a fixed loss (as well as a proportional loss). When operating with a fat zero, the fixed portion of the loss due to surface states is eliminated as well as any fixed loss from other sources. It should be noted that the proportional loss due to surface states (edge effect trapping) is only weakly affected by a fat zero.

3) Nonlinear Loss: Nonlinear loss is a more general phenomenological description of transfer loss which includes the dependence of loss on signal amplitude, which can be nonlinear. Although this general model actually includes proportional and fixed losses as special cases, they will be specifically excluded in this definition. In order to describe this loss, the inefficiency parameter, ϵ, introduced in (1), will be generalized to include a dependence on the signal amplitude, N_{sig}, so that the loss per

Reprinted from *IEEE Trans. Electron Devices*, vol. ED-22, pp. 40–46, Feb. 1975.

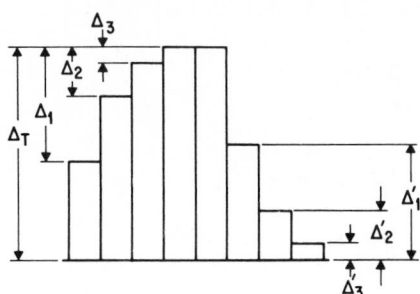

Fig. 1.   The output of a CCD which has a proportional loss when the input is five clock periods long.

transfer is given by

$$N_{\text{Loss}} = \epsilon(N_{\text{sig}})N_{\text{sig}}. \qquad (3)$$

With this generalization, it is usually not possible to obtain simple expressions for the dispersion of a pulse train as it transfers along the device, and computer simulation of CCD operation must be used.

B. Measurement Techniques

The usual method of measuring the inefficiency parameter ϵ is presented in Fig. 1, which shows a train of uniform signal charges (five shown here) in the midst of a long series of fat zeros. Δ_T is the difference between the signal charge and the fat-zero charge. The differences Δ_1, Δ_2, etc., are the amounts of charge missing from the first, second, etc., pulses in the train; and Δ_1', Δ_2', etc., are the amounts of charge in excess of the fat-zero charge which emerge in the first, second, etc., pulses trailing the pulse train. The normalized total loss in the leading edge (L_L) is defined by

$$L_L = \frac{1}{\Delta_T} \sum_i \Delta_i \qquad (4)$$

and the normalized total loss in the trailing edge (L_T) is defined by

$$L_T = \frac{1}{\Delta_T} \sum_i \Delta_i'. \qquad (5)$$

If the operation of a CCD with a proportional loss is analyzed step by step by keeping track of the amount of charge in every well as a function of time, it is possible to infer analytic expressions for the leading- and trailing-pulse amplitudes. It is found by this procedure that the loss in a given leading-edge pulse is equal in amplitude to the corresponding trailing pulse, i.e., $\Delta_1 = \Delta_1'$, $\Delta_2 = \Delta_2'$, etc.

An expression for Δ_i is given in Appendix 2 of Carnes and Kosonoky [2]. The results of Carnes and Kosonoky are obtained for the case of a nondestructive readout, i.e., after the charge is read out; ϵ times the signal charge is allowed to remain in the readout position and is added to the next signal packet [3]. An example of a nondestructive readout is a floating gate output [4]. However, a more common output technique is through an output diode

which is a destructive output, because there is nothing left at the readout position after the signal charge has been sampled [3]. An expression for this type of output is

$$\Delta_i = \Delta_i' = 1 - \epsilon - \sum_{k=0}^{i-1} \binom{N-k-2}{k} \epsilon^k (1-\epsilon)^N \qquad (6)$$

where N is the number of transfers and $\binom{i}{j}$ is a binomial coefficient. This expression also has a computational advantage over the form given by Carnes and Kosonoky because of a vastly decreased number of terms in the summation (to determine Δ_1 or Δ_1', there is only one term compared to N terms for Carnes and Kosonoky's expression.) Using (4) and (6) it can be shown that [5]

$$L_L = L_T = \frac{(N-1)\epsilon}{1-\epsilon} \qquad (7)$$

which is independent of signal amplitude. For a nondestructive output, the form of (6) is slightly changed, but the summation result of (7) is still valid except that $(N-1)$ is replaced by N. The result given in (7) differs from that given by Carnes and Kosonoky (namely $L_T = N\epsilon$). They summed the trailing pulses at a fixed instant of time, whereas in (7), the sum is taken as the pulses appear at the output of the device. If ϵ is small compared to one (as it usually is), (7) reduces to the expression of Carnes and Kosonoky. Equation (7) is a very useful result and is the basis for the step-response method of evaluating the inefficiency parameter ϵ. To measure ϵ, it is only necessary to introduce a sufficiently long pulse train such that the input signal level Δ_T can be determined. The leading-edge (L_L) or trailing-edge (L_T) loss is then measured and ϵ is calculated by evaluating (7).

Another technique commonly used for measuring a proportional loss, which is somewhat more difficult to implement (but is sometimes necessary in CCD's which do not have means for an electrical input) is to optically introduce a signal packet into a single well. The amplitude of the first pulse out of the register is then proportional to $(1 - \epsilon)^N$. To determine the value of ϵ, the light spot is moved along the register and the amplitude of the first pulse as a function of the number of transfers is recorded. When plotted on semilog paper, the slope of this amplitude as a function of the number of transfers will then reveal the value of ϵ. This technique has the advantage of demonstrating whether the loss occurs at selected points along the register or whether it is uniformly distributed. Unfortunately, it is not possible to obtain a simple expression such as (7) for calculating ϵ with a single pulse input, but it is still possible to determine ϵ by comparing the output with plots of the following expression [3], [6]:

$$P_i = \binom{N+i}{i} \epsilon^i (1-\epsilon)^N \qquad (8)$$

where $i = 0$ is the first pulse out, $i = 1$ is the first trailing pulse, etc.

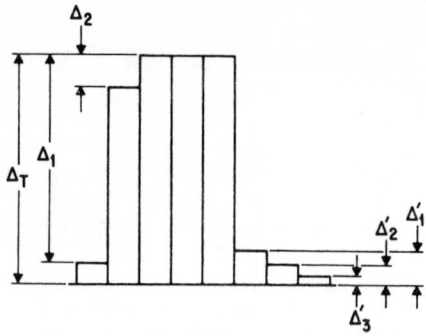

Fig. 2. The output of a CCD with a fixed loss when the input is five clock periods long.

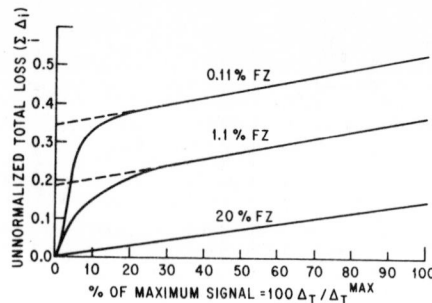

Fig. 3. A plot of unnormalized total loss versus signal level which shows that the fixed loss part of the total loss can be eliminated by a fat zero.

It should be noted that both fixed and nonlinear losses are not properly characterized by the previous considerations. Both of these losses will almost always give rise to a nonsymmetric-looking output from a pulse train at the input. The effect of a fixed loss on the leading edge of a pulse train is to increase Δ_1 (see Fig. 2). If the loss is large enough so that Δ_1 is almost equal to Δ_T as in Fig. 2, then Δ_2 increases and so on. On the other hand, the emission of the charge after the trailing edge of the pulse train is usually relatively slow and it takes many pulses for the charge to be reemitted. Since the amount of charge is conserved $L_L = L_T$, but in general $\Delta_i \neq \Delta_i'$.

Nonlinear loss, however, is even more complicated. Very little information can be gained about a nonlinear loss by use of (4) and (5). A new technique will therefore be presented which makes possible an evaluation of the parameters involved in modeling a nonlinear loss.

III. SURFACE STATE LOSS

A. Fixed Loss

The transfer loss due to surface states results in both a fixed and proportional type of loss. The proportional part of the loss is due to trapping under the edges of the transfer gate (edge-effect trapping) whereas, the fixed loss is due to trapping directly underneath the gates [7]–[9]. A fat zero is only effective in filling the surface states directly under the transfer gate and thus only the fixed portion of the loss can be eliminated.

In determining the amount of fat zero required to eliminate this fixed loss, one can simply increase the fat zero until the loss in the first pulse is equal to the charge in the first trailing pulse. At this point, the loss is proportional, and the fixed loss due to surface states has been eliminated.

A more accurate determination of fat zero required may be obtained by measuring the sum of the losses in the leading edges ($\sum \Delta_i$) versus the input signal level (Δ_T). In Fig. 3, this unnormalized total loss is measured on a 64-bit device as a function of signal level for a fat-zero level of 0.11, 1.1, and 20 percent. The proportional loss (slope of the lines) is nearly independent of the fat-zero level (actually, it decreases slightly with increasing fat

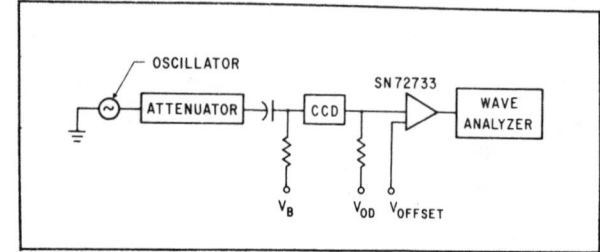

Fig. 4. Circuit for measuring frequency response of a CCD as a function of signal level.

zero), but the dramatic part of the measurement is the shift along the vertical axis. The 0.11-percent fat-zero curve extrapolates to a fractional fixed loss of 0.34 of a full well, which, for 193 transfers, gives 1.8×10^{-3} per transfer. The 1.1-percent fat-zero curve extrapolates to a fixed loss of 1.0×10^{-3} per transfer, whereas the 20-percent fat-zero curve extrapolates to zero fixed loss. (On this particular device, the fixed loss disappears at about 10-percent fat zero and the curves shift very little with further increases in the fat-zero level.)

In this measurement, the determination of the fat-zero level was made by monitoring the average current at the output of the device using a picoammeter. Measurement of the current can be used to obtain a very accurate determination of a full well as well as fat-zero levels and makes it possible to calibrate the signal from the output amplifiers in terms of charge.

B. Verification That Fixed Loss is Eliminated By Fat Zero

The results of the previous section indicate that if a sufficiently large fat zero is used, fixed loss can be eliminated and thus very small signal levels can be transferred with acceptable loss. This result is very important in the operation of surface-channel CCD's in applications involving low-signal levels, such as low-light level imaging. Because of this importance, an experimental verification of this result was undertaken. In order to eliminate problems with noise in the measurement, the loss was determined by measuring the frequency response of a CCD with the experimental apparatus shown in Fig. 4. A sinusoidal input of frequency f was capacitively coupled through an attenuator to the input of a CCD. The fat-zero level as

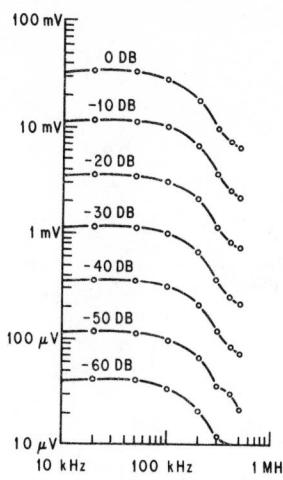

Fig. 5. Measured frequency response for signal levels spanning 60 dB.

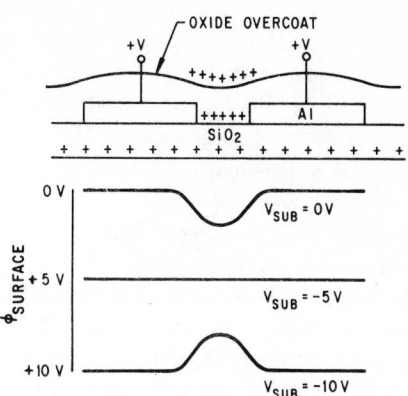

Fig. 6. The formation of barriers and wells by varying the substrate bias.

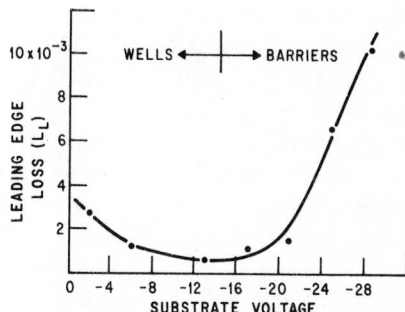

Fig. 7. Loss per transfer as a function of substrate voltage.

determined by V_B was set at approximately half a full well so that a maximum ac signal could be achieved. The CCD output was amplified using a wideband video amplifier (SN72733) and fed to the wave analyzer. The frequency response was measured at the maximum signal level, and then the signal was attenuated in 10-dB increments down to −60 dB. The limitation at −60 dB was due to the noise in the external circuitry (primarily the video amplifier); if care were taken to eliminate these external noise sources, much lower signal levels could be attained.

Measurements of the frequency response $H(f)$, were made and compared with the expression of $H(f)$ for a proportional loss, ϵ,

$$H(f) = \exp\left[-n\epsilon(1 - \cos 2\pi f/f_c)\right] \quad (9)$$

where n is the number of transfers and f_c is the clock frequency [3], [10]. The experimental results are shown in Fig. 5 for a device having 300 transfers at a clock frequency of 1 MHz. When ϵ is chosen to be 2.7×10^{-3}, the curves calculated using (9) agree with the experimental curves to within the experimental error. The transfer loss for this device is far from optimum, but one thing is strikingly clear: the loss is independent of signal amplitude over at least a 60-dB range of signal level. By extrapolation, we believe this is true over an even wider range.

This measurement, therefore, proves that even though a high density of fast surface states is detrimental from a noise standpoint, the effect of these states on transfer efficiency can be considerably decreased provided a sufficiently large fat zero is used (which is usually 10–15 percent).

IV. EFFECT OF THE INTERELECTRODE GAPS

A. Charge in the Gap Region

The interelectrode gaps in CCD's which are fabricated photolithographically with a single level of metallization are limited to 2–3 μm. If these gaps are not covered by means of a "resistive sea" [11], it is possible for potential

barriers and wells to develop in this region. In the gap, the only capacitance is the small depletion-layer capacitance so that a very small amount of charge in this region results in a large surface potential. In an exposed gap, the surface potential for an n-channel device tends toward the free surface potential, ϕ_{fs}, which is given by

$$\phi_{fs} = \frac{Q_{TOTAL}^2}{2q\epsilon_{si}N_A} \quad (10)$$

where Q_{TOTAL} is the total amount of charge per unit area in the gap region. This total charge is composed of the fixed positive charge which exists at the oxide–silicon interface as well as all of the impurity charge which exists in and above the gate oxide. (See Fig. 6(a).) For a sufficiently wide gap, the surface potential would attain this value, but for gaps of interest (≈3 μ), the surface potential in the gap region is a complicated function of the gap width and substrate doping. For this case, it is necessary to numerically solve the two-dimensional Poisson equation in order to determine the size of the barriers or wells.

The effect of the interelectrode potential on transfer efficiency is shown in Fig. 7. The device used in this figure was a 3-phase, n-channel, 100-bit serial shift register with 0.4- by 1-mil electrodes and 0.1-mil gaps. The device was clocked at 1 MHz with drivers which operated between 0 and 15 V, and 15-percent fat zero. The substrate was biased with a negative voltage which was then varied, and measurements of leading edge loss, L_L, were made. This variation of loss with substrate voltage can be understood by referring back to Fig. 6. For this figure, the free surface potential as calculated by (10) is $+5$ V. For this value of ϕ_{fs} with a gate to substrate voltage, V_{gs}, on both gates of $+5$ V, there will be no barriers or wells in the gap between these gates. However, if V_{gs} is decreased to 0 V, a well will appear in the gap. On the other hand, if V_{gs} is increased to 10 V, a barrier will arise. The actual mechanism of the loss resulting from the barriers and wells requires a more complicated model than that shown in Fig. 6, but the essential point to be obtained from this figure is that there is an optimum gate-to-substrate voltage for a given charge in the gap region which minimizes the effect of the barriers and wells. Unfortunately, since the charge above the gate oxide is extremely dependent on environmental conditions, the optimum voltage for operation is also variable and thus CCD's with exposed oxide have unstable operating characteristics. This widely reported [3], [11]–[13], behavior of CCD's with exposed oxide is closely related to the charge-spreading phenomena seen in planar diodes [14].

It has been reported that there is a wide range of interface charge for which complete transfer across a gap can be achieved, regardless of electrode separation [15]. This analysis did not properly include the effect of wells which arise in the gaps at high levels of interface charge (or low-substrate voltages) which can result in a considerable loss. If the size of the gap is increased, then the charge in the gap, Q_{TOTAL}, will be more effective in controlling the potential in the gap region (due to smaller fringing fields from the adjacent gates) and the loss from the barriers and wells will become even more acute. Our results, therefore, show that transfer across gaps depends on the relative values of Q_{TOTAL} and the substrate voltage, and becomes even more critical for large gaps. This critical dependence that results for the electrode structure shown in Fig. 6, imposes severe constraints on device passivation and threshold stability. Therefore, a resistive sea [11] or an overlapping gate structure [16] is needed to control the gap potential and thus eliminate the barrier and well problem. A more detailed analysis of how the barriers and wells result in loss follows.

B. Potential Wells

For an n-channel device, when the "off" gate-to-substrate voltage is too small (or equivalently the substrate voltage is too low relative to ϕ_{fs} due to the charge in the gap region, Q_{TOTAL}), the device will suffer a charge-transfer loss which is due to wells. These wells exist between the transferring electrode and the one preceding it as shown in Fig. 8. The loss which results from the wells can be considered to consist of two parts: a fixed loss and a proportional loss. The fixed loss is simply the retention of charge in the wells and can be eliminated by use of a fat zero in much the same way as the fixed loss from surface states is eliminated. The proportional loss results from the variation in the amount of trapped charge with a change in the size of the signal-charge packet. Unfortunately, this proportional loss does not decrease with use of fat zero. In fact, for very large wells, the loss becomes nonlinear

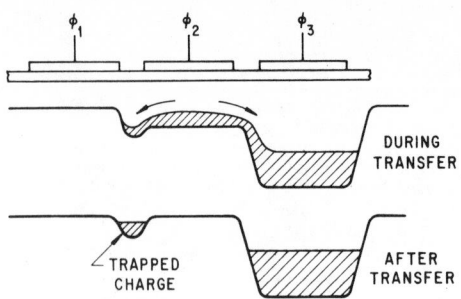

Fig. 8. Illustration of the trapped charge due to wells which results in both a fixed and proportional loss.

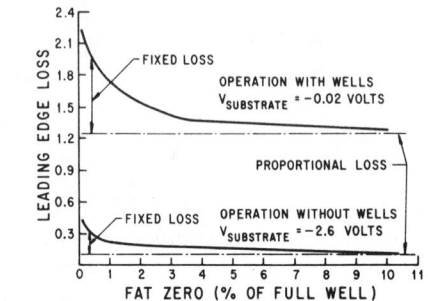

Fig. 9. Loss per transfer versus fat zero for a device with and without wells.

and the transfer inefficiency actually degrades with an increase in fat zero [17]. This nonlinear loss due to the wells is in many respects similar to the barrier loss which will be discussed in the next section.

In Fig. 9, the leading-edge loss, L_L, is shown as a function of increasing fat zero for a 100-bit serial register (3-phase, 0.4- by 5-mil electrodes) for two different substrate voltages. The curve labeled "Operation without Wells" is, for the substrate bias, set for optimum operation. The other curve is, with the substrate bias, set at a more positive voltage so that wells appear in the gap regions. The dashed lines in these two curves represent the proportional loss which remains when sufficient fat zero is used to eliminate all fixed loss (approximately 15 percent). The difference between the measured curves and the dashed line is the fixed loss. For operation without wells, the fixed loss is due to surface states; however, for operation with the wells, the fixed loss is much greater and is caused by the trapping of the signal charge in the potential wells in the gaps. It should also be noted that the proportional loss for operation with wells is over an order of magnitude higher than the proportional loss for operation with optimum substrate bias, as mentioned previously. This increase in loss is due to the dependence of the amount of charge in the wells on the size of the transferring charge packet.

Some insight into the mechanism of this loss can be obtained from the observation that the amount of charge retained in the wells decreases with increasing fall time of the clock drivers. This dependence is expected, because, with longer fall times, less charge remains under the transfer electrode as the transfer-clock voltage turns off.

DYNAMIC RANGE LIMITATION

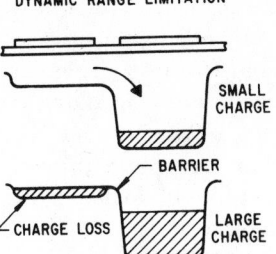

Fig. 10. Illustration of how a small charge can transfer without barriers while barriers will appear for a large charge.

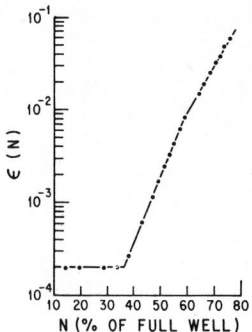

Fig. 11. The inefficiency parameter as a function of signal charge level.

Since the wells do not form until the transfer clock is nearly off, less charge is available to be trapped in the wells [18].

C. Potential Barriers

When the substrate voltage is too negative, it is possible for potential barriers to appear in the gap region. The loss which results from these barriers has several distinctive characteristics, which can be attributed to the signal amplitude dependence of the barrier height [1], [19] as shown in Fig. 10. For input signals lower than a certain threshold level, the loss due to barriers is very small, because the potential difference between adjacent wells is large enough to suppress the barriers. However, as the signal is increased above this level, the barriers appear and the transfer efficiency decreases drastically. The output signal then spreads dramatically with any further increase in input level. In many respects, this dispersion makes it appear as if the device has only a very small dynamic range.

The analysis of loss due to the potential barriers is interesting because it is an example of a nonlinear loss. Since a nonlinear loss has a dependence on signal amplitude, the usual method of measuring loss (leading- and trailing-edge loss) is inadequate, and an incremental technique was therefore employed. This technique linearizes the loss at a given signal amplitude, N_o, by measuring the loss in an incremental increase, ΔN, in charge above its background level, N_o.

Since the background charge, N_o, is continually being transferred through the device, an amount of charge $\epsilon(N_o)N_o$ is always left behind. When the incremental signal, ΔN, is added to the background level, the signal charge, $N_o + \Delta N$, has a loss $\epsilon(N_o + \Delta N)(N_o + \Delta N)$. The net loss from the incremental signal, $L_{\Delta N}$, is the difference between the loss with and without the signal charge which is

$$L_{\Delta N} = \epsilon(N_o + \Delta N)(N_o + \Delta N) - \epsilon(N_o)N_o. \quad (11)$$

If $\epsilon(N)$ is expanded in a series in ΔN and only lowest order terms are kept, $(\Delta N \ll N_o)$ the loss, $L_{\Delta N}$, can be written as

$$L_{\Delta N} \approx \epsilon(N_o)\Delta N. \quad (12)$$

Therefore, by measuring the loss in an incremental signal, ΔN, in excess of a background charge on N_o, the fractional loss per transfer, $\epsilon(N_o)$, of a charge packet of size N_o can

be obtained. The subscript on N_o will be dropped in remaining considerations so that the nonlinear-loss parameter is given by $\epsilon(n)$.

In Fig. 11, $\epsilon(n)$, as determined by the incremental technique, is plotted versus signal amplitude, N, where N is a fraction of a full well. This curve was taken when the device was operating with the substrate at a voltage more negative than the optimum value so that potential barriers existed in the gaps between the electrodes. It can be seen from this plot that, as the bias level increases above 38 percent, a barrier forms and grows which drastically decreases the transfer efficiency. If the bias level had been decreased much below the percentage shown in this figure (15 percent), the loss would also have increased due to trapping in surface states.

With this determination of $\epsilon(N)$, it is possible to calculate the dispersion of a signal charge as it transfers down a device. However, to make this calculation, it is necessary to simulate the operation of a CCD with a computer program. For a transfer (at time t_0) from the ith gate to the $i + 1$ gate, the amount of charge remaining in the ith well after a transfer (at time $t_1 = t_0 + 1/3f_c$) is

$$N_i(t_1) = \epsilon[N_i(t_0) + N_{i+1}(t_0)]N_i(t_0). \quad (13)$$

In (13), the inefficiency parameter $\epsilon(N)$ is evaluated at $N = N_i(t_0) + N_{i+1}(t_0)$ because it is the total amount of charge in the $i + 1$ well at the end of transfer which determines the height of the barrier. This expression is correct as long as $\epsilon(N)$ remains small. For very large signals $(N > 0.8)$, because of the difficulty in measuring such a large loss in the differential measurement $(N\epsilon > 30)$, it is not even possible to evaluate $\epsilon(N)$. However, because the loss is so large, a large signal will be quickly attenuated to a lower value and only a few stages near the input will be involved in transferring large charge levels. A calculation of the dispersion is therefore essentially independent of the modeling of the very high loss transfers. The dashed lines of Fig. 12 are the calculated output after 300 transfers for input signals of seven pulses with three different amplitudes, 0.27, 0.57, and 1.0. Also shown in this figure as solid lines are the experimentally obtained results for the same conditions used in the calculation. The agreement is excellent and thus demonstrates the accuracy of the incremental method of obtaining the nonlinear loss parameter. It is important to note that all

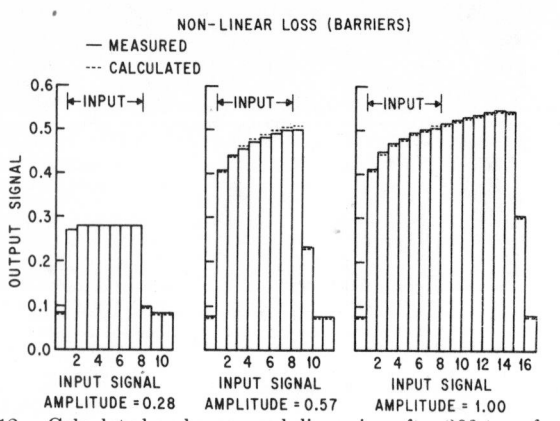

NON-LINEAR LOSS (BARRIERS)
— MEASURED
--- CALCULATED

Fig. 12. Calculated and measured dispersion after 300 transfers for a CCD with barriers.

of the characteristics of a barrier loss are predicted by the $\epsilon(N)$ of (13), and, in particular, demonstrates that the dynamic range limitation is only due to the nonlinear characteristic of the transfer efficiency.

The description of a loss by means of an inefficiency parameter $\epsilon(N)$ has wide applicability. Another situation where the nonlinear description of a loss is necessary is the characterization of the loss resulting from the interaction of the "buried" signal charge with surface states in a buried channel device. This interaction occurs if the clock voltage is too high or the signal charge is too large.

V. CONCLUSIONS

To characterize properly the transfer loss in a charge-coupled device, it is necessary to determine more than the transfer inefficiency, ϵ, at one signal level. This applies even for devices for practical applications in which, normally, the transfer efficiency is sufficiently high so that the proportional description would be adequate. For, if these devices are operated in a nonoptimum mode (e.g. too low fat zero or too large a signal), the more comprehensive techniques presented here for determining the dispersion of the input signal are useful.

The CCD-loss mechanism can be grouped into three basic types of loss: fixed, proportional, and nonlinear. Methods of measurement and examples of all three types of loss have been given. The more detailed characterization techniques which have been presented made possible a deeper understanding of the limitations of a given device. In using these techniques, several interesting results have been obtained. The fixed loss from the surface states has been shown to be eliminated entirely by use of a fat zero and a method for obtaining the size of this fat zero has been given. Also, when operated at high-substrate voltages, a dynamic range limitation, found in devices with exposed gaps, has been attributed to barriers which give rise to a nonlinear loss. A phenomenological model has been devised which predicts the dispersion due to this barrier loss and comparison with experiment yields excellent agreement. Finally, in contrast with previous analysis, it has been found that at a given substrate bias, zero-loss transfer across gaps is possible only for a very

narrow range of charge density in the gap region. This requirement places unrealistic constraints on device passivation and threshold stability. Therefore, these results demonstrate the necessity of controlling the gap potential using a technique such a "resistive sea" or an overlapping gate structure. If they have been properly designed, the performance in these types of devices is limited by surface states in surface-channel devices [9], [20] or in bulk states in buried-channel devices [21], [22].

ACKNOWLEDGMENT

The authors express their appreciation to M. Gosney, D. Brown, B. Barton, and B. Hewes for helpful discussions and to F. Wall for his assistance in the measurements.

REFERENCES

[1] C. N. Berglund, "Analog performance limitations of charge-transfer dynamic shift registers," *IEEE Trans. J. Solid-State Circuits*, vol. SC-6, pp. 391–394, Dec. 1971.
[2] W. F. Kosonocky and E. J. Carnes, "Two-phase charge-coupled devices with overlapping polysilicon and aluminum gates," *RCA Rev.*, vol. 34, p. 164, 1973.
[3] W. B. Joyce and W. J. Bertram, "Linearized dispersion relation and Green's function for discrete-charge transfer devices with incomplete transfer," *Bell Syst. Tech. J.*, vol. 50, no. 6, p. 1741, 1971.
[4] D. D. Wen and P. J. Salsbury, "Analysis and design of a single-stage floating gate amplifier," *ISSCC Digest Tech. Papers*, pp. 154–155, Feb. 1973.
[5] H. H. Hosack and J. B. Barton, private communication.
[6] C. H. Sequin, "Experimental investigation of a linear 500-element 3-phase CCD," *Bell Syst. Tech. J.*, vol. 53, no. 4, pp. 581–610, April 1974.
[7] A. M. Mohsen, T. C. McGill, Y. Daimon, and C. A. Mead, "The influence of interface states on incomplete charge transfer in overlapping gate charge-coupled devices," *IEEE Trans. J. Solid-State Circuits*, vol. SC-8, pp. 125–138, Apr. 1973.
[8] M. F. Tompsett, "The quantitative effects of interface states on the performance of charge-coupled devices," *IEEE Trans. Electron Devices*, vol. ED-20, pp. 45–55, Jan. 1973.
[9] J. E. Carnes and W. F. Kosonocky, "Fast interface state losses in CCD's," *Appl. Phys. Lett.*, vol. 20, no. 7, pp. 261–263, April 1972.
[10] M. F. Tompsett and E. J. Zimany, "Use of charge-coupled devices for delaying analog signals," *IEEE Trans. J. Solid-State Circuits*, vol. SC-8, pp. 151–157, April 1973.
[11] C. K. Kim and E. H. Snow, "p-channel CCD's with resistive gate structures," *Appl. Phys. Lett.*, vol. 20, no. 12, pp. 514–515, June 1972.
[12] C. H. Sequin, D. A. Sealer, W. J. Bertram, Jr., M. F. Tompsett, R. R. Buckley, T. A. Shankoff, and W. J. McNamara, "Charge-coupled image sensor and frame store," *IEEE Trans. Electron Devices*, vol. ED-20, p. 244, Mar. 1973.
[13] M. F. Tompsett, B. B. Kosicki, and D. Kahng, "Measurements of transfer inefficiency of 250-element undercut-isolated charge-coupled devices," *Bell Syst. Tech. J.*, vol. 52, no. 1, p. 1, 1973.
[14] W. Schroen, "Ion transport on and through thin-oxide layers of MOS structures," in *Proceedings of the International Symposium on Basic Problems in Thin Film Physics*. R. Niedermayer and H. Mager, Eds. Gottingen, Holland: Vandenhoek and Reprecht, 1966.
[15] R. H. Krambek, "Zero loss transfer across gaps in a CCD," *Bell Syst. Tech. J.*, vol. 50, no. 10, 1971.
[16] W. F. Kosonocky and J. E. Carnes, "Charge-coupled digital circuits," *IEEE Trans. J. Solid-State Circuits*, vol. SC-6, pp. 314–326, Nov. 1971.
[17] M. Gosney and D. Brown, private communication.
[18] G. F. Amelio, "Computer modeling of charge-coupled device characteristics," *Bell Syst. Tech. J.*, no. 3, p. 705, 1972.
[19] N. A. Patrin, "Performance of very high density charge-coupled devices," *IBM J. Res. Dev.*, vol. 7, May 1973.
[20] W. J. Bertram *et al.*, "3-phase CCD using 3 levels of poly-silicon," *IEDM Wash. Late News.*, Dec. 1973.
[21] R. W. Brodersen and A. F. Tasch, Jr., "A comparison of two different types of buried channel devices," *IEEE Device Res. Conf.*, June 1974.
[22] A. M. Mohsen and M. F. Tompsett, "The effect of bulk states on buried channel performance," *IEEE Device Res. Conf.*, June 1974.

Charge-Control Method of Charge-Coupled Device Transfer Analysis

HSING-SAN LEE AND LAWRENCE G. HELLER

HSING-SAN LEE AND LAWRENCE G. HELLER

Abstract—An analysis of charge transfer based on the "charge-control" approach has been made for charge-coupled devices (CCD's). A general closed-form equation for the charge transfer efficiency has been obtained that includes the major mechanisms of 1) charge-gradient induced drift, 2) thermal diffusion, 3) an external fringing field, and 4) charge loss due to traps or recombination. When the charge loss and fringing field terms are neglected, the results are in close agreement with the numerical solutions by Strain and Schryer. With the fringing field term included, the closed-form solution compares well with the numerical results by Heller, Chang, and Lo. The effect of charge loss on the transfer efficiency is studied and the temperature dependence of the efficiency, including the temperature dependent surface mobility, is discussed. The effect of a "fat" zero on the diminution of a digital one is discussed with and without charge loss to surface states.

It is believed that the charge-control approach not only simplifies the mathematics involved, but also provides practical charge-coupled device and circuit design guides.

I. INTRODUCTION

IT IS well known [1] that the operation of a charge-coupled device (CCD) depends on the storage and the absence of minority carriers in semiconductor surface depletion regions created by a series of MOS structures, and the controlled transport of these charges along the surface by unidirectional movement of the potential minima. The transient behavior of the charge transfer from one potential well to the next well has been analyzed by many authors [2]–[6]. These analyses are based on the numerical solution of the nonlinear transport equation. While the results of numerical analyses have provided considerable information on the various charge transport mechanisms, it is not very practical to use numerical results in circuit analyses programs and device design tradeoffs. In this work, we develop a compact closed-form solution for CCD transfer characteristics.

In the following paragraphs a "charge-control" [7]–[9] approach for the analysis of the transfer characteristics of CCD's is presented. This approach, similar to that proposed by Beaufoy and Sparkes [8] and by Koehler [7] for bipolar transistors, has not been heretofore applied to the transient behavior of CCD's.

The applications of the closed-form equation for the charge transfer efficiency are demonstrated in the estimations of the effect of fringing fields, the charge loss to surface states during transfer, the temperature dependence, and a "fat" zero in the shift-register operation.

Manuscript received January 25, 1972; revised June 7, 1972.

The authors are with the IBM Components Division Laboratory, Essex Junction, Vt. 05452.

II. BASIC CHARGE-CONTROL MODEL

A schematic of a section of a CCD is shown in Fig. 1. The length definition, the initial condition, and the boundary conditions on which the following analysis is based are also illustrated. It is assumed that no variation in the charge distribution at the semiconductor surface occurs in the direction perpendicular to the charge transfer.

Following the classical approach of charge-control theory, the starting point is the integration of the continuity equation over the length L of the discharging potential well (Fig. 1). We thus obtain

$$\frac{d}{dt} \int_0^L q_n(x, t)\, dx$$

$$= -\frac{1}{\tau_s} \int_0^L q_n(x, t)\, dx + \int_0^L \frac{\partial}{\partial x} J_n(x, t)\, dx \quad (1)$$

where $q_n(x, t) = qn(x, t)$ is the magnitude of the electron charge distribution per unit area, $J_n(x, t)$ is the electron current density per unit width, and τ_s is the effective time constant for charge loss. Since $q_n(x, t)$ has been defined as the magnitude of the electron charge density, all our final results apply to both n- and p-channel devices. The first term on the right-hand side of (1) accounts for the charge lost to surface state traps during charge transfer to the next well. A more detailed discussion on charge trapping is given in Section V. By introducing the total charge $Q_w(t)$ per unit width in the discharging potential well, the functional dependence on the spatial coordinate x is removed from (1). This results in the following basic charge-control equation:

$$\frac{d}{dt} Q_w(t) = -\frac{Q_w(t)}{\tau_s} + J_n(L, t) - J_n(0, t) \quad (2)$$

where

$$Q_w(t) = \int_0^L q_n(x, t)\, dx.$$

Thus, the transient behavior of the total charge in the discharging potential well is determined by the current boundary conditions and the model used to represent trapping or the recombination of charges by surface states during charge transfer to the next potential well. Note, aside from the current boundary conditions at the ends of the discharging potential well, which determines the mode of operation of the device, no restrictions on the specific charge distributions within the

Reprinted from *IEEE Trans. Electron Devices*, vol. ED-19, pp. 1270–1279, Dec. 1972.

77

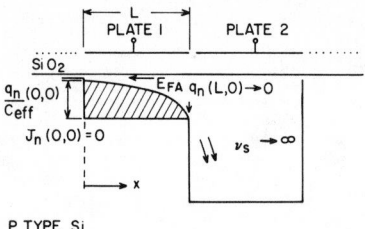

Fig. 1. Schematic of a section of CCD with the length definition, initial condition, and boundary conditions for which a closed-form solution for the charge transfer efficiency is obtained.

well are implied in (2). This point will be illustrated later.

We assume that the current at any instant at the left end of the discharging potential well is zero, namely, $J_n(0, t) = 0$. This implies that an infinite potential barrier exists that eliminates the back flow of charges and assures unidirectional flow. This is one of the necessary conditions for the successful operation of CCD's. The current boundary condition at the right end of the discharging potential well has to be finite and nonzero. In charge-transfer devices, the gate producing the next potential well is driven into deep depletion such that in the direction of charge flow there is no potential barrier that impedes the flow of information bearing carriers. Mathematically, we may assume an ideal charge sink starting at the right edge of the discharging potential well. The charge condition at $x = L$ is such that the charges are flowing at the saturation limited velocity $v_s = 10^7$ cm/s with a finite current density. However, in the analysis to follow we may assume that the charge at $x = L$ is approaching zero but moving at an infinite velocity such that the current is finite. Thus, the boundary conditions for the discharging potential well can be expressed as

$$J_n(0, t) = 0$$
$$J_n(L, t) = -\lim_{x \to L} v_s q_n(x, t) \qquad (3)$$

where $v_s \to \infty$ as $x \to L$ and

$$\lim_{x \to L} q_n(x, t) = 0. \qquad (4)$$

The boundary condition given by (4) has been justified by Strain and Schryer [2] who numerically studied the effect of a finite limiting velocity v_s and concluded that the effect is negligible.

In general, for $0 \leq x \leq L$, the current equation is given by

$$J_n(x, t) = D_n \frac{\partial}{\partial x} q_n(x, t) + \mu_n q_n(x, t) E \qquad (5)$$

where the electric field E can be expressed as a combination of the external fringing field E_{ext} and a contribution due to the gradient of the surface potential $\psi_s(x, t)$.

The surface potential ψ_s based on the one-dimensional analysis [1] can be expressed as

$$\psi_s = V_g - V_{FB} - (q_n + q_B)/C_{ox}$$
$$\simeq V_g - V_{FB} - q_n/C_{eff} \qquad (6)$$

where q_B is the bulk depletion charge per unit area and C_{eff} is the effective surface capacitance per unit area. The linear approximate relationship between ψ_s and q_n that is indicated by (6) has been known [1]–[4] to be a good approximation, especially for thin oxide thickness and low substrate doping. Using (2) and (3), the charge-control or transport equation becomes

$$\frac{d}{dt} Q_w(t) = -\frac{Q_w(t)}{\tau_s} + J_n(L, t) \qquad (7)$$

which upon substitution of (4)–(6) results in

$$\frac{d}{dt} Q_w(t) = -\frac{Q_w(t)}{\tau_s} + \left[D_n \frac{\partial}{\partial x} q_n(x, t) \right.$$
$$+ \frac{\mu_n}{C_{eff}} q_n(x, t) \frac{\partial}{\partial x} q_n(x, t)$$
$$\left. + \mu_n q_n(x, t) E_{ext} \right]_{x=L}. \qquad (8)$$

Note that the first drift term in (8) cannot be neglected just because the charge $q_n(x, t)$ at $x = L$ is approaching zero. To take into account this finite drift current, the product $q_n(x, t) \partial/\partial x q_n(x, t)$ at $x = L$ must be evaluated in terms of a definite quantity. As a first-order approximation, we equate the product at $x = L$ to the average value of the product within the discharging potential well. Physically, this is equivalent to setting the charge-gradient induced drift current at $x = L$, at any instant, equal to its space average within the discharging potential well.

$$\left[q_n(x, t) \frac{\partial}{\partial x} q_n(x, t) \right]_{x=L}$$
$$= \frac{1}{L} \int_0^L q_n(x, t) \frac{\partial}{\partial x} q_n(x, t) \, dx$$
$$= -\frac{1}{2L} [q_n(0, t)]^2. \qquad (9)$$

The approximation for the charge-gradient induced drift current, in terms of charge, given by (9) is similar to that, in voltage form, often found in FET analyses [10], [11].

To solve the transport equation, a knowledge of the charge distribution is required. The numerical solution of the nonlinear transport equation in the differential form [2], [3] has revealed the transient behavior of the charge distribution. However, an analytical solution of (8) can only be obtained by transforming the equation to an ordinary differential equation with known time-independent coefficients. To facilitate the solution of the transport equation, we now apply the important assumption of the charge-control method [7]–[9]—the postulate of instantaneous charge redistribution during transient decay!

It is generally accepted [12] that the response time

for electron redistribution is just the dielectric relaxation time of the inversion layer. According to Hofstein and Warfield [12], a typical value for this response time in $10\ \Omega \cdot$cm p-type silicon with a heavy inversion layer is 10^{-13} s. Thus, it seems to be reasonable to postulate instantaneous charge redistribution. Mathematically, this results in a separation of variables of the form

$$q_n(x, t) = f(t)\phi(x). \qquad (10)$$

This implies that the shape function $\phi(x)$ of the charge distribution remains unchanged during the transient, i.e., the mobile charge (electrons) always redistribute themselves instantaneously to follow the shape of the initial distribution—the magnitude being weighted by the function $f(t)$ that satisfies the integral form of the transport equation.

In the following section we develop a generalized formula for the charge transfer efficiency that is based on (10) and examples are shown for cosine, ellipsoidal, and higher order shape functions.

III. GENERAL DERIVATION OF THE CHARGE TRANSFER EFFICIENCY

Based on the postulate of instaneous charge redistribution, the essential equations in the time-space separated form are the following:

1) The initial condition at $x = 0$ is

$$q_n(0, 0) = f(0)\phi(0) = q_{n0}. \qquad (11)$$

2) Using (3), (4), and (9), the boundary conditions with no fringing fields are

$$J_n(0, t) = 0 \Leftrightarrow \phi'(0) = 0 \qquad (12)$$

$$q_n(L, t) = 0 \Leftrightarrow \phi(L) = 0 \qquad (13)$$

$$q_n(L, t) \frac{\partial}{\partial x} q_n(x, t)\big|_{x=L} = -\frac{1}{2L}[f(t)\phi(0)]^2. \qquad (14)$$

3) The total charge in the discharging potential well is

$$Q_w(t) = f(t) \int_0^L \phi(x)\ dx = f(t)N \qquad (15)$$

where

$$N = \int_0^L \phi(x)\ dx.$$

4) By substituting (10), (14), and (15) into (8) in the absence of a fringe field we obtain

$$\frac{d}{dt} Q_w(t) = -\frac{Q_w(t)}{\tau_s} + \frac{D_n}{N}\phi'(L)Q_w(t)$$

$$-\frac{\mu_n}{2LC_{\text{eff}}N^2}\phi^2(0)Q_w^2(t). \qquad (16)$$

It is interesting to note that the form of (16) is similar to that of the rate equation for the bimolecular re-

combination processes in photoconductivity [13]. Since (16) is an ordinary differential equation with constant coefficients, the following solution is readily obtained

$$Q_w(t) = A\frac{C_1 \exp[C_1 t]}{C_2(1 - A\exp[C_1 t])}. \qquad (17)$$

where

$$C_1 = -\left[\frac{1}{\tau_s} - \frac{D_n}{N}\phi'(L)\right] \qquad (18)$$

and

$$C_2 = -\frac{\mu_n}{2LC_{\text{eff}}N^2}\phi^2(0). \qquad (19)$$

Using (11), (15), and (17), the integration constant A can be found from

$$Q_w(0) = Nf(0) = N\frac{q_{n0}}{\phi(0)} = Lq_{\text{av}}$$

$$= \frac{AC_1}{C_2(1 - A)} \qquad (20)$$

where q_{av} is the average initial charge level per unit area. After some algebraic manipulations, we obtain

$$A = \frac{\mu_n q_{\text{av}}\phi^2(0)}{\mu_n q_{\text{av}}\phi^2(0) + 2N^2 C_{\text{eff}}[1/\tau_s - (D_n/N)\phi'(L)]}. \qquad (21)$$

Combining (17)–(21), we find the fraction $F(t, \tau_s)$ of the mobile charge remaining in the discharging potential well to be

$$F(t, \tau_s) = \frac{Q_w(t)}{Q_w(0)} = \frac{(1 - A)\exp[C_1 t]}{(1 - A\exp[C_1 t])}. \qquad (22)$$

Using the Einstein relation $D_n/\mu_n = kT/q$ we obtain the final form

$$F(t, \tau_s) = \frac{K\exp[-Kt/t_{\text{tr}}]}{K + (q_{\text{av}}L^2\phi^2(0)/2N^2 C_{\text{eff}})[1 - \exp[-Kt/t_{\text{tr}}]]} \qquad (23)$$

where

$$t_{\text{tr}} = \frac{L^2}{\mu_n}$$

is the discharging potential well transit time per unit voltage drop and

$$K = \begin{cases} \left[\frac{t_{\text{tr}}}{\tau_s} - \frac{D_n L^2 \phi'(L)}{\mu_n N}\right], & \text{for } \tau_s \text{ finite} \\ -\frac{D_n L^2 \phi'(L)}{\mu_n N}, & \text{for no surface state loss.} \end{cases} \qquad (24)$$

It should be noted that (23) and (24) are explicitly related to the device parameters. The form of (23) is the same as that of the decay equation for the bimolecular process [13] mentioned previously.

The charge transfer efficiency, when there is no surface state loss, is simply one minus the fraction of the charge remaining:

$$\eta(t) = 1 - F(t, \tau_s \rightarrow \infty). \qquad (25)$$

If surface state loss cannot be neglected, the transfer efficiency has to be developed from (7). This results in

$$\eta(t, \tau_s) = \frac{-1}{Q_w(0)} \int_0^t J_n(L, t) \, dt = - \int_0^t \frac{d}{dt} \frac{Q_w(t)}{Q_w(0)} \, dt$$

$$- \frac{1}{\tau_s} \int_0^t \frac{Q_w(t)}{Q_w(0)} \, dt = 1 - F(t, \tau_s)$$

$$- \frac{1}{\tau_s} \int_0^t F(t, \tau_s) \, dt$$

$$= 1 - F(t, \tau_s) - F_{\text{loss}}(t, \tau_s) \qquad (26)$$

where $F_{\text{loss}}(t, \tau_s)$ is the fraction of the charge lost during transfer due to trapping or recombination and can be obtained by integrating (23). We obtain

$$F_{\text{loss}}(t, \tau_s) = \frac{2(N/L)^2}{(\tau_s/t_{\text{tr}})(q_{\text{av}}/C_{\text{eff}})\phi^2(0)}$$

$$\cdot \left[- K \frac{t}{t_{\text{tr}}} + \ln \frac{1}{F(t, \tau_s)} \right] \qquad (27)$$

where K is given by (24).

The transient decay due to various transfer mechanisms can be readily obtained from (17)–(21) by using L'Hospital's rule. When there is no fringing field, we have obtained the following special cases.

Case 1: By neglecting the drift term in (16) we obtain for the diffusion mechanism acting alone

$$F(t, \tau_s) = \exp \left[- \left[\frac{1}{\tau_s} - \frac{D_n}{N} \phi'(L) \right] t \right]. \qquad (28)$$

Case 2: For the drift mechanism acting alone

$$F(t, \tau_s) = \frac{\exp \left[-t/\tau_s \right]}{\left[1 + (\mu_n \tau_s q_{\text{av}} \phi^2(0)/2N^2 C_{\text{eff}})(1 - \exp \left[-t/\tau_s \right]) \right]} \cdot \qquad (29)$$

Case 3: When there is no surface state loss and transfer is by drift alone

$$F(t, \tau_s) = \frac{1}{\left[1 + (\mu_n q_{\text{av}} \phi^2(0)/2C_{\text{eff}} N^2)t \right]} \cdot \qquad (30)$$

Equations (23)–(30) are general expressions that can be used for any given initial charge distribution within the validity limit of the postulate of instantaneous charge redistribution.

The choice of an initial charge distribution that satisfies boundary conditions (12) and (13) is not as difficult as it may seem. One of the choices is the cosine function, $\cos \pi x/2L$. This is the first term in the series expansion solution of the linear partial differential equation of continuity and also close to the charge decay distribution obtained in the numerical solution of the nonlinear transport equation by Strain and Schryer [2, Fig. 12]. Another choice for the charge de-

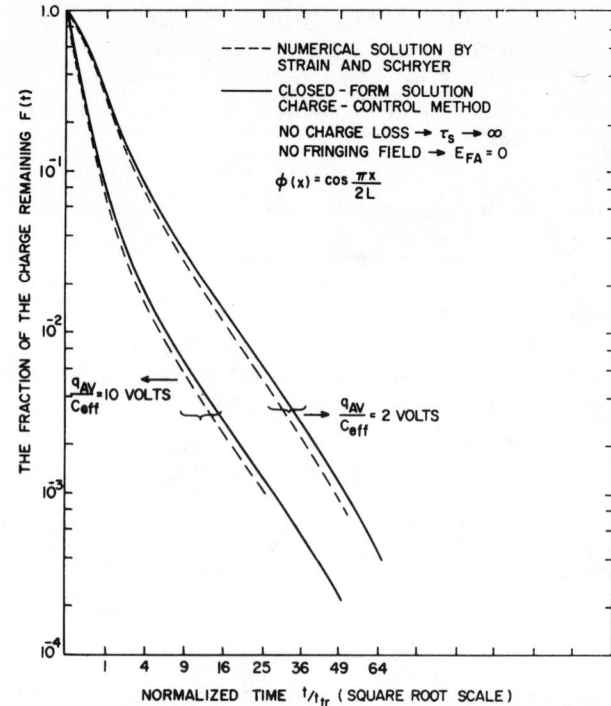

Fig. 2. The fraction of the charge remaining under the first plate as a function of the normalized time t/t_{tr}. A comparison of the closed-form analytical solution and the numerical solution by Strain and Schryer.

cay distribution is the form $1 - (x/L)^m$. Examples using these choices of $\phi(x)$ are shown and discussed in the following paragraphs.

The Cosine Distribution $\phi(x) = \cos \pi x/2L$

Using the cosine distribution results in

$$\phi(0) = 1$$

$$\phi'(L) = - \frac{\pi}{2L}$$

and

$$N = \int_0^L \cos \frac{\pi x}{2L} \, dx = \frac{2L}{\pi} \cdot \qquad (31)$$

Substituting (31) into (23) yields

$$F(t, \tau_s) = \frac{K \exp \left[-Kt/t_{\text{tr}} \right]}{K + (\pi^2 q_{\text{av}}/8C_{\text{eff}})(1 - \exp \left[-Kt/t_{\text{tr}} \right])} \qquad (32)$$

where

$$K = \left[\frac{\pi^2 D_n}{4\mu_n} + \frac{t_{\text{tr}}}{\tau_s} \right].$$

By comparing (32) when $\tau_s \rightarrow \infty$ with the numerical solution by Strain and Schryer, using the same set of constants, we find the close result shown in Fig. 2. Since the comparison is made on the same normalized time scale with the same initial charge levels, it shows good agreement over a wide range of parameters. The effects of the drift and the diffusion mechanisms are also ap-

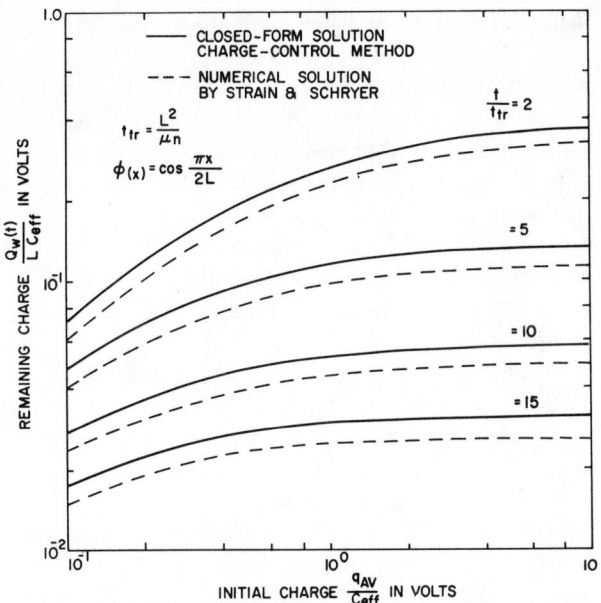

Fig. 3. The remaining charge at several normalized times after the initiation of transfer versus the initial charge q_{av}/C_{eff}. The comparison with the numerical solution is made under the conditions that no charge is lost to surface states and no fringe field exists.

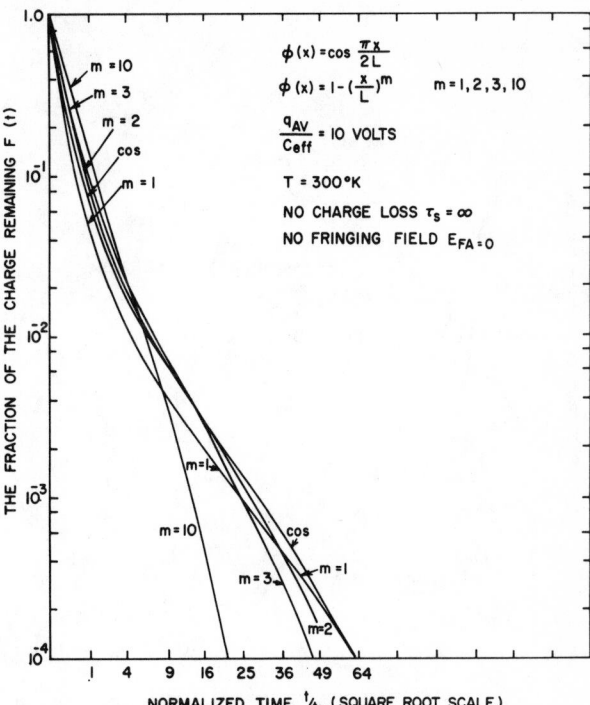

Fig. 4. A comparison of the fraction of the charge remaining F, using various charge distributions of a given total charge (subject to the validity limit of the charge-control method for large values of m).

parent from the two different slopes in the decay curves. Fig. 3 shows the remaining charge at several normalized times after the initiation of transfer versus the initial charge q_{av}/C_{eff}. Again, the comparison between (32) and the numerical solution by Strain and Schryer is made under the conditions that no charges are lost to surface states and no fringe field exists in the potential well. The charge transfer efficiency for one transfer can be obtained simply by substituting (31) and (32) into (27).

The Polynomial Distribution $\phi(x) = 1 - (x/L)^m$

Using the following substitutions in (23),

$$\phi(0) = 1$$

$$\phi'(L) = -\frac{m}{L}$$

and

$$N = \int_0^L \left[1 - \left(\frac{x}{L}\right)^m \right] dx = \frac{mL}{m+1}$$

results in

$$F(t, \tau_s)$$

$$= \frac{K \exp\left[-Kt/t_{tr}\right]}{K + \frac{1}{2}(1 + 1/m)^2 (q_{av}/C_{eff})\left[1 - \exp\left(-Kt/t_{tr}\right)\right]} \quad (33)$$

where

$$K = \left[(m+1)\frac{D_n}{\mu_n} + \frac{t_{tr}}{\tau_s} \right].$$

To test the validity limit of the postulated instantaneous charge redistribution or the charge-control

method, a comparison of the fraction of charge remaining $F(t, \tau_s \rightarrow \infty)$ for the above two distribution functions with the same total initial charge is illustrated in Fig. 4. We find the fraction of charge remaining $F(t, \tau_s \rightarrow \infty)$ for the ellipsoidal distribution ($m = 2$) is very close to the cosine distribution and Strain and Schryer's result. For high values of m in the polynomial distribution, the deviation becomes large due to the assumptions of (9) and (10) and the deviation of the diffusion decay constant away from the analytical result by Kim and Lenzlinger [3]. In the following sections, the results based on the cosine charge distribution will be exclusively used.

IV. Effects of the Fringing Field

A first order calculation of the fringing field effect due to adjacent gate plates can be incorporated into the general expression obtained previously. Recently, Carnes *et al.* [5] have formulated the surface fringing field distribution within the potential well and the carrier transit time corresponding to the average fringing field in the discharging potential well. The value of the average fringing field can be calculated using the equation given by Carnes *et al.*,

$$E_f = -13 \frac{x_{ox}V}{L^2}\left[\frac{5x_{d/L}}{1 + 5x_{d/L}}\right]^4 \quad (34)$$

where the following nomenclature is used.

x_{0x} Oxide thickness.

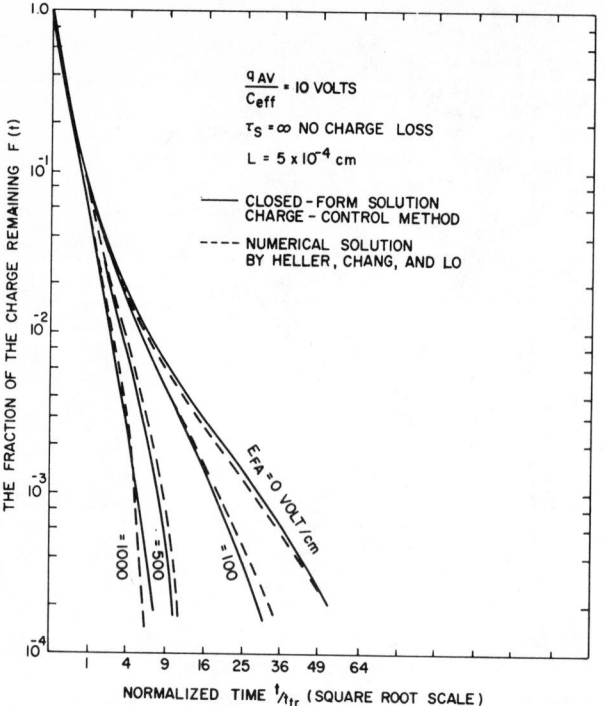

Fig. 5. The effect of a fringing field on the fraction of the charge remaining in the discharging potential well. The parameter is the fringing field in μ/volts per centimeter.

L Electrode center to center spacing.
x_d Depletion depth at the electrode.
V $\frac{1}{2}$ the total pulse voltage.

In the following paragraphs, we estimate a fringing field enhanced effective diffusion constant and evaluate this constant at the center of the discharging well. We lump the fringing field term and the diffusion term in (8) to obtain the fringing field enhanced diffusion constant D_f. We obtain

$$D_n\phi'(x) + \mu_n\phi(x)E_f = \left[D_n + \mu_n E_f \frac{\phi(x)}{\phi'(x)}\right]\phi'(x) \quad (35)$$

$$= D_f\phi'(x).$$

By using $\phi(x) = \cos \pi x/2L$, and evaluating D_f at $x = L/2$, we obtain

$$D_f = D_n - \frac{2\mu_n L E_f}{\pi} \quad (36)$$

where $E_f = E_{fR} > 0$ for a retarding field for an electron and $E_f = -E_{fA} < 0$ for an aiding field for an electron. The coefficient K in (32) now becomes

$$K = \frac{\pi^2 D_f}{4\mu_n} + \frac{t_{tr}}{\tau_s} = \frac{\pi^2 D_n}{4\mu_n} + \frac{\pi}{2}LE_{fA} + \frac{t_{tr}}{\tau_s} . \quad (37)$$

Fig. 5 illustrates the effect of the fringing field on the fraction of the charge remaining in the discharging potential well as a function of the decay time normalized to a carrier transit time L^2/μ_n. From this figure it is

readily seen that the later part of the decay, which is predominately due to the diffusion mechanism, is substantially accelerated by the fringing field, resulting in a faster transfer and an improved transfer efficiency.

Numerical solutions of the fringe field enhanced nonlinear transport equation, due to Heller, Chang, and Lo [6], are indicated by the broken-line plots in Fig. 5 for comparison with the charge-control results. In all the numerical and closed-form solutions, it is assumed that the fringe field is constant in both time and space within the discharging potential well. Under these assumptions the numerical and closed-form solutions compare well, as indicated in the figure. Since the fringe field affects only the last part of charge transfer, when the electric field is slowly approaching a static configuration, it seems reasonable to approximate the actual fringe field by a constant field equal to the space average given by (34). However, (34) should be used to calculate the fringe field near the end of transfer. In other words, the pulse voltage V used in (34) may have to be adjusted to include the effect of the transferred charge in the charging potential well.

V. THE EFFECT OF CHARGE LOSS TO SURFACE STATES

The existence of charge loss to surface states has been described first by Strain [14] and subsequently by Tompsett [15]. The surface states are characterized by their density N_s, capture cross section σ, distribution in energy E_s, and by the mean thermal velocity of surface carriers ν [16]. Electronic transitions or the filling and emptying of states occur between the states and the semiconductor bands at the surface. Thus, from the rigorous standpoint, we should solve the nonequilibrium rate equations for electron trapping in the surface states [13]–[18], in addition to the continuity equation in the charge-control form. However, in order to obtain a tractable solution we carry out our analysis based on a hypothetical single-level interface state. Such an approach is still preliminary in nature but the details of surface-state effects can be considered more readily. The coupled differential equations for a single-level interface state are

$$\frac{dQ_w(t)}{dt} = -\frac{dQ_s(t)}{dt} + J_n(L, t) \quad (38)$$

$$\frac{dQ_s(t)}{dt} = \frac{Q_w(t)}{\tau_c} - \frac{Q_s(t)}{\tau_e} \quad (39)$$

where

$$\tau_c = \frac{1}{(\sigma\nu/x_i)[N_s - Q_s(t)/qL]} \quad (40)$$

and

$$\tau_e = \frac{1}{\sigma\nu N_c} \exp[E_s/kT]. \quad (41)$$

Equation (39) implies that the net rate of change of trapped carriers $Q_s(t)$ is equal to the difference between the rate at which mobile electrons are captured by surface state traps and the rate at which the trapped electrons leave the surface states [16]–[18]. Equation (40) indicates [17] that τ_c, the life time of a mobile carrier, is inversely proportional to the density of empty traps $N_s - Q_s(t)/qL$. The quantities N_s and $Q_s(t)$ are the total density of surface states per unit area and the total charge trapped in the surface states per unit device width in the discharging potential well, respectively. The emission rate (the probability per unit time that an occupied surface state will lose its electrons to the conduction band) is $1/\tau_e$ [13], [16]. In other words, τ_e is the average time that an electron stays in a trap [17]. From (41) we observe that τ_e is exponentially dependent on the temperature and the energy depth of the surface states with respect to the conduction band. The effective density of states in the conduction band is N_c and x_i is the depth of the inversion layer.

A transient solution of the coupled nonlinear differential equations (38) and (39) has been solved numerically and the result is shown in Fig. 6. The net surface state density used is 10^{10} states/cm² and a capture cross section σ_n equal to 5.9×10^{-16} cm² has been used in the plot. The parameter shown is the energy depth of the single trap levels normalized to kT. The figure indicates that as the trap level becomes deeper, the capture process dominates over the emission process during the charge transfer. Within the validity limit of the hypothetical single-level interface state analysis, there seems to exist a critical range of $E_s/kT \approx 8$ in which surface traps degrade signal transfer.[1] For $E_s/kT < 8$ charge is released sufficiently rapidly that surface states are of no consequence. For $E_s/kT > 8$ the traps cannot empty sufficiently rapidly; thus using fat zeros, the traps remain nearly full and are again of no consequence.

A simplified approach may be taken if a closed-form solution is desired. If charge loss to surface states exists during charge transfer from one well to the next well through the interelectrode region, due to trapping and some possible recombination by the subsequent capture of majority carriers (a rate equation [19] describing the transition between the surface states and the valence band is required in addition to the previously discussed coupled nonlinear differential equations), we may lump all the complicated mechanisms into a constant effective charge loss parameter τ_s and replace the right-hand side of (39) by $Q_w(t)/\tau_s$. This reduces (38) to (7). The charge released from the traps after transfer is then included in the charge remaining in the discharged potential well. Based on the simplified model, the effect of the charge loss time constant on the fraction of the charge remaining, in the discharging potential well, and

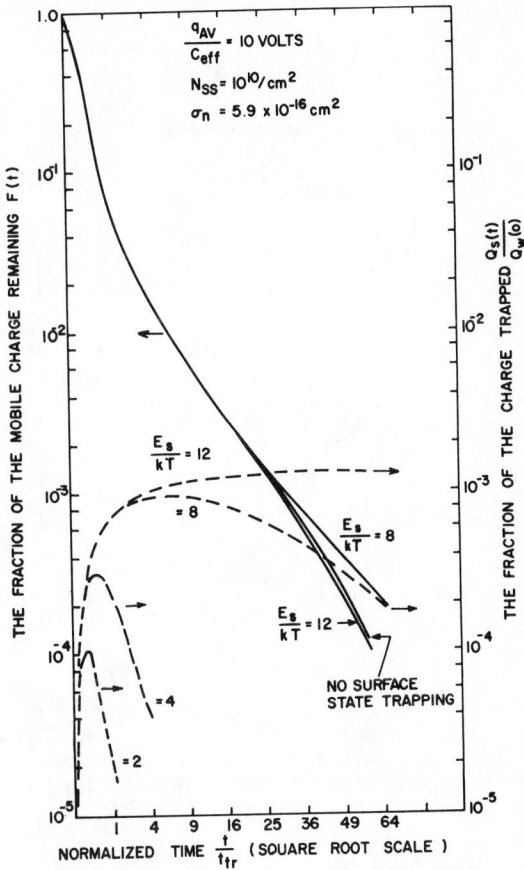

Fig. 6. The effect of the charge trapping on the fraction of the charge remaining in the discharging potential well and the fraction of the charge trapped in the surface states versus normalized time.

the fraction of the charge lost is shown in Fig. 7. The loss time constant is normalized to the carrier transient time $t_{tr} = L^2/\mu_n$. It is clear from Fig. 7 that when the loss time constant becomes comparable to the carrier transit time, the charge loss to surface states increases significantly. On comparing Figs. 5 and 7, we find that for a fringing field strength of 0.01 V/μ, the fraction of the charge remaining in the discharging potential well is approximately the same as that for the case when the time constant τ_s approaches 10 t_{tr}. However, the drift-aiding fringing field accelerates the charge transfer process and thereby improves the transfer efficiency, while the trapping or recombination process degrades the transfer efficiency by losing charges to the interface states. We have assumed the charge loss time constant τ_s to be independent of the amount of charge trapped in the surface states. However, the following time dependent relationship follows from (39):

$$\frac{1}{\tau_s} = \frac{\sigma \nu}{x_i}\left[N_c - \frac{Q_s(t)}{qL} \right] - \frac{Q_s(t)}{\tau_e Q_w(t)}. \tag{42}$$

Note that the loss time constant τ_s increases when the surface states are filled by charges $Q_s(t)$. A practical application of this concept to reduce the surface states effect is the use of a background charge [15] or the fat

[1] The authors would like to thank the reviewer for emphasizing this point.

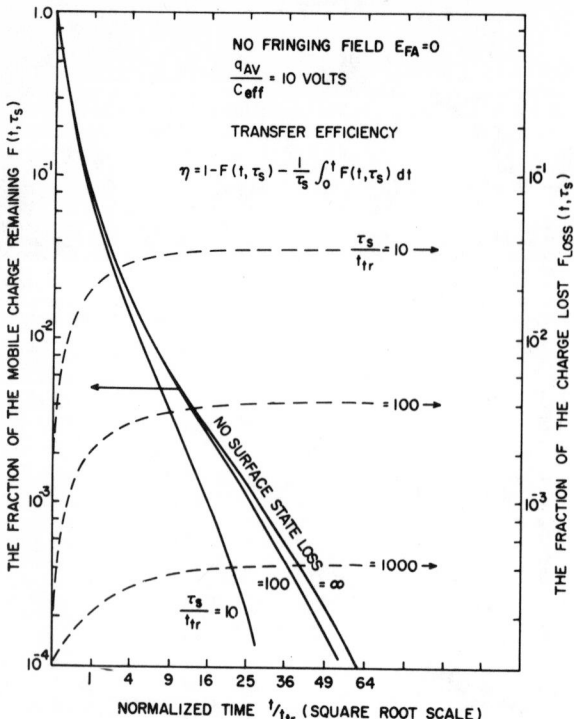

Fig. 7. The effect of the charge loss time constant τ_s on the fraction of the charge remaining in the discharging potential well and the fraction of the charge lost versus normalized time.

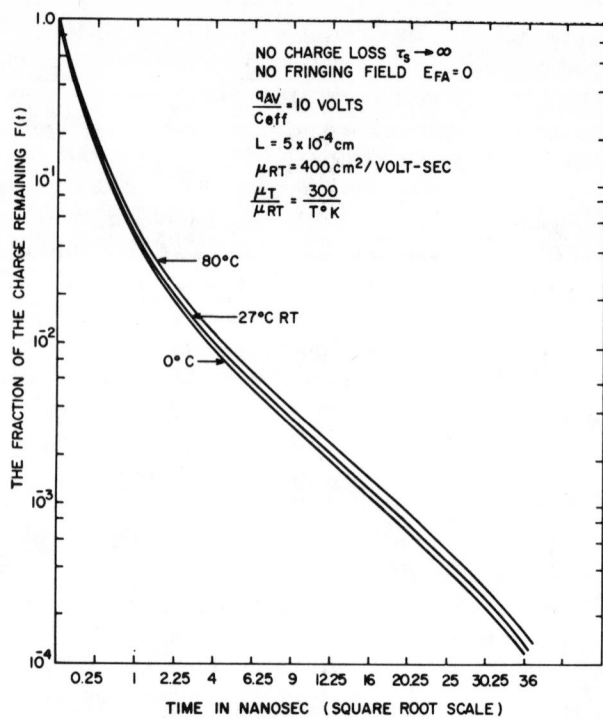

Fig. 8. The temperature dependence of the fraction of the charge remaining in the discharging potential well with temperature as the parameter. The temperature dependence of the surface mobility is included.

zero mode of operation which is discussed in Section VII.

VI. TEMPERATURE DEPENDENCE

One of the important questions concerning charge-transfer devices is the effect of temperature on their transfer efficiency. If charge loss to surface states is neglected, the only temperature sensitive parameter is the surface mobility. In the temperature range of interest, 0–100°C, it is known that the surface mobility follows the $1/T$ dependence [20], where the temperature T is in degrees Kelvin. Using (32), in the absence of charge loss to surface states, the explicit expression for the fraction of the charge remaining in the discharging potential well in terms of the temperature is

$$F(t, \tau_s \to \infty)$$

$$= \frac{(kT/q)\exp\left[-(\pi^2 kT/4L^2 q)\mu_n t\right]}{kT/q + (q_{av}/2C_{eff})(1 - \exp\left[-(\pi^2 kT/4L^2 q)\mu_n t\right])} \quad (43)$$

where

$$\mu_n = \frac{300}{T}\mu_{RT}$$

and μ_{RT} is the surface mobility at room temperature. Note that the exponential term is no longer dependent on the absolute temperature T; the transfer efficiency becomes a very weak function of temperature. For an increase in temperature, it is seen from (43) that the transfer efficiency decreases only slightly. Fig. 8 shows the fraction of the charge remaining versus the transfer

time for the temperature range of actual interest. The n-channel mobility has been taken to be 400 cm²/V·s at room temperature. The plate length and the initial charge level are taken to be 5 μ and 10 V, respectively.

If charge trapping or recombination is not neglected, the transfer efficiency may further depend on the temperature through the electronic transitions between the surface states and the conduction and valence bands. Since the time constant τ_e, given by (41), associated with the emptying of a trap level decreases exponentially with an increase in temperature, the release of trapped charges may affect the latter part of the discharging process. At higher temperatures the thermal generation of charges from the field-induced depletion regions contribute background charges and may reduce the effect of surface states.

VII. FAT ZERO OPERATION

The effect the digital zero charge level has on CCD operation is discussed in this section. For simplicity, we analyze the case when a digital one follows a digital zero in a shift register and the one is transferred from the wth potential well to the $(w+1)$th potential well. Only one transfer will be considered here. However, our results can be readily extended to the analysis and simulation of the signal transfer characteristics of CCD shift registers. The quantities $Q_w(0)$ and $Q_w(T)$ are used to represent the charge, in the wth well, that is associated with a digital bit at the beginning and at the end of transfer, respectively, where T is the transfer pulse width. A superscript is used to indicate whether

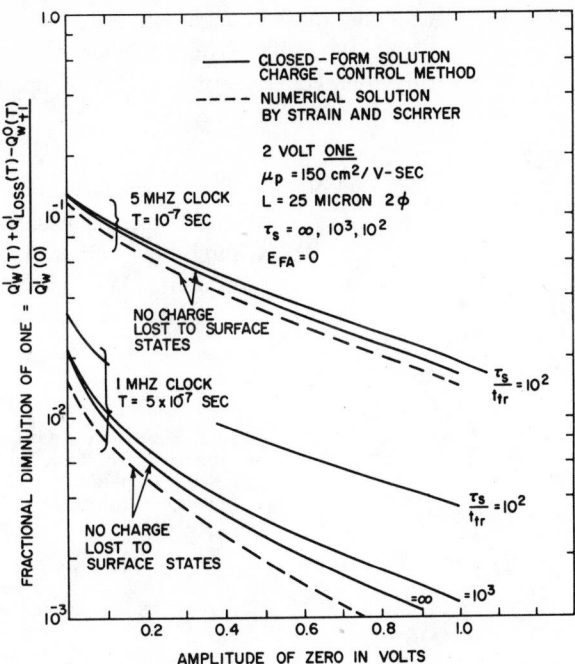

Fig. 9.   The fractional diminution of a one as a function of the amplitude of a fat zero. A comparison of the closed-form solution and the numerical solution by Strain and Schryer is shown for the conditions when no charge is lost to surface states and no fringe field exists. Also shown is the fractional diminution of a one with various charge loss time constants.

Q_w represents a digital one or zero. $Q_{\text{loss}}(T)$ is the amount of charge lost to surface states during the transfer of the one from the wth to the $(w+1)$th potential well. Using the simplified approach discussed in Section V, the qualitative aspects of charge loss will be studied by characterizing $Q_{\text{loss}}(T)$ in terms of the effective charge loss time constant τ_s rather than numerically solving the coupled differential equations (38) and (39). $Q_{w+1}^0(T)$ is the charge left behind in the $(w+1)$th well by the zero. We neglect the effect of any thermally generated leakage during storage. Prior to the transfer of the two adjacent bits, the signal "window" above the amplitude of the zero is $Q_w^1(0) - Q_{w+1}^0(0)$. In transferring from the wth well to the $(w+1)$th well, the one loses the charge $Q_w^1(T) + Q_{\text{loss}}(T)$ and gains the charge left in the $(w+1)$th well by the zero $Q_{w+1}^0(T)$. Thus, the fractional diminution of the one, resulting from the transfer, may be defined as

$$F_{\text{diminution}} = \frac{Q_w^1(T) + Q_{\text{loss}}(T) - Q_{w+1}^0(T)}{Q_w^1(0)} \quad (44)$$

where

$$Q_{\text{loss}}(T) = Q_w^1(0) F_{\text{loss}}(T, \tau_s). \quad (45)$$

Fig. 9 shows the fractional diminution of a one as a function of the amplitude of a fat zero. A comparison of the closed-form solution, based on (32) and (27), and the numerical solution by Strain and Schryer is shown for the case when no charge is lost to surface states and

the fringing field is negligible. The parameters used in the closed-form solution are the same as those used by Strain and Schryer. For two-phase operation, we have related the clock frequency f to the transfer pulse width by $2T = 1/f$. Again, good agreement between the two solutions is obtained. Also shown in the same figure are the cases with various constant charge loss time constants. It is seen that even in the presence of surface state losses, the signal transfer characteristics improve with the amplitude of the fat zero.

If the charge loss time constant τ_s is not assumed a constant but varies with the mobile and trapped charge levels, the fractional diminution of a one will be even less for a given amplitude of a fat zero, as mentioned in Section V. The fractional diminution of a one with charge loss versus fat zero amplitude should decrease faster than is shown in Fig. 9.

Two important performance parameters for charge-transfer dynamic shift registers have been introduced by Berglund [21] and discussed by Thornber [22] and also by Joyce and Bertram [23]. These parameters are the fraction of the charge lost per transfer $F_{\text{loss}}(T, \tau_s)$ and the fraction of the incomplete signal transferred per transfer α. In Section III, $F_{\text{loss}}(T, \tau_s)$ was shown to be related to the integral of the fraction of the charge remaining $F(t, \tau_s)$. A closed-form expression for $F_{\text{loss}}(T, \tau_s)$ is given by (27). However, the reader is cautioned that the use of (27) for evaluating $F_{\text{loss}}(T, \tau_s)$ requires the numerical solution of the coupled equations

(38) and (39) so that the effective loss time constant τ_s can be estimated.

When charge loss is small, α can be usefully defined by the form

$$\alpha \equiv \frac{Q_w{}^1(T) - Q_{w+1}{}^0(T)}{Q_w{}^1(0) - Q_{w+1}{}^0(0)} . \quad (46)$$

Substituting (32) into (46) results in

$$\alpha = \frac{\exp\left[-KT/t_{\text{tr}}\right]}{K_1{}^2 Q_w{}^1(0) Q_{w+1}{}^0(0) + K_1 Q_w{}^1(0) + K_1 Q_{w+1}{}^0(0) + 1} \quad (47)$$

where

$$K_1 = \frac{(1 - \exp\left[-KT/t_{\text{tr}}\right])\pi^2}{8 K L C_{\text{eff}}} . \quad (48)$$

When diffusion of carriers is neglected ($K \to 0$), it is easy to show that

$$\alpha \text{ goes as } \begin{cases} 1/T, & \text{for } Q_{w+1}{}^0(0) = 0 \\ 1/T^2, & \text{for } Q_{w+1}{}^0(0) > 0. \end{cases}$$

In the small-signal limit [21]

$$\alpha \to \frac{dQ_w(T)}{dQ_w(0)} = \frac{\exp\left[-KT/t_{\text{tr}}\right]}{K_1{}^2 Q_w{}^2(0) + 2 K_1 Q_w(0) + 1} . \quad (49)$$

$dQ_w(T)/dQ_w(0)$ is indicated by the slopes of the curves in Fig. 3. Note that the small-signal α decreases rapidly with $Q_w(0)$.

VIII. SUMMARY

We have presented an analysis of the charge transfer in charge-coupled devices using the charge-control approach and obtained a practical closed-form solution for device design tradeoffs and circuit analysis programs. The results were first compared with the numerical solutions by Strain and Schryer and good agreement was obtained over a wide range of parameters. The charge-control analysis is then extended to include the effect of an assumed constant fringing field and the solution is checked by comparing it with previous results obtained by the numerical approach. The effect of surface states is also analyzed by presenting the rate equation governing the electronic transitions between a surface state trap level and the conduction band, in addition to the continuity equation in the charge-control form. In order to extend the closed-form solution to include surface state effects, an effective charge loss time constant τ_s is introduced, and the effect of this charge loss time constant is evaluated in comparison with the effect of a fringe field. Temperature dependence of the charge transfer efficiency is also analyzed. Finally, the effect of a fat zero on the signal transfer from one potential well to the next, in light of charge loss to surface states, is reassessed. It is believed that the charge-control method of analysis can permit deeper physical understanding of the operation of charge-coupled devices and provide possible extensions to the analysis of other surface transport devices in various modes of operation.

ACKNOWLEDGMENT

The authors wish to thank Dr. W. H. Chang for his stimulating discussions and Drs. B. Agusta and J. J. Chang for their support of this work and their useful comments. We also thank Prof. A. W. Lo of Princeton University for his contribution to the initial phases of this study.

REFERENCES

[1] W. S. Boyle and G. E. Smith, "Charge coupled semiconductor devices," *Bell Syst. Tech. J.*, vol. 49, pp. 587–593, 1970.
[2] R. J. Strain and N. L. Schryer, "A nonlinear diffusion analysis of charge-coupled-device transfer," *Bell Syst. Tech. J.*, vol. 50, pp. 1721–1740, 1971.
[3] C. K. Kim and M. Lenzlinger, "Charge transfer in charge-coupled devices," *J. Appl. Phys.*, vol. 42, pp. 3586–3594, 1971.
[4] W. E. Engler, J. J. Tieman, and R. D. Baertsch, "Surface charge transport in silicon," *Appl. Phys. Lett.*, vol. 17, pp. 469–472, 1970.
[5] J. E. Carnes, W. F. Kosonocky, and E. G. Ramberg, "Drift-aiding fringing fields in charge-coupled devices," *IEEE Trans. Solid-State Circuits (Special Issue on Semiconductor Memories and Digital Circuits)*, vol. SC-6, pp. 322–326, Oct. 1971.
[6] L. G. Heller, W. H. Chang, and A. W. Lo, "Generalized model for surface-charge-transfer devices," presented at the Device Research Conf., Ann Arbor, Mich., June 1971.
[7] D. Koehler, "The charge-control concept in the form of equivalent circuits, representing a link between the classic large signal diode and transistor models," *Bell Syst. Tech. J.*, vol. 46, pp. 523–576, 1967.
[8] R. Beaufoy and J. J. Sparkes, "The junction transistor as a charge-controlled device," *ATE J.*, vol. 13, no. 4, pp. 310–324, 1957.
[9] H. K. Gummel, "A charge control relation for bipolar transistors," *Bell Syst. Tech. J.*, vol. 49, pp. 115–120, 1970.
[10] J. S. T. Huang, "Charge control approach to the small signal theory of field-effect devices," *IEEE Trans Electron Devices*, vol. ED-16, pp. 775–781, Sept. 1969.
[11] S. M. Sze, *Physics of Semiconductor Devices*. New York: Wiley, 1969, pp. 517–518.
[12] S. R. Hofstein and G. Warfield, "Physical limitations on the frequency response of a semiconductor inversion layer," *Solid-State Electron.*, vol. 8, pp. 321–341, 1965.
[13] R. H. Bube, *Photoconductivity of Solids*. New York: Wiley, 1960, pp. 277–278.
[14] R. J. Strain, "Power and surface state loss analysis of charge coupled devices," presented at the IEEE Electron Devices Meeting, Washington, D. C., Oct. 1970.
[15] M. F. Tompsett, "The quantitative effects of interface states on the performance of charge coupled devices," presented at the IEEE Electron Devices Meeting, Washington, D.C., Oct. 1971.
[16] P. V. Gray, "The silicon–silicon dioxide system," *Proc. IEEE*, vol. 57, pp. 1543–1551, Sept. 1969.
[17] T. J. O'Reilly, "Effect of surface traps on characteristics of insulated-gate field-effect transistors," *Solid-State Electron.*, vol. 8, pp. 267–274, 1965.
[18] R. R. Haering, "Theory of thin film transistor operation," *Solid-State Electron.*, vol. 7, pp. 31–38, 1964.
[19] D. R. Frankl, *Electrical Properties of Semiconductor Surfaces*. New York: Pergamon, 1967, pp. 58.
[20] L. Vadasz and A. S. Grove, "Temperature dependence of MOS transistor characteristics below saturation," *IEEE Trans. Electron Devices*, vol. ED-13, pp. 863–866, Dec. 1966.
[21] C. N. Berglund, "Analog performance limitations of charge-transfer dynamic shift registers," *IEEE Trans. Solid-State Circuits (Special Issue on Analog Integrated Circuits)*, vol. SC-6, pp. 391–394, Dec. 1971.
[22] K. K. Thornber, "Incomplete charge transfer in IGFET bucket-brigade shift registers," *IEEE Trans. Electron Devices*, vol. ED-18, pp. 941–950, Oct. 1971.
[23] W. B. Joyce and W. J. Bertram, "Linearized dispersion relation and Green's function for discrete-charge-transfer devices with incomplete transfer," *Bell Syst. Tech. J.*, vol. 50, pp. 1741–1759, 1971.

Noise Sources in Charge-Coupled Devices

J. E. Carnes and W. F. Kosonocky

RCA Laboratories, Princeton, N. J.

Abstract—Potential noise sources in charge-coupled devices are analyzed and their contributions to the rms fluctuation in the number of charge carriers in a signal packet are evaluated. The noise sources considered are (1) transfer loss noise, (2) background charge generation noise, (3) output amplifier noise, and (4) fast interface state trapping noise. The comparison of the relative contribution of these sources shows that for a reasonable number of transfers (> 100) fast state trapping should dominate the noise behavior of conventional interface channel CCD's.

1. Introduction

To date, most of the analytical and experimental work on charge-coupled devices (CCD's)[1] has centered around maximum achievable transfer efficiency (η), since this performance parameter is the most important in determining the practical maximum number of stages between signal refreshing. However, recent announcements of experimentally observed η values[2] of up to 99.99% at 1 MHz have demonstrated CCD practicality, confirmed transfer efficiency analyses, and, in addition, resulted in increased attention to other CCD performance characteristics such as noise, which so far has received only minimal coverage in the CCD literature.[3] The assessment of CCD

Reprinted with permission from *RCA Rev.*, vol. 33, pp. 327–343, June 1972.

noise is important in establishing the minimum size of CCD memory elements, the noise figure of CCD analog delay lines, and the dynamic range and low-light-level sensitivity of CCD image sensors.

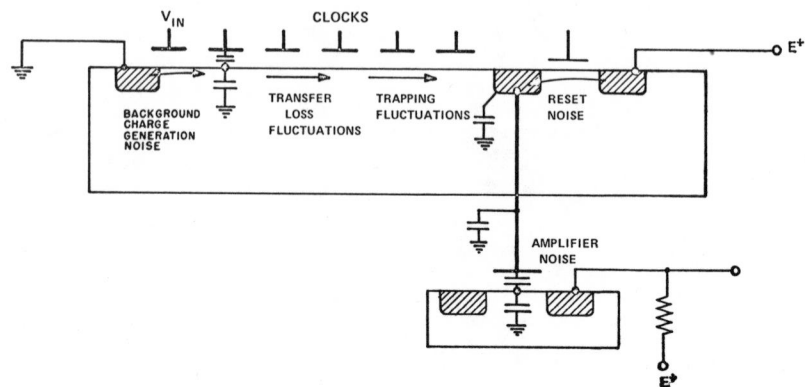

Fig. 1—Cross-sectional view of charge-coupled device showing origin of various noise sources.

This paper, which is essentially analytical in nature, discusses the various noise sources expected in typical CCD operation. Every effort has been made to use simple physical ideas, probability concepts, and noise models, and to avoid complicated mathematical procedures. Since CCD's are basically charge-packet shift registers with no gain mechanism (so that the same carriers remain essentially intact in the same packets throughout the entire transfer process), the rms fluctuation in the number of carriers in each packet has been used to quantify and compare the various noise sources. This quantity is denoted by $\overline{N}_n$, the rms fluctuation in the total number of carriers in each charge packet being transferred by the CCD.

Fig. 1 is a schematic cross-section of a CCD labeled with the various noise sources we have considered. These sources are (1) transfer loss fluctuations; (2) background charge generation noise; (3) output amplifier noise, including reset noise for floating diffusion and floating gate voltage-sensing and MOS field-effect transistor (MOSFET) noise; and (4) fast interface state noise. It is shown that the largest fluctuation will result from fast state trapping in conventional surface channel CCD's. This mechanism should limit the sensitivity of CCD image sensors and the minimum gate area and packing density of CCD digital memories.

2. CCD Noise Sources

2.1 Transfer Loss Fluctuations

For CCD's operating in the "complete" charge-transfer mode a small fraction ϵ of the charge is left behind at each transfer. If N_s is the total number of carriers in the signal packet, then on the average ϵN_s will be left behind at each transfer. There will, however, be fluctuations about this average with mean-squared value of $2\epsilon N_s$, i.e., shot noise introduced into the signal packet once upon entering and a second time upon leaving the potential well. If ϵ is independent of the amount of charge being transferred, as it is for incomplete free charge transfer and small signal fast interface state trapping losses, the fluctuations at each transfer will be independent of one another and the mean-squared fluctuations will add.

Both signal carriers and background carriers will fluctuate because of this mechanism. Thus,

$$\overline{N}_{n,\ \text{Transfer Loss}} = [2\epsilon N_g\ (N_s + N_{s,o})]^{1/2}, \tag{1}$$

where ϵ is the fractional loss per gate or transfer

N_g is the number of gates or transfers

N_s is the number of signal carriers per charge packet

$N_{s,o}$ is the number of background carriers ("fat zero") per charge packet.

2.2 Background Charge Generation Noise

Background charge may either be electrically generated at the input as fat zeros to suppress fast interface state losses or it may be generated thermally or optically via bias light. In any case, this background charge will be noisy. If it is introduced thermally or optically (both processes are stochastic with Poisson-distributed arrival times), the variance in the average number of carriers in each packet will be that average number, i.e., shot noise.

$$\overline{N}_{n,\text{Background Charge, Thermal}} = (N_{s,o})^{1/2} \tag{2}$$

If the device is cooled, thermal generation can be suppressed and the background charge or fat zero may be introduced electrically at the input with less than shot noise. If a MOSFET is used to introduce background charge into the first potential well once during each cycle, then the analysis of Appendix 1 indicates that the rms fluctuation would be:

$$\overline{N}_{n,\text{ Background Charge, Electrical}} = 400 \ (C_{pf})^{1/2} \qquad [3]$$

where C_{pf} is the capacitance of the potential well in picofarads.

2.3 Output Amplifier Noise

In an amplifier whose bandwidth is limited by the RC time constant at the input, the thermal noise associated with the input resistance is the major noise source. This resistance is in parallel with the input capacitance of the amplifier, and, as shown in Appendix 1, the rms carrier fluctuation at the input is given by

$$\overline{N}_{n,\text{ Amp } RC} = \frac{1}{q} \ (kTC)^{1/2} = 400 \ (C_{pf})^{1/2} \qquad [4]$$

However, the above analysis applies only for the case where the bandwidth is determined by the input RC time constant. In a CCD this is not necessarily the case. In the voltage-sensing amplifier shown in Fig. 1, the voltage on the input gate is proportional to the signal charge in the floating diffusion. Since the floating diffusion is reset to a reference potential once each clock period, the bandwidth is determined by the Nyquist criterian ($\Delta B = f_c/2$) and is not RC limited. Thus, the input resistance can be arbitrarily large and the noise will be determined by the resetting of the floating diffusion.

a. *Floating Diffusion Reset Noise*

In this voltage-sensing scheme, the floating diffusion is directly connected to the gate of the output transistor, and the potential of floating diffusion controls the output current. The floating diffusion is also connected to a fixed voltage source E_o via an MOS channel, which is periodically turned on to reset the floating diffusion to E_o. Alternately stated, the signal charge is removed from the floating diffusion once each cycle by the MOS channel. Several methods for resetting the floating diffusion are discussed in Ref. [4]. The noise introduced by this resetting will be the thermal noise of the MOS channel resistance in parallel with the floating diffusion capacitance. As shown in Appendix 1, the rms carrier fluctuation will be

$\sim 400\ \sqrt{C_{pf}}$, where C_{pf} is the total capacitance associated with the floating diffusion and amplifier input gate in picofarads.

$$\overline{N}_{n,\ \text{Reset Diffusion}} = 400\ (C_{pf})^{1/2}. \qquad\qquad [5]$$

b. *Floating Gate Reset Noise*

In the floating gate sensing scheme, a gate above the channel oxide can be set to a fixed potential E_o through a large resistance, or reset to E_o through an MOS transistor at a frequency less than the clock frequency. Since the capacitance between the substrate and this floating gate C_g is the series combination of the oxide capacitance C_o and depletion layer capacitance C_d, it will increase when signal charge is present. Because this gate is floating and has fixed charge, its voltage will decrease as C_g increases, i.e., when signal charge is present at the Si-SiO$_2$ interface. The sensitivity of the floating gate method can be comparable to that of the floating diffusion. The floating gate is connected directly to the gate of the MOSFET amplifier.

The noise introduced by this scheme will again be the thermal noise of the resistance through which the floating gate is reset. This resistance can be large enough so that the time constant RC (where C is the sum of the floating gate and amplifier gate) is large compared to the operating period of the CCD. The current through R need only be large enough to restore charge lost by leakage. Thus, while the thermal noise of the resetting resistor will be large up to frequencies of approximately $(\pi RC)^{-1}$, it will not be large in the bandwidth of interest, and appropriate filtering will eliminate it. Alternatively, if the floating gate is reset once each frame time by an MOS transistor, the reset noise will again be below the video bandpass and can be removed by filtering.

Thus, by this alternate approach to voltage sampling, the resetting noise of the output amplifier can be essentially eliminated.

c. *MOSFET Noise*

There are, however, other noise sources in a MOSFET amplifier, namely thermal channel noise[5] and $1/f$ surface noise.[6] Both these sources create fluctuations in the drain current that can be referred back to the input as fluctuations in the number of electrons on the input gate. As proposed by Klaassen and Prins,[5] this thermal noise of the channel is equivalent to an input noise resistance R_n given by

$$R_n = \frac{\alpha}{g_{mo}},$$ [6]

where g_{mo} is the transconductance of the device and α varies between $\tfrac{2}{3}$ and 10 and depends upon gate and drain voltage, oxide capacitance, and substrate doping.

The other well-known source of noise in MOSFETs is the $1/f$ noise usually associated with fast interface state trapping. $1/f$ noise has a noise power spectrum proportional to $1/f$ and can be related to the thermal noise if the frequency where they are equal f' is known.

$$\overline{v_n{}^2}_{\text{MOSFET}} = \widehat{\overline{v_n{}^2}}_{\text{THERMAL}} \left(1 + \frac{f'}{f}\right).$$ [7]

Klaassen[6] reports f' values ranging between 10^4 Hz for low current levels to 10^5 Hz at high current levels. Work at this Laboratory on low-noise silicon-gate MOSFET's has also produced cross-over frequencies of 10^4 Hz[7]. For many CCD applications, appropriate filtering of the output signal can, therefore, eliminate most $1/f$ noise, and the input-referred mean-squared thermal noise voltage of the channel will be thermal white noise:

$$\overline{v_n{}^2} = \frac{4kT\Delta B\alpha}{g_{mo}}.$$ [8]

Thus, the rms carrier fluctuation at the input will be

$$\overline{N}_{n,\text{ MOSFET}} = \frac{C_{\text{gate}}}{q} \, \overline{v}_n$$ [9]

$$\overline{N}_{n,\text{ MOSFET}} = 60\, C_{\text{gate}} \left[\left(\frac{\Delta B}{5\text{MHz}}\right)\left(\frac{1000\ \mu\text{mho}}{g_{mo}}\right)\right]^{1/2}$$ [10]

where C_{gate} is measured in pf.

Thus for a C_{gate} value of 0.1 pf or less, the fluctuation introduced by the output amplifier MOSFET itself is anticipated to be negligible compared with other sources.

2.4 Fast Interface State Noise

Fast interface states[8] will generate noise in CCD's even in the absence of signal charge. Suppose we have a CCD that is operating either in the bias charge* or the complete transfer with fat zero mode with no signal input. On the average, the fast state occupation will either remain constant for bias charge operation or fill and empty to the same level during each period for complete transfer. On the average, then, there will be no net loss of charge into fast states. However, there will be fluctuations in the total number of carriers trapped at any instant of time. These fluctuations will be reflected in the free charge in the potential well, and hence in the output. If the mean-squared fluctuation $N_t{}^2$ in the total trapped charge per unit area is independent of the free background charge levels, as it will be for background charge levels high enough to effectively suppress fast state losses, then the mean-squared fluctuations a particular charge packet encounters at each gate as the charge packet transits through the device must add. Further, the fluctuation in the number of carriers in the charge packet will reflect the fluctuations in the trapped charge twice at each gate—once when the packet is transferred in and once when it is transferred out. As shown in Appendix 2, the $N_t{}^2$ for bias charge operation is simply kTN_{ss}; while as shown in Appendix 3, for complete transfer operation with fat zero, it is $0.7\ kTN_{ss}$.

Thus, after N_g transfers, in the case of complete transfer with fat zero, the rms fluctuations will be:

$$\overline{N}_{n,\ \text{Trap}} = [1.4\ kTN_{ss}N_gA_g]^{1/2} = 670 \left[\frac{N_{ss}}{10^{10}} \frac{N_g}{10^3} \frac{A_g}{1.2 \times 10^{-6}} \right]^{1/2}, \quad [11]$$

where A_g is the area of one gate in cm^2 and N_{ss} is in units of $(\text{cm}^2\text{-}eV)^{-1}$.

3. Discussion

The degree to which this analysis of CCD noise agrees with actual CCD noise performance will depend upon the thoroughness with which we have identified all of the noise sources. For this reason a dis-

---

* The bias charge mode occurs when a certain amount of charge (bias charge) is held in each potential well continuously. The transfer-loss noise analysis discussed earlier, however, considered only the complete transfer mode of operation.

93

cussion of why certain potential noise sources have not been included is in order.

First, switching noise is expected to be minimal in CCD sensors and delay lines because all switching noise will be at least twice the maximum signal frequency and can be filtered out by various techniques.[9]

Another possible candidate for a CCD noise source is $1/f$ noise in the CCD channel. $1/f$ noise is a generic term used to describe noise with a $1/f$ or close to $1/f$ power spectrum. It appears in many different electronic systems. $1/f$ noise in MOSFET transistors is often attributed to some type of interface trapping[10,11] and is observed to be proportional to the fast state density[6] and, therefore, might be expected to contribute in CCD's. There is no general agreement as to the exact mechanism which gives rise to $1/f$ noise in MOSFET's. The $1/f$ spectrum can be described by a set of nonconducting electronic states, equally-populated, that can communicate with the conducting channel with a wide range of response times τ. Carriers from the conducting channel are trapped in these states and detract from the output charge in the ratio τ/T_{tr} where T_{tr} is the transit time of carriers in the channel. The $1/f$ spectrum then arises because fluctuations in the long time constant states (low frequencies) contribute a greater fluctuation in the output than short time constant states (high frequencies) because of the τ/T_{tr} gain effect. However, in CCD's (assuming complete charge transfer), the trapped charge is part of the signal charge itself and there is no amplification or gain mechanism. Thus, long time constant states count no more than short time constant states and the $1/f$ mechanism should be absent in CCD's.

Table 1 shows the four basic noise sources considered, the expression for N_n, and finally the actual number for the rms fluctuation for typical values of the parameters involved. Clearly, fast interface state trapping will dominate the noise characteristics for a reasonable number of transfers (> 100). Transfer loss noise will not be a problem provided the ϵN_g product remains low. Even if ϵN_g were 1 or 2, the fluctuations would only increase to about 150 and still lie below the contribution from fast state trapping. For large signal levels ($N_s \sim 10^6$), the transfer loss noise approaches fast state noise, but the signal-to-noise ratio is correspondingly higher at the large signal levels. Likewise output-amplifier-associated noise and background-charge-generations noise are relatively small compared to trapping noise, provided capacitance levels remain in the 0.1 pf range.

The degree to which the fast state trapping will limit performance of CCD devices will depend upon the details of the specific application.

However, the present analysis suggests that other CCD-type devices in which transport does not occur at the interface and is not subject to fast interface state trapping (buried channel devices[12]) may be desirable where low-noise performance is at a premium.

Table 1—Charge-Coupled-Device Noise Sources

Source	$\overline{N_n}$	Typical Values for $\overline{N_n}$	
Transfer Loss	$[2\epsilon N_g(N_s + N_{s,o})]^{1/2}$	70	
Background Charge Generation			
Thermal or Optical	$\sqrt{\overline{N_{s,o}}}$	100	
Electrically-introduced	$400\sqrt{C_{pf}}$	40	$C_{pf}=0.01$
Output Amplifier			
RC-limited bandwidth	$400\sqrt{C_{pf}}$	120	$C_{pf}=0.1$
Floating diffusion reset	$400\sqrt{C_{pf}}$	120	$C_{pf}=0.1$
Floating gate reset	filterable	negligible	
MOSFET	$60C_{pf}\sqrt{\left(\dfrac{\Delta B}{5\text{MHz}}\right)\left(\dfrac{1000\mu\text{mho}}{g_{mo}}\right)}$	6	$C_{pf}=0.1$ $\Delta B=5\text{MHz}$ $g_{mo}=1000\mu\text{mho}$
Fast Interface State Trapping	$670\left[\dfrac{N_{ss}}{10^{10}}\dfrac{N_g}{10^3}\dfrac{A_g}{1.25\times10^{-6}}\right]^{1/2}$	670 for 1000 transfers	

$N_{s,o} = 1.25 \times 10^4$ $N_s = 1.25 \times 10^4$ $\epsilon N_g = 0.1$

$A_g = 1.25 \times 10^{-6}\text{cm}^2$ $N_{ss} = 10^{10}\,(\text{cm}^2-\text{eV})^{-1}$

Acknowledgments

The authors are grateful to A. Rose for his continuing interest and guidance in the development of the noise concepts reported here. We are also pleased to acknowledge R. Martinelli for his suggestion to use the Bernoulli trials approach and E. Ramberg for his help in evaluating the resultant integral. Stimulating discussions with B. Williams and R. Ronen also contributed to this work. The careful manuscript typing by Mrs. Mary Frances Pennington is also appreciated.

Appendix 1—Carrier Fluctuations for Parallel RC Circuit

Several of the noise sources associated with the CCD involve the thermal noise of a resistance that is in parallel with a capacitance.

Of interest is the rms fluctuation in the number of carriers on the capacitance. Fig. 2 shows the circuit diagram of the configuration to be analyzed along with the noise equivalent circuit. The resistance R is equivalent to a noise current source $\overline{i_n^2} = 4kT\Delta B/R$ in parallel

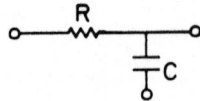

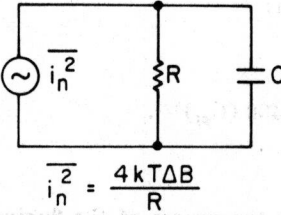

$$\overline{i_n^2} = \frac{4\,k\,T\Delta B}{R}$$

Fig. 2—RC circuit and noise equivalent circuit.

with R. The voltage fluctuations across the RC network will depend upon its frequency characteristics. The mean-squared noise voltage per unit bandwidth is given by:

$$\overline{v_n^2} = \frac{\overline{i_n^2}}{|G|^2} \tag{12}$$

where

$$G = \frac{1}{R} + j\omega C \tag{13}$$

$$|G|^2 = \frac{1}{R^2}\,(1 + \omega^2 R^2 C^2). \tag{14}$$

The total mean-squared noise voltage will be the integral over frequency of $\overline{v_n^2}$:

$$\overline{V_n{}^2} = \int_0^\infty \overline{v_n{}^2}\,df = \frac{4kT}{2\pi C}\int_0^\infty \frac{\dfrac{1}{RC}}{\dfrac{1}{R^2C^2}+\omega^2}\cdot \qquad [15]$$

Since the integral is just $\pi/2$,

$$\overline{V_n{}^2} = \frac{kT}{C}\,. \qquad [16]$$

The rms carrier fluctuation is then:

$$\overline{N_n} = \frac{1}{q}\,C\overline{V_n} = \frac{1}{q}\,(kTC)^{1/2} = 400\,(C_{pf})^{1/2}. \qquad [17]$$

Note that while the resistance is the source of the fluctuations, its value does not affect the total carrier fluctuations. This is so because, while larger values of R increase the mean-squared noise voltage per unit bandwidth, they decrease the effective bandwidth by the same factor. A large capacitance reduces the bandwidth by $1/\sqrt{C}$, thereby reducing the noise voltage, but increases the number of carrier fluctuations for a given rms noise voltage by C; thus the $\sqrt{C}$ dependence.

Appendix 2—Steady-State Mean-Squared Fluctuations in Number of Carriers Trapped in Fast Interface States

A. van der Ziel[13] has derived the probability distribution of the number of carriers in the conduction band of a semiconductor $P(N)$, where the probability of an electron being generated in the time dt is $g(N)\,dt$ and the probability of an electron recombining in time dt is $r(N)\,dt$ (dt is small enough that multiple electron processes are negligible). The derivation is quite general and assumes that g and r are functions only of N, not of time. Thus the results apply for any collection of particles whose number changes only via the probabilities $g(N)$ and $r(N)$.

Van der Ziel shows that the most probable value N_o will satisfy the following condition:

$$g(N_o) = r(N_o+1) \cong r(N_o), \text{ for large } N_o. \tag{18}$$

Also, the mean-squared fluctuation about N_o is given by

$$\overline{\delta N^2} = -\left[\frac{g'(N_o)}{g(N_o)} - \frac{r'(N_o+1)}{r(N_o+1)}\right]^{-1} \cong \frac{g(N_o)}{r'(N_o)-g'(N_o)}, \tag{19}$$

where the primes denote differentiation with respect to N.

To apply these results to find the fluctuations of the total number of trapped carriers under steady-state conditions, we must identify $g(N)$ and $r(N)$ for incremental energy intervals $d\mathscr{E}$ and then integrate Eq. [19] over the bandgap to determine the total mean-squared fluctuations.

The kinetic equation that has been assumed to describe changes in occupation of fast states in energy interval $d\mathscr{E}$ located $\mathscr{E}$ below the conduction band is given by[8]

$$\frac{d}{dt}(n_{ss}d\mathscr{E}) = k_1 n_s (N_{ss}-k_2 n_{ss}d\mathscr{E} \exp n_{ss}) d\mathscr{E} - \left\{-\frac{\mathscr{E}}{kT}\right\}, \tag{20}$$

where n_{ss} is the number of occupied fast states in $(cm^2-eV)^{-1}$

N_{ss} is the density of fast states in $(cm^2-eV)^{-1}$

k_1 and k_2 are constants $(k_1 \approx 10^{-2}, k_2 \approx 10^{11})$

n_s is the density of free carriers in the conduction band (assumed constant)

Eq. [20] assumes that n_s is large enough that the number of carriers removed by trapping is small compared to n_s. Thus $n_{ss}d\mathscr{E}$ corresponds to N and

$$g(n_{ss}d\mathscr{E}) = k_1 n_s (N_{ss} - n_{ss}) d\mathscr{E} \tag{21}$$

$$r(n_{ss}d\mathscr{E}) = k_2 n_{ss}d\mathscr{E} \exp\left\{-\frac{\mathscr{E}}{kT}\right\}. \tag{22}$$

Here, generation corresponds to the trapping of a free carrier from the conduction band, and recombination corresponds to emission of a trapped carrier into the band.

The most probable value $n_{ss}{}^o$ is given by

$$n_{ss}{}^o = \frac{N_{ss}}{1 + \exp\left\{\dfrac{\mathcal{E}_Q - \mathcal{E}}{kT}\right\}}, \qquad [23]$$

where

$$\mathcal{E}_Q = kT\ln\left(\frac{k_2}{k_1 n_s}\right). \qquad [24]$$

Using Eq. [19], the mean squared fluctuation in each energy interval $d\mathcal{E}$ is given by

$$d(\overline{\delta N^2}) = \frac{k_1 n_s d\mathcal{E}(N_{ss} - n_{ss}{}^o)}{k_2 \exp\left\{-\dfrac{\mathcal{E}}{kT}\right\} + k_1 n_s} = \frac{N_{ss} d\mathcal{E} \exp\left\{\dfrac{\mathcal{E}_Q - \mathcal{E}}{kT}\right\}}{\left[1 + \exp\left\{\dfrac{\mathcal{E}_Q - \mathcal{E}}{kT}\right\}\right]^2} \qquad [25]$$

Eq. [25] expresses the fluctuations in each interval $d\mathcal{E}$ only, and the total mean squared fluctuations $N_t{}^2$ is the integral over the forbidden gap:

$$\overline{N_t{}^2} = \int_0^{\mathcal{E}_g} d(\overline{\delta N^2}) = \int_0^{\mathcal{E}_g} \frac{N_{ss} \exp\left\{\dfrac{\mathcal{E}_Q - \mathcal{E}}{kT}\right\} d\mathcal{E}}{\left[1 + \exp\left\{\dfrac{\mathcal{E}_Q - \mathcal{E}}{kT}\right\}\right]^2}. \qquad [26]$$

Assuming N_{ss} is uniform in energy, $\mathcal{E}_Q/kT \gg 1$ and $\mathcal{E}_g - \mathcal{E}_Q/kT \gg 1$, and using the substitution $u = \exp\left\{(\mathcal{E}_Q - \mathcal{E})/kT\right\}$, we have

$$\overline{N_t{}^2} = kTN_{ss} \int_0^\infty \frac{du}{(1 + u)^2} = kTN_{ss}. \qquad [27]$$

Thus, for steady-state conditions, the mean-squared fluctuation in the number of trapped carriers is simply kTN_{ss} per unit area.

Appendix 3—Mean-Squared Fluctuation of Trapped Carriers
in Complete-Charge Transfer Mode with Fat Zero

For the complete charge transfer mode of CCD operation, the mean-squared fluctuation in the total number of trapped carriers $\overline{N_t^2}$ is no longer given by the steady-state value kTN_{ss}. A different approach must be used to find the fluctuations in the number of trapped carriers. During the transfer-in-period, $-1/2f < t < 0$, with the fat zero present, the fast states fill to the level

$$n_{ss}(0) = n_{ss}\left(-\frac{1}{2f}\right) + \left[n_{ss}(\infty) - n_{ss}\left(-\frac{1}{2f}\right)\right](1 - \exp\{-f_o/f\}) \qquad [28]$$

where $n_{ss}(t)$ is the density of occupied fast interface states in (cm^2 eV)$^{-1}$ and $f_o = k_1 n_{s,o}/2$. In the time from 0 to $1/(2f)$, the transfer-out period, the occupation decreases toward $n_{ss}[1/(2f)]$ which must be on the average the same as $n_{ss}[-1/(2f)]$.

The probability that any state will empty in time dt is dt/τ times the probability that it is still filled, $\exp\{-t/\tau\}$ [τ is the empty time given by $(1/k_2)\exp\{\mathcal{E}/kT\}$]. Thus the probability that a state will empty in time $1/(2f)$ is given by

$$p = \int_0^{\frac{1}{2f}} \frac{dt}{\tau}\exp\left\{-\frac{t}{\tau}\right\} = 1 - \exp\left\{-\frac{1}{2f\tau}\right\}. \qquad [29]$$

So, at $t = 0$, at each energy interval $d\mathcal{E}$, there are $n_{ss}(0)d\mathcal{E}$ states filled. They each have probability $p = 1 - \exp\{-1/(2f\tau)\}$ of emptying by time $1/(2f)$ and probability $q = 1 - p$ of not emptying. This is an example of Bernoulli trials[14] with $n = n_{ss}(0)d\mathcal{E}$, $p = 1 - \exp\{-1/(2f\tau)\}$, and $q = \exp\{-1/(2f\tau)\}$. The average number of states that have emptied at $t = 1/(2f)$ is

Average Number of States to Empty $= pn$
$$= [1 - \exp\{-1/(2f\tau)\}]n_{ss}(0)d\mathcal{E}.$$

The variance, or mean-squared fluctuation about this average, is

Variance $= pqn = n_{ss}(0)d\mathcal{E}\,[1 - \exp\{-1/(2f\tau)\}]\exp\{1/(2f\tau)\}.$ $\qquad [30]$

The total mean-squared fluctuations in the number of trapped carriers at $t = 1/(2f)$ is the integral of Eq. [30] over energy

$$\overline{N_t^2} = \int_0^{\mathcal{E}_g} n_{ss}(0)\, d\mathcal{E}\; [1 - \exp\{-1/(2f\tau)\}]\exp\{1/(2f\tau)\}$$

$$= \int_0^{\mathcal{E}_g} n_{ss}(\infty)\, d\mathcal{E}\; \frac{[1 - \exp\{1/(2f\tau)\}][1 - \exp\{-f_o/f\}]}{\exp\{1/(2f\tau)\} - \exp\{-f_o/f\}}. \qquad [31]$$

$n_{ss}(\infty)$, the steady-state occupation with the fat zero present, can be replaced by N_{ss} in this integral because the factor $[1 - \exp\{-1/(2f\tau)\}]\exp\{-1/(2f\tau)\}$ is a sharply-peaked function at energy $\mathcal{E}_p = kT\ln[k_2/(2f\ln 2)]$. $n_{ss}(\infty)$ is just N_{ss} times the Fermi function with quasi-Fermi level $kT\ln[k_2/(k_1 n_{s,o})]$. Only for $n_{s,o}$ values less than approximately 10^9 will $n_{ss}(\infty)$ have a value less than N_{ss} at $\mathcal{E}_p$.

Even with this approximation, the integral in Eq. [31] cannot be evaluated in closed form, but a series solution is possible. By making appropriate substitutions, Eq. [31] can be written

$$\overline{N_t^2} = kTN_{ss}(1 - \exp\{-f_o/f\})\int_{l_2}^{l_1} \frac{(1 - x)\, dx}{[1 - x\exp\{-f_o/f\}]\ln x} \qquad [32]$$

where

$$l_1 = \exp\left\{\frac{-k_2}{2f}\right\}$$

$$l_2 = \exp\left\{-\frac{k_2}{2f}\left[\exp\left(-\frac{\mathcal{E}_g}{kT}\right)\right]\right\}.$$

By expanding $(1 - x\exp\{-f_o/f\})^{-1}$ as a series, we have

$$\overline{N_t^2} = kTN_{ss}(1 - \exp\{-f_o/f\})\int_{l_2}^{l_1} dx\left[\sum_{j=0}^{\infty} \frac{(-1)^j\left(x\exp\left\{-\dfrac{f_o}{f}\right\}\right)^j}{\ln x}\right.$$

$$\left. - \sum_{j=0}^{\infty} \frac{(-1)^j\left(\exp\left\{-\dfrac{f_o}{f}\right\}\right)^j x^{j+1}}{\ln x}\right]. \qquad [33]$$

This can be further reduced to a sum of logarithmic integrals li(y) where

$$\mathrm{li}(y) = \int_0^{e^y} \frac{dx}{\ln x}, \qquad [34]$$

which results in the following form:

$$\overline{N_t^2} = kTN_{ss}(1 - e^{-f_o/f}) \sum_{j=1}^{\infty} (-1)^{j-1} (e^{-f_o/f})^{j-1} \ln\left(\frac{j+1}{j}\right). \quad [35]$$

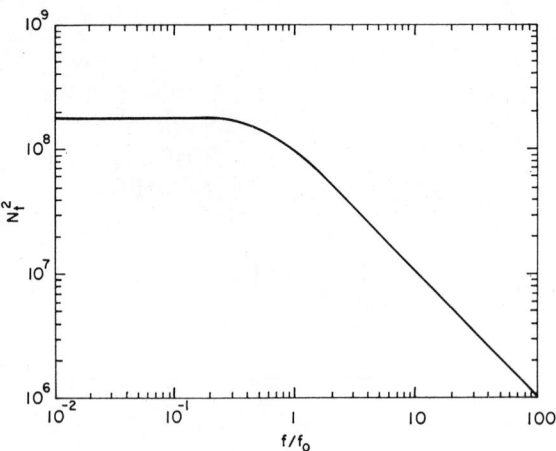

Fig. 3—Mean-squared fluctuation in total number of trapped carriers versus normalized frequency for complete charge transfer mode with fat zero.

Fig. 3 shows $\overline{N_t^2}$ as a function of f/f_o for $N_{ss} = 10^{10}$. Above $f/f_o = 1$ the fluctuations fall off as f/f_o. However, below $f/f_o = 1$, they are constant at $kTN_{ss} \ln 2$. Since as discussed earlier, f/f_o values must be less than 1 in order to suppress interface state losses, the mean-squared fluctuations due to fast state trapping in the complete charge transfer mode will be $0.7kTN_{ss}$.

References:

[1] W. S. Boyle and G. E. Smith, "Charge-Coupled Semiconductor Devices," **Bell Syst. Tech. J.,** Vol. 49, p. 587, (1970).
[2] W. F. Kosonocky and J. E. Carnes, "Two-Phase Charge-Coupled Shift Registers," **Digest Tech. Papers, IEEE International Solid State Circuit Conf.,** Philadelphia, Pa., p. 132, Feb. 1972.

[3] M. F. Tompsett, G. F. Amelio, W. J. Bertram, R. R. Buckley, W. J. McNamara, J. C. Mikkelsen, and D. A. Sealer, "Charge-Coupled Imaging Devices: Design Considerations," **IEEE Trans. Elec. Dev.,** Vol. ED-18, p. 992, Oct. 1971.

[4] W. F. Kosonocky and J. E. Carnes, "Charge-Coupled Digital Circuits," **IEEE J. Solid-State Circuits,** Vol. SC-6, p. 314, (1971).

[5] F. M. Klaassen and J. Prins, "Thermal Noise of MOS Transistors," **Phillips Res. Repts,** Vol. 22, p. 505, 1967.

[6] F. M. Klaassen, "Characterization of Low 1/f Noise in MOS Transistors," **IEEE Trans. Elec. Dev.,** Vol. ED-18, p. 10, Oct. 1971.

[7] R. Ronen, private communication.

[8] J. E. Carnes and W. F. Kosonocky, "Fast Interface State Losses in Charge-Coupled Devices," **Appl. Phys. Ltrs.,** Vol. 20, April 1, 1972, p. 261.

[9] M. G. Kocac, W. S. Pike, F. V. Shallcross and P. K. Weimer, "Solid State Imaging Emerges from Charge Transport," **Electronics,** p. 72, Feb. 28, 1972.

[10] S. Christensson, I. Lundstrom, and C. Svensson, "Low Frequency Noise in MOS Transistors," **Solid-State Electron.,** Vol. 11, p. 813, 1968.

[11] F. Berz, "Theory of Low Frequency Noise in Si MOST's," **Solid-State Electron.,** Vol. 13, p. 631, 1970.

[12] G. E. Smith, et. al., "A Buried Channel Charge Coupled Device," IEEE Device Research Conf., Ann Arbor, Mich., June, 1971.

[13] A. van der Ziel, "Semiconductor Noise," in **Noise in Electron Devices** ed. by L. D. Smullin and H. A. Haus, Wiley, New York, 1959, p. 320.

[14] William Feller, **An Introduction to Probability Theory and Its Applications,** Vol. 1, p. 146, Wiley, New York (1950).

One-Dimensional Study of Buried-Channel Charge-Coupled Devices

HAMDI EL-SISSI, STUDENT MEMBER, IEEE, AND RICHARD S. C. COBBOLD, MEMBER, IEEE

Abstract—An analysis leading to some basic relations is performed on a one-dimensional model of a buried-channel charge-coupled device (CCD). Expressions for the charge distribution, potential, channel thickness, and location are obtained. These enable the effects of varying the device structural parameters, as well as the gate voltage and signal charge, to be examined very simply. The maximum charge-carrying capacity is discussed and compared to that for a surface-type CCD. Furthermore, the analysis is extended to a device with a nonuniformly doped semiconductor layer.

LIST OF SYMBOLS

It should be noted that unnormalized quantities are represented by upper case symbols, while normalized quantities employ lower case symbols.

C_o	Oxide capacitance.
C_s	ϵ_s/X_{m0}, semiconductor capacitance.
k	$\epsilon_s x_o/\epsilon_o$, a constant that is proportional to the oxide thickness.
k_B	Boltzmann's constant.
$\mathcal{L}$	Effective Debye length.
l	Length of a unit cell in a three-phase device.
N_D	Donor impurity density in the substrate.
$N_A(x)$	Acceptor impurity density in the surface region.
q	Magnitude of the electronic charge.
q_s	Signal charge per unit area normalized by $qN_D\mathcal{L}$.
$\hat{q}_s$	Maximum signal charge-carrying capacity (per unit area).
$\hat{q}_{sm}$	Value of $\hat{q}_s$ for $x_1 = x_{1q}$.
$\hat{q}_{sf}$	Value of $\hat{q}_s$ for $x_1 = x_{1f}$.
T	Absolute temperature.
u	Electrostatic potential in thermal volts.
u_m	Electrostatic potential minimum.
u_0, u_1	Values of u at $x = 0^+$ and $x = x_1$.
u_0', u_1'	Electrostatic field at $x = 0^+$ and $x = x_1$.
V_G	Gate-substrate voltage.
v_g	Effective gate voltage in thermal volts.
v_s, v_t, v_b	Storage, transfer, and blocking voltages in a three-phase device.
w	Channel thickness normalized by $\mathcal{L}$.
w_1, w_2	Channel thickness measured from x_m to the surface and substrate sides ($w = w_1 + w_2$).

X	Distance from the SiO₂–Si interfac
x	$X/\mathcal{L}$, normalized distance.
x_o	Normalized oxide thickness.
x_1	Location of the p-layer to n-substrate metallurgical junction.
x_m	Location of the potential minimum.
x_{m0}	Value of x_m for zero signal charge.
x_{1q}, x_{1f}	Value of x_1 that makes q_s a maximum and the fringing field a maximum.
$\alpha(x)$	$N_A(x)/N_D$, normalized doping.
α_m	$\alpha(0)$, peak normalized surface doping.
ϵ_o, ϵ_s	Oxide and semiconductor permittivities.
ρ_s	Signal charge density normalized by qN_D.
ρ_{sm}	Maximum value of ρ_s.

I. INTRODUCTION

THE most recent addition to the family of charge-coupled devices (CCD's) is the buried-channel arrangement wherein the mobile charge is stored and transported in an induced channel contained within the bulk of a thin semiconductor layer [1]–[3]. Several advantages result from the charge being located in the bulk rather than at the oxide–silicon interface, as is the case for a surface-type CCD [4], [5]. First, surface states and trapping effects associated with the interface are avoided, thereby improving the charge-transfer efficiency. Second, the absence of any interface carrier scattering increases the mobility. And finally, the higher fringing field penetration under the storage electrode of a buried-channel CCD enhances the speed of charge transport, thereby resulting in higher operating frequencies [6]. On the other hand, as we shall see, the charge-carrying capacity is substantially smaller than that for a surface CCD of comparable dimensions.

Although CCD's are basically two-dimensional in nature, a one-dimensional analysis, because of its relative simplicity, enables certain physical insights to be obtained that are obscured by the more exact approach. Furthermore, such insights form a valuable starting point for the more difficult static and dynamic two-dimensional analyses, since they enable one to see what approximations are reasonable.

Using numerical methods, Kent [7] has recently studied some of the problems considered in this paper.[1]

Manuscript received November 19, 1973; revised March 12, 1974. This work was supported in part by a National Research Council of Canada Grant to R. S. C. Cobbold and in part by a National Research Council Postgraduate Scholarship to H. El-Sissi.

The authors are with the Department of Electrical Engineering and Institute of Biomedical Engineering, University of Toronto, Toronto, Ont., Canada M5S 1A4.

[1] In addition, Gunsager et al. [8] and Esser et al. [9] have reported that they used one-dimensional calculations in studying charge transfer dynamics.

Reprinted from *IEEE Trans. Electron Devices*, vol. ED-21, pp. 437–447, July 1974.

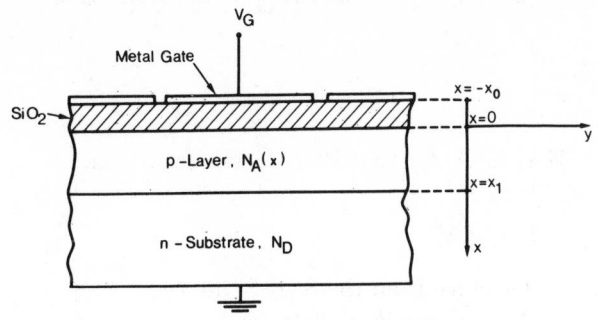

Fig. 1. Sketch of a buried-channel CCD showing the normalized coordinate system and structure assumed for the analysis.

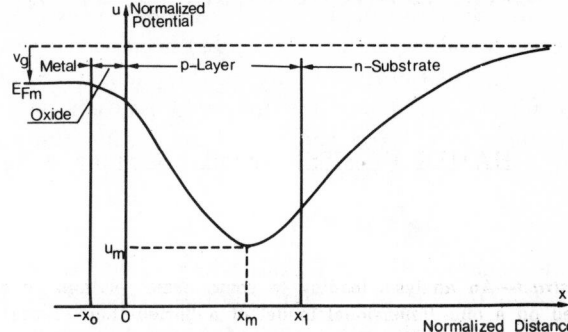

Fig. 2. Sketch of the potential variation in a buried-channel CCD. The location and value of the potential minimum depend on the total signal charge.

While such work is obviously very useful, it fails to give much physical insight to device operation. Moreover, such methods are rather expensive in computer time, especially when examining the effects of varying various parameters. We shall show, using some well justified approximations, that analytical solutions for the mobile carrier distribution, the channel potential, location, and thickness, can be obtained as functions of the device geometry, gate voltage, and total stored mobile charge. From these expressions we shall estimate the device charge-carrying capacity, and will qualitatively discuss the nature of the charge transfer in two dimensions. Where possible, our results are compared with the numerical computations of Kent [7]; discrepancies between the two are pointed out and discussed.

The organization of this paper is as follows. In Section II the basic equations and approximations necessary for obtaining expressions for the mobile charge density contained within a nonuniformly doped layer are obtained. A solution for the case of a uniformly doped layer is obtained in Section III. In addition, expressions for the channel position, potential, and thickness are also derived. These enable the charge-carrying capacity and the effects of various parameters on device performance to be studied. Finally, in Section IV, we proceed on a stronger basis to analyze the more general case of a nonuniformly doped layer device in a manner similar to that employed in Section III.

II. BASIC MODEL AND APPROXIMATIONS

The device to be analyzed is shown in Fig. 1. It consists of four layers: a metal gate, an insulator (SiO_2), a p-type layer, and an n-type substrate.[2] Since the analysis is one-dimensional, it is implicitly assumed in this first-order treatment that the gate is sufficiently wide and long so that the influence of adjacent gates and fringing effects can be neglected. As a result, the analysis can be expected to be reasonably accurate in the central region beneath the gate of a CCD whose gate length is greater than the surface layer thickness. A second assumption is that the hole and electron currents, whether they be due to charge

transport or generation recombination processes, are assumed to be zero. This corresponds to the quasi-static situation existing during the storage phase of operation.

Assuming the validity of Einstein's relation, it follows from the fact that the hole and electron current densities are zero, that the electron density is proportional to e^{+u} and the hole density is proportional to e^{-u}, where u is the normalized potential in units of $k_B T/q$, measured with respect to the potential deep within the substrate.

Under normal operating conditions, the p-layer is reverse biased, and its thickness and doping are such that in the absence of any signal charge, both this layer and part of the n-substrate are fully depleted. As illustrated in Fig. 2, the electrode potentials are such as to create a potential minimum in the p-layer, which is capable of storing holes delivered to it from an adjacent region during the transfer phase of operation. The normalized signal charge density in the neighborhood of this minimum can be written as

$$\rho_s(x) = \rho_{sm} \exp [u_m - u], \tag{1}$$

where ρ_{sm} is the peak value.

The potentials in the various regions can be found from Poisson's equation. When expressed in normalized form for the three regions of interest, they become the following.

Oxide:

$$\frac{d^2u}{dx^2} = 0. \tag{2}$$

p-*layer*:

$$\frac{d^2u}{dx^2} = \alpha(x) - \rho_s(x). \tag{3}$$

n-*substrate*:

$$\frac{d^2u}{dx^2} = e^u - 1 - \rho_s(x). \tag{4}$$

In the preceding equations, x is a normalized distance given by $x = X/\mathcal{L}$, in which the effective Debye length $\mathcal{L}$ is given by $\mathcal{L} = (k_B T \epsilon_s / q^2 N_D)^{1/2}$. Furthermore, $\alpha(x) = N_A(x)/N_D$, where N_A and N_D are the net ionized acceptor

[2] The results for an n-layer on a p-substrate can be obtained by simply reversing the appropriate signs.

and donor densities in the p-and n-regions, respectively, and the normalization factor for $\rho_s(x)$ is qN_D.

For the potential distribution in the oxide, (2) can be very simply solved and the following relation between the potential (u_0) and field (u_0') just within the semiconductor ($x = 0^+$) can be obtained.

$$v_g = u_0 - ku_0'. \qquad (5)$$

In this equation $k = x_o(\epsilon_s/\epsilon_o)$, where ϵ_s and ϵ_o are the semiconductor and oxide permittivities, respectively, and x_o is the normalized oxide thickness.[3] Furthermore, v_g is related to the gate voltage V_G by

$$v_g = \{V_G - \Phi_{MS} + (Q_{ss}/C_o)\} \Big/ \left(\frac{k_BT}{q}\right), \qquad (6)$$

where C_o is the oxide capacitance per unit area, Φ_{MS} is the metal-semiconductor work function difference, and Q_{ss} is the equivalent oxide interface charge per unit area.

In the n-substrate, u and u' are zero at $x = \infty$, so that by integrating (4) from x to ∞ we find that for $x \geq x_1$

$$u' = \sqrt{2}\{\exp(u) - u - 1$$
$$+ \rho_{sm}\exp(u_m)[\exp(-u) - 1]\}^{1/2}. \qquad (7)$$

At the p-n metallurgical junction x_1, the field and potential are related by

$$u_1' = \sqrt{2}\{\exp(u_1) - u_1 - 1$$
$$+ \rho_{sm}\exp(u_m)[\exp(-u_1) - 1]\}^{1/2}. \qquad (8)$$

Equations (5) and (8) furnish the nonlinear mixed boundary conditions necessary to solve for the potential in the p-layer. If one wishes to pursue the calculations numerically from this point on, the calculations are greatly simplified by using these equations. In fact, what we have done is to confine the solution to just the p-layer, as compared to the original problem that covered the whole half space.

Further simplification of the boundary conditions can be made if it is realized that, under practical operating conditions, $v_g \simeq -200$, u_1 and u_m are in the order of -1000, and that $u_m - u_1 \ll 0$. Hence (8) reduces to

$$u_1' = (-2u_1)^{1/2}, \qquad (9)$$

which is the well-known depletion approximation.

III. ANALYSIS FOR A UNIFORMLY DOPED LAYER DEVICE

A. Signal Charge Distribution and Channel Thickness

The potential about the minimum can be expanded in a Taylor series

$$u - u_m = (x - x_m)u_m' + \tfrac{1}{2}(x - x_m)^2 u_m'' + \cdots, \qquad (10)$$

where x_m is the position of the minimum. Now $u_m' = 0$;

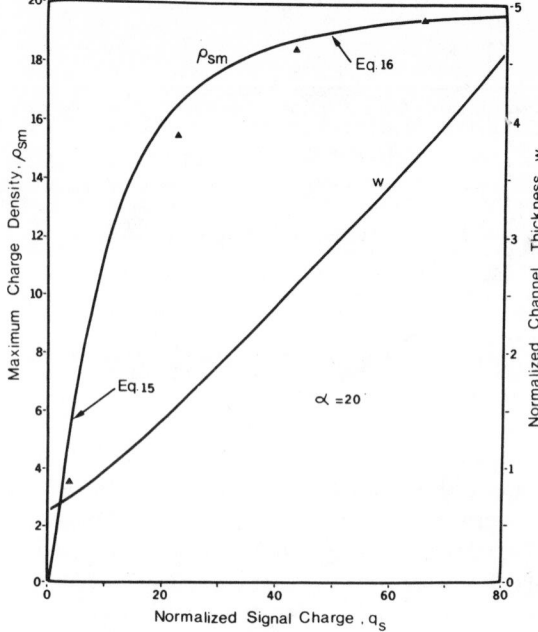

Fig. 3. Variation of the normalized signal charge density at the potential minimum and normalized channel thickness with signal charge for a uniformly doped p-layer device. The results obtained by Kent [7] are shown as points on the graph.

furthermore, it can easily be shown from (3) that all odd derivatives are zero at x_m and the even derivatives are given by

$$u_m'' = \alpha - \rho_{sm}, u_m'''' = \rho_{sm}(\alpha - \rho_{sm}), \text{ etc.} \qquad (11)$$

Hence from (1) and (10) the normalized signal charge density can be expressed as

$$\rho_s(x) = \rho_{sm}\exp\left\{-\frac{1}{2!}(x - x_m)^2(\alpha - \rho_{sm})\right.$$
$$\left. -\frac{1}{4!}(x - x_m)^4(\alpha - \rho_{sm})\rho_{sm}\cdots\right\}, \qquad (12)$$

which is a Gaussian-like curve with a peak value at x_m. It should be noted from (12) that the charge is symmetrically distributed about x_m with half the total charge lying on either side. If the higher order terms can be neglected, corresponding to the condition $|x - x_m| < (12/\rho_{sm})^{1/2}$, then the total normalized signal charge per unit area q_s can be obtained by integrating (12):

$$q_s = \int_0^\infty \rho_s(x)\,dx \simeq \rho_{sm}\left(\frac{2\pi}{\alpha - \rho_{sm}}\right)^{1/2}. \qquad (13)$$

Thus the charge density at the potential minimum is given by

$$\rho_{sm} = \frac{q_s}{(2\pi)^{1/2}}\left\{\left(\alpha + \frac{q_s^2}{8\pi}\right)^{1/2} - \frac{q_s}{(8\pi)^{1/2}}\right\}, \qquad (14)$$

which is plotted in Fig. 3. It should be noted that for $q_s \ll (8\pi\alpha)^{1/2}$, (14) reduces to

$$\rho_{sm} \approx q_s \left(\frac{\alpha}{2\pi} \right)^{1/2} \left(1 - \frac{q_s}{(8\pi\alpha)^{1/2}} \right), \qquad (15)$$

while for larger stored charges, such that $q_s \gg (8\pi\alpha)^{1/2}$,

$$\rho_{sm} \simeq \alpha - (2\pi\alpha^2/q_s). \qquad (16)$$

From (15) it follows that the charge density is proportional to the total signal charge when the latter is small. In addition, it will be noted from (16) that the peak charge density saturates at a value equal to α. This last observation also follows from (3), by noting that the condition for a potential minimum is that $u'' > 0$, i.e., that $\alpha > \rho_{sm}$.

It is convenient to characterize the spatial distribution of the signal charge by defining the channel thickness to be the separation of those points where the charge density falls to $e^{-\eta}$ of the peak value, i.e., where the potential is $u_m + \eta$. From (10) and (11), neglecting higher order terms, the thickness w can be expressed as

$$w = 2 \left(\frac{2\eta}{\alpha - \rho_{sm}} \right)^{1/2}$$

which, with the help of (14), can be re-expressed in terms of the signal charge, yielding

$$w \simeq \frac{(8\eta)^{1/2}}{(\alpha + q_s^2/8\pi)^{1/2} - q_s/(8\pi)^{1/2}}. \qquad (17)$$

Now the fraction of charge within the channel is simply erf (η), so that if we arbitrarily choose $\eta = 1$ (the value assumed in the remainder of this paper), we find that 84 percent of the charge resides in the channel. If we had chosen $\eta = 2$, this figure increases to just over 99 percent. In passing, it should be noted that the higher order terms in (12) tend to localize the charge rather more than for a simple Gaussian function, so that the channel thickness will be slightly less than that given by (17).

In Fig. 3, both the channel thickness and peak signal charge density are shown as functions of the signal charge density for $\alpha = 20$. The points on the ρ_{sm} curve, taken from the numerical results of Kent [7], are seen to be in reasonable agreement. The discrepancies that do occur appear to arise from Kent's assumption that the potential is zero at a normalized distance of $x = 42$. It can be shown that for all practical signal charges such an assumption is rather poor; only for very large signal charges can this assumption yield accurate results.

B. Channel Potential and Position

The spatial variation of the signal charge density has so far been discussed in terms of the quantities u_m, x_m, and the potential profile. To obtain values for these quantities it is necessary to solve Poisson's equation for the p layer. Therefore, integrating (3), using the boundary condition $u' = u_0'$ at $x = 0^+$, we find that

$$u' = u_0' + \alpha x - \int_0^x \rho_s(\xi)\, d\xi. \qquad (18)$$

Noting that $u_m' = 0$, this enables the position of the potential minimum to be expressed as

$$x_m = (q_s - 2u_0')/2\alpha. \qquad (19)$$

In addition, it also follows from (18) that for $x < x_m - (w/2)$

$$u' = u_0' + \alpha x, \qquad (20)$$

and that for $x > x_m + (w/2)$

$$u' = u_0' + \alpha x - q_s. \qquad (21)$$

Thus, as expected from the application of Gauss' law, the field changes by $\alpha(x_a - x_b) + q_s$ from a location x_a on one side of the channel to a location x_b on the other side.

By integrating (18) and simplifying we find that

$$u = u_0 + u_0'x + \frac{\alpha x^2}{2} - (x - x_m) \int_0^x \rho_s(\xi)\, d\xi$$
$$+ \int_0^x (\xi - x_m)\rho_s(\xi)\, d\xi. \qquad (22)$$

Thus for $x < x_m - (w/2)$,

$$u \simeq u_0 + u_0'x + \frac{\alpha x^2}{2} \qquad (23)$$

and, for $x > x_m + (w/2)$,

$$u \simeq u_0 + u_0'x + \frac{\alpha x^2}{2} - q_s(x - x_m). \qquad (24)$$

In addition, for the Gaussian form of $\rho_s(x)$, with the higher order terms in (12) neglected, the potential minimum becomes

$$u_m = u_0 + u_0'x_m + \frac{\alpha x_m^2}{2} - \frac{\rho_{sm}}{\alpha - \rho_{sm}}. \qquad (25)$$

Indeed, because the last term in (22) is small compared to the others, the use of the Gaussian form for $\rho_s(x)$ in deducing (25) is well justified.[4] By making use of (13) and (19), (24) and (25) can be expressed in the more convenient forms

$$u = u_0 + u_0'\left(x - \frac{q_s}{\alpha} \right) + \alpha \left(x - \frac{q_s}{\alpha} \right)^2 \Big/ 2 \qquad (26)$$

$$u_m = u_0 - \frac{(u_0')^2}{2\alpha} + q_s^2 \left(\frac{1}{8\alpha} - \frac{1}{2\pi\rho_{sm}} \right) \qquad (27)$$

or,

$$\simeq u_0 - (u_0)^2/2\alpha.$$

It is perhaps of interest to note that by comparing (26) to (23), it can be seen that the change in normalized potential from one side of the channel to the other appears as a change in the normalized distance from x to $x - (q_s/\alpha)$.

---

[4] In point of fact, it can be shown that (23) and the subsequent equations of this subsection are generally true, provided $\rho_s(x)$ is a symmetrical function.

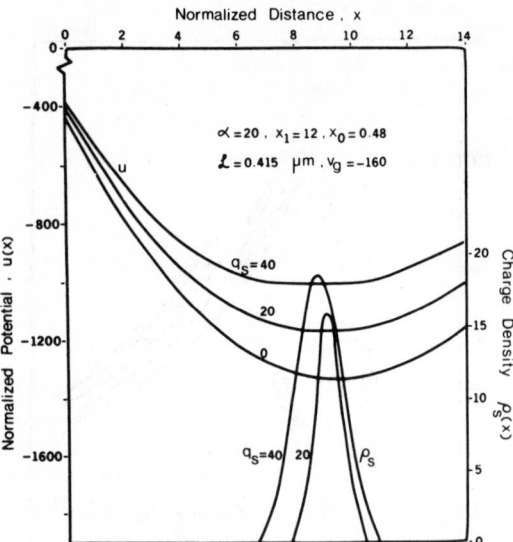

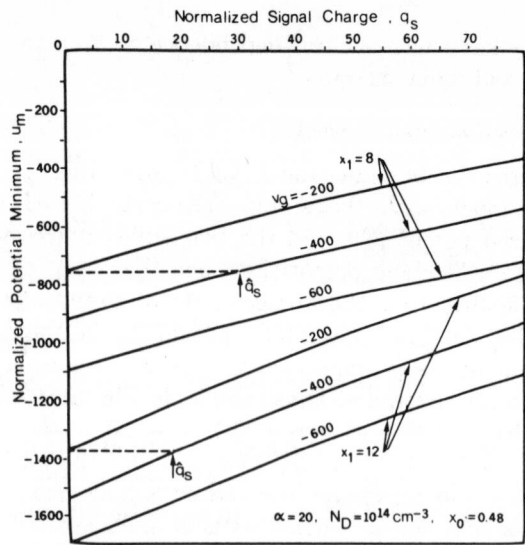

Fig. 4. Potential and charge density variation for different values of the signal charge, for a uniformly doped device with a bulk doping of $N_D = 10^{14}$ cm^{-3}, a surface layer doping of $N_A = 2 \times 10^{15}$ cm^{-3}, and an oxide thickness of 2000 Å. Note that the charge density ρ_s is in units of 10^{14} cm^{-3} while q_s is in units of 0.415×10^{10} cm^{-2}.

Fig. 5. Dependence of the potential minimum on the signal charge for p-layer thicknesses of 5 μm and 3.32 μm, and various gate voltages. The other device parameters and the normalization constants are the same as for Fig. 4. Also indicated are the normalized maximum charge-carrying capacities for a normalized gate voltage difference of -200.

The boundary potential at $x = x_1$ can be obtained from (26) as

$$u_1 = u_0 + u_0'\left(x_1 - \frac{q_s}{\alpha}\right) + \alpha\left(x_1 - \frac{q_s}{\alpha}\right)^2 \Big/ 2. \quad (28)$$

By making use of the other boundary conditions given by (9) and (21) the normalized field at $x = 0^+$ can be explicitly expressed in the form

$$u_0' = \left\{k^2 - 2v_g + (\alpha + 1)\left(x_1 - \frac{q_s}{\alpha}\right)\left(x_1 - \frac{q_s}{\alpha} + 2k\right)\right\}^{1/2}$$
$$- \left\{k + (\alpha + 1)\left(x_1 - \frac{q_s}{\alpha}\right)\right\}. \quad (29)$$

Thus by using either (22), (23), or (24), the potential throughout the region $x_1 \geq x \geq 0$ can be computed in a very simple manner. The results presented in Fig. 4 for two different signal charges illustrate the influence of the charge on both the potential distribution and the signal charge density. The parameters of this graph were chosen to reflect typical operating conditions.

Of more practical significance is the variation of the potential minimum and its spatial location as a function of the signal charge. From (19), (27), and (29), it can easily be shown that

$$x_m = \frac{1}{\alpha}\left[k + x_1(1+\alpha) - q_s\left(\frac{2+\alpha}{2\alpha}\right) - \left\{k^2 - 2v_g\right.\right.$$
$$\left.\left. + (\alpha+1)\left(x_1 - \frac{q_s}{\alpha}\right)\left(x_1 - \frac{q_s}{\alpha} + 2k\right)\right\}^{1/2}\right] \quad (30)$$

$$u_m = -\frac{(\alpha+1)}{2\alpha}\left[\left(k + x_1 - \frac{q_s}{\alpha}\right) - \left\{k^2 - 2v_g\right.\right.$$

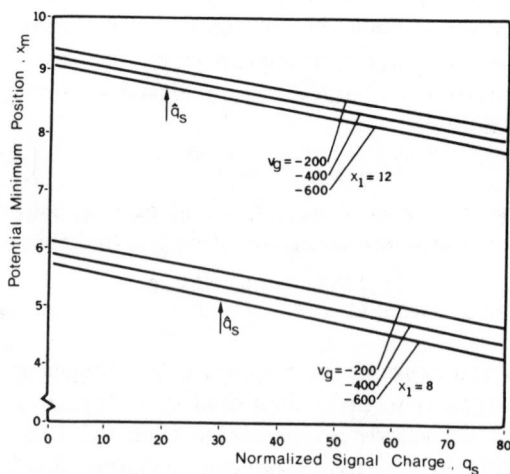

Fig. 6. Position of the potential minimum for a uniformly doped device as a function of the signal charge for p-layer thicknesses of 5 μm and 3.32 μm. The other device parameters are the same as assumed for Fig. 4.

$$\left. + (\alpha+1)\left(x_1 - \frac{q_s}{\alpha}\right)\left(x_1 - \frac{q_s}{\alpha} + 2k\right)\right\}^{1/2}\right]^2$$
$$+ q_s^2\left(\frac{1}{8\alpha} - \frac{1}{2\pi\rho_{sm}}\right) \quad (31)$$

where ρ_{sm} is given by (14). It will be noted from these equations that both x_m and u_m depend on the stored charge in such a way that as the charge is increased so the potential minimum moves towards the surface and the depth of the potential well is reduced. Both these effects are illustrated in Fig. 4. More specifically, the dependence of u_m and x_m on q_s for various gate voltages is illustrated in Figs. 5 and 6, from which it will be noted that each quantity has a nearly linear dependence both on the gate voltage and signal charge.

C. Charge-Carrying Capacity

Consider a buried-channel CCD operating in the storage mode with the storage electrode biased to a normalized potential v_s and the two adjacent electrodes biased to a blocking potential v_b such that $|v_b| < |v_s|$. Charge spillage from the storage electrode to the adjacent electrodes can be prevented, provided the potential minimum of the storage electrode is less than that of the adjacent electrodes. For example, in Fig. 5, it will be noted that a device with $x_1 = 12$, $\alpha = 20$, $v_b = -200$, $v_s = -400$,[5] and zero signal charge under the adjacent electrodes, the maximum normalized signal charge per unit area that can be carried without spillage is 20. For $N_D = 1 \times 10^{14}$ cm^{-3}, this corresponds to 1.32×10^{-8} C/cm^2. If the p-layer thickness is decreased so that $x_1 = 8$, and if the other conditions remain the same, then the maximum charge $\hat{q}_s$ is increased to 30.

Since the practical charge-carrying capacity is relatively small, the dependence of u_m on q_s and v_g can be represented by linear equations (see Figs. 5 and 6), which, as will be seen, enables an approximate expression to be obtained for the device charge-carrying capacity.

The maximum charge-carrying capacity corresponds to the condition

$$u_m \big|_{\hat{q}_s, v_s} = u_m \big|_{q_s=0, v_b}. \tag{32}$$

From the expansion of $u_m(q_s, v_g)$, and making use of (6) together with the preceding condition, we find that

$$\hat{q}_s = (v_b - v_s) \frac{\partial u_m}{\partial v_g} \bigg/ \frac{\partial u_m}{\partial q_s}, \tag{33}$$

where v_b and v_s are the normalized gate voltages and the partial derivatives are evaluated at $q_s = 0$ and $v_g = v_b$. When the derivatives are evaluated from (31), the maximum normalized signal charge can be expressed as[6]

$$\hat{q}_s = \frac{\alpha(v_b - v_s)}{-[-2v_b + k^2 + (\alpha + 1)(x_1 + 2k)x_1]^{1/2} + (\alpha + 1)(x_1 + k)} \tag{34a}$$

which is plotted in Fig. 7 as a function of x_1. For reasonable device design parameters it turns out that $x_1 \gg k$ and that $\alpha \gg 1$, so (34a) reduces to

$$\hat{q}_s \simeq \frac{v_b - v_s}{x_1(1 - \alpha^{-1/2})} \tag{34b}$$

i.e., $\hat{q}_s$ is inversely proportional to the p-layer thickness.

If one tries to increase $\hat{q}_s$ by decreasing x_1, two penalties are incurred. First a decrease in $\partial u_m/\partial q_s$ occurs corresponding to a loss in the self-induced field, and therefore a slower transfer of signal charge. Second, the fringing field

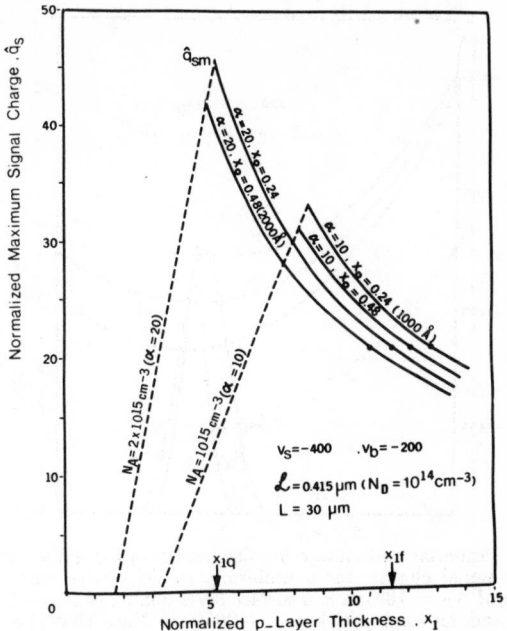

Fig. 7. Maximum charge-carrying capacity as a function of the normalized p-layer thickness. The broken line corresponds to the minimum thickness as determined by the requirement that the channel should not penetrate to the Si–SiO$_2$ interface. Values of $\hat{q}_{sf}$ for the various graphs are shown as dots. The values of x_{1f}, x_{1q} and $\hat{q}_{sm}$ are indicated for the $\alpha = 20$, $x_o = 0.24$ curve.

is reduced and this also increases the signal transfer time [6]. Hence, a compromise is necessary between increasing the maximum charge storage and the loss of transfer speed.

It should be noted that as the p-layer thickness is reduced so the signal charge gets closer to the Si–SiO$_2$ interface. Evidently, when $w/2 = x_m$ some of the charge will be subject to interface trapping effects. For a CCD to behave as a buried channel rather than a mixed-mode device, we can use the approximate criterion $w/2 \leq x_m$. This enables us to obtain an upper limit value for the maximum signal charge that can be used for a given p-layer thickness.

Substituting (30) into $x_m = w/2$, we find that, corresponding to a normalized p-layer thickness of x_{1q}, the upper limit value for the maximum signal charge $\hat{q}_{sm}$ is given by

$$\alpha x_{1q} = \hat{q}_{sm} + \beta + \left[-2v_g + \beta\left(2k + \frac{\beta}{\alpha}\right) \right]^{1/2} \tag{35}$$

where $\beta = (\alpha w/2) + (\hat{q}_{sm}/\alpha)$. This equation is plotted in Fig. 7 as a broken line for two values of α. It will be noted that for $x_o = 0.24$ and $\alpha = 20$, $x_{1q} = 5.3$ and $\hat{q}_{sm} = 46$, while for the same value of x_o and $\alpha = 10$, $x_{1q} = 8.6$ and $\hat{q}_{sm} = 33$. Such limit values are shown in Fig. 8 as a function of the normalized doping.

It is most important to note that by substituting (30)

---

[5] These normalized gate voltages correspond to $V_B = -5$ V and $V_S = -10$ V when $\Phi_{MS} = Q_{ss}/C_o$.

[6] Note that since the charge under the edges of the storage electrode is less than that at the center, the average charge per unit area will be somewhat less than this.

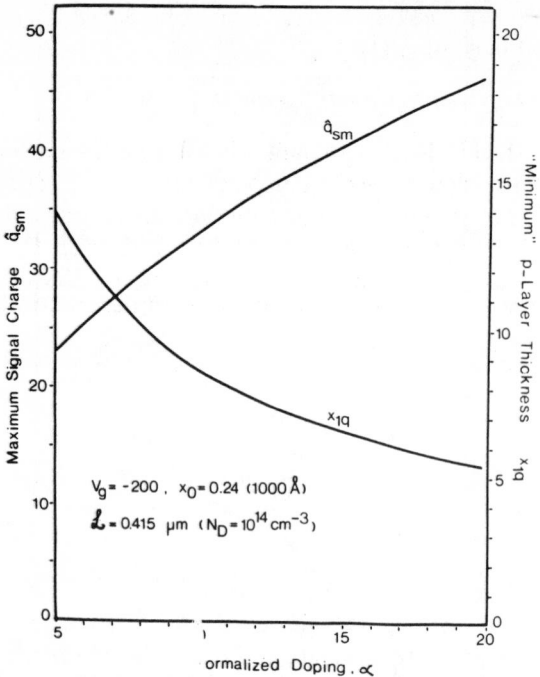

Fig. 8. Minimum p-layer thickness and the corresponding limiting values of the maximum signal charge-carrying capacity as functions of the normalized doping α ($= N_A/N_D$). The minimum thickness corresponds to the conditions $x_m = w/2$.

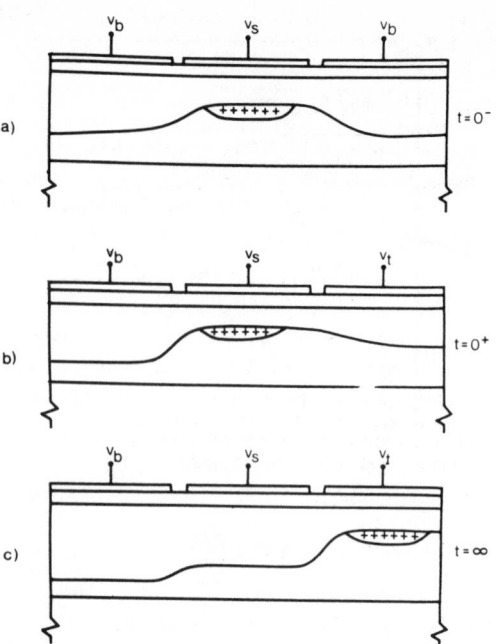

Fig. 9. Illustrating the variation of the potential minimum position during the transfer of signal charge. (a) Steady-state conditions prior to transfer. (b) Situation immediately following the application of a transfer voltage to the receiver electrode. (c) Final steady-state situation—all the charge transferred.

into (34a) the maximum normalized signal charge can be expressed very simply in terms of the position of the potential minimum for $q_s = 0$:

$$\hat{q}_s = \frac{v_b - v_s}{k + x_{m0}} \qquad (36)$$

where $x_{m0} = x_m(q_s = 0)$. When this expression is denormalized it is found that

$$\frac{\text{Maximum signal charge per unit area}}{\text{Voltage difference between electrodes}} = 1 \Big/ \left(\frac{1}{C_o} + \frac{1}{C_s} \right),$$

$$(37)$$

where $C_s = \epsilon_s/x_{m0}$ is the capacitance per unit area of a semiconductor layer of thickness equal to the position of the potential minimum when $q_s = 0$. Exactly the same result would have been obtained if it were assumed that all the signal charge was located at x_{m0}.

Equation (36) enables the maximum charge-carrying capacity of a buried-channel CCD to be compared to that of the surface CCD. Thus,

$$\frac{\text{Maximum charge for surface CCD}}{\text{Maximum charge for buried-channel CCD}} = 1 + \frac{x_{m0}}{k},$$

which typically turns out to be in the range from 10 to 30. Hence, a surface CCD compared to a buried-channel device of the same dimensions can carry more than an order of magnitude more charge. However, this is a small price to be paid for the superior high-frequency capability of the buried-channel device [9].

Returning to the question of the best choice of x_1, it should be pointed out that the results of a two-dimensional analysis [10] have shown that there exists an optimum value for x_1 (denoted by x_{1f}), which makes the minimum fringing field under the storage electrode a maximum. For a three-phase buried-channel CCD whose unit cell normalized length is l and which is biased such that $v_t - v_s = v_s - v_b$, it is found that this optimum occurs when $x_m \simeq 0.133 \, l - k$. From (36), this means that for a normalized p-layer thickness of x_{1f}, the maximum normalized charge is given by

$$\hat{q}_{sf} \simeq 7.5(v_s - v_b)/l,$$

which, together with x_{1f}, is indicated in Fig. 7. It therefore appears that, for a given v_g and α, there may be an optimum choice for x_1 that lies between x_{1q} and x_{1f}, depending on the particular application.

D. Nature of the Charge Transfer Process

From (30) and Fig. 6 it will be noted that the x position of the potential minimum depends on the signal charge as well as the gate voltage. Hence, during the process of signal-charge transfer from one gate to an adjacent gate, the charge transport occurs along a channel whose position varies in the x direction. This is illustrated in Fig. 9. An additional complication arises from the change in potential profile with signal charge. The transfer of charge causes the local charge density to change, and this affects the potential. Such involved interactions make the analysis rather difficult; however, some simplification is possible by noting that the variation of x_m with both v_g and q_s is relatively small. This can be seen by evaluating the

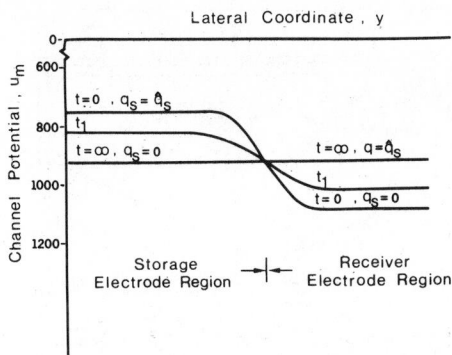

Fig. 10. Sketch showing the variation of the potential minimum under the storage and receiver electrodes during the transfer process. The conditions correspond to a signal charge equal to the maximum charge-carrying capacity.

partial derivatives of x_m using (30):

$$\frac{\partial x_m}{\partial q_s} \simeq \frac{1}{2\alpha} \; ; \frac{\partial x_m}{\partial v_g} \simeq \frac{1}{\alpha} \left[-2v_g + x_1(\alpha + 1)(x_1 + 2k) \right]^{-1/2},$$

$$\text{for } q_s \ll \alpha x_1. \quad (38)$$

For reasonable values of α, v_g, and q_s, both derivatives are small. This is also evident from Fig. 6 when it is remembered that the maximum normalized charge is in the neighborhood of 20.

The influence of the signal charge on the lateral profile of the potential minimum is illustrated in Fig. 10 at various times following the application of a transfer voltage step to the receiver electrode. It will be noted that the maximum lateral field[7] in the neighborhood of the electrode boundary occurs at the start before the signal charge has had time to be transported. As the charge is depleted from the storage electrode and as charge builds up in the receiver electrode, so the electric field is reduced, reaching its minimum value when all the charge has been transported. As a result it can be expected that the final emptying of the storage electrode will be a dominant factor in controlling the maximum frequency of operation. Furthermore, it is clear that the magnitude of the signal charge in relation to the maximum charge-handling capacity plays an important role in determining the decay of the final fraction of charge left under the storage electrode. The larger the signal charge the longer will be the time taken to transfer the final fraction.

IV. ANALYSIS FOR A NONUNIFORMLY DOPED LAYER

Having gained some useful insights into the properties of a uniformly doped layer CCD, we are now in a position to examine the more general case of a nonuniformly doped layer and to study the effects of profile on the characteristics. As will be seen, the basic features concerning the charge distribution and potential profile are qualitatively similar to the uniformly doped case considered in Section III.

A. Charge and Potential Distribution

The highly localized charge distribution found for the uniformly doped case again suggests that a series expansion for the potential distribution about the minimum should yield fairly accurate results. Following the procedure used in obtaining (12), but noting that α is a function of x, we obtain for the normalized signal charge density

$$\rho_s(x) = \rho_{sm} \exp \left\{ \frac{-1}{2!} (x - x_m)^2 (\alpha(x_m) - \rho_{sm}) \cdot \right.$$
$$\left. - \frac{1}{3!} (x - x_m)^3 \alpha'(x_m) \cdots \right\} \quad (39)$$

which differs from (12) by the presence of odd terms in the expansion, making the charge density an asymmetrical function of x. For example, if the p-layer doping is a Gaussian function centered at the SiO_2–Si interface, the charge density will have a longer and more gradual tail on the substrate side of x_m.

By integrating (39), treating the third-order term as perturbation, and neglecting higher order terms, the total normalized signal charge can be found. This enables the following expression for the maximum charge density to be obtained

$$\rho_{sm} = \frac{q_s}{(2\pi)^{1/2}} \left\{ \left(\alpha(x_m) + \frac{q_s{}^2}{8\pi} \right)^{1/2} - \frac{q_s}{(8\pi)^{1/2}} \right\} \quad (40)$$

which is the same as (14) but with α replaced by $\alpha(x_m)$.

To obtain an expression for the channel thickness w, we again consider the third-order term as a perturbation, yielding[8]

$$w = 2 \left(\frac{2\eta}{\alpha(x_m) - \rho_{sm}} \right)^{1/2}. \quad (41)$$

However, unlike the uniformly doped case, the channel is not symmetrical about x_m, but extends to distances of

$$w_{1,2} = \frac{w}{2} \pm \frac{w^4}{192\eta} \alpha'(x_m) \quad (42)$$

on the surface side (+ sign) and substrate side (− side), i.e., to positions where the normalized potential has changed from u_m to $u_m + \eta$.

Calculation of the potential distribution follows essentially the same method as that used for the uniformly doped device, and details are given in Appendix A.

B. An Example

To illustrate how the device is influenced by a non-

---

[7] The lateral field is approximately the sum of the fringing field and the field induced by the charges under the storage and receiving electrodes.

[8] Equations (41) and (42) are only accurate when the signal charge is appreciably less than the maximum charge-carrying capacity.

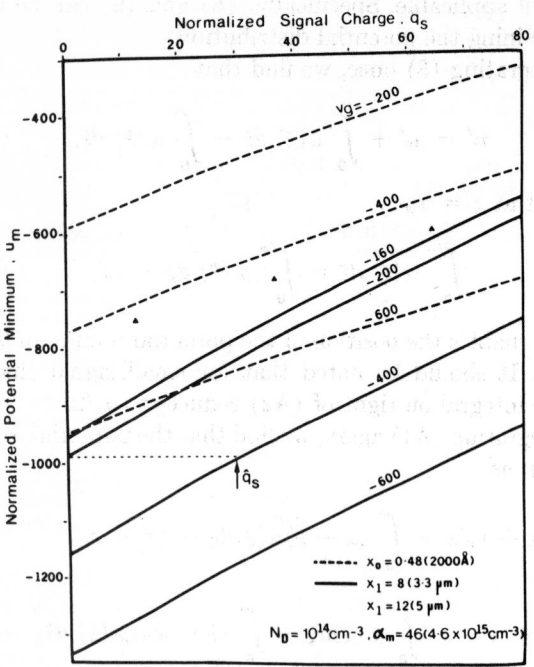

Fig. 11. Variation of the normalized potential minimum with signal charge for a device with a Gaussian p-layer doping given by (43). The results obtained by Kent [7] for $v_g = -160$, and $x_1 = 12$ are shown as triangles. Note that the normalization constant for q_s is 0.415×10^{10} cm^{-2}.

uniform impurity profile we shall assume that

$$\alpha(x) = \alpha_m \exp\left[-x^2/2\sigma^2\right] - 1 \qquad (43)$$

corresponding to the profile obtained during a drive-in diffusion. The standard deviation σ is related to x_1 by $\sigma = x_1/(2 \ln \alpha_m)^{1/2}$, thereby making α zero at $x = x_1$. Strictly speaking, since the theory developed assumed a constant doping of N_D beyond x_1, this choice for x_1 is improper. Ideally, one should choose x_1 such that $\alpha(x_1) \simeq -1$ in order that the charge density for $x > x_1$ be nearly uniform and equal to qN_D. We feel that the choice of $\alpha(x_1) = -0.9$, i.e., $\sigma = x_1/[2 \ln (10\alpha_m)]$ is sufficient for our purposes. Indeed the difference in the results for the two choices of x_1 becomes insignificant when x_1 is not too small.

To insure reasonable accuracy in the numerical calculations, six terms in the expansion of the potential u were used [see (39)]. Provided the signal charge was limited to less than one half the maximum charge-carrying capacity, the results obtained in this way agreed well with those obtained using the simple Gaussian form for u.

In Fig. 11 the variation of the channel potential with signal charge is shown. While qualitatively the results are similar to those for the uniformly doped layer (Fig. 5), quantitatively, when the same average doping[9] is used as the basis of comparison, two differences may be noted. First, the potential minimum for the Gaussian-doped layer is appreciably less than that for the uniformly doped layer. Second, the maximum charge-carrying capacity is significantly greater for the Gaussian-doped layer.

[9] The Gaussian profile with $\alpha_m = 46$, and $x_1 = 12$, has an average normalized doping of approximately 20.

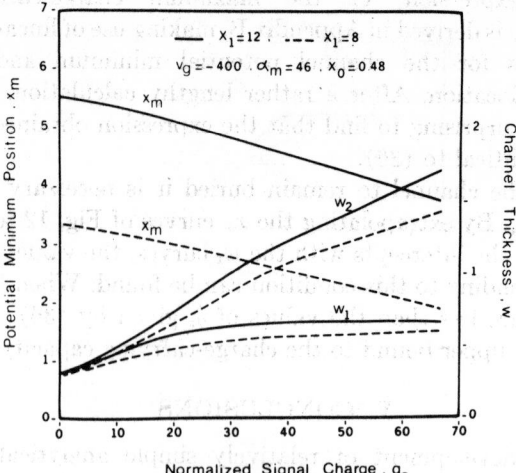

Fig. 12. Channel dimensions and position of the potential minimum as functions of the signal charge for a Gaussian p-layer doping and two p-layer thicknesses. The device parameters and normalization constants are the same as for Fig. 11, and the gate voltage is -10 V.

This arises from the increased doping close to the Si–SiO$_2$ interface, which causes the channel to be displaced toward the surface, thereby leading to a higher charge-carrying capacity.

Also shown in Fig. 11 are some points taken from the numerical computations of Kent [7]. It will be noted that for small signal charges the agreement is rather poor, while for larger charges the two results converge. This discrepancy, like that discussed in Section II-A, is believed to arise from Kent's assumption that $u = 0$ at $x = 42$.

In Fig. 12 the position of the channel and its widths w_1 and w_2 are shown as functions of q_s. It will be noted that the channel width on the surface side w_1 tends to become constant at higher charges, while w_2 continues to increase. This behavior is physically reasonable, since higher signal charges can be expected to cause less penetration of the more highly doped surface region. In addition, by comparing Figs. 12 and 6, it will be noted that the Gaussian-doped device, compared to a uniformly doped device with the same average doping, has a shallower channel. This also is a direct result of the more heavily doped surface region.

C. Charge-Carrying Capacity

The arguments used in Section III-C to arrive at the condition governing the maximum signal charge-carrying capacity of a uniformly doped device are equally valid when the doping is nonuniform. Hence, (32) and (33) can be used for this case. Equation (32) and Fig. 11 enable the maximum charge-carrying capacity to be found for the case $x_1 = 12$, $v_s = -400$, $v_b = -200$. This value, namely $\hat{q}_s \simeq 30$, when compared to that obtained from Fig. 5 for the same gate voltages and p-layer thickness, shows that the maximum charge-carrying capacity for the Gaussian-doped device is approximately 50 percent greater than that of a uniformly doped device having the same average doping.

An expression for the maximum charge-carrying capacity is derived in Appendix B, making use of linearized relations for the channel potential minimum and its spatial location. After a rather lengthy calculation, it is hardly surprising to find that the expression obtained for $\hat{q}_s$ is identical to (36).

For the channel to remain buried it is necessary that $w_1 \leq x_m$. By extrapolating the x_m curves of Fig. 12 to determine the intercepts with the w_1 curves, the values of q_s corresponding to this condition can be found. When these values are less than the values of $\hat{q}_s$ given by (36), they form an upper bound to the charge-carrying capacity.

V. CONCLUSIONS

The development of relatively simple analytical expressions for the signal charge distribution and potential profile in a buried-channel CCD has led directly to better physical insights of the device properties and behavior, and in addition has provided useful design equations and criteria. Furthermore, by eliminating the need for numerically integrating the basic equations, the computation time has been greatly reduced. More specifically, we have developed simple expressions for the channel thickness, location, and potential minimum, which have enabled us to examine the manner in which they are influenced by the gate voltage, total signal charge, and the structural parameters. It appears that over the range of practical interest, both the channel position and potential can be expressed as linear functions of the signal charge and gate voltage. This led to a simple estimate of the device charge-carrying capacity, and enabled the surface and buried-channel devices to be compared.

A most important aspect of buried-channel CCD design concerns the proper choice of the p-layer thickness and doping. It has been shown that a minimum thickness exists for a given doping, governed by the requirement that the channel charge must lie below the oxide/semiconductor interface. It has also been pointed out that the need to achieve a high lateral fringing field calls for a thicker p-layer. This is in conflict with the requirements for a large charge-carrying capacity. In practice, depending on the application, a compromise is necessary.

Examination of the effects of the p-layer doping profile suggests that substantial changes in the profile do not have a major influence on the device characteristics when the average doping throughout the layer remains the same. This suggests that, with the range of profiles that can be achieved in practice, it may be difficult to effect any major improvements in the device performance.

APPENDIX A

POTENTIAL DISTRIBUTION IN A NON-UNIFORMLY DOPED DEVICE

Since the conditions exterior to the p layer for the non-uniformly doped case are the same as for the uniformly doped layer, the boundary equations previously derived are still applicable. Specifically, (5) and (9) can be used in obtaining the potential distribution.

Integrating (3) once, we find that

$$u' = u_0' + \int_0^x \alpha(\xi) \, d\xi - \int_0^x \rho_s(\xi) \, d\xi, \qquad (A1)$$

so that at $x = x_m$

$$\int_0^{x_m} \alpha(\xi) \, d\xi = \int_0^{x_m} \rho_s(\xi) \, d\xi - u_0', \qquad (A2)$$

which enables the position of the potential minimum to be found. It should be noted that for small signal charges q_s, the integral on right of (A2) reduces to $q_s/2$.

Integrating (A1) again, we find that the potential can be written as

$$u = u_0 + u_0'x + \int_0^x (x - \xi)\alpha(\xi)d\xi - (x - x_m)$$

$$\cdot \int_0^x \rho_s(\xi)d\xi + \int_0^x (\xi - x_m)\rho_s(\xi) \, d\xi. \quad (A3)$$

This equation enables the potential minimum to be expressed as

$$u_m = u_0 + u_0'x_m + \int_0^{x_m} (x_m - \xi)\alpha(\xi) \, d\xi$$

$$+ \int_0^{x_m} (\xi - x_m)\rho_s(\xi) \, d\xi. \quad (A4)$$

From (A1) and (A3), the boundary conditions at x_1 can be written as

$$u_1' = u_0' + \int_0^{x_1} \alpha(\xi)d\xi - q_s$$

$$u_1 = u_0 + u_0'x_1 + \int_0^{x_1} (x_1 - \xi)\alpha(\xi)d\xi - (x_1 - x_m)q_s$$

$$+ \int_0^{x_1} (\xi - x_m)\rho_s(\xi) \, d\xi \quad (A5)$$

which, together with the boundary conditions (5) and (9), can be solved to yield the normalized field at $x = 0$:

$$u_0' = \{-2v_g + k^2 + 2\int_0^{x_1} (\xi + k)(\alpha(\xi) + 1) \, d\xi$$

$$- 2(x_m + k)q_s - 2\int_0^{x_1} (\xi - x_m)\rho_s(\xi) \, d\xi\}^{1/2}$$

$$- \{k + \int_0^{x_1} (\alpha(\xi) + 1) \, d\xi - q_s\}. \quad (A6)$$

It should be noted that, for small signal charges, the integral

$$\int_0^x (\xi - x_m)\rho_s(\xi) \, d\xi,$$

which appears in (A3) to (A6), can be approximated by $\rho_s(x)/\{\alpha(x_m) - \rho_{sm}\}$. This greatly simplifies the calculations.

To obtain numerical values for u_0' and x_m, (A6), (A2), and (39) must be solved iteratively. Once these quantities have been obtained, (A3) can be numerically evaluated to obtain the potential profile and other quantities such as u_m, $\rho_s(x)$, and ρ_{sm} can then be calculated using the appropriate equations.

APPENDIX B

CHARGE-CARRYING CAPACITY FOR A NON-UNIFORMLY DOPED DEVICE

It will be noted from Figs. 11 and 12 that, over the range of practical interest, x_m and u_m have a nearly linear variation with the signal charge. Consequently, we can write

$$u_m \simeq u_{m0} + q_s \left.\frac{\partial u_m}{\partial q_s}\right|_{q_s=0}$$

$$x_m \simeq x_{m0} + q_s \left.\frac{\partial x_m}{\partial q_s}\right|_{q_s=0}, \qquad (B1)$$

where the zero signal charge values, denoted with a subscript 0, can be found from (A4) and (A2). It can be shown from the differential forms of (5), (9), and (A5), together with (A2) and (A4), that the partial derivatives are given by

$$\left.\frac{\partial u_m}{\partial q_s}\right|_{q_s=0} = \frac{(k + x_{m0})(u_{10}' + x_1 - x_{m0})}{u_{10}' + x_1 + k}$$

$$\left.\frac{\partial x_m}{\partial q_s}\right|_{q_s=0} = \frac{1}{\alpha(x_{m0})}\left(\frac{x_{m0} + k}{u_{10}' + x_1 + k} - \frac{1}{2}\right). \qquad (B2)$$

By substituting (B2) into (B1) we obtain the required linear relations. Furthermore, an expression for the maximum charge-carrying capacity can be obtained from (33) by evaluating the partial derivatives $\partial u_m/\partial v_g$ and $\partial u_m/\partial q_s$ at $q_s = 0$. This can be performed with the help of (A2), (A4), (A6), and (B2), and leads to

$$\hat{q}_s = (v_b - v_s)/(k + x_{m0}), \qquad (B3)$$

which is identical to (36).

REFERENCES

[1] R. H. Walden, R. H. Krambeck, R. J. Strain, J. McKenna, N. L. Schryer, and G. E. Smith, "The buried channel charge coupled device," *Bell Syst. Tech. J.*, vol. 51, pp. 1635–1640, Sept. 1972.
[2] L. J. M. Esser, "Peristaltic charge-coupled device: A new type of charge-transfer devices," *Electron. Lett.*, vol. 8, pp. 620–621, Dec. 14, 1972.
[3] C. K. Kim, J. M. Early, and G. F. Amelio, "Buried channel charge coupled devices," presented at NEREM, Nov. 1972.
[4] W. S. Boyle and G. E. Smith, "Charge-coupled semiconductor devices," *Bell Syst. Tech. J.*, vol. 49, pp. 587–593, Apr. 1970.
[5] M. F. Tompsett, "Charge transfer devices," *J. Vac. Sci. Technol.*, vol. 9, pp. 1166–1181, July–Aug. 1972.
[6] J. McKenna and N. L. Schryer, "The potential in a charge coupled device with no mobile minority carriers and zero plate separation," *Bell Syst. Tech. J.*, vol. 52, pp. 669–696, May–June 1973.
[7] W. H. Kent, "Charge distribution in buried-channel charge-coupled devices," *Bell Syst. Tech. J.*, vol. 52, pp. 1009–1024 July–Aug. 1973.
[8] K. C. Gunsager, C. K. Kim, and J. D. Philips, "Performance and operation of buried channel charge coupled devices," in *Tech. Dig. 1973 Int. Electron Devices Meeting*, pp. 21–23.
[9] L. J. M. Esser, M. G. Collet, and J. G. Van Santen, "The peristaltic charge coupled device," *Tech. Dig. 1973 Int. Electron Devices Meeting*, pp. 17–20.
[10] H. El-Sissi and R. S. C. Cobbold, unpublished work.

The Effects of Bulk Traps on the Performance of Bulk Channel Charge-Coupled Devices

AMR M. MOHSEN, MEMBER, IEEE, AND MICHAEL F. TOMPSETT, MEMBER, IEEE

Abstract—The effects of bulk traps on the transfer efficiency and transfer noise in bulk channel charge-coupled devices (BCCD's) are calculated for different charge packet sizes and operating frequencies. These predictions are compared with experimental results and the distribution and density of bulk states in actual devices are thereby measured. The measured low transfer inefficiency of 10^{-4} per transfer with no intentionally introduced background charge and low transfer noise are shown to be due to a low bulk state density of $2 \times 10^{11}/cm^3$. A detailed comparison of estimated noise in both surface and bulk channel versions of an image sensor and an analog delay line show that BCCD's are very attractive for low-light level imaging but not as attractive for analog signal processing.

I. INTRODUCTION

BULK channel charge-coupled devices (BCCD's) were first introduced [1] with the name buried channel CCD in order to avoid the transfer inefficiency limitations expected in surface channel CCD's (SCCD's) due to the interaction of the signal charge with interface states

Manuscript received June 6, 1974; revised July 22, 1974.
The authors are with Bell Laboratories, Murray Hill, N. J. 07974.

Reprinted from *IEEE Trans. Electron Devices*, vol. ED-21, pp. 701–712, Nov. 1974.

[2]–[4]. The use of a circulating background charge minimizes these limitations and, practically extremely low interface state densities of 1×10^9 cm$^2 \cdot$eV have now been measured [5], [6]. Interface states also give rise to transfer noise which is unaffected by the background charge. As we shall see in this paper, BCCD's are also subject to similar but normally smaller effects, which can be attributed to the interaction of the signal charge with bulk states. The BCCD was also proposed with the name peristaltic CCD [7] as a means of increasing the fringing fields under the electrodes and thereby obtaining very high-frequency operation.

Whereas our measurements [5] on surface channel CCD's indicate that the interface states are distributed continuously in the bandgap, we expect bulk traps to exist at discrete energy levels. In this paper, we consider the effects of these traps on the performance of the BCCD's. In particular, we shall calculate the transfer inefficiency and transfer noise as a function of frequency and charge packet size in Sections II, III, and IV and compare these parameters with experimental results in Section V, followed by a discussion in Section VI.

II. INCOMPLETE CHARGE TRANSFER WITHOUT BACKGROUND CHARGE

We shall assume an n-channel BCCD with a uniform density N_t of bulk traps at an energy level E below the conduction band. The density of filled states is described by the Shockley–Read–Hall [8] rate equation

$$\frac{dn_t}{dt} = \frac{(N_t - n_t)}{\tau_t} - \frac{n_t}{\tau_e} \qquad (1)$$

where τ_t and τ_e are the trapping and emission time constants of the trap center and are given by

$$\tau_t = 1/\sigma v_{\text{th}} n_e \qquad \tau_e = 1/\sigma v_{\text{th}} N_c \exp (E/kT) \qquad (2)$$

where n_e is the volume density of mobile electrons, σ is the trap capture cross section for electrons (cm^2), v_{th} is the average thermal velocity of the mobile electrons (cm/s), N_c is the effective density of states per unit volume (cm^{-3}) in the conduction band, and kT is the electronvolt equivalent of temperature.

Since for most traps of importance the trapping time constant is much less than the reemission time constant, the charge trapped in bulk traps from a signal charge stored under an electrode and occupying a volume V_s in the bulk at moderate and low clock frequencies is given by

$$q_t = \int_{V_s} \frac{eN_t \, dV}{1 + (N_c/n_e) \exp (-E/kT)} \approx eN_t V_s. \qquad (3)$$

At high clock frequencies the traps do not have enough time to capture any carriers. In this case (3) should be multiplied by a filling probability F which will be defined later in (7). We consider first a multiphase CCD where all electrodes are equivalent, such as a three-phase CCD. When the signal charge is transferred to the next electrode, the density of the signal charge under the first electrode

drops rapidly during the transfer period, and the emission of the trapped charge becomes dominant after a time t'. In the next period of time $T_0 - t'$ the emitted carriers move forward to join the main signal packet. The amount of trapped charge which cannot join the signal charge packet is then given by

$$\Delta q = eN_t V_s \exp \left[-(T_0 - t')/\tau_e\right]$$

where T_0 is the transfer time and is related to the *clk* frequency f_c, the clock period T_c, and number of phases m by $T_0 = 1/mf_c = T_c/m$. Since $t' \ll T_c$ for good transfer efficiency the preceding expression becomes

$$\Delta q \simeq eN_t V_s \exp (-T_0/\tau_e). \qquad (4)$$

If a given charge packet is followed by a string n_z of empty charge packets, the filled bulk states continue to emit trapped charge into the empty charge packets. When the next signal charge packet comes along the bulk states are filled again by capturing charge ΔQ_s per transfer from this packet. This charge appears as a charge loss $n\Delta Q_s$ from the leading charge packet of a string of charge packets transferred through a device with n electrodes when preceded by n_z empty charge packets. ΔQ_s is given by (5) as the difference between the total number of filled traps given by (4) and the number of traps that remain filled after time $(n_z mT')$ so that

$$\Delta Q_s \cong eN_t V_s \exp (-T_0/\tau_e)[1 - \exp (-n_z mT_0/\tau_e)]. \qquad (5)$$

In a similar manner we can calculate the charges $\Delta Q_{(1)}$, $\Delta Q_{(2)}$, and $\Delta Q_{(i)}$ emitted, respectively, into the first, second, and ith trailing empty charge packets, respectively, after a string of signal charge packets.

$$\Delta Q_{(1)} = eN_t V_s \exp (-T_0/\tau_e)[1 - \exp (-mT_0/\tau_e)]$$

$$\Delta Q_{(2)} = eN_t V_s \exp \left[-(m + 1)T_0/\tau_e\right]$$
$$\cdot [1 - \exp (-mT_0/\tau_e)]$$
$$= \Delta Q_{(1)} \exp (-mT_0/\tau_e)$$

$$\Delta Q_{(i)} = eN_t V_s \exp \{-[(i - 1)m + 1]T_0/\tau_e\}$$
$$\cdot [1 - \exp (-mT_0/\tau_e)]$$
$$= \Delta Q_{(1)} \exp [-(i - 1)mT_0/\tau_e]. \qquad (6)$$

In the case where several bulk states with different densities and emission time constants interact with the signal charge, the right-hand sides of (5) and (6) will be the sum of similar terms corresponding to each of the bulk states.

In other types of devices, such as a two-phase BCCD [9], half the electrodes act as transfer gates and the effect of the partial filling of the bulk states under these gates as the charge moves by must be considered. Following the same approach used for the surface channel overlapping gate two-phase CCD [4] the filling probability F_s of the traps as the free charge sweeps by can be shown to be

$$F_s = 1 - \exp (-\Delta T/\tau_c), \qquad 1/\tau_c = 1/\tau_e + \sigma v_{\text{th}} n_{\text{av}} \qquad (7)$$

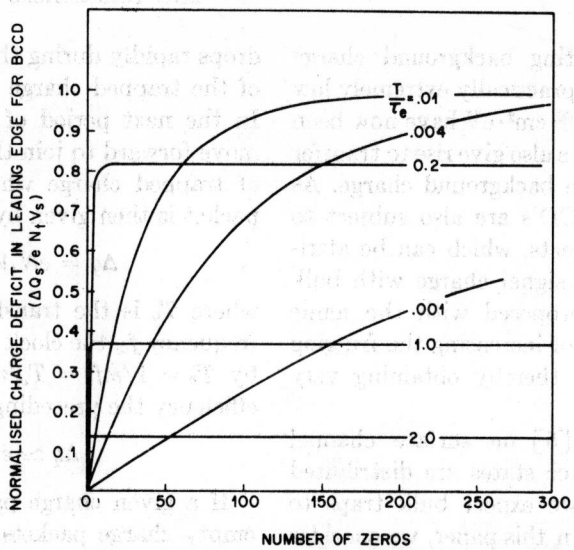

Fig. 1.   Normalized charge deficit ($\Delta Q_s/eN_tV_s$) in the leading edge versus the number of "zeros" for different ratios of the transfer time T to the emission time constant τ_e.

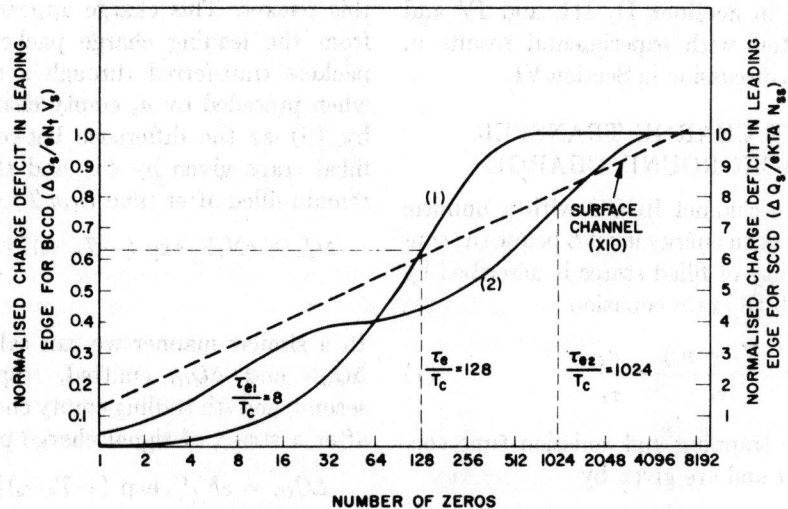

Fig. 2.   Normalized charge deficit in the leading edge of a series of charge packets versus the number of zeros for CCD's. Curve 1 for the case of a BCCD with one bulk state of emission time constant $\tau_e = 128T_c$. Curve 2 for a BCCD with two bulk states of emission time constants $\tau_{e1} = 8T_c$ and $\tau_{e2} = 1024T_c$. The dashed curve is for a surface channel device.

where n_{av} is the average carrier concentration under the transfer gate during the time interval ΔT and can be obtained from free charge transfer calculations [10], [11], σ is the capture cross section of the bulk state under consideration, and v_{th} is the thermal velocity of the carriers. In this case the charge deficit ΔQ_s from the leading edge and the charge excess ΔQ_1 in the trailing edge are also asymmetrical and are given by

$$\Delta Q_s = eN_t(V_{st} + F_sV_{Tr})$$
$$\cdot \exp(-T_0/\tau_e)[1 - \exp(-2n_zT_0/\tau_e)]$$

$$\Delta Q_1 = eN_t(V_{st} + F_sV_{Tr})$$
$$\cdot \exp(-T_0/\tau_e)[1 - \exp(-2T_0/\tau_e)] \quad (8)$$

where V_{st} is the volume occupied by the signal charge

under the storage gate, V_{Tr} is the volume swept by the signal charge during its transfer under the transfer gate.

In Fig. 1, using (5) for a three-phase BCCD we have plotted the normalized charge deficit $\Delta Q_s/eN_tV_s$ in the leading signal charge packet Q_s per transfer versus the number of empty charge packets preceding it assuming a single bulk state for different values of T_0/τ_e. The normalized charge deficit per transfer is plotted in Fig. 2 on a semilog scale in the cases where one and two bulk states interact with the signal charge. The normalized charge deficit in a SCCD is also plotted in Fig. 2 for comparison. The exponential saturation of the charge loss with the increasing number of empty charge packets n_z in BCCD's in Figs. 1 and 2 differs from the continuously (logarithmically) increasing value of the charge loss in SCCD's in Fig. 2. This is because in BCCD's the trap centers are localized

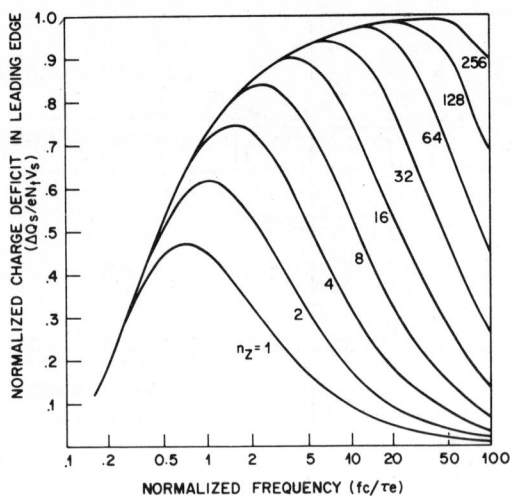

Fig. 3. Normalized charge deficit from leading edge ($\Delta Q_s/eN_tV_s$) versus normalized clock frequency ($f_c\tau_e$) for different number of "zeros" n_z.

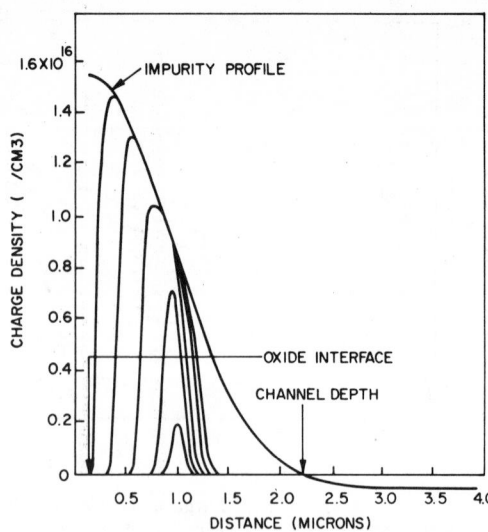

Fig. 4. The distribution of charge in the buried channel for different sizes of charge packets. The buried channel depth is 2.2 μ with a Gaussian doping profile of surface concentration $C_s = 1.6 \times 10^{16}/\text{cm}^3$.

in the bandgap. When the number of empty charge packets n_z becomes greater than the ratio of the emission time constant τ and the clock period T_c for a particular bulk state.

$$n_z > \tau_e/T_c \qquad (9)$$

most of the trapped charge in this bulk state is emitted. However, in SCCD's, there is a continuum of interface states across the bandgap, and as the number of empty charge packets n_z increases, more interface states empty their trapped charge. In that case the charge loss per transfer is given by [2]–[4]

$$\Delta Q_s = eAkTN_{ss} \ln{(n_z m + 1)} \qquad (10)$$

where $N_{ss}(E)$ is the interface state density (cm$^{-2}\cdot$eV^{-1}) assumed constant across the bandgap and A is the area covered by the signal charge at the interface. In Fig. 3 we have plotted the normalized charge deficit per transfer in the leading edge $\Delta Q_s/eN_tV_s$ for a three-phase CCD using (5) in the case of a single bulk state for different numbers of empty charge packets n_z versus $f_c\tau_e$. It is clear that the charge deficit ΔQ_s, from the leading charge packet has a maximum value when the emission time constant τ_e is of the order of the time between successive strings of charge packets. The maximum charge deficit occurs at a frequency $f_{\max}$, where

$$f_{\max} = \frac{1}{\tau_e} \frac{n_z}{\ln{(mn_z + 1)}} . \qquad (11)$$

For clock frequencies $f_c \ll f_{\max}$ most of the trapped charge is emitted during the transfer of the signal charge and therefore the charge deficit decreases exponentially with the inverse of the clock frequency. For $f_c \gg f_{\max}$ the bulk states do not have enough time to emit the trapped charge in the time period between successive signal charge packets, therefore they do not have to be refilled and the charge deficit decreases with the inverse of the clock frequency.

III. INCOMPLETE CHARGE TRANSFER WITH CIRCULATING BACKGROUND CHARGE

A background charge may be inserted at the input end of the device so that charge packets are passed along the device in the absence of signal charge. As in the case of surface channel devices the use of the background charge will reduce the incomplete charge transfer effects. In this section we consider the effect of the background charge on the filling and emptying of the localized bulk states in the buried channel and the resulting transfer inefficiencies.

Fig. 4 shows the distribution of charge normal to the surface in the buried channel for different charge packet sizes obtained by numerically solving the one-dimensional Poisson equation [12]. The doping profile in the buried channel is assumed Gaussian with a peak concentration at the surface of $1.6 \times 10^{15}/\text{cm}^3$ and channel depth of 2.1 μ. As the size of the charge packet increases, it spreads to cover a larger volume in the buried channel. Using the data in Fig. 4, we have plotted in Fig. 5 the distance over which the charge packet spreads versus the charge packet size for gate voltages of 0 and 10 V. It can be seen that the volume of bulk silicon per electron in which the bulk states can be filled is greatest for small charge packets but rapidly decreases as the bottom of the well is filled and then the volume of the charge packet increases almost linearly with increasing charge. Thus a little background charge or slim zero is expected to improve the performance once the linear part of the curve is reached. As the gate voltage increases the charge packet moves closer to the surface where the implanted ion density is greater and hence the volume occupied by the charge packet decreases.

In treating the trapping and release of charges by the bulk states in the buried channel with a background charge one must consider the ways in which the different volumes of the buried channel interact with the charge packets. Fig. 6 defines the two kinds of edges bounding

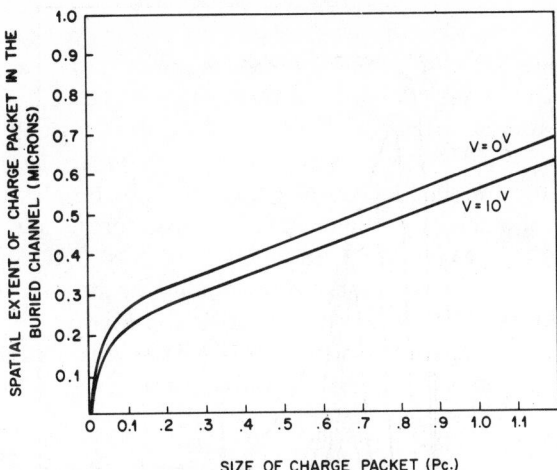

Fig. 5. The spatial extent of the charge in the buried channel for different charge packet sizes at gate voltages of zero and 10 V. The doping profile is Gaussian with a surface concentration of $1.6 \times 10^{16}/cm^3$ and channel depth of 2.1 μ.

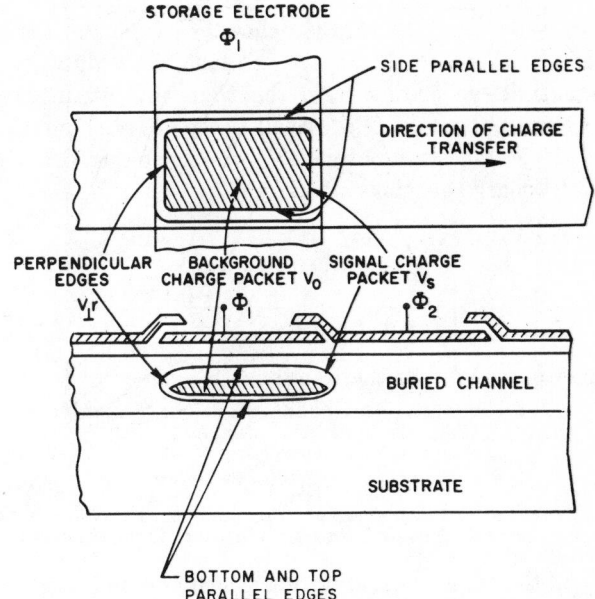

Fig. 6. Volumes occupied by a background charge and a signal charge packet in a buried channel charge-coupled device.

a charge packet, perpendicular and parallel [4]. The perpendicular edges are the regions of the buried channel perpendicular to the channel, which are not occupied by a background charge but are occupied by a larger signal charge packet when stored under a gate electrode. The background charge flows through these edges during the transfer cycle. Thus the bulk states in the perpendicular edges can capture charge from a large signal charge packet during storage and from a background charge during transfer. The parallel edges are the regions of the buried channel parallel to the channel which are not occupied by a circulating background charge but are occupied by a large signal charge packet. The bulk states in the parallel edges cannot capture charge from the background charge during tranfer. Fig. 6 shows the side parallel edges (as in SCCD's) [4] and the top and bottom parallel edges over which the charge packet spreads with increasing size. The parallel edges are residual volumes in the buried channel which the background charge cannot reach and hence the trapping and release of charge by the bulk states there is similar to an empty channel without background charge. We shall calculate the effects of charge trapping in these volumes on the incomplete charge transfer.

Trapping Under the Storage Gates

We consider first the volume V_0 occupied by both the background and signal charge under the storage gates as shown in Fig. 6. The bulk states in this volume will be filling and emptying every cycle and for a sufficiently large background charge, almost the same net trapping occurs. Thus the background charge is expected to reduce the effects of the trapping in these bulk states. However, the dependence of the trap occupation on the time available for emission $(T - t')$ and the size of the charge packet results in a residual component of incomplete charge transfer. Using (3)–(6), the charge deficit ΔQ_s from the leading edge and the charge excess in the trailing edge of a string of large charge packets Q_s after n_z back-

ground charge packets is given by

$$\Delta Q_s = \frac{eN_tV_0 \exp\left[-(T_0 - t')/\tau_e\right]}{1 + 1/\tau_e\sigma v_{th}n_s}$$

$$- \frac{eN_tV_0 \exp\left[-(T_0 - t' + \Delta t')/\tau_e\right]}{1 + 1/\tau_e\sigma v_{th}n_0}$$

$$\Delta Q_s \approx eN_tV_0 \exp\left(-T_0/\tau_e\right)$$

$$\cdot \left[\frac{1}{\tau_e n_s \sigma v_{th}} \frac{n_s - n_0}{n_0} + \frac{\Delta t'}{\tau_e}\right] \quad (12)$$

where n_s and n_0 are the average carrier concentration over the common volume V_0 of the signal and background charges and $\Delta t'$ is the difference in the time available for emission. Thus any charge deficit in the leading edge and charge excess in the trailing edge arising from traps under the storage gate are equal and independent of the number of background charge packets preceding the string of large charge packets.

In deriving (12), it has been assumed that the filling probability of the bulk states is unity. However, at very high frequency and low carrier densities, this may not be valid and (12) should be modified accordingly. For example, the trapping time constant τ_t for $n_e = 10^{15}$, $\sigma = 10^{-15}$ cm², and $v_{th} = 10^7$ cm/s is 0.1 μs. Thus for high frequencies (>10 MHz in this case) the trap occupation would be unable to follow the carrier density variations in the charge packets and the effective number of traps would decrease. Therefore, any charge deficit or charge excess due to bulk states trapping will also decrease.

Trapping in the Perpendicular Edges

The bulk states in the perpendicular edges will capture charge from the large signal charge packet during the storage time and from the background charge during charge transfer. The charge deficit ΔQ_s from the leading charge packet and the charge excess in the trailing background charge arising from the traps are symmetrical. For a multiphase CCD where all electrodes are equivalent and the storage volumes under successive electrodes overlap the charge deficit is given by

$$\Delta Q_s = eN_t V_{\text{tr}} \exp\left(-T_0/\tau_e\right)$$

$$\cdot \left[\frac{F_s}{1 + 1/\tau_e \sigma v_{\text{th}} n_s} - \frac{F_0 \exp\left(-T_0/\tau_e\right)}{1 + 1/\tau_e \sigma v_{\text{th}} n_0}\right] \quad (13)$$

where F_s and F_0 are the fractional occupancy of the bulk states for the signal and background charge packets and are given by

$$F_s = \left[1 - \exp\left(-T_0/\tau_{cs}\right)\right]$$

$$F_0 = 1 - \exp\left(-\Delta T_0/\tau_{c0}\right). \quad (14)$$

ΔT_0 is the time interval during which the bulk states in the perpendicular edges can capture charge from the background charge. τ_{cs} and τ_{c0} are the effective capture time constants for the signal and background charge packets and are given by

$$\frac{1}{\tau_{cs}} = \frac{1}{\tau_e} + \sigma v_{\text{th}} n_s \quad \text{or} \quad \frac{1}{\tau_{c0}} = \frac{1}{\tau_s} + \sigma v_{\text{th}} n_0 \quad (15)$$

n_s and n_0 are the average carrier concentration for the signal and background charge, respectively, in the perpendicular edge volume.

In the case of two-phase buried channel CCD's [8], the emptying and filling of the bulk states under the transfer gates is similar to that under the perpendicular edges. The charge deficit ΔQ_s from the leading edge and the charge excess in the trailing edge arising from these traps are also symmetrical and given by

$$\Delta Q_s = eN_t V_{\text{Tr}} \exp\left(-T_0/\tau_e\right)$$

$$\cdot \left[\frac{F_s}{1 + 1/\tau_e v_{\text{th}} \sigma n_s} - \frac{F_0}{1 + 1/\tau \sigma v_{\text{th}} n_0}\right]. \quad (16)$$

Trapping in the Parallel Edges

Since the parallel edges are residual volumes in the buried channel which the background charge cannot reach, the trapping and release of charge by the bulk states is similar to an empty channel without background charge. The charge deficit from the leading edge after n_z background charge packets and the charge excess in the trailing edge are given by (5) and (6) with the volume V_S replaced by the volume of the parallel edges V_{PL}. Similar to the empty channel case, the charge deficit and charge excess are asymmetrical, and their dependence on the clock frequency and number of zeros is illustrated in Figs. 1–3.

Thus we see that the relative contributions of the bulk traps in the different volumes of the buried channel considered earlier to the charge deficit at the leading edge and charge excess at the trailing edge of a group of ones with a background charge will depend on several factors. These factors include the clock frequency, the size of the charge

packet relative to the background charge packet, the number of slim zeros between the group of "ones," the length of the electrodes and the channel width. For example, for large charge packets and relatively small slim zeros in BCCD's with a deep buried layer, long electrodes and narrow channel widths trapping in the bulk states in the parallel edges is dominant. However, for signal charge packets and fat zeros in BCCD's with shallow buried layers and large channel width trapping in the bulk states in the perpendicular edge volumes and storage volumes becomes more important, with the latter dominating for long electrodes.

IV. INTRODUCTION OF TRANSFER NOISE BY BULK STATES

Similar to the effects of interface states in SCCD's [2], the constant filling and emptying of the bulk states as a signal charge moves along the buried channel in a BCCD can introduce fluctuations in the number of carriers in the signal charge packet. These fluctuations will then appear as transfer noise in the output signal of the device. It is clear that when the bulk states are completely empty or completely full, the resulting fluctuations are zero since the number of carriers emitted from these states is then exactly determined. However, when the time available for the bulk states to emit the trapped carriers is of the order of the emission time constant τ_e, the resulting fluctuations are maximum. An expression for the fluctuations introduced by trapping in a bulk state may be derived in the following way.

The trap center is considered to be full and starts to emit the trapped carriers at $t = 0$ according to (1) with the mobile carrier density $n = 0$. It is assumed that the trapped carriers, which are emitted, are transferred each cycle and do not accumulate under the storage electrodes or become retrapped; i.e., for the mobile carrier density n_e to be small enough that the emission time constant $\tau_e \ll$ trapping time constant τ_t that is $n_e \ll N_c \exp\left(-E_t/kT\right)$ or the quasi-Fermi level describing the electron density in the conduction band $E_{PQ} < E_t$. The probability P of a carrier trapped in a bulk state of emission time constant τ_e being emitted in time t is given by

$$P \equiv 1 - \exp\left(-t/\tau_e\right). \quad (17)$$

The distribution of full and empty bulk states will be binomial with a variance V_n per unit volume of the buried channel given by

$$V_n = N_t \exp\left(-t/\tau_e\right)\left[1 - \exp\left(-t/\tau_e\right)\right]. \quad (18)$$

Thus, as expected, the variance of the fluctuations in the carriers emitted from a localized bulk state is dependent on the time available for emission t. The variance V_n has a maximum equal to

$$V_{n\,\text{max}} = 0.25 N_t \quad (19)$$

when the time t available for emission is given by

$$t = \tau_e \ln 2 = 0.69 \tau_e. \quad (20)$$

As the charge packets are transferred along the device, an excess of carriers reemitted into the charge packet leads to an equal deficit in the succeeding charge packet. This is because the bulk states are filled completely by each charge packet. Hence, a charge fluctuation in one charge packet is always accompanied by a fluctuation of the opposite sign in the succeeding packet. Since these pairs of fluctuations are independent, the total variance of the number of carriers in a charge packet at each transfer cycle is equal to the sum of the variances of the fluctuations in the emitted carriers into that charge packet and into the preceding charge packet at the preceding transfer cycle.

In SCCD's the mean square fluctuation F^2 in the number of charges introduced into a signal charge packet will not depend on the information of the preceding charge packets as the interface states are rather uniformly distributed across the bandgap. But in BCCD's the signal charges interact with bulk traps which exist at discrete energy levels, therefore the mean square fluctuation F^2 in a signal charge packet will depend in general on the information content of the preceding charge packets. For example, for a multiphase CCD where the electrodes are equivalent and the case of a string of full charge packets separated by a string of n_z empty charge packets, the mean square fluctuations F^2 in the number of charges per transfer in the first leading charge packet is given by

$$F^2 = V_s N_t \exp\left(-T_c/m\tau_e\right)\left[1 - \exp\left(-T_c/m\tau_e\right)\right]$$
$$+ V_s N_t \exp\left[-(n_z + 1/m)T_c/\tau_e\right]$$
$$\cdot\{1 - \exp\left[-T_c(n_z + 1/m)/\tau_e\right]\} \quad (21)$$

where V_s is the volume occupied by the charge packet, m is the number of phases, and T_c is the clock period. The mean square fluctuation F^2 in the number of charges in any of the other full charge packets is given by

$$F^2 = 2V_s N_t \exp\left(-T_c/m\tau_e\right)\left[1 - \exp\left(-T_c/m\tau_e\right)\right]. \quad (22)$$

It follows that the maximum mean square fluctuation F is given by

$$F^2 = 0.5 V_s N_t. \quad (23)$$

which is approximately equal to the shot noise in the trapped carriers emitted from the bulk state during the available transfer time. From (20) and (22) we see that this maximum noise will occur at a transfer frequency given by

$$f_0 = \frac{1.45}{m}\frac{1}{\tau_e}. \quad (24)$$

For clock frequencies $f_c \gg f_0$, the mean square fluctuation due to trapping in a bulk state with emission time constant τ_e is given by

$$F^2 \cong 2V_s N_t/m f_c \tau_e, \quad \text{for } f_c \gtrsim 5/m\tau_e. \quad (25)$$

For clock frequencies $f_c \ll f_0$, then the mean square fluctuation is given by

$$F^2 \cong 2V_s N_t \exp\left(-1/m\tau_e f_c\right), \quad \text{for } f_c \lesssim \frac{1}{2.3\,\mathrm{m}}\frac{1}{\tau_e}. \quad (26)$$

For two-phase BCCD's where the volume occupied by the signal charge under the storage and transfer gates is V_{st} and V_{Tr} and the filling probability of the traps under the transfer gates is F_s, the mean square fluctuation is also given by (18)–(26) with V_s replaced by $(V_{st} + V_{Tr}E_s)$. In the case that several bulk states with different densities and emission time constants interact with the signal charge, the total mean square fluctuations will be the sum of the mean square fluctuations resulting from each bulk state. After n_T transfers, neglecting the suppression of the compounding of the transfer noise [14] (for $n_T\epsilon < 1$ where ϵ is the transfer inefficiency), the total mean square fluctuations is given by

$$F_T^2 = n_T \sum_i F_i^2 \quad (27)$$

where the summation is over the bulk states interacting with the signal.

The noise due to trapping in the bulk states between neighboring charge packets is correlated, since a charge excess in one packet results in an equal deficit in the trailing packet. Therefore, the current noise spectrum $S(f)$ in the output signal will depend on the frequency f and is given by [13]

$$S(f) = 2f_c F_T^2[1 - \cos\left(2\pi f/f_c\right)] \quad (28)$$

where f_c is the clock frequency.

V. EXPERIMENTAL RESULTS

We have measured the transfer properties of a 256-bit three-phase three-level metallization buried channel CCD [6]. The polysilicon electrodes length is 10 μm and the channel width is 200 μm. The buried channel was obtained by implanting 1.5×10^{12} phosphorous ions in a p-substrate of doping $4.5 \times 10^{14}/\mathrm{cm}^3$. After several anneal and oxide growth cycles, the doping profile in the buried channel was found [15] to be almost Gaussian with peak concentration at the surface of $1.6 \times 10^{16}/\mathrm{cm}^3$ and channel depth of 2.1 μm.

Fig. 7 shows the output from a buried channel device operated at a clock frequency of 6 MHz with two groups of eight "ones" separated by 18 and 4067 "zeros," respectively. The leading charge packet of each group shows a charge deficit as it has to replenish all the charge emitted from the bulk states in the buried channel since the last passage of a "one" charge through the device. The charge deficit in the leading one and the charge excess in the trailing zero are not equal. By adding a small amount of background charge the charge deficit in the leading "one" and the charge excess in the trailing "zero" of the two groups of "ones" are both decreased. However, they remain asymmetrical. This indicates that with a circulating background charge, charge trapping in the

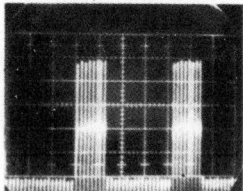

Fig. 7. Appearance of two groups of eight "ones" separated by 18 and 4071 "zeros" at the output of a three-phase 256-bit buried channel CCD operated at 6-MHz clock frequency without and with 10-percent background charge.

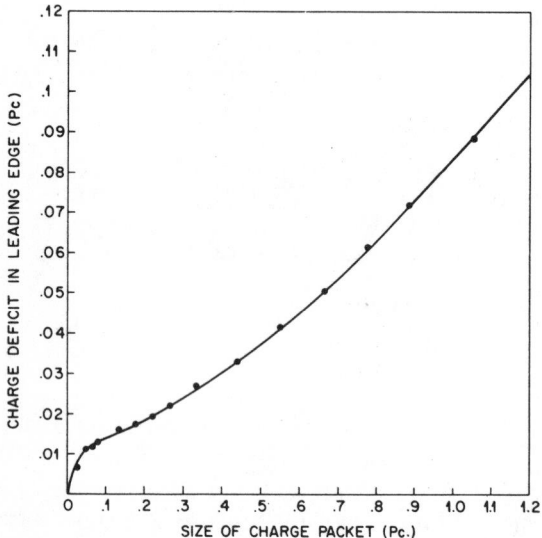

Fig. 8. Charge deficit in picocoulombs at the leading edge of a group of eight "ones" separated by 4090 "zeros" versus the size of the charge packet in picocoulombs for a three-phase buried channel CCD operated at 1-MHz clock frequency.

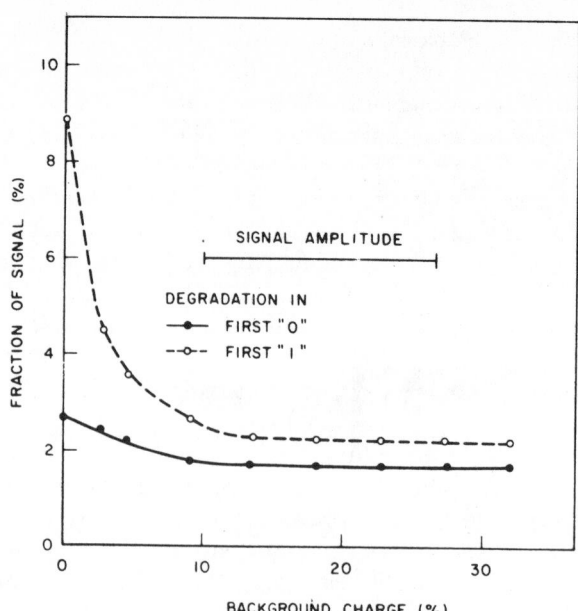

Fig. 9. Charge deficit (o) of first "one" after 4962 "zeros" and charge excess (●) in first "zero" following a group of "ones" in a buried channel device operated at 1-MHz clock frequency as a function of background charge.

bulk states of the parallel edges as defined in Fig. 6 are dominant as we would expect.

Fig. 8 illustrates the increase of the charge deficit in the leading charge packet of a group of "ones" following 4090 empty charge packets in a buried channel device operated at 1 MHz, as the size of the charge packet increases. The shape of this curve, particularly the sharp rise at small charge packets, corresponds to the shape of the spatial extent of the charge packet in the buried channel plotted in Fig. 5. The data in Fig. 8 fitted to (5) indicate an average effective density of states of $2 \times 10^{11}/cm^3$ in the bulk. The nonlinearity at large packet sizes can be explained by an increase of bulk states towards the silicon–oxide interface, perhaps explained as un-annealed ion implant damage.

Fig. 9 illustrates quantitatively the effect of a background charge on the charge deficit of the leading station (·O) and the charge excess in the trailing station (●) of a group of eight "ones" separated by 4088 "zeros." The amount of charge in the ONE corresponded to 16 percent of a full well. Even with no intentionally introduced background charge, the charge deficit from the leading charge packet after 4088 "zeros" is 8.9 percent. The charge excess in the trailing edge is only 2.6 percent. This charge deficit in the leading edge with no intentionally introduced background charge corresponds to a transfer inefficiency of 10^{-4} per transfer. By adding a background charge, the charge deficit at the leading edge and the charge excess at the trailing edge decreases rapidly and reaches a constant value of 2.2 percent and 1.7 percent, respectively, after about 10-percent background charge. This is equivalent to a transfer inefficiency of about 3×10^{-5} per transfer, which is close to the limit of measurement for a single passage through a 256-bit device.

In Fig. 10 the charge deficit from the leading "one" in picocoulombs is plotted versus the number of "zeros' between the groups of "ones" for a buried channel device at 1-MHz clock frequency. As expected according to the model described above based on two localized bulk states interacting with the signal charge, the charge deficit shows the exponential saturation behavior illustrated in Fig. 2. The data presented in Fig. 10 when fitted to (5) in addition to other data taken at 6-MHz clock frequency indicate that a bulk state of emission time constant equal to 275 μs and density of $1.2 \times 10^{11}/cm^3$ and another bulk state of emission time constant approximately 0.3 μs and effective density of $1.8 \times 10^{11}/cm^3$ are interacting with the signal charge. Since the volume that a signal charge packet of 0.17 pC occupies in the buried

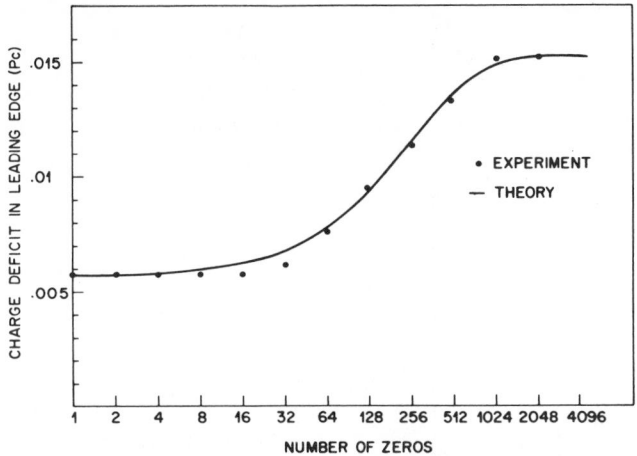

Fig. 10. Charge deficit in the first "one" in picocoulombs for a buried channel device operated at 1-MHz clock frequency as a function of the number of "zeros" between the group of "ones." The solid curve is calculated from (5).

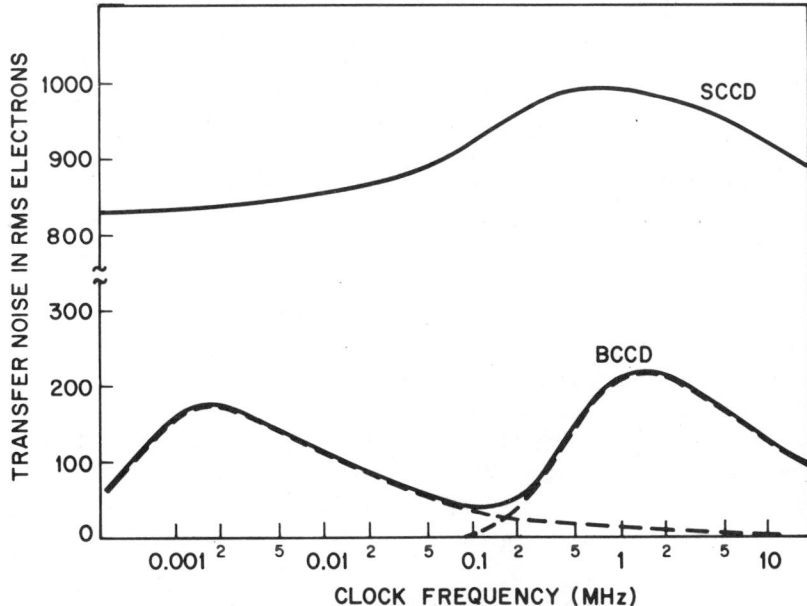

Fig. 11. Transfer noise in rms electrons due to trapping in bulk states in BCCD's and trapping in interface states in SCCD's of a three-phase 256-bit device versus clock frequency.

channel of the device used in the above measurements is about 6×10^{-10} cm³, the effective number of bulk states per storage site which interact with the charge packet at 1-MHz clock frequency is only about 180. By repeating the measurements in Fig. 10 at various temperatures, one can obtain the variation of the emission time constants of the bulk states with temperature. Then using (2) one can determine their location in the bandgap and their capture cross section.

Noise measurements have been carried out on the three-phase BCCD's [5], but the limit of measurement sensitivity was 250 rms electrons per charge packet. Our measurements indicated that the BCCD measured had a transfer noise component less than this limit for full charge packets.

Using the bulk trap densities and distribution obtained from the charge deficit calculations in Fig. 10 the mean square fluctuation in a charge packet is calculated from (22 for a charge packet of 0.17 pC after transfer along) the 256-bit three-phase BCCD and plotted in Fig. 11 versus clock frequency. The transfer noise peaks at 1.7 kHz and 1.5 MHz are due to localized bulk states with emission time constants of 275 μs and 1/3 μs, respectively. At 1-MHz clock frequency the transfer noise is mainly due to the second bulk state. As described in (24)–(26), the transfer noise due to any bulk states decreases linearly and exponentially for frequency above and below the peak frequency, respectively. Using the data in Figs. 5 and 11 we have also plotted the transfer noise versus the size of the charge packet in Fig. 12. As the size of charge packet increases, it occupies a larger volume in the channel and interacts with more traps. Therefore, the transfer noise

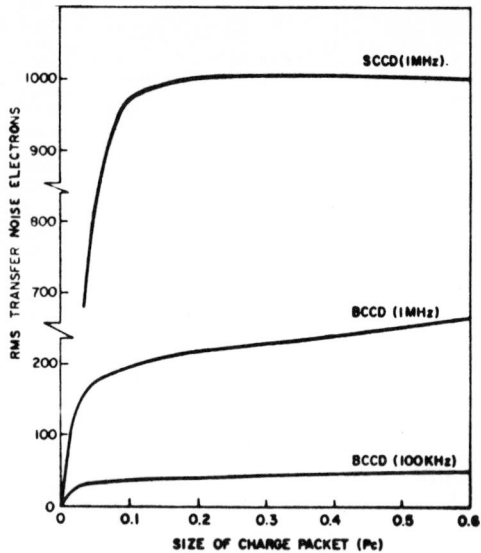

Fig. 12. Transfer noise in rms electrons due to trapping in bulk states in BCCD's and in interface states in SCCD's of a three-phase 256-bit device versus charge packet size.

increases directly with the square root of the size of the charge packets. For comparison, we have also plotted the measured transfer noise due to interface state trapping in a similar 256-bit three-phase SCCD [5] as a function of clock frequency and charge packet size in Figs. 11 and 12, respectively.

VI. DISCUSSION

We have described in the foregoing sections the effects of bulk trapping on the transfer inefficiency and transfer noise in bulk channel devices. In this section we will discuss the differences between bulk trapping in BCCD's and interface states trapping in SCCD's and present a detailed comparison of estimated noise in both surface and bulk channel versions of an image sensor and an analog delay line.

It should be pointed out that in the analysis of the transfer properties of BCCD's in Sections II–IV we have assumed that the capture time constant of the bulk traps is small enough that they are completely filled with carriers trapped from the signal charge. However, due to the lateral spreading of the charge packets in the buried channel, this assumption may not hold at high clock frequency and small signal charges and the trap occupation would be unable to follow the carrier density variations in the successive signal charge packets. Thus the effective number of bulk traps contributing to transfer noise and incomplete charge transfer in the equations derived above would be less than the actual number of the bulk states in the buried channel.

One of the assumptions of our calculations is that the volume of the bulk exposed to the free electrons is given by electrostatic considerations. This is valid for large charge packets. However, for very small charge packets (less than 1000 electrons for a 200 μm^2 electrode area) the kinetic energy of the electrons must be considered. As the number of electrons decreases the volume of bulk exposed

to the charge packet tends to a constant value. Values of transfer inefficiency and transfer noise calculated from our model would lead to optimistic predictions for the case of a few electrons only. If this volume is appropriately calculated, the method of calculation of the effects of the bulk trapping given in this paper is still statistically valid so long as the number of active traps in this volume times the number of transfers is significantly less than the number of electrons in the packet.

The dependence of the transfer inefficiency and transfer noise on the clock frequency and the size of the charge packet in the BCCD differs from that of the SCCD. This is mainly due to the localized energy levels of the bulk traps and the spatial spreading of the signal charge packet in the buried channel in BCCD. In this SCCD the interface states are distributed continuously across the bandgap and the signal charge is localized at the interface.

The charge deficit at the leading edge of a group of ones due to bulk trapping in BCCD's exhibits a maximum at a clock frequency f_{max} related to τ_e and n_z by (11) and drops considerably at higher and lower frequencies as is shown in Fig. 3. It increases directly with the size of the charge packet as the volume occupied by the charge packet increases. The charge deficit at the leading edge also exhibits exponential saturation characteristics with increasing numbers of zeros as shown in Figs. 1, 2, and 10, whereas the charge deficit at the leading edge of a group of ones due to surface trapping in SCCD's is almost constant over several decades of clock frequencies, and is almost independent of the size of the charge packet [6]. Also it increases almost logarithmically with the number of zeros preceding it.

The transfer noise due to bulk trapping in BCCD's varies with the clock frequency and exhibits distinctive maxima and minima due to the localized energy levels of the bulk traps. Also, the transfer noise in BCCD's increases with the size of the charge packet due to the signal charge spreading in the buried channel. However, transfer noise due to interface state trapping in SCCD's is almost constant for different clock frequencies and charge packet sizes due to the continuous distribution of the interface states across the bandgap and the localization of the charge in the inversion layer at the interface.

In view of the results presented above, it is worthwhile to consider the advantage of bulk transfer devices in some of the applications of CCD's. BCCD's do not require the use of a circulating background charge to achieve high transfer efficiency ($\epsilon \lesssim 10^{-4}$ per transfer), whereas SCCD's require a fat zero of at least 10 to 20 percent. This is of utmost importance for low light level imaging applications. With SCCD's a background charge must be used and this results in electrical insertion noise [5] in addition to the interface states transfer noise of the signal and background charge. This considerably increases the minimum detectable signal. By cooling the SCCD, the minimum detectable signal will decrease with the square root of the temperature only. But with BCCD's, no background charge should be used and the signal

TABLE I
NOISE LEVELS IN 500-LINE CCAIS AT LOW LIGHT LEVELS
(Active Element Area 20 × 30 μm²)

Noise Source	Noise Equivalent Signal in SCCD	Noise Equivalent Signal in BCCD
Electrical insertion noise of background charge	200 (70[a])	not required
Trapping noise assuming $N_{ss} \cong 1 \times 10^6/cm^9$ eV and $N_t \cong 2 \times 10^{11}/cm^3$	450	20–80
On-chip amplifier noise $C_0 = 0.2$ pF	180	180 (few electrons[b])
Dark current noise (10 nA/cm²)	100	100 (10[c])
Total noise equivalent	540 (≈500[a])	≈220 (≈20[b,c])

[a] Theoretical limit given by $[\frac{2}{3}(kTC/q)]^{1/2}$, number outside parenthesis is that obtained experimentally.
[b] Use of ideal amplifier, i.e., floating gate [16] or reset detection noise with double sampling [17].
[c] Cooling by 60°C.

TABLE II
NOISE LEVELS IN 256 ELEMENTS DELAY LINE
(Active Element Area 200 × 30 μm²)

Noise Source	Noise Equivalent Signal in SCCD	Noise Equivalent Signal in BCCD
Electrical insertion noise of background charge	600–900 (220[a])	not required
Electrical insertion noise of the signal	600–900 (200[a])	600–900 (220[a])
Trapping noise assuming $N_{ss} = 1 \times 10^9/cm^2$ eV and $N_t = 2 \times 10^{11}/cm^3$	720	70–370
On-chip amplifier noise $C_0 = 0.2$ pF	180	180 (few electrons[b])
Dark current noise for 1-ms delay (10 nA/cm²)	60	60
Total noise	960 (770[a])	730 (230[a,b])
Maximum signal for $V_p = 14$ V	40 × 10⁶	20 × 10⁶
Dynamic range	94 dB[a]	99 dB[a,b]

[a] Theoretical limit given by $[-(kTC/q)]^{1/2}$, number outside parenthesis is that obtained experimentally.
[b] Use of ideal amplifier, i.e., floating gate [16] or reset detection node with double sampling [17].
[c] Cooling by 60°C.

transfer noise due to bulk state trapping is proportional to the size of the signal. Thus small signals will have proportionally small transfer noise. Also by cooling the BCCD's, one can considerably reduce the transfer noise. This can be explained with Fig. 11. By cooling the device, the emission time contant of the bulk traps will increase exponentially and the peaks of the transfer noise in Fig. 11 will shift to lower frequencies. Thus for a particular clock frequency, one can avoid the peaks of the transfer noise and operate the device at one of the transfer noise minima.

Table I summarizes the major noise sources in a 500 × 500 element charge-coupled array image sensor at low light levels assuming the active area per element is 20 × 30 μm². In this table the pulser noise, which is due to the random fluctuations of the voltage levels of the clock pulses coupled to the output, is neglected [5]. The dominant noise source in the SCCAIS is the interface state trapping noise which limits the minimum noise equivalent number of electrons per charge packet to 500 in the best case. In the BCCAIS the dominant noise sources are the dark current shot noise and the on-chip preamplifier reset noise, which limit the minimum detectable signal to about 220 electrons. However, using a floating gate on-chip amplifier [16] or a reset detection node with correlated double sampling [17] and appropriate cooling, the minimum detectable signal could be very small indeed on the order of a few electrons only. It should be pointed out that using an ideal amplifier with the SCCAIS is of little value and cooling would reduce the dominant trapping noise with the square root of the temperature only.

For delay line and signal processing applications, the low noise and the no-background-charge operation of the BCCD are very attractive. However, the reduction of the signal handling of a BCCD as compared to that of SCCD due to the lateral spreading of the charge in the buried channel instead of being localized at the interface, should also be considered. For the particular device whose doping profile is shown in Fig. 4 the signal handling is reduced by about a factor of two compared to a similar SCCD. This signal handling reduction could be decreased by using a

shallow buried channel [1], [9] or modifying the doping profile [18]. Values of the rms noise electrons, due to electrical inserrtion, transfer noise, and output amplifier for the 256-bit three-phase device already described above are listed in Table II. The maximum signal charge with a clock voltage of 14 V for both SCCD and BCCD is also shown. Again for surface channel devices the dominant noise sources are the interface state trapping noise and electrical insertion noise and the use of an ideal amplifier is of little value. For buried channel devices the electrical insertion noise and the bulk state trapping noise are the dominant noise sources. From the data presented in Table II we see that if the theoretical electrical insertion noise is achieved, the reduced signal handling capability of the buried channel device will be more than compensated by the reduced noise and the expected overall dynamic range of the buried channel device will be about 5 dB better than that of the surface channel device. It should be pointed out that the noise associated with the electrical insertion of the signal is a major component in the titai noise and even if the 2/3 KTC limit is achieved it is still the dominant component in the buried channel device.

VII. CONCLUSIONS

In conclusion, we have described the effects of bulk trapping on the transfer efficiency and transfer noise in BCCD's. We have measured the transfer properties of a 256-bit three-phase buried channel CCD with 200-μm wide channel. At 1-MHz clock frequency, the transfer noise for full charge packets is less than our measurement limit of 250 rms electrons. The transfer inefficiency for half full packets is 10^{-4} per transfer with no background charge and 3×10^{-5} with 10-percent background charge.

We have shown that by plotting the deficit in the leading edge of a group of "ones" versus the number of "zeros" preceding them, we may determine the density and distribution of the bulk traps in the buried channel. Operating the device with a higher clock frequency permits a better resolution of the distribution of the traps. More information about the location of the bulk states in the bandgap and their capture cross sections can be extracted by taking these measurements at various temperatures. Two bulk traps with a density of $1.2 \times 10^{11}/cm^3$ and $1.8 \times 10^{11}/cm^3$ and emission time constants of $275 \ \mu s$ and $0.3 \ \mu s$ at room temperature were identified.

We have presented a comparison of estimated noise in both surface and bulk channel versions of an image sensor and an analog delay line. The low transfer noise and n-background charge operation make BCCD's very attractive for low light level imaging.

ACKNOWLEDGMENT

The authors are grateful to K. K. Thornber and C. H. Séquin for useful comments and discussions on the manuscript.

REFERENCES

[1] R. H. Walden et al., "A buried channel charge coupled device," Bell Syst. Tech. J., vol. 51, pp. 1635–1640, 1972.
[2] M. F. Tompsett, "The quantitative effects of interface states on the performance of charge coupled devices," IEEE Trans. Electron Devices, vol. ED-20, pp. 45–55, Apr. 1973.
[3] J. E. Carnes and W. F. Kosonocky, "Fast interface states losses in charge coupled devices," Appl. Phys. Lett., vol. 20, pp. 261–263, 1972.
[4] A. M. Mohsen, T. C. McGill, Y. Daimon, and C. A. Mead, "The influence of interface states on incomplete charge transfer in overlapping gates charge coupled devices," IEEE J. Solid-State Circuits, vol. SC-8, pp. 125–138, Apr. 1973.
[5] A. M. Mohsen, M. F. Tompsett, C. H. Séquin, "Noise measurements in charge-coupled devices," to be published.
[6] W. J. Bertram et al., "A three-level metallization three-phase CCD," to be published.
[7] L. J. M. Esser, "The peristaltic charge coupled device: a new type of charge-transfer device," Electron. Lett., vol. 8, pp. 620–621, 1972.
[8] W. Shockley and W. T. Read, "Statistics of recombination of holes and electrons," Phys. Rev., vol. 87, p. 835, 1952.
[9] A. M. Mohsen, R. W. Bower, T. C. McGill, and T. A. Zimmerman, "Overlapping gates buried channel and charge-coupled devices," Electron. Lett., vol. 9, pp. 396–397, 1973.
[10] A. M. Mohsen, T. C. McGill, and C. A. Mead, "Charge transfer in overlapping gate charge coupled devices," IEEE J. Solid-State Circuits, vol. SC-8, pp. 191–207, June 1973.
[11] Y. Daimon, A. M. Mohsen, and T. C. McGill, "Charge transfer in buried channel charge coupled devices," ISSC Digest of Technical Papers, Feb. 1974, pp. 146–147.
[12] W. H. Kent, "Charge distribution in buried channel charge-coupled devices," Bell Syst. Tech. J., vol. 52, pp. 1009–1023, 1973.
[13] K. K. Thornber and M. F. Tompsett, "Spectral density of noise generated in CCD's," IEEE Trans. Electron Devices, vol. ED-20, p. 456, Apr. 1973.
[14] K. K. Thornber, "Noise suppression in charge transfer devices," Proc. IEEE, vol. 60, pp. 1113–1114, Sept. 1972.
[15] A. M. Mohsen and F. J. Morris, "Characterization of bulk transfer charge-coupled devices from MOS Capacitors and MOS transistors," to be published.
[16] D. D. Wen and P. J. Salsbury, "Analysis and design of a single-stage floating gate amplifier," ISSCC Digest of Technical Papers, Feb. 1973, pp. 154–155.
[17] M. H. White, D. R. Lampe, I. A. Mack, and B. C. Blaha, "Characterization of charge-coupled device line and area array imaging at low light levels," ISSCC Digest of Technical Papers, Feb. 1973, pp. 134–135.
[18] L. J. Esser, "The peristaltic charge-coupled device for high speed charge transfer," ISSCC Digest of Technical Papers, Feb. 1974, pp. 28–29.

The Peristaltic Charge-Coupled Device for High-Speed Charge Transfer

Leonard J. M. Esser

Philips Research Laboratories

Eindhoven, Netherlands

THE PCCD, like all bulk[1,2,3,4] Charge Transfer Devices (CTDs), offers the possibility of sole bulk-mode operation. The influence of surface states on the charge transfer-efficiency is then eliminated.

It has been demonstrated[5,6] that:

(1) A PCCD with a very shallow and relatively high doped N-layer (Figure 1(a)) has a high charge handling capability but operates only at a low speed . . . .

(2) A PCCD with a thick homogeneously doped N-layer (Figure 1(b)) has a low charge handling capability, but operates at a high speed.

The profiled PCCD (Figure 2) combines the two advantages; e.g., high charge handling capability and high transfer speed.

These features can be noted in the calculated results; Figure 3. These calculations were carried out on a simplified one-dimensional model, with very high ohmic substrate, by using the gradual channel approximation introduced by Schockley. Three PCCD structures were compared for transfer of the same amount of charge. The main part of the charge packet is transferred by self-induced fields, the last charge fraction is transferred either by diffusion (solid curves) or by externally induced (fringing) fields (broken curve).

Comparison of curves 1 and 2 shows that for transfer deeper in the bulk, a larger fraction of the charge is transferred by self-induced fields[4,5,6]. Deep in the bulk externally induced fields are of increased importance. However, the effective distance between the stored charge and the electrode in the case of curve 2 is about 9 times larger than in the case of curve 1. Thus, a corresponding increase for the clock-voltage is needed for the transfer of the same amount of charge, or a similar decrease in charge handling capability for the same clockvoltage; curve 2 of Figure 3. Curve 1 of Figure 3 shows that the transfer time is almost completely determined by the transfer of the last charge fraction. So one might expect that it is sufficient to improve the transfer of this last vestige of charge. This occurs in the profiled PCCD where for the calculated case (Figure 3, curve 3) 80% of the stored charge is contained in the very thin top layer. When the bulk charge in the top layer dominates the transfer process, curve 3 follows approximately curve 1. After a few time constants, when the charge in the top layer has been transferred, curve 3 almost follows curve 2. Hence the transfer time needed to attain a given residual charge level is almost the same for the profiled PCCD and the homogeneously-doped thick layer PCCD.

As a result of the profiled structure, the effective distance of the stored charge to the electrode is now about 3.5 times smaller than for the thick homogeneously-doped layer PCCD. This causes a corresponding reduction of the necessary clockvoltage for equal amounts of charge, or a similar increase in charge handling for equal clockvoltages as shown in Figure 4, curve 3.

Figures 3 and 4 show how the profiled PCCD combines high speed and high charge handling capability by transferring only the last vestige of charge deep in the bulk.

In bulk CTDs one expects bulk trapping centers to increase the charge transfer inefficiency. Similar to the surface CCD, one would like to eliminate the influence of these centers by using a fat zero. However in a homogeneously-doped layer bulk CTD the volume occupied by a charge packet is proportional to its magnitude. When the trapping centers are homogeneously distributed, the number of traps occupied in the presence of a packet is proportional to the magnitude of the packet. Obviously a fat zero will not be useful.

In the profiled PCCD most of the volume occupied by a charge packet is covered by a small fraction of the total charge. A fat zero packet of this magnitude will eliminate the major part of the effect of bulk traps and an improved transfer will be found.

New experimental results have been obtained from both types of PCCDs:

(1) A homogeneously doped layer PCCD[6] (Figure 1(b)) with an epitaxially grown 4.5 μm thick N-type layer and a dope level of 7 x 10^{14} cm^{-3}. The P-type substrate has a dope level of about 5 x 10^{14} cm^{-3}. The oxide thickness is 0.15 μm.

(2) The profiled PCCD (Figure 2) with an epitaxial layer thickness of 4.5 μm and dope level of about 3 x 10^{14} cm^{-3}.

[1] Walden, R. H., Krambeck, R. H., Strain, R. J., McKenna, J., Schrijer, N. L. and Smith, G. E., "The Burried Channel Charge Coupled Device", *BSTJ*, p. 1635-1640; Sept., 1972.

[2] Kim, C. K., Early, J. M. and Amelio, G. F., "Burried Channel Charge Coupled Devices", *NEREM*; Nov., 1972.

[3] Takemoto, I., Sunami, H., Ohba, S., Aoki, M. and Kubo, M., "Bulk Charge Transfer Device", Japan, *Intl. Conf. on Solid State Devices*, Tokyo; Aug., 1973.

[4] Esser, L. J. M. "Peristaltic Charge-Coupled Device: A New Type of Charge Transfer Device", *Electronics Letters*, Vol. 8, p. 620-621; Dec., 1972.

[5] Esser, L. J. M., "The Peristaltic Charge Coupled Device", *Charge Coupled Device Applications Conference*, San Diego, p. 269-277; Sept., 1973.

[6] Esser, L. J. M., Collet, M. G. and van Santen, J. G., "The Peristaltic Charge Coupled Device", *Intl. Electron Device Meeting*; Dec., 1973.

Reprinted from *1974 IEEE Int. Solia-State Circuits Conf., Dig. Tech. Papers*, Session II, Feb. 13-15, 1974, pp. 23–29.

It is provided with an implanted top layer of about 5×10^{11} cm^{-2} phosphor atoms. The thickness of the top layer is about 0.5 μm. The oxide thickness is 0.1 μm.

Both devices have a 128-stage 4-ϕ electrode structure (512 gates) with a double layer metalization of aluminum and polycrystalline silicon. The electrode length in the transfer direction is about 7.5 μm. The channel width 240 μm. The clockvoltages used are 10-V peak-to-peak and the devices show an almost constant transfer efficiency up to at least 135 MHz (the highest available measuring frequency).

The homogeneously-doped layer PCCD shows transfer efficiencies of about 0.99995 per transfer for real-zero operation. Fat zero operation does not influence the transfer efficiency. The charge handling capability is about 1.5×10^{11} electrons cm^{-2}. At higher charge values the transfer inefficiency starts to increase rapidly.

The profiled PCCD shows transfer efficiencies of 0.9998 per transfer for real zero operation. Fat zero operation of 1% and higher provides an improvement to 0.9999. Charge handling has 3×10^{11} electrons cm^{-2}. Above this value an increase of the transfer inefficiency is found.

Both types of PCCDs are expected to operate satisfactorily up to 1 GHz.

It is obvious that due to its high frequency operation and large charge-handling capability the PCCD will greatly extend the range of CTD-applications.

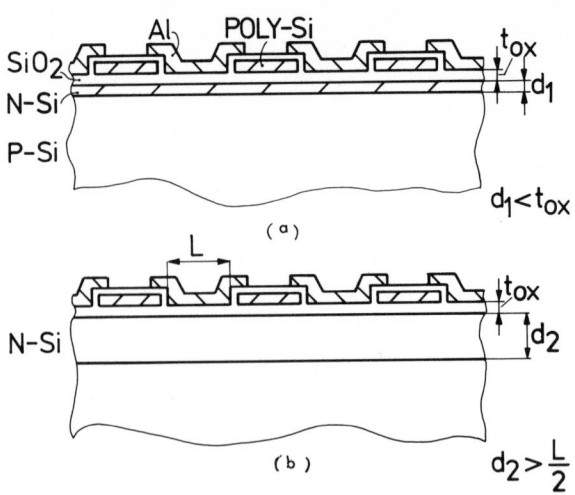

FIGURE 1—The homogeneously doped layer PCCD: (a)—the shallow layer PCCD with high storing capability; (b)—the thick layer PCCD for high-speed operation.

[See page 219 for Figures 4 and 5.]

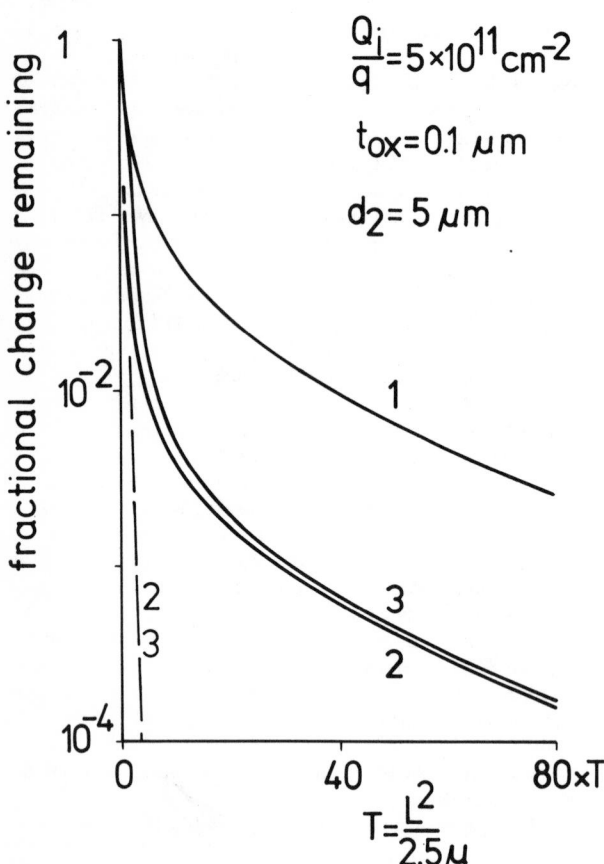

FIGURE 3—Charge decay. Curve 1 = shallow layer PCCD; Figure 1(a). Curve 2 = thick layer PCCD; Figure 1(b), N = 10^{15} cm^{-3}. Curve 3 = profiled PCCD; Figure 2, $N_2 = 2 \times 10^{14}$ cm^{-3}.

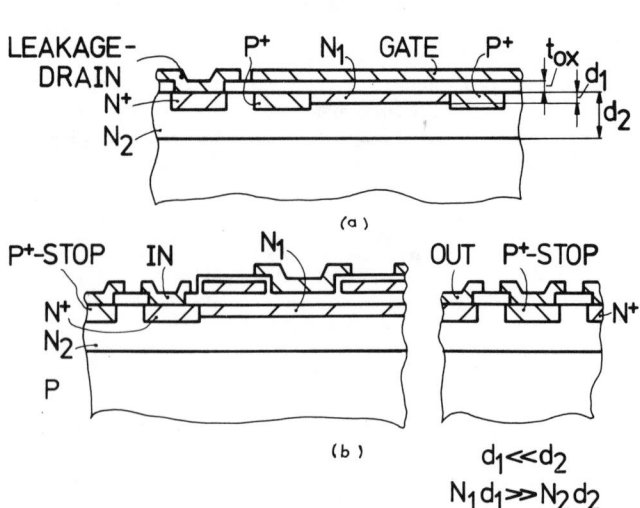

FIGURE 2—The profiled layer PCCD: (a)—Cross section perpendicular to the transfer direction showing the P$^+$ channel stop diffusions; (b)—Cross section in transfer direction.

Part III
Imaging

Some of the first experiments with CCD's involved the imaging of visible light. The first CCD built had no diffused regions with which to inject charge, leaving oxide-avalanche generation (often destructive to the device) and optically excited carrier generation techniques as the two popular methods of injecting the signal charge into the first experimental devices.

At the outset, CCD image sensors were proposed as offering many potential advantages over other semiconductor candidates for future compact, low-power, solid-state replacements of the vidicon-type tube for video camera applications. These proposed advantages included

1) high resolution,
2) low thermal noise,
3) low light level performance,
4) flat spectral response (high sensitivity to blue as well as red light), and
5) low cost.

These factors are reviewed by Barbe in the first reprint paper in this part. This is followed by paper 2 by Amelio *et al.* which contains design considerations for CCD image sensors. In paper 3 by White *et al.*, CCD image sensor noise performance and spectral response are discussed. An important concept reviewed in this paper is that the Johnson/Nyquist thermal noise associated with charging the capacitor of a sample-and-hold circuit (the so-called kTC noise) may be eliminated by "correlated-double sampling."

Following this is a series of papers by authors from the laboratories of Bell Telephone (paper 4), General Electric Company (paper 5), RCA Corporation (paper 7), and Fairchild Semiconductor (paper 9), presenting devices which have potential commercial application. In all cases, except the Bell Telephone device, commercial product specification sheets follow the respective technical papers. These commercial papers and products are intended to reveal the latest in present-day state of the art in a rapidly improving technology, and to introduce to the reader the performance levels of some of the first devices to become commercially available.

Special note should be made of the Michon and Burke paper as their device, the "CID," differs from other CCD's. The CID differs from the other charge-coupled sensor designs presented in this part in that there is only one transfer between storage potential wells before the charge is detected. Since the array is XY-addressed by MOS shift register circuits on the same silicon substrate, this device represents a combination of both CCD circuit and conventional circuit concepts.

In the final paper in this part, Melen notes that many of the performance capabilities of CCD's lend themselves well to moderate resolution television-oriented applications, and he compares them with MOS-scanned diode array sensors for industrial control applications.

Although not yet widely used in commercial broadcast cameras, the CCD concept has had a dramatic impact on the design of television-resolution sensors and the miniature cameras which contain them.

Imaging Devices Using the Charge-Coupled Concept

DAVID F. BARBE, MEMBER, IEEE

Invited Paper

Abstract—A unified treatment of the basic electrostatic and dynamic design of charge-coupled devices (CCD's) based on approximate analytical analysis is presented. Clocking methods and tradeoffs are discussed. Driver power dissipation and on-chip power dissipation are analyzed. Properties of noise sources due to charge input and transfer are summarized. Low-noise methods of signal extraction are discussed in detail. The state of the art for linear and area arrays is presented. Tradeoffs in area-array performance from a systems point of view and performance predictions are presented in detail. Time delay and integration (TDI) and the charge-injection device (CID) are discussed. Finally, the uses of the charge-coupled concept in infrared imaging are discussed.

I. INTRODUCTION

CHARGE COUPLING is a simple but extremely powerful concept. Basically, a charge-coupled device (CCD) is a metal–oxide–semiconductor (MOS) structure, as shown in Fig. 1, which can collect and store minority carrier charge packets in localized potential wells at the Si–SiO$_2$ interface [1]. The CCD can transfer charge packets in discrete-time increments via the controlled movement of potential wells. The charge packets can then be detected at the output via capacitive coupling. Thus a CCD acts as an analog shift register composed of three sections. 1) The input section which contains a diffusion, which is the source of minority carriers, and whose potential can be controlled, and an input gate which can be turned on and off to control the flow of charge from the source diffusion into the first potential well. 2) The transfer section, containing a series of electrodes which control the potential at the Si–SiO$_2$ interface. When the voltages on the electrodes are properly manipulated, the potential wells are moved toward the output and the charge packets follow. 3) The output section is a reverse-biased p-n junction capacitance whose voltage is changed when a charge packet is transferred into it. The diode is then reset via a reset switch to prepare for the next packet to be transferred into it. This node is typically connected to an MOS amplifier.

Charge can be entered into the device in a series manner via the input diffusion or via the absorption of photons near the potential wells. Suppose that a charge configuration has been entered into the device, by one of these methods, as shown in Fig. 1(c). In a device having a planar oxide and uniform substrate doping, three phases are required for unidirectional charge transfer, i.e., if a barrier is maintained behind the charge packet while a deeper well is formed in front of the packet, then charge will flow into the deeper well. The clocking diagram is shown in Fig. 1(d). At $t = t_1$, charge resides in the wells under the ϕ_1 electrodes. At $t = t_2$, the potential on ϕ_2 is made positive forming wells under the ϕ_2 electrodes. Charge will then flow from the ϕ_1 wells into the ϕ_2 wells. At

Manuscript received July 16, 1974; revised August 19, 1974.

The author is with the U. S. Naval Research Laboratory, Washington, D.C. 20375.

Fig. 1. Three-phase CCD. (a) Cross-sectional view showing input section, transfer section, and output section. A primitive electrode structure having unprotected gaps is shown for simplicity. (b) Surface-potential profile showing potential wells under the ϕ_1 electrodes. (c) Surface-potential profiles showing progression of charge transfer during one clock period. (d) Clocking waveforms used to drive the CCD during transfer.

$t = t_3$, the potential on the ϕ_1 electrodes is reduced to a low value and the remaining charge in the ϕ_1 wells will be pushed into the ϕ_2 wells. This sequence repeats with the result that the charge configuration moves from one cell to the next every clock period. Thus the clock rate is equal to the data rate.

Since the charge packets follow the potential wells which are controlled by external voltages, transfer can be achieved in two dimensions. Information can be entered in analog or digital form. The device can have multiple inputs and multiple nondestructive outputs via floating gates. The CCD is inherently a low-power device requiring low voltages for its operation, and has a high packing density. As a result, there are many applications for CCD's in imaging, analog and digital signal processing, and memory.

An attempt is made in this paper to present a unified treatment of the fundamental design parameters of the device, discuss the technological and design tradeoffs, and consider the state of the art. Section II discusses the electrostatic design considerations; Section III, the dynamic design considerations; Sections IV and V, the input and output design considerations;

Reprinted from *Proc. IEEE*, vol. 63, pp. 38–67, Jan. 1975.

Section VI, noise; Section VII, the design and performance of linear imaging arrays; Section VIII, the tradeoffs in area-array design; Section IX, radiation effects on CCD's; and Section X, the uses of the CCD concept in IR imaging.

II. Electrostatics

A. Surface Channel

When a voltage is applied to the conducting electrode of an MOS capacitor with respect to the substrate, the energy bands in the semiconductor bend. If the applied field is in the direction to repel majority carrier from the surface of the semiconductor, i.e., from the semiconductor–insulator interface, the bands in p-type silicon will bend as shown in Fig. 2(a). The effective voltage across the capacitor, $V_G - V_{FB}$, will be divided between the semiconductor, as indicated by the band bending in the semiconductor, and the oxide, as indicated by the "tilt" of the energy bands in the oxide. The electron potential at the semiconductor–insulator interface under the electrode is lower than that in the bulk of the semiconductor by amount ϕ_{s_0}; thus a "potential well" of depth ϕ_{s_0} is formed at the semiconductor–insulator interface under the electrode. If charge is collected in the potential well as a result of photon absorption, injection from an input diffusion, or thermal generation, the potential across the insulator and the semiconductor will be redistributed as shown in Fig. 2(b). Thus the potential well has been partially filled to ϕ_s. The equation relating the surface potential or potential-well depth to the factors affecting it is discussed in the following paragraphs.

The basic electrostatic design equation for a surface-channel CCD relating surface potential ϕ_s to doping density N_A, oxide thickness d, and gate voltage V_G is obtained by solving Kirchhoff's voltage equation for an MOS capacitor:

$$V_G - V_{FB} = \phi_s + V_{ox} \tag{1}$$

where V_{FB} is the flatband voltage and V_{ox} is the voltage drop across the oxide. But

$$V_{ox} = \frac{1}{C_{ox}} \text{ [mobile charge density + fixed charge density]}$$

$$= \frac{1}{C_{ox}} [\quad eN \quad + N_A eW \quad] \tag{2}$$

where $C_{ox} = \epsilon_{ox}/d$, $e = 1.6 \times 10^{-19}$C, and N is the number of mobile electrons per unit area.

Using the depletion approximation, the width of the depletion region, W, is

$$W = \left(\frac{2\epsilon_s \phi_s}{eN_A} \right)^{1/2} \tag{3}$$

where ϵ_s is the dielectric constant of the semiconductor.
Combining (1), (2), and (3) gives

$$V_G - V_{FB} - \frac{eN}{C_{ox}} = \phi_s + \frac{1}{C_{ox}} (2eN_A \epsilon_s \phi_s)^{1/2}. \tag{4}$$

Solving (4) for ϕ_s gives

$$\phi_s = V_{FB} - V_{FB} - \frac{eN}{C_{ox}} + \frac{eN_A \epsilon_s}{C_{ox}^2} - \frac{1}{C_{ox}}$$

$$\cdot \left[2eN_A \epsilon_s \left(V_g - V_{FB} - \frac{eN}{C_{ox}} \right) + \left(\frac{eN_A \epsilon_s}{C_{ox}} \right)^2 \right]^{1/2}. \tag{5}$$

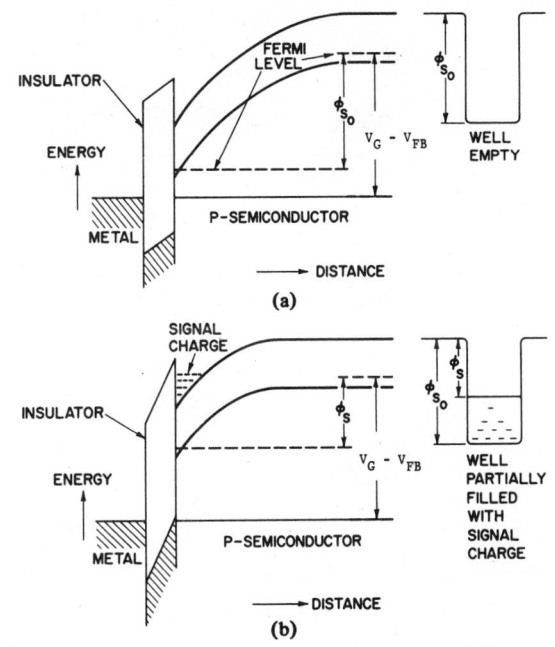

Fig. 2. Energy-band diagrams for a surface-channel MOS capacitor. (a) Band-bending at deep depletion and the empty potential-well representation. (b) Band-bending with mobile charge at the Si–SiO$_2$ interface and the partially filled potential-well representation.

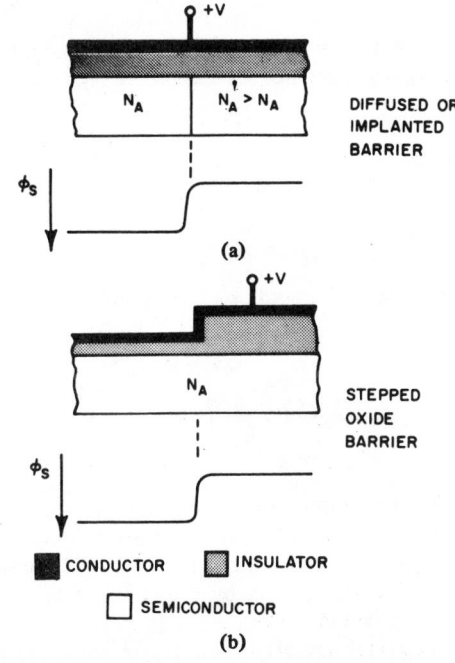

Fig. 3. Methods of forming fixed potential barriers. (a) Surface-potential barrier due to nonuniform doping. (b) Surface-potential barrier due to nonuniform oxide thickness.

Equation (5) shows how barriers can be controlled by proper choice of gate voltage, doping density, and oxide thickness [2]. For example, (5) shows that ϕ_s decreases when N_A is increased and when C_{ox} is decreased. This is illustrated in Fig. 3, which shows qualitatively the two methods of channel confinement (channel stops)—high doping or thick oxide. These two methods are also used to build in charge flow directionally along the channel in two-phase CCD's.

Equation (5) can also be used to calculate the maximum electron density that can be stored on an MOS capacitor, i.e.,

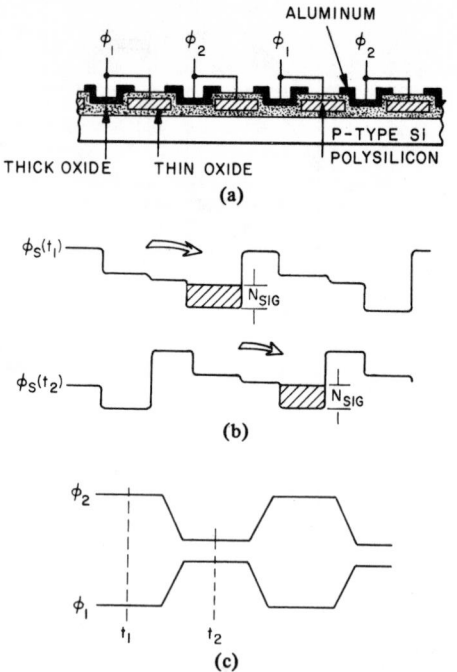

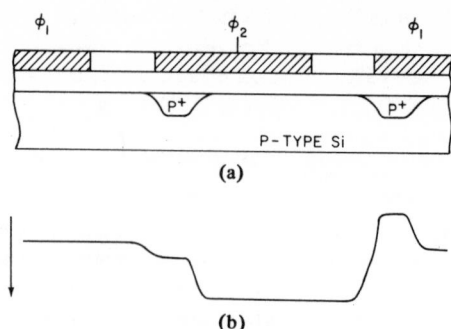

Fig. 5. Implanted-barrier two-phase CCD. (a) Cross-sectional view of the structure. (b) Surface-potential profile showing built-in potential barriers due to nonuniform doping.

Fig. 4. Stepped-oxide two-phase CCD. (a) Cross-sectional view of the structure. (b) Surface-potential profiles showing charge transfer. (c) Clocking waveforms used to drive the CCD during transfer.

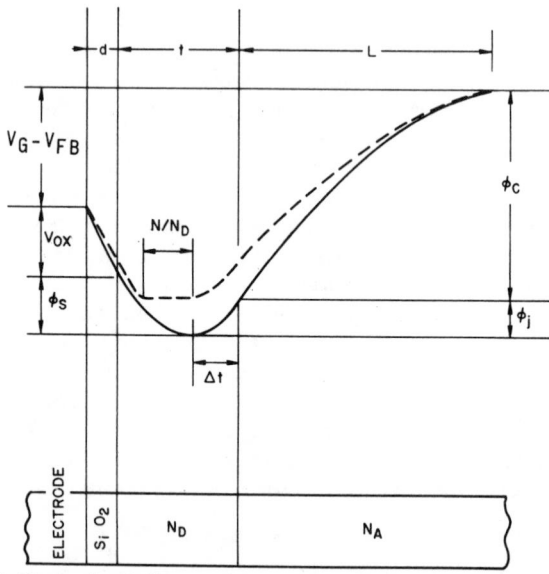

Fig. 6. Energy-band diagram for a buried-channel MOS capacitor showing the potential minimum in the semiconductor.

$N = N_{\max}$ when $\phi_s = 2\phi_B = 2kT/e \ln (N_A/n_i)$, and n_i is the intrinsic density of carriers. Solving (5) for $N_{\max}$ gives

$$N_{\max} = \frac{C_{ox}}{e} [V_G - V_{FB} - 2\phi_s - V_B] \qquad (6)$$

where

$$V_B = (4eN_A \epsilon_s \phi_s)^{1/2}/C_{ox}.$$

Typically, V_{FB}, $2\phi_s$, and V_B are much smaller than V_G; therefore,

$$N_{\max} \simeq C_{ox} V_G/e. \qquad (7)$$

Thus for $d = 1000$ Å and $V_G = 10$ V,

$$N_{\max} \simeq 2 \times 10^{12} \text{ electron/cm}^2.$$

This is the maximum electron density in an inversion layer with $V_G = 10$ V. In CCD's, the saturation level is usually less than this value because the full well is determined by the height of a barrier formed by a difference in electrode potentials, an implanted barrier, or oxide step.

Fig. 4 illustrates the stepped-oxide method for building in potential barriers that allow the device to be clocked with two phases. In a typical device, the thin oxide is 1000 Å. Doped and patterned polysilicon provides the set of electrodes on the thin oxide. Subsequent oxidation provides the thick (2400 Å) oxide onto which aluminum electrodes are formed. This oxidation also isolates the polysilicon and aluminum electrodes from each other. Thus the oxide steps (1400 Å) formed in this way provide barriers that prevent charge from spilling backward. The electrodes are connected as shown schematically in Fig. 4(a) to form a two-phase CCD. The electrode interconnections for this stepped-oxide device are made off chip, thus four on-chip buses are required. Surface-potential profiles illustrating charge transfer are shown in Fig. 4(b), and the clocking waveforms are shown in Fig. 4(c).

An alternative method for achieving two-phase operation, the implanted barrier technique, is shown in Fig. 5. Basically, nonuniform doping is used to achieve the required barrier. This technique uses planar oxides and only two buses are required. Barriers of about 4 V are typical for both of these techniques. The full-well capacity is

$$N_{full} = C_{ox} V_{barrier} A_{storage}/e$$

where $A_{storage}$ is the area of the electrode under which charge is stored. For two-phase stepped-oxide or implanted-barrier designs, $A_{storage} \simeq 1/4 A_{cell}$. Thus the number of electrons at saturation is approximately 2×10^6 electron/cell.

B. Buried Channel

The profile of potential versus distance into silicon for a buried-channel CCD element is shown in Fig. 6. As for the surface-channel CCD, the basic electrostatic design equation relating the minimum electron potential in the channel, $\phi_{\min}$; to the doping density of the substrate, N_A; doping density of the channel, N_D; oxide thickness d; thickness of the donor layer, t; and gate voltage V_G is calculated from Kirchhoff's equation [3]:

$$\phi_s + V_G - V_{FB} + V_{ox} = \phi_j + \phi_c. \qquad (8)$$

Since $V_{ox} = 1/C_{ox} \times$ (mobile charge density + fixed charge density), it follows that

$$V_{ox} = \frac{eN_D(t - \Delta t - N/N_D)}{\epsilon_{ox}/d}. \tag{9}$$

Using the depletion approximation gives

$$\phi_s = \frac{eN_D(t - \Delta t - N/N_D)^2}{2\epsilon_s} \tag{10}$$

$$\phi_c = \frac{eN_A L^2}{2\epsilon_s} \tag{11}$$

$$\phi_j = \frac{eN_D(\Delta t)^2}{2\epsilon_s} \tag{12}$$

and

$$\phi_{min} = \phi_c + \phi_j. \tag{13}$$

Combining these six equations gives

$$V_G - V_{FB} + V_I - eN\left(\frac{d}{\epsilon_{ox}} + \frac{t}{\epsilon_s} - \frac{N/N_D}{2\epsilon_s}\right)$$

$$= \left[2e\epsilon_s\left(\frac{N_D N_A}{N_D + N_A}\right)\phi_{min}\right]^{1/2}\left(\frac{d}{\epsilon_{ox}} + \frac{t}{\epsilon_s} - \frac{N/N_D}{\epsilon_s}\right)$$

$$+ \frac{N_D}{N_D + N_A}\phi_{min} \tag{14}$$

where

$$V_I \equiv eN_D t\left(\frac{d}{\epsilon_{ox}} + \frac{t}{2\epsilon_s}\right) \tag{15}$$

and V_I is the voltage that must be applied across the p-n junction to deplete the channel. The quadratic formula gives the solution:

$$\phi_{min} = \left\{-\left(\frac{e\epsilon_s}{2}\frac{N_A(N_D + N_A)}{N_D}\right)^{1/2}\left(\frac{d}{\epsilon_{ox}} + \frac{t}{\epsilon_s} - \frac{N/N_D}{\epsilon_s}\right)\right.$$

$$+ \left[\frac{e\epsilon_s}{2}\frac{N_A(N_D + N_A)}{N_D}\left(\frac{d}{\epsilon_{ox}} + \frac{t}{\epsilon_s} - \frac{N/N_D}{\epsilon_s}\right)^2\right.$$

$$+ \frac{N_D + N_A}{N_D}\left(V_G - V_{FB} + V_I - eN\right.$$

$$\left.\left.\left.\left(\frac{d}{\epsilon_{ox}} + \frac{t}{\epsilon_s} - \frac{N/N_D}{2\epsilon_s}\right)\right)\right]^{1/2}\right\}^2 \tag{16}$$

Fig. 7 shows ϕ_{min} versus $V_G - V_{FB}$, with $N/N_D t$ as a parameter for a shallow buried-channel device. For a voltage ($V_G - V_{FB}$) swing from -2 V to $+5$ V, a full well corresponds to $N/N_D t \simeq 0.75$, which gives $N \simeq 6.75 \times 10^{11}$ cm^{-2}. Assuming that the area of the well is 10^{-6} cm^2, the number of electrons in the full well is approximately 6.75×10^5.

III. DYNAMICS

A. Free-Charge Transfer

The dynamics of charge transfer are governed by the continuity equation and Poisson's equation with the appropriate boundary conditions. In general, both drift and diffusion contribute to charge transfer; however, in well-designed devices, drift processes are dominant. Drift current is caused either by

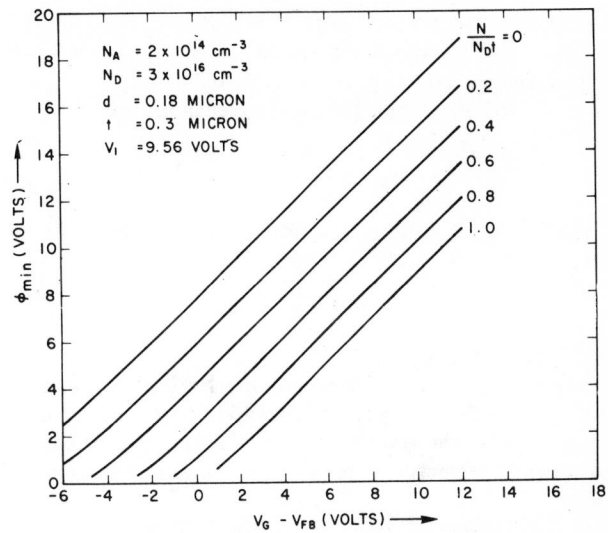

Fig. 7. Minimum potential versus effective gate voltage, with density of carriers in the potential well, N, as a parameter. Curves are calculated from (16).

the electrical field due to a nonuniform distribution of electrons under two electrodes at the same potential, or by a field due to the potential difference between electrodes. These processes are called self-induced drift and fringe-field drift, respectively. The detailed solution of these equations for real device structures requires numerical calculations [4]. However, insight can be gained from approximate analytical solutions [5]. The approximate equation for the fraction of charge remaining in a given position after transfer time t due to self-induced drift is

$$\frac{N_r(t)}{N_0} \simeq \left(1 + \frac{t}{t_{SI}}\right)^{-1} \tag{17}$$

with

$$t_{SI} = \frac{2L^2 C_{eff}}{\pi\mu_n e N_0} \tag{18}$$

where

N_r charge remaining after time t,
L center-to-center electrode spacing,
μ_n electron mobility,
N_0 number of electrons per unit area in the packet to be transferred at $t = 0$,
C_{eff} effective storage capacitance per unit area.

The approximate equation for the fraction of charge remaining in a given position after transfer time t due to fringe-field drift is

$$\frac{N_r(t)}{N_0} \simeq \exp(-t/\tau_{FF}) \tag{19}$$

with

$$\tau_{FF} \leqslant \frac{L}{\mu_n E_{min}} \tag{20}$$

and

$$E_{min} \simeq \frac{2}{3}\frac{\Delta V}{L^2}\frac{\pi\epsilon_s}{C_{eff}} \tag{21}$$

where ΔV is the voltage difference between the electrode under which charge was stored and the receiving electrode, and ϵ_s is the dielectric constant of the semiconductor.

From (5), it follows that for a surface-channel device,

$$C_{\text{eff}} = C_{\text{ox}} = \frac{\epsilon_{\text{ox}}}{d} \qquad (22)$$

and from (16), it follows that for a shallow buried-channel device,

$$C_{\text{eff}} = \left(\frac{d}{\epsilon_{\text{ox}}} + \frac{t}{\epsilon_s} - \frac{N/N_D}{2\epsilon_s} \right)^{-1}. \qquad (23)$$

Self-induced drift is the dominant mechanism in the initial part of the transfer process and fringe-field drift is dominant in the final part of the transfer process. Therefore, for shallow buried-channel structures, it is appropriate to use $N = N_0$ in (18):

$$C_{\text{eff}} = \left(\frac{d}{\epsilon_{\text{ox}}} + \frac{t}{\epsilon_s} - \frac{N_0/N_D}{2\epsilon_s} \right)^{-1} \qquad (24)$$

and $N = 0$ in (21):

$$C_{\text{eff}} = \left(\frac{d}{\epsilon_{\text{ox}}} + \frac{t}{\epsilon_s} \right)^{-1} \qquad (25)$$

According to (18), (20), and (21), it follows that the transfer time for a shallow buried-channel device (0.5 μm) is not significantly smaller than that of a surface-channel device with relatively thick oxide ($d \simeq 0.2$ μm). However, deep buried-channel devices (5 μm) have significantly shorter transfer times because C_{eff} is much smaller and, as a result, the fringe field is larger. The physical reason for the fringe fields being larger in a deep buried-channel device can be seen from Fig. 8. Because the interface is close to the electrodes, the surface potential under a given electrode is determined almost totally by the potential of that electrode, except near the edges. Because the minimum potential in a deep buried-channel device is relatively far from the plane of the electrodes, the potential under a given electrode is influenced not only by the potential of that electrode but also by the potential of adjacent electrodes. Therefore, in a deep buried-channel device, the potential profile has more slope under the transferring electrode, i.e., the fringe field is larger. Deep buried-channel devices have been operated at data rates up to 130 MHz [6].

B. Interface-State Trapping in Surface-Channel Devices

Another mechanism that limits the efficiency of charge transfer is trapping at the semiconductor–insulator interface. The distribution of interface states in the silicon energy gap is shown in Fig. 9 [7]. The emission time of interface states is proportional to exp $[(E_c - E_t)/kT]$. The particular states that contribute to charge-transfer inefficiency depend on the clock frequency. States near the conduction-band edge emit trapped electrons quickly after the charge-transfer process is initiated and these electrons rejoin the main packet; therefore, traps with emission times $\ll 1/f_c$ do not contribute to charge-transfer inefficiency. States having emission times $\gg 1/f_c$ result in an almost permanent trapping and thus do not contribute to charge-transfer inefficiency on a steady-state basis. Therefore, it is states having emission times $\approx 1/f_c$ which trap charge from a packet and emit charge into trailing packets, thus contributing to charge-transfer inefficiency. The transfer inefficiency due to interface-state trapping can be minimized

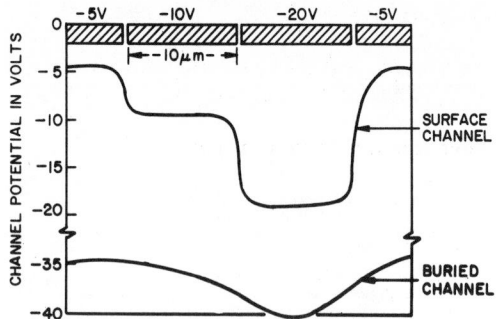

Fig. 8. Channel-potential profile (surface potential in surface-channel device and minimum potential in buried-channel device) for a surface-channel device and a deep buried-channel device. These curves illustrate the greater potential gradient under the transferring electrode in the buried-channel case.

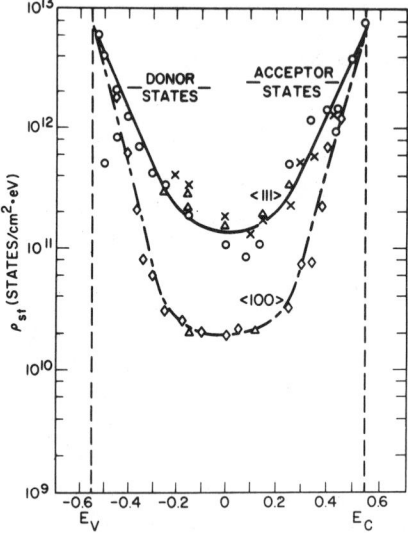

Fig. 9. Si–SiO$_2$ interface-state density versus energy in the silicon bandgap. Data were compiled by several investigators using different techniques.

by filling each well to a fixed level, typically 10–20 percent of saturation [8]. This effectively sets a bias charge level such that a ZERO in the digital sense is not an empty well but rather a well that is 10–20 percent filled; thus the term "fat zero" is used for this bias charge. The effect of the fat zero is to keep the interface states under the gates filled so they do not trap signal charge, and each charge packet will receive about the same number of electrons from preceding packets as it loses to trailing packets. The transfer inefficiency due to interface-state trapping can be written as [9]

$$\epsilon_{\text{FIS}} = \frac{1}{n_s + n_{s,0}} kTN_{\text{FIS}} \ln \left(1 + \frac{2f_c}{k_1 n_{s,0}} \right) \qquad (26)$$

where

$n_{s,0}$	number of fat-zero electrons per unit area,
n_s	number of signal electrons per unit area,
n_{FIS}	fast interface-state density in cm^{-2} (eV)$^{-1}$,
f_c	clock frequency,
k_1	a constant parameter that depends on the trapping cross section ($k_1 \simeq 10^{-2}$ cm^2/s).

Interface states at the edges of the electrodes are not "covered" by the fat-zero charge because the edges of the potential wells are not vertical [10]. In these regions, the trapping is not reduced by the fat zero. This is called the "edge effect."

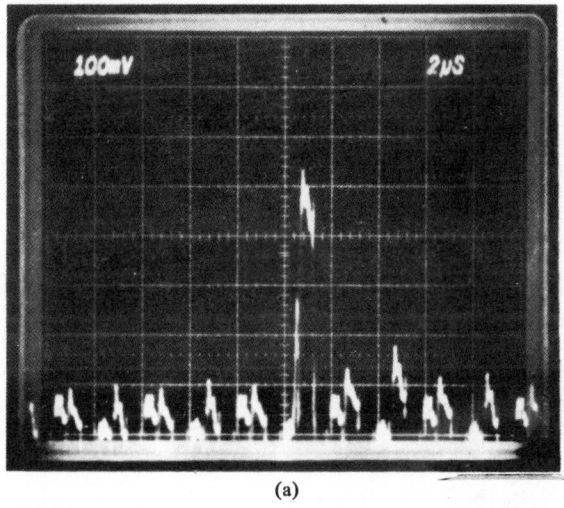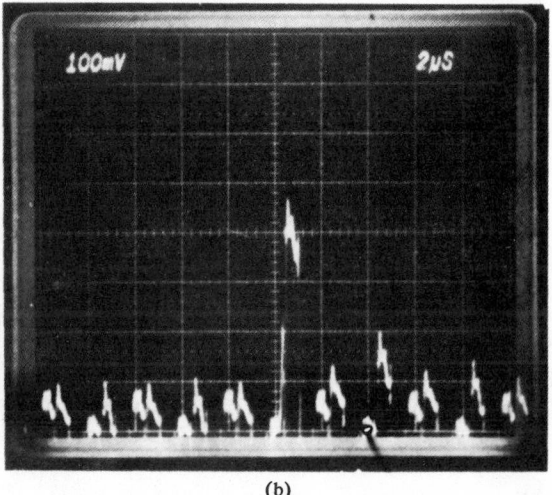

Fig. 10. Oscilloscope traces showing output voltage versus time. (a) Output voltage when a cell was illuminated near the output diode. (b) Output voltage when a cell was illuminated far from the output diode. The number of transfers between the two illuminated cells is 126.

A typical curve of ϵ versus clock frequency has two sections. 1) At low frequencies, ϵ is limited by trapping and is nearly independent of clock frequency. 2) At high frequencies, ϵ is limited by free-charge transfer mechanisms and increases sharply with increasing f_c.

C. Transfer Inefficiency Measurement

Fig. 10 shows an example of transfer inefficiency measurement for a shift register. The shift register is clocked to provide an integration period with subsequent readout via shift-register transfer. Charge is entered into a cell by means of a small light spot. The trace on the left side of Fig. 10 shows the output voltage V_N when the light spot is near the output end, and the trace on the right side shows the output voltage V_F when the light spot is far from the output end. If these two cells are n transfers apart, then the transfer inefficiency is

$$\epsilon = \frac{V_N - V_F}{V_N n}. \tag{27}$$

For the example shown, $V_N = 465$ mV, $V_F = 435$ mV, and $n = 126$; therefore, $\epsilon \simeq 5 \cdot 10^{-4}$.

D. Effect of Transfer Inefficiency on an Impulse

In the discussion of the effects of transfer inefficiency on waveforms, the following definitions will be used:

ϵ fractional loss per elemental transfer,
n number of elemental transfers,
P number of clocking phases,
N number of cell or bit transfers,
α fractional loss incurred by a charge packet in moving from one *cell* to the next.

Then it follows that $n = PN$ and $\alpha = P\epsilon$.

If a single packet of charge containing charge Q is placed into a CCD well ($i = 0$), then after N cell transfers, the distribution of charge in the cells $i = 0, 1, 2, \cdots$ is given by [11]

$$D(i,N) = \frac{Q_i}{Q} = \frac{N!}{(N-i)! \, i!} (1 - \alpha)^i \, \alpha^{N-i}. \tag{28}$$

The dispersion caused by the transfer inefficiency is shown for the first few transfers in Fig. 11. The fractional loss from

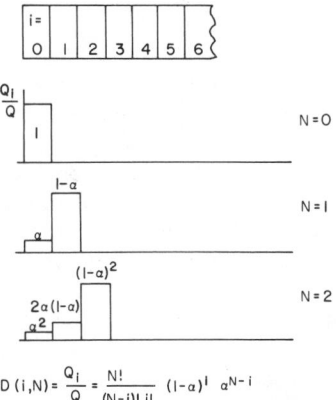

$$D(i,N) = \frac{Q_i}{Q} = \frac{N!}{(N-i)! \, i!} (1-\alpha)^i \, \alpha^{N-i}$$

Fig. 11. Schematic diagram of dispersion in a single-charge packet (impulse) as it is transferred through a CCD.

the charge packet after N cell transfers is given by $1 - D(N,N)$:

$$\text{loss} = 1 - (1 - \alpha)^N \simeq N\alpha, \qquad \text{for } N\alpha \ll 1. \tag{29}$$

The equation $D(N, N) = D(N - 1, N)$ gives the value of N for which the first trailing packet and the leading packet have equal amounts of charge. The solution is $N = 1/\alpha$. It can be shown in general that for every $1/\alpha$ cell transfers, the peak of the charge distribution shifts back by one unit, e.g., after $2/\alpha$ transfers, the second trailing packet contains the largest fraction of the charge. This constitutes a delay in addition to the delay time N/f_c required to clock a packet out of a CCD, i.e., the total delay is

$$\tau = N/f_c \, (1 + \alpha). \tag{30}$$

E. Effect of Transfer Inefficiency on Sinusoids

The effect of transfer inefficiency on a sinusoid of frequency f transferred through n elemental transfers at frequency f_c is characterized by a gain G and a phase shift $\Delta\phi$, where [10]

$$G = \exp \left\{ -n\epsilon \left[1 - \cos \left(2\pi f/f_c \right) \right] \right\} \tag{31}$$

and

$$\Delta\phi = -n\epsilon \left[2\pi f/f_c - \sin \left(2\pi f/f_c \right) \right]. \tag{32}$$

The sampling frequency is equal to the clock frequency; there-

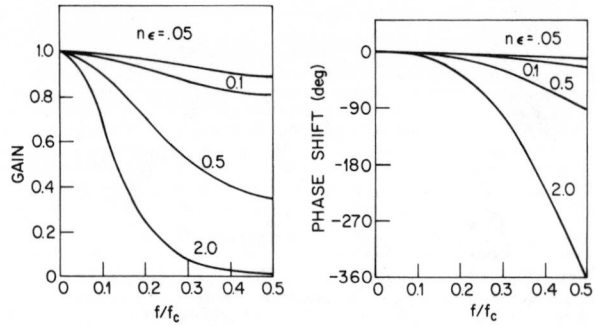

Fig. 12. Effect of transfer inefficiency on the propagation of sinusoids through a CCD shift register. Left curves show gain versus signal frequency (normalized to clock frequency). Right curves show phase shift versus signal frequency. The parameter is the product of number of transfers and transfer inefficiency.

fore, the Nyquist limit is $f_c/2$. In Fig. 12, the gain and phase shift are plotted versus frequency up to the Nyquist limit, with $n\epsilon$ as a parameter. From Fig. 12, it should be noted that if $n\epsilon \leqslant 0.1$, then $0.8 \leqslant G \leqslant 1.0$ and $0 \leqslant \Delta\phi \leqslant \pi/10$. These limits are acceptable for most applications.

F. Dark Current

Since a CCD operates by controlling depletion volumes, there is a continual generation of hole–electron pairs due to the thermal vibration of the silicon lattice, and minority carriers tend to flow to the point of minimum potential where they are collected. The three components of dark current are thermal generation at the $Si–SiO_2$ interface, thermal generation in the depleted volume, and thermal generation in the neutral bulk within a diffusion length of the interface.

The dark current imposes three limitations on the operating characteristics of a CCD—finite storage time, fixed-pattern "noise," and temporal noise. Each of these effects will be discussed in the following paragraphs.

The length of time required for the dark current to fill a well is called the storage time T_s, and is given by

$$T_s \simeq \frac{C_{\text{eff}} \Delta V}{J_d} \qquad (33)$$

where C_{eff} is the effective storage capacitance per unit area of the well, J_d is the dark current density, and ΔV is the height of the barrier isolating the wells. Storage times of 100 s have been achieved in some devices; however, values of 0.1–10 s are more typical.

The dark current is not completely uniform throughout a device. The dark-current signature of a device is obtained by integrating in the dark. A dark-current signature for a short linear device is shown in Fig. 13. The effect of this dark-current nonuniformity is to impose a fixed-pattern "noise" on the signal. If the element-to-element nonuniformity is ΔN_d, then the minimum signal that can be detected will be limited to $N_s \simeq \Delta N_d$. If the element-to-element variation of N_d is 10 percent, then at room temperature this will limit the minimum detectable signal to about 1000 electrons.

Since the generation process is random, there is temporal noise associated with it. Since the process obeys Poisson statistics, the temporal noise is the square root of the mean. This noise source will also limit the imagery at room temperature.

Since the states that contribute to thermal generation are near the center of the forbidden gap, the temperature de-

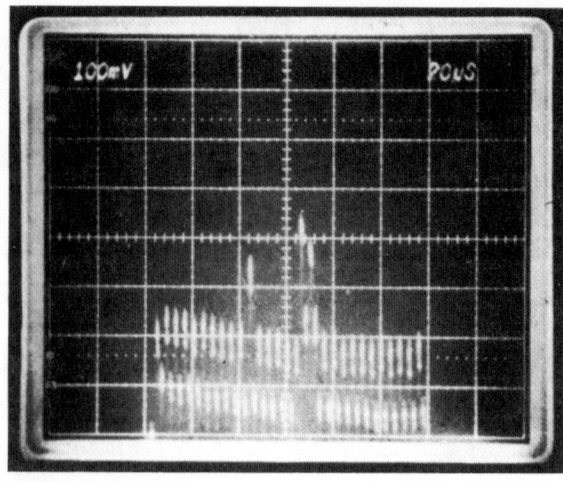

Fig. 13. Oscilloscope trace of a CCD after integration in the dark, showing nonuniformity in the dark current from cell to cell.

pendence of the dark current is

$$J_d \sim \exp\left(-E_g/2kT\right) \qquad (34)$$

where E_g is the energy gap. Thus the dark current is a strong function of temperature, decreasing by a factor of 2 for every $10°C$ decrease in T. The dark-current spikes have the same temperature dependence as the average dark current. If a device having $J_D = 10$ nA/cm^2 at $20°C$ is cooled to $-50°C$, then the number of electrons per cell due to the dark current is approximately 80. A 10-percent nonuniformity will impose a detection limit of 8 electrons and the temporal noise is 9 electrons.

G. Square-Wave Driver Power Dissipation

The equivalent circuit of a square-wave clock driver is simply a series RC circuit, where R is the internal resistance of the clock driver and C is the capacitance on a clock line. Assume that the voltage swing is V and the clock frequency is f_c. If $f_c^{-1} \gg RC$, the energy dissipated in R during charging or discharging C is

$$E_R = \int_0^{f_c/2} i^2 R \, dt = \int_0^{f_c/2} \frac{V^2}{R} e^{-2t/RC} \, dt = \frac{1}{2CV^2}. \qquad (35)$$

Each CCD bit is charged and discharged once per cycle. Thus the average power dissipated in the driver per CCD bit is

$$P_{\text{avg}} = C_{\text{bit}} V^2 f_c. \qquad (36)$$

H. Sinusoidal Drivers

The power dissipated in a sinusoidal driver per CCD phase is

$$P_{\text{avg}} = V^2 R \left(R^2 + \frac{1}{(2\pi f_c C)^2}\right)^{-1} \qquad (37)$$

where V is the peak voltage at the CCD electrodes, and f_c is the frequency of the driver. For practical cases, $R \ll (2\pi f_c C)^{-1}$, and the average power dissipated per CCD bit is

$$P_{\text{avg}} = \frac{\pi}{Q} C_{\text{bit}} V^2 f_c \qquad (38)$$

where Q is the quality factor of the circuit: $Q = 2\pi f_c RC$.

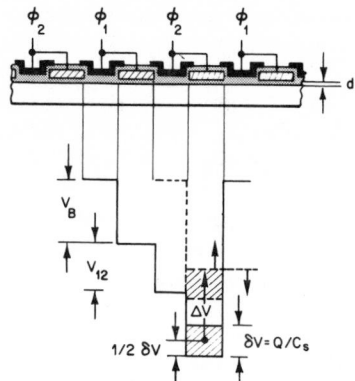

Fig. 14. Potential profile illustrating power dissipation in a two-phase device.

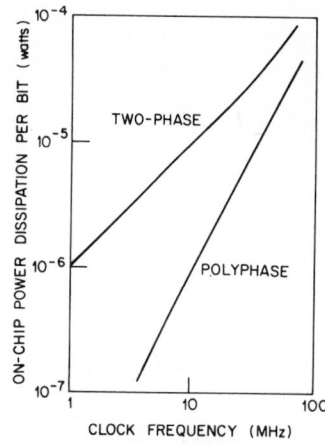

Fig. 15. On-chip power dissipation versus clock frequency for two-phase and polyphase CCD's.

For tuned sinusoidal drivers, Q can be quite large thereby reducing the driver power requirements considerably.

I. On-Chip Power Dissipation

For any CCD there is power dissipated on the chip per CCD bit due to the current flow [12]:

$$P = (J \cdot \mathcal{E})V = (\rho v \cdot \mathcal{E})V$$
$$= \rho V(v \cdot \mathcal{E}) = \rho V v^2 / \mu$$
$$= ne(f_c L)^2 / \mu \qquad (39)$$

where

J	current density along the channel,
$\mathcal{E}$	electric field along the channel,
V	bit volume,
v	average bit velocity,
μ	effective mobility,
L	bit length,
n	number of electrons per bit,
f_c	clock rate.

For two-phase CCD's that have built-in barriers, there is an additional power dissipation term resulting from the need to "lift" the charge packets over the built-in barriers to allow transfer to occur [13]. This is illustrated in Fig. 14. The energy required to lift a charge packet containing n electrons over a potential barrier ΔV is the $ne\Delta V$. This is done twice per clock period; thus the average power dissipated per bit due to the mechanism is

$$P = 2ne\Delta V f_c. \qquad (40)$$

The factor ΔV can be calculated with reference to Fig. 14, which shows the potential profile for a two-phase stepped-oxide device. V_B is the built-in barrier. V_{12} is the surface potential difference between the polysilicon electrode when the clock is low and the aluminum electrode when the clock is high. The average potential of the charge packet is $\frac{1}{2}\delta V = ne/C_p$, where C_p is the capacitance of the polysilicon MOS capacitor. From Fig. 14, it then follows that

$$\Delta V = V_B + \frac{1}{2} V_{12} - \frac{1}{2} \frac{ne}{C_p} \qquad (41)$$

$$P_{\text{avg}} = 2nef_c \left(V_B + \frac{1}{2} V_{12} - \frac{1}{2} \frac{ne}{C_p} \right). \qquad (42)$$

Therefore, the total average power dissipation per bit for a two-phase CCD is

$$P_{\text{avg}} = ne(f_c L^2 / \mu) + 2nef_c \left(V_B + \frac{1}{2} V_{12} - \frac{1}{2} \frac{ne}{C_p} \right). \qquad (43)$$

Fig. 15 shows typical power dissipation curves for two-phase and polyphase CCD's at saturation.

From (36) and Fig. 15, it can be shown that for a two-phase device at saturation driven by a 10-MHz square-wave driver, the *driver* power dissipation per bit is about a factor of 30 greater than the *on-chip* power dissipation per bit.

J. Clocking Tradeoffs

Fig. 16 shows a four-phase CCD structure. There are two methods for clocking four-phase devices—normal clocking and double clocking—as shown in Fig. 17(a) and (b), respectively. In the normal-clocking mode, charge is stored under only one electrode, as shown in Fig. 17(c). In the double-clocking mode, charge is stored under two adjacent electrodes, as shown in Fig. 17(d) [14].

From stability and speed considerations, practical CCD structures have overlapping electrodes. Table I compares the characteristics of two-, three-, and four-phase CCD's having overlapping electrodes.

It is assumed that the minimum geometry along the channel (electrode length or implantation dimension) is L; therefore, the cell or bit length of two-phase and four-phase devices is $4L$ while the cell length of a three-phase device is $3L$. Recall that each time a capacitance C (corresponding to one electrode of length L) is charged or discharged, the power dissipation in a square-wave driver is $\frac{1}{2}CV^2$. To move a charge packet from one cell to the next in two-phase and four-phase devices, electrodes have to be charged or discharged a total of eight times in f_c^{-1} s. Thus the driver power dissipation is $8\frac{1}{2}CV^2 f_{\text{bit}} = 4CV^2 f_c$. To move a charge packet from one cell to the next in a three-phase device, electrodes have to be charged or discharged a total of six times in f_c^{-1} s. Thus the driver power dissipation is $3CV^2 f_c$. Since the capacitance of two- and four-phase cells is $4C$ and the capacitance of a three-phase cell is $3C$, the general expression for square-wave driver dissipation is $C_{\text{cell}} V^2 f_c$.

The two-phase and four-phase double-clocking schemes have an inherent charge pushing action, whereas this must be built into the three-phase clocking pulses, as shown in Fig. 1.

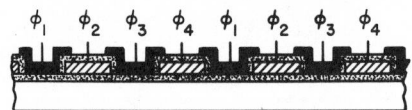

Fig. 16. Cross-sectional view of a four-phase CCD.

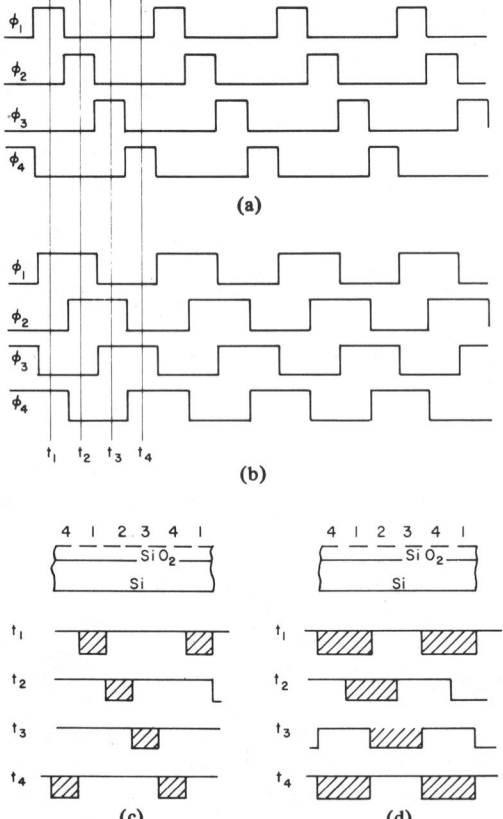

Fig. 17. Four-phase clocking. (a) Normal four-phase clocking wave-forms. (b) Double-clocking waveforms. (c) Length of potential wells with normal clocking. (d) Length of potential wells with double clocking.

TABLE I
CLOCKING TRADEOFFS

	2-PHASE	3-PHASE	4-PHASE
CELL LENGTH	4 L	3 L	4 L
DRIVER POWER	4 CV²f_c	3 CV²f_c	4 CV²f_c
SENSITIVITY TO CLOCK WAVEFORMS?	NO	YES	NO
SIGNAL HANDLING CAPACITY	1 × 4 VOLTS	1 × 6 VOLTS	2 × 6 VOLTS[a]
TRANSFER PER BIT	2	3	4
FABRICATION DIFFICULTY	2-LEVELS	3-LEVELS[b]	2-LEVELS
SELF ALIGNED?	YES	YES	YES

L = MIMIMUM LENGTH OF ELECTRODE OR IMPLANTATION ALONG CHANNEL
[a] DOUBLE CLOCKING
[b] BELL LABS 3-PHASE, 3-LEVEL POLYSILICON

Typical built-in barriers in two-phase devices have heights of 4 V, whereas about 6 V can be achieved with barriers controlled by gate voltages. Therefore, the signal-handling capacity (in terms of barrier height) of two-phase devices is about 4 V, and for three-phase devices it is 6 V. Since the four-phase device operated in the double-clocking mode stores charge under two electrodes and has barriers of about 6 V, its

signal-handling capacity is equivalent to that of a device storing charge under a single electrode having a 12-V barrier.

Typical two- and four-phase CCD's with overlapping gates utilize polysilicon–aluminum or anodized aluminum technology. Overlapping-gate, three-phase devices utilize triple-layer polysilicon technology. All three designs can be self-aligned.

In summary, the two-phase CCD requires the fewest transfers, the three-phase CCD has the highest packing density, and the four-phase CCD operated in the double-clocking mode has the largest signal-handling capability.

IV. LOW-NOISE INPUT CIRCUITS

A means of introducing a low-noise charge packet into a CCD is illustrated in Fig. 18 [15]. The receiving well is filled with electrons to a level higher than the largest expected signal. This can be accomplished by pulsing the input diode to a low voltage. With the receiving well full, the input gate is set to the desired level, and the input diode is pulsed positive. This removes electrons from the receiving well until the potential of the receiving well increases to the point where the channel pinches off. This process can be modeled as that of charging a capacitor through a resistor having a Johnson–Nyquist spectral noise density $\overline{V_n^2}(f) = 4kTR$ V²/Hz. If the charge Q on the capacitor C_{RW} is treated as a random variable, the variance of Q is kTC_{RW} (C)², where C_{RW} is the capacitance of the receiving well. For $C_{RW} \simeq 0.1$ pF, the noise in a charge packet introduced into the receiving well is approximately 130 electrons.

This technique can be extended to further reduce the input noise by presetting a small floating diffusion and then transferring the charge from the floating diffusion into the first CCD well [15]. The structure and the timing diagram for this technique are shown in Fig. 19(a) and (b), respectively. The advantage of this technique over that of injecting from a source diffusion into the receiving well via a gate is that the capacitance of the floating diffusion can be made smaller than C_{RW}. Thus the noise is less.

In surface-channel area arrays, it is necessary to insert a fat zero into each column. The techniques described in the foregoing can be used to insert the fat zero with low temporal noise; however, the charge entered from column to column is not uniform due to geometrical tolerances. For example, due to photolithographic tolerances, the uncertainty in the capacitance of a receiving well or floating diffusion is about one percent. Thus, for a ten-percent fat zero, the nonuniformity in fat zero from column to column is about 500 electrons. This imposes an obvious limitation in low-light-level imaging applications.

V. LOW-NOISE OUTPUT CIRCUITS

A. Correlated Double Sampling

Consider a capacitor C with an initial charge Q_0 and corresponding voltage V_0. At $t = 0$, a noiseless ideal switch is closed, charging the capacitor to the voltage V_{DR} through a resistor as shown in Fig. 20. The mean charge on the capacitor, Q_m, is

$$Q_m(t) = CV_{DR}(1 - e^{-t/RC}) + Q_0. \quad (44)$$

It is shown in the Appendix that the mean-square deviation from the mean value of $Q(t)$ is

$$\overline{(Q(t) - Q_m(t))^2} \equiv \overline{Q(t)^2} = kTC(1 - e^{-2t/RC}) \quad (45)$$

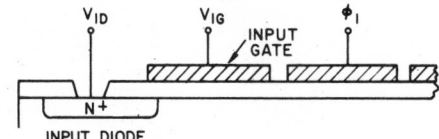

Fig. 18. Input structure.

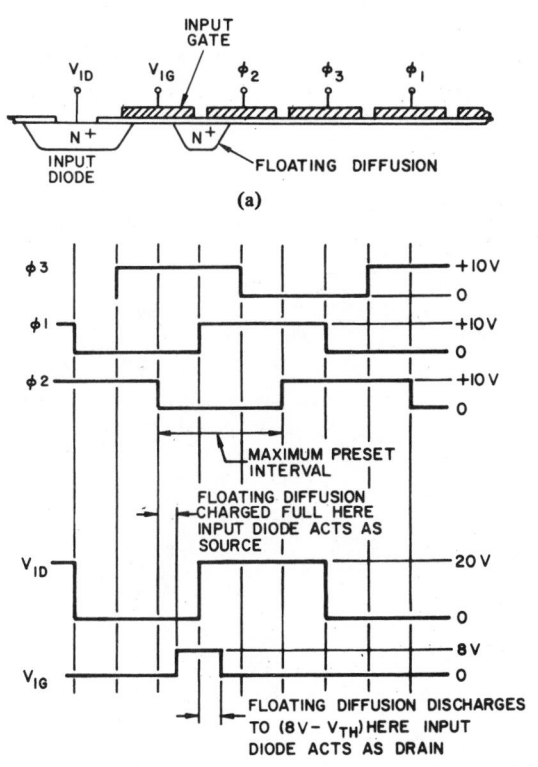

(a)

(b)

Fig. 19. Low-noise input structure utilizing a low-capacitance floating diffusion.

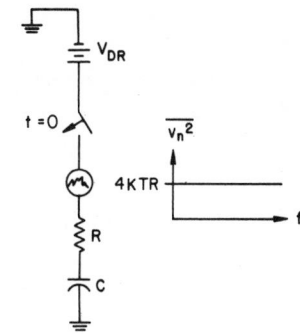

Fig. 20. Noise-equivalent circuit for the charging of a capacitor through a resistor having thermal noise.

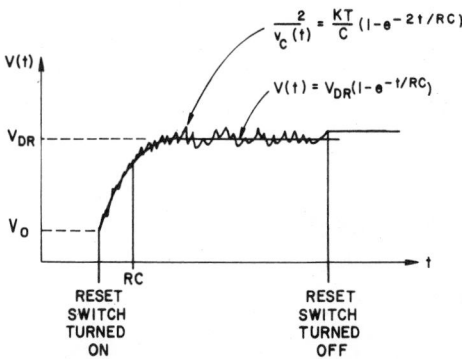

Fig. 21. Curve of mean voltage and noise voltage across a capacitor as a function of time when the capacitor is charged through a resistor having thermal noise.

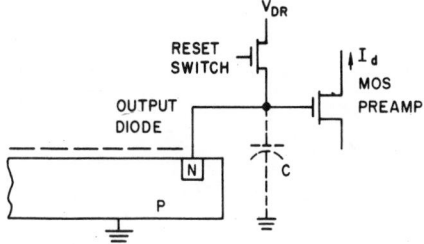

Fig. 22. Conventional reset amplifier output circuit for a CCD.

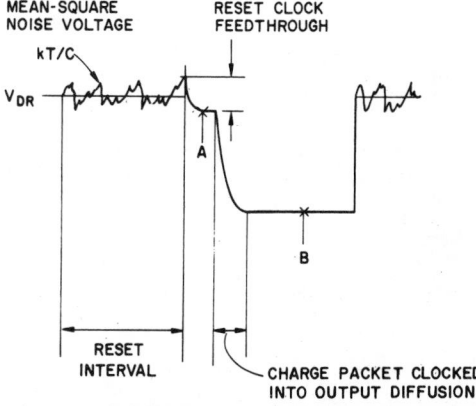

Fig. 23. Output voltage waveform of reset amplifier output circuit.

or in terms of voltage

$$\overline{V_c(t)^2} = \frac{kT}{C}(1 - e^{-2t/RC}). \qquad (46)$$

The complete charging curve is shown in Fig. 21. Thus, when charging a capacitor through a resistor, the variance of the voltage across the capacitor is small for $t \ll \frac{1}{2}RC$ and is kT/C for $t \gg RC$. The significance of this is that the correlation time for $V_c(t)$ is $\frac{1}{2}RC$, which forms the basis for correlated double sampling.

The conventional CCD output circuit is shown in Fig. 22, and the voltage across C is shown in Fig. 23. The readout sequence is as follows. The output diode is reset to some large reverse-bias voltage V_{DR} through a MOSFET switch. The RC time constant is

$$R_{on}C = (10^4 \ \Omega \times 10^{-13} \ \text{F}) = 10^{-9} \ \text{s}.$$

When the switch is turned off, there will be a droop in the voltage waveform across C due to reset clock feedthrough (Miller effect) in the reset transistor.

The RC time constant is $R_{off}C = (10^{12} \ \Omega \times 10^{-13} \ \text{F}) = 10^{-1}$ s. Next, the charge packet is clocked into the output diode, reducing its reverse bias. The cycle then repeats itself. Typically, the voltage is sampled at B with respect to some fixed reference. In this case, the noise voltage in this sample is

$(kT/C)^{1/2}$. However, if two samples are taken, one at A just after the reset switch is turned off and the other at B after the signal charge has been clocked into the output diode, and if the time between A and B is $\ll \frac{1}{2}R_{off}C$, then the noise on these time samples is correlated and can be removed by subtracting V_A from V_B [16]. This is called correlated double sampling. The subtraction process is achieved by inverting and storing the reset voltage for subsequent comparison with the signal voltage.

Since the kTC noise is removed by correlated double sampling, the remaining noise is due to the amplifiers. The noise in the on-chip MOS preamplifier referred to the CCD channel is

$$N_{n_1}^2 = \left(\frac{C_{O-N}}{e}\right)^2 \frac{8kTB}{3g_m} \quad \text{(electron)}^2 \quad (47)$$

where C_{O-N} is the CCD output-node capacitance, B is the bandwidth, and g_m is the transconductance of the MOS preamplifier. For $C_{O-N} = 0.1$ pF, $B = 4$ MHz, and $g_m = 250$ μmho, $N_{n_1}^2 \simeq 70$ (electron)^2.

At the output of the preamplifier, the signal is still at a low level. If an off-chip amplifier having an equivalent input noise current i_n (amperes per root hertz) is used, this noise referred to the CCD channel is

$$N_{n_2}^2 = \left(\frac{C_{O-N}}{g_m e}\right)^2 Bi_n^2 \quad \text{(electron)}^2. \quad (48)$$

For $i_n = 1$ pA/(Hz)$^{1/2}$, $N_{n_2}^2 = 250$ (electron)^2. Therefore, the total noise referred to the CCD channel is $N_n \simeq 18$ electrons.

B. Distributed Floating-Gate Amplifier (DFGA)

The distributed floating-gate amplifier (DFGA), shown schematically in Fig. 24, makes use of repeated nondestructive sensing of charge packets by floating gates that are over the CCD channel [17]. A schematic diagram of a floating-gate amplifier stage is shown in Fig. 25. The floating gate is biased to the desired operating point via capacitive coupling to a bias electrode (not shown) above the floating gate. When a charge packet containing S_i electrons transferring along the DFGA input register moves under a floating gate, a potential change ΔV_{FG} is induced on the floating gate.

The floating-gate responsivity $\mathfrak{R}$ is defined as

$$\mathfrak{R} \equiv \frac{\Delta V_{FG}}{\Delta S_i} \quad \text{(volt/electron)}. \quad (49)$$

The calculation of $\mathfrak{R}$ follows directly from the capacitance equivalent circuit shown in Fig. 26, and $\mathfrak{R}$ is

$$\mathfrak{R} = \frac{eC_1}{C_1(C_2 + C_{GSUB} + C_{CS}) + C_3(C_1 + C_2 + C_{GSUB} + C_{CS})}. \quad (50)$$

The voltage on the floating gate modulates the flow of electrons from a source diffusion into the adjacent potential well of the DFGA output CCD register. The charge gain G between a position in the DFGA input register and the corresponding position in the DFGA output register depends on the amount of time τ that the charge packet is under the floating gate and also on $\mathfrak{R}$:

$$G = \mathfrak{R}g_m \tau/e \quad (51)$$

where g_m is the amplifier transconductance. It is important to clock the DFGA in such a way that τ is maximized. Clocking schemes have been used which give

$$\tau \simeq 0.4/f_c \quad (52)$$

where f_c is the clocking frequency. If $g_m = 10$ μmho, $\mathfrak{R} = 5$ μV/electron, and $f_c = 8$ MHz, then $G \simeq 16$.

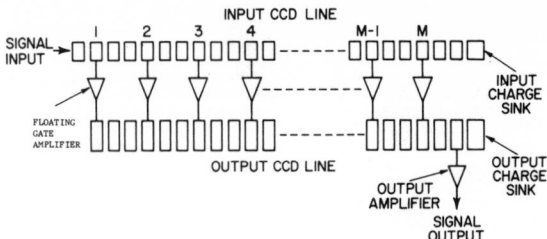

Fig. 24. Schematic diagram of charge-coupled distributed amplifier.

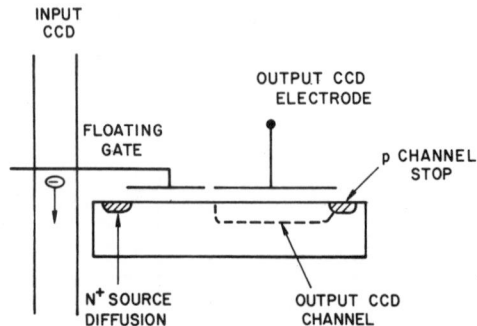

Fig. 25. Schematic diagram of a single distributed floating-gate amplifier (DFGA) stage.

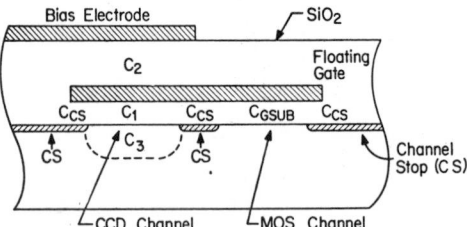

Fig. 26. Capacitance equivalent circuit of a single floating-gate amplifier stage.

The operation of the DFGA is as follows. When a charge packet S_i in the DFGA input register is transferred under the first floating gate, an amplified charge packet $[(0.4/f_c) I_o - GS_i]$ is injected into the first cell of the DFGA output register, where I_o is the bias current. The two registers are clocked synchronously so that when S_i is under the second floating gate, $[(0.4/f_c) I_o - GS_i]$ is in the second cell of the DFGA output register, and another amplified charge packet is injected into the second cell of the DFGA output register. Thus, after M DFGA stages, the charge at the output of the DFGA is $M[(0.4/f_c) I_o - GS_i]$, and the output signal S_o is $S_o = MGS_i$. The noise in the charge packets in the DFGA input register, N_i, is amplified and added in the same way. Furthermore, noise is introduced at each DFGA stage due to the amplification and injection processes; let N_A denote the noise charge introduced at each stage referred to the DFGA input register. The noise charges introduced at each stage, $(N_A)_1$, $(N_A)_2$, and $(N_A)_M$, are uncorrelated. Therefore, the total noise (in number of electrons) at the output of the DFGA, N_o, is

$$N_o = (M^2 G^2 N_i^2 + M N_A^2 G^2)^{1/2}. \quad (53)$$

Then the output signal-to-noise ratio is

$$\frac{S_o}{N_o} = \frac{S_i}{N_i} \left(1 + \frac{N_A^2/M}{N_i^2}\right)^{-1/2} \quad (54)$$

Thus the condition required to preserve the signal-to-noise

ratio inherent in the charge packets in the DFGA input register is

$$M \geqslant 5 \left(\frac{N_A}{N_i} \right)^2. \tag{55}$$

Analysis of the noise sources in the DFGA indicates that the dominant noise is shot noise in the floating-gate amplifiers. The number of noise electrons referred to the DFGA input CCD channel is

$$N_A = \frac{1}{G} \left[\left(I_o - \frac{\Re g_m}{e} S_i \right) \frac{0.4}{f_c} \right]^{1/2}. \tag{56}$$

For $f_c = 8$ MHz, $I_o = 10^{-6}$ A, $g_m = 10^{-5}$ mho, $\Re = 5 \cdot 10^{-6}$ V/electron, and $S_i = 0$, $N_A = 36$ electrons. For $N_i = 20$ electrons, the required number of stages to preserve the signal-to-noise ratio in the DFGA input CCD channel according to (55) is $M = 16$.

At the output of the DFGA, the signal has been amplified by the factor MG; thus the signal is at a high level when it is taken off the chip. The obvious way to reduce the noise of the DFGA is to increase $\Re$ and g_m.

VI. Noise

A. Types

The types of noise expected to be present in a CCD can be separated into three categories: those associated with the input, those associated with integration and transfer, and those associated with the output. The types of noise expected from these three operations are defined as follows.

1) Input

Photon noise: The emission of photons from any source is a random process. The number of photoelectrons collected in a potential well in time Δt is therefore a random variable. The standard deviation of this random variable is the photon noise. Since the statistics for the emission process are Poisson, the standard deviation equals the square root of the mean.

Electrical input noise: The injection of charge from a source diffusion into a potential well is a random process because of thermal noise in the resistance of the input circuit. The number of electrons injected into a potential well from a source diffusion is therefore a random variable whose standard deviation is the electrical input noise. When a bias charge is injected in this way, the noise is sometimes called fat-zero noise.

2) Integration and Transfer

Fast interface-state noise [18], *slow interface-state noise* [19], *and bulk-state noise:* The transfer of electrons from one site to another is a random process because of trapping and emission by fast interface states, slow interface states (oxide states), and bulk states. The standard deviation of the number of electrons transferred from one well to the next is the noise associated with each of the trapping processes. Fast interface-state noise should be present in surface-channel CCD's during transfer, slow interface-state noise may be present in surface-channel CCD's during integration, and only bulk-state noise should be present in buried-channel CCD's.

Dark-current noise: The thermal generation of hole-electron pairs in the semiconductor is a random process that contributes charge to the CCD potential wells. Therefore, the number of electrons in a charge packet due to thermal generation is a random variable whose standard deviation is the dark-current noise. Like photon emission, this is a Poisson process; thus the standard deviation is the square root of the mean.

3) Output

Reset noise: A common technique for CCD readout involves the charging of a capacitance through a switch (reset). Unless special techniques are used, e.g., correlated double sampling, the noise in this reset process which is due to thermal noise in the reset circuit is directly reflected in the output signal. Thus the reset level is a random variable whose standard deviation is the reset noise.

MOS amplifier noise: Due to various noise sources in an MOS amplifier, the output voltage is a random variable. The standard deviation of the output voltage referred back to the input (as a charge on the gate) is the amplifier noise. By characterizing the noise as a charge on the gate of the MOS preamplifier, it can be directly compared with other noise sources in the CCD.

Signal processing noise: It is assumed that any appreciable noise introduced by off-chip signal processing should be referred back to the gate of the preamplifier and included in the noise analysis.

B. Bulk-Trapping Noise

Models exist for all of the noise sources except bulk-state trapping. Therefore, it is appropriate to discuss a model for bulk-state trapping in some detail. Bulk-trapping states are characterized by discrete energy levels. According to the Shockley–Read theory, the emission time characteristic of a bulk-trapping state at energy ΔE below the conduction-band edge is proportional to $\exp (\Delta E / kT)$. If a charge packet is located in a volume V for time τ and the emission time of the bulk state is τ_e, then the probability of occupation is

$$P = 1 - e^{-\tau/\tau_e}. \tag{57}$$

Since the random variable of interest is the density of traps occupied at time τ, the mean value of n_t is

$$n_t = P N_t \tag{58}$$

and the variance is

$$\overline{n_t^2} = (1 - P) P N_t. \tag{59}$$

Consider a buried-channel area imager. The operation can be separated into three distinct steps: 1) integration, with effective volume V_I and integration time τ_I; 2) vertical transfer, with effective volume V_v, time τ_v, and number of transfers M_v; and 3) horizontal transfer, with effective volume V_H, time τ_H, and number of horizontal transfers M_H. It follows that the variance of the complete sensing and readout operation is

$$\begin{aligned} \overline{n_t^2} \simeq\ & N_t V_I \left(1 - e^{-\tau_I/\tau_e}\right) \left(e^{-\tau_I/\tau_e}\right) \\ &+ N_t V_v M_v \left(1 - e^{-\tau_v/\tau_e}\right) \left(e^{-\tau_v/\tau_e}\right) \\ &+ N_t V_H M_H \left(1 - e^{-\tau_H/\tau_e}\right) \left(e^{-\tau_H/\tau_e}\right). \end{aligned} \tag{60}$$

The effective volume is calculated as follows.

$V =$ (area occupied by the charge packet under an electrode)
$\quad \times$ (thickness of the region occupied by the charge packet)
$\quad = A \times t$.

At high signal levels, the thickness from (9) is

$$t \simeq \frac{N}{N_D} = \frac{n/A}{N_D} \tag{61}$$

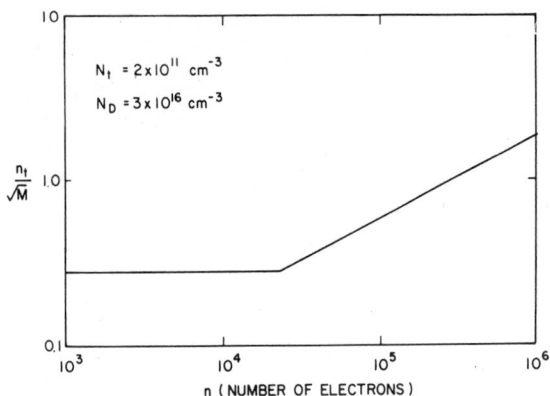

Fig. 27. Bulk trapping noise (in number of electrons) per transfer versus number of electrons in a charge packet for shallow buried-channel device.

where n is the number of electrons in the charge packet. Thus $V \simeq n/N_D$.

Then the number of noise electrons due to bulk trapping can be written as

$$n_t \equiv (\overline{n_t^2})^{1/2} = \left\{ n\frac{N_t}{N_D}\left[(1-e^{-\tau_I/\tau_e})e^{-\tau_I/\tau_e} \right.\right.$$
$$+ M_v(1-e^{-\tau_v/\tau_e})e^{-\tau_v/\tau_e}$$
$$\left.\left. + M_H(1-e^{-\tau_H/\tau_e})e^{-\tau_H/\tau_e}\right]\right\}^{1/2} \quad (62)$$

The factor $(1-e^{-\tau/\tau_e})e^{-\tau/\tau_e}$ has a maximum value of $\frac{1}{2}$ when $\tau = \tau_e \ln(2)$ and drops off exponentially when τ increases or decreases.

Equation (62) does not apply for n small enough to make $n/N_D \approx L_D$, where L_D is the extrinsic Debye length. For small n, the formulation is as follows [20]. According to (10),

$$\Delta\phi_{min} = \frac{eN_D t^2}{2\epsilon_s}\frac{N_D + N_A}{N_D}. \quad (63)$$

Using the conservative criterion that most of the electrons have energies $\leqslant 6kT$, then it follows that when $(\Delta\phi_{min})_{max} = 6kT/e$,

$$t_{min} = \left(\frac{12\epsilon_s kT}{e^2(N_A + N_D)}\right)^{1/2} = \sqrt{6}L_D. \quad (64)$$

The effective area is the area within which electrons having energy less than $6kT$ are confined. Due to the curvature of the potential well, this area is clearly less than the area of the storage electrode. An accurate determination of this area would require the precise knowledge of the shape of the potential wells from computer calculations. Using the quadratic approximation for the potential well gives $A_{eff} \simeq 8.8 \times 10^{-8}$ cm². Using $L_D = 4 \times 10^{-6}$ cm gives $V_{eff} = 8.4 \times 10^{-13}$ cm³. Using $N_t = 2 \times 10^{11}$ cm⁻³ gives $0.28\sqrt{M}$ electrons as the worst case noise level at low signal levels. The bulk-state noise versus signal level is plotted in Fig. 27.

Table II gives expressions and numerical estimates for the variances of the processes which contribute to noise in surface-channel CCD's at room temperature and in buried-channel CCD's at room temperature and at $-50°$C. From this noise analysis, it is concluded that the number of noise electrons at the output of a cooled buried-channel CCD could be as low as 10 electrons.

An additional limitation on surface-channel area arrays is the "fixed-pattern" noise arising from the inability to insert a fat

TABLE II
SUMMARY OF CCD NOISE SOURCES

OPERATION	PROCESS	VARIANCE (Number of electrons)²			
			SURFACE CHANNEL	BURIED CHANNEL (20°C)	BURIED CHANNEL (-50°C)
INPUT	PHOTON NOISE	n_s			
	FAT ZERO INPUT	$\frac{kT}{e^2}C_{I-N}$	1.6×10^4	0	0
INTEGRATION AND TRANSFER	FAST INTERFACE STATES	$mkT N_{FIS}A$	4×10^5	0	0
	SLOW INTERFACE STATES	$mkT N_{SOS}A \ln(\tau/\tau_s)$ if $\tau \gg \tau_s$	1300	0	0
		$m\frac{2}{\pi}\frac{kT}{e}N_{SOS}A(\tau/\tau_s)$ if $\tau \ll \tau_s$			
	BULK STATES	$mn_s\frac{N_t}{N_D}(1-e^{-\tau/\tau_e})e^{-\tau/\tau_e}$ if $\frac{14 n_s}{AN_D} > L_D$			
		$\frac{m}{14}AL_D(1-e^{-\tau/\tau_e})e^{-\tau/\tau_e}$ if $\frac{14 n_s}{AN_D} < L_D$		<85	<85
	DARK CURRENT	n_{dark}	10^4	10^4	80
OUTPUT	RESET AMPLIFIER	$\frac{kT}{e^2}C_{O-N}+\left(\frac{C_{O-N}}{e}\right)^2\frac{8kTB}{3g_m}$	1.6×10^4	1.6×10^4	1.6×10^4
	CORRELATED DOUBLE SAMPLING	$\left[\left(\frac{C_{O-N}}{e}\right)^2\frac{8kTB}{3g_m}+\left(\frac{C_{O-N}i_n}{g_m e}\right)^2\right]B$	320	320	320
	DISTRIBUTED FLOATING GATE AMPLIFIER	$\frac{ef_c i_0}{0.4(g'_m)^2 \mathscr{R}^2 M}$	100	100	100

n_s = number of signal electrons collected in a CCD cell,
m = number of positions involved in the process; i.e., for integration, m = 1; and for transfer m is the number of transfers,
A = area of a CCD gate,
N_{FIS} = fast interface-state density,
N_{SOD} = area density of slow oxide states,
τ = time a charge packet is under a gate,
τ_s = time constant of slow oxide states located at the Si-SiO₂ interface,
N_t = volume density of bulk traps in the silicon,
N_D = doping density of the buried channel,
τ_e = emission time constant of the bulk traps,
L_D = extrinsic Debye length in the channel,
n_{dark} = number of electrons added to a charge packet during integration and transfer due to dark current,
C_{O-N} = output node capacitance,
B = bandwidth of the output circuit,
g_m = transconductance of the output amplifier, and
i_n = equivalent noise current of the off-chip amplifier in amps/(Hz)$^{1/2}$,
M = number of distributed amplifier stages,
$\mathscr{R}$ = responsivity of the floating gate in volts per electron,
g'_m = transconductance of the floating gate amplifier,
i_0 = bias current of each floating-gate amplifier stage.

C_{I-N} = 0.1 pf,
m = 1000,
N_{FIS} = 10^{10} cm⁻² (eV)⁻¹,
A = 10^{-6} cm²,
N_{SOS} = 2×10^{11} cm⁻² (eV)⁻¹,
$\tau_{integrate}$ = 1/30 s,
$\tau_{transfer}$ = 10^{-6} s,
τ_s = 10^{-6} s,
N_t = 2×10^{11} cm⁻³,
N_D = 3×10^{16} cm⁻³,
L_D = 4×10^{-6} cm,
C_{O-N} = 0.1 pf,
B = 4 MHz,
g_m = 250 micromhos,
i_n = 1 pa/(Hz)$^{1/2}$,
M = 12,
$\mathscr{R}$ = 5 microvolts per electron,
g'_m = 10 micromhos,
i_0 = 10^{-6} amps.

zero uniformly into each cell. The basic problem arises from the nonuniform capacitance of each source providing each column with a fat zero. If the variation is 1 percent, then the fixed-pattern noise for a 10-percent fat zero is ~500 electrons. Thus, without special off-chip processing, the surface-channel area imager will be limited to a noise-equivalent signal (NES) of about 500 electrons. Nonuniform dark current has the same effect; however, it can be reduced substantially by cooling the array.

VII. IMAGING ARRAYS WITH MECHANICAL SCAN

A. Linear Arrays

Linear imaging arrays can be designed in three ways. 1) A simple CCD shift register can be used if it is clocked in such a way that the shift-out time is very much less than the integration time. This condition reduces image smear caused by shifting pixels through light-sensitive regions. 2) An imager can also be designed with separate sensors and shielded readout register as shown in Fig. 28. After integration in the sensors, the charge configuration is shifted into the shift register, i.e., a parallel-to-series transformation is effected. The line of video

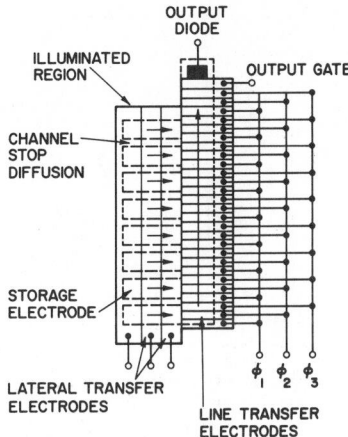

Fig. 28. Linear CCD imaging array with single readout register.

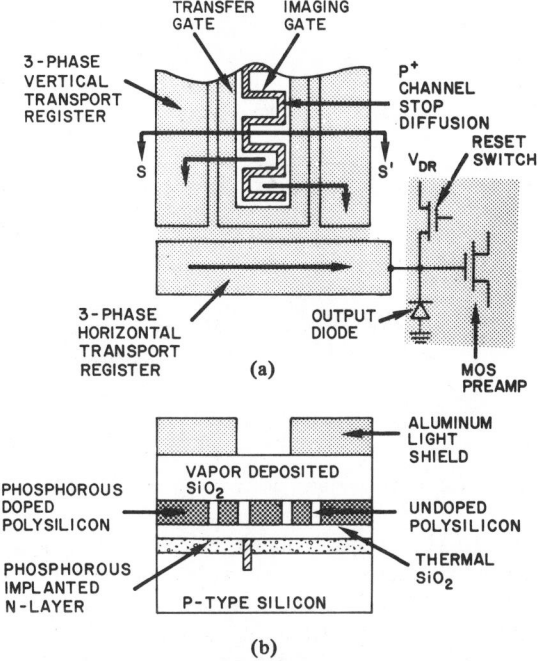

Fig. 29. Linear CCD imaging array with double readout register.

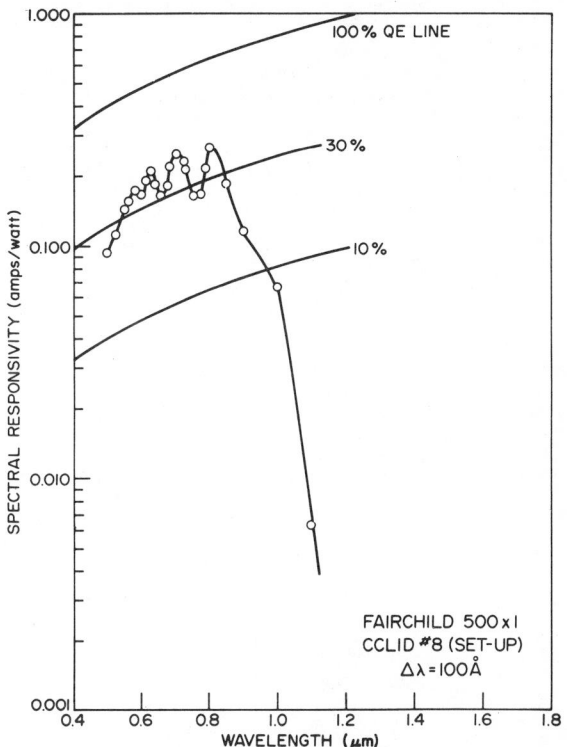

Fig. 30. Spectral responsivity versus wavelength for frontside-illuminated linear array.

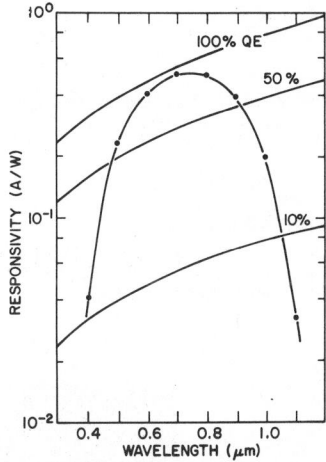

Fig. 31. Spectral responsivity versus wavelength for thinned backside-illuminated linear array.

is then shifted out via the shift register while a new line is being integrated. This design effectively eliminates the image smear problem. 3) The third design, shown in Fig. 29, uses a line of sensors and two shielded readout registers. After integration, odd-numbered pixels are shifted into one readout register and even-numbered pixels are shifted into the other readout register. The information in the two vertical registers is clocked into a two-bit horizontal register thus reforming the pixels in the order in which they were formed. The number of bits in each vertical register is half the total number of pixels. Thus, for two-phase vertical registers, the number of transfers required to clock out the pixel farthest from the output is equal to the number of pixels. The primary advantages of 3) are higher sensor packing density and fewer transfers to read out a given pixel.

Linear arrays can be excited through frontside illumination if polysilicon, which is nearly transparent, is used for the electrodes. Fig. 30 shows the spectral responsivity versus wavelength for a 500-element linear frontside-illuminated device [21]. The structure in the response is caused by inter-

ference effects at the layer boundaries shown in the cross section of Fig. 29. If the device is thinned in the sensor area to about half the center-to-center pixel spacing, and if the backside is accumulated to minimize recombination at the back surface, the array can be backside illuminated. In this case, losses due to multiple reflections at layer boundaries which occur in frontside-illuminated devices are eliminated; also, the structure in the response curve is eliminated. The spectral response curve for a thinned backside-illuminated device is shown in Fig. 31 [22].

Table III gives the parameters of linear imagers which have been fabricated. The primary application of long linear imaging chips is "reading" printed pages.

Fig. 32 shows the noise-equivalent power (NEP) versus power for a buried-channel frontside-illuminated 500-element

TABLE III
LINEAR IMAGING CHIP PARAMETERS

Manufacturer	Number of Resolution Elements	Cell length (mils)	Technology	Design
Bell Labs	1600	0.63	SC	S + SRR
Fairchild	1728	0.53	BC	S + DRR
RCA	500	1.2	SC	SR
TI	500	1.2	SC	SR

SR = shift registers
S + SRR = sensors plus single readout register
S + DRR = sensors plus double readout registers

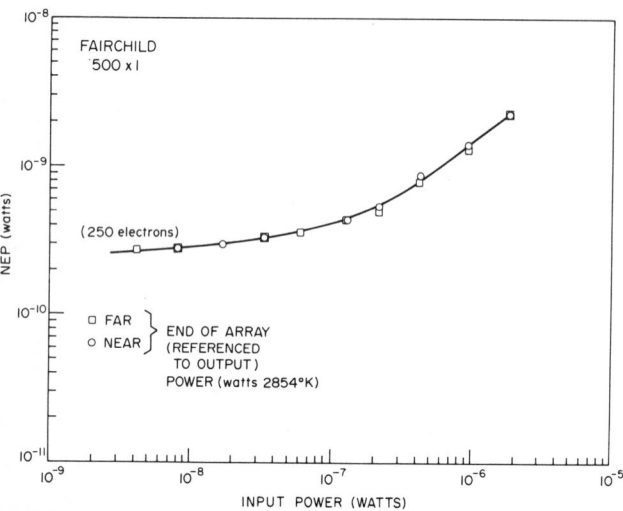

Fig. 32. Noise-equivalent power (NEP) versus input power for 500-element linear buried-channel array.

Fig. 33. Image of facsimile chart using a 1728-element linear CCD.

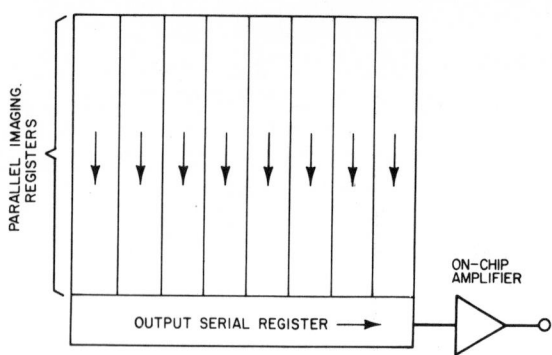

Fig. 34. Schematic diagram of a multicolumn time delay and integration (TDI) chip.

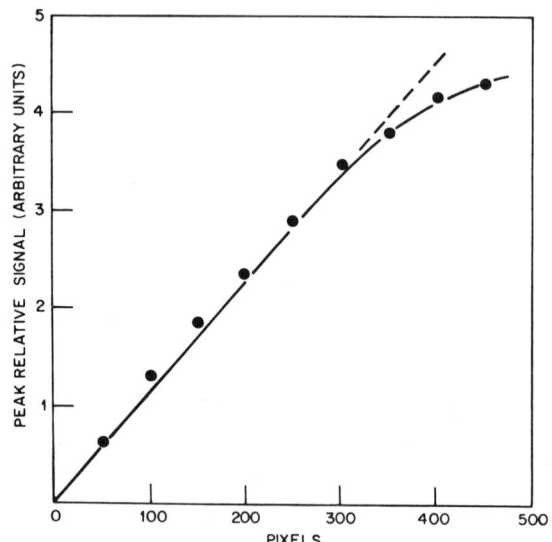

Fig. 35. Plot of peak relative signal versus number of TDI pixels.

array [21]. This array produced the first CCD imaging which was limited by temporal noise. At low light levels, the noise per pixel per frame is 250 electrons, due to a noise level of approximately 200 electrons in the reset amplifier and 150 electrons of shot noise in the dark current. Fig. 33 shows an example of imaging from a 1728-element high-resolution buried-channel array to be used in facsimile applications.

B. Time Delay and Integration (TDI) Arrays

In applications where the CCD imaging chip has a velocity relative to the object to be imaged, the CCD can be used in the time delay and integration (TDI) mode to enhance the signal-to-noise ratio. In such applications which lend themselves to this mode of operation, a CCD composed of N_x columns each containing N_y bits is oriented in such a direction and clocked at such a rate that the transfer of pixels down the CCD columns compensates for the movement of the image along the CCD columns due to the relative velocity of the chip and the object. The array organization for this mode of operation is shown in Fig. 34. Fig. 35 presents a plot of the relative signal achieved experimentally versus the number of pixels integrated [23]. After N_y bits of TDI, the signal is $N_y S_p$, where S_p is the signal representing a single bit without TDI. The noise accumulated during transfer, i.e., shot noise in the dark current and trapping noise, adds incoherently, and after N_y bits of TDI, the noise is $(N_y)^{1/2} N_s$. Thus the signal-to-noise ratio with TDI is greater than that without TDI by the factor $(N_y)^{1/2}$.

VIII. AREA ARRAYS

A. Tradeoffs

There are two basic designs for CCD low-light-level imaging chips—backside-illuminated frame transfer (BIFT) and frontside-illuminated interline transfer (FIIT).

To obtain the best possible performance from a CCD array, it is necessary to maximize the signal-to-noise ratio. The noise is basically determined by the technology used (surface channel and buried channel) and by the noise characteristics of the amplifier. On the other hand, the responsivity is largely determined by array design.

The purpose of this section is to provide a framework for comparing CCD arrays of different design and to compare the BIFT and FIIT designs in detail. To make the comparison, we assume that the same CCD technology and amplifier technology are used for both designs.

The general framework for comparison is formulated in terms of the overall system responsivity as a function of spatial frequency. The CCD chip design parameters that affect system responsivity are then used to calculate the system responsivity for the BIFT and FIIT designs. Finally, curves of system responsivity versus spatial frequency are plotted for comparison.

1) System Considerations

a) Chip responsivity at zero spatial frequency: The sensor chip responsivity at zero spatial frequency, R_{chip}, is defined as

$$R_{chip} = \frac{\text{peak signal current out of sensor chip}}{\text{irradiance incident on chip}} = \frac{I_s}{H} \quad (65)$$

where the units of H are W/m^2.

i) Narrow-band excitation: For narrow-band excitation between the wavelengths λ and $\lambda + d\lambda$, the signal current is

$I_s(\lambda) =$ (incident irradiance per unit wavelength)

    × (incremental width of the excitation band)

    × (active area of resolution element)

    × (integration time)

    × (energy per photon)$^{-1}$

    × (number of electrons collected per incident photon)

    × (charge per electron)

    × (output current per electron collected)

$$= H_\lambda \, d\lambda \, \Delta x \Delta y \, t_{integ} \, (hc/\lambda)^{-1} \, \eta(\lambda) \, e(g_m/C) \quad (66)$$

where

H_λ	irradiance per unit wavelength,
$\Delta x \Delta y$	active area of resolution element,
t_{integ}	integration time,
hc/λ	photon energy,
η	$\dfrac{\text{no. of electrons collected in a resolution element}}{\text{no. of photons incident on a resolution element}}$,
e	electronic charge,
g_m	transconductance of on-chip preamplifier,
C	effective capacitance at preamplifier input node.

ii) Wide-band excitation: For wide-band excitation, the signal current is obtained from (66) by integrating from λ_1 to λ_2:

$$I_s = \frac{g_m}{C} \Delta x \Delta y \, t_{integ} \int_{\lambda_1}^{\lambda_2} e H_\lambda \, \eta(\lambda) \, \frac{\lambda}{hc} \, d\lambda. \quad (67)$$

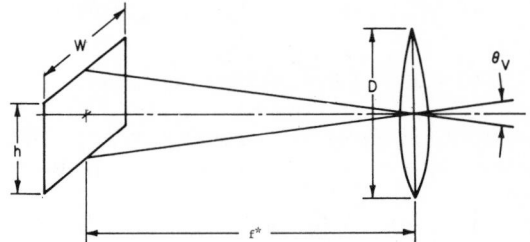

Fig. 36. Schematic diagram of imaging system.

The wide-band irradiance is

$$H = \int_{\lambda_1}^{\lambda_2} H_\lambda \, d\lambda. \quad (68)$$

Combining (65), (67), and (68) gives

$$R_{chip} = \frac{g_m}{C} \frac{\Delta x \Delta y \, t_{integ}}{\int_{\lambda_1}^{\lambda_2} e H_\lambda \, \eta(\lambda) \, \frac{\lambda}{hc} \, d\lambda}{H}. \quad (69)$$

The ratio

$$R_{element} = \frac{\int_{\lambda_1}^{\lambda_2} e H_\lambda \, \eta(\lambda) \, \frac{\lambda}{hc} \, d\lambda}{H} \quad (70)$$

is the responsivity of the sensor resolution element, i.e., the charge collected per joule of incident energy. Then the chip responsivity can be written as

$$R_{chip} = \frac{g_m}{C} \Delta x \Delta y \, t_{integ} \, R_{element}. \quad (71)$$

b) System responsivity at zero spatial frequency: In practical applications, a lens must be used with the sensor chip, and it is the performance of the overall system which is important. The lens–sensor system is shown in Fig. 36, in which the symbols used are defined as follows:

W	width of the chip,
h	height of the photosensitive part of the chip,
D	absolute lens aperture,
f^*	lens focal length,
θ_v	vertical field of view.

The focal ratio is defined as

$$F = f^*/D \quad (72)$$

and the relation between chip height, focal length, and field of view is

$$\tan(\theta_v/2) = \tfrac{1}{2} h/f^*. \quad (73)$$

The system responsivity at zero spatial frequency R_{system} is defined as

$$R_{system} = I_s/N \quad (74)$$

where N is the scene radiance in W/m$^2 \cdot$ sr.

The sensor irradiance and scene radiance are related by

$$H = NT\pi/4F^2 \quad (75)$$

where T is the lens transmission.

Combining (72)–(75) gives the system responsivity

$$R_{\text{system}} = \frac{\pi T}{4F^2} \frac{g_m}{C} \Delta x \Delta y \, t_{\text{integ}} R_{\text{element}}. \quad (76)$$

Equation (76) can be written more explicitly by considering the narrow-field-of-view case and the wide-field-of-view case separately.

i) Narrow-field-of-view case: For narrow fields of view, the limiting factor is the absolute lens aperture; thus the focal ratio used in (76) is set by the field of view and the maximum practical lens aperture. Combining (72), (73), and (76) gives

$$R_{\text{system}} = \pi T D^2 \tan^2 (\theta_v/2) \left(\frac{g_m}{C} \right)$$
$$\times \frac{\Delta x \Delta y}{h^2} t_{\text{integ}} R_{\text{element}}. \quad (77)$$

ii) Wide-field-of-view case: For wide fields of view, the focal ratio is the limiting factor; thus the focal ratio is chosen as the minimum value determined by practical lens design. The system responsivity for wide fields of view is given by (76).

c) System response at nonzero spatial frequencies: Since most of the information content in a scene is contained in high spatial frequencies, the system response should include a factor which takes into account the rolloff of system response with increasing spatial frequencies. This is accomplished by multiplying (76) by the modulation transfer function (MTF) of the sensor chip and the MTF of the lens:

$$R_{\text{system}}(f) = \frac{\pi T}{4F^2} \frac{g_m}{C} \Delta x \Delta y \, t_{\text{integ}} R_{\text{element}}$$
$$\times \text{MTF}_{\text{chip}} \text{MTF}_{\text{lens}}. \quad (78)$$

d) Basis for comparing sensor chips: The basic figure of merit of a sensor chip is the signal-to-noise ratio. In charge-coupled imagers, the noise can be attributed broadly to two sources—noise sources within the CCD proper, and noise due to the preamplifier. The output noise current is then

$$I_n = \left[\left(\frac{q_n g_m}{C} \right)^2 + I_{npa}^2 \right]^{1/2} \quad (79)$$

where q_n is rms noise charge per chip resolution element due to noise sources in front of the preamplifier, and I_{npa} is rms noise current due to preamplifier noise sources.

The basic figure of merit of a chip is the signal-to-noise ratio

$$\frac{I_s}{I_n} = \frac{\frac{g_m}{C} \Delta x \Delta y \, t_{\text{integ}} R_{\text{element}} \text{MTF}_{\text{chip}} H}{\left[\left(\frac{q_n g_m}{C} \right)^2 + I_{npa}^2 \right]^{1/2}}. \quad (80)$$

If noise sources in front of the preamplifier dominate, then the signal-to-noise ratio is independent of g_m/C. If the preamplifier dominates, then a large g_m/C ratio is desirable. Another common way of specifying noise performance is noise-equivalent irradiance (NEI), i.e., NEI is that value of H which gives $I_s/I_n = 1$.

2) Chip Considerations

In this subsection we will discuss chip design parameters in terms of their effect on the figure of merit.

a) Vertical interlace-integration time: Fig. 37 illustrates the operation of the backside-illuminated frame-transfer (BIFT)

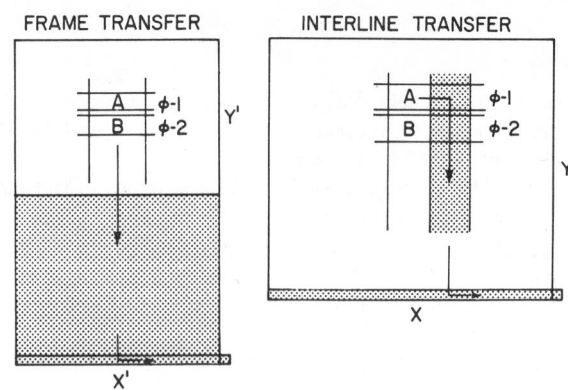

Fig. 37. Operation of frame-transfer and interline-transfer chips with 2:1 interlace in the vertical direction.

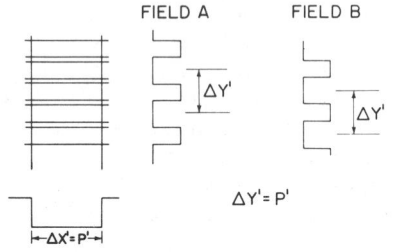

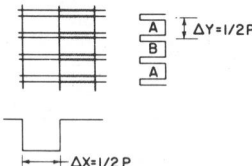

Fig. 38. Basic integration cells for backside-illuminated frame-transfer (BIFT) and frontside-illuminated interline-transfer (FIIT) chips.

structure and the frontside-illuminated interline-transfer (FIIT) structure. In both cases, two-phase structures are shown with 2:1 interlace in the vertical direction. In the frame-transfer structure, the top half of the chip is photosensitive. If a frame rate of 30 frames per second is assumed, then field A is formed by collecting photoelectrons under the ϕ-1 electrodes for $\frac{1}{60}$ s. This charge configuration is shifted into the shielded storage register in typically $\frac{1}{600}$ s. Field A is then read out a line at a time while field B is being formed by collecting photoelectrons under the ϕ-2 electrodes.

In the interline-transfer structure, the shielded vertical readout registers are interdigitated with the photosensitive lines. Potential wells are formed in the photosensitive regions by applying voltages to the vertical polysilicon stripes. The horizontal polysilicon stripes are used to clock the vertical shielded register. Because the integrating cells and shift-out cells are separate, the effective integration time for both fields A and B is $\frac{1}{30}$ s. The operation is as follows. After collecting photoelectrons in field A for $\frac{1}{30}$ s, the charge configuration is shifted into the shielded registers and down, a line at a time, into the horizontal output register. When field A has been completely read out ($\frac{1}{60}$ s), field B is shifted into the shielded registers and out. It is important to note that the effective integration time for the interline-transfer structure is twice that of the frame-transfer structure.

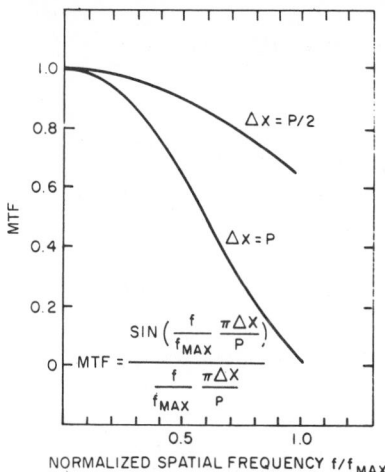

Fig. 39. Integration MTF versus normalized spatial frequency for $\Delta x = \frac{1}{2}$ and $\Delta x = p$.

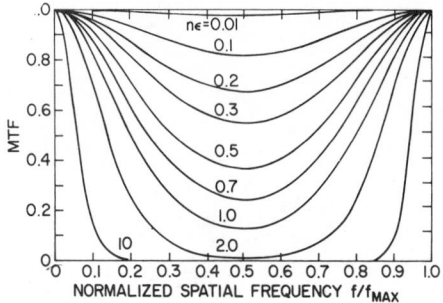

Fig. 40. Transfer MTF versus normalized spatial frequency.

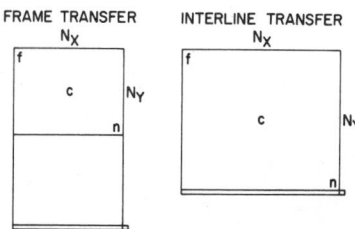

Fig. 41. Frame-transfer and interline-transfer chips, with points labeled to indicate all positions relative to output diode.

Using 2:1 interlace in both structures effectively doubles the vertical spatial sampling frequency, and the resulting Nyquist limit is $1/p$, where p is the center-to-center distance between adjacent resolution elements in a given field on the chip (pitch), as shown in Fig. 38.

b) Modulation transfer function (MTF): The MTF describes the rolloff of imager response with increasing spatial frequency. The overall MTF of the chip is composed of three factors: 1) the loss of frequency response due to the geometry of the integrating cell (MTF_{integ}), 2) the loss of frequency response due to transfer inefficiency ($MTF_{transfer}$), and 3) the loss of frequency response due to the diffusion of charge between photon absorption and photoelectron collection (MTF_{diff}). Each of these factors will be discussed separately.

i) Integration MTF: The integration MTF is given by the Fourier transform of the basic integration cell. For a rectangular cell of length Δx repeated with periodicity p, the MTF is

$$MTF_{integ} = \frac{\sin\left(\frac{f}{f_{max}}\frac{\pi\Delta x}{p}\right)}{\frac{f}{f_{max}}\frac{\pi\Delta x}{p}} \qquad (81)$$

where $f_{max} = 1/p$ for 2:1 interlace. For $\Delta x = p$, the first zero in the MTF occurs at $f = f_{max} = 1/p$. For $\Delta x = p/2$, the first zero occurs at $f_{max} = 2/p$. Fig. 39 gives the integration MTF versus normalized spatial frequency. As shown in Fig. 38, $\Delta x = \Delta y = p$ for the BIFT chip and $\Delta x = \Delta y = p/2$ for the FIIT chip.

ii) Transfer MTF: During the transfer of a sampled sinusoid along a CCD shift register, a fraction of the charge ϵ is lost from each of the samples at each transfer, and this charge is added to trailing samples. The effect of this dispersion effect on MTF is given by

$$MTF_{transfer} = \exp\left\{-n\epsilon\left[1 - \cos\left(\pi\frac{f}{f_{max}}\right)\right]\right\}. \qquad (82)$$

Fig. 40 shows the transfer MTF versus normalized spatial frequency with the $n\epsilon$ product as the parameter. The symmetry of the curves about $f/f_{max} = 0.5$ can be explained as follows. In a 2:1 interlaced array, the Nyquist frequency is

$1/p$. Also, the sampling in any one field occurs at spatial frequency $1/p$. If the signal to be sampled is at spatial frequency f, then the frequency carried in the field is f for $f \leqslant \frac{1}{2}p$. However, if $\frac{1}{2}p \leqslant f \leqslant 1/p$, the frequency carried in the field [24] is $(1/p) - f$. Thus the maximum frequency carried in a field is $\frac{1}{2}p$, and the minimum frequency carried in a field occurs when $f = 0$ and $f = 1/p$.

Fig. 41 shows the frame-transfer and interline-transfer chips. N_x and N_y are the number of geometrical resolution cells in the x and y directions. Points are marked to denote cells—the farthest from the output diode (f), the center of the photosensitive array (c), and the nearest to the output diode (n).

Table IV gives the number of transfers required to read out the f, c, and n cells and the rates at which the charge is transferred. The number of transfers in the x direction is the same for both frame-transfer and interline-transfer arrays. Therefore, the horizontal MTF degradation due to transfer would be the same for both arrays. The number of transfers in the y direction is greater for the frame-transfer chip by the amount PN_y, where P is the number of phases. Therefore, the vertical MTF for the frame-transfer chip is less than that for the interline-transfer chip as given by (82).

iii) Diffusion MTF: If photons are absorbed within the depletion regions, then we assume that the collection is 100 percent efficient. However, if photons are absorbed away from the depletion regions, then the charge configuration will spread as it diffuses toward the depletion regions with a resulting decrease in MTF.

If photons are absorbed a distance d from the depletion regions and if the diffusion length in the silicon is L_0, then the MTF due to the diffusion of charge is given by [25], [26]

$$MTF_{diff} = \frac{\cosh(d/L_0)}{\cosh(d/L)} \qquad (83)$$

where $L^{-2} = L_0^{-2} + (2\pi f)^2$.

Fig. 42 shows the diffusion MTF versus normalized spatial frequency with d as the parameter.

c) Photoelement responsivity ($R_{element}$): The photoelement responsivity is determined by the efficiency with which

TABLE IV
NUMBER OF TRANSFERS FOR FRAME-TRANSFER AND INTERLINE-TRANSFER AREA ARRAYS

	FRAME			INTERLINE		
	f	c	n	f	c	n
AT VIDEO RATE	PN_x	$1/2PN_x$	—	PN_x	$1/2PN_x$	—
AT LINE RATE	PN_y	$1/2PN_y$	—	PN_y	$1/2PN_y$	—
AT INTERMEDIATE RATE	PN_y	PN_y	PN_y	—	—	—
TOTAL	$P(N_x + 2N_y)$	$1/2P(N_x + 3N_y)$	PN_y	$P(N_x + N_y)$	$1/2P(N_x + N_y)$	—

P = NUMBER OF PHASES

$(N_x)_{FT} = (N_x)_{IT}$

$(N_y)_{FT} = (N_y)_{IT} + PN_y$

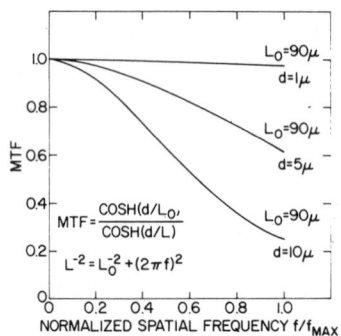

Fig. 42. Diffusion MTF versus normalized spatial frequency ($p = 1.2$ mils is assumed).

TABLE V
ASSUMPTIONS USED IN TRADEOFF EXAMPLE

PARAMETER	INTERLINE TRANSFER	FRAME TRANSFER
TOTAL CHIP AREA	$XY = A$	$X'Y' = A$
ASPECT RATIO	r	r
VERTICAL FIELD OF VIEW	θ_v	θ_v
SAMPLING DISTANCE IN OBJECT PLANE	G_{SD}	G_{SD}
RANGE	R	R
NUMBER OF SENSOR CELLS	N_x, N_y	N_x, N_y
FOCAL RATIO	F	F

photons are absorbed and the resulting photoelectrons are collected. Basically four mechanisms act to reduce $R_{element}$: 1) reflection at layer interfaces before the photons reach the silicon, 2) absorption in these layers before the photons reach the silicon, 3) recombination at the Si–SiO$_2$ interface after hole–electron generation, and 4) absorption too far away from potential wells for the photoelectrons to be collected. Mechanisms 2) and 3) cause a reduction in $R_{element}$ in the blue, mechanism 4) causes a reduction in $R_{element}$ in the infrared, and mechanism 1) causes interference fringes throughout the spectrum. Mechanism 1) is mainly responsible for $R_{element}$ being different for BIFT and FIIT structures. Computer analysis [27] indicates that if the layer thicknesses are chosen properly, 50 percent of the incident photons in the 0.4–1.0 μm band are transmitted into the silicon for the FIIT structure. Therefore, $R_{element}$, averaged over the 0.4–1.0-μm band, for the FIIT structure is one-half that for the BIFT structure.

3) Example

Suppose that an area array for a wide-field-of-view low-light-level application is to be designed with the following constraints. 1) Square elements are assumed, i.e., $p_x = p_y = p$; 2) the distance from the object plane to the image plane, R, is specified; 3) the vertical field of view θ_v is specified; 4) the sampling distance in the object plane, G_{SD}, is to be minimized; 5) the aspect ratio is specified; and 6) the minimum practical focal ratio is to be used. The problem is to determine the relative performance of the BIFT and the FIIT chips designed for these requirements. We assume that 7) the largest practical chip for either design has total area A, and that 8) the number of elements for either chip is N_xN_y. Table V summarizes the assumptions. The image format for the BIFT chips is X′ by Y′/2 and the image format for the FIIT chip is X by Y. From 7) and 5), it follows that $Y' = 1.414Y$ and $X' = 0.707X$. From

(73), it follows that $f^{*'} = 0.707f^*$. From the magnification relation

$$\frac{P}{G_{SD}} = \frac{f^*}{R} \qquad (84)$$

it follows that $p' = 0.707p$. From (72) and constraint 6), it follows that $D = 1.414D'$.

To compare the performance of the BIFT and FIIT chips, we assume equal-noise performance. Therefore, the figure of merit reduces to the system response given by (78). We also assume equal-lens MTF's. Table VI summarizes the values used in (78) and (81). The results of the comparison are given in Fig. 43. At zero spatial frequency, the response of the BIFT chip is twice that of the FIIT chip. However, the response of the BIFT chip rolls off sharply and crosses the response of the FIIT chip at approximately $0.6/G_{SD}$.

4) Alternate Formulation of R_{system}

An alternate definition of system responsivity is [21]

$$R_{system} = \frac{I_{avg}}{N} = \frac{\pi S A^* T}{4F^2} \qquad (85)$$

where S is time average current out of the CCD per watt input, and

$A^* =$ total image format area of the chip
$= XY$ for FIIT
$= X'Y'/2$ for BIFT.

If a CCD array is uniformly illuminated so that every charge packet contains n electrons, the average current out of the de-

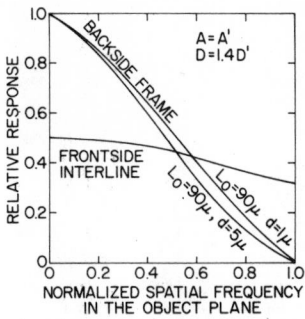

Fig. 43. Relative systems response versus normalized spatial frequency for strongly absorbed light.

TABLE VI
CHIP PARAMETERS USED IN TRADEOFF EXAMPLE

PARAMETER	FIIT	BIFT
Δx	0.5 p	$p' = 0.707\ p$
Δy	0.5 p	$p' = 0.707\ p$
RELATIVE t_{integ}	2	1
RELATIVE $R_{element}$	1	2

TABLE VII
SYSTEM RESPONSIVITY TRADEOFFS

	η_A	η_t	RELATIVE $R_{element}$	S(ma/W)	RELATIVE FORMAT AREA	RELATIVE R_{system}
FIIT	1/4	2	1/2	23	2	1
BIFT	1	1	1	90	1	2

ASSUMPTIONS: A = A', F = F', T = T'

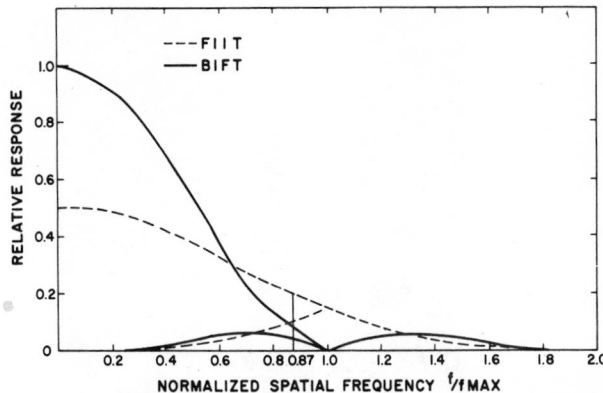

Fig. 44. Relative response versus spatial frequency in the vertical direction for BIFT and FIIT designs showing aliased spectra (2:1 interlace is assumed).

vice (no amplification or signal processing) is

$$I_{avg} = \frac{N_x N_y n e}{t_{field}} \qquad (86)$$

where t_{field} is the field time. The peak current out of the device with on-chip preamplifier is

$$I_p = \frac{n e g_m}{C} \qquad (87)$$

where g_m is the transconductance of the on-chip preamplifier, C is the capacitance on the output node of the CCD, and $N_x N_y$ is the number of elements read out in a field time. Then it follows that

$$I_{avg} = \frac{N_x N_y}{t_{field}} \frac{C}{g_m} I_p. \qquad (88)$$

Combining (85), (88), and (76) gives

$$S = \left(\frac{N_x N_y \Delta x \Delta y}{A^*} \right) \left(\frac{t_{integ}}{t_{field}} \right) R_{element} = \eta_A \eta_t R_{element}. \qquad (89)$$

Thus S is the product of the area utilization factor η_A, the time utilization factor η_t, and the element responsivity $R_{element}$.

Table VII summarizes the FIIT versus BIFT tradeoffs in terms of the "average-current" formulation. Again it is con-

cluded that the zero spatial-frequency response of the BIFT chip is twice that of the FIIT chip of equal total size.

5) Aliasing versus MTF

Fig. 44 shows the relative response curves using only the integration MTF curves. The curves are reflected about the Nyquist frequency f_{max} to show the amplitude of the aliased spectra. Using the criterion that aliasing occurs when the reflected branch equals half of the fundamental branch, it can be shown that for both the FIIT and the BIFT chips aliasing in the vertical direction occurs at frequencies $> 0.87\,f_{max}$. Thus both designs are equally susceptible to aliasing in the vertical direction. In the horizontal direction, the MTF is quite high at f_{max} for the FIIT design, and such chips are highly susceptible to aliasing. However, in cases where aliasing is an important consideration, two FIIT chips can be boresighted with a $\frac{1}{2} P$ offset. This doubles the horizontal sampling frequency and effectively eliminates aliasing in the horizontal direction.

Fig. 45 illustrates the tradeoff between aliasing and MTF [28]. For the conditions shown in Fig. 45(a)–(c), the horizontal and vertical MTF's [MTF$_H$ and MTF$_V$] at the respective Nyquist frequencies were [50 and 25 percent] for Fig. 45(a), [35 and 15 percent] for Fig. 45(b), and [5 and 0 percent] for Fig. 45(c). These conditions were achieved by using an FIIT array with high MTF [Fig. 45(a)] and defocusing the lens to simulate the prefiltered conditions [Fig. 45(b) and (c)].

In Fig. 45(a), there is considerable high-frequency detail; however, moiré can be seen in the flag. It should be noted that the moiré in the flag has no effect on the remainder of the image, i.e., the effects of aliasing are limited to the extent of the object causing the aliasing. In Fig. 45(c) much of the high-frequency detail is lost; however, the moiré is also not present. These pictures illustrate the nature of the tradeoff between MTF and aliasing. From this simulation, it is concluded that the effects of aliasing are not as severe as might be predicted from frequency-domain analysis. Since Fig. 45(a) corresponds *roughly* to the FIIT design and Fig. 45(c) corresponds *roughly* to the BIFT design, the results of the simulation provide some qualitative insight into the tradeoffs between FIIT and BIFT designs.

B. The Charge-Injection Device (CID)

The charge-injection device (CID) is an MOS structure in which surface-potential wells are created by means of external voltages applied to an array of gate electrodes [29]. When photons are absorbed near the potential wells, minority carriers

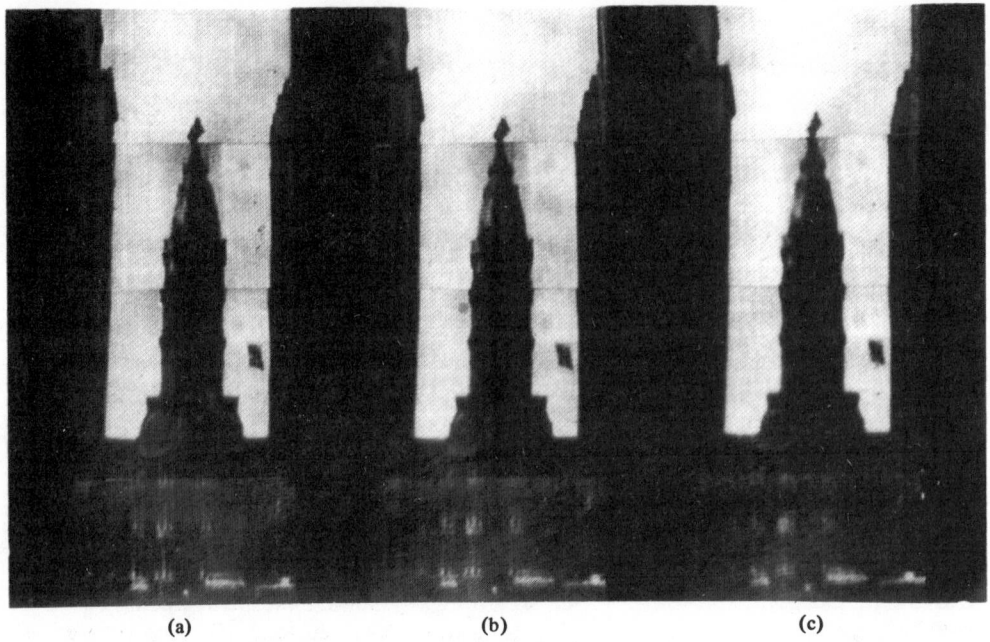

Fig. 45. Photographs showing CCD imagery. (a) Noncontiguous horizontal integration cells. (b) Simulated contiguous horizontal integration cells. (c) Simulated overlapping horizontal integration cells.

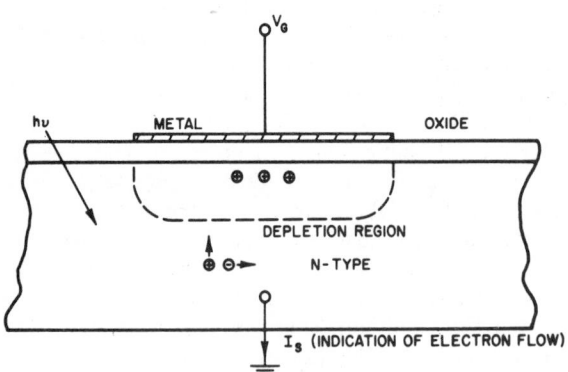

Fig. 46. Integration cell for charge-injection device (CID) showing the collection of photon-generated holes in the potential well.

Fig. 47. Injection of the collected charge from the integration cell into the substrate by collapsing the potential well.

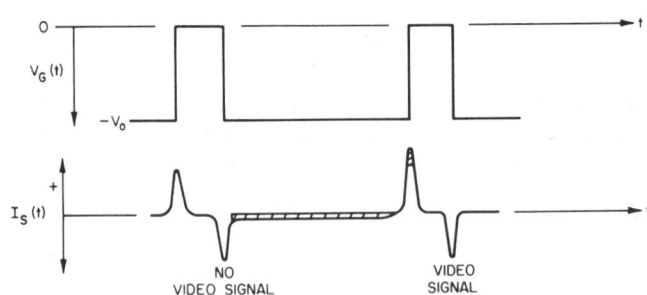

Fig. 48. CID output-current waveforms during integration and injection. Negative shaded area = positive shaded area = video signal.

are collected in the wells, as shown in Fig. 46. The charge configuration in the potential wells is a point-by-point sampling of the light falling on the image plane. Thus the integration process in a CID is the same as that in a CCD.

For video signal readout, the charge stored in a potential well is injected into the bulk at the end of the integration time by removing the gate voltage, as shown in Fig. 47. These minority carriers recombine at the substrate contact thereby causing a current flow in the external circuit. The details of the current flow during integration and readout are described in the following paragraphs.

As a drive voltage pulse is applied to the gate, a depletion region is formed, creating majority carriers, electrons in this case, flowing through the substrate. Because of capacitance coupling both positive and negative pulses of substrate current will result in the waveform, shown in Fig. 48. During integration, photon-generated minority carriers, holes in this case, will be collected at the surface in the potential well and stored as shown in Fig. 46, while the photon-generated electrons will continuously flow through the substrate, as shown by the negative shaded area of the substrate current curve in Fig. 48. At the end of the frame time, the injection of the stored charges will increase the positive pulse height, as shown in the second

pulses of Fig. 48, which constitutes the desired video information. The integration of this current waveform results in a net signal proportional to the injected video charge. The net voltage is sampled to provide the video output signal.

For two-dimensional CID arrays, the structure requires two separate metal electrodes per unit cell, in order to scan the array elements in the x–y direction. The two-dimensional operation of this device is illustrated in Fig. 49. Each sensor element consists of two separate electrodes for two storage capacitors.

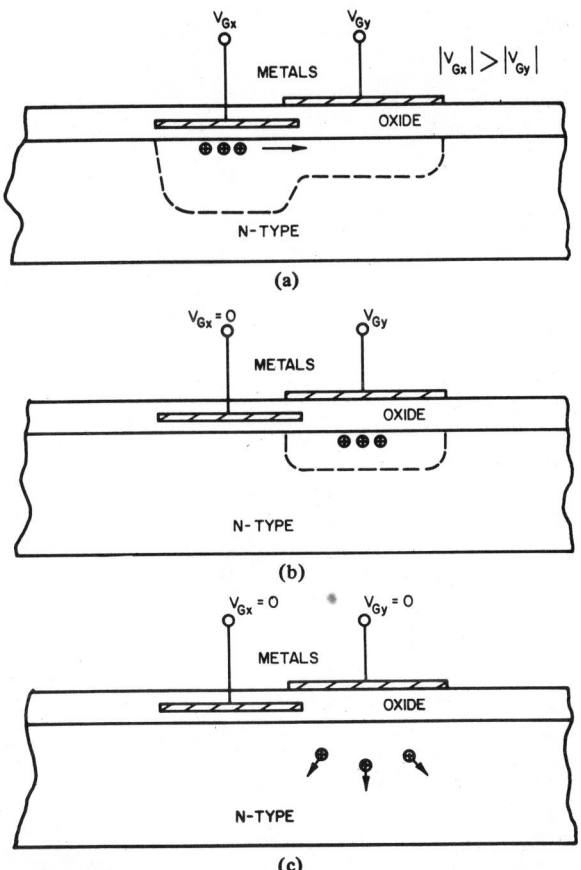

Fig. 49. Basic cell of a CID area array. (a) Charge storage. (b) Charge transfer. (c) Charge injection.

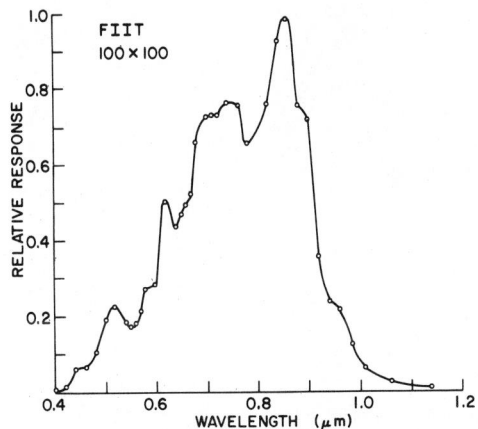

Fig. 50. Relative response versus wavelength for an FIIT area array.

TABLE VIII
AREA IMAGING CHIP PARAMETERS

Manufacturer	Number of Resolution Elements*	Image Format (mils)^2	Technology	Design	Amplifier
Bell Labs	256 X 220	200 X 260	SC	FT	Reset
Fairchild	244 X 190	170 X 220	BC	IT	DFGA and DDA
GE	244 X 188	340 X 450	SC	CID	OFF CHIP
RCA	512 X 320	310 X 380	SC	FT	Reset
TI	150 X 100	90 X 120	SC	FT	Reset

*Includes 2:1 interlace in the vertical direction
SC = surface channel   IT = interline transfer
BC = buried channel   DFGA = distributed floating gate amplifier
FT = frame transfer   DDA = dual differential amplifier

One of the electrodes is connected to a horizontal access line, the other to a vertical line.

Fig. 49(a) shows the normal charge storage in the potential well; as the one electrode in the particular cell shown is turned off, the stored charges under that electrode are transferred to the other potential well, as shown in Fig. 49(b). The injection of the stored charge in the particular sensor cell will occur only when both electrode voltages are switched off, as demonstrated in Fig. 49(c); therefore, a sensor element is sampled only when both the electrode voltages are removed from the sampling cell.

The CID is simpler to fabricate than the CCD. The CID is x–y addressed and therefore lends itself to random access. The CCD readout is serial. The output capacitance of a CID is the capacitance of one row and one column if all other rows and columns are floated during the readout of a given cell. The output capacitance of a CCD is the capacitance of a reverse-biased diode. Therefore, the process of reading out a CID is noisier than that of reading out a CCD.

C. Further Tradeoffs and State of the Art

There are other comparisons that can be made between BIFT, FIIT, and CID imaging arrays. A thinned and backside-passivated frame-transfer chip can be operated in the electron-bombarded induced-conductivity (EBIC) mode. This provides a preintegration gain; however, high voltages (in the kilovolt range) are required. Also, it is not clear that a CCD can withstand bakeout in vacuum and long-term operation in vacuum without degradation. The interline-transfer design is incompatible with backside-illumination or EBIC mode operation. It

is not clear whether the CID will be efficient in the EBIC mode of operation due to the conflicting constraints that high recombination is required at the backside of the chip for efficient CID readout whereas low recombination is required for efficient collection of carriers generated by photoelectron absorption. The non-EBIC approach to low-light-level imaging is cooled buried-channel photon-excited arrays with low-noise amplifiers.

Another factor in comparing area arrays is the spectral response. The response of a backside photon-illuminated linear device is shown in Fig. 31. The response of a frontside-illuminated interline-transfer area array is shown in Fig. 50. The structure is caused by interference effects in the layered frontside structure. The shape of the response can be altered in the frontside-illuminated case by choice of layer thicknesses during fabrication.

Most of the area arrays that have been fabricated are front-side illuminated, although Texas Instruments has imaged with a backside-illuminated 64 by 64 array. Table VIII summarizes the area arrays that have been fabricated.

D. Performance Predictions

In the noise-limited region, the noise-limited resolution R is proportional to the elemental signal-to-noise ratio. Using the "average-current" formulation, the signal is $CHSA\, t_{integ}/e$, where C is the contrast. The photon noise is $(HSA\, t_{integ}/e)^{1/2}$ and the total CCD noise is N. Then [21]

$$R = K \frac{CHSA\, t_{integ}/e}{(HSA\, t_{integ}/e + N^2)^{1/2}}. \tag{90}$$

It has been shown empirically with low-light-level imaging tubes that patterns having bar lengths extending over one-third of the picture height can be resolved at mean photoelec-

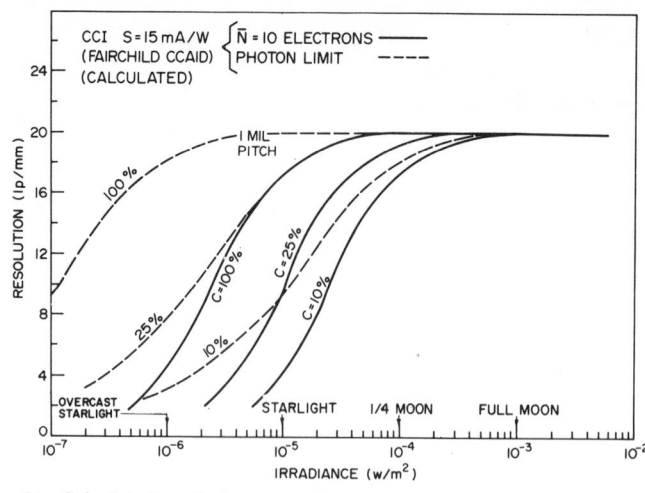

Fig. 51. Calculated resolution versus irradiance for FITT array based on (91).

T/I.4; 25mm DIAGONAL SENSOR; 2854 °K RESPONSE $\begin{cases} \text{4 MA/W, I-SIT} \\ \text{20 MA/W, CCI} \end{cases}$

INTENSIFIED SIT CAMERA

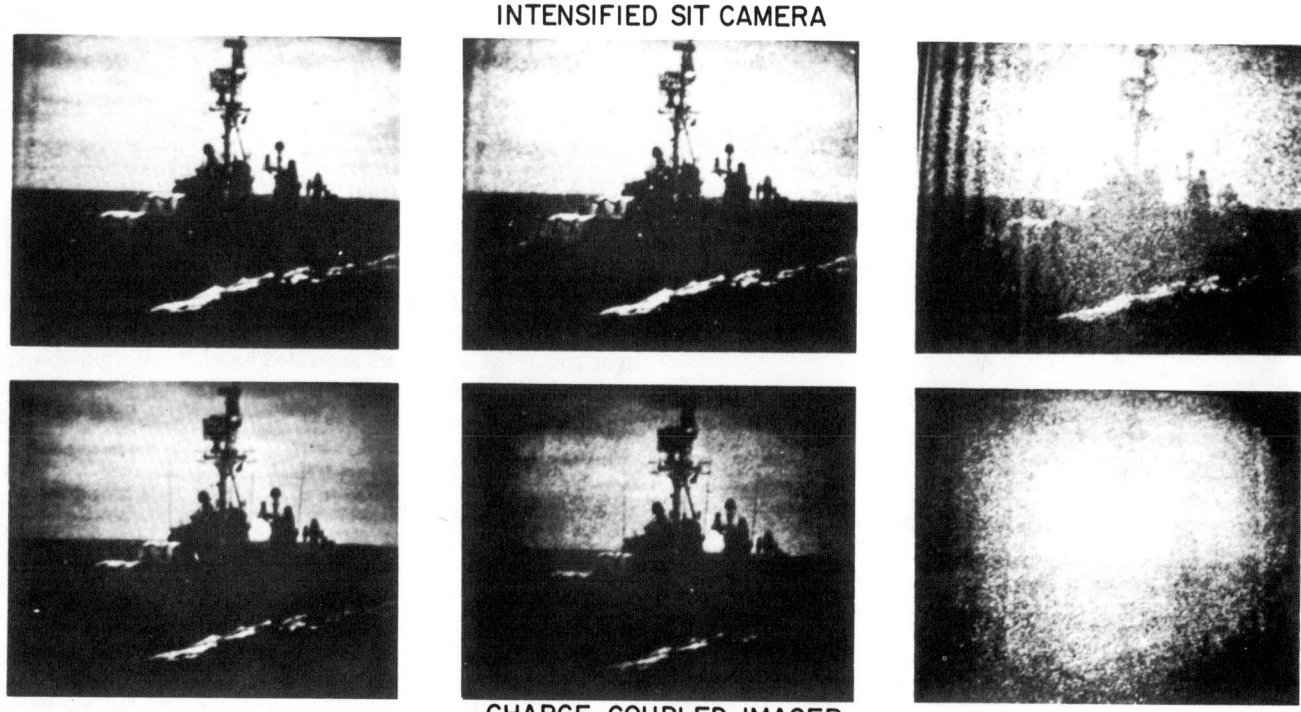

CHARGE-COUPLED IMAGER

Fig. 52. Actual I-SIT imagery versus simulated CCD imagery (with noise-equivalent charge per pixel per frame = 10 electrons) at quarter-moon, no moon (starlight), and overcast starlight conditions.

tron rates lower than 30 per second in a square resolution cell defined by the bar width. Using this fact, together with the condition that $R = R_{geom}$ when $N = 0$ and $C = 1$, gives $K = R_{geom}$. Thus (90) becomes

$$R = R_{geom} \frac{CHSA\, t_{integ}/e}{(HSA\, t_{integ}/e + N^2)^{1/2}} \qquad (91)$$

where R_{geom} is the maximum resolution determined by the geometry of the device. This equation can be used to predict resolution versus irradiance curves for CCD's. Using the values

$$S = 15 \text{ mA/W},$$
$$t_{integ} = 1/30 \text{ s},$$

$$A = 6.45 \cdot 10^{-10} \text{ m}^2,$$
$$R_{geom} = 20 \text{ line pairs (lp)/mm}$$

gives the curves shown in Fig. 51 for contrasts of 100, 25, and 10 percent. The solid curves correspond to a total CCD noise of 10 electrons per pixel per frame and the dashed curves correspond to photoelectron noise-limited performance, i.e., $N = 0$. This model is verified qualitatively by the CCD simulation shown in Fig. 52. In the simulation, the signal-to-noise ratio of an I-SIT camera tube was controlled by adjusting the input irradiance and the gain to simulate a CCD having $S = 20$ mA/W and $N = 10$. The upper three pictures of Fig. 52 show actual I-SIT imagery and the lower three pictures show the CCD simulation.

IX. RADIATION EFFECTS

The effects of space radiation on CCD's are important because the CCD is ideal in terms of size, weight, and power for space applications.

The CCD structures that have been evaluated include: 1) three-phase surface-channel structures with planar aluminum electrodes [30]; 2) two-phase stepped-oxide surface-channel structures with polysilicon and aluminum electrodes [31]; and 3) three-phase buried-channel structures with planar doped polysilicon electrodes and undoped polysilicon interelectrode isolation [32]. Device 1) is a 500-element shift register, 2) is 64-element shift register, and 3) is composed of 60 photosensitive cells and 2 readout registers. Each structure was evaluated by irradiation with cobalt 60 gamma radiation and observed for the effects on charge-transfer inefficiency, dark current, and charge-handling capacity.

The 500-element three-phase surface-channel structures with planar aluminum electrodes were irradiated while operating as shift registers at a 1-MHz clock rate with no intentionally introduced fat zero. The p-type substrate was held at -7 V and the clock-voltage swing was from -5 V to $+5$ V. The flat-band voltage shift versus dose is shown in Fig. 53. At 10^3 rad, the shift in flat-band voltage was sufficient to cause devices operated with an input-gate voltage set for 20-percent preirradiation fat zero to increase the bias charge to 100 percent of full-well capacity. The transfer inefficiency versus radiation dose is shown in Fig. 54. Roughly an order-of-magnitude increase in the transfer inefficiency was observed at 3×10^4 rad with no fat zero and also with 20-percent fat zero. Increasing the fat zero from 20 to 40 percent at 3×10^4 rad produced only a small decrease in the inefficiency (2.8×10^{-2} to 2.3×10^{-2}). The increased transfer inefficiency is attributed to increased interface trapping. The CV curves of the electrode-substrate capacitance show severe distortion in addition to the flat-band voltage shift.

The fast interface-state density, as calculated from the periodic-pulse technique, and the measured dark-current density are shown as a function of dose in Fig. 55. Note the increase in the fast interface-state density from the preirradiation value of $\sim 10^{10}$ cm^{-2} (eV)$^{-1}$ to $\sim 10^{11}$ cm^{-2} (eV)$^{-1}$ at 3×10^4 rad. An order-of-magnitude increase was also observed for the dark current (14 nA/cm$^2 \rightarrow 150$ nA/cm^2). The gain of the on-chip output amplifier began to degrade for total doses greater than 10^4 rad. At 10^5 rad, the output signal for a full-well charge packet was <15 percent of its preirradiation value.

The 64-element, surface-channel, stepped-oxide structures were operated as shift registers with a 1-MHz clock rate and a fat zero (bias charge) of 20 percent of the full-well capacity. In these devices, for total dose up to 10^5 rad (Si), the primary cause of changes in the CCD operation was a negative shift in the flat-band voltage, with a larger shift for the polysilicon gates than for the aluminum gates, as shown in Fig. 56. With fixed applied voltages, the flat-band shifts cause 1) an increase in the transfer inefficiency shown in Fig. 57 (from a preirradiation value of 10^{-4} to a value of 4×10^{-3} after 10^5 rad) due to the cut-off of the fat-zero injection and 2) a decrease in the full-well capacity (down to 20 percent of the preirradiation value after 3×10^5 rad) due to the modification of the surface-potential profile shown in Fig. 58. By adjusting the operating voltages to compensate for the flat-band shifts, preirradiation transfer inefficiency and most of the preirradiation full-well capacity were recovered for total doses up to 10^5 rad. Improved schemes for fat-zero injection have been devised to

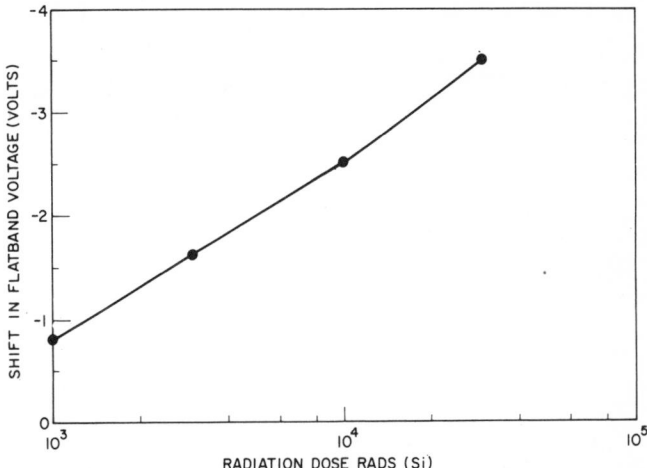

Fig. 53. Shift in flat-band voltage versus radiation dose for 500 × 1 aluminum-electrode n-channel planar-oxide surface-channel CCD shift register.

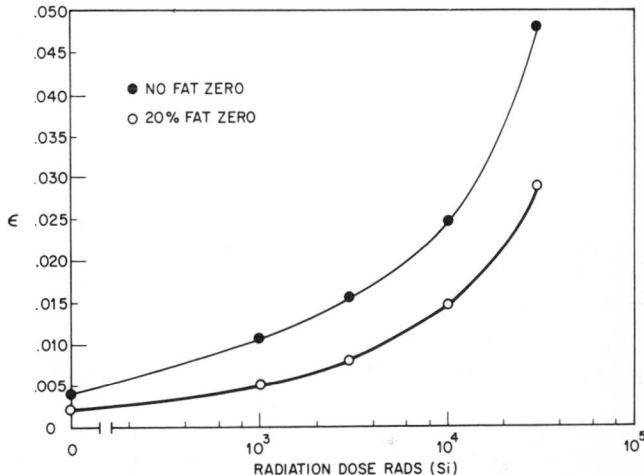

Fig. 54. Transfer inefficiency versus radiation dose for three-phase 500 × 1 aluminum-electrode n-channel planar-oxide surface-channel CCD shift register.

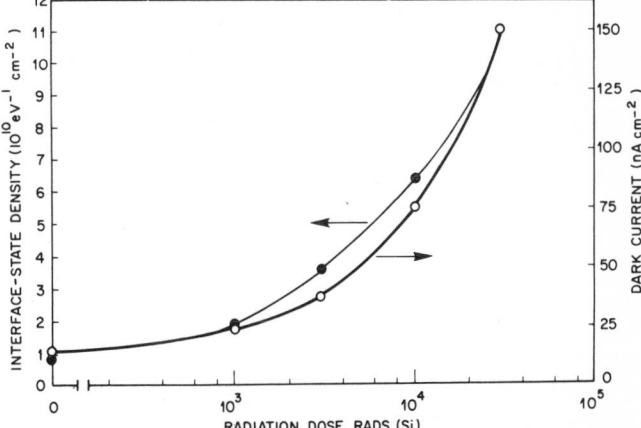

Fig. 55. Interface-state density (determined from periodic-pulse measurements) and dark current versus radiation dose for aluminum-electrode n-channel planar-oxide surface-channel CCD shift register.

eliminate the problem of fat-zero cutoff due to flat-band shifts. At a total dose of 3×10^5 rad, however, the formation of interface states (measured by the periodic-pulse method) results in an increase in the transfer inefficiency and an increase

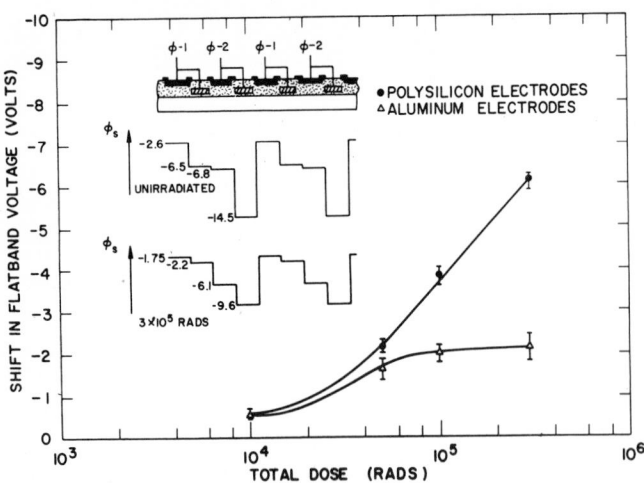

Fig. 56. Shift in flat-band voltage versus radiation dose for two-phase 64 × 1 stepped-oxide p-channel surface-channel CCD shift register.

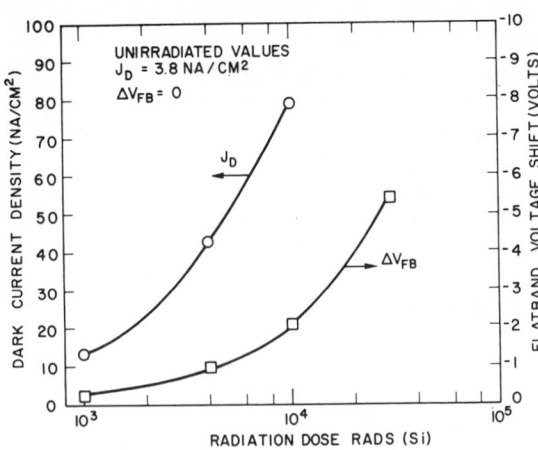

Fig. 59. Flat-band voltage shift and dark-current density versus radiation dose for three-phase 60-element buried-channel imager with doped polysilicon electrodes and undoped polysilicon interelectrode isolation.

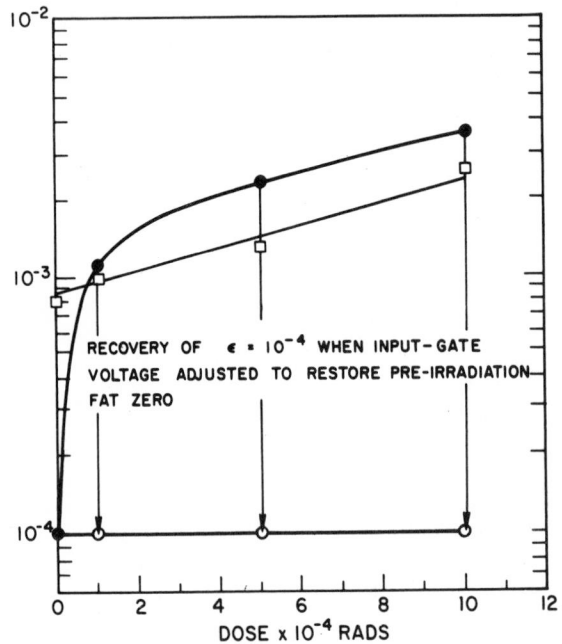

Fig. 57. Transfer inefficiency versus radiation dose for stepped-oxide p-channel surface-channel CCD shift register.

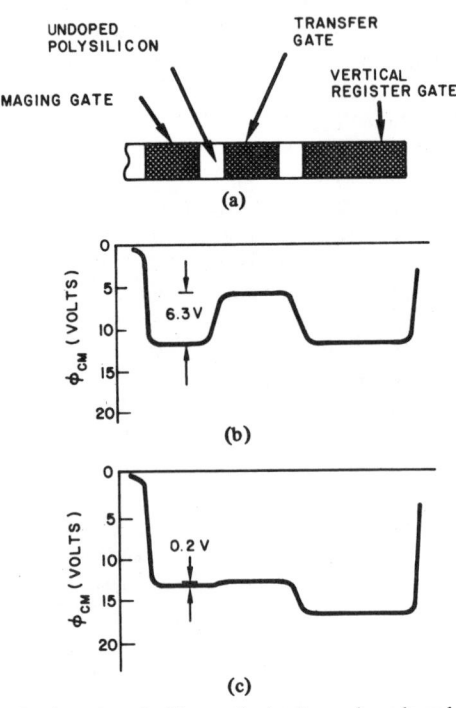

Fig. 60. CCD doped polysilicon electrode and undoped polysilicon interelectrode CCD structure and potential profiles. (a) Cross-sectional view of gate structure. (b) Preirradiation profile of minimum channel potential. (c) Profile of minimum channel potential after 3 × 10⁴ rad.

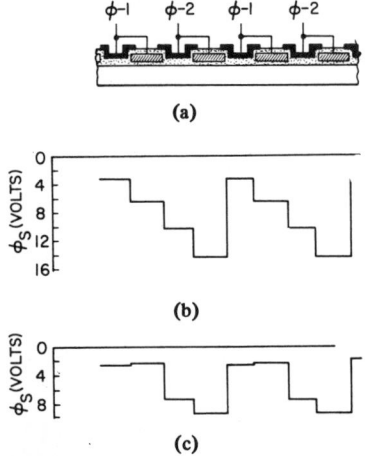

Fig. 58. Stepped-oxide structure and surface-potential profiles. (a) Cross-sectional view of the structure. (b) Preirradiation surface-potential profile. (c) Surface-potential profile after 3 × 10⁵ rad.

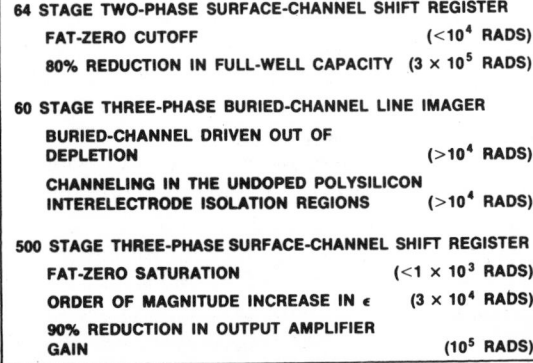

TABLE IX
FAILURE MODES OF CCD'S IN RADIATION ENVIRONMENT

64 STAGE TWO-PHASE SURFACE-CHANNEL SHIFT REGISTER	
FAT-ZERO CUTOFF	(<10^4 RADS)
80% REDUCTION IN FULL-WELL CAPACITY	(3 × 10^5 RADS)
60 STAGE THREE-PHASE BURIED-CHANNEL LINE IMAGER	
BURIED-CHANNEL DRIVEN OUT OF DEPLETION	(>10^4 RADS)
CHANNELING IN THE UNDOPED POLYSILICON INTERELECTRODE ISOLATION REGIONS	(>10^4 RADS)
500 STAGE THREE-PHASE SURFACE-CHANNEL SHIFT REGISTER	
FAT-ZERO SATURATION	(<1 × 10^3 RADS)
ORDER OF MAGNITUDE INCREASE IN ε	(3 × 10^4 RADS)
90% REDUCTION IN OUTPUT AMPLIFIER GAIN	(10^5 RADS)

in the dark current which cannot be compensated for by adjusting the operating voltages.

The 60-element buried-channel structures, with doped polysilicon electrodes and undoped polysilicon interelectrode isolation, were operated as optical imagers with a 60-bit photosensitive region, two 30-bit vertical registers, and a 2-bit horizontal register. The device structure is shown in Fig. 29. The vertical registers were operated at a 250-kHz clock rate and the horizontal register at a 500-kHz rate. In these devices, for total dose up to 10^4 rad, the transfer inefficiency remains nearly constant at the preirradiation value of 10^{-4}. The dark current and the flat-band voltage shift are shown as a function of dose in Fig. 59. At a total dose of 3×10^4 rad, the devices were inoperative due to two effects. The first effect was that the flat-band voltage shift became large enough so that the original buried-channel drain bias was insufficient to deplete the channel completely. This problem was corrected by increasing the drain bias voltage. The second effect was a reduction in the resistance of the undoped polysilicon isolating regions by a factor of 20. This drop in the isolation resistance allowed a mixing of the various clock voltages, the net effect being that almost no potential barrier could be formed to store photogenerated charge under the imaging gates. This situation is shown in Fig. 60. The failure of the isolation was due to induced channeling in the undoped polysilicon as a result of trapped positive charge in the oxide. Altering the applied voltages could not correct this problem, leaving the devices functionally inoperative at 3×10^4 rad. Any devices using undoped polysilicon for isolation regions could potentially suffer from this failure mechanism. Table IX summarizes the radiation failure modes determined for the three CCD designs tested. Recommendations for a hardened design should include the following. 1) Use a buried-channel structure, which eliminates fat zero and interface-state trapping. The n-channel reduces ΔV_{FB}. 2) Use a planar insulator and a single gate material. 3) Avoid undoped polysilicon isolation. 4) Control surface potential between gates.

X. INFRARED DEVICES

There are basically three ways in which the charge-coupled concept can be used in infrared imaging. 1) A silicon CCD can be used to multiplex an array of IR detectors, 2) a silicon CCD can be used to provide time delay and integration for an array of IR detectors, and 3) a CCD or CID can possibly be fabricated in IR-sensitive semiconductors to provide monolithic infrared charge-coupled devices (IRCCD's).

The multiplexer implementation is shown in Fig. 61. The output of the detectors is a voltage. Once per line time, the CCD, via the capacitively coupled input circuits, samples the detector output voltages, obtains charge in each CCD cell proportional to the corresponding detector output voltage, and shifts this charge configuration out. The object of this approach is to perform the multiplexing within the dewar with the least amount of power dissipation. In this way, the number of leads from the dewar to the outside world will be minimized and the heat load will be minimized. There are two problems associated with this approach—crosstalk between channels at the CCD output due to transfer inefficiency, and the low-noise injection of charge packets into the CCD which are proportional in charge to the voltage at the output of the detectors [33]. The crosstalk problem is largely solved by injecting into alternate CCD cells instead of injecting into each CCD cell. Equation (24) can be used to show that

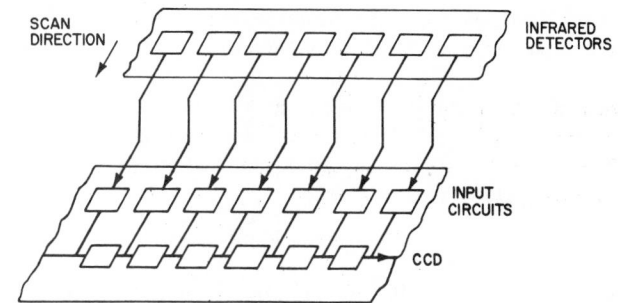

Fig. 61. Schematic diagram of the use of a silicon CCD to multiplex an IR detection array.

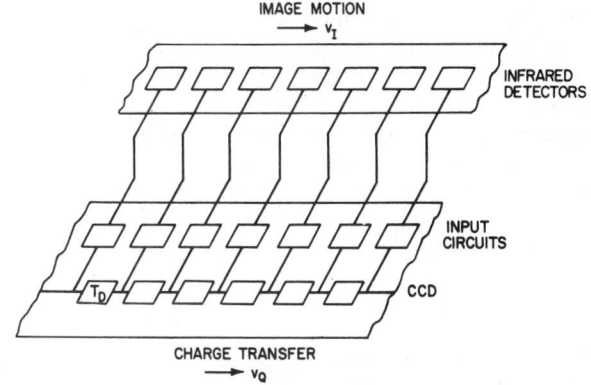

Fig. 62. Schematic diagram of the use of a silicon CCD array to provide TDI for an IR detector array.

injecting into alternate CCD cells reduces the interchannel crosstalk at the CCD output by the factor $[2/(N - 1)]$ $[(1 - \alpha)/\alpha]$ compared with the interchannel crosstalk when the signal is injected into each cell. For $N = 100$ and $\alpha = 10^{-4}$, this factor is 200. The larger problem is to introduce charge into the CCD in response to low-level voltages. Since the input signal is a voltage, it is the noise voltage $(kT/C_{I-N})^{1/2}$ which must be minimized. This requires that C_{I-N} should be large.

The time delay and integration (TDI) implementation is shown in Fig. 62. TDI is achieved by making a 1:1 interconnection between each detector element and the corresponding CCD element via the appropriate low-noise input circuits. The charge injected into a CCD cell at a given point is proportional to the detector output voltage. The transfer of a charge packet along the CCD is synchronous with the velocity V_Q of the corresponding point in the image plane along the detector column. Thus the effective integration is M times longer than the single-detector integration time, where M is the number of detectors in the column. For an N-column by M-element image plane, the number of interconnections between the detector array and the CCD array is M times N. Reliable fabrication of this large number of interconnections is a major concern for this approach [34].

The third approach is monolithic IRCCD, where the CCD or the CID is fabricated in a narrow-bandgap semiconductor. The major problems with this approach are that 1) MOS technologies in suitable semiconductors are not well developed, 2) the high-background photon flux in the IR saturates storage cells quickly, and 3) the combination of high-background photon flux and low contrast imposes severe limits on the tolerable amount of nonuniformity of response (from cell to cell). Each of these problems will be discussed more fully in the following paragraphs.

Some progress has been made in the development of an MOS technology in InSb using deposited silicon oxynitride as the insulator and nichrome as the metal. Storage times (with no excitation) of 0.1 s at 77 K and interface-state densities of 10^{12} cm^{-2} (eV)$^{-1}$ have been reported [35].

The full-well capacity (in carrier/cm^2) of an MOS capacitor is approximately given by

$$N_{\max} = C_{ox} V_G / e. \qquad (92)$$

Using $C_{ox} = 3.5 \times 10^{-8}$ F/cm^2 and $V_G = 2$ V, $N_{\max} = 2 \times 10^{12}$ carrier/cm^2. If the photon flux is 6×10^{14} photon/cm^2 · s, then the well will fill in 3 ms. Thus the MOS cell (CCD or CID) is limited to short integration times for wide-band thermal imaging due to the finite storage capacity. The most fundamental problem with the use of an integrating-type sensor in the IR is the nonuniformity problem. If the element-to-element fractional nonuniformity of response in an integrating array is ΔR, and if the background flux is F_B and the scene contrast is C, then the number of carriers N_T collected in the cells after integrating for t_{integ} seconds is

$$N_T = (R + \Delta R) F_B (1 + C) \, t_{integ}$$
$$= \underset{(a)}{RF_B t_{integ}} + \underset{(b)}{RF_B C t_{integ}} + \underset{(c)}{\Delta R F_B t_{integ}}.$$

Term (a) is a constant number of carriers in each cell and has no effect other than using dynamic range. Term (b) is the desired signal. Term (c) is the cell-to-cell variation in the number of carriers collected due to spatial nonuniformities in response. When (c) is larger than (b), the signal will not be detectable. For example, in the 3–5-μm range, the contrast is 3.7 percent/K. Thus, in order to detect a temperature differential of 1 K, the maximum nonuniformity tolerable is 3.7 percent. This imposes severe constraints on material homogeneity and photolithographic tolerances.

Figs. 63–69 illustrate recent results from a single CID cell fabricated in InSb [36]. Fig. 63 shows the CID readout and signal processing circuit. The substrate current is fed into an integrator and the two current pulses are integrated during the charge-injection period. The final value of the signal voltage is amplified and then sampled and held. Finally, the integrator is reset to zero for the next period. The optical measurement setup is shown in Fig. 64. The blackbody temperature is 500 K representing the signal and the background temperature is 300 K. Fig. 65 shows the output signal as a function of integration time. The output voltage is a linear function of integration time up to saturation. Fig. 66 shows the output voltage versus blackbody irradiance, and Fig. 67 gives data on the same device with narrow-band excitation ($\lambda_0 = 4.5\ \mu$m, $\Delta\lambda = 0.2\ \mu$m). The sensitivity is approximately a factor of 3 lower than BLIP.

Fig. 68 shows the saturation characteristic of the device. The saturation corresponds closely to the calculated value of 2×10^{11} cm^{-2}. Finally, Fig. 69 shows the measured spectral response of the CID cell.

It is concluded that InSb MOS structures can be operated in the integration mode via the CID implementation with values of D^* approaching the BLIP limit.

A scheme for background charge rejection can be used to skim off the signal Q_s from the background Q_B and dump the background charge subsequently by injecting it into the substrate. Such a circuit is shown in Fig. 70. The left well is photosensitive and collects charge. The barrier ϕ_B is con-

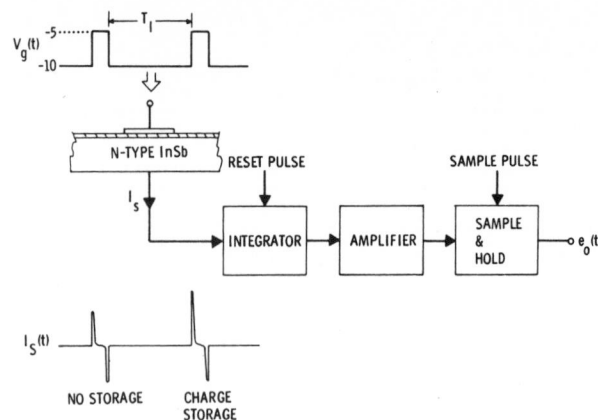

Fig. 63. Charge-injection operation of InSb MIS structure.

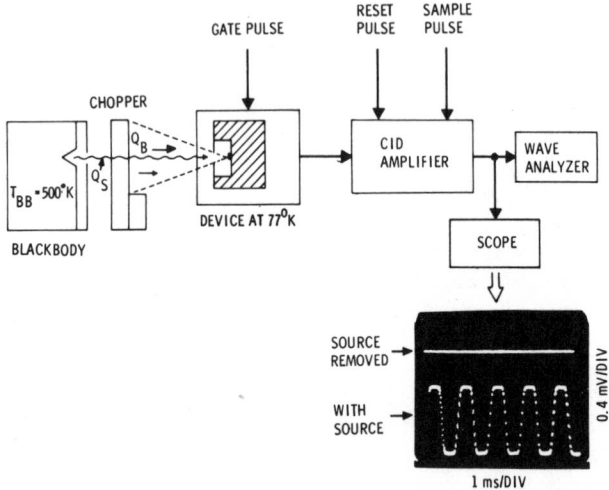

Fig. 64. Block diagram of experimental apparatus to test the InSb CID structure.

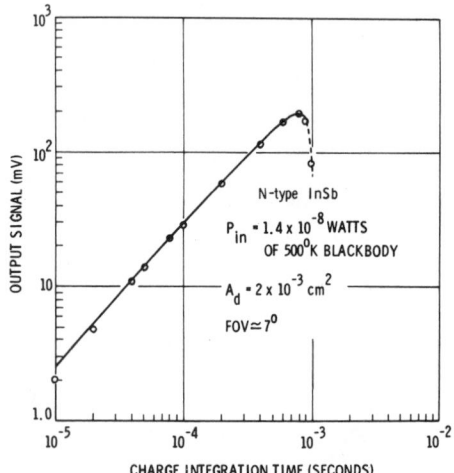

Fig. 65. Output voltage versus integration time for InSb CID cell with 1.4×10^{-8} W input power from 500 K blackbody.

trolled by the voltage on the center electrode and is set to the level corresponding to the background charge Q_B. Therefore, any signal charge in excess of Q_B will spill into the right well. The charge remaining in the left well, Q_B, can be extracted by injecting it into the substrate or by transferring it to a drain region. The charge remaining in the right well is Q_s. This technique can be used to reduce the dynamic range re-

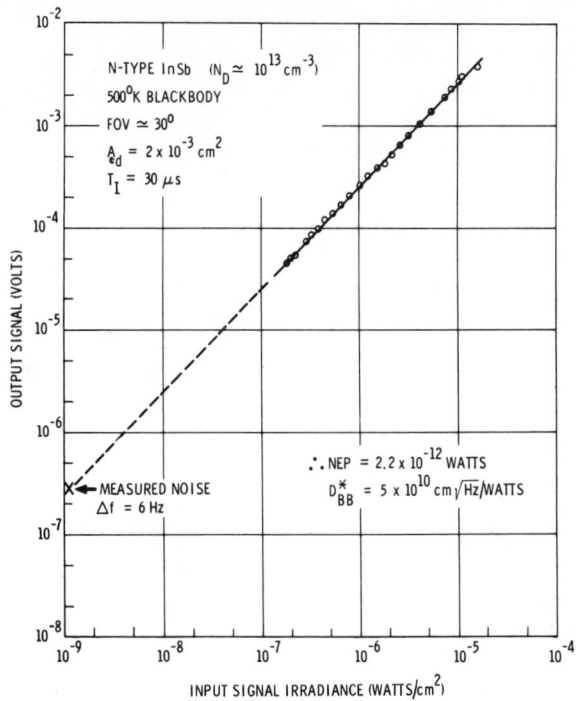

Fig. 66. Output voltage versus wide-band input irradiance for InSb CID cell.

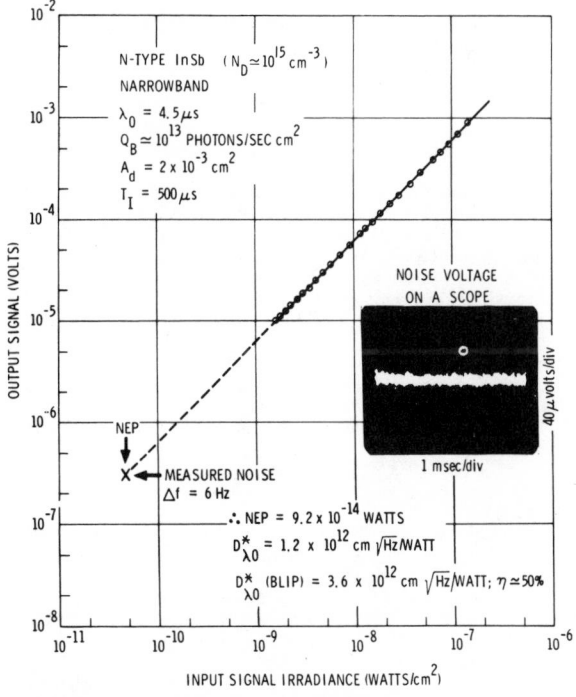

Fig. 67. Output voltage versus narrow-band (0.2 μm at $\lambda_0 = 4.5$ μm) irradiance for InSb CID cell.

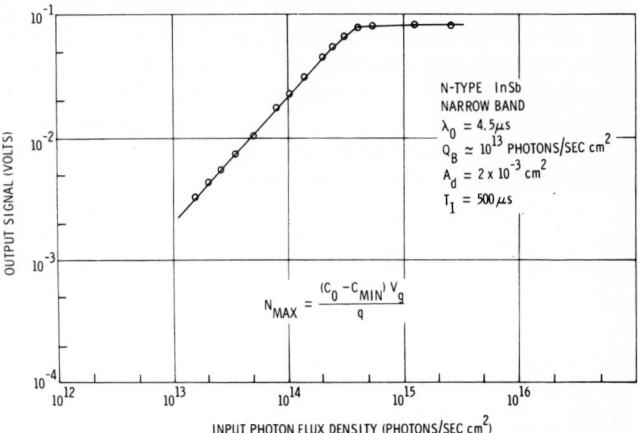

Fig. 68. Output voltage versus input photon flux density showing saturation level.

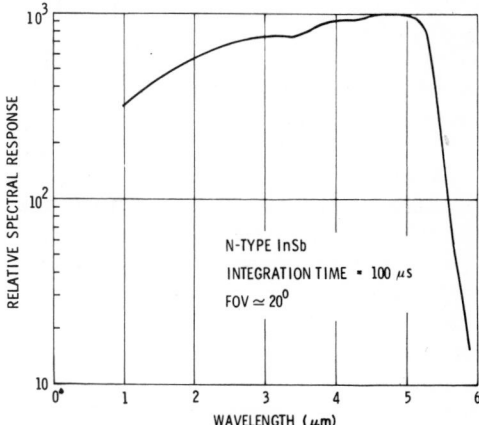

Fig. 69. Relative spectral response of InSb CID cell.

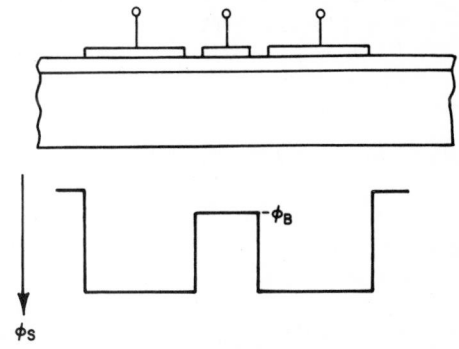

Fig. 70. Schematic diagram of background rejection or antibloom circuit.

quired of the amplifier. The same structure can be used in a slightly different way to provide antiblooming and exposure control.

XI. Conclusions

From the survey of CCD technology and design presented here, the following general conclusions can be drawn.

1) Buried-channel CCD's have been shown to be superior to surface-channel CCD's in terms of transfer inefficiency, spatial and temporal noise, and speed. Some sacrifice of charge-handling capacity is made to achieve speeds in excess of 20 MHz.

2) The CCD is unique in that signal charge can be sensed nondestructively. This makes ultralow-noise signal amplification via distributed amplifiers possible.

3) Long linear arrays in excess of 1700 elements have been fabricated and evaluated. Such devices are viable candidates for document reading and surveillance applications.

4) Area arrays having approximately half of conventional TV resolution have been fabricated. The fabrication of large arrays having full TV resolution appears to be feasible.

5) The CCD concept is especially useful for the TDI function; the CID concept is especially useful for applications requiring random accessibility.

6) Cooled (−50°C), frontside-illuminated, photon-excited area arrays having NES's of ≤ 10 electrons appear to be achievable. The performance of such arrays would approach that of photoelectron-excited arrays of equal area.

7) The use of two boresighted interline-transfer area arrays with $\frac{1}{2}P$ horizontal optical offset and $\frac{1}{2}T$ electrical offset appears to be a practical technique for increasing the horizontal sampling density in the focal plane, thereby increasing the responsivity by a factor of 2 and decreasing the susceptibility to aliasing.

8) The use of silicon CCD's at 77 K for focal-plane signal processing (TDI and multiplexing) of information from IR detector arrays has been demonstrated. The key problem in these approaches is associated with the interface between the IR detector arrays and the CCD arrays.

9) The primary effects of ionizing radiation are the shift in flat-band voltage and the increase in interface-state density. Buried-channel CCD's having planar oxides and overlapping electrodes would appear to be the least susceptible to total-dose ionizing radiation environments.

10) The basic operation of the InSb CID has been demonstrated. The simplicity of this approach makes it attractive for low-background applications. The key problem with the use of this device for high-background applications is the severe restriction on the allowable nonuniformity of response from element to element when the device is used in the integrating mode. Therefore, InSb CID arrays are best utilized when used in a scanned mode.

APPENDIX
NOISE CORRELATION TIME FOR RC CIRCUIT

Consider an RC circuit with charge Q on the capacitor. The differential equation (A-1) completely determines the response of the circuit to any forcing function $e(t)$.

$$\frac{dQ}{dt} + \frac{Q}{RC} = \frac{e(t)}{R}. \tag{A-1}$$

This equation can be rewritten as

$$\frac{d}{dt}(Qe^{t/RC}) = \frac{1}{R}e(t)e^{t/RC} \tag{A-2}$$

Then

$$Q(t) = \frac{1}{R}e^{-t/RC}\int_0^t e(\tau)e^{\tau/RC}\,d\tau \tag{A-3}$$

and

$$\overline{(Q(t) - \overline{Q(t)})^2} = \overline{Q(t)^2} = \frac{1}{R^2}e^{-2t/RC}$$

$$\cdot\int_0^t\int_0^\tau \langle e(\tau_1)e(\tau_2)\rangle e^{(\tau_1+\tau_2)/RC}\,d\tau_1\,d\tau_2. \tag{A-4}$$

If $e(t)$ is white noise arising from the thermal fluctuations of the resistor, then

$$\langle e(\tau_1)e(\tau_2)\rangle = 2kTR\delta(\tau_1 - \tau_2). \tag{A-5}$$

Making this substitution into (A-4), and performing the integrations, gives

$$\overline{Q(t)^2} = kTC(1 - e^{-2t/RC}) \tag{A-6}$$

and it follows that

$$\overline{V_c(t)^2} = \frac{kT}{C}(1 - e^{-2t/RC}).$$

Thus the correlation time is $\frac{1}{2}RC$.

ACKNOWLEDGMENT

The author wishes to thank L. W. Sumney of the Naval Electronics Systems Command, S. B. Campana of the Naval Air Development Center, and W. D. Baker, J. M. Killiany, N. S. Saks, and J. Freeman of the Naval Research Laboratory for many helpful discussions.

REFERENCES

[1] W. S. Boyle and G. E. Smith, "Charge coupled semiconductor devices," *Bell Syst. Tech. J.*, vol. 49, pp. 587–593, 1970.
[2] G. F. Amelio, W. J. Bertram, Jr., and M. F. Tompsett, "Charge-coupled imaging devices: Design considerations," *IEEE Trans. Electron Devices*, vol. ED-18, pp. 986–992, Nov. 1971.
[3] C. K. Kim, J. M. Early, and G. F. Amelio, "Buried-channel charge-coupled devices," in *Proc. 1972 NEREM Conf.*, pp. 161–164.
[4] A. M. Mohsen, T. C. McGill, and C. A. Mead, "Charge transfer in overlapping gate charge-coupled devices," *IEEE J. Solid-State Circuits*, vol. SC-8, pp. 191–207, June 1973.
[5] M. G. Collet and L. J. M. Esser, "Charge transfer devices," *Advan. Solid State Phys.*, vol. 13, pp. 337–358, 1973.
[6] L. J. M. Esser, "The peristaltic charge-coupled device for high speed charge transfer," in *Proc. 1974 IEEE Solid State Circuits Conf.*, pp. 28–29.
[7] M. H. White and J. R. Cricchi, "Characterization of thin-oxide MNOS memory transistors," *IEEE Trans. Electron Devices*, vol. ED-19, pp. 1280–1288, Dec. 1972.
[8] J. E. Carnes and W. F. Kosonocky, "Fast interface-state losses in charge-coupled devices," *Appl. Phys. Lett.*, vol. 20, pp. 261–263, 1972.
[9] J. E. Carnes and W. F. Kosonocky, "Charge-coupled devices and applications," *Solid State Technol.*, vol. 17, pp. 67–77, 1974.
[10] M. F. Tompsett, "Charge transfer devices," *J. Vac. Sci. Technol.*, vol. 9, pp. 1166–1181, 1972.
[11] W. B. Joyce and W. J. Bertram, "Linearized dispersion relation and Green's function for discrete-charge-transfer devices with incomplete transfer," *Bell Syst. Tech. J.*, vol. 50, pp. 1741–1759, 1971.
[12] R. J. Strain, "Properties of an idealized traveling-wave charge-coupled device," *IEEE Trans. Electron Devices*, vol. ED-19, pp. 1119–1130, Oct. 1972.
[13] "Advanced scanner and imaging systems for earth observations," NASA Ref. SP-335, pp. 239–264, 1972.
[14] D. R. Collins *et al.*, "Electrical characteristics of 500-bit Al-Al$_2$O$_3$–Al CCD shift registers," *Proc. IEEE* (Lett.), vol. 62, pp. 282–284, Feb. 1974.
[15] S. P. Emmons and D. D. Buss, "Techniques for introducing a low noise fat zero in CCD's," presented at the 1973 IEEE Device Res. Conf., Boulder, Colo.
[16] M. H. White, D. R. Lampe, F. C. Blaha, and I. A. Mack, "Characterization of surface channel CCD imaging arrays at low light levels," *IEEE Trans. Solid-State Circuits*, vol. SC-9, pp. 1–13, Feb. 1974.
[17] D. D. Wen and P. J. Salsbury, "Analysis and design of a single stage floating gate amplifier," presented at the 1973 IEEE Solid State Circuits Conf., Philadelphia, Pa.
[18] J. E. Carnes and W. F. Kosonocky, "Noise sources in charge-coupled devices," *RCA Rev.*, vol. 33, pp. 327–343, 1972.
[19] D. F. Barbe, "Noise and distortion considerations in charge-coupled devices," *Electron. Lett.*, vol. 8, pp. 207–208, 1972.
[20] J. M. Early, "Theory of bulk trapping in buried-channel CCD's near −50°C for levels below 100 electrons," presented at the 1974 Device Res. Conf., Santa Barbara, Calif.
[21] S. B. Campana, "Charge coupled devices for low light level imaging," in *Proc. 1973 CCD Applications Conf.*, pp. 235–245.
[22] S. R. Shortes, W. W. Chan, W. C. Rhines, J. B. Barton, and D. R. Collins, "Characteristics of thinned backside-illuminated charge-coupled device imagers," *Appl. Phys. Lett.*, vol. 24, pp. 565–567, 1974.
[23] "Moving target sensors," Texas Instruments, Navy Contract

N00039-73-C-0070, Final Rep., 1973.

[24] C. H. Séquin, "Interlacing in charge-coupled devices," *IEEE Trans. Electron Devices*, vol. ED-20, pp. 535–541, June 1973.

[25] M. H. Crowell and E. F. Labuda, "The silicon diode array camera diode," *Bell Syst. Tech. J.*, vol. 48, pp. 1481–1528, 1969.

[26] D. H. Seib, "Carrier diffusion degradation of modulation transfer function in charge coupled imagers," *IEEE Trans. Electron Devices*, vol. ED-21, pp. 210–217, Mar. 1974.

[27] "Low-light level charge-coupled imaging devices," Fairchild, Navy Contract N00039-73-C-0015, Final Rep. on Phase I, pp. B-1–B-20, 1973; also DDC AD 915544.

[28] S. B. Campana and D. F. Barbe, presented at the CCD Applications Conf., Edinburgh, Scotland, Sept. 1974.

[29] J. M. Hooker, "A sampling mode surveillance system using the CID imager," presented at the IEEE Int. Conf., New York, N.Y., Mar. 1974.

[30] J. M. Killiany, N. S. Saks, W. D. Baker, and D. F. Barbe, "Effects of gamma radiation on charge-coupled devices," presented at the GOMAC Conf., Boulder, Colo., June 1974.

[31] D. F. Barbe, J. M. Killiany, and H. L. Hughes, "Effects of gamma radiation on charge-coupled devices," *Appl. Phys. Lett.*, vol. 23, pp. 400–402, 1973.

[32] J. M. Killiany, N. S. Saks, W. D. Baker, and D. F. Barbe, "Effects of gamma radiation on buried-channel CCD's with doped polysilicon gates and undoped polysilicon interelectrode isolation," *Appl. Phys. Lett.*, vol. 24, pp. 506–508, 1974.

[33] T. F. Cheek, A. F. Tasch, J. B. Barton, S. P. Emmons, and J. E. Schroeder, "Design and performance of charge-coupled device time division multiplexers," in *Proc. 1973 CCD Applications Conf.*, pp. 127–139.

[34] D. M. Erb and K. Nummedal, "Buried-channel charge coupled devices for infrared applications," in *Proc. 1973 CCD Applications Conf.*, pp. 157–167.

[35] J. C. Kim, "Interface properties of InSb MIS structures," *IEEE Trans. Parts, Hybrids, and Packag. (Special Issue on Materials)*, vol. PHP-10, pp. 200–207, Dec. 1974.

[36] J. C. Kim, W. E. Davern, and T. Shepelavy, "InSb surface charge-injection imaging devices," presented at the 22nd IRIS Meeting.

Charge-Coupled Imaging Devices:
Design Considerations

GILBERT F. AMELIO, MEMBER, IEEE, WALTER J. BERTRAM, JR., MEMBER, IEEE,
AND MICHAEL F. TOMPSETT

Abstract—In this paper some of the parameters relevant to the design of charge-coupled imaging devices are considered. Among these are charge storage capability, transfer efficiency, charge conservation, dark current, and the anticipated signal-to-noise ratio. Each is discussed, and the resultant effects on the performance of imaging devices are investigated.

I. INTRODUCTION

THE idea of charge coupling in MIS structures first described by Boyle and Smith [1] leads to some inherently simple designs of self-scanned imaging devices. The basic concept consists of storing electrons (or holes) in potential wells created at the surface of a doped semiconductor and moving this charge as a packet across the surface by translating the potential minima. The minima are produced by applying a voltage to conducting electrodes formed on the surface of the insulator covering the semiconductor and driving the surface into depletion. This is shown graphically in Fig. 1(a), which is a plot of potential energy versus distance perpendicular to the surface for an MIS structure on a p-type conductivity substrate for which a voltage sufficient to cause depletion is instantaneously applied. Available minority carriers, in this case electrons, may then be collected at the surface, forming an n-type inversion layer and resulting in a decrease in depth of the potential well originally produced in the semiconductor, as shown in Fig. 1(b). The minority carriers may originate in many ways. In this paper our concern will be centered on those generated by incident photons. Movement of this localized inversion layer may be accomplished by the three-phase scheme depicted in Fig. 2(a). The geometry of this charge-coupled device is chosen so that the potential minimum of the more positive electrode couples across the interelectrode gap, resulting in the lateral transfer of the charge packet. Other configurations for obtaining surface charge transport are possible, such as the two-phase scheme [5] illustrated in Fig. 2(b), in which the thicker oxide under one of a pair of electrically connected electrodes produces a potential hill at the silicon surface and causes charge to flow in one direction only.

For optical imaging purposes, photons incident on the semiconductor generate the necessary minority carriers, which are then stored in and moved through

Manuscript received May 7, 1971; revised June 1, 1971.
The authors are with Bell Telephone Laboratories, Inc., Murray Hill, N. J. 07974.

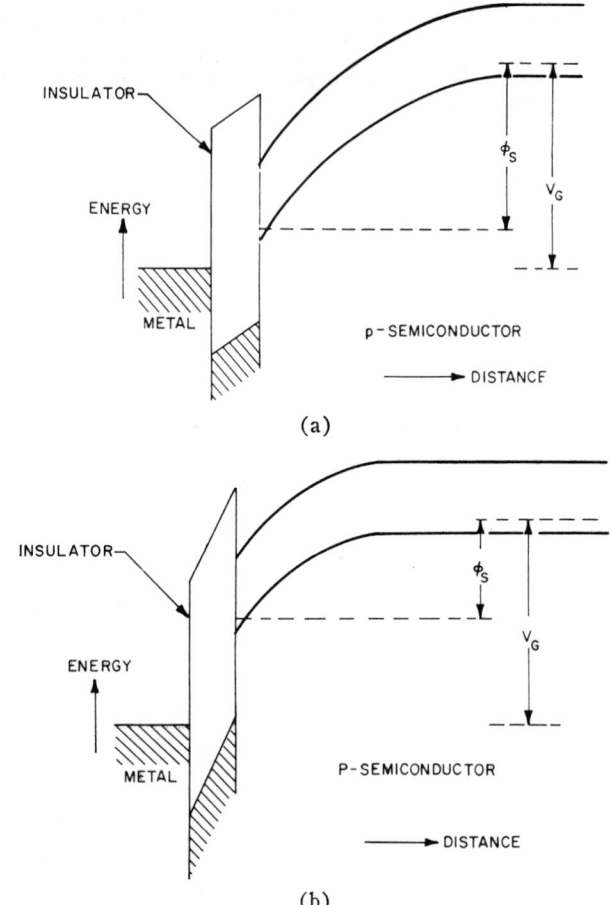

Fig. 1. Energy level diagram of a metal–oxide p-type semiconductor combination. (a) When a voltage V_G has just been applied to the metal electrode. (b) After the accumulation of some negative carriers at the oxide–silicon interface.

the MIS capacitors in the manner described. At an appropriate location these packets may be detected, for example, by removing them sequentially from the semiconductor by means of a single reverse-biased diode coupled to one of the transfer electrodes. Self-scanning arrays for both line and area imaging applications thereby become possible. Among the advantages of such an approach is the very low capacitance presented by the reverse-biased output diode.

An eight-bit charge-coupled device (CCD) has been described [2] and simple line scan imaging demonstrated. This device used 24 transfer electrodes connected in a three-phase configuration. A transfer efficiency of 99.9 percent was measured at a bit frequency of 160 kHz.

Reprinted from *IEEE Trans. Electron Devices*, vol. ED-18, pp. 986–992, Nov. 1971.

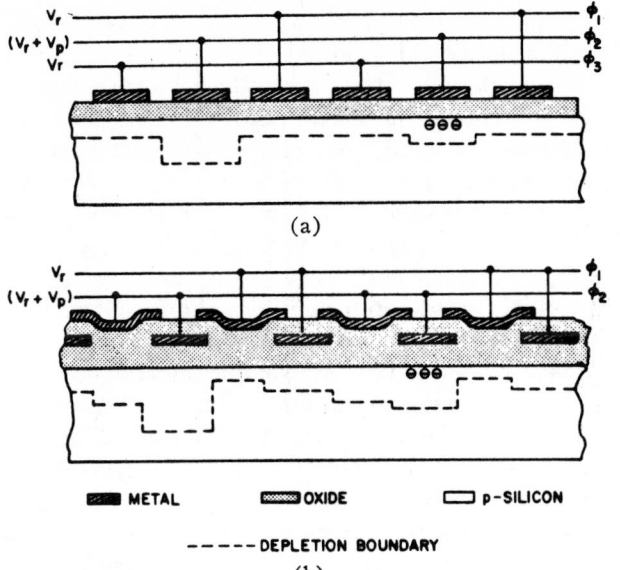

METAL   OXIDE   p-SILICON

- - - - - DEPLETION BOUNDARY

(b)

Fig. 2. Schematic cross section of charge-coupled device structures. Charge transfer is to the right if the electrodes are pulsed in the sequence $\phi_1\phi_2\phi_3$. (a) Three-phase device. (b) Two-phase structure with stepped oxide to provide directionality and overlapping electrodes.

The transfer efficiency is the percentage of charge remaining in the original packet after transfer along the device. The fact that transfer efficiency is in general less than 100 percent is a key consideration in the design of charge-coupled self-scanned imaging devices and limits the total number of transfers. The transfer efficiency is determined principally by the rate at which the minority carriers in the packet move from one MIS region to the next. The rate is governed by both diffusion processes and the tangential electric field at the surface [3]. The tangential electric field is determined by the electrode structure and will dominate the transfer rate in a well-designed configuration. Use of the minority carriers with the highest mobility, electrons in the case of silicon, speeds the charge motion.

In Section II the relationship of charge density and surface potential in a charge-coupled device is presented. This is followed in Sections III and IV by a discussion of the salient features of transfer efficiency and charge conservation. Methods of improving performance are discussed. The presence of unwanted minority carriers, or dark current, is considered in Section V followed by a treatment of sensitivity and signal-to-noise ratio in Sections VI and VII.

II. Charge–Potential Relations in a CCD

As in all electrostatic problems related to semiconductor devices, the relationship between potential and charge is obtained by solving Poisson's equation

$$\nabla^2 V = \frac{-\rho}{\epsilon} = \frac{q}{\epsilon}(N_D - N_A + n - p) \qquad (1)$$

where N_D and N_A are the donor and acceptor concen-

trations and n and p are the electron and hole densities. Typically, in approaching an MIS problem the three-dimensional Laplacian is replaced by a one-dimensional second derivative. The justification for such a simplifying reduction is that the capacitor is "large" and the variations in potential along the surface may be ignored. In order to present the basic ideas of charge and potential in a CCD we shall follow this course; however, we indicate that for regions near the perimeter of the capacitor the one-dimensional formalism is likely to be quantitatively in error. In addition, we employ the depletion approximation which calls for strict application of the Poisson equation in the depleted region and the Laplace equation in the neutral bulk. Electron and hole densities are ignored except for any surface charge present at the interface.

Upon the instantaneous application of a voltage V_G to the electrode a depletion region is formed in the semiconductor. The gate voltage V_G is then related to the potential at the surface φ_s by

$$V_G - V_{FB} = \varphi_s + \frac{1}{C_0}(2\epsilon_s q N \varphi_s)^{1/2} \qquad (2)$$

where ϵ_s is the semiconductor dielectric constant, N is the doping density, and C_0 is the insulator capacitance per unit area. The flat-band voltage V_{FB} is given by

$$V_{FB} = \phi_{MS} - \frac{1}{C_0 d}\int_0^d x\rho(x)dx$$

$$= \phi_{MS} - \frac{Q_{ss}}{C_0} \qquad (3)$$

where we have used Q_{ss} as an effective surface charge density to account for the volume charge density $\rho(x)$ distributed throughout the insulator of thickness d, and ϕ_{MS} is the metal–semiconductor contact potential difference. Now, if in addition to this effective oxide charge, we add a charge density q_0 at the interface resulting from the formation of an inversion layer, the potential of (2) is altered by a term q_0/C_0. Including this term and solving for the surface potential gives

$$\varphi_s = V_0 + V - (V_0{}^2 + 2VV_0)^{1/2} \qquad (4)$$

where

$$V = V_G - V_{FB} - \frac{q_0}{C_0}$$

$$V_0 = \frac{\epsilon_s q N}{C_0{}^2}.$$

The depletion depth is in turn given by

$$w = \left(\frac{2\epsilon_s\varphi_s}{qN}\right)^{1/2}. \qquad (5)$$

The surface potential φ_s is of overriding importance in a discussion of CCD operation and will be referred to frequently. For convenience, (4) is plotted as a function

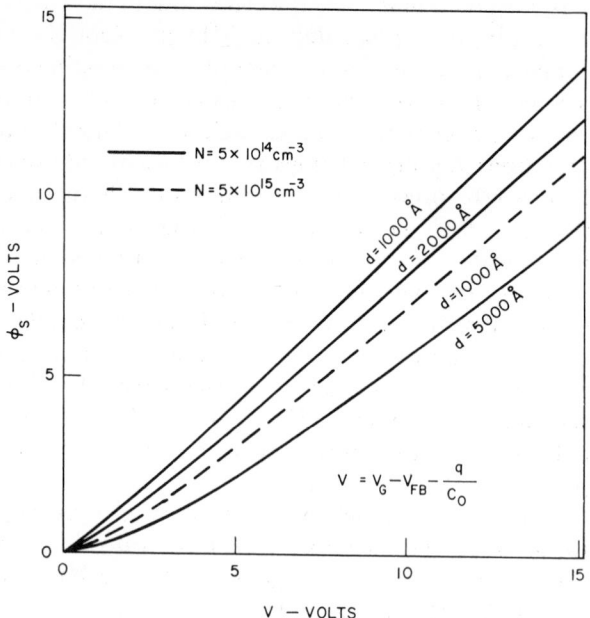

Fig. 3. Silicon surface potential versus applied electrode voltage corrected for fixed charge in the oxide and free charge at the interface, with oxide thickness d and semiconductor doping N as parameters.

of V in Fig. 3 with N and d as parameters. Note that for large V (corresponding to a large depletion depth and small depletion capacitance) the curves become linear with a slope of one.

III. TRANSFER EFFICIENCY

A fundamental property of an imaging device is that it exhibit an analog (preferably linear) behavior; i.e., that the charge measured in the output and attributable to a given light-sensitive element is directly related to the integrated light intensity incident on that light-sensitive element. This is inherently the case in a charge-coupled imaging device unless charge is left behind or lost during transfer. In these cases signal will be attenuated and spatial resolution will be degraded. In the eight-bit charge-coupled device reported earlier [2], it was observed that part of the charge packet appeared to transfer rapidly, whereas some lesser charge density, say a fraction f of the maximum charge density, transferred more slowly with an exponential time constant τ. Since, at the frequencies on the order of a few megahertz required in imaging devices, the time available for transfer is much greater than the duration of the rapid transfer behavior, the efficiency η is essentially governed by the fraction f and the slow time constant τ. Thus, after time t

$$\eta \approx 1 - f e^{-t/\tau}. \qquad (6)$$

This kind of behavior has been confirmed not only by experiment but also by analytical studies and by computer modeling [4]. Ignoring for a moment the effects of tangential surface fields, the behavior is qualitatively easy to understand. At time $t=0$, the

beginning of the transfer, the packet is very dense and localized, and the gradient of the carrier density at the edges of the packet is large. When the surface potential of the adjacent electrode becomes more positive than the local potential, namely, when the adjacent MIS structures couple, a considerable fraction of the packet transfers very quickly due to the strong repulsive forces felt by these electrons. As time evolves, this repulsive force decreases and becomes asymptotic to a value given by considerations of thermal diffusion and surface electric field intensity. This determines the time constant τ in (6), and f is an empirically defined quantity reflecting the density below which the electrons do not "feel" the presence of the other electrons. A detailed presentation of this behavior has been given by Strain and Schryer [3]. Their results indicate, for example, that a transfer efficiency of 99.99 percent per transfer can be expected from a device with 10 μm electrodes operated at 1 MHz.

One important impediment to charge transfer is inadequate coupling across the region below the inter-electrode gap. This condition results from too large an electrode separation for the operating voltage used and/or from a detrimental concentration of oxide charge in the gap region possessing the same sign as the minority carriers. One obvious way of circumventing these difficulties is to effectively eliminate the gap altogether by using overlapping electrodes [5], [6] with oxide insulation as in the two-phase arrangement shown in Fig. 2. This approach, however, complicates the structure. It is not necessary to use overlapping electrodes to bypass this problem so long as the interelectrode gaps and the oxide charge of the same polarity as the stored charge are kept "small." Since silicon dioxide tends to charge positively, use of a p-type silicon substrate is most advantageous. In this case the surface tends to be positive with respect to the neutral bulk and aids in the depletion of the surface. Similarly, when the gaps are large there is more charge in this region to be swept out in order to achieve depletion and, consequently, coupling. Empirically, it has been observed that for interelectrode gaps on the order of 3 μm and oxide charge densities on the order of 10^{11} cm^{-2}, satisfactory coupling can be achieved using peak voltages of less than 10 V. A detailed analysis of the effects of gap spacing and oxide charge is beyond the scope of the present discussion but will be reported in a future publication [4].

Another mechanism whereby transfer efficiency may be reduced is the temporary trapping and re-emission of the minority carriers at a later time by surface states. The carriers may be trapped in either the lower or the upper half of the bandgap for p- and n-channel devices, respectively. The probability of a state being occupied by a minority carrier depends on the energy level of the state, its relative density, and the density of carriers. When a charge packet moves to the next element, the minority carrier density falls several orders of magnitude and any trapped carriers are then re-emitted with a characteristic time depending on the energy difference to the nearest band [7]. If this time is very short compared to the transfer time, the charge is re-emitted into the correct packet and there is no net effect on transfer efficiency, but if it is of the same order of magnitude as the time of transfer or longer, some of the trapped charge is re-emitted into a trailing packet of charge. Since the statistics of trapping is in part a function of the charge density, the addition of a small background charge may reduce the effects of surface states.

The effect of inefficiency in charge transfer is to deteriorate image quality by reducing the modulation transfer function (MTF) or sine wave response and by producing a phase shift for different spatial frequencies. The interpretation of the transfer efficiency in terms of a sine wave input enables the quantitative degradation to be calculated [8].

Suppose the length of a light-sensitive element (three transfer electrodes long for a three-phase device) is taken as p, then the angular spatial frequency of the optical input may be written as k/p where $k/2\pi$ is the number of line pairs per light-sensitive element imaged on the device. If the transfer inefficiency is $\epsilon = 1 - \eta$, then after n transfers the phase shift of the signal relative to its correct position is given by

$$\Delta\phi = -n\epsilon(k - \sin k).\tag{7}$$

The maximum spatial frequency resolvable corresponds to $k = \pi$, and for a three-phase device $n = 3N$, where N is the number of light-sensitive elements, so that for this worst case

$$\Delta\phi = -3N\epsilon\pi.$$

The frequency-dependent attenuation is given by

$$\frac{\text{voltage MTF}(k)}{\text{voltage MTF}(0)} = \exp\left(-n\epsilon(1 - \cos k)\right).\tag{8}$$

Again, the worst case is for $k = \pi$, and with $n = 3N$

$$\frac{\text{voltage MTF}(\pi)}{\text{voltage MTF}(0)} = \exp\left(-6N\epsilon\right).$$

Hence, for example, if $N = 10^3$ and $\epsilon = 10^{-4}$, the MTF ratio in the worst case is only degraded by 5 dB.

IV. CHARGE CONSERVATION

Thus far we have limited our discussion of nonideal charge-coupled device performance to transfer inefficiency; however, actual loss or gain of charge is also possible. In normal operation the central region of the transfer channel is permanently depleted (or inverted), and due to a lack of majority carriers, loss of charge is

extremely unlikely. If the regions between transfer electrodes are allowed to enter accumulation, however, majority carriers can recombine with any trapped minority carriers, thereby depopulating these surface states. When the next charge packet passes this surface area, these states will be reoccupied only to be subsequently depopulated when accumulation reoccurs. Thus, some minority carriers are being effectively "pumped" into the substrate at every transfer. This effect was observed in the eight-bit charge-coupled device reported earlier [2] as the resting potential on the electrodes is reduced below the threshold potential for channel formation along the structure. It is prevented by ensuring that the transfer regions of the device are always depleted.

Similar arguments apply for the regions adjacent to the transfer region where the channel may be defined by a heavy majority carrier doping, and the surface is always in accumulation. As the electrodes are pulsed, there is a small region at the edges of the device where the quasi-Fermi level for majority carriers is perpetually sweeping across the band gap resulting in, first, accumulation and then depletion. If minority carriers can enter this region when a charge packet is under the electrode in question, some can be pumped by the process described above. Except for very full packets, however, the minority carriers tend to localize in the center of the channel where their potential energy is minimized, and this effect is not expected to be significant. A possibly more serious problem is the reversal of this effect, where a packet gains charge from the bulk because of minority carriers moving into the packet at these edge regions. For a sufficiently well-doped channel-stop region, however, the edge area of possible difficulty can be made very small, thereby minimizing the effect.

Channel isolation may also be accomplished by using a metal electrode edge guard separated from the transfer electrodes by a small gap. This electrode is dc biased so as to produce a shallow depletion region within the semiconductor and to effectively clamp the surface to a fixed potential. The method has the disadvantage of coupling more dark current into the channel region than the channel-stop doping approach. It also requires multilayer metallization capability and an additional voltage source.

V. DARK CURRENT

In normal operation the capacitance elements of a CCD are not saturated with charge; i.e., they are not in equilibrium. With the evolution of time, however, the system will tend to equilibrium by means of the thermal generation of minority carriers. Thus, even in the absence of light or other means of charge injection into the device, an unwanted dark current will be present. As is well known, a dark current contribution is characteristic of most imaging devices. It is an important criterion that this current be only a small fraction of the expected video current, particularly if it is nonuniform over the imaging area.

Whether this is likely to be the case for a charge-coupled imaging device can be determined by an investigation of the various generation current contributions in a semiconductor relative to the incident light. There are three principal dark current components: one arising from carriers generated in the bulk depletion region, one from the neutral bulk, and one from the semiconductor insulator interface. The generation current I_{gd} per unit area in the bulk depletion region is given by [9]

$$I_{gd} = \frac{1}{2} \frac{q n_i w}{\tau_{p,n}} \tag{9}$$

where n_i is the intrinsic carrier concentration $= 1.6 \times 10^{10}$ cm^{-3}; w is the depletion width; and $\tau_{p,n}$ is the carrier lifetime in the neutral material $= \sigma v_{th} N_t$, in which σ is the carrier capture cross section $\approx 2.2 \times 10^{-16}$ cm^2, v_{th} is the thermal velocity of carriers $\approx 10^7$ cm/s, and N_t is the concentration of recombination–regeneration centers in the bulk.

Since the depletion width may be expressed as

$$w = \left(\frac{2\epsilon_s \varphi_s}{qN}\right)^{1/2} \tag{10}$$

then

$$I_{gd} = \frac{n_i}{\tau_{p,n}} \left(\frac{\epsilon_s q \varphi_s}{2N}\right)^{1/2} \tag{11}$$

Minority carriers generated in the neutral bulk have the possibility of diffusing to the depletion region and also being collected. The generation current I_{gb} per unit area stemming from this cause is given by [9]

$$I_{gb} = \frac{q n_i^2}{N \tau_{p,n}} \cdot L_{p,n} = \frac{6.6}{N} \left(\frac{\mu}{\tau_{p,n}}\right)^{1/2} \quad \text{A/cm}^2 \tag{12}$$

where $L_{p,n}$ is the diffusion length of minority carriers,

$$L_{p,n} = \sqrt{D\tau_{p,n}} = \sqrt{\mu(kT/q)\tau_{p,n}},$$

in which μ is the minority carrier mobility.

The surface generation current I_{gs} per unit area is produced by those surface states within a few kT of midband. These surface recombination–regeneration centers of density N_{st} per unit area produce a current I_{gs} given by [9]

$$I_{gs} = \frac{1}{2} q n_i S_0 = 2.8 \times 10^{-18} N_{st} \tag{13}$$

where S_0 is the surface recombination velocity,

$$S_0 = \sigma v_{th} N_{st}.$$

For quality bulk material with lifetimes on the order of 100 μs, the dark current is essentially surface dominated and from the above equations can be expected to be on the order of 30 nA/cm^2.

A further small source of dark current is the minority carriers collected from the changing depletion volume. This effect becomes important only at very high frequencies.

A useful experimental figure of merit for the device is that of storage time. We may define this storage time τ_s for the MIS capacitor electrodes on the device as the time it takes for equilibrium to become reestablished following the application of a voltage step ΔV pulsing the device from inversion into depletion. Thus τ_s may be related to the average dark current by the expression

$$\tau_s \approx \frac{\Delta V C_0}{\langle \Sigma I_g \rangle} \tag{14}$$

where C_0 is the oxide capacitance. A more detailed study of the transient response of the basic MOS capacitor has been given by Heiman [10] who used the method to infer effective lifetime. For our purpose it is sufficient to have τ_s very much greater than the integration time of the optical image.

These data must be interpreted with some care for a charge-coupled imaging device in a given application. For example, in integrating charge in a three-phase device only one electrode is held at a deep depletion potential, but dark current appropriate to the shallow depletion condition is collected from the two neighboring electrodes.

VI. Sensitivity

The charge-coupled device has essentially the same light sensitivity as any other silicon device under corresponding conditions with allowance for any geometrical factors. Assume a scene illumination of L fc, an object reflectivity of unity, and a lens aperture of $f{:}x$. Then the illumination in the image is given by

$$E = 2.8 L x^{-2} \, \text{lnm}^{-2}. \tag{15}$$

The sensitivity of a silicon detector corrected to have photopic response is 500 μA ln^{-1}. If all the generated carriers were collected, the signal current is given by

$$I_L = 1.4 \times 10^{-7} L x^{-2} \, \text{A cm}^{-2}. \tag{16}$$

VII. Signal-to-Noise Ratio in the Video Signal

Noise in the video signal arises from several possible sources: 1) thermal noise in the preamplifier; 2) shot noise in the signal current; 3) coherent noise or pickup; 4) unspecified noise, which may be inherent in charge-coupled devices.

Thermal noise generated in the input resistance of the preamplifier will first be considered. Since the output diode of the device acts as a current source, a high input impedance voltage amplifier is appropriate. The minimum spatial wavelength that the device can resolve will be two elements long, so that the bandwidth required of the preamplifier in imaging applications is just $f/2$, where f is the frequency of the pulses applied to the transfer electrodes. This defines the amplifier input resistor R in the presence of feedback βA and input capacitance C by

$$\pi RC = (1 + \beta A)f^{-1} \qquad (17)$$

for the response to be 3 dB down at $f/2$. The equivalent thermal noise current i_N generated by the input resistor in the $f/2$ bandwidth, which would be defined at a later stage in the amplifier, is given by

$$\overline{i_N^2} = \frac{4kT}{R}\frac{f}{2}. \qquad (18)$$

Now the peak video signal current is just Qf where Q is the charge in a full packet. Hence, using (17) and (18), the signal-to-noise ratio that would be obtained by the ideal amplifier is given by

$$\frac{(\text{peak signal})^2}{(\text{rms noise})^2} = \frac{(Qf)^2}{\overline{i_N^2}} = \frac{Q^2(1 + \beta A)}{2\pi kTC}$$

$$= 4 \times 10^{19} Q^2 C^{-1} (1 + \beta A). \qquad (19)$$

For most practical values of Q and C, this ratio is very favorable. The limiting value of A occurs when the noise of the next stage of amplification becomes comparable with that of the input stage. The details of low noise video preamplifiers are described elsewhere [11]. The important conclusion is that for a CCD the signal-to-noise ratio on the basis of thermal noise only is independent of frequency.

Shot noise in the signal current Qf in the bandwidth $f/2$ is given by

$$\overline{i_s^2} = 2qQff/2. \qquad (20)$$

Hence, the signal-to-noise ratio from this source is

$$\frac{(Qf)^2}{\overline{i_s^2}} = \frac{Q}{q}. \qquad (21)$$

The variation of signal-to-noise ratio as a function of Q is plotted in Fig. 4. The figure shows the fundamental performance limits of the device. The value of C for a well-designed CCD should be at least one order of magnitude smaller than in any other currently available imaging device.

Coherent noise or pickup from the transfer pulses will always be present. However, this will be at frequencies f and higher and may be filtered out since the upper signal frequency is $f/2$. Techniques of pulse cancellation and neutralization are useful in preventing preamplifier overload. Nonetheless, careful design of the imaging device to ensure minimum capacitive pickup to the output is required.

There will be other sources of noise inherent in charge-coupled devices related to the statistics of transfer, charge trapping, and edge effects. However, if it is assumed that these effects are small, then the noise associated with them will be even smaller and should

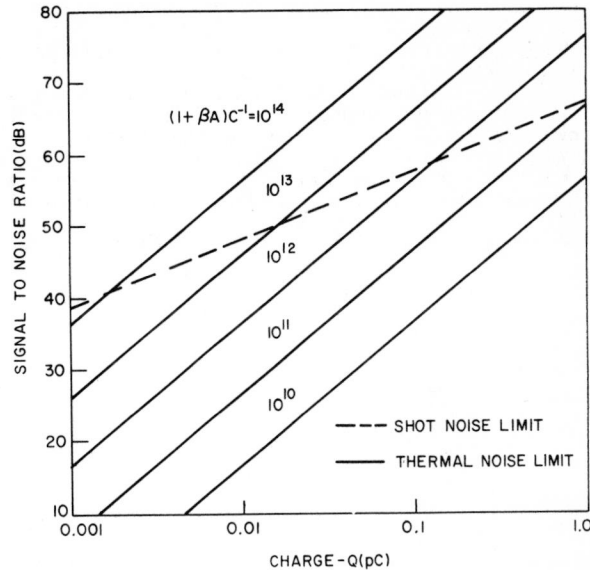

Fig. 4. The theoretical signal-to-noise ratios in the output signal of a charge transfer device as a function of charge Q per element.

not be significant in providing any further limitations than these effects might already have on performance.

VIII. Conclusions

In this paper some of the factors that must be considered in the design of a charge-coupled imaging device have been indicated and relevant parameters evaluated. Probably the most critical parameter of such a device is its transfer efficiency. For most imaging applications, adequate transfer efficiency should be achieved provided the geometry of the structure is properly designed and the processing optimized in regard to the bulk, interface, and oxide properties. Since none of these requirements appear out of line with the existing silicon MOS technology, the practical fabrication of charge-coupled imaging devices for a wide range of applications can be anticipated.

It should be emphasized that this paper is not intended as a complete design study for charge-coupled imaging devices. Many of the design considerations involve more detailed studies than those presented here. These studies include modeling of charge transport [4] and surface state effects [13].

A following paper [12] will describe possible design configurations of charge-coupled imaging devices and experimental results obtained with a 96-element linear structure. Detailed comparison between theoretical predictions and the experimental measurements of performance will be made in a later paper [13].

References

[1] W. S. Boyle and G. E. Smith, "Charge coupled semiconductor devices," *Bell Syst. Tech. J.*, vol. 49, pp. 587–593, 1970.
[2] M. F. Tompsett, G. F. Amelio, and G. E. Smith, "Charge coupled 8-bit shift register," *Appl. Phys. Lett.*, vol. 17, pp. 111–115, 1970.
[3] R. J. Strain and N. L. Schryer, "A nonlinear diffusion analysis of charge coupled device transfer," *Bell Syst. Tech. J.*, vol. 50, July/Aug. 1971.
[4] G. F. Amelio, *Bell Syst. Tech. J.*, to be published.

[5] D. Kahng and E. H. Nicollian, private communication.
[6] W. E. Engeler, J. J. Tiemann, and R. D. Baertsch, "Surface charge transport in silicon," *Appl. Phys. Lett.*, vol. 17, pp. 460–472, 1970.
[7] E. H. Nicollian and A. Goetzberger, "The Si–SiO₂ interface-electrical properties as determined by MIS conductance technique," *Bell Syst. Tech. J.*, vol. 46, pp. 1055–1133, 1967.
[8] W. B. Joyce and W. J. Bertram, "Linearized dispersion relation in Green's function for discrete charge transfer devices with incomplete transfer," *Bell Syst. Tech. J.*, vol. 50, July/Aug. 1971.
[9] See for example, A. S. Grove, *Physics and Technology of Semi-conductor Devices*. New York: Wiley, 1967, pp. 298–302.
[10] F. P. Heiman, "On the determination of minority carrier lifetime from the transient response of an MOS capacitor," *IEEE Trans. Electron Devices*, vol. ED-14, pp. 781–784, Nov. 1967.
[11] H. Breimer, W. Holm, and S. L. Tan, "A color television camera with Plumbicon camera tubes," *Philips Tech. Rev.*, vol. 28, pp. 336–351, 1967.
[12] M. F. Tompsett, G. F. Amelio, W. J. Bertram, R. R. Buckley, W. J. McNamara, J. C. Mikkelsen, and D. A. Sealer, "Charge-coupled imaging devices: Experimental results," this issue, pp. 992–996.
[13] M. F. Tompsett, in preparation.

Characterization of Surface Channel CCD Image Arrays at Low Light Levels

MARVIN H. WHITE, senior member, ieee, DONALD R. LAMPE, member, ieee,

FRANKLYN C. BLAHA, member, ieee, AND INGHAM A. MACK, member, ieee

Abstract—The characterization of surface channel charge-coupled device (CCD) line imagers with front-surface imaging, interline transfer, and 2-phase stepped oxide, silicon-gate CCD registers is presented in this paper. The analysis, design, and evaluation of 1 × 64 CCD line arrays are described in terms of their performance at low light levels. The signal-to-noise(S/N) is formulated in terms of charge at the collection diode. A dynamic range of 80 dB and a noise equivalent signal (NES), where S/N = 1, of 135 electrons is achieved with a picture element time of 20 μs and an integration time of 1.32 ms in the absence of a fat zero. A unique CMOS readout circuit, which uses correlated double sampling within a picture element time window, removes the Nyquist noise of the reset switch, eliminates switching transients, and suppresses low-frequency noise to provide low-noise analog signal processing of the video signals. This paper describes the responsivity, resolution, spectral, and noise measurements on silicon-gate CCD sensors and CCD interline shift-registers. The influence of transfer inefficiency and electrical fat-zero insertion on resolution and noise is described at low light levels.

I. Introduction

CHARGE-COUPLED devices (CCD's), since their invention [1] and experimental investigation [2], have promised low dispersive, analog delay lines with low-noise signal propagation and high dynamic range. The CCD analog delay line may be coupled with a monolithic photosensor, such as a "semitransparent" silicon-gate CCD or diffused photodiode, and a low-noise monolithic preamplifier to provide low light level imaging. The transfer inefficiency ϵ, in practical *surface-channel* CCD structures, is determined by surface-state trapping [3] which limits the free-charge transfer process and introduces dispersion [4] into the analog delay line. The CCD analog delay line should have a $NP\epsilon \leq 0.1$ (N = number of bits, P = number of phases) to prevent loss in image resolution or modulation transfer function (MTF) degradation.

The CCD imager uses the flow of minority carriers to transfer the video signal to a low capacitance collection diode. The collection diode is typically 0.25 pF in the CCD imager, whereas, the readout line capacitance in the x-y addressed image arrays is at least an order of magnitude larger in value. The low capacitance provides a reduction in noise and an increase in voltage swing at the gate electrode of an on-chip MOS electrometer amplifier. Thus, with the CCD principle a photon-generated signal charge may be transported over long distances within the silicon and amplified at low input noise charge levels. Although the clock and video signal levels are noninteracting within the CCD imager, there is an interaction at the collection diode. We have developed a method of signal processing called *correlated double sampling* [5], [6] to remove the switching transients, eliminate the Nyquist noise associated with the reset switch/node capacitance combination, and suppress "1/f" surface-state noise contributions. With this technique we have realized the intrinsic noise equivalent signal (NES) of the CCD imager which is set by the thermal "shot" noise of the leakage current. In the CCD imager the video signal is processed within the array by an analog CCD shift register, whereas, in non-CCD arrays the signal is transferred from the sensor to the video preamplifier by the closure of an address switch with a gate pulse from a digital shift register [7], [8]. Thus, in a CCD imager we require analog characteristics from both the sensor and the shift register, and this dual requirement places limitations on yield due to the presence of nonuniformities within the array.

II. CCD Line Arrays

General Considerations

Fig. 1 illustrates the line array functional block diagram. The *interline transfer* approach requires a transfer pulse ϕ_T, at the start of a line time, to transfer the stored photocharge from the individual sensor locations to a corresponding bit in the parallel-to-serial CCD shift register. In our case, a 2-phase clock system transfers the charge along the shift register to a CMOS readout circuit. The shift register is a 2-phase stepped-oxide geometry with a surface field and potential profile as shown in Fig 2. Fig. 3 is an enlargement of the CCD line array after the definition of the shift register with an aluminum interconnection. The sensors are constructed with transparent, conductive, polycrystalline silicon-gate electrodes with an n+ stopper diffusion (n-type, ⟨100⟩ substrates) on three sides. This diffusion reduces the interaction between adjacent elements and inhibits blooming in the array. The sensor bias voltage is adjusted such that the maximum collected charge at any sensor location cannot overflow the storage well of the shift register. The fourth side of

Manuscript received May 8, 1973; revised July 23, 1973.
The authors are with the Advanced Technology Laboratories, Westinghouse Electric Corporation, Baltimore, Md. 21203.

Reprinted from *IEEE J. Solid-State Circuits*, vol. SC-9, pp. 1–12, Feb. 1974.

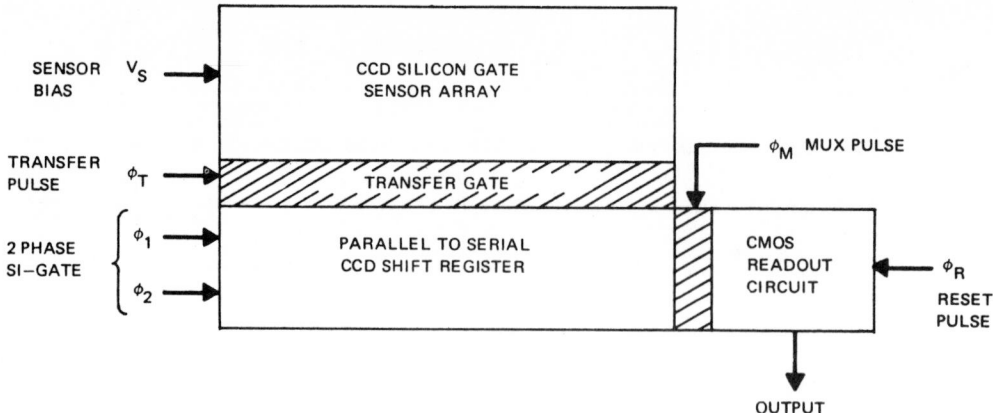

Fig. 1. CCD line array functional block diagram illustrating interline transfer.

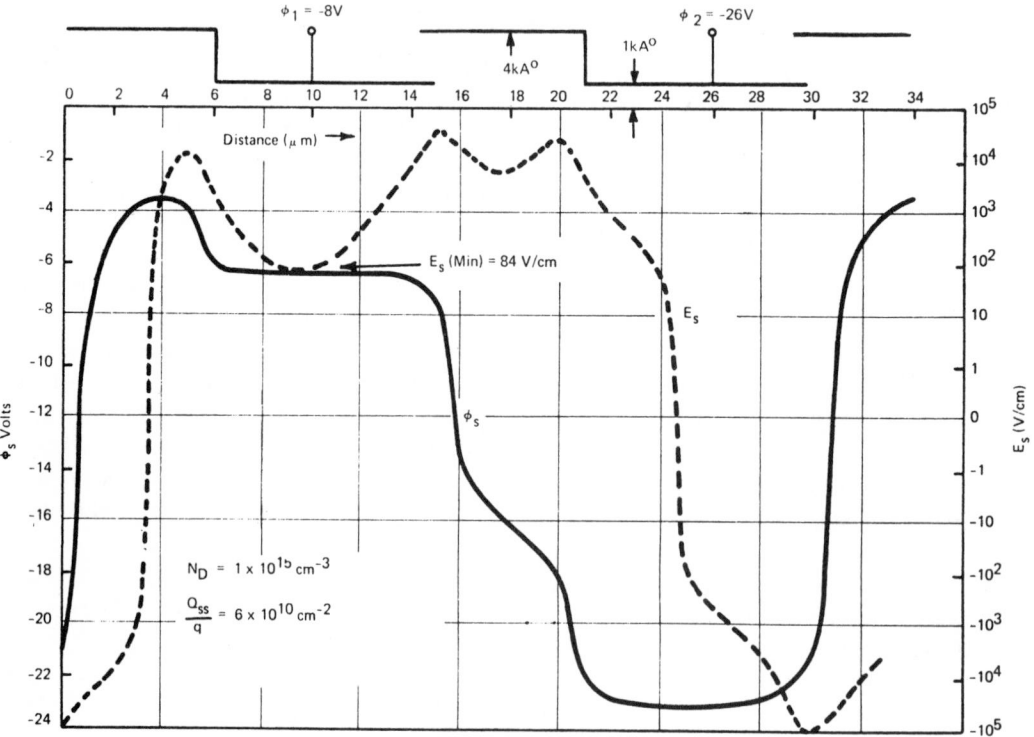

Fig. 2. Two-phase stepped oxide CCD unit cell with curves of periodic surface potential and Longitudinal electric field relative to electrode cross section at top of figure.

the CCD sensor is "stoppered" by holding the transfer gate at a positive potential with respect to the substrate during the integration period to form an n⁺ accumulation layer. The sensors are defined by and aluminum light shield which also serves to cover the CCD shift-register and CMOS readout circuits. The CCD line array is constructed of 128 CCD sensors with a $2P$ offset in the along-track direction ($P = 15$ μm, $\Delta x = 22$ μm, $\Delta y = 18$ μm). The dimension Δx is in the across-track electron-scan direction and Δy is in the along-track mechanical-scan direction.

Fig. 4 illustrates the CMOS readout circuit which consists of a multiplex gate (i.e., muxgate ϕ_M), a reverse-biased collection diode, n-channel MOS reset switch (i.e., reset gate ϕ_R), and p-channel MOS electrometer ampli-

fier. The voltage waveform on the gate of the electrometer is also displayed in Fig. 4. Below the waveform four distinct timing intervals are labeled for discussion. These four timing intervals comprise a *pixel* (i.e., picture element) time and form the basis of a signal processing method called *correlated double sampling*. The node capacitance at the collecting diode is 0.25 pF and is not influenced by the parasitic n⁺/p⁻ diode of the reset switch which is reverse-biased to prevent discharge of the collecting node by the reset feedthrough "pedestal" when the reset switch is turned off. The aluminum light shield provides about 0.03 pF of the node capacitance since it rests over 1.3 to 1.5 μm of deposited SiO_2 and forms a ground plane to shield the sensitive output collection circuit from pickup.

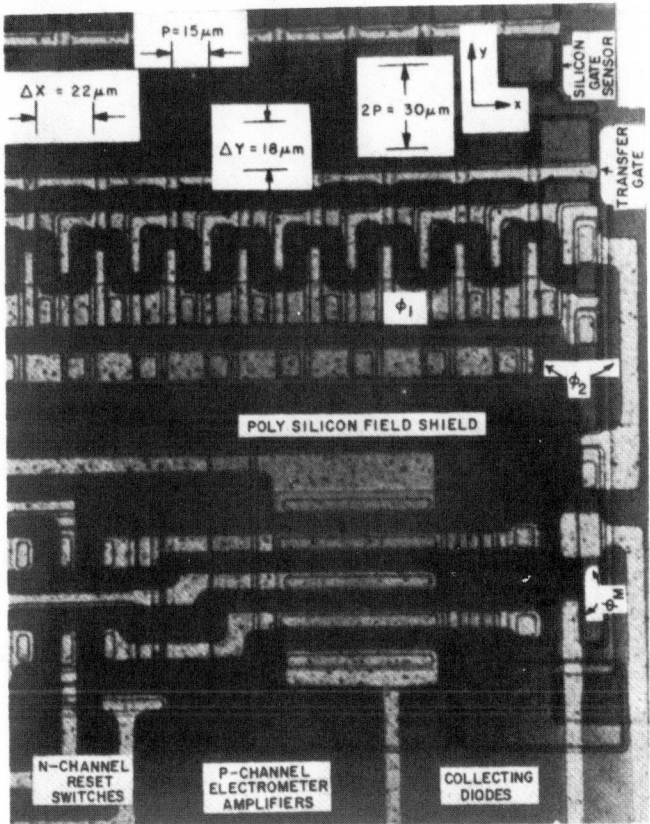

Fig. 3. CCD "interline-transfer" line array with CMOS readout circuit (prior to aluminum light shield).

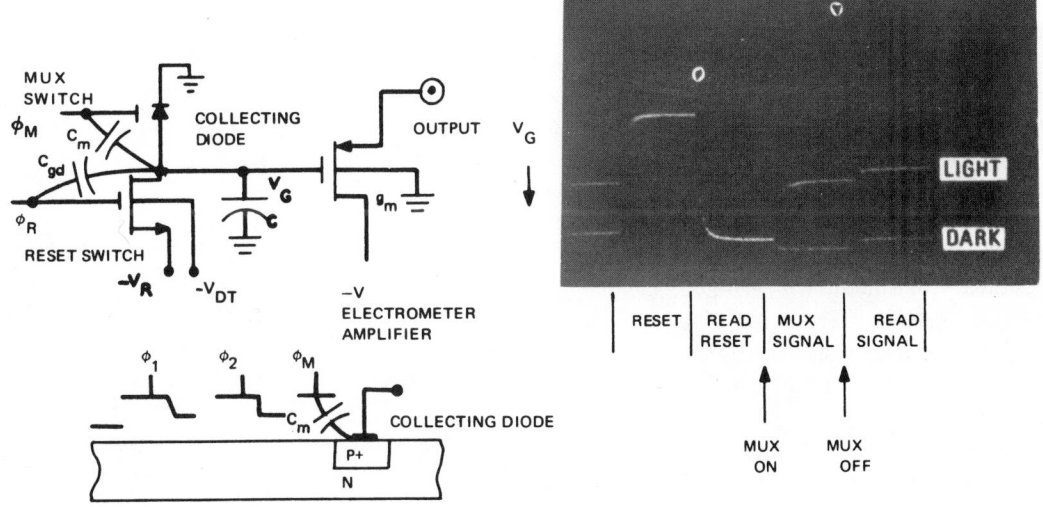

Fig. 4. "On-chip" correlated double sampling (CDS) readout circuit with gate voltage waveform comprised of separate sequential steps.

III. CORRELATED DOUBLE SAMPLING ANALOG SIGNAL PROCESSOR

In the use of a CCD sensor array the natural output is charge integration, and sampling techniques are advantageous to provide on-chip signal processing and reconstruction. In the introduction we mentioned a major limitation to x-y addressed image arrays was the interaction of the clock pulses with the video signal charge to be detected and amplified. In addition to this problem, the random rise and fall times of the clocks give rise to frequency components which fall in the passband of the video preamplifier. The minimum detectable signal or noise equivalent signal (NES) after preamplification was set by the system noise in these image arrays. It seemed as though all of the effort was placed on sensor development with little emphasis on the problem of signal detection and video processing. With the advent of CCD

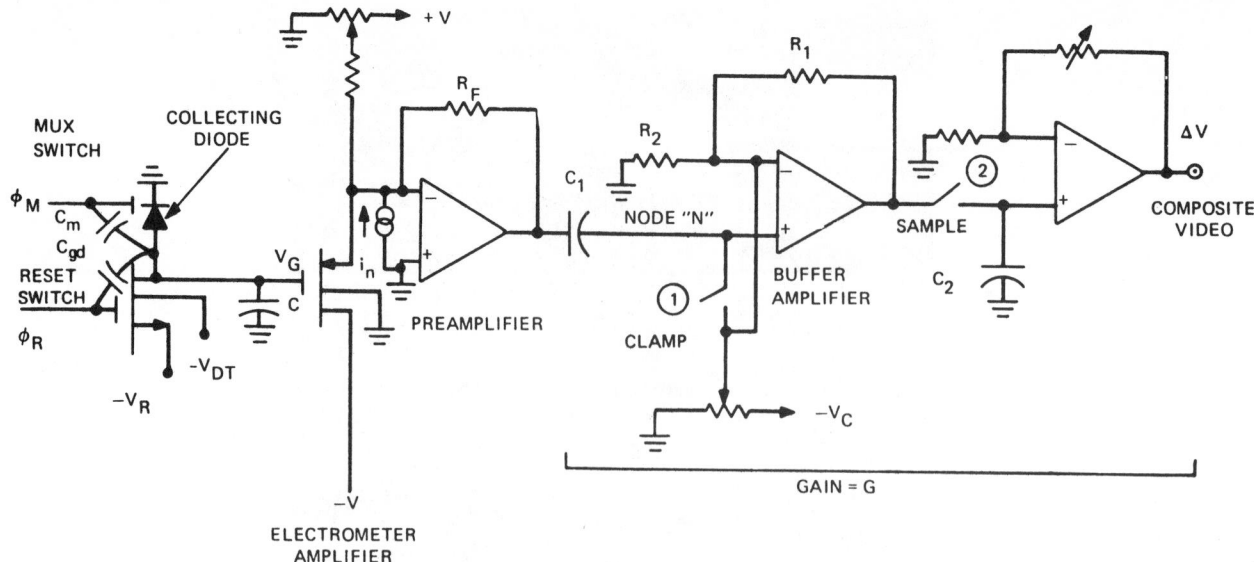

CMOS READOUT CIRCUIT

Fig. 5. Schematic diagram of a CDS processor with critical capacitances, noise sources, and signal nodes.

imagers, the integrity of the signal charge is maintained until signal detection occurs at a common collection diode. The importance of this point cannot be emphasized too strongly since the problem of signal and clock interaction is localized at a common readout circuit. All of the pixel information passes through a common collection diode, integrating capacitance, and electrometer amplifier. This is true for the line and area CCD imagers. We also mentioned in the introduction the low output capacitance ($C = 0.25$ pF) at the collection diode. For a well designed 10-MHz video amplifier, as used with a vidicon, the shunt capacitance may be 20–30 pF. For a noise current i_n, which is shown at the input to the pre-amplifier in Fig. 5, the equivalent noise charge at the collection diode is $C\, i_n/g_m$. Thus, the small C/g_m of the electrometer increases the signal-to-noise charge ratio at the collection diode for such noise currents as shot noise, preamplifier noise, surface-state noise, and supply noise.

Let us examine the four distinct timing intervals employed in the readout circuit of Fig. 4.

1) Reset: The n-channel MOSFET reset switch is turned on and the voltage V_G across the capacitor C is reset to the reference voltage V_R with a noise uncertainty V_n. This noise voltage may be introduced through inadequate filtering of the reference supply voltage and the Nyquist noise contribution of the reset switch, [9], [10] where the latter is given by $V_n = (kT/C)^{1/2}$ or in terms of noise charge $Q_n = (kTC)^{1/2}$. The full Nyquist voltage appears across C when the electrical time constant formed by the series resistance of the reset switch and integration capacitance C is much less than the time the reset switch is on. The p-channel electrometer is connected to an operational amplifier which is the preamplifier in the CCD signal processor circuit shown in Fig. 5.

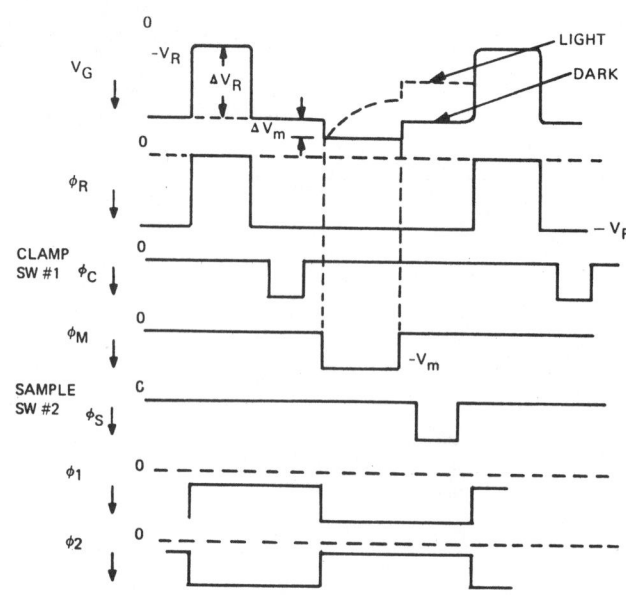

Fig. 6. Timing diagram for CCD line array with CDS analog signal processor. The four steps are separated in sequence to correspond to explanation in the text.

The timing diagram for the signal processing is illustrated in Fig. 6. At the start of the reset interval the pixel charge is in transit to the last well (i.e., see phase ϕ_2 and Fig. 4) of the electrical bit adjacent to the muxgate ϕ_M and the collection diode.

2) Read Reset: After the n-channel reset switch is turned off, the voltage present on the gate of the electrometer consists of a feedthrough pedestal ΔV_R and a noise voltage V_n. With the reset switch off, the gate voltage is holding on a high impedance point with a time constant of seconds. In the read reset interval, the clamp switch 1 is turned on and C_1 is charged to a voltage indicative of the voltage on the gate of the electrometer. Switch 1 is turned off and one side of the capacitor, node N, is *clamped* or dc restored to a reference voltage V_c, while the other side of the capacitor represents the instantaneous sample of the gate voltage. The instantaneous voltage across the clamp switch from this moment on is the differential or incremental charge caused by a change in the on-chip gate or collection diode voltage. With the clamp switch turned off, the measured reset level is holding on the high impedance node N formed by the clamp capacitor C_1, and the noninverting input of the buffer amplifier.

3) Mux Signal: At the start of the mux signal interval the pixel charge is raised in the storage well (ϕ_2 goes high) and ϕ_M goes low to transfer the pixel charge to the collection diode. The collection of pixel charge (minority carriers) discharges the voltage V_G as shown in the signal waveform of Fig. 4. If we assume, for the moment, there is no pixel charge, then the only charge transferred to the gate electrode is the feedthrough pedestal $\Delta V_m = V_m C_m / C$, where V_m is the mux voltage swing and C_m the feedthrough capacitance from the muxgate to the collection diode. The charge is removed, however, when the muxgate is turned off as shown in Fig. 4. A Nyquist noise of $Q_n{}^2 = 2KTC_m$ is introduced, which may be minimized for $C_m < 0.01$ pF for the case where the muxgate does not overlap the collection diode. Alternatively, an overlapping muxgate ϕ_m may be held at a fixed d-c potential with the clock ϕ_2 transferring charge to the collection diode (see Fig. 4). In the absence of any optical pixel charge we would collect the leakage current from the sensor and the shift register wells.

4) Read Signal: After the mux signal is turned off, the running output voltage on node N is the *time difference* between the previously clamped reset level and the hame reset level plus signal increment introduced by the closure of the mux switch (i.e., there is negligible leakage of the reset level between read reset (clamp) and read signal (sample) intervals). *Thus, the reset noise, which includes Nyquist noist and V_R power supply noise, is correlated within a pixel time window.* The signal increment, which consists of sensor and shift register leakage current added to photocharge, is amplified and passed to the output of the signal processor by the closure of the sample switch 2. The output video stream is a sequence of pixel element responses free from reset noise and proportional to the minority carrier signal increment introduced by closure of the mux switch.

The correlated double sampling (CDS) method removes switching transients similar to an earlier technique [11] which used a gated charge integrator in lieu of storing (clamping) the actual diode reset level for subsequent subtraction from the reset level plus signal increment to give the signal increment without reset noise. The Nyquist noise of the reset switch has been removed

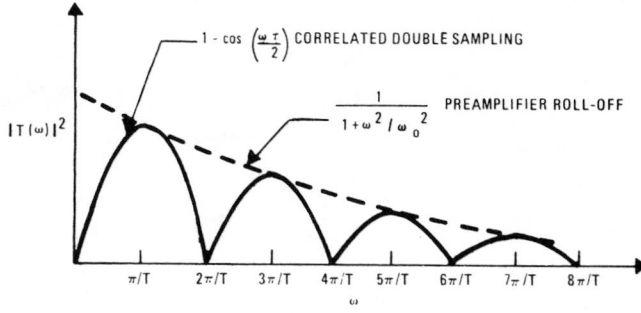

Fig. 7. Filter characteristic of CDS analog processor with $\tau = T/2$.

since it is correlated within a pixel time, and this means a removal of a noise charge

$$Q_n = \frac{(kTC)^{1/2}}{q} = 200e^- \qquad (1)$$

for a 0.25 pF capacitor. The "$1/f$" surface-state noise is also suppressed by the filter characteristic of the analog signal processor which is shown as follows.

The transfer function, which acts on any time-varying components of the signal between clamp and sample intervals, may be written as

$$T(s) = T_0 \frac{(1 - e^{-s\tau})}{1 + s/\omega_0} \qquad (2)$$

where T_0 is the signal gain and τ is the delay time between the end of the clamp pulse and the end of the sample pulse, and ω_0 is the bandwidth of the front-end preamplifier. Fig. 7 illustrates a plot of the filter characteristic for a value of $\tau = T/2$, where T is the clock period. The important features of this filter are the "double zeros" of $|T(\omega)|^2$ at $\omega = 2N\pi/T$ ($N = 0, 1, 2, \cdots$). The "double zero" at the origin ($\omega = 0$) serves to suppress $1/f$ and low-frequency noise arising from power supplies, pulse jitter, etc., and the double zero at even harmonics of the fundamental clock frequency also suppresses surface-state noise generation in the p-channel MOS electrometer amplifier. Thus, the signal processing does not degrade but enhances the qualities of the sensor element by removing the Nyquist noise and filtering the $1/f$ noise while amplifying the signal. In addition to these features, we have automatic dark level subtraction to increase the dynamic range since the video signal is clamped by the reference voltage V_c. There is no need to filter out the clock fundamental and higher order harmonics; and the video output is already in a format for image display or further data processing (see Fig. 8).

IV. RESPONSIVITY AND NOISE CONSIDERATIONS

Analytical Formulation

We can begin by an examination of what we mean by responsivity. We will define this quantity to be R in pA/

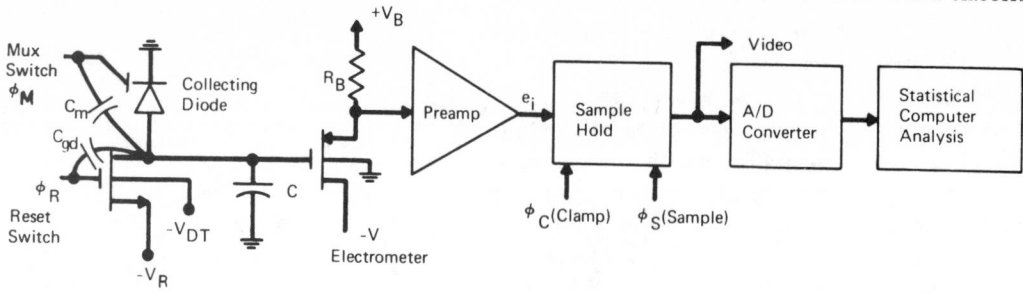

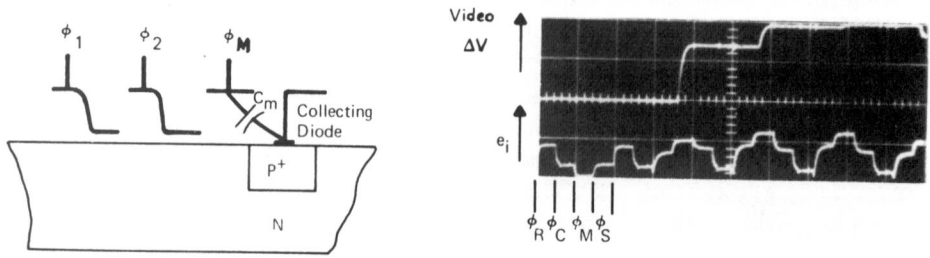

Fig. 8. Test apparatus for measurement of responsivity and noise with typical output waveforms.

mW/m² or current output per input irradiance,[1] which makes the responsivity a function of sensor area. Suppose we attempt to formulate the responsivity by reference to an ideal blackbody source of temperature T_s (e.g., $T_s = 6000$ K). Thus, we can write

$$R = \frac{\int_{\lambda_1}^{\lambda_2} R_\lambda H_s(\lambda)\, d\lambda}{\int_{\lambda_1}^{\lambda_2} H_s(\lambda)\, d\lambda} = \text{individual sensor responsivity} \quad (3)$$

and

$$R_\lambda = \text{spectral responsivity} = \frac{q\eta(\lambda)A}{hc/\lambda}$$

where H_s is the specified source irradiance, λ_2 and λ_1 the reference spectral band (e.g., $\lambda_1 = 400$ nm, $\lambda_2 = 800$ nm), A the area of the sensor, and $\eta(\lambda)$ the effective quantum efficiency. Integration of (3) for the specified temperature source and wavelength interval yields

$$R = 0.186\eta \left[\frac{A}{18\ \mu\text{m} \times 22\ \mu\text{m}} \right] \frac{\text{pA}}{\text{mW/m}^2} \quad (4)$$

where η has been assumed to be constant over the wavelength interval. *The experimental responsivity may be referred to the gate capacitance C of the electrometer in Fig. 8 and expressed in terms of quantities in Fig. 5,*

$$R \cdot (1 - \epsilon)^{2N} = \frac{C\Delta V}{g_m R_F G \Delta E} = \frac{\Delta Q}{\Delta E}$$

= responsivity measured at collection diode   (5)

where C is the node capacitance at the gate of the electrometer, g_m the electrometer transconductance, R_F the pre-

amplifier feedback resistance, G the gain of the signal processor following the preamplifier, and ΔV the change in output voltage for a change in input exposure density ΔE. The responsivity of each sensor is degraded by a factor $(1 - \epsilon)^{2N}$ where $2N$ is the number of transfers the signal undergoes before it reaches the collection diode. In practice, the integral in the numerator of (3), which represents current output, is obtained with a test source whose spectral irradiance profile is known. The irradiance supplied by the test source, while effective over the entire spectral response of the sensor (i.e., 200 nm to 1200 nm for silicon), is converted to an effective irradiance from a 6000 K blackbody source in the 400 nm to 800 nm window.

The measurement procedure involves the use of a 10-b A/D converter, which provides an accuracy of 1024 b. In practice, the signal output from each sensor element in the array is sampled 1024 times at each irradiance level and recorded in terms of A/D bits. The mean and variance are calculated in A/D bits for each signal. The mean represents the signal while the variance corresponds to the uncertainty or noise as determined over 1024 samples. With these quantities we can determine linearity, streaking, responsivity, noise, etc. The measurement circuit is illustrated in Fig. 8 which shows the timing sequence (discussed in Section III) at the preamplifier output and video output signal. The noise is converted from rms A/D bits to an equivalent input exposure density (microjoules/meter²), called the noise equivalent signal (NES), by multiplying the rms A/D bits by the reciprocal slope of the transfer curve at the particular irradiance level (exposure density). The transfer curve is essentially a plot of signal A/D bits versus input exposure density. The reciprocal slope of the transfer curve is called the quantizing interval and is given as

[1] R may easily be converted to electrons/microjoule/meter².

$$Q_I = \frac{\Delta E}{\Delta(\text{A/D bits})} \left(\frac{\mu\text{J}}{\text{m}^2 \cdot \text{bit}}\right). \qquad (6)$$

Thus, if we have B_{rms} bits variance at a specified irradiance level, then the NES becomes

$$\text{NES} = Q_I B_{\text{rms}} \; (\mu\text{J/m}^2). \qquad (7)$$

The NES, which is measured by the above procedure, consists of 4 principal terms:

1) *System noise* from analog signal processor, power supplies, pulse jitter, mechanical vibrations, etc.
2) *Chip noise* from the CCD sensor array. This noise is determined by geometrical design and fabrication processes for the chip.
3) *Radiation shot noise* from the fluctuation in arriving signal photons and which represents a background limited performance.
4) *Quantization noise* from the uncertainty associated with the finite size of the quantizing interval Q_I.

The measured or total NES is given as

$$\text{NES}_T{}^2(\text{total}) = \text{NES}_{\text{syst}}{}^2 + \text{NES}_{\text{chip}}{}^2 + \frac{qE}{R} + \frac{Q_I{}^2}{12} \qquad (8)$$

where the radiation shot noise term involves the exposure density and the quantization noise assumes an error which varies linearly with time [12]. The system noise can be reduced through painstaking debugging of the voltage logic waveforms, examination of grounding, heavy filtering on supplies, careful layout of the analog signal processor, low noise preamplifier, and high source biasing resistor R_B for a given R_F to reduce Nyquist noise contribution. The quantization interval is determined by the dynamic range requirements and the resolution or accuracy in the measurements. We can determine the system noise by measurement, and see the effect of radiation shot noise through measurements at different exposure densities.

The chip NES is determined by the remaining noise in (8). We can formulate analytically the chip NES at the collection diode in Fig. 4. There are 3 sources of noise on the CCD chip:

1) *Nyquist noise* at the reset switch and output capacitance $Q_n{}^2 = kTC$.
2) *Thermal shot noise* associated with the sensor and shift register $Q_n{}^2 = q[I_{LR} + I_{LS} (1 - \epsilon)^{2N}]\tau$, where here $\tau = $ line readout time.
3) *Surface state noise* associated with the transfer of charge to and from Si-SiO$_2$ interface states within the sensor, shift register and electrometer amplifier

$$Q_n{}^2 = kTC_{st}$$

where C_{st} is an effective surface state capacitance determined by the number of transfers, the clock frequency, the effective bandwidth of the signal processor, etc.

We have mentioned that the correlated double sampling

TABLE I
EXPERIMENTAL PARAMETERS OF CCD READOUT CIRCUIT AND ANALOG SIGNAL PROCESSOR

$g_m = 250 \; \mu\text{mho}$	$G = 2.57$	$\Delta V = 3.56 \text{ V}$
$R_F = 75 \; \Omega$	$\Delta E \; 290 \; \mu\text{J/m}^2$	$C = 0.25 \text{ pF}$
$A = 18 \; \mu\text{m} \times 22 \; \mu\text{m}$	$\tau = 1.32 \text{ ms}$	$\epsilon = 10^{-4}$
$2N = 5 \text{ to } 133$	$V_R = -12 \text{ V}$	

removes the *Nyquist noise* of the reset switch; however, to be general we will include this term in the formulation of signal-to-noise which may be written as

$$\frac{\text{S}}{\text{N}} = \frac{\text{signal charge}}{\text{noise charge}}$$

$$= \frac{R(1 - \epsilon)^{2N}E}{(kT(C + C_{st}) + q[I_{LR} + I_{LS}(1 - \epsilon)^{2N}]\tau)^{1/2}}. \qquad (9)$$

If we set the S/N $= 1$ in (5)–(7), then the chip NES becomes

$$\text{NES}_{\text{chip}} = \frac{(kT(C + C_{st}) + q[I_{LR} + I_{LS}(1 - \epsilon)^{2N}]\tau)^{1/2}}{R(1 - \epsilon)^{2N}}. \qquad (10)$$

Measurements

Table I lists the experimental measurements of parameters associated with a 1×64 element CCD line array. The experimental responsivity calculated from these parameters [see (5)] is

$$R(\text{exp}) = \frac{0.065 \text{ pA}}{\text{mW/m}^2} \left(\frac{406e^-}{\mu\text{J/m}^2}\right) \qquad (11)$$

and substitution of (11) into (4) yields

$$\eta(\text{effective}) = 0.35 \qquad (12)$$

for the CCD sensor. The effective quantum efficiency for this CCD sensor must be taken with caution since the relative spectral response profile displays interference fringes due to the $2k$ Å polysilicon electrodes over a $1k$ Å gate oxidation. The relative spectral responses of the CCD sensor compared with a diffused p$^+$/p$^-$/n$(x_j = 12 \; \mu\text{m})$ photodiode are shown in Fig. 9. The photodiode response illustrates a constant quantum efficiency $\eta = 0.65$ between 400 nm and 800 nm and the responsivity for an equivalent area size is 0.124 pA/(mW/m^2). Thus, it is clear that a photodiode sensor would have a definite advantage over its CCD sensor counterpart. Both sensors are overcoated with approximately 3 μm of SiO$_2$.

The response of the CCD sensor array to input irradiance (responsivity) may be determined in another manner through the radiation shot noise. Fig. 10 illustrates the total (NES)^2 plotted versus exposure density $E(\mu\text{J}/\text{m}^2)$. We notice the straight line obtained which indicates the CCD becomes background limited at exposure densities above $50 \mu\text{J/m}^2$. The quantizing interval for these measurements was $Q_I = 0.30 \; \mu\text{J/m}^2 \cdot$ bit, and the re-

173

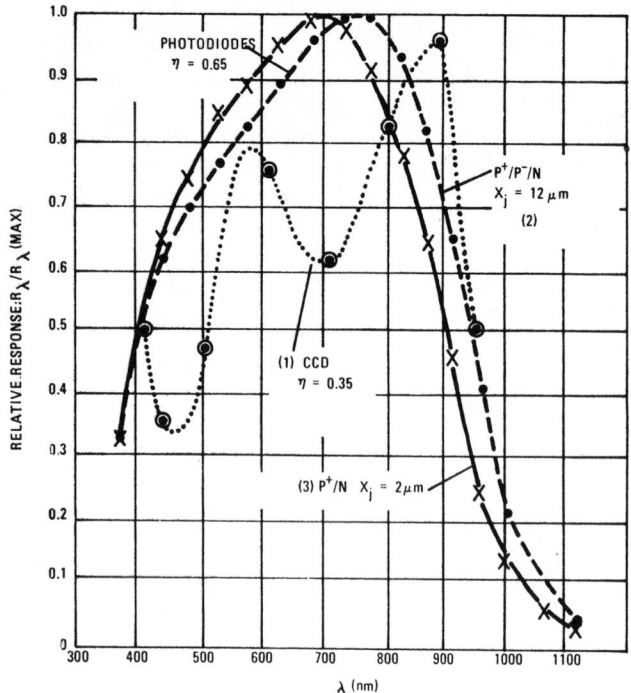

Fig. 9. Relative spectral responses of (1) a CCD sensor with a 0.2 μm silicon gate over a 0.1 μm SiO₂ and of two diffused photodiodes; (2) a p⁺/p⁻/n diode 12 μm deep; and (3) a p⁺/n diode 2 μm deep.

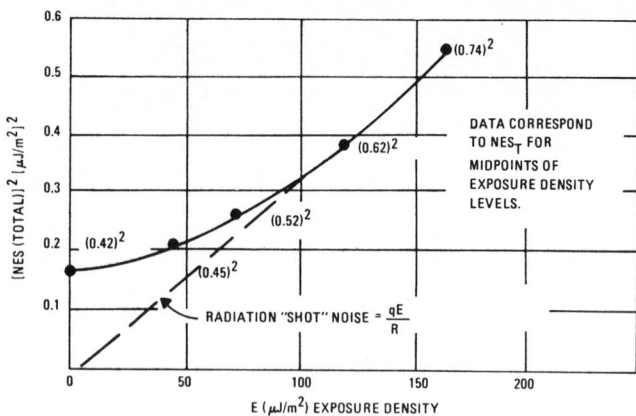

Fig. 10. Effect of radiation shot noise on the total noise-equivalent signal (NES) with a quantizing interval, $Q_I = 0.30$ μJ/m²·bit.

sponsivity obtained from the slope of the curve in Fig. 10 [see (8)] is

$$R = \frac{q\Delta E}{\Delta(\text{NES})^2} = \frac{1.6 \times 10^{-19} \times 140 \times 10^{-6}}{[(0.72)^2 - (0.42)^2] \times 10^{-12}} \cong \frac{0.065 \text{ pA}}{\text{mW/m}^2} \tag{13}$$

in agreement with the responsivity obtained through (5) as calculated in (11).

The dynamic range of a CCD sensor line array (1 × 64 element) is shown in Fig. 11. This particular CCD array had a high quantizing interval $Q_I = 0.80$ μJ/m² · bit and a leakage current of 7.0 pA/well as measured by the offset between clamp and sample intervals (see Fig. 4). The leakage current, to a first approximation, is assumed to buildup linearly down the N-b CCD shift register such that the last well in the register carries approximately $2N$ times the leakage charge as the first well. This leakage current is affected severely by the choice of clock voltages which are -8 and -26 V (substrate grounded). A reduction of the clock voltage amplitudes will reduce the leakage current substantially, however, the resolution will be impaired due to inadequate charge transfer. Fig. 11 illustrates the variance or noise measured for the correlated and uncorrelated sampling case. To obtain the uncorrelated noise we reversed the clamp and sample sequence such that they were not performed in the same pixel time window. This is proof of the correlated double sampling technique in action. Measurements had been taken on another CCD line array with a lower leakage current (1.5 pA/well) and the total NES

measured was 0.40 μJ/m² (160e^-) for a quantizing interval of 0.30 μJ. This total NES is less than the Nyquist noise contribution which is

NES (Nyquist)

$$= \frac{(2kTC)^{1/2}}{R} = \frac{0.70 \text{ }\mu\text{J}}{\text{m}^2}(284e^-) \tag{14}$$

where the factor of 2 is used to illustrate the uncorrelated case of 2 independent readings of clamp and sample. The NES values shown in Fig 11 are for the basic sensor chip. The system NES = 0.15 μJ/m² which is referred to the input of the electrometer amplifier. As Fig. 11 illustrates, the dynamic range is about 75 dB for this particular CCD array with a chip NES = 260e^-; however, measurements on lower leakage devices indicate a chip NES = 135e^- and a dynamic range of 80 dB. All of the measurements were performed under uniform irradiance (line time τ = 1.32 ms with the limitation as the "shot" noise associated with the leakage current on the chip. For example, the device measured in Fig. 11 had a leakage current I_{LR} = 7.0 pA in the register (the sensor leakage current I_{LS} was negligible compared with the register leakage current because of the low sensor bias voltage $V_S = -8$ V) and the shot noise NES contribution.

$$\text{NES(shot)} = \frac{(qI_{LR}\tau)^{1/2}}{R} = 0.591 \frac{\mu\text{J}}{\text{m}^2}(240e^-) \tag{15}$$

which accounts for most of the noise. In the low leakage devices the measured chip NES = [0.33 μJ/m² (135e^-)] can be attributed to the leakage current of 1.5 pA/well or a NES = [0.275 μJ/m² (112e^-)]. The remaining noise is attributed to the surface-state noise and in the low leakage device this is a NES = [0.18 μJ/m² (74e^-)]. The noise measurements do not appear to change across the 1 × 64 element line array and the values quoted are representative of the average noise along the array. One reason for the noise to remain constant across the array is that the major surface-state noise contribution is from the electrometer amplifier. There is also a suppression of

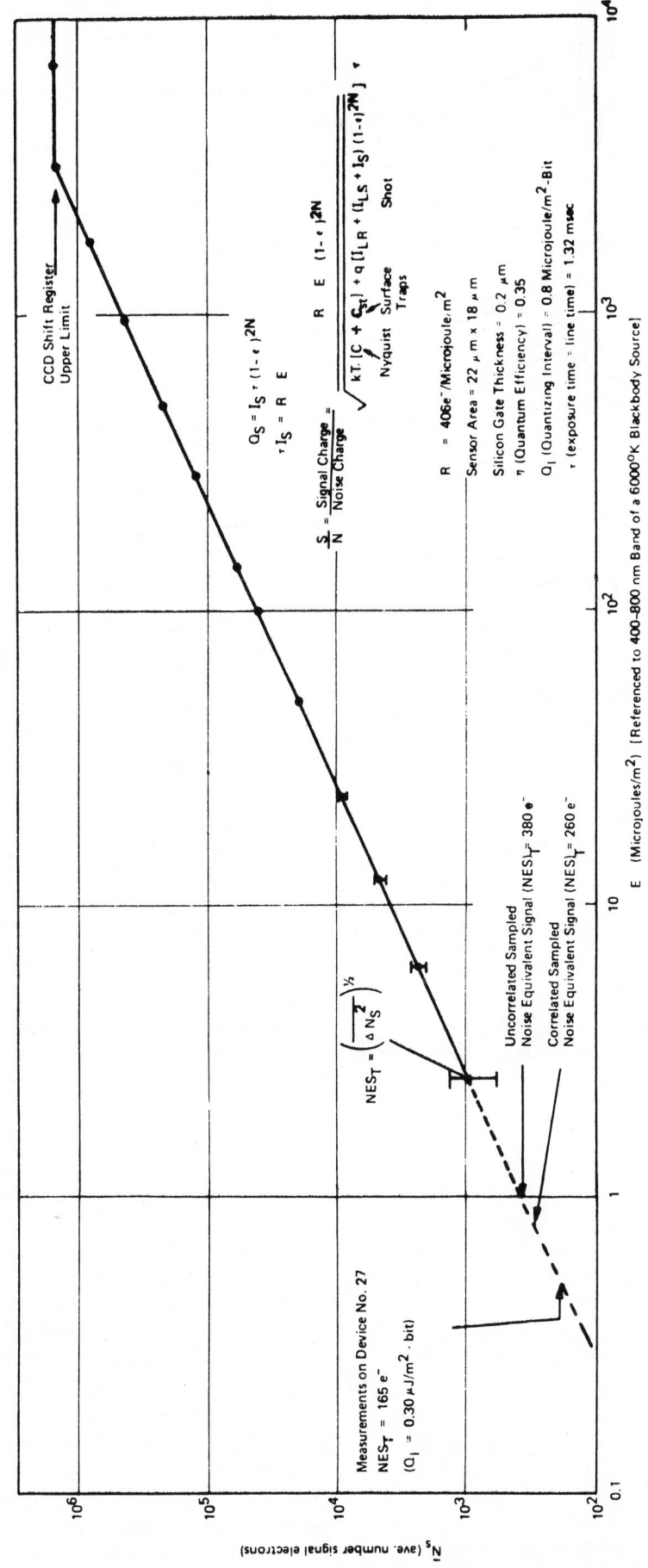

Fig. 11. Dynamic range and noise in a CCD sensor array (1 × 64 elements) with uniform irradiance.

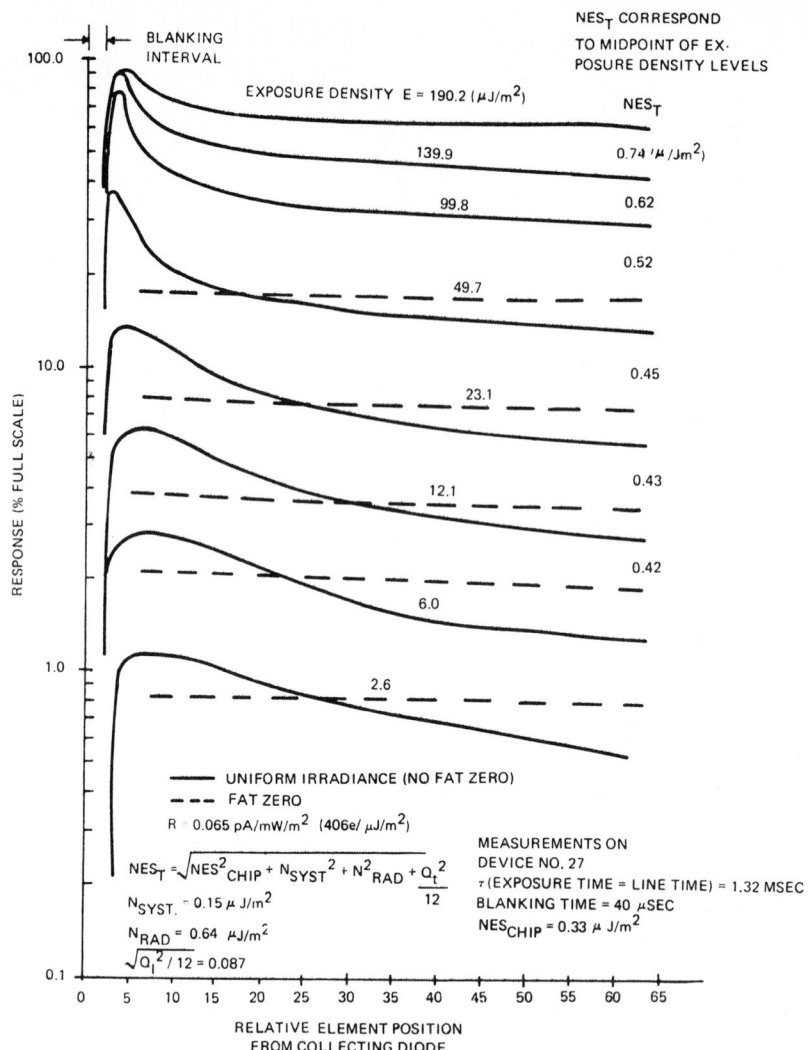

Fig. 12. Low light level response uniformity and noise performance of a 1 × 64 element CCD line array with and without a 20 percent fat zero. (CDS analog signal processing at a clock frequency $f_c = 50$ kHz.)

the surface-state noise within the array by the correlation of noise between adjacent wells [13]. The surface state capacitance may be written as

$$C_{st} \simeq e^2 N_{st} A \qquad (16)$$

where N_{st} is the interface state density[2] (states/cm²·eV) and A the area of the gate electrode of the MOS electrometer. Typical values of $N_{st} \simeq 4 \times 10^{10}$ cm⁻² ev⁻¹ near the band edge at 50 kHz and $A = 6.2 \times 10^{-6}$ cm² yield a $C_{st} = 0.04$ pF and a noise contribution of

$$\text{NES (surface state)} = \frac{(kTC_{st})^{1/2}}{R} = 0.20 \frac{\mu J}{m^2} (80e^-) \quad (17)$$

which is close to the observed value.

Fig. 12 illustrates the responsivity and noise performance of the 1 × 64 element CCD line array discussed in the preceding paragraphs. Table II summarizes the noise

[2] N_{st} is an effective surface-state density modified by the transfer function of the signal processor and the distribution of surface states.

performance and noise sources which influence the performance of the CCD array. The line array is "blanked" or held still for approximately 2 pixel times between lines. In the absence of a fat zero, the video response (response minus dark) varies across the array: high for the first pixels (i.e., pixels nearest collection diode) after the blanking interval and low for the last pixels. Inefficient transfer of excess charge, which accumulates during the blanking interval in the larger potential wells nearest the collecting diode (see Fig. 3), contributes only to the high response. Inefficient transfer of photocharge along the CCD register causes the video output from the last pixels to be diminished by residual charge which is added to the first pixels in the next video line. If we inject an electrical fat zero (20 percent full CCD well) into the shift register, the output is uniform [see dashed lines in Fig. (12)] because of improved transfer efficiency. An electrical fat zero injected over the entire pixel time window produced a high, nonuniform noise at the output. The output noise for a 10 percent full CCD well under these conditions was approximately [4.0 $\mu J/m^2$

TABLE II
SUMMARY OF NOISE PERFORMANCE OF CCD IMAGING ARRAY
(NO ELECTRICAL FAT ZERO)

$R = 406e^{-}/\mu\mathrm{J/m^2}$

Total NES (measured) at $166 \frac{\mu\mathrm{J}}{\mathrm{m^2}} =$ ... $0.74 \frac{\mu\mathrm{J}}{\mathrm{m^2}}$

Calculated radiation shot noise $\left(\frac{qE}{R}\right)^{1/2} =$ ... $0.64 \frac{\mu\mathrm{J}}{\mathrm{m^2}}$

Signal processor noise (measured system noise) $=$ ... $0.15 \frac{\mu\mathrm{J}}{\mathrm{m^2}}$

Quantizing noise $\left(\frac{QI^2}{12}\right)^{1/2} =$ ... $0.087 \frac{\mu\mathrm{J}}{\mathrm{m^2}}$

NES (chip) $= ((0.74)^2 - (0.64)^2$
$\qquad - (0.15)^2 - (0.087)^2)^{1/2} = 0.33 \frac{\mu\mathrm{J}}{\mathrm{m^2}}$

$\qquad = (kTC_{st} + qI_{LR}\tau)^{1/2}$ (theoretical) ... $0.34 \frac{\mu\mathrm{J}}{\mathrm{m^2}}$

NES (shot) $= \frac{(qI_{LR}\tau)^{1/2}}{R} =$ ... $0.275 \frac{\mu\mathrm{J}}{\mathrm{m^2}}$

NES (surface) $= \frac{(kTC_{st})^{1/2}}{R} =$ ... $0.20 \frac{\mu\mathrm{J}}{\mathrm{m^2}}$

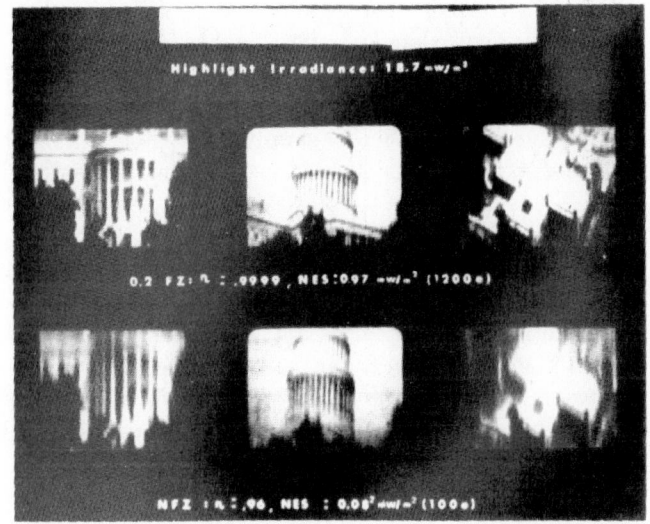

Fig. 13. Low light level imaging with and without a 20 percent fat zero. (Collection diode at photo-bottom for end pictures, at photo-top for middle pictures.)

$(1600e^{-})$] and consisted of irregular "burst" noise. When the time window for electrical injections was narrowed to about 1 percent of the pixel window, the output noise for a 20 percent full CCD well was reduced to approximately [$3.0 \ \mu\mathrm{J/m^2} \ (1200e^{-})$] with no evidence of irregular noise. The method of electrical fat zero injection requires further investigation to produce a truly low noise injection process.

V. LOW LIGHT LEVEL IMAGING

Transfer Inefficiency at Low Light Levels in Surface-Channel CCD Imagers

The *modulation transfer function* (MTF) [4] is determined by the charge transfer inefficiency ϵ. The ability of the CCD to transfer charge from one potential well to another is described by the transfer inefficiency ϵ, which in a practical CCD is limited by surface-state trapping [3]

$$\epsilon = \frac{kTN_{st}}{N_{sig}} \ln \left(f/f_0\right) \qquad (18)$$

where N_{st} is the interface state density (typically $\simeq 1 \times 10^{10}/\mathrm{cm^2 \cdot eV}$), N_{sig} the signal charge in the CCD shift register well (charges/cm^2), and f_0 the characteristic surface-state escape frequency (typically $f_0 \simeq 1 - 10$ kHz). A small amount of background charge or *fat zero* is required to fill the interface states permanently and reduce the transfer inefficiency. Our transfer inefficiency measurements, with a 10 μm optical profile slit placed in the center of the various sensor elements, indicate a variable transfer inefficiency across the CCD line array. The transfer efficiency improves as leakage accumulates to fill the surface states along the shift register. With an electrical fat zero the transfer inefficiency $\epsilon \simeq 1 \times 10^{-4}$ from low frequencies to 2 MHz, which is sufficient to operate both line and area CCD imagers. We have made measurements on the response uniformity with a fat-zero injection and there is less than $\pm$ 5 percent variation in response across a 1×64 element array with $\pm$ 5 percent variation in NES across the array. The deviation from straight line linearity over the entire dynamic range (> 60 dB with an electrical fat zero) is within $\pm$ 2 percent.

Imaging Evaluation

In order to evaluate the CCD line arrays we employed a rotating drum to provide the along track image motion while the line array was scanned electronically in the across track direction. A transparency was mounted on the drum and irradiated by a calibrated source with all irradiance levels referenced to a blackbody source at 6000 K in the 400 nm to 800 nm band. Fig. 13 illustrates the low light level performance of the CCD 1×64 element line array with 3 ms exposure time (0.75 ms line readout time) and a highlight irradiance of 18.7 mW/m^2, which is about 120 pA of signal current. Our minimum detectable signal with a 20 percent fat zero is 0.97 mW/m^2 or about 6 pA. The minimum detectable signal, if we could introduce the fat zero with negligible noise,[3] is 0.08 mW/m^2 or about 0.4 pA. Notice the loss in resolution if we remove the electrical fat zero. In the scene of the White House the elements furthest from the collecting diode are at the top of the picture. It is apparent that the long number of empty CCD wells (as denoted by the black "tree" area), affects the resolution severely since these wells do not have a background "radiation fat zero" to provide low transfer inefficiencies.

[3] A method for low-noise, electrical injection of a fat zero [14] can give a NES less than the simple shot noise associated with the fat zero.

Since a CCD imager uses an analog shift register we must have low leakage register elements in addition to the requirements of low leakage sensor elements. With the interline transfer method we can separate the causes of leakage in an imager by simply turning on and off the transfer gate. In general, we encounter more leakage in the shift register than the sensor elements. One advantage of the interline transfer approach is that the shift register can be operated "free-running" to distribute the leakage throughout the register. If the register is "blanked" or held stationary for a period of time, then the local dark-current spots will begin to saturate the analog signal processor. The pictures illustrated in Fig. 13 illustrate the high quality imaging that is possible with CCD imagers.

VI. Conclusions

We have developed a method of signal processing called *correlated double sampling* which removes the switching transients at the output collection diode, eliminates, the Nyquist noise of the reset switch-output capacitance combination, provides dc restoration and increases dynamic range, and suppresses surface state and $1/f$ noise contributions. The analog signal processor, which uses this technique has been operated with clock frequencies from 800 Hz to over 3 MHz. We have measured the intrinsic noise-equivalent-signal (NES) or minimum detectable input exposure density of the CCD imager with this technique. The primary limitation to the NES is the shot noise associated with thermal leakage current generated in the sensor and shift register. The surface channel CCD imager requires a 20 percent fat zero for transfer inefficiency $\epsilon \leq 10^{-4}$ to obtain geometrical resolution. The excess noise associated with the electrical fat zero limits the sensitivity of the surface channel CCD and attention must be directed toward a method of low noise electrical injection. For an electrical fat zero introduced by a gated diode, our measurements indicate a noise charge of 1000 to 1200 electrons; however, a redesign of the input electrical injection circuit [14] should enable surface channel CCD's to operate with an NES limited by thermal leakage shot noise and a dynamic range in excess of 80 dB. For our simple gated diode injection circuit with a 20 percent fat zero, we have achieved greater than 60 dB dynamic range with less than ± 2 percent deviation from linearity over this range and less than ± 3 percent variation in responsivity and NES across a 1×64 element line array. The dark current variation was less than ± 3.4 percent across the array. We have made measurements without an electrical fat zero and under uniform exposure to indicate a noise charge of 135 electrons at an integration time of 1.32 ms. The noise was associated primarily with the thermal leakage current in the CCD shift register. The thermal leakage current on the better arrays was about 50 nA/cm².

Spectral response and responsivity measurements indicate a nonuniform spectral response due to interference fringes caused by this silicon gate thickness and the underlying SiO_2. The effective quantum efficiency of the silicon gate CCD sensor in the range from 400 nm to 800 nm was $\eta = 0.35$, compared with a photodiode $\eta = 0.65$, which is constant. Consideration should be given to the design of the CCD imaging array with the use of a diffused photodiode for the sensor. The g_m/C ratio should be increased at the output collection diode to overcome system noise limitations; and the quantizing interval Q_I, if digital signal processing is used, should be decreased in accordance with dynamic range and data rate limitations.

Acknowledgment

The authors wish to express their appreciation to R. M. McLouski for the diffusions and oxidations, P. R. Reid for the polysilicon depositions, J. Grossman for the chemical processing, C. J. Taylor for the photolithography, and D. S. Herman for the special aluminum/silicon evaporations. They would also like to thank D. H. McCann for his valuable assistance and technical discussions, and to express appreciation to W. S. Corak (Manager), M. N. Giuliano and R. C. Gallagher (Supervisors), of the Solid State Systems Technology Laboratory, and to G. Strull, Manager of the Advanced Technology Laboratories, for their encouragement and support. A special thanks is accorded to C. Lesniewski and S. Viscomi for preparation of the manuscript.

References

[1] W. S. Boyle and G. E. Smith, "Charge-coupled semiconductor devices," *Bell Syst. Tech. J.*, vol. 49, p. 587, 1970.
[2] G. F. Amelio, M. F. Tompsett, and G. E. Smith, "Experimental verification of the charge coupled device concept," *Bell Syst. Tech. J.*, vol. 49, p. 593, 1970.
[3] M. F. Tompsett, "The quantitative effects of interface states on the performance of CCD's," presented at the Int. Electron Devices Meeting, Washington, D. C., Oct. 1971; also *Tech. Papers Abstracts*, p. 70, and *IEEE Trans. Electron Devices*, vol. ED-20, pp. 45–55, Jan. 1973.
[4] W. B. Joyce and W. J. Bertram, "Linearized dispersion relation and Green's function for discrete charge transfer devices with incomplete transfer," *Bell Syst. Tech. J.*, vol. 50, p. 1741, 1971.
[5] M. H. White, D. R. Lampe, F. C. Blaha, and I. A. Mack, "Charge coupled device (CCD) imaging at low light levels," presented at the IEEE Int. Electron Devices Meeting, Washington, D. C., 1972.
[6] ——, "Characterization of charge coupled device line and area-array imaging at low light levels," presented at the IEEE Int. Solid-State Circuit Conf., Philadelphia, Pa., 1973.
[7] R. H. Dyck, and G. P. Weckler, "A new self-scanned photodiode array," *Solid-State Technol.*, p. 37, July 1971.
[8] M. Ashikawa, N. Koike, T. Kamiyama, and S. Kubo, "A new spike noise elimination technique for photosensitive arrays," presented at the IEEE Int. Solid-State Circuits Conf., Philadelphia, Pa., 1973.
[9] J. E. Carnes and W. F. Kosonocky, "Noise sources in charge-coupled devices," *RCA Rev.*, vol. 33, p. 327, 1972.
[10] D. F. Barbe, "Noise and distortion considerations in charge-coupled devices." *Electron. Lett.*, vol. 8, p. 207, 1972.
[11] J. D. Plummer and J. D. Meindl, "MOS electronics for a portable reading aid for the blind," *IEEE J. Solid-State Circuits*, vol. SC-7, pp. 111–120, Apr. 1972.
[12] J. A. Betts, *Signal Processing, Modulation and Noise*. New York: Elsevier, 1971.
[13] K. K. Thornber and M. F. Tompsett, "Spectral density of noise generated in charge transfer devices," *IEEE Trans. Electron Devices*, vol. ED-20, p. 456, Apr. 1973.
[14] S. P. Emmons and D. D. Buss, "Techniques for introducing a low noise fat zero in CCD's," presented at the Device Res. Conf., Boulder, Colo., June 1973.

All-Solid-State Camera for the 525-Line Television Format

CARLO H. SÉQUIN, MEMBER, IEEE, EDWARD J. ZIMANY, JR., MICHAEL F. TOMPSETT, SENIOR MEMBER, IEEE, AND E. N. FULS

Abstract—A charge-coupled image sensor of the vertical frame transfer type has been fabricated with three-phase three-level polysilicon electrodes. The device has 496 vertically interlaced rows of elements and 475 resolution elements/line. The imaging area, measuring 12.8 × 9.6 mm², corresponds to that of a 1-in vidicon. Defect-free devices have not yet been fabricated, but with an appropriate effort such devices seem feasible.

The device is operated in a self-contained camera, measuring 6 × 6 × 15 cm, containing the countdown circuitry, and implemented with commercially available TTL. The clock line drivers are built with discrete transistors. The camera produces a suitably filtered black-and-white video signal in the standard 525-line television format, which includes the necessary blanking periods and synchronization signals. Large-scale integration (LSI) of these circuits could readily lead to a camera which could be smaller by more than a factor of two.

I. INTRODUCTION

CHARGE-COUPLED devices (CCD's) employing three separate levels of polysilicon to form a three-phase overlapping semitransparent electrode structure have been described previously [1]. The advantage of this electrode structure for the fabrication of large area image sensors lies in the reduced probability of fatal interphase shorts. The realization of a 220 × 256-element area image sensor operating in the Picturephone® format [2] demonstrated that indeed a large number of operational devices could be readily obtained.

The same basic approach has now been used in the design and fabrication of a device with the imaging format of the 1-in vidicon and with an output that matches the 525-line television format. A compact self-contained camera, which produces the complete video signal with the proper blanking and synchronization periods, has been built for operating and evaluating the devices.

The basic operation of CCD's as well as alternative approaches to the design of solid-state image sensors are presented and discussed elsewhere [3].

II. DEVICE DESIGN

The organization of the device is shown schematically in Fig. 1. As in the earlier device [2], a surface channel frame transfer approach is employed. The imaging area of the device measures 12.825 by 9.672 mm², so that the actual sensing area is about the same as the area scanned on a 1-in vidicon camera tube. In the 525-line television format two interlaced

Manuscript received July 30, 1975; revised September 19, 1975.
The authors are with Bell Laboratories, Murray Hill, NJ 07974.
®Registered service mark of the American Telephone and Telegraph Company.

fields are displayed at a frame rate of 30 Hz, where each field contains from 242 to 244 active lines. To provide some extra margin, the imaging area of the device consists of 248 rows of vertical transfer cells, measuring 39 μm in the vertical direction. The storage area contains the same number of cells. However, in order to conserve active silicon area, its transfer cells are only 30 μm long and have an active electrode length of 10-μm and 2-μm mutual overlap.

Black-and-white television displays normally require at least 450 picture elements/line, to be read out in an active line time of 52.5 μs. The readout rate of the serial register has been chosen to be equal to 8.946 MHz, or 2.5 times the subcarrier of 3.578 MHz used in transmitting color television signals. This would lead to 470 cells in the horizontal direction. In order to obtain an integer number of μm for the cell size, the horizontal dimension of the imaging area is actually composed of 475 cells, each 27 μm long. Even for a surface channel device, the resulting 9-μm-long transfer electrodes in the serial register should not then limit transfer efficiency at the required clock rate of almost 9 MHz, since the low substrate doping level of 5×10^{14} cm^{-3} leads to substantial fringe fields which assist charge transfer.

The serial register is extended several elements beyond the bus lines of the storage area to minimize noise in the output amplifier. Furthermore, a short serial dummy register has been added so that the video signal can be sensed differentially between the two registers and some cancellation of the pulse pickup obtained. In both registers a resettable floating gate is employed to sense the signal and typically the potential on the floating gate is reset once every line time during the horizontal retrace interval.

III. DEVICE FABRICATION

Photomasks for the devices were made on an electron beam exposure system (EBES) [4] that permitted single step exposure of a master mask at unity magnification. This avoided step-and-repeat errors, the introduction of defects, and the large number of reticles that would have been required with a conventional optical step-and-repeat camera. The size of the devices was such that only two could be fitted onto a 5-cm wafer.

The devices are fabricated on p-type substrates doped with 5×10^{14} cm^{-3} boron. The gate oxide of 100 nm thickness is thermally grown and annealed in H$_2$ at 500°C for 1 hour. It is then covered with a 100-nm-thick layer of silicon nitride. This layer is part of a dual-dielectric-gate insulator and protects the gate oxide in the transfer areas during all subsequent processing steps. The nitride layer is patterned and the under-

Reprinted from *IEEE Trans. Electron Devices*, vol. ED-23, pp. 183–189, Feb. 1976.

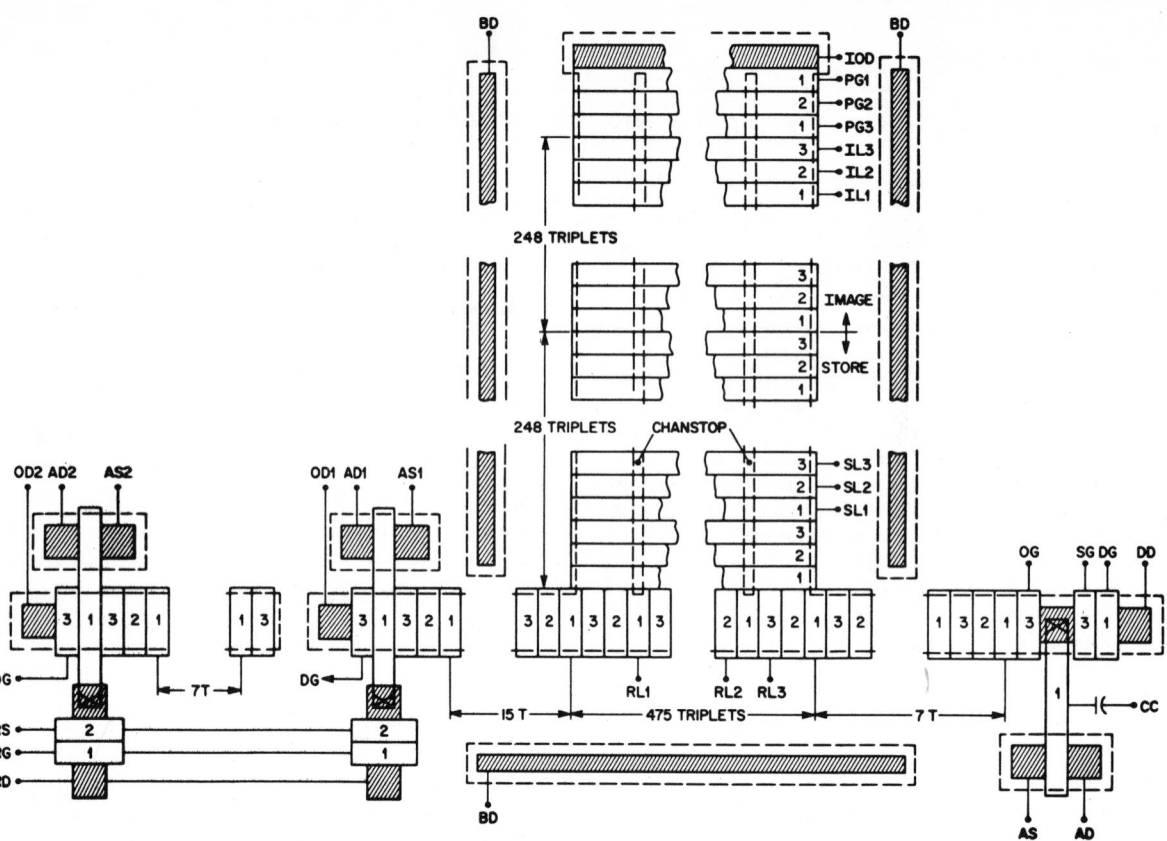

Fig. 1. Schematic layout of the 475 × 496 element charge-coupled area image sensor.

lying oxide etched to allow the diffusion of the n^+ source and drain features. The transfer channels are subsequently defined with a 4×10^{14} cm^{-2} boron ion implant which is performed at 110 keV through the double gate insulator using the photoresist as a mask. The three polysilicon electrode levels are subsequently deposited, patterned, and thermally oxidized. No intermediate oxide etching steps are required. Because of the silicon nitride layer, all three sets of electrodes will automatically be formed on the same gate insulator structure.

The whole device is then covered with 400 nm of a deposited oxide. Two separate window masks are used to open the contact windows through the nitride and the oxide layer, respectively, so as to avoid undercuts which might cause breaks in the top metal pattern. Annealing steps are then performed to improve transfer efficiency and reduce dark current. An aluminum metallization provides bus lines of high conductivity, interlevel contacts, and bonding pads. Devices in wafer form are first operated on a probe test station and functioning devices are then mounted in a 30-pin dual-in-line package, as shown in Fig. 2.

IV. Experimental TV Camera

A self-contained television camera, shown in Fig. 3, which uses the standard 525-line television format, was built to operate the device and to verify the circuit designs required in the construction of compact solid-state cameras. In order to minimize device heating by the 3.4 W of power dissipated in the logic and drive circuitry, the sensor device was placed in

a separate compartment which was thermally isolated from a rear chamber containing the circuit boards. A solid lens mount, consisting of aluminum and brass, allows the attachment of any standard C-mount lens, while providing a heat sink to ambient room temperature. The overall camera size, excluding the lens, is $6 \times 6 \times 15$ cm (Fig. 3). The circuitry compartment is 11 cm deep and contains four plug-in boards for the countdown logic, the driving interface, and the video processing.

V. Timing Logic

The main considerations in the design of the countdown logic were to minimize package count and power consumption. Complementary MOS circuitry was first considered, but since many of the logic blocks have to switch at speeds corresponding to frequencies in the MHz range, the low-power Schottky-clamped TTL family was chosen, yielding a suitable compromise between power and speed. The present design uses 21 dual-in-line TTL packages, and a commercially available p-MOS synchronization generator package. Although the total gate count in all these packages is higher, it is estimated that all the necessary functions could be generated using a customer-designed large-scale integration (LSI) chip containing not more than 250 gates.

A block diagram of the camera logic circuit is shown in Fig. 4. The master clock, operating at three times the chosen element rate of 8.946 MHz, is built with a crystal and two high-speed Schottky-clamped gates. It drives a triplet genera-

Fig. 2. Illustration of the 16×20 mm^2 chip mounted in a 30-pin dual-in-line package. (Background has mm grid.)

Fig. 3. Illustration of the compact all-solid-state black-and-white television camera operating with the 525-line television format.

tor, built with a shift register, which produces the three driving phases for the serial register. A 10-bit element counter defines the line time equal to 568 elements. It also produces, through a logic decoding tree, the auxiliary pulses necessary to gate the driving pulses of the serial register, to control the line steps in the storage area, and to control the sync generator package. Furthermore, a signal at twice the line frequency f_L is used as the input to the 10-bit line counter, which controls the vertical format. Another signal at $\frac{1}{8}$ of the element rate is the master clock for the vertical triplet generator during frame transfer and during line steps.

The 10-bit line counter counts 525 half-lines to define each field. It controls the vertical triplet generator to produce 248 line steps in the storage area and subsequently 249 consecutive triplets for frame transfer lasting 10.5 line times or about 667 μs. This vertical triplet generator uses an internal feedback so that it will always stop in the same state, i.e., with the phase-2 storage area driver high. To obtain the proper interlacing format, an additional flip-flop, which is driven by the line counter, switches the drivers of the imaging area to the proper integration potentials. The two fields are thus integrated under electrodes 1 and 2 + 3, respectively. Total current drawn from the 5-V logic power supply is 170 mA. The 11 dual-in-line packages occupy two of the four plug-in circuit boards in the camera.

VI. DRIVERS

Operation of the image sensor in the frame transfer mode requires nine driving interfaces between the logic circuitry and the device, three each for the imaging area, the storage area, and the serial register. The image sensor requires serial readout drivers which are capable of driving a worst case load capacitance of 240 pF at 9 MHz between two levels, while the storage and image section drivers have to drive 7000 and 9000 pF at 0.373 MHz, respectively. In addition, all drivers must be capable of driving and holding the required voltage levels in the presence of large interphase capacitances caused by the mutual overlap of adjacent electrodes. For maximum signal handling and best transfer efficiency, subsequent clock pulses must also overlap. This result can be obtained by pulse stretching in the driver itself, although the required overlap could alternatively have been produced in the logic triplet generators.

This requirement for insensitivity against interphase capacitance coupling eliminated any of the commercially available integrated drivers. Separate drivers built from discrete components were thus used. They have to fulfill the following additional constraints: the design should be simple to give low component count, and power dissipation must be minimized since nine of them have to fit on a small board. The final design is shown in Fig. 5. All drivers basically consist of one or two switches, which drive a complementary push-pull pair of emitter followers. This configuration can provide the current necessary to drive the capacitive loads and can also supply and sink the currents caused by the interphase capacitances. Pulse stretching, and thus mutual overlap, is obtained by capacitively loading the logic drive signal with C_{OL}. The resistor diode network R_1, D_1 causes the positive logic pulse edge (trailing edge of the output pulse) to experience greater delay.

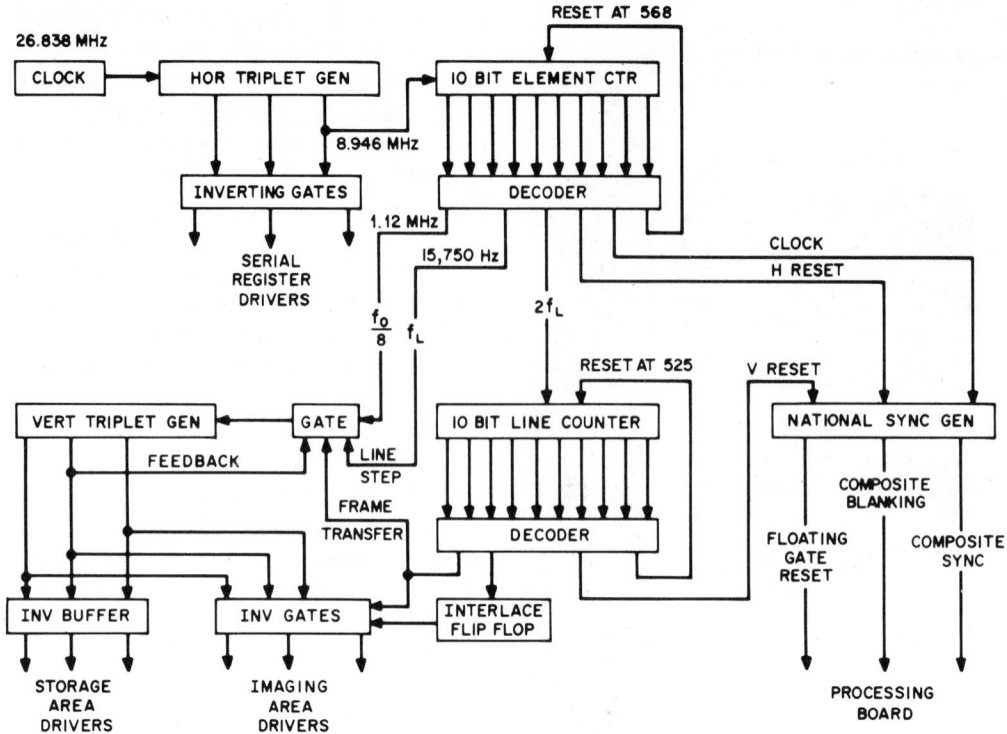

Fig. 4. Block diagram of the camera control logic.

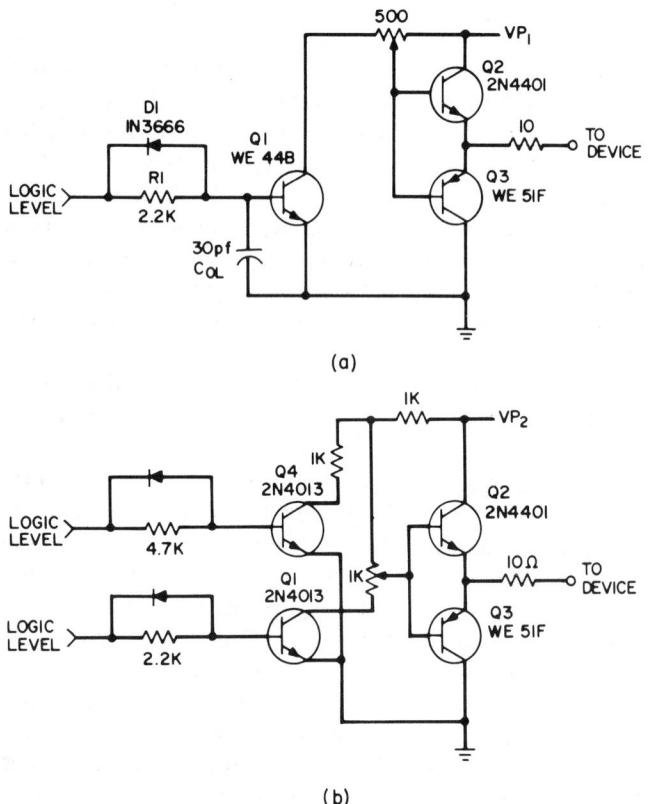

(a)

(b)

Fig. 5. (a) Circuit diagram of the two-level driver. (b) Circuit diagram of the four-level driver.

For simplicity the same design was used for all nine drivers, with the component values and the number of switch transistors being varied to match the driving requirements. While the drivers for the serial register and for the storage area [Fig. 5(a)] operate between only two voltage levels (V_R, V_P) the drivers for the image section are somewhat more complicated [Fig. 5(b)]. An additional switch transistor Q_4 has been added to allow switching between four voltage levels. During the frame transfer the potential on the electrodes is pulsed regularly between V_R and V_P. But during the integration time both of these potentials are lowered. Charge integration at a voltage $V_A < V_P$ reduces the amount of dark current collected. At the same time the nonintegrating electrodes are held at a lower potential, which biases the interface potential slightly into accumulation and thereby reduces preferential blooming along the vertical transfer channels [5]. The total power consumption of the nine drivers, which all fit on one 4 × 11 cm board, with the device operating at a clock rate of 9 MHz and with a V_P of 12 V, is 1.9 W.

VII. VIDEO PROCESSING

The image sensor incorporates a dummy register shown in Figs. 1 and 6 and a differential amplifier to minimize pulse pickup by on-chip preamplification and cancellation. The signals are sensed in the two registers by floating gates, which are reset to the reference potential applied to diode RD when a logic pulse turns on the gate RG during the horizontal retrace interval. During the active line interval, each time that phase 3 turns off, a signal charge packet is transferred under the floating gate. The corresponding change in potential is amplified differentially by the on-chip amplifier consisting of transistors Q_3 and Q_4 and the load resistors R_L, which are set to provide a voltage gain of four. The difference signal is then amplified off-chip by a further factor of five using

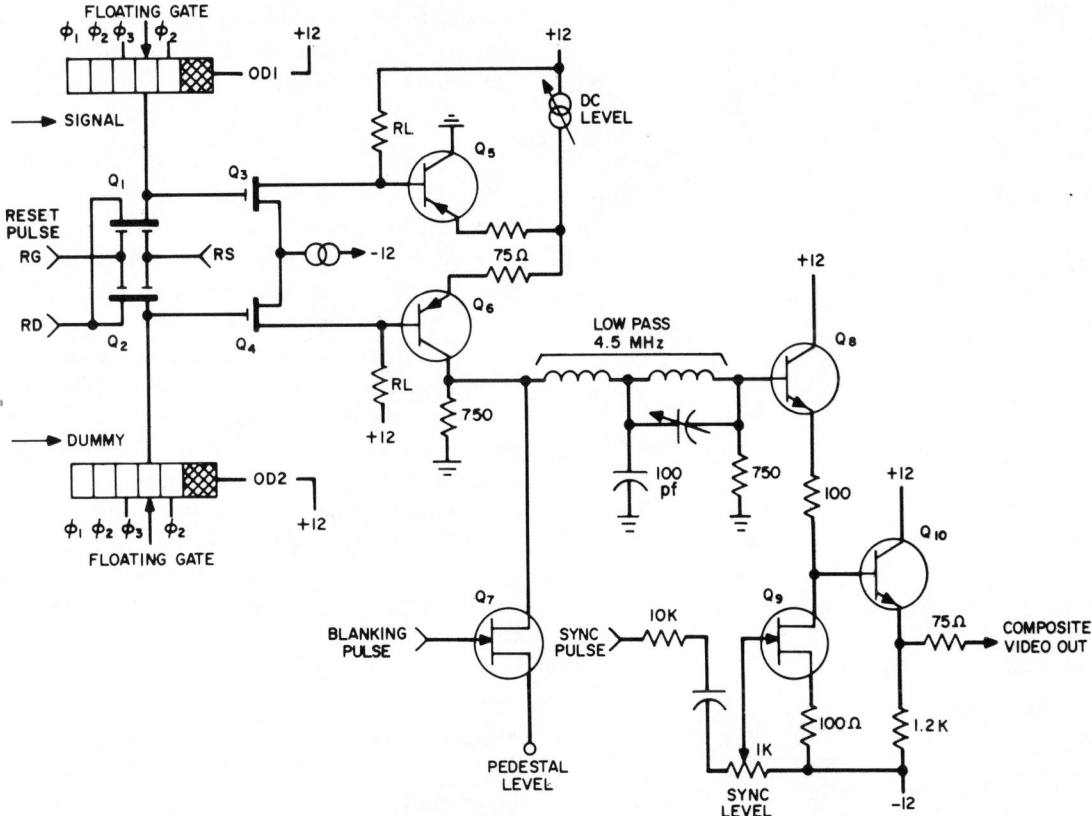

Fig. 6 Circuit diagram of the video processor.

Fig. 7. Illustration of a television display produced by the all-solid-state 525-line television camera.

transistors Q_5 and Q_6 on a video processing board (Fig. 6). Transistor Q_7 is used to insert the blanking pedestal at the output of Q_6 and to gate out unwanted transients occurring during the retrace intervals. The signal is then filtered by a constant-K T-section low-pass filter with an output arm resonant at 8.946 MHz to further reduce clock pickup. Q_9 adds the composite synchronization signal to the emitter follower output of Q_8. The composite video signal is then put

onto a 75-Ω coaxial line by Q_{10}. The output voltage is approximately + 0.7 V for a saturated white level. Blanking pedestal, sync pulse height, and output dc level are all adjustable. The bandwidth of this circuit is determined by the low-pass filter and is set to have its −3-dB point at 4.5 MHz. This processing circuitry is built on part of a 4 × 11 cm board with enough room to spare for the future incorporation of other circuitry, for example, a modulated RF transmitter for remote operation. Overall power consumption is 25 mA at ±12 V.

VIII. RESULTS

Very few wafers of this device have been fabricated owing to a limited processing effort. However, the number of functioning devices is remarkably high (on the order of 25 percent) considering their unusual size. Unfortunately, these devices did not have what we now consider to be an optimum hydrogen anneal at the end of the processing sequence, and extra surface state densities that were formed during the ion implantation through the nitride and gate oxide were not removed. Thus values of transfer inefficiency and dark current were not as good as those obtained on other devices having the same electrode structure[1], [2]. An example of the overall picture quality is shown in Fig. 7. The lack of sharpness in the lower part of the picture is due to limited transfer efficiency in the vertical direction.

Other devices, however, which had some strong localized defects, demonstrated that the spatial resolution can be made to meet the design specifications when it is not degraded by

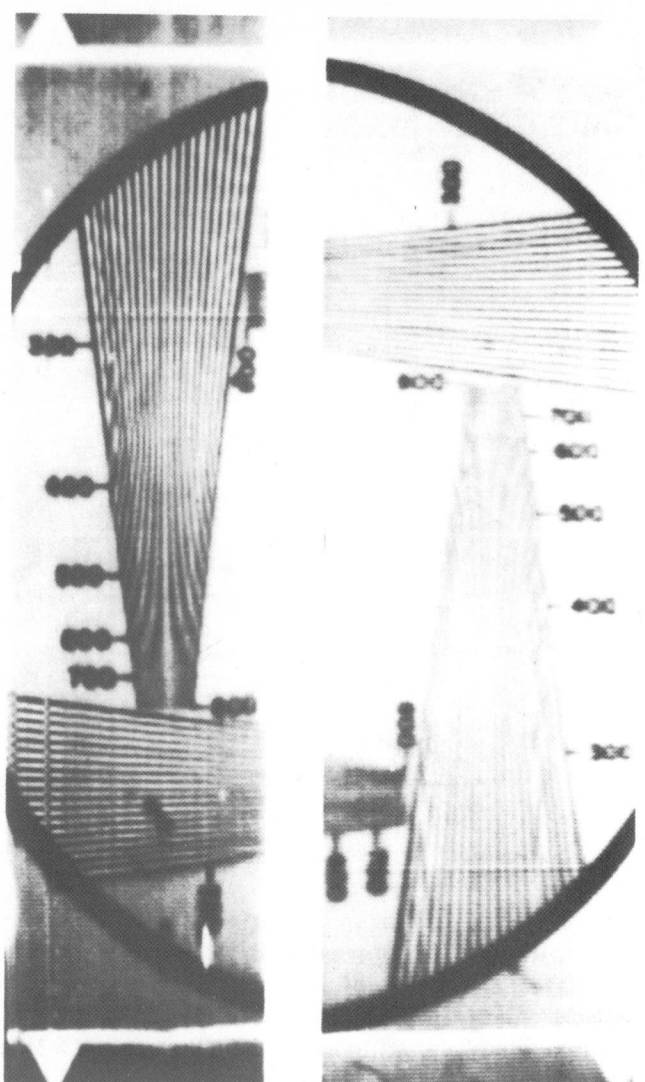

Fig. 8. Resolution wedges showing horizontal and vertical resolution at the left and right extremes of the imaging area of the device.

aliasing is observed since it is suppressed by the substantial overlap of the collection areas of the cells corresponding to the two interlaced fields.

IX. DISCUSSION

At this time we have not made high-quality solid-state television image sensors which were at the same time free of defects, had low transfer inefficiency, and had low dark current. All the necessary steps to that goal, however, have been shown to be feasible by this and other work, and it remains for a dedicated development effort to bring everything together. Large area photomasks virtually free of defects have been fabricated. A high percentage of operating devices has been found, verifying the suitability of the basic design. The three-level polysilicon electrode structure has again proven to be very tolerant of minor photolithographic defects introduced during device fabrication. Low dark current densities (<5 nA $\cdot$ cm^{-2}) and low surface state densities (1×10^9 cm^{-2}) have been demonstrated on other devices with this electrode structure, but a sustained effort is required in order to obtain them routinely.

In order to obtain uniform spectral response in such a device with polysilicon electrodes, it is necessary to use substrate side illumination. The backside surface of the thinned-down device can then be properly treated so as to minimize light reflection and carrier recombination. This has been demonstrated on small devices [6] and the technique could be extended to larger ones.

About three quarters of the volume of the experimental camera described in this paper is used for the countdown circuitry and the MOS pulse drivers, which are built with standard TTL logic components and discrete transistors, respectively. Integration of these circuits into custom-made medium-scale integration (MSI) or LSI circuits should reduce the size of the camera by more than 50 percent. Ultimately one would arrive at the point where the size of the camera is dominated by the optical components.

X. CONCLUSIONS

Although the use of CCD's with three levels of polysilicon is not the only approach which may lead to commercially viable television image sensors, it has permitted, with only a limited effort, the realization of a 475 × 496-element sensor for the standard 525-line television format. Considerable development effort is still required to obtain inexpensive high-quality devices and a fully integrated implementation of the associated circuitry on only one or two chips. Nevertheless, we believe that this work demonstrates the feasibility of very compact solid-state TV cameras.

ACKNOWLEDGMENT

The authors would like to thank J. R. Barner for his expert assistance in laying out the device, R. F. W. Pease, W. J. McNamara, and J. P. Ballantyne for their support in the fabrication of the photomasks on EBES, D. O'Shea and J. L. Statile

transfer inefficiency. Fig. 8 demonstrates this fact by showing resolution wedges reproduced with the extreme left- and right-hand parts of the imaging area. In both cases the horizontal resolution is limited by the number of vertical transfer columns. The corresponding Nyquist limit occurs at 475 × 3/4 TV lines, since the number of lines is referred to the picture height. Indeed the displayed wedges merge towards the point corresponding to 356 TV lines. Beyond that point aliasing occurs, which shows that the resolution is not seriously degraded by the cell geometry or by the lateral diffusion of the generated carriers. Furthermore, comparing corresponding wedges at left and right extremes of the imaging area shows that the transfer efficiency in the serial register, even at the 9-MHz clock rate, produces negligible loss in horizontal resolution.

In the vertical direction the expected resolution is 496 TV lines reduced by the Kell factor of about 75 percent owing to the discrete display lines in the monitor. Indeed the corresponding resolution wedges fade out around 350 lines. No

IEEE TRANSACTIONS ON ELECTRON DEVICES, VOL. ED-23, NO. 2, FEBRUARY 1976

for help in testing and screening of the devices, and D. A. Sealer for many helpful discussions.

REFERENCES

[1] W. J. Bertram, A. M. Mohsen, F. J. Morris, D. A. Sealer, C. H. Séquin, and M. F. Tompsett, "A three-level metallization three-phase CCD," *IEEE Trans. Electron Devices*, vol. ED-21, pp. 758–767, Dec. 1974.

[2] C. H. Séquin, F. J. Morris, T. A. Shankoff, M. F. Tompsett, and E. J. Zimany, "Charge-coupled area image sensor using three levels of polysilicon," *IEEE Trans. Electron Devices*, vol. ED-21, pp. 712–720, Nov. 1974.

[3] C. H. Séquin and M. F. Tompsett, *Charge-Transfer Devices*, suppl. 8 for *Advances in Electronics and Electron Physics*. New York: Academic, 1975.

[4] D. R. Herriott, R. J. Collier, D. S. Alles, and J. W. Stafford, "EBES: A practical electron lithographic system," *IEEE Trans. Electron Devices*, vol. ED-22, pp. 385–392, July 1975.

[5] C. H. Séquin, "Blooming suppression in charge coupled area imaging devices," *Bell Syst. Tech. J.*, vol. 51, pp. 1923–1926, Oct. 1972.

[6] G. A. Hartsell and A. R. Kmetz, "Design and performance of a three phase double level metal 160 × 100 element CCD imager," in *Tech. Dig., Int. Electron Devices Meeting*, 1974, pp. 59–62.

Charge-Injection Imaging: Operating Techniques and Performances Characteristics

HUBERT K. BURKE, MEMBER, IEEE, AND GERALD J. MICHON, MEMBER, IEEE

Abstract—The charge-injection device (CID) imaging technique employs intracell transfer and injection to sense photon-generated charge at each sensing site. Sites are addressed by an X–Y coincident-voltage technique that is not restricted to standard scanning. Free-format (random) site selection is possible.

An epitaxial structure provides a buried collector to prevent recollection of the injected charge. In sequential injection, the charge is injected into the substrate and the resulting displacement current sensed. In parallel injection, the functions of signal charge detection and injection have been separated. The injection operation is used to reset (empty) the charge storage capacitors after line readout has been completed. Nondestructive readout (NDRO) is possible by deferring the injection operation. Low-loss NDRO operation has been achieved using a cooled imager.

High sensitivity, low dark current, high modulation transfer function (MTF), and low blooming are some additional advantages of sensing signal charge levels within the array. Compared to the charge-coupled device (CCD), the CID approach results in a relatively high output capacitance; however, this is not considered to be a performance-limiting factor for most imaging applications.

Manuscript received July 30, 1975; revised September 15, 1975. CID imager development was sponsored in part by the United States Army Electronics Command, Night Vision Laboratory, The Advanced Research Projects Agency (ARPA), and the Air Force Systems Command, United States Air Force.

The authors are with Corporate Research and Development, General Electric Company, Schenectady, NY 12301.

I. INTRODUCTION

IN CONTRAST to charge-coupled devices (CCD's), in which the signal charge is transferred to the edge of an array for sensing, the charge-injection device (CID) approach [1]–[3] confines this charge to an image site during sensing. Site addressing is done by an X–Y coincident-voltage technique, not unlike that used in digital memory structures. In early structures, readout was effected by injecting the charge from individual sites into the substrate and detecting the resultant displacement current. A number of variations of this technique are possible, some of which will be described below. The CID technique offers a number of basic advantages, among which are simple mechanization, tolerance to processing defects, avoidance of charge transfer losses, and minimized blooming. In this paper we describe the basic CID structure, several site selection and readout techniques, and CID fabrication methods. Performance data for various CID image structures are given and discussed.

II. BASIC DESCRIPTION

CID image sensors use an X–Y addressed array of charge storage capacitors which store photon-generated charge in MOS inversion regions. Readout of the first self-scanned

Reprinted from *IEEE Trans. Electron Devices*, vol. ED-23, pp. 189–195, Feb. 1976.

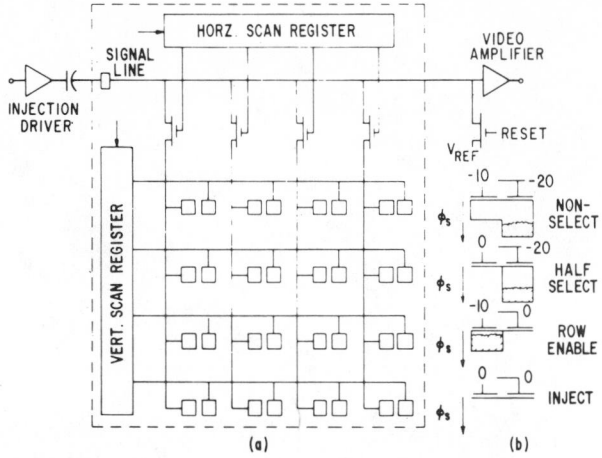

Fig. 1. Diagram of basic X–Y accessing scheme for a CID imager. (a) Schematic diagram of a 4 × 4 array. (b) Sensing site cross section showing silicon surface potentials and location of stored charge for various operating conditions.

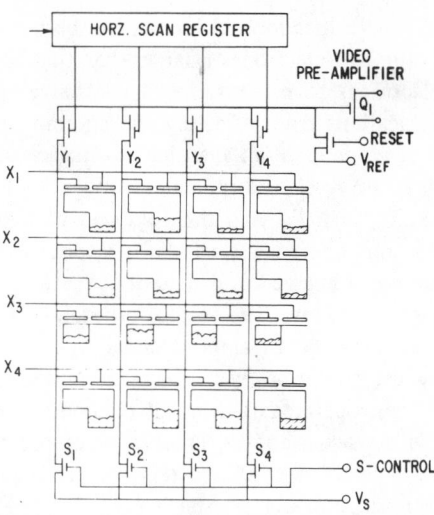

Fig. 2. Schematic diagram of a 4 × 4 CID array designed for parallel-injection readout. Silicon surface potentials and signal charge locations are included.

arrays was effected by injecting the stored charge into the substrate and detecting the resultant displacement current to create a video signal [1]–[3]. This readout method has been termed "sequential injection."

Readout can also be implemented by measuring charge transfer between the two storage capacitors that are used at each sensing site in an array. In an X–Y addressable array, the transfer can be performed on all sensing sites along a row in parallel. Each row can also be cleared of signal charge by performing the injection operation in parallel at all sites in the addressed row. This readout technique has been termed "parallel injection" [4].

A. Sequential Injection

An array designed for progressive scan which includes integral shift registers is shown in Fig. 1(a). Each sensing site consists of two MOS capacitors with their surface inversion regions coupled such that charge can readily transfer between the two storage regions. A larger voltage is applied to the row-connected electrodes so that photon-generated charge collected at each site is stored under the row electrode, thereby minimizing the capacitance of the column lines. The sensing site cross sections, Fig. 1(b), illustrate the silicon surface potentials and locations of stored charge under various applied voltage conditions.

For convenience, the array drive voltages will be described as being switched to 0 V during charge transfer and injection operations. In practice, this would result in some charge loss, since switching into accumulation would allow some surface states to empty and lose charge through recombination. Signal charge would then be consumed in refilling these states when voltage is reapplied. A bias voltage, slightly larger than the threshold voltage, can be added to the drive voltages to avoid this loss. A bias charge will then be maintained at each site.

A row is selected for readout by setting its voltage to zero by means of the vertical scan register. Signal charge at all sites

of that row is transferred to the column capacitors, corresponding to the row enable condition shown in Fig. 1(b). The charge is then injected into the substrate by driving each column voltage to zero, in sequence, by means of the horizontal scan register and the signal line. Charge in the unselected rows remains under the row-connected electrodes during the injection pulse time (column voltage pulse). This corresponds to the half select condition of Fig. 1(b).

The basic signal is contained in the majority-carrier displacement current that flows upon injection of the stored charge. This signal can be detected anywhere in the loop composed of the substrate, driver circuit, and the driven array lines. In general, sensing on the array lines provides the lowest capacitance environment. Array line sensing can be mechanized by precharging and floating the signal line and applying the injection pulse to the signal line through a series capacitor. The change in signal line voltage from its precharge level is proportional to the net injected charge.

The injection-detection process used for each sensing site is sequential and consists of the following operations: 1) charge injection; 2) amplifier overdrive recovery; 3) signal level sample-hold; and 4) integrating capacitance voltage reset (precharge).

Because of the time required for charge injection and overdrive recovery, sequential injection is generally not used for high-speed readout applications such as for large-area TV-compatible imagers.

B. Parallel Injection

In this readout technique, the functions of signal charge detection and injection are separated. The level of signal charge at each sensing site is detected by intracell transfer [5] during a line scan and, during the line retrace interval, all charge in the selected line can be injected.

A diagram of a 4 × 4 array designed for parallel injection is illustrated in Fig. 2 with the relative silicon surface potentials and signal charge locations included. As before, the voltage applied to the row electrodes is larger than that applied to the column electrodes to prevent the signal charge stored at unaddressed locations from affecting the column lines. At the beginning of a line scan, all rows have voltage applied and the column lines are reset to a reference voltage V_S by means of switches S_1 through S_4, and then allowed to float. Voltage is removed from the row selected for readout (X_3 in Fig. 2) causing the signal charge at all sites of that row to transfer to the column electrodes. The voltage on each floating column line then changes by an amount equal to the signal charge divided by the column capacitance. The horizontal scanning register is then operated to scan all column voltages and deliver the video signal to the on-chip preamplifier Q_1. The input voltage to Q_1 is reset to a reference level prior to each step of the horizontal scan register.

At the end of each line scan all charge in the selected row can be injected simultaneously by driving all column voltages to zero through switches S_1 through S_4. Alternatively, the injection operation can be omitted and voltage reapplied to the row after readout, causing the signal charge to transfer back under the row electrodes. This action retains the signal charge and constitutes a nondestructive readout (NDRO) operation.

The parallel-injection approach permits high-speed readout and is thus well adapted to TV-scan formats, and offers optional NDRO. A 244-line by 248-element imager, employing this technique and including an on-chip preamplifier, has been designed, fabricated, and evaluated in both the normal and NDRO modes.

For TV-compatible operation, a line time interval of 63 μs (5-MHz element rate) is used and the vertical scan rate is 60 fields/s. The imager is completely read out during each interlaced field of the standard TV frame such that video is displayed on all 488 active lines of the 525-line system, but with two identical fields.

C. Charge Injection

The CID imaging technique requires that the collected photon-generated charge be ultimately disposed of by injection into the substrate. Upon injection, this charge must either recombine or be collected to avoid interference with subsequent readouts. For the high lifetime material usually required for image sensors, recombination is not a suitable method of charge disposition, since recollection of this charge can give rise to objectionable image lag and crosstalk. For this reason, most CID imagers are fabricated on epitaxial material. The epitaxial junction, which underlies the imaging array, acts as a buried collector for the injected charge. If the thickness of the epitaxial layer is comparable to the spacing between sensing sites, most of the injected charge will be collected by the reverse-biased epitaxial junction and injection crosstalk is minimized. The rate at which charge injected at the surface is removed from the epitaxial layer can be determined analytically by solving the one-dimensional diffusion equation [6, pp. 121-

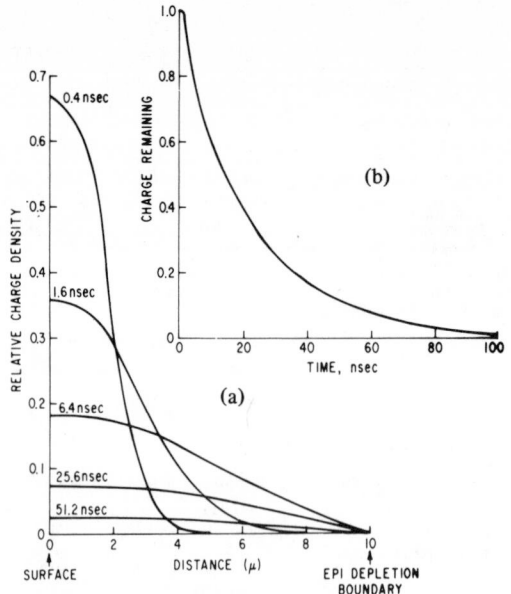

Fig. 3. Plot of the calculated charge-collection characteristics of an epitaxial junction. (a) Distribution of the injected charge in the epitaxial layer at various times after injection. (b) Relative amount of charge remaining as a function of time.

124] with appropriate boundary conditions. Using p-channel terminology, this equation is

$$\frac{\partial p_n}{\partial t} = D_p \frac{\partial^2 p_n}{\partial x^2} - \frac{p_n - p_{n0}}{\tau_p},$$

with the following boundary conditions for $p_n(x, t)$:

$$p_n(x, 0) = p_{n0}, \quad \text{for } 0 < x < X$$

$$\int_0^X p_n(x, 0)\, dx = P_{ns}$$

$$p_n(X, t) = 0, \quad \text{for any } t$$

where

P_{ns} minority carriers/unit area prior to injection
p_n minority carrier concentration
p_{n0} equilibrium minority carrier concentration
D_p hole diffusivity
t time from injection
x distance from surface
X distance from surface to epi depletion layer.

The results of a numerical solution of this equation are shown in Fig. 3. Diffusion of a quantity of charge from the surface to the epitaxial depletion region boundary has been calculated, assuming low-level injection and no carrier recombination in the epi layer. Fig. 3(a) shows the charge distribution in the epitaxial layer at various times after injection, while Fig. 3(b) shows the fraction of injected charge remaining in the epitaxial layer as a function of time from injection.

It can be seen from Fig. 3(b) that essentially all of the injected charge can be collected in about 100 ns for the operating conditions typically used in the CID.

The use of the buried charge collector also modifies the spectral response and modulation transfer function (MTF) characteristics in a manner that will be described below.

For image-intensification applications, where it is required to thin the silicon substrate for backside illumination by electrons, the epitaxial structure can be replaced by charge collectors on the front surface. These take the form of a grid of diffused conductors interconnected along the rows and columns. CID imagers operated in this fashion exhibit injection characteristics quite similar to epitaxial devices.

D. Random Selection

Certain imaging techniques, such as those involving bandwidth compression, could be much more effectively implemented if the primary image information could be made available in other than the progressive scan format employed by most currently available imaging devices. The X–Y coincident-voltage configuration of the CID is easily adaptable to special scan formats, including completely "random" addressing, in which each sensing site is directly accessible without regard to its time or spatial sequence with respect to other sites in the array.

In many respects, image site selection in the CID resembles MOS memory addressing, and many of the decoding techniques developed for addressing MOS memories can be applied to random-access imaging. Charge can be injected during readout in the manner of sequential injection, or deferred to a later time, as is done in the parallel-injection mode of operation. Image site selection differs from memory site selection in two important aspects. These are 1) a much higher dynamic range is required of the readout signal, and 2) any variation in integration time between sensing sites due to nonsequential readout order must be accommodated. For these reasons, the type of decoder used for random-access imaging may be different from those for memory selection.

E. Array Fabrication

Most CID processing to date has employed the standard p-channel silicon-gate technology. The advantages of using a well-understood widely available MOS integrated circuit (IC) technology has helped speed the development of the CID. The single modification to the standard silicon gate process required for imager manufacture is the addition of an n+ diffusion step to provide a top-contact when an epitaxial structure is utilized.

In order to allow charge transfer between the two polysilicon storage capacitors at each sensing site, a p-type diffusion is used in the interelectrode gap. One of the storage capacitors at each site must be connected to the upper aluminum conductor level in order to construct the X–Y addressable array. These requirements are troublesome in the design of high density arrays; the junction capacitance of the coupling p-diffusion presents a load to the storage region that results in a reduction of charge storage capacity and the interlevel contact requires valuable silicon area. Both of these problems can

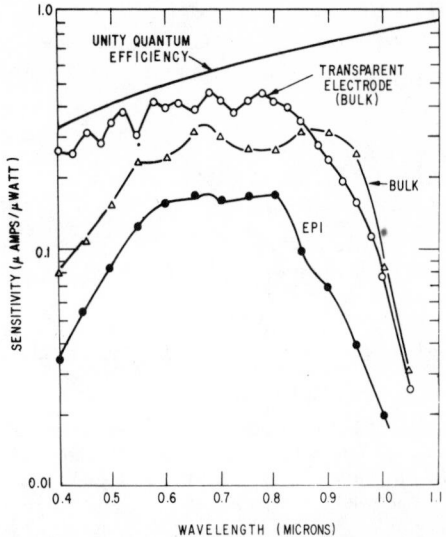

Fig. 4. Sensitivity plotted as a function of wavelength of incident radiation for three types of CID imagers; 100 × 100 bulk (triangles), 100 × 100 epitaxial (closed circles), and 32 × 32 bulk with transparent electrodes (open circles).

be circumvented by a design in which a portion of the storage capacitor electrodes overlaps to provide charge transfer in a manner analogous to that used in CCD's. Such a technique has been developed and used in the manufacture of high-density CID imagers. This technique permits the simultaneous fabrication of active devices for array scanning and can include highly transparent upper level electrodes [7].

III. PERFORMANCE

Imaging performance has been measured on devices of various sizes and configurations; while the results presented below were not all derived using similar devices, they are considered to be representative of performance of CID imagers. Wherever measured data are presented, the type and size of the measured device are given, and extrapolation to other size devices is made if considered relevant.

A. Sensitivity

Sensitivity as a function of wavelength for both the bulk and epitaxial 100 × 100 structures is plotted in Fig. 4 (lower curves). The sensitivity maximum for the bulk structure occurs at a wavelength of 0.67 μm. At this point, the overall efficiency in conversion of incident photon energy to output current is 66 percent of the theoretical maximum.

The sensitivity of the epitaxial structure is about half that of the bulk imager in the visible region of the spectrum. This sensitivity reduction reflects collection by the epitaxial junction of some of the signal charge generated in the space between sensing sites. Thus, some of the indicated sensitivity loss represents loss of charge that was contributing to crosstalk in the bulk sample. The somewhat greater loss in the infrared region is to be expected because of the deeper photon penetration at these wavelengths.

The upper curve shows the sensitivity improvement resulting from the use of transparent metal oxides for the upper level conductors. Since these data were taken using a bulk imager,

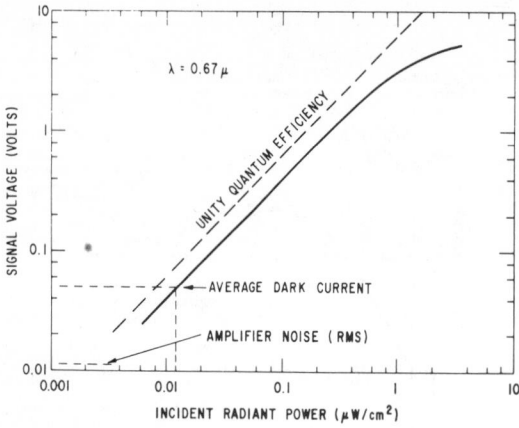

Fig. 5. Transfer characteristics of bulk and epitaxial 100 × 100 sequential-injection imagers. Measurements were made at λ = 0.67 μm. The rms value of device thermal noise and typical dark current at 25°C are shown for comparison.

Fig. 6. Image produced by a 244 × 188 CID imager exposed to a high-contrast scene. Blooming resistance is indicated by the image of the headlamps.

they should be compared to standard bulk results (middle curve). Of particular significance is the greatly increased blue response which results from the substitution of the transparent material for a large portion of the polysilicon gate-electrode material. The apparent loss of response above 8500 Å for this device can be attributed to low carrier lifetime for the particular device tested. Note that the sensitivities presented are independent of site area and array size. Thus these data, although measured on differing devices, can be directly compared.

The transfer characteristics are shown in Fig. 5. Measurements were made at the point of maximum spectral sensitivity (λ = 0.67 μm), 30 frames/s. A curve representing 100 percent quantum efficiency for silicon at this wavelength is shown for comparison. The nonlinearity near saturation results from depletion capacitance loading on the injected charge signal. In other words, the signal current measured upon injection is reduced by the displacement current needed to reestablish the depletion region under the storage electrode when the electrode voltage is reapplied. This is a characteristic of the sequential injection mode of operation.

B. Blooming

The epitaxial CID structure is resistant to image blooming since each sensing site is electrically isolated from its neighbors. Charge spreading in the substrate is minimized by the underlying charge collector.

In sequential injection, blooming of the displayed image occurs if charge is injected from a half-selected site during readout of another site on the same column. This excess injected charge is detected as signal and adds to the displayed video. This results in brightening of the affected column upon overload of a single site.

The image displayed in the parallel injection approach exhibits relatively little blooming as a result of sensing site overload. This is because the half-select and injection operations occur during the horizontal blanking interval. While excess charge can accumulate during a line scan interval and cause column brightening for overloads occurring in the right-hand

portion of the image field, this effect is attenuated by the line-to-frame integration time ratio.

During NDRO, virtually no blooming occurs, since the charge is not injected. The affected sites simply fill to capacity and cease collecting charge.

In all cases, radial spreading of excess charge is prevented by the underlying charge collector.

The image shown in Fig. 6 was produced by a 244 × 188 CID imager exposed to a high-contrast scene. This particular camera uses a modified parallel-injection readout method that is highly resistant to image blooming, as indicated by the image of the automotive head lamps.

C. MTF and Lag

The array cells are designed to allow charge transfer between the two storage capacitors at each sensing site but not between adjacent sites. Two mechanisms exist, however, which can result in signal crosstalk between adjacent sites. Charge generated in the undepleted silicon between sensing sites will divide between the sites as long as the distance to be traveled is less than the minority carrier diffusion length. This effect can be attenuated in epitaxial imagers if the distance between the surface and the buried collector is made less than, or comparable to, the spacing between sensing sites.

The second crosstalk mechanism is the migration of injected charge to neighboring sites. Proper placement of the epi collecting junction also attenuates this effect. In-phase square wave response greater than 90 percent at the Nyquist limit has been measured on 100 × 100 imagers in which the effective epitaxial layer thickness was 10 μm and the storage region separation was 15 μm. The horizontal cell pitch was 82.5 μm. There is a lower measured sensitivity for the epitaxial structure, as illustrated in Figs. 4 and 5.

Lag can occur in CID imagers through two mechanisms: 1) recapture of the injected charge by the injecting site upon reapplication of the storage electrode voltage; and 2) migration of injected charge to adjacent sites that are not to be read out until the following field. Image lag resulting from the

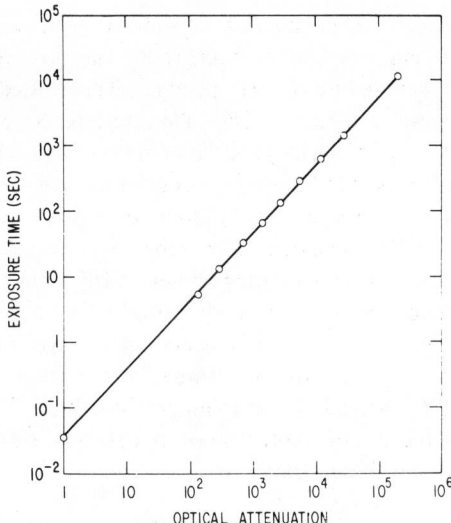

Fig. 7. Plot showing exposure time required to accumulate a given quantity of signal charge as a function of incident light level. Time is plotted as a function of optical attenuation. Measurements were made on a 244 × 248 array, cooled to 200 K and operated at 30 frames/s, NDRO.

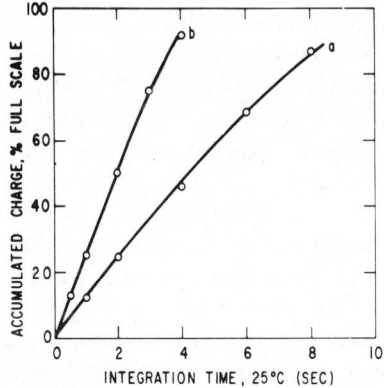

Fig. 8. Thermal charge buildup versus time in 100 × 100 CID imager at 25°C, for curve *a*, both storage regions inverted, and for curve *b*, one storage region depleted.

second mechanism occurs to the same degree as crosstalk (loss of MTF). Recapture of charge by the injecting site is a function of the time allowed for injection (see Fig. 3). This can vary from virtually zero recapture to full recapture of the injected charge (NDRO, for example). In practice, the injection pulse width is adjusted to insure that image lag is consistent with system requirements.

D. Nondestructive Readout

The NDRO characteristic of a 244 × 248 CID array has been evaluated by operating the device at low temperature to minimize dark current and thereby achieve long storage time intervals.

Two experiments were performed to identify the limiting factors in nondestructive image readout. First, a charge pattern of an image was generated and stored by momentarily opening a shutter, and then the image was read out continuously at 30 frames/s, until image degradation was noted. At a chip temperature of 200 K, images were read out for three hours (324 000 NDRO operations) with no detectable charge loss. The charge lost during each NDRO operation was, on the average, much less than one carrier/pixel/frame.

The second experiment was performed to insure that charge could be generated and stored at very low light levels under continuous (30 frames/s) NDRO conditions. A series of exposures was made at successively lower light levels and the time required to reach a given level of signal voltage was measured. The results, Fig. 7, show that the exposure time is inversely proportional to light level with no measurable reciprocity loss for exposure times up to three hours. The light level used for the three-hour exposure resulted in the accumulation of approximately 600 000 carriers/site in the highlight regions of the image. Since the imager was read out more than

300 000 times during this time exposure, the NDRO loss must, on the average, be much less than two carriers/site/readout.

E. Dark Current

The CID approach permits significantly more silicon area to be used for photon charge generation than for charge storage. This results in an advantageous dark current situation because the thermal charge generation rate in nondepleted bulk silicon is orders of magnitude less than the generation rate in the depleted storage region [6, pp. 298–302]. Consequently, each image sensing site collects and stores photon-generated charge from essentially the total site area but generates dark current only in the storage area. Also, no separate storage area is required for image readout, so that a dark current contribution from this source is avoided.

The use of bias charge in the storage area results in an additional reduction in dark current since the surface thermal generation rate in MOS structures is much smaller under inversion conditions than under depletion conditions [8]. Measured thermal charge generation rates for a 100 × 100 imager are shown in Fig. 8. Minimum dark current results from biasing the array such that charge is stored under both electrodes, Fig. 8(a). Fig. 8(b) shows the effect of operating with one storage region depleted. These data were obtained by operating the device in a "burst" mode in which the sensing sites were allowed to integrate charge for the appropriate time after which the entire array was read out in $\frac{1}{30}$ s. The average accumulated charge for the entire array was recorded.

The 100 × 100 imager is normally operated with one storage region depleted, so that the conditions of Fig. 8(b) apply. The peak signal-to-dark current ratio exceeds 100:1 under these conditions at 30 frames/s. For lower frame rates, it is feasible to operate under the conditions of Fig. 8(a), thereby doubling the useful integration time.

F. Noise Sources

Noise in CCD devices has been extensively investigated and reported upon [9], [10]. (In contrast, little discussion of CID noise sources has appeared in the literature.) The CCD and CID approaches have many similarities in the manner by which the signal charge is detected, and so some of the same considerations apply. The primary thermal noise sources in the CID are preamplifier noise, capacitor reset noise, and dark current shot noise. Unlike the CCD, the CID has negligible charge transfer noise.

An on-chip MOS transistor is generally used as the first stage of the video amplifier. This on-chip preamplifier stage is advantageous to minimize load capacitance and induced noise (pickup). This transistor should be designed for the lowest possible thermal noise. Johnson noise in the transistor channel is proportional to the channel resistance; therefore, a high g_m is desirable to minimize this noise component. MOS transistors exhibit a relatively large amount of excess low-frequency noise ($1/f$ noise). It has been shown, both theoretically and experimentally [11], that the magnitude of this noise component is proportional to the surface-state density in the MOS channel and is inversely proportional to the channel area. Since g_m is proportional to channel width-to-length ratio, and $1/f$ noise varies inversely with channel area, the S/N ratio is best for an on-chip transistor that has minimum channel length and a channel width such that the transitor input capacitance equals the source (array) capacitance. It is not always necessary to optimize the on-chip amplifier for minimum $1/f$ noise, since the correlated double sampling technique can be used to attenuate noise components that are lower in frequency than the sampling frequency.

If a switch is used to set the voltage across a capacitor, thermal noise in the resistance of the switch results in an uncertainty in the final capacitor voltage. The magnitude of this uncertainty [7] is

$$V_n = (KT/C)^{1/2}$$

or

$$Q_n = (KTC)^{1/2}$$

where K = Boltzmann's constant = 1.38×10^{-23} W·s/K and T = temperature in degrees Kelvin. Sequential injection CID imagers are not limited by this noise component because it is possible to reference the net injected charge signal to the input capacitor voltage after reset has been completed. This technique, called correlated double sampling, results in the substitution of KTC noise on a clamping capacitor for KTC noise on the array output capacitance. The level of KTC noise referred to the array can be made arbitrarily small, however, since gain can be used between the array output and the clamp capacitor. The parallel-injection technique does not allow complete elimination of KTC noise. The column reset transistors introduce KTC noise that is not rejected. Voltage noise at the input of the preamplifier results in an equivalent input charge that is directly proportional to the array output capacitance ($q = cv$). Theoretical preamplifier noise levels of a few hundred carriers result from array output capacitance levels in the 10-pF region. KTC noise can be either negligible or the predominant temporal noise source, depending upon the specific array design and readout method.

In addition to thermal noise, solid-state image sensors exhibit a fixed nonuniform spatial background in the reproduced image that can reduce image quality. The sources of this "pattern noise" are many: drive line interference, patterning irregularities, and bias charge variations, to name a few. These effects tend to become more pronounced as scan rates are increased. To the extent that differential sensing can be performed, the CID technique offers relief from these annoying and often performance-limiting effects. Differential sensing allows rejection of common mode drive line interference and minimizes the effects of patterning and bias charge differences.

Dark current nonuniformity can be an important source of pattern noise, particularly at room temperature. The inherently low dark current performance of CID imagers can be an advantage under these conditions.

IV. CONCLUSIONS

The charge-injection approach to image sensing leads to many desirable performance characteristics that make this technique attractive as a method for the realization of high quality solid-state imaging systems. These include high sensitivity over a wide spectral range, low dark current, high MTF, and resistance to blooming—in a structure that affords ease of fabrication, tolerance to defects, and flexibility of design. NDRO of the stored image and random image site addressing are options available to the system designer.

ACKNOWLEDGMENT

The authors would like to thank M. Ghezzo for his assistance in array design and circuit analysis, K. Keller, D. Meyer, and P. Salvagni for sensor fabrication, and S. Dworak for test circuit fabrication and device testing.

REFERENCES

[1] E. Arnold, M. H. Crowell, R. D. Geyer, and D. P. Mathur, "Video signals and switching transients in capacitor-photodiode and capacitor-phototransistor image sensors," *IEEE Trans. Electron Devices*, vol. ED-18, pp. 1003–1010, Nov. 1971.

[2] G. J. Michon and H. K. Burke, "Charge injection imaging," in *Dig. Tech. Papers, Int. Solid-State Circuits Conf.*, 1973, pp. 138–139.

[3] ——, "Operational characteristics of CID imager," in *Dig. Tech. Papers, Int. Solid-State Circuits Conf.*, 1974, pp. 26–27.

[4] G. J. Michon, H. K. Burke, and D. M. Brown, "Recent developments in CID imaging," in *Proc. Symp. Charge Coupled Device Technology for Scientific Imaging Applications*, 1975, pp. 106–115.

[5] J. J. Tiemann, W. E. Engeler, R. D. Baertsch, and D. M. Brown, "Intracell charge-transfer structures for signal processing," *IEEE Trans. Electron Devices*, vol. ED-21, pp. 300–308, May 1974.

[6] A. S. Grove, *Physics and Technology of Semiconductor Devices.* New York: Wiley, 1967.

[7] D. M. Brown, M. Ghezzo, and M. Garfinkel, "Transparent metal oxide electrode CID imager," this issue, pp. 196–200.

[8] D. J. Fitzgerald and A. S. Grove, "Surface Recombination in Semiconductors," *Surface Sci.*, vol. 9, pp. 347–369, 1968.

[9] J. E. Carnes and W. F. Kosonocky, "Noise sources in charge coupled devices," *RCA Rev.*, vol. 83, pp. 327–343, June 1972.

[10] A. M. Mohsen, M. F. Tompsett, and C. H. Séquin, "Noise measurements in charge-coupled devices," *IEEE Trans. Electron Devices*, vol. ED-22, pp. 209–218, May 1975.

[11] S. Christensson, L. Lundstrom, and C. Svensson, "Low frequency noise in MOS transistors," *Solid-State Electron.*, vol. 11, pp. 797–812, 1968.

CID (CHARGE INJECTION DEVICE) IMAGING SYSTEMS

Now available from General Electric is a fully TV-compatible, all solid state camera featuring an advanced Charge Injection Device (CID) sensor.

This device, having 244 rows with 188 elements-per-row, produces a TV image that provides video information on all unblanked raster lines of a 525-line, 30 frame-per-second monitor.

Data brochure TPD 6235 provides operational theory, lists applications, and illustrates typical performance.

Data pertinent to the 244x188 sensor is shown in figures 4, 5 & 6. Photographs of monitor displays (figures 1, 2 & 3) provide illustration of typical performance. Note the ability to accept small area exposure overload conditions without the resultant blooming or streaking that occurs in other imagers. This is a feature of the General Electric Type Z7892 CID Camera that allows viewing of scenes having wide dynamic range, allows identification in close proximity to highlight conditions, and minimizes the need for continuous illumination compensation.

The Z7892 camera employs standard "C" mount lensing, thereby making available a full complement of lens selection. Provided with the camera is a 25mm, f-1.4 lens. Also offered as an option is a lens system that provides light compensation automatically.

```
+-------------------------------------+
|                                     |
|       KEY FEATURES                  |
|                                     |
|                                     |
|          ANTI-BLOOMING              |
|             *  *  *                 |
|            HIGH MTF                 |
|             *  *  *                 |
|             NO LAG                  |
|             *  *  *                 |
|        WIDE SPECTRAL RANGE          |
|                                     |
+-------------------------------------+
```

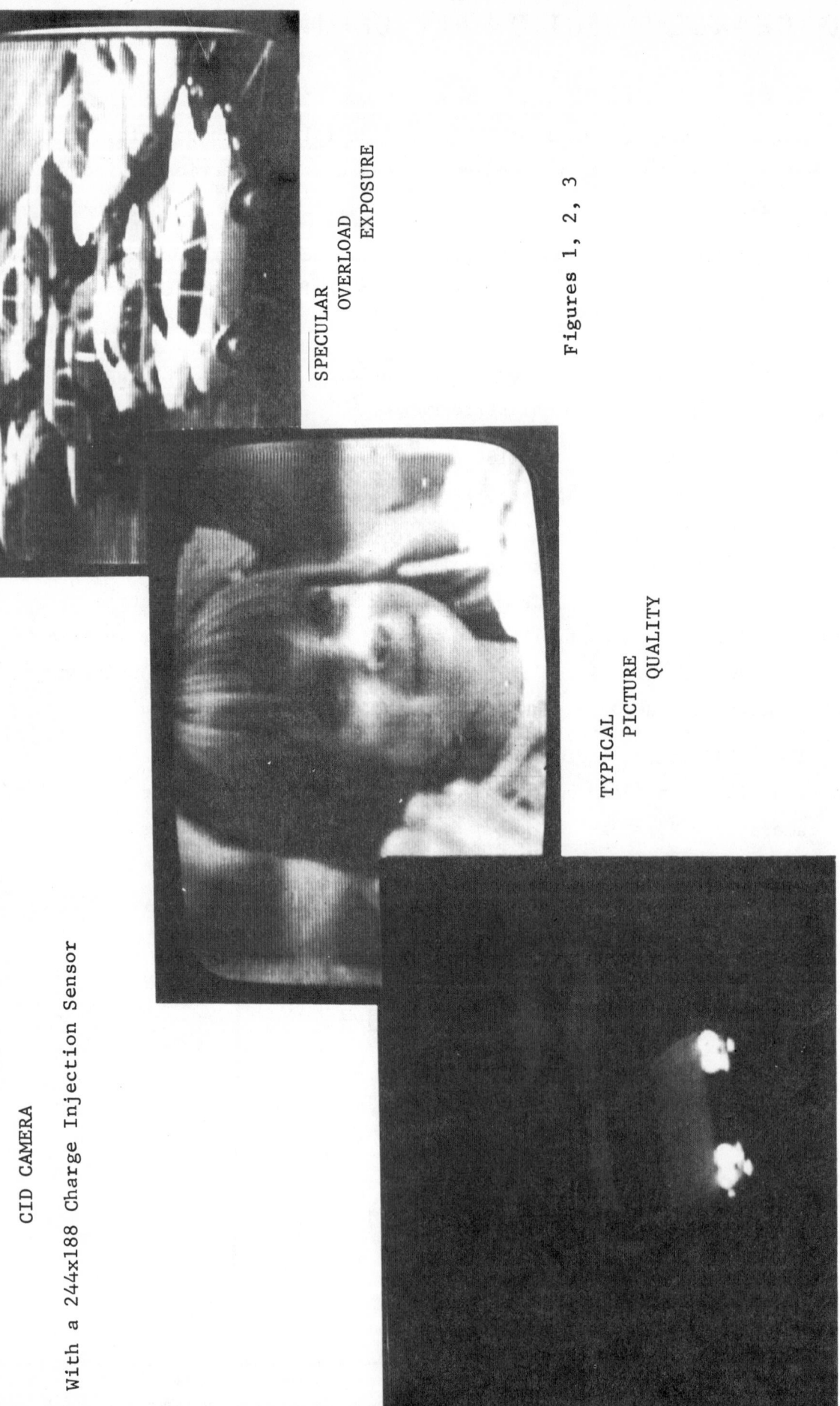

TYPICAL PERFORMANCE

CID CAMERA

With a 244x188 Charge Injection Sensor

SPECULAR
OVERLOAD
EXPOSURE

TYPICAL
PICTURE
QUALITY

NIGHT SCENE

WIDE DYNAMIC RANGE

Figures 1, 2, 3

194

HOW THE CID IMAGER SYSTEM PERFORMS

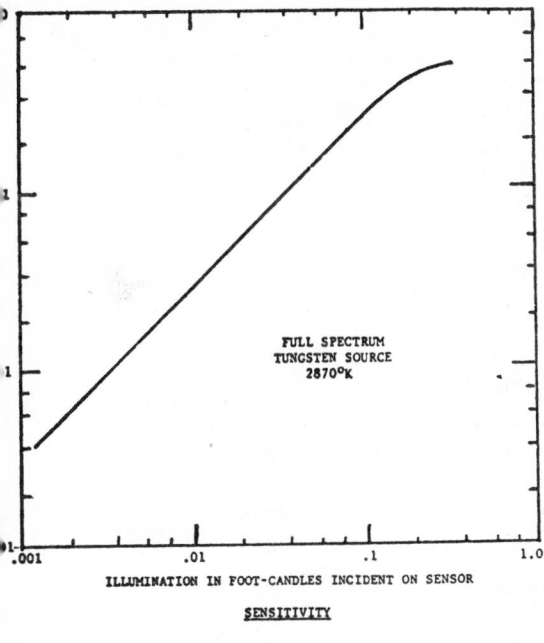

FULL SPECTRUM
TUNGSTEN SOURCE
2870°K

ILLUMINATION IN FOOT-CANDLES INCIDENT ON SENSOR

SENSITIVITY

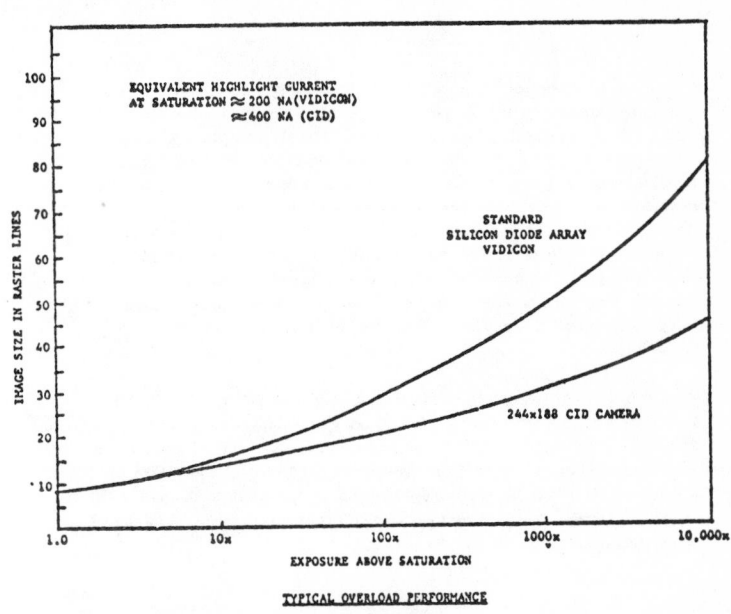

EQUIVALENT HIGHLIGHT CURRENT
AT SATURATION ≈ 200 NA (VIDICON)
≈ 400 NA (CID)

STANDARD
SILICON DIODE ARRAY
VIDICON

244x188 CID CAMERA

IMAGE SIZE IN RASTER LINES

EXPOSURE ABOVE SATURATION

TYPICAL OVERLOAD PERFORMANCE

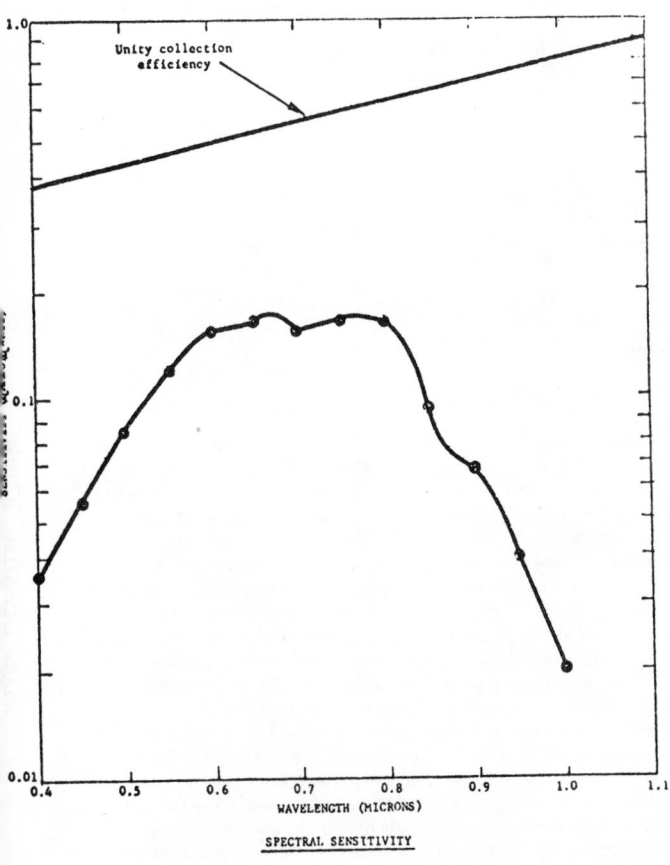

Unity collection
efficiency

WAVELENGTH (MICRONS)

SPECTRAL SENSITIVITY

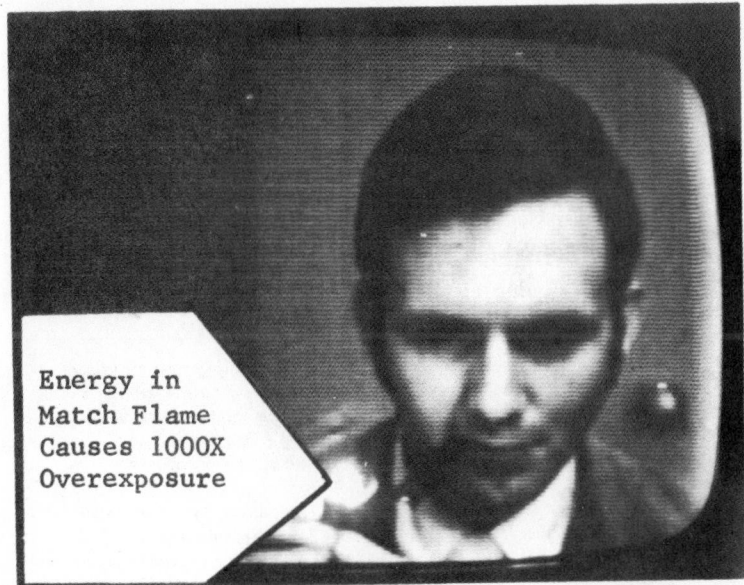

Energy in
Match Flame
Causes 1000X
Overexposure

CONTACT INFORMATION:

General Electric Company
Optoelectronic Systems Operation
Building 3, Electronics Park
Syracuse, New York 13201

ATTN: Fred Sachs
Phone (315) 456-2832

Figs. 4, 5, 6 and 7

CHARGE-COUPLED IMAGER FOR 525-LINE TELEVISION

R. L. Rodgers, III
RCA Corporation
Electro-Optic Products
Lancaster, PA 17604

INTRODUCTION

Considerable effort went into the development of a TV system that would make an acceptable picture utilizing all of the eye-brain visual perception principles. Factors such as the visual acuity of the eye and critical flicker frequency effects were considered. This effort led to the development of the U.S. Standard 525-line television system. For a solid-state image sensor to obtain its greatest market potential it should be compatible with and meet the minimum performance parameters of this standard system. Many approaches have been taken toward achieving this goal. The most promising approach thus far is a charge-coupled imager which has the potential for meeting both performance and manufacturing requirements. The production experience with silicon vidicon targets will provide a base for the manufacture of solid-state charge-coupled image sensors[1].

A 512 x 320 element CCD imager containing 163,840 individual analog storage sites has been designed and fabricated for the requirements of the standard 525-line television system. In this paper, the requirements for this system are defined and the resultant imager performance is described. Several smaller size vertical frame transfer CCD imagers were developed prior to the 512 x 320 imager[2,3,4]. These devices demonstrated that the required device performance parameters could be obtained.

IMAGE SENSOR REQUIREMENTS FOR 525-LINE TELEVISION COMPATIBILITY

The television system used for almost all broadcast and closed circuit television in the United States is the system described by the National Television System Committee (NTSC). This standard is described in EIA standards RS-170 for broadcast and RS-330 for closed circuit industrial applications. For the purpose of the following discussion, the requirements are the same.

The standard RS-170 system consists of two 262.5 line fields each 1/60 second long (1/59.939 second for color), interlaced 2:1 to form one complete frame every 1/30 second. Approximately 23% of each frame is devoted to blanking time for the display. This time is used to transfer charge from one register to another in a CCD.

Vertical Cell Count

The RS-170 vertical blanking interval is .075 ± .005 of the field interval yielding 241.5 to 244.125 (243 nominal) active display lines per field. This places a requirement of a minimum of 242 lines to be generated by the imager for each field. The maximum number of lines in each field is 262 to enable all of the cells for that field to be read out before the next field starts. One row of cells is required for each line in the display. The RS-170 sync pulses cause the display to interlace each field with the preceding field. This causes a display with approximately 486 active lines per frame. The same 243 line positions generated on one field may be repeated on the next field giving a 486 active line display, with the resolution of 243 lines, or a new interlaced set of 243 elements may be supplied by the sensor each field for a total of 486 lines of vertical resolution per frame.

Horizontal Cell Count

The RS-170 standards call for a horizontal line frequency of 15,750 lines/second (15,734 lines/second for color). The line time is therefore 63.49 microseconds. The nominal horizontal blanking interval is 17% of the horizontal line time leaving 52.7 microseconds active display time. In closed circuit applications, the horizontal video bandwidth is unlimited. In monochrome and color broadcast, the luminance bandwidth is 4.2 MHz. Filling this bandwidth requires approximately 450 cells. A composite color signal has the color information modulated on a 3.58 MHz subcarrier. Unless the video monitor or receiver uses an expensive comb filter to remove the chroma information from the luminance signal (not commercially used), the color subcarrier beats with the luminance

signal causing an annoying "edge creep". It is therefore common practice to limit the luminance video bandwidth to 3 MHz. Most present-day video tape recorders are also limited to 3 MHz video bandwidths for luminance information. This leads to a practical minimum number of cells for the 3 MHz bandwidth of 320 cells. The corresponding video data rate is 6 MHz. Figure 1 summarizes the 525-line system requirements.

SUMMARY OF 525 LINE TELEVISION SYSTEM REQUIREMENTS (RS—170)

Vertical Parameters

Frame Time (1/30 Second)

2 Interlaced Fields (1/60 Second Each)

Total Number of Lines Per Field (262.5)

Vertical Blanking Interval (.075 ± .005 V)

Active Scan Lines Per Field (Nominal/243)

Required Vertical Cell Count:
  242-262 Each Field (Repeated or Interlaced)

Horizontal Parameters

Luminance Electrical Bandwidth (4.2 MHz)

Color Receiver and VTR Practice (3 MHz Bandwidth)

Required Cell Count:
  4.2 MHz (Approx. 450 Cells)
  3 MHz (Approx. 320 Cells)

Format Parameters

Picture Aspect Ratio (4:3)

Figure 1

Picture Format Considerations

There are several factors which affect the choice of picture format. Figure 2 lists the most popular formats used in television, movie film and still photography. Most movie film and TV formats have the 4:3 aspect ratio which is required for the standard TV system. There has been a general trend towards smaller formats through the years in all three media. This has been the result of improvements in the resolution of lenses and photosurfaces as well as the need for more compact and lightweight equipment. The most popular lenses are the "C-mount" variety. There are two basic subdivisions of C-mount lenses. These are the designs for the 16 mm format of one-inch vidicons and those designed for the 11-12 mm formats of 2/3 inch vidicons and 16 mm movie film. Any of the 16 mm vidicon lenses can however be used for the smaller formats. The 11-12 mm formats seem very attractive because of their small size and ready availability of inexpensive interchangeable lenses. Lenses for 8 mm and Super 8 mm have low MTF's in the TV line ranges of interest and the small format requires extremely small CCD cell dimensions beyond the state of the art for fabrication and device operation considerations to achieve standard TV resolution (240 television lines per picture height). An additional drawback of very small formats is the inability to achieve selective focus special effects with normal fields of view due to the inherent very large depth of focus.

Reprinted from *1974 IEEE Int. Conv. and Exposition, Tech. Program Papers*, Session 2, Mar. 26–29, 1974, pp. 1–4.

POPULAR IMAGE FORMATS

Movie Film	Diag. (mm)	Aspect Ratio
8 mm	5.46	4:3
Super 8 mm	6.68	4:3
16 mm	12.0	4:3
35 mm	25.5	4:3
Television		
2/3″ Vidicon	11.0	4:3
1″ Vidicon	15.9	4:3
1.2″ Vidicon	21.2	4:3
1.5″ Vidicon	25.0	4:3
Still Film		
110	21.2	4:3
126	39.2	1:1
135	43.3	3:2

Figure 2

512 x 320 CCD IMAGER PERFORMANCE PARAMETERS

A 512 x 320 element CCD array has been designed and fabricated for the standard 525-line TV system as described above. It is capable of generating the full resolution requirements of broadcast color receivers and tape recorders. Figure 3 shows the layout of this vertical frame transfer imager. The imager is composed of three sections. The lower section contains 256 x 320 cells which can be interlaced on alternate fields to generate 512 x 320 picture elements per frame[5]. In actual use, only 486 lines (243 per field) are displayed as described above. The extra elements are provided to allow variations in system blanking and timing and to avoid nonuniformities at the picture edges. The middle section contains a 256 x 320 cell storage area to provide format conversion to a sequential horizontal readout. The top section shows the 320-element readout register which shifts the video out at 6 MHz data rate. The cell size is 1.2 x 1.2 mils resulting in a 12 mm image format. The overall chip size is 500 mils x 750 mils.

512 x 320 ELEMENT CCD IMAGER

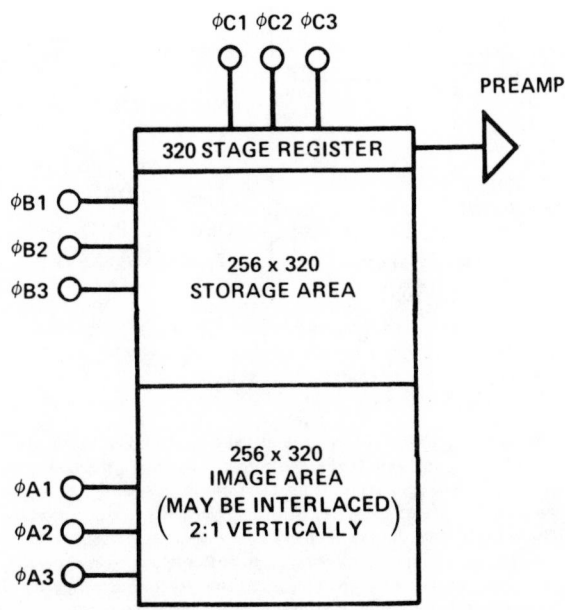

Figure 3

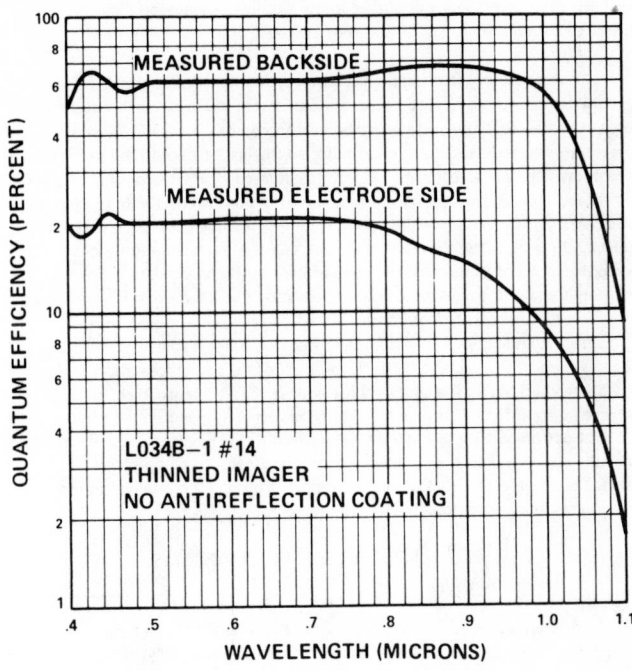

CCD IMAGER SPECTRAL RESPONSE

Figure 4

Spectral Response

The spectral response of a CCD is similar to the spectral response of a silicon vidicon. The spectral response extends from a .4 micrometers to 1.1 micrometers. Figure 4 shows the spectral response for a thinned area imager test device. Absorption of light in the transfer electrodes reduces the available light to the substrate when an image is formed on the electrode side. When the light is incident on the backside (side opposite transfer electrodes), the deviation in sensitivity from 100% QE is caused primarily by reflection losses and incomplete absorption in the infrared. The magnitude of light loss in electrode side imaging varies with process variations and wavelengths, and can differ somewhat from that shown in Figure 4 particularly in the blue end of the spectrum. The backside illumination is basically independent of electrode details.

SIT-CCD

A vertical frame transfer CCD imager processed for backside illumination that has good blue response may be fabricated into a new type of high-sensitivity CCD imager. This device is the CCD version of the Silicon Intensifier Target (SIT) Camera Tube[6]. The SIT-CCD imager consists of a photoemissive cathode, electron-optical focusing section, and CCD imager in an evacuated envelope. The photoelectron image is injected into the silicon at high energy forming many secondary hole-electron pairs. This current gain greatly increases the sensitivity of a CCD by raising the output signal above background signal variation limitations. Interline transfer CCD imagers cannot readily be operated with backside illumination, and therefore cannot use this method of low-light-level sensitivity enhancement.

Resolution

The most basic limitation on limiting resolution is the number of cells contained in the imager. In the horizontal direction, the number of cells is determined by the number of channels formed by the channel stop diffusions (320 for the 512 x 320 device). In the vertical, the number of resolution elements is determined by the number of different storage cell configurations formed by the transfer electrodes in the image area (256 actual cells electronically interlaced 2:1 to achieve 512 sample positions in the vertical direction). Resolution is usually defined in terms of Television Lines/Picture Height (TVL/PH). In the horizontal direction, this number is 3/4 of the horizontal cell count or 240 TVL/PH due to the 4:3 aspect ratio. In the vertical direction, only 486 resolution samples are displayed out of the 512 possible, and the finite line structure in the display is often quoted as reducing the effective useful resolution to approximately 350 TVL/PH due to the "Kell factor" even though actual resolution may be seen to the 486 sample limit with proper pattern phasing. In the horizontal, the video is smoothed into a continuous waveform and the finite cell aliasing (moire) effects are not as severe.

Another factor affecting picture sharpness is the reduction of MTF (sine-wave response) due to resolution loss mechanisms in the image formation and readout processes. These losses are tabulated in Figure 5 for a 1.2 mil cell dimension. The lens is the first contributor to the loss of resolution in image formation. The lateral diffusion of electrons in the imager before collection is the second mechanism. It is a Sech function of the optical spatial frequency and field-free diffusion length in the bulk. The numbers shown are representative of typical geometries. The third loss is the Sin X/X loss due to the finite cell sampling process. All of these effects are also present in silicon vidicons[1]. The resolution losses in the readout process are given by the equation in Figure 5. $N\epsilon$ is the transfer loss and f_o is the frequency of one cell. This resolution loss mechanism is analogous to the resolution loss of a scanning electron beam in a silicon vidicon. The overall MTF is the product of the image formation MTF and the readout MTF. The squarewave response (CTF) is larger than the sine-wave response (MTF). The actual measured responses depend on the phasing of the light pattern with respect to the cell pattern. If the black-and-white bars are lined up with the cells in an in-phase condition, the response is greatest. If the bars are lined up with half a white bar and half a black bar contained in each cell, the response at the cell limit will be zero.

RESOLUTION LOSS MECHANISMS

Image Formation Losses

Mechanism	1/2 Cell Limit (2 Cells/TVL)	Cell Limit (1 Cell/TVL)
(1) Lens MTF	.94	.88
(2) Sin X/X Cell Sampling MTF	.90	.63 $(\frac{2}{\pi})$
(3) Lateral Diffusion MTF	.95	.90
MTF Product 1,2,3	.80	.50

Image Readout Losses

Readout MTF $= \exp\left[-N\epsilon\left(1 - \cos\ (2\pi f/f_o)\right)\right]$

Figure 5

Transient Response

In the vertical frame transfer design, the picture is integrated for one field (16.66 milliseconds). The picture is then read out completely each field. There are therefore no smearing effects due to "lag" during panning except for resolution loss due to image motion during exposure as in any movie film camera. Some other types of solid-state sensors integrate for a full frame (33.3 milliseconds) and have twice the panning resolution loss because of the longer exposure time.

Dark Current

During the operation of a CCD imager, a two-dimensional charge pattern replica of the light pattern is formed inside the CCD. Thermally generated charge partially fills the potential wells and generates a background signal known as dark current. There are three basic dark-current generating mechanisms[1]. These sources are illustrated in Figure 6. The first source is diffusion current from the substrate. This current source can be neglected. The second source of dark current is from the depleted layer between the SiO_2 interface and the substrate. A simplified equation for this current density is $J_{Dep} = e\ N_i\ D/2\ \tau$, where τ is the effective lifetime and N_i is the intrinsic carrier concentration. The effective lifetime is a complex function of the actual hole and electron lifetimes and the energy level of the generation centers. If CCD's are processed to remove most of the generation centers near the center of the bandgap, this component of dark current is less than the third dark-current source. The third dark-current source is generation current from surface states at the silicon-SiO_2 interface. A simplified equation for the surface-generation current density is $J_s = e\ N_i\ S/2$. S is the effective surface generation velocity and is proportional to the surface-state density N_{ss} near the middle of the bandgap. The value of N_{ss} near the center of the band may differ significantly from the density of surface states nearer the band edges which affect transfer efficiency. Actual dark current measurements have yielded dark current densities of 5 nanoamperes/cm[2]. This results in a dark current of about 4 nanoamperes in the image sensing area of a 512 x 320 cell device.

The second and third dark-current sources depend on the temperature variation of N_i. This results in the standard silicon dark-current variation of a factor of 2 for every 9°-10° C temperature change around room temperature.

DARK CURRENT SOURCES IN CCD IMAGERS

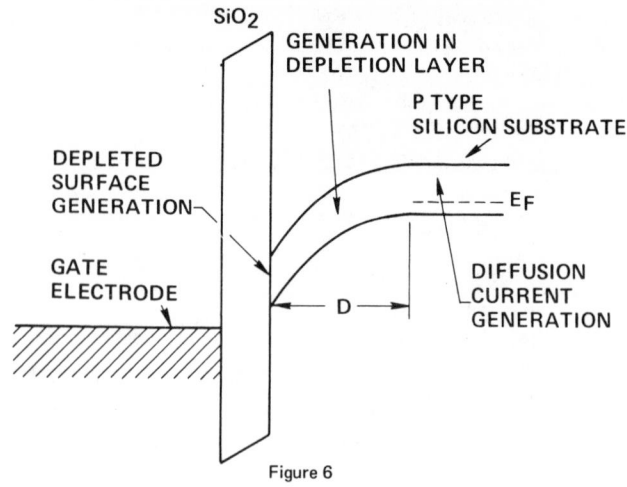

Figure 6

Blooming

CCD's exhibit unique problems of charge containment when overloaded with light. Charge is fairly well confined from spreading sideways by the channel stop diffusions. However, excess charge can spill up and down the channel. A special mode of operation has been developed to greatly reduce the spreading of charge down the channel under overload conditions. This low-blooming mode is accomplished by maintaining the phase fingers adjacent to a charge integrating phase finger at a voltage that maintains the surface under these fingers in light accumulation. This forms a temporary extension of the channel stop around each charge integration collecting site minimizing charge spread down the channel.

ON-CHIP VIDEO PROCESSING

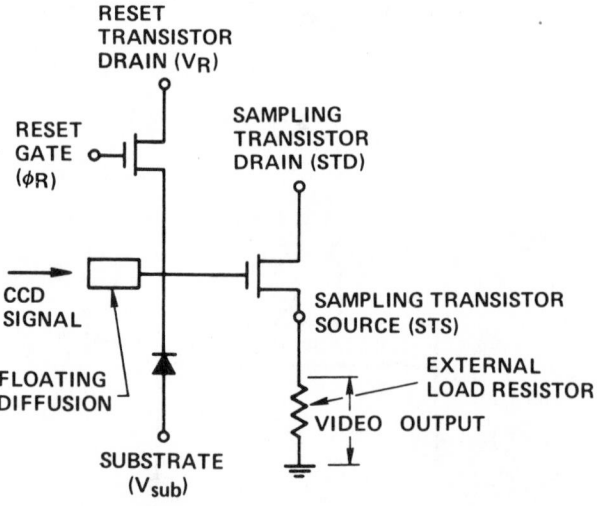

Figure 7

On-Chip Video Processing

The charge signal in a CCD is manipulated around without touching a finite electrode until it is extracted at the output. A floating diffusion is used as a charge detector. The voltage on the floating diffusion is reset to a fixed potential once each clock period by a reset transistor. When each charge packet reaches the floating diffusion, the charge changes the voltage on the capacity at that node. This voltage is sampled by a sampling transistor operating as a source follower. These circuits are shown in Figure 7. The very low node capacity of the floating diffusion results in a significant improvement in signal-to-noise ratio over a silicon vidicon operating at the same light level. In fact, at normal signal levels, no noise is visible in a displayed picture generated by the CCD imager.

Developmental Camera

A developmental black-and-white camera has been fabricated for the 512 x 320 CCD imager. Figure 8 shows a picture of this camera. The camera is approximately the size of a pack of cigarettes and has a "C-mount" for lens interchangeability. Figure 9 and Figure 10 show 525-line monitor pictures made by this camera with a 512 x 320 CCD imager.

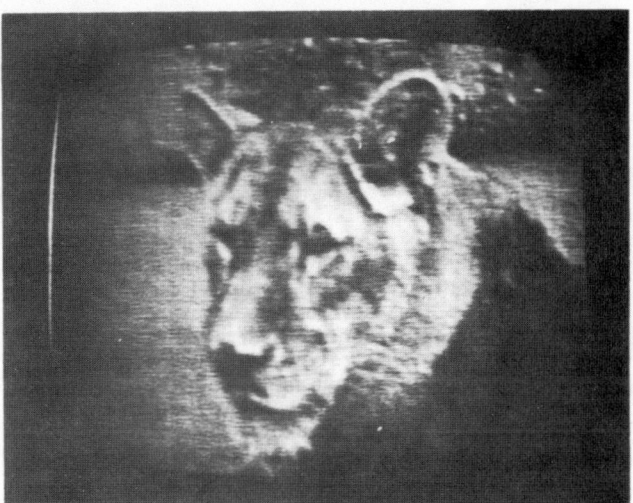

Figure 10

Figure 8

SUMMARY OF DEVELOPMENTAL 512 x 320 CCD IMAGER PERFORMANCE PARAMETERS

(1) Spectral Response Similar to Silicon Vidicon
(2) Operable in SIT-CCD Mode
(3) 500 Nanoamperes DC Maximum Signal
(4) 4 Nanoamperes DC Dark Current at 25° C
(5) Compatible With Standard 525 Line TV
(6) Supplies Full Resolution of Color TV (240 TVL/PH)
(7) No Lag — Picture Erased in One Field

Notes:

3φ, N-Channel Vertical Frame Transfer
163, 840 1.2 x 1.2 Mil Cells
6 MHz Data Rate, 16.66 ms Integration
12 mm Image Format
500 x 750 Mil Chip Size

Figure 11

Figure 9

SUMMARY AND CONCLUSIONS

A CCD imager has been developed that will offer an attractive alternative to camera tubes for many 525-line television applications. It is capable of supplying the full resolution of color TV (240 TVL/PH). The features of CCD imagers that are superior to camera tubes are signal to noise ratio, freedom from lag, and absence of microphonics. The small size, light weight, low-power consumption, and precision image characteristics are additional benefits. Figure 11 summarizes the performance parameters for this developmental 512 x 320 CCD imager.

REFERENCES

1. R. L. Rodgers, III, "Beam Scanned Silicon Targets for Camera Tubes," Paper presented at IEEE Intercon, New York, NY, March 1973.

2. P. K. Weimer, W. S. Pike, M. G. Kovac, and F. V. Shallcross, "The Design and Operation of Charge-Coupled Image Sensors," Paper presented at IEEE Solid State Circuit Conference, Philadelphia, PA, February 1973.

3. M. G. Kovac, F. V. Shallcross, W. S. Pike, P. K. Weimer, "Design, Fabrication, and Performance of a 128 x 160 Element Charge-Coupled Image Sensor," Paper presented at CCD Applications Conference, San Diego, California, September 1973.

4. M. F. Tompsett, G. F. Amelio, W. S. Bertram, R. R. Buckley, W. J. McNamara, J. C. Mikkelsen, Jr., and D. A. Sealer, "Charge-Coupled Imaging Devices: Experimental Results," IEEE Trans. Electron Devices, Vol. ED-18, pp. 992–996, November 1971.

5. C. H. Sequin, "Interlacing in Charge-Coupled Imaging Devices," IEEE Trans. Electron Devices, Vol. ED-20, pp. 535–541, June 1973.

6. R. L. Rodgers, 3rd, et al, "Silicon Intensifier Target Camera Tube," Paper presented at IEEE International Solid State Circuits Conference, Philadelphia, PA, February 1970.

Silicon Imaging Device

SID 51232

512 x 320 Element Sensor for Very Compact TV Cameras

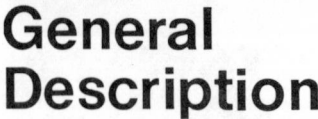

Features

- High Resolution
- Standard Interlaced 525-Line TV Output
- Ultra Low Blooming
- No Lag
- Precision Image Geometry
- No Microphonics
- Low Voltage and Power Requirements
- Small Size
- Highly Resistant to Image Burn-In
- Stable Sealed Silicon Gate Structure
- No Aliasing

L-644

General Description

RCA SID51232* is a self-scanned, charge-coupled device (CCD), solid-state image sensor intended primarily for use in generating standard interlaced 525-line television pictures. The device contains 512 x 320 elements (163,840 individual storage sites). The device is constructed with a 3-phase, N-channel, Vertical Frame Transfer organization using a sealed silicon gate structure. This device features high resolution combined with ultra-low blooming characteristics. Overall picture performance is comparable to that of 2/3-inch vidicon camera tubes but undesirable characteristics such as lag and microphonics are eliminated.

The SID51232 is supplied in a hermetic, edge-contacted, 24-connection ceramic dual in-line package. The package contains an optical glass window which allows an image to be focused onto the sensor's 12.2 mm image diagonal.

*Two developmental variants are available. See page 7 for grade classifications and ordering information.

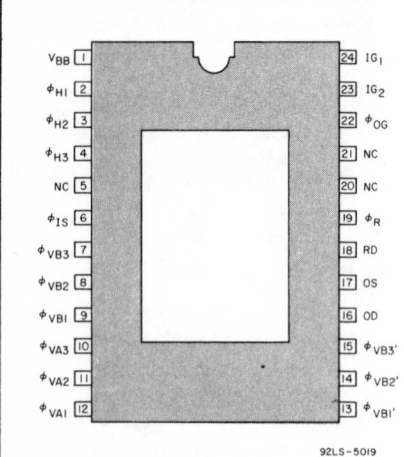

92LS-50I9

Figure 1 — Connection Diagram (Top View)

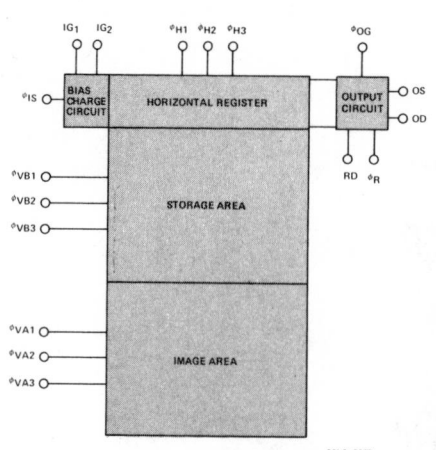

92LS-5023

Figure 2 — Block Diagram

Connection Names

V_{BB}	Substrate Bias Voltage
ϕ_{H1} ϕ_{H2} ϕ_{H3}	Horizontal Register Clocks
ϕ_{IS}	Output Register Source Clock
ϕ_{VA1} ϕ_{VA2} ϕ_{VA3} ϕ_{VB1} ϕ_{VB2} ϕ_{VB3} ϕ_{VB1}' ϕ_{VB2}' ϕ_{VB3}'	Vertical Register Clocks
OD	Output Transistor Drain
OS	Output Transistor Source
RD	Output Reset Transistor Drain
ϕ_R	Output Reset Transistor Gate Clock
ϕ_{OG}	Output Gate Clock
IG$_1$ IG$_2$	Input Gates

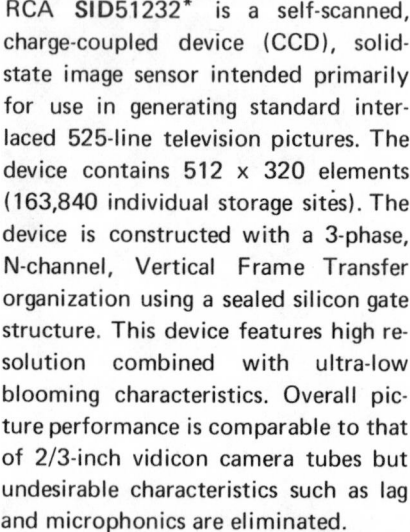

For further information or application assistance on this device, contact your RCA Sales Representative or write Charge-Coupled Device Marketing, Lancaster, PA 17604.

Developmental-type devices or materials are intended for engineering evaluation. The type designation and data are subject to change, unless otherwise arranged. No obligations are assumed for notice of change or future manufacture of these devices or materials.

Information furnished by RCA is believed to be accurate and reliable. However, no responsibility is assumed by RCA for its use; nor for any infringements of patents or other rights of third parties which may result from its use. No license is granted by implication or otherwise under any patent or patent rights of RCA.

Printed in U.S.A./ **1-75**
SID51232

Trademark(s) Registered ®
Marca(s) Registrada(s)

Reprinted with permission from a specification sheet of the RCA Corporation, Lancaster, PA, pp. 1–7, Jan. 1975.

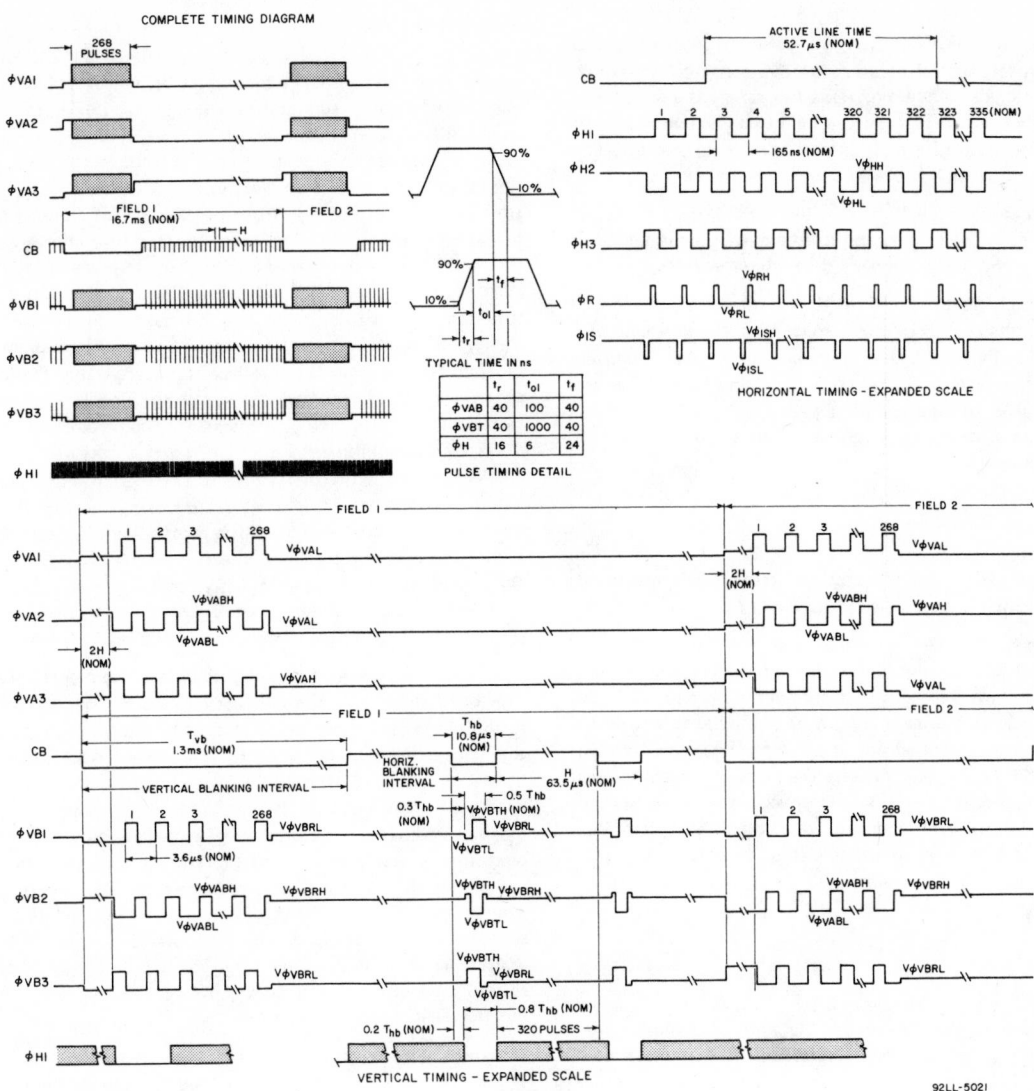

Figure 3 — Timing Diagram

Functional Description

The **SID51232**, shown in the block diagram of **Figure 2**, consists of the following subsections:

Image Area

The image area is an array of analog CCD shift registers containing 320 parallel vertical columns of 256 sensing cells. An elemental cell is defined by each grouping of three adjacent polysilicon gate electrodes in the vertical direction and adjacent channel stops in the horizontal direction. The three gates in each cell are connected in parallel with the corresponding gates in the other cells. These three connections are called vertical register clocks ϕ_{VA1}, ϕ_{VA2}, and ϕ_{VA3}.

When these vertical register clocks are pulsed with the voltage waveforms shown in **Figure 3**, a light image focused on this register is integrated into a charge pattern of electrons during the normal

active TV display time and transferred to the storage area during the vertical blanking interval. To prevent blooming of the image during picture integration, the electrodes which are not storing charge are pulsed slightly negative with respect to the substrate to drive the surface into accumulation. This operational method electronically extends the channel stops completely around each sensing site to prevent vertical charge spreading. The effective number of vertical resolution elements is interlaced to 512 by alternation of ϕ_{VA2} and ϕ_{VA3} as the average sensing location in the vertical direction on successive fields. In a normal display approximately 482 lines are displayed.

The transfer electrodes are made of polysilicon and are transparent to most wavelengths. The entire image sensing area is therefore light sensitive and has no opaque areas to cause "aliasing" due to small picture details being obscured and not contributing to the video output.

Storage Area

The storage area has the same construction as the image area and also contains 320 parallel vertical columns of 256 sensing cells which line up with the image area columns. This area serves as a temporary storage site for the previous TV picture field to allow conversion of the charge pattern image into a sequential horizontal readout. The storage area must normally be covered by an external light shield.

There are three storage area clock drives. The first is ϕ_{VB1} connected in parallel with ϕ_{VB1}'. The second and third are: ϕ_{VB2} connected in parallel with ϕ_{VB2}', and ϕ_{VB3} connected in parallel with ϕ_{VB3}'.

The storage area is clocked in unison with the image area during vertical blanking to transfer the complete image from the image area to the storage area (See **Figure 3**). During the horizontal blanking interval the entire charge pattern in the storage area is advanced toward the horizontal register by one complete cell transfer. The line adjacent to the horizontal register is loaded into the horizontal register to allow it to be read out.

Horizontal Register

The output register has the same 3-phase structure as the image and storage areas. The clock drives are ϕ_{H1}, ϕ_{H2}, and ϕ_{H3}.

The horizontal register receives one line of picture information from the storage area during each horizontal blanking interval (See **Figure 3**). The register contains 320 cells corresponding to each of the 320 columns in the image and storage areas plus two additional cells at the output. Each line is clocked out at a 6.1 MHz picture element rate so that the 320 active elements are read out in the active line time of 52.7 μs. Typically the register is "over-clocked" by several pulses to provide a clean dark signal for video black level clamping. (Total number of pulses is approximately 335.) A bias charge is introduced into the input of the register and the signal is extracted by the output circuit.

Bias Charge Circuit

Bias charge is introduced electrically into the input of the horizontal register by the IG_1, IG_2, and ϕ_{IS} connections. When these connections are operated as shown in **Figure 3**, a uniform low-noise bias charge (often referred to as "Fat Zero") is introduced into the output register to maximize the horizontal resolution. This circuit provides a constant amount of charge during each clock cycle. The amount of charge is determined primarily by the difference in voltage between IG_1 and IG_2.

It should be noted that a bias charge is required in the image and storage areas. This is provided by illuminating the image area with a uniform background light source (i.e., a couple of LEDs placed in a position to act as a uniform diffuse light source).

Output Circuit

The CCD signal is extracted from the horizontal register by the circuit shown in **Figure 4**. ϕ_{OG} is the last CCD gate in the horizontal register. The CCD signal charge is collected at the floating diffusion. This diffusion is reset to a positive potential once each clock cycle by RD and ϕ_R. The voltage change on the floating diffusion is the signal which is sensed by the output transistor. The signal may be taken from OS as a source follower or from OD as an inverter.

For measurement purposes, the photocurrent delivered by the CCD in response to a light stimulus can be measured at the drain of the reset transistor — RD.

Waveform Description

The drive signals required to operate the SID51232 are shown in reference to the composite blanking (CB) waveform in **Figure 3**. The charge pattern is integrated in the image area during the active frame time under the electrode which is held at $V_{\phi VAH}$. This position changes between ϕ_{VA2} and ϕ_{VA3} on alternate fields to interlace the image. The other two electrodes are held at $V_{\phi VAL}$ during this time to control blooming from overloads. The transfer between the image area and storage area takes place during the vertical blanking interval. There should be approximately a 2H delay (127 μs) for the start of these pulses as shown. The 268 transfer pulses are applied simultaneously to the image and storage area. All pulses swing from $V_{\phi ABL}$ to $V_{\phi ABH}$ during this time. The sequence is slightly different on each field so that the interlace is preserved and the image charge pattern comes to rest under ϕ_{VB2} on each field. The image is transferred one line at a time into the horizontal register during the active horizontal blanking intervals. During this transfer the horizontal clocks are stopped with ϕ_{H2} resting at $V_{\phi HH}$ and ϕ_{H1} and ϕ_{H3} at $V_{\phi HL}$. The first transfer must occur during the first full horizontal blanking pulse as shown to match the display interlace. While a horizontal line is being read out, the storage area stores the image with ϕ_{VB1} and ϕ_{VB3} resting at $V_{\phi VBRL}$ and ϕ_{VB2} at $V_{\phi VBRH}$. During transfer the pulses swing from $V_{\phi VBTL}$ to $V_{\phi VBTH}$. The transfer pulses must be completed within the horizontal blanking time.

The horizontal readout begins just before the display is unblanked. The third pulse will contain picture information and should occur in the active line time. There should be a total of 320 pulses within the active time to display the full image. Additional horizontal pulses (approx. 13) should be continued into the beginning of the next blanking interval to provide a black reference for clamping the video signal. The horizontal clocks must come to rest before the next line is transferred from the storage area into the horizontal register as shown. All ϕ_H pulses swing between $V_{\phi HL}$ and $V_{\phi HH}$. The ϕ_R pulses occur in the center of every ϕ_{H3} pulse as shown and swing from ϕ_{RL} to ϕ_{RH}. The ϕ_{IS} pulse occurs while ϕ_{H1} is at its ϕ_{HL} level and swings from ϕ_{ISL} to ϕ_{ISH}. The horizontal clocks should continue running during the vertical blanking time to read-out the dark charge buildup.

All pulses within a group should have overlap as defined by **Figure 3**. Typical overlaps, rise and fall times are given in the figure.

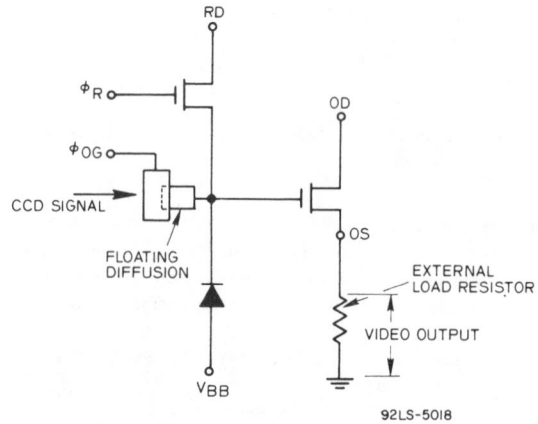

Figure 4 — Typical Output Circuit

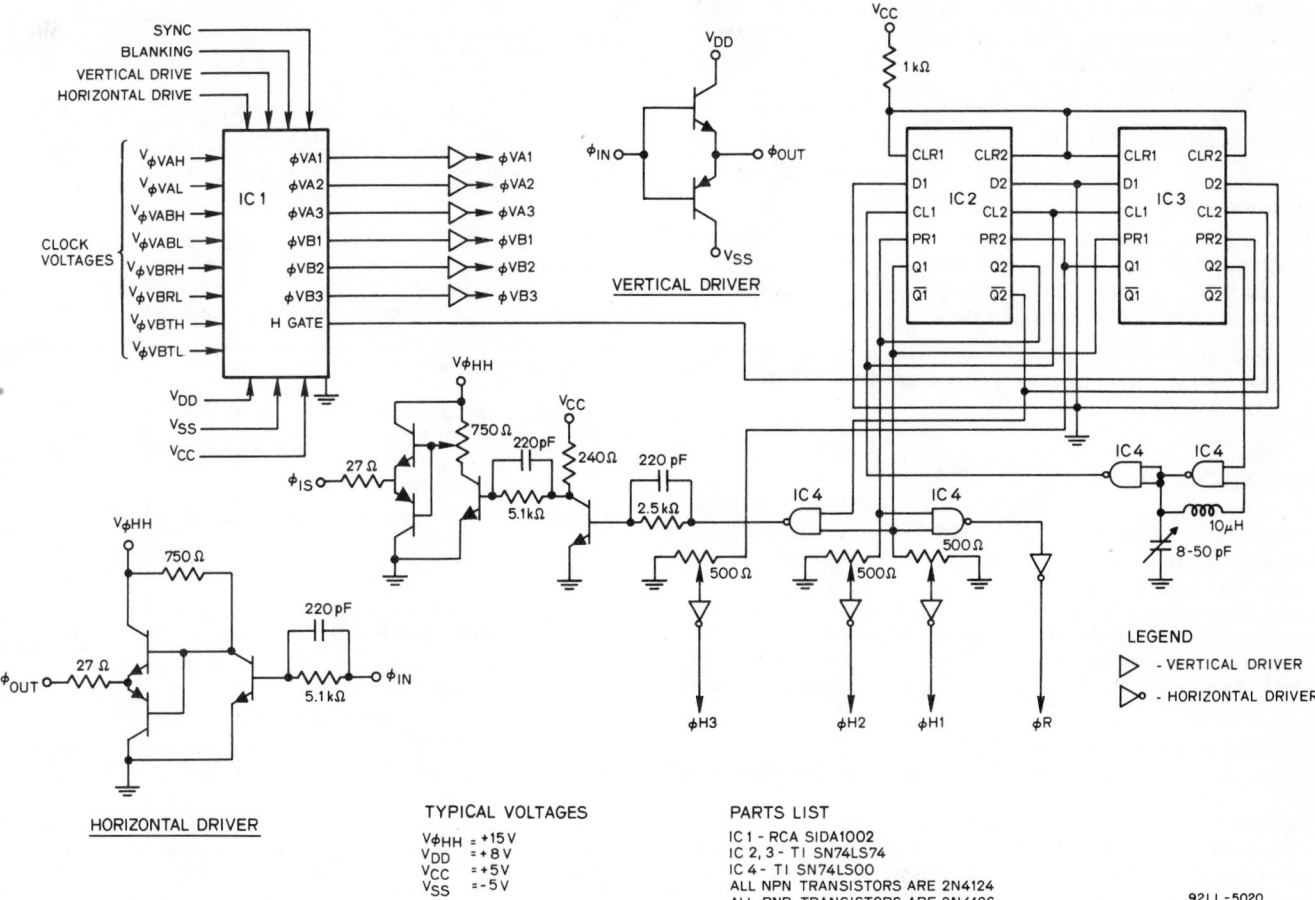

Figure 5 — Typical Drive Circuitry

Video Processing

The unprocessed video output (at OS) during the active line time consists of the light induced voltage output changes plus horizontal register clock pickup and reset pulses. The clock and reset pulse voltage swings have a fundamental frequency of 6.1 MHz. This frequency is twice the maximum video generation rate due to the number of horizontal cells (sampling limit). The clock pulse amplitude should be suppressed by limiting the video bandwidth to 3 MHz with a low pass filter having good rejection at 6 MHz and above.

The video output should be clamped during the horizontal register "over-clocking" period (**Figure 3**) to establish a black level which is not light sensitive and does not depend on the electrical bias charge setup conditions.

Drive Circuitry

Figure 5 shows typical circuitry required to drive the **SID**51232 in a standard 525-line television system. IC1 generates the six vertical timing waveforms plus a horizontal gate pulse which is used to synchronize the horizontal line circuitry. IC1 is a custom CMOS LSI circuit and thus has very low power consumption.

The vertical timing is derived in IC1 from the four standard sync signals (ref. EIA Standard RS-170). The voltage levels of the pulses are set by the eight DC-clock voltage inputs through the use of multiplexers in the IC. These voltages are typically supplied by pots connected from V_{DD} to V_{SS}.

The horizontal timing is generated by three low-power Schottky TTL ICs. One-half of IC4 is the gated horizontal oscillator whose frequency is adjusted by the variable capacitor. The oscillator runs at three times the horizontal clock frequency (approximately 18.3 MHz). Symmetrical three-phase pulses are generated by the three D flip-flops of IC2 and one-half of IC3. This circuit is a modified ring counter where the Q output of each flip-flop is connected to preset of the previous flip-flop in the ring. The propagation delay associated with the preset generates the overlap required for proper CCD transfer. The fourth flip-flop (half of IC3) assures that the three-phase clocks stop and start in proper sequence every horizontal line. The other two NAND gates of IC4 are used to decode the ϕ_R and ϕ_{IS} pulses.

The bipolar drivers provide the necessary current drive capability. The horizontal drivers also convert the TTL levels to the proper drive levels. The three 500-ohm pots for the ϕ_H clocks control the fall time of the output pulses. These are adjusted for the maximum fall time without pulse distortion.

203

Maximum Ratings, Absolute-Maximum Values

Storage Temperature (non-operating) . . . . −40°C to +100°C
Operating Temperature . . . . −40°C to +65°C
Voltages (with respect to Connection 1):
  Connections 2, 3, 4, 7, 8, 9, 10, 11, 12,
           13, 14, 15, 19, 22, 23, 24 . . . . −5 to +20 volts
  Connections 6, 16, 17, 18 . . . . −0.6 to +20 volts

Caution

1. Static discharge to any connection may cause permanent damage. Store in shorting clip or in conductive foam. Use grounded soldering irons and tools. Personnel should wear grounding bracelets and avoid synthetic smocks and gloves.

2. When placing CCD into socket (**SIDA1001**), press only on ceramic or metal. Do not exert pressure on glass window to avoid possible glass breakage.

Optical Characteristics

Image Cell Size . . . . 1.2 x 1.2 mils (0.03 x 0.03 mm)
Image Size (3 x 4 aspect ratio). . . . 288 x 384 mils (7.31 x 9.75 mm)
Image Diagonal . . . . 12.2 mm
Effective Optical Distance
from Socket Top (See **Figure 9**) . . . . approx. 100 mils (2.54 mm)
Minimum Useful Wavelength . . . . 420 nm
Maximum Useful Wavelength . . . . 1100 nm
Saturation Exposure . . . . 2.67×10^{-3} fc·s 2856 K

Typical Performance Data

Conditions: Standard EIA RS-170 525-line TV format operating at 25° C.

Electrical

Parameter		Unit	Note
Horizontal Clock Rate	6.1	MHz	
Video Bandwidth	3.0	MHz	
Vertical Transfer Rate	0.28	MHz	
Light Integration Time	16.67	ms	
Image Area Dark Current	4	nA	1
Image Area Light Bias Current	30	nA	1
Peak-to-Peak Signal Current	250	nA	1
Peak-to-Peak Signal Voltage	12.5	mV	2
Saturation Peak-to-Peak Signal Current	400	nA	1
Horizontal Register Bias Current	300	nA	1
Horizontal Limiting Resolution	240	TVL/PH	3
Vertical Limiting Resolution	480	TVL/PH	3
Contrast Transfer Function (CTF)	0.4	—	4
Gamma	1	—	5
Signal-to-Noise Ratio	50	dB	6

Optical

Parameter		Unit	Note
Sensitivity (Luminous)	3250	μA/lm	1,7
Sensitivity (Radiant)	65	mA/W-2856 K	1,7
Faceplate Illumination (250 nA signal)	0.1	fc	7

Notes

1. DC current from image area measured at RD.
2. Developed across 3 kilohm load resistor at OS.
3. Observed with EIA Resolution Chart and good quality monitor.
4. Measured at 200 TVL/PH with an infrared blocking filter (2 mm HA-11). See **Figure 7**.
5. Measured between 4 and 400 nA signal current.
6. Video peak signal (250 nA DC) to RMS noise.
7. Measured with 2856 K illumination.

Forcing Functions

Symbol	Parameter	Range			Units
		Min.	Typ.	Max.	
$f_{\phi V}$	Vertical Transfer Frequency		0.28		MHz
$f_{\phi H}$	Horizontal Transfer Frequency		6.1		MHz
$V_{\phi VAH}$	Image Area Voltage High		1.5	8	V
$V_{\phi VAL}$	Image Area Voltage Low	−5	−4.5		V
$V_{\phi VABH}$	Image-to-Storage Voltage High		8.0	12	V
$V_{\phi VABL}$	Image-to-Storage Voltage Low	−5	−2.0		V
$V_{\phi VBRH}$	Storage Area Rest Voltage High		7.5	12	V
$V_{\phi VBRL}$	Storage Area Rest Voltage Low	−5	−1.5		V
$V_{\phi VBTH}$	Storage Area Transfer Voltage High		8.0	12	V
$V_{\phi VBTL}$	Storage Area Transfer Voltage Low	−5	−2.0		V
$V_{\phi HH}$	Horizontal Register Clock Voltage High		15.0	16	V
$V_{\phi HL}$	Horizontal Register Clock Voltage Low	0	0		V
$V_{\phi RH}$	Reset Transistor Clock Voltage High		15.0	16	V
$V_{\phi RL}$	Reset Transistor Clock Voltage Low	0	0		V
$V_{\phi ISH}$	Input Source Clock Voltage High		15.0	16	V
$V_{\phi ISL}$	Input Source Clock Voltage Low	0	1.0		V
V_{RD}	Reset Transistor Drain Bias Voltage	0	12.0	16	V
V_{IG1}	Input Gate Bias Voltage	0	4.0	16	V
V_{IG2}	Input Gate Bias Voltage	0	6.0	16	V
V_{BB}	Substrate Bias Voltage	−5	−3.0	0	V

Notes

1. ϕ_{OG} is normally connected to ϕ_{H1}.
2. ϕ_{VB} connections should be connected to corresponding ϕ_{VB}' connections.
3. $C_{\phi VA1} = C_{\phi VA2} = C_{\phi VA3} = C_{\phi VB1} = C_{\phi VB2} = C_{\phi VB3} \approx 2500$ pF.
4. $C_{\phi H1} = C_{\phi H2} = C_{\phi H3} \approx 66$ pF.

Mechanical Data

The **SID51232** is packaged in a hermetic, edge-contacted, 24-connection ceramic dual in-line package. **Figure 8** shows the mechanical dimensions of the sensor package. **Figure 9** shows the sensor mounted in its socket (**SIDA1001**) and the position of the center of the image area. The package contains an optical glass window which allows an image to be focused onto the sensor's 12.2 mm image diagonal.

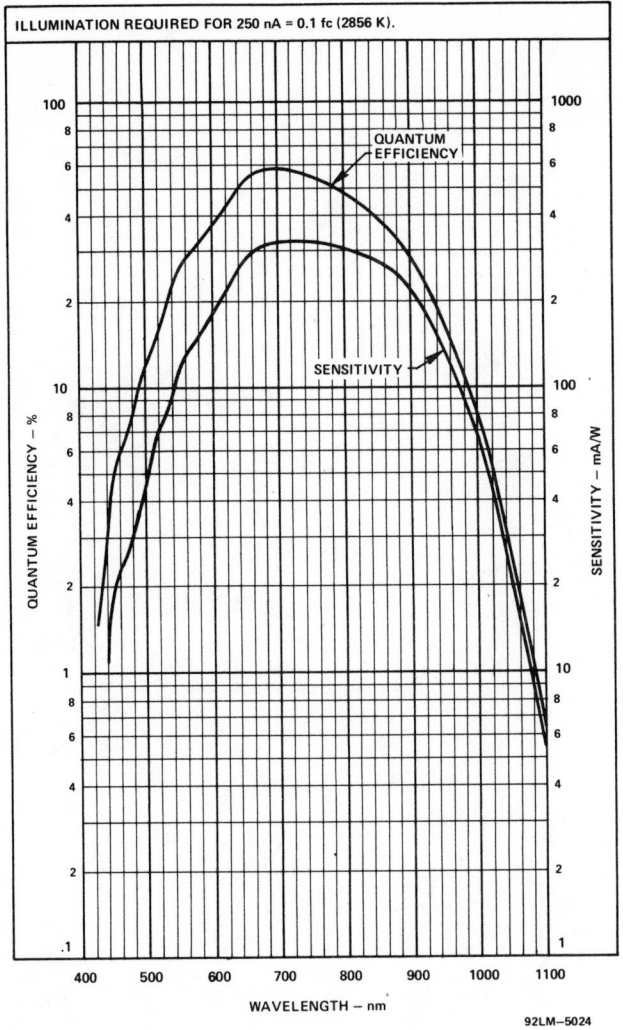

ILLUMINATION REQUIRED FOR 250 nA = 0.1 fc (2856 K).

Figure 6 — Typical Spectral Response Characteristics

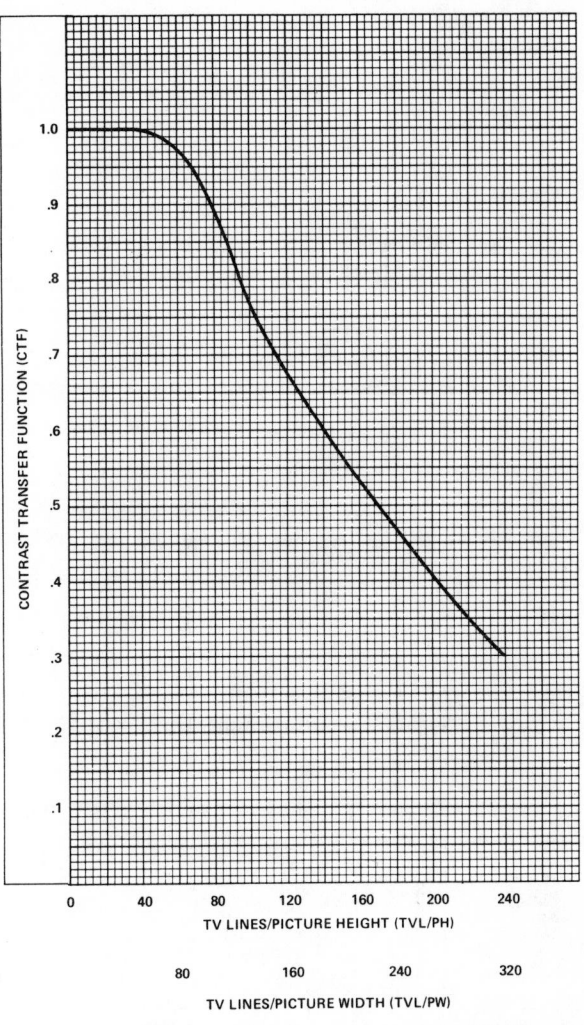

92LM–5022

Figure 7 — Typical Contrast Transfer Function (CTF) Characteristic

Optical

RCA **SID51232** utilizes a 1.2 x 1.2 mil cell size in the image sensing area. The image area consists solely of imaging cells with no additional registers separating cells. A light image is focused onto the image area after first passing through an optical glass window. The light image is absorbed in the silicon substrate and converted into an electron charge pattern after passing through the polysilicon gate electrodes which are transparent to most wavelengths. The useful spectral response range extends from 420 nm to 1100 nm (See **Figure 6**). Due to the absence of any structure blocking picture details in the image area, there is no "aliasing" in the reproduced picture from small picture segments not contributing to the video output. A small amount of "moiré" may however appear in the reproduced picture from picture details having spatial frequencies higher than the cell frequency. This effect is directly equivalent in the vertical direction to a beam scanned camera tube which also has approxi-

mately 482 scan lines. The effect in the horizontal direction is equivalent to one field in the vertical direction.

The image diagonal is 12.2 mm which allows the use of most "C" mount and other lenses intended for 16-mm film or 1-inch and 2/3-inch vidicons. The position of the image area is shown in **Figure 9**. The storage area and output register must normally be covered by an external light shield placed midway between connections 6-7 and 18-19 extending toward connections 1 and 24. This shield prevents light from striking these areas.

To provide a bias charge for the image and storage areas, a uniform background illumination of the image area is required. A light source (i.e., two or more LEDs) must be placed in a position to uniformly illuminate this area and not block the image from the lens.

The normal orientation of the **SID51232** is with connections 1 and 24 facing downward and a lens between the scene and the sensor.

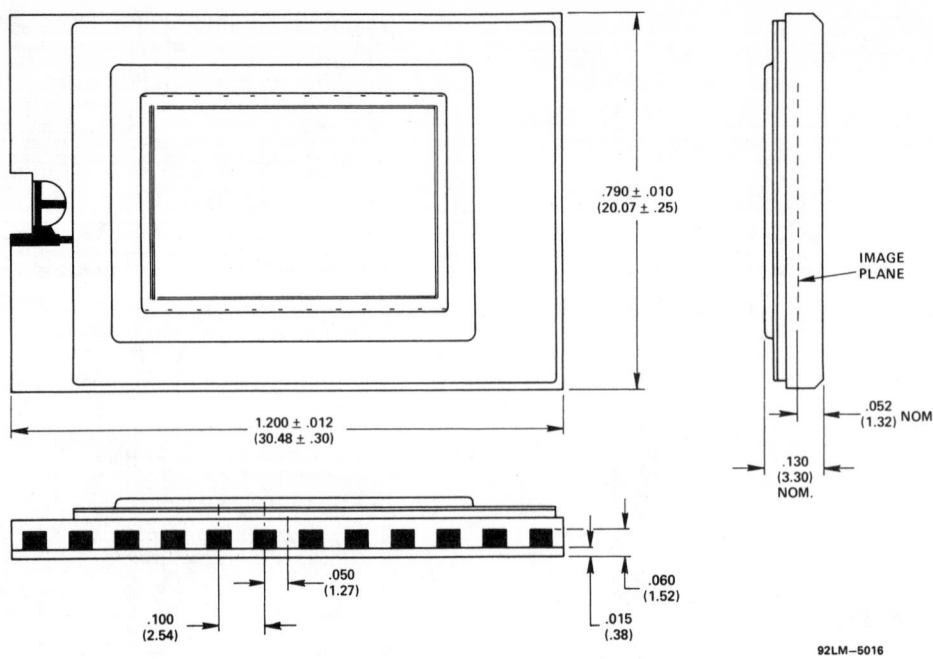

Figure 8 — Dimensional Outline of Sensor (SID51232)

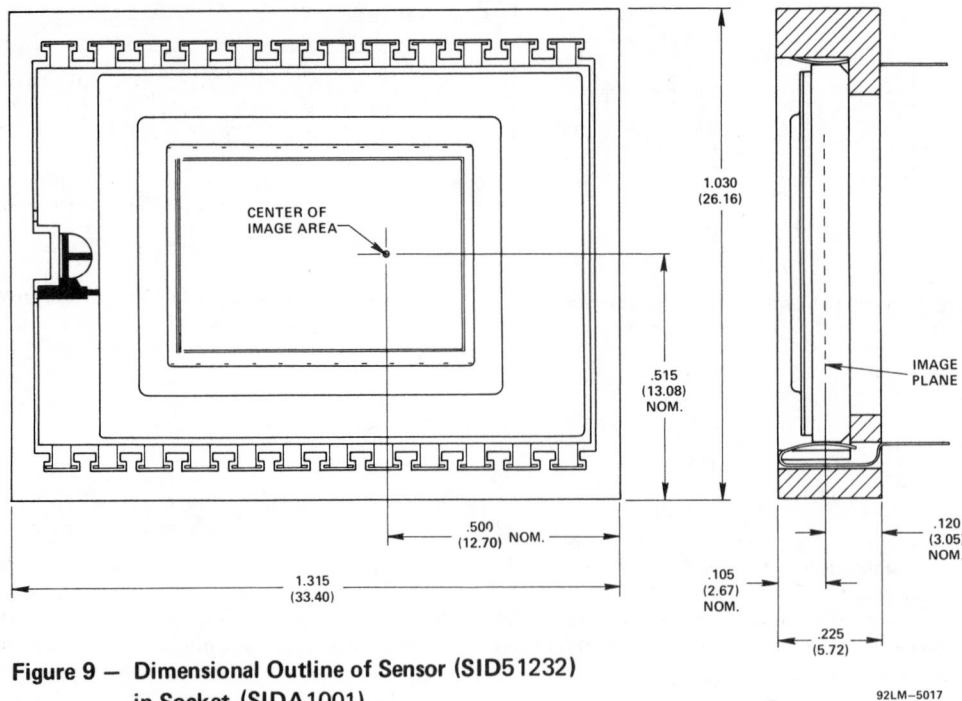

Figure 9 — Dimensional Outline of Sensor (SID51232) in Socket (SIDA1001)

Dimensions in parentheses are in millimetres.

Classification and Ordering Information

SID51232 is available in two grades. **SID**51232BD is the higher performance device and is intended for the more demanding applications. **SID**51232AD is intended for those applications where less stringent electrical and blemish criteria are permissible. The A and B in the type designations indicate the grade difference and the suffix D indicates a developmental status.

LOW LIGHT LEVEL PERFORMANCE OF CCD IMAGE SENSORS

David D. Wen

Fairchild Research & Development, 4001 Miranda Ave., Palo Alto, Ca. 94304

ABSTRACT   This paper reports the low light level performance of CCD image sensors. Theoretical limitations of transfer efficiency for small charge packets in CCD shift registers are reviewed. Low noise charge detection techniques are discussed. Specifically, the operation of a distributed floating-gate amplifier (DFGA) is described. A DFGA employs several charge amplifiers which repeatedly sense a signal charge packet in a CCD. The outputs of the charge amplifiers are coherently summed in a second CCD shift register. Signal-to-noise ratio is improved so that extremely small charge packets can be detected.

Low light level imaging performance of both linear and area arrays are reported. The linear array contains 1728 photoelements and uses a single-stage floating-gate amplifier (FGA) as the on-chip detector. The area array has 244 x 190 photoelements and contains a twelve-stage DFGA and an FGA. Both arrays employ two-phase, buried channel CCD shift registers. Low light level images at 50 and 25 electron levels have been achieved with the linear and area sensors, respectively.

INTRODUCTION

The low light level performance of CCD image sensors is reported in this paper. The basic requirement of transferring a small charge packet in a CCD shift register is considered first. It is well known[1] that in a surface channel CCD shift register, a background charge (fat zero) is required at all times to suppress charge-trapping effects of surface states. The noise[2] associated with the generation of the fat zero makes the surface channel CCD sensor inadequate to perform low light level imaging functions.

In a buried-channel CCD register[3], the signal charge packets are stored and transferred in the bulk of the semiconductor so that surface state trapping can be avoided. However, the signal charge packets may be trapped by the crystalline imperfections in the bulk of the semiconductor. The effect of these bulk traps on charge transfer efficiency plays a dominant role in the low light level performance of buried-channel CCD image sensors. This effect has been analyzed and reported previously by J. Early.[4] Some of his principal assumptions and conclusions are repeated in the first part of this paper. It is shown that signal charge packets of approximately 10 electrons can be transferred in a buried-channel CCD register.

In order to fully exploit the low light level capability of buried-channel CCD image sensors, a special low-noise, high-gain amplifier must be used. Such an amplifier is the distributed floating-gate amplifier (DFGA)[5]. A DFGA employs several charge amplifiers with floating gate inputs to sense repeatedly a signal charge packet in a CCD register. The outputs of the charge amplifiers are summed using a second CCD register. Since the random noise of the charge amplifiers is uncorrelated, the resulting signal-to-noise voltage ratio is enhanced by a factor equal to the square root of the

Reprinted from *Proc. 1975 Naval Electron. Lab. Center Int. Conf. on the Application of Charge-Coupled Devices*, Oct. 29–31, 1975, pp. 109–119.

207

number of charge amplifiers used. Signal charge packets of the order of 10 electrons can thus be detected. Operation and experimental results of a twelve-stage DFGA are discussed.

Low light level imaging performance of both linear and area arrays are reported. The linear array contains 1728 photoelements. Charge packets in the photosites are transferred in parallel into an adjacent opaque, two-phase, buried-channel, implanted-barrier CCD shift register. A single-stage floating-gate amplifier is positioned at the end of the register to detect the signal charge packets. The area array has 244 x 190 photoelements and employs the interline transfer organization. Signal charge packets from each column are transferred into an adjacent opaque two-phase vertical shift register. The vertical shift register in turn transfers each row of signal charge packets into a two-phase horizontal shift register. The signal charge packets are then transferred along the horizontal register and detected by a twelve-stage DFGA and an FGA. Image resolution and noise performance are examined at different light levels.

LOW LIGHT LEVEL CHARGE TRANSFER CONSIDERATIONS

In buried-channel CCD image sensors, the signal charge packets may be trapped by crystalline imperfections in the bulk of the semiconductor. The effect of bulk traps and dark current on charge transfer efficiency and signal-to-noise ratio for a hypothetical 500 x 500 element CCD area image sensor, operating in the standard NTSC mode has been analyzed by J. Early. His analysis revealed the following:

1) A reduction in operating temperature reduces dark charge and thereby increases the signal-to-noise ratio.

2) An increase in dark current improves charge transfer efficiency because bulk

traps can be filled by dark charges.

3) Under worst-case conditions, for an average dark charge of 10 electrons per pixel and a signal of 10 electrons per pixel, the overall transfer efficiency is approximately 0.8, and the signal-to-noise voltage ratio is 1.4.

4) Bulk trapping is of consequence only in long registers operating at conventional television horizontal line repetition rates of 15.75 KHz.

5) Transfer efficiency and signal-to-noise ratio are excellent at the high horizontal transfer rate of 5MHz.

Charge transfer efficiency of a 15-electron charge packet in a two-phase buried-channel CCD was measured, using statistical methods.[6] The device was cooled to obtain an average dark charge of 4 electrons per pixel. It was observed that after 238 transfers at a 15.7 KHz clock rate, approximately one electron was lost. This data confirms the theoretical analysis that in a buried-channel CCD shift register there is useful transfer efficiency after 500 transfers in a signal charge packet of approximately 10 electrons.

DISTRIBUTED FLOATING-GATE AMPLIFIER (DFGA)

In a CCD register, the signal charge packet can be detected by a sensing "floating-gate" electrode as shown in Figure 1. It can be seen that the floating gate provides a capacitive coupling between the signal charge packet in the CCD register and the current in the MOS channel without making physical contact with either of them. Since the signal charge packet is not destroyed by the floating gate, it can be transferred along the CCD register and be detected repeatedly by similar structures. In a DFGA[5], the outputs of several floating-gate structures are summed with a

second CCD register. The signal-
to-noise ratio of the summed output is enhanced so that extremely
small charge packets can be detected.

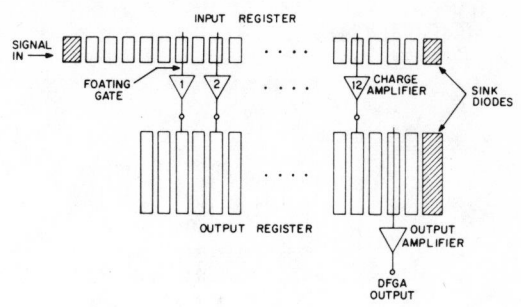

FIGURE 2   Schematic of DFGA.

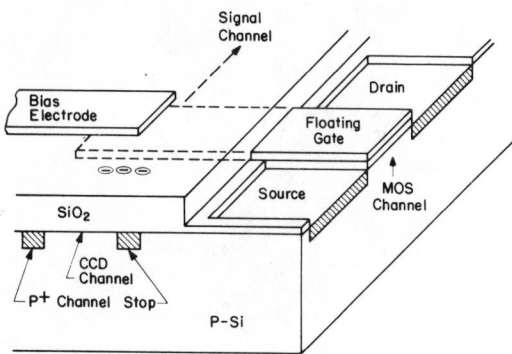

FIGURE 1   Schematic of FGA.

A schematic diagram of a twelve-
stage DFGA is shown in Figure 2.
It consists of an input register,
a bank of twelve charge amplifiers
with floating-gate inputs, an output CCD register, and an output
amplifier. A signal charge packet
is sensed by the floating gates
when it is transferred in the
input register. The operation
of the charge amplifier is illustrated in Figure 3. During
the period when the signal charge
packet in the input register is
under the floating gate, the control gate is pulsed "on" so that
a small amount of charge flows
into the output register. The
magnitude of this charge is determined by the signal charge in
the input register through the
coupling of the floating gate.
This is a charge inverting amplifier in the sense that the larger
the signal charge in the input register, the lower the floating
gate potential, and the less the
charge that flows into the output
register. A dc gate is used to
eliminate clock coupling from the
control gate to the floating gate.

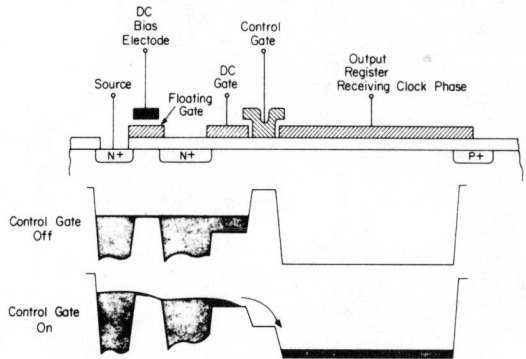

FIGURE 3   Charge amplifier in DFGA.

The operation of a DFGA can be
illustrated by considering a hy-
pothethical three-stage structure.
Figure 4(a) illustrates the initial charge distribution in the
DFGA after a long series of empty
charge packets have been transferred along the input register.
Figure 4(b) shows that one clock
cycle later ($t = t_c$) a large charge
packet D has been transferred under
the floating gate of the first
charge amplifier stage. After
the control gate has been pulsed
"on" a small amount of charge D'
is injected into the output register. Figure 4(c) shows that at
$t = 2t_c$, charge packet D has been
transferred under the floating gate
of the second charge amplifier
stage while charge packet D' in
the output register has also been

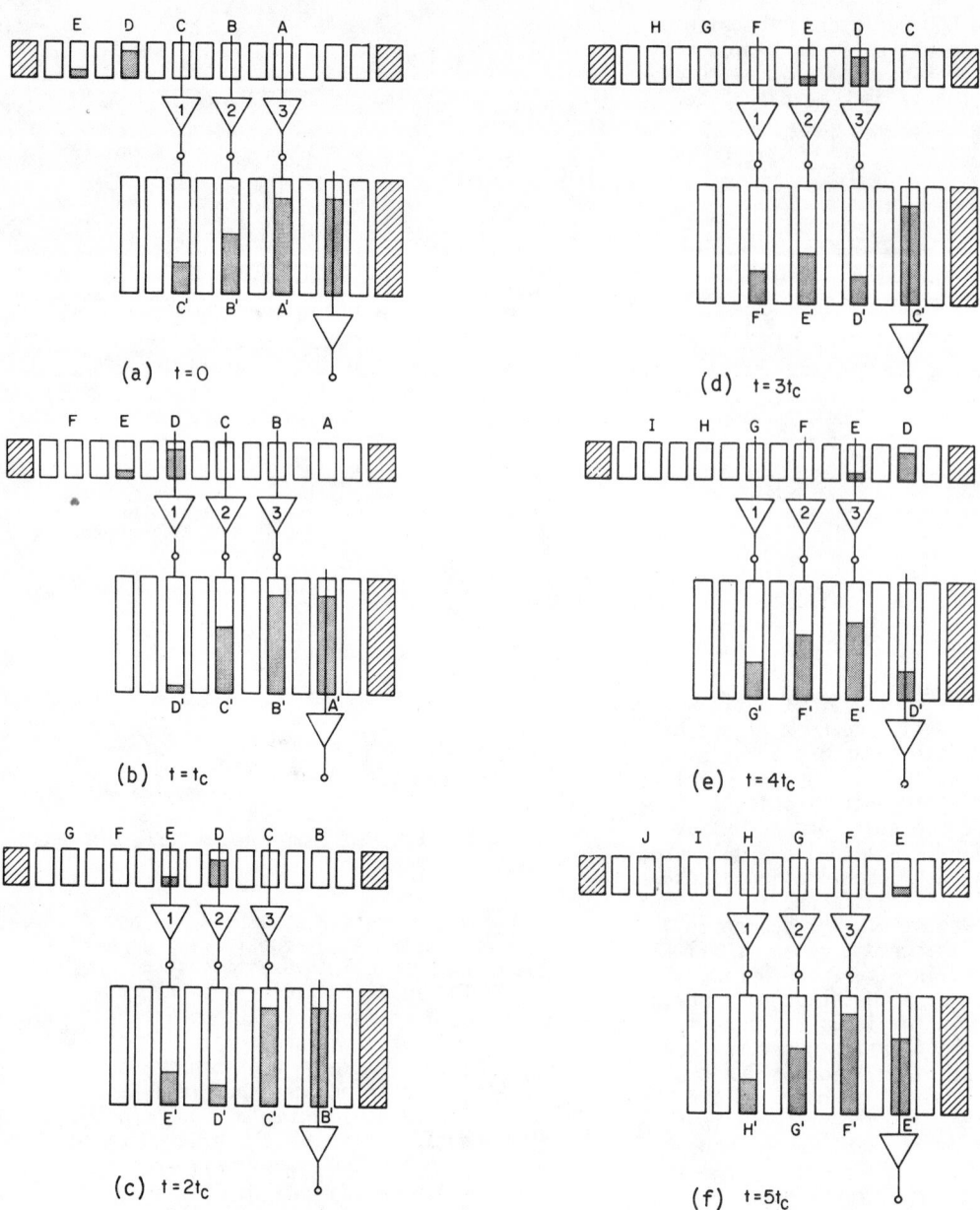

FIGURE 4   Charge distribution in DFGA.

transferred to the corresponding
position. After the control gate
is pulsed "on", another small
amount of charge is added to the
charge packet D'. Figure 4(d)
illustrates the charge distribution at $t = 3t_c$. At $t = 4t_c$,
the charge packet D' has been
transferred to the output amplifier where it produces the final
DFGA output. Charge packets E
and E' in the same sequence
illustrate the situation when a
small charge packet is transferred along the input register.
At $t = 5t_c$, charge packet E' is
detected by the output amplifier.

The DFGA output waveform is illustrated in Figure 5. The DFGA output
at $t = 4t_c$ corresponds to the initial
charge packet D. For zero initial
charge in the input register, a maximum
amount of charge is injected into the
output register. The corresponding output is V_{BIAS}. The signal output for
charge packet D which is the difference
between the DFGA output and V_{BIAS}, is
designated by V_S. The DFGA output at
$t = 5t_c$ corresponds to the small charge
packet E in the input register.

Twelve-stage DFGA test structures have
been built and tested. A typical transfer characteristic curve for a 50nsec
control gate "on" time, t_{on}, at 3 MHz
bandwidth and room temperature is plotted
in Figure 6. The smallest signal level
measured is approximately 30 electrons.
The RMS Noise measured is 10 to 20
electrons.

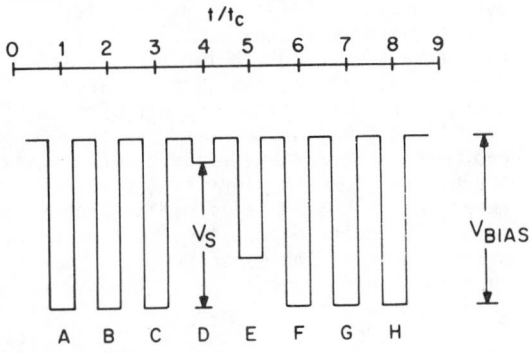

FIGURE 5   DFGA output waveform.

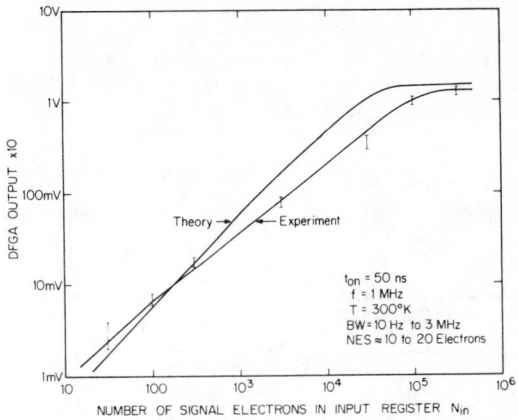

FIGURE 6   DFGA characteristics.

EXPERIMENTAL RESULTS OF
CCD IMAGE SENSORS

Linear Image Sensor

A block diagram of the 1728-element
interlaced linear image sensor is shown
in Figure 7. It consists of an array
of 1728 photosites, a two-phase CCD
shift register, and a single-stage
floating-gate amplifier. An opaque aluminum layer is deposited over the device to block incident light except in
the photogate ϕ_p area. A positive dc
voltage is applied to the photogate to
collect the signal electrons in the
potential wells formed. The 1728 photosites under the photogate are defined
by the p-type channel-isolation
diffusion shown in this figure. The
center-to-center spacing of these
photosites is 13μm.

At the end of an integration period,
the transfer gate ϕ_X is pulsed "high"
to transfer the signal electrons in two
fields into the neighboring two phase
CCD shift register. The signal
electrons are then transferred along the
shift register and detected by the single-
stage floating-gate amplifier. A sink
diode and exposure control gate ϕ_{EC} are
incorporated to provide exposure control
and antiblooming functions[7]. The CCD

shift register is constructed using two layers of polysilicon with self-aligned ion-implanted barriers[8] as shown in Figure 8. The buried channel is accomplished with the ion-implanted N-layer.

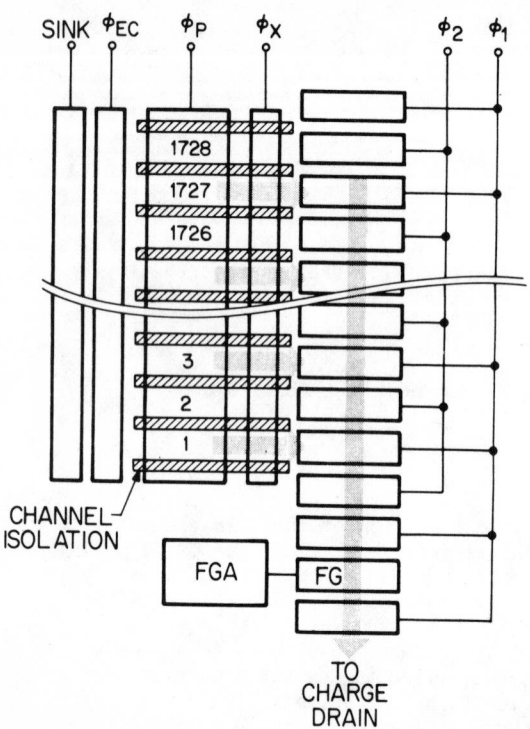

FIGURE 7   A 1728-element linear sensor.

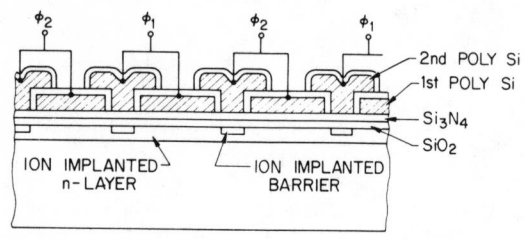

FIGURE 8   Two phase CCD structure.

A CRT monitor display of the IEEE Facsimilie Test Chart is shown in Figure 9. The image was horizontally scanned by the image sensor, while vertical scanning was obtained by mechanical rotation of the test chart. A portion of this displayed image is also shown with an expanded monitor sweep. The maximum resolution obtained is approximately 36 line parts/mm.

Low light level performance of this image sensor is illustrated by the photographs in Figure 10. This series of single-frame photographs show the display of approximately 700 photosites at illumination levels successively reduced from near saturation. The ambient temperature was $25^\circ C$ and the clock rate was 1.5MHz. The high-light area in Figure 10(a) represents an illumination of $200\mu W/cm^2$ with a maximum charge per photosite of approximately 500,000 electrons. At a 1/1000 reduction in light intensity, a high-quality image is retained although some dark current spikes appear as vertical streaks. At a 1/10000 reduction in light intensity, the brightest area in the picture represents approximately 50 electrons per photosite. The dark charge per pixel is approximately 800 electrons, resulting in a dark charge noise of 28 electrons. The noise-equivalent-signal per pixel in Figure 10(e) is approximately 100 electrons. The low light level performance of this sensor is therefore limited by the noise in the amplifier.

Area Image Sensor

A photograph of the 244 x 190 area image sensor [9] is shown in Figure 11. This device employs ihe interline transfer organization where the signal charge packets are read out in two successive fields. In operation, signal charge packets are generated and stored under the 190 vertical photogates. At the end of an integration period the photogates are pulsed "low" to transfer the signal charge packets from half of the photosites into an adjacent opaque two-phase vertical shift register. The vertical shift register in turn transfers each row of the signal charge packets into a two-phase horizontal shift register. The signal charge packets are then transferred along the horizontal register and detected by the output amplifiers. After all the signal charge

FIGURE 9   Image of linear sensor.

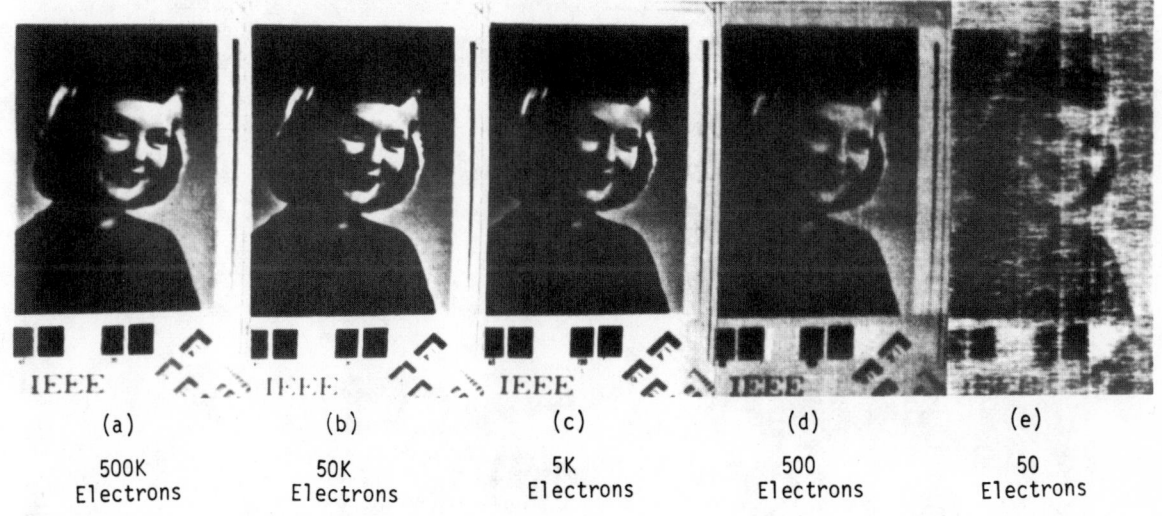

(a)	(b)	(c)	(d)	(e)
500K Electrons	50K Electrons	5K Electrons	500 Electrons	50 Electrons

FIGURE 10   Imaging performance of linear sensor.

packets have been detected, the process is repeated to read out the signal charge packets from the remaining photosites. Both the vertical and horizontal shift retisters are two-phase, buried-channel structures identical to that shown in Figure 8. The photoelement center-to-center spacing is 30μm horizontally and 18μm vertically.

This device employs a twelve-stage DFGA, and a single-stage floating-gate amplifier similar to that used in the linear image sensor described earlier. A photograph of the DFGA area is shown in Figure 12. This DFGA is identical to the twelve-stage DFGA test structure described in Section III, except that the structure of the output register has been modified.

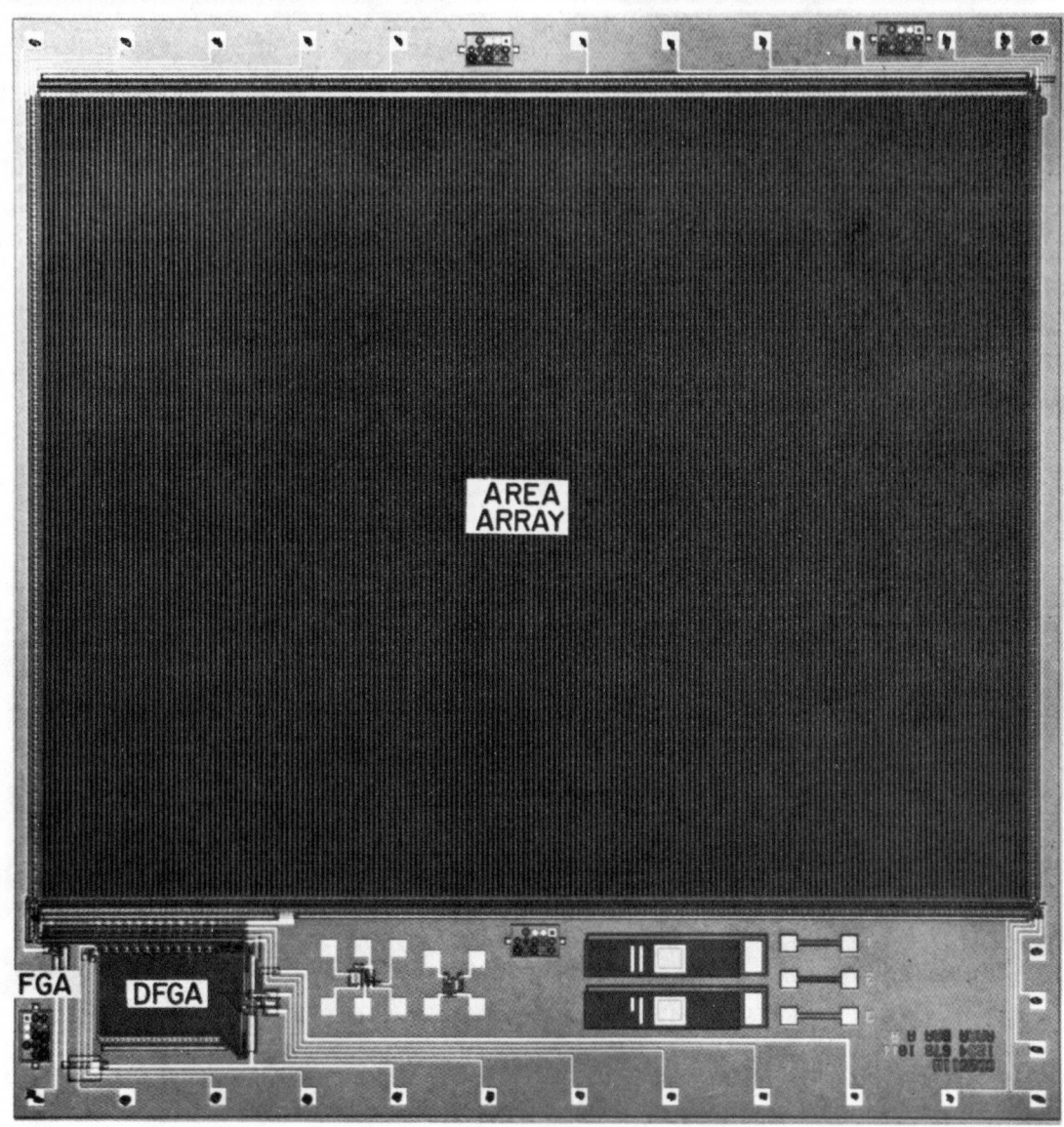

FIGURE 11 Photograph of the 244 x 190 area sensor.

It can be seen that the output register separates into two parallel registers after the twelfth charge amplifier stage. An output amplifier is provided at the end of each register, one being delayed from the other by one half horizontal clock period. By summing these two outputs off chip, the large output swing V_{BIAS} can be cancelled, as illustrated in Figure 13 and the resulting waveform can be much more easily processed. A theoretical analysis indicates that the noise equivalent signal of this DFGA is approximately 17 electrons at room temperature and 3 MHz bandwidth.

A block diagram of the imaging test set-up is shown in Figure 14. The regular and delayed DFGA outputs are combined to remove V_{BIAS} in the output waveform. The output is amplified and then dc restored during each horizontal blanking period with the line clamp circuit. It is amplified again and displayed on a monitor.

Low light level imaging performance of this sensor is illustrated in Figure 15. These photographs show DFGA images at $-10^{\circ}C$. The horizontal clock frequency was 2MHz, resulting in a frame rate of 23 frames/sec. The light integration time was 43 msec. The width of the coarse bars is 120μm, and the width of the fine bars is 60μm on the CCD image plane. Figure 14(a) shows the image at a near saturation exposure of approximately 4.7 μW/cm². The bright areas in this photograph contain approximately 200,000 electrons per photoelement. Figure 15(b) and (c) show the same image when the

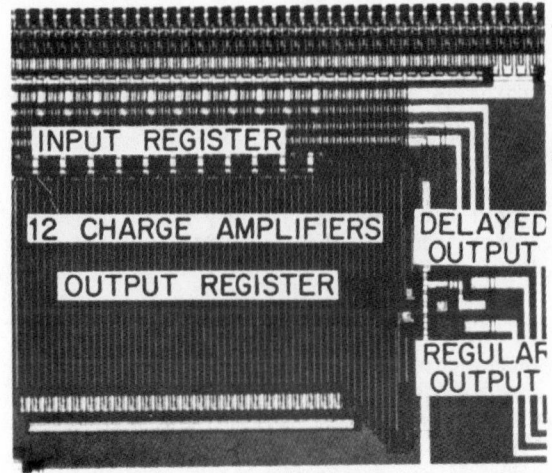

FIGURE 12   DFGA on the area sensor.

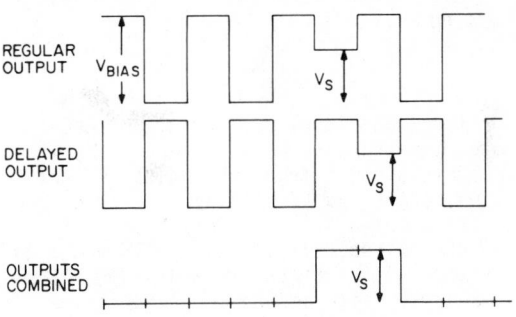

FIGURE 13   Modified DFGA output waveform.

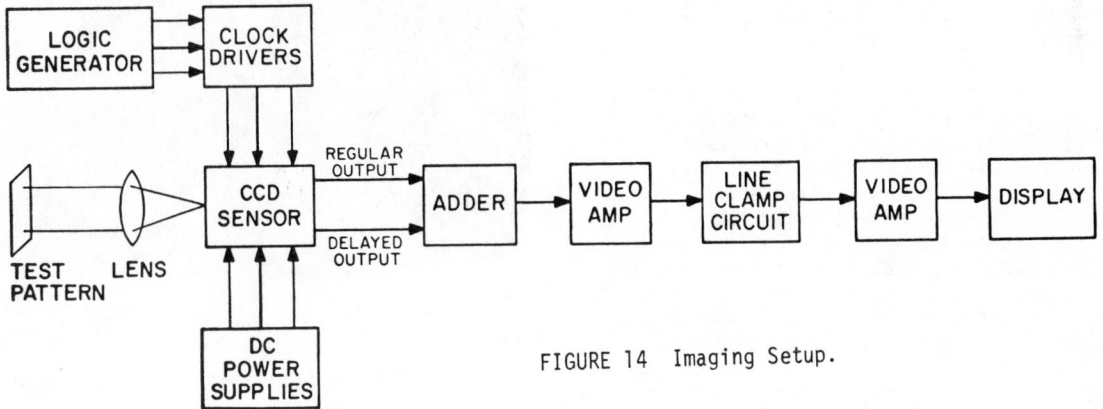

FIGURE 14   Imaging Setup.

215

light level is reduced by a factor of 1000 and 8000, respectively. At the 1/8000 reduction in light level, there are approximately 25 electrons per photoelement. A good transfer efficiency is maintained. The coarse (1/4 Nyquist) bars are visible and some of the fine (1/2 Nyquist) bars can also be recognized. Noise-equivalent-signal per pixel is probably 2 to 3 times higher than the 10 to 20 electron level which is predicted and measured on separate DFGA test structures.

At -10°C, the dark charge per pixel is approximately 1000 electrons. The dark charge noise should be approximately 32 electrons. The dominant noise source in Figure 15(c) is not the dark charge noise since it can not be reduced by cooling the array further. The limiting factor of the low light level performance of this array is not known. The most probable sources are the noise on the clock drivers, power supplies, and signal processing circuits.

CONCLUSION

The low light level performance of CCD image sensors have been examined. An analysis of bulk trapping indicates that signal charge packets of ten electrons can be transferred in large buried-channel CCD sensors operating at the NTSC mode. 1728-element linear arrays and 244 x 190-element area arrays both showed excellent transfer efficiency at signal levels well below 100 electrons. Using a twelve-stage DFGA, the area image sensor has demonstrated half Nyquist limit (60μm width) bar images at a signal level of approximately 25 electrons at -10°C. The corresponding irradiance was 6×10^{-4} μW/cm^2, and the integration time was 43 msec.

Separate measurements on twelve-stage DFGA test structures have shown noise level of 10 to 20 electrons. Actual images obtained with the DFGA on the area array, however, showed 2 to 3 times higher noise per pixel. The dominant noise source is suspected in the external supplies and signal processing circuits rather than in the sensor itself. It is expected that a factor of 2 to 3 improvements in the lwo light level imaging performance can be achieved with the present CCD sensors by improving the external supplies and signal processing circuits.

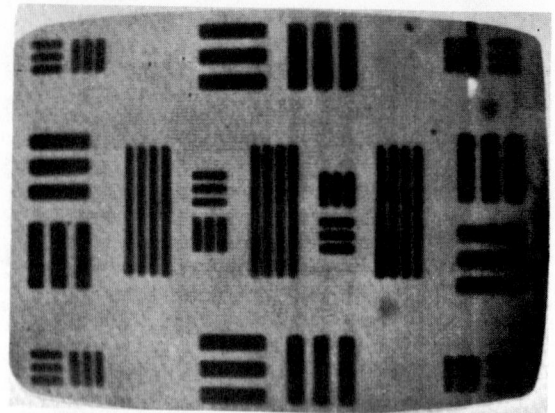
(a)   200K Electrons

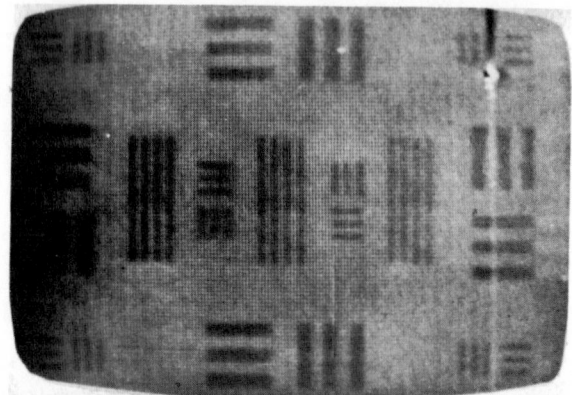
(b)   200 Electrons

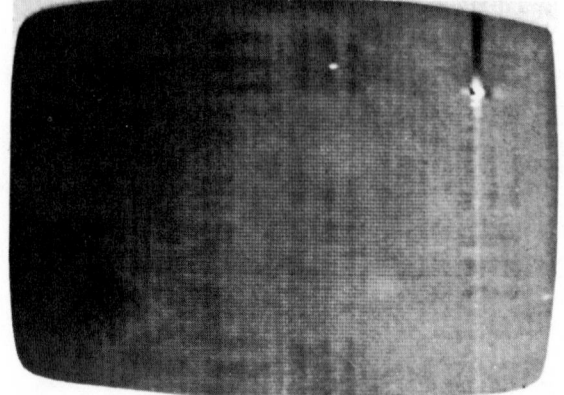
(c)   25 Electrons

FIGURE 15   Imaging performance of area sensor.

ACKNOWLEDGEMENT

The author wishes to express his gratidude to J. Early, G. Amelio, C. K. Kim, and W. Steffe for many helpful discussions. Thanks also go to M. Sklar for fabricating the devices, G. Lee for testing the linear sensors, and H. Dean and L. Walsh for providing the low light level images of the 244 x 190 area sensor. The support from D. F. Barbe of the office of Naval Research Laboratory and L. W. Sumney of the Naval Electronic Systems Command is also appreciated.

BIBLIOGRAPHY

1. M. H. White, D. R. Lampe, F. C. Blaha and I. A. Mack, "Characterization of Surface Channel CCD Image Arrays at Low Light Levels," Proceedings of CCD Applications Conference, pp. 23-25, 1973.

2. David F. Barbe, "Imaging Devices using the Charge-Coupled Concept," Proceedings of IEEE, Vol. 63, No. 1, January, 1975.

3. C. K. Kim, J. M. Early and G. F. Amelio, "Buried Channel Charge-Coupled Devices," presented at NEREM, November, 1972.

4. J. M. Early, "Theory of Bulk Trapping in in Buried-Channel CCD's near -50°C for Levels Below 100 Electrons," presented at the 1974 Device Research Conference, Santa Barbara.

5. D. D. Wen, J. M. Early, C. K. Kim and G.F. Amelio, "A Distributed Floating-Gate Amplifier in Charge-Coupled Devices," Digest of Tech. Papers, pp. 24-25, ISSCC, February, 1975.

6. M. D. Jack and R. H. Dyck, "Charge Transfer Efficiency in a Buried Channel CCD at Very Low Signal Levels," to be publishes.

7. D. D. Wen, C. K. Kim and G. F. Amelio, "The Latest in CCD Image Sensor Technology," presented at SEMICON/WEST, May, 1975.

8. C. K. Kim, "Two-Phase Charge-Coupled Imaging Device with Self-Aligned Implanted Barrier," Digest of Tech. Papers, IEDM, 1974.

9. W. Steffe, L. Walsh and C. K. Kim, "A High Performance 190 x 244 CCD Area Imager," CCD Applications Conference, San Diego, Ca., October, 1975.

CCD211
244 × 190 ELEMENT AREA IMAGE SENSOR
CHARGE COUPLED DEVICE

GENERAL DESCRIPTION – The CCD211 is a solid state self-scanned area image sensor suitable for use as the sensor in camera type applications. The device is organized in an array of 244 horizontal lines by 190 vertical columns. The 46,360 image sensing elements are 14 μm horizontally by 18 μm vertically and are located on 30 μm horizontal centers and 18 μm vertical centers. The dimensions of the image sensing area are 4.4 mm x 5.7 mm with a diagonal dimension of 7.2 mm.

The X-Y format of the array was selected to provide a 4 x 3 image aspect ratio and to approximate the image size of the Super 8 movie lenses. The highly precise location of the photosites allows precise identification of each component of the image signal. This feature allows the device to be used in applications requiring precise dimensional measurement. The device is also intended to be used in video cameras that require low power, small size, high sensitivity and high reliability. The device is packaged in a 24-pin Dual In-line Package with an optical glass window.

- 46,360 IMAGE SENSING ELEMENTS ON A SINGLE CHIP
- COLUMN ANTI-BLOOMING
- NO LAG, NO GEOMETRIC DISTORTION
- GAMMA CHARACTERISTIC OF APPROXIMATELY 1.0
- ON-CHIP VIDEO PREAMPLIFIER PROVIDING MORE THAN 200 mV OF OUTPUT SIGNAL
- HIGH DYNAMIC RANGE – TYPICALLY: 300:1
- LOW LIGHT LEVEL CAPABILITY, LOW NOISE EQUIVALENT EXPOSURE
- WIDE RANGE OF VIDEO DATA RATES UP TO 100 FRAME/s
- ALL OPERATING VOLTAGES UNDER 20 V
- LOW POWER DISSIPATION – TYPICALLY 75 mW
- HIGH RELIABILITY
- DESIGNED TO MATCH SUPER 8 MOVIE LENSES

CONNECTION DIAGRAM
DIP (TOP VIEW)

AB	1	24	ϕH1
ϕV1	2	23	ϕH2
ϕV2	3	22	TP11
TP1	4	21	ϕBE
TP2	5	20	TP10
TP3	6	19	TP9
VSS	7	18	TP8
TP4	8	17	VDD
ϕP	9	16	TP7
ϕV2	10	15	TP6
ϕV1	11	14	TP5
SF	12	13	VIDEOOUT

PIN NAMES

AB	Anti-blooming Bias
SF	Source of Floating Gate Amplifier
VIDEO$_{OUT}$	Video Output
ϕ_P	Photogate Clock
ϕ_{V1}, ϕ_{V2}	Vertical Analog Transport Register Clocks
ϕ_{H1}, ϕ_{H2}	Horizontal Analog Transport Register Clocks
BE	Bias Electrode
ϕ_{BE}	Bias Electrode Clock
V$_{DD}$	Power Supply
V$_{SS}$	Ground
TP1 – TP11	Test Points

BLOCK DIAGRAM

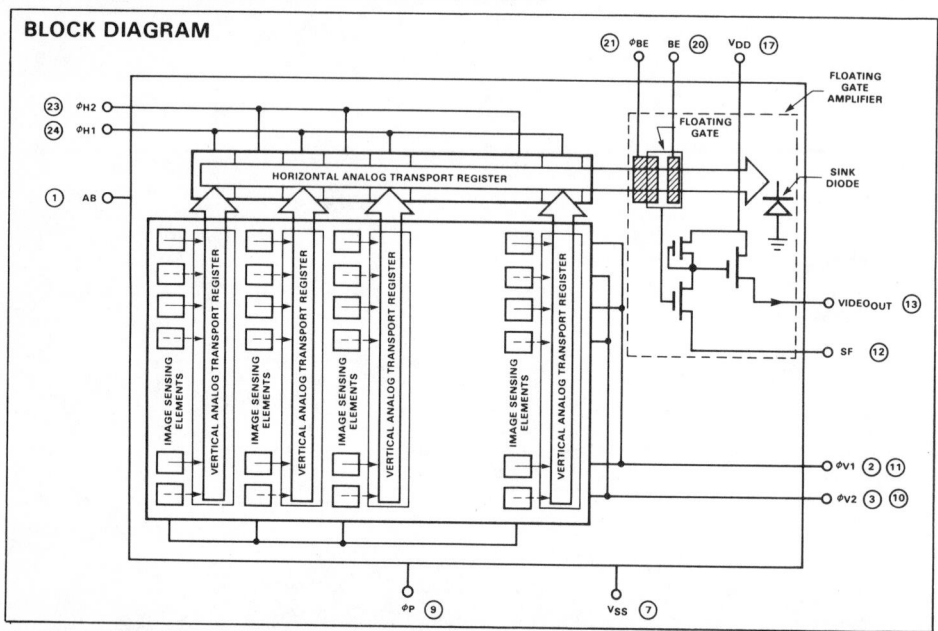

464 ELLIS STREET, MOUNTAIN VIEW, CALIFORNIA 94042 (415) 962-5011/TWX 910-379-6435

Manufactured under one or more of the following U.S. Patents: 2981877, 3015048, 3025589, 3064167, 3108359, 3117260; other patents pending.

Reprinted with permission from a preliminary data sheet of Fairchild Semiconductor, Mountain View, CA, pp. 1-6, Mar. 1976.

ABSOLUTE MAXIMUM RATINGS

Operating Temperature $-25°C$ to $55°C$
Storage Temperature $-25°C$ to $100°C$

VOLTAGES

Pins 1, 4, 12, 13, 14, 17 -0.6 V to $+15$ V
Pins 2, 3, 6, 8, 9, 10, 11, 21, 23 and 24 -10 V to $+15$ V
Pins 5, 7, 15, 16, 18, 19, 20, 22 $V_{SS} = 0$ V

Caution: The device has limited built-in gate protection. Static charge build-up should be minimized.

FUNCTIONAL DESCRIPTION – The CCD211 consists of the following subsections as illustrated in the Block Diagram on the previous page.

1. 46,360 image sensing elements in a 244 x 190 array.
2. 190 columns of 122-element 2-phase vertical analog transport registers.
3. A 200 element 2-phase horizontal analog output shift register charge coupled to the output of each of the 190 column shift registers.
4. A two stage low noise floating gate preamplifier which detects and converts the charges delivered from the horizontal analog transport register to the output terminal VIDEO$_{OUT}$.

Light energy incident on the image sensor elements generates a packet of electrons at each sensing element. Electrical clocking of the photogate, the vertical analog transport registers, and the horizontal analog output register sequentially delivers the charge packets to the preamplifier. Detailed descriptions of the functional subsections follow.

Image Sensing Elements – Image photons pass through a transparent polycrystalline silicon gate structure and are absorbed in the single crystal silicon by hole-electron pair production. The resulting photoelectrons are collected in the photosites during the HIGH state of the photogate. The amount of charge accumulated is a linear function of the incident illumination intensity and of the integration period. The output signal voltage ranges from a thermally generated background level in the absence of illumination to a maximum at saturation.

Vertical Analog Transport Registers – At the end of an integration period, the charge packets are transferred out of the array in two sequential fields of 122 lines each. When the photogate voltage is lowered, alternate lines of charge packets are transferred to their corresponding sites in the vertical registers (i.e., the odd numbered photoelements in the block diagram). Clocking of the vertical register at ϕ_{V1} and ϕ_{V2} delivers the charge packets from the 190 vertical registers to the horizontal analog transport register. A minimum of 124 vertical transfers (124 clock cycles) are required to transport each field of charge packet out of the vertical registers. Subsequent to the removal of one field of charge packets, a second field cycle is initiated to receive the information from photosites corresponding to the other field (i.e., the even numbered photoelements). Clocking of the register transports the charge packets in a similar fashion to the output.

Horizontal Analog Transport Register – The horizontal register is a 200-element 2-phase register that receives the charge packets from the vertical transport registers line by line. As each row of information is received from the vertical registers it is serially moved to the output amplifier by the horizontal clocks ϕ_{H1} and ϕ_{H2}. A minimum of 200 horizontal clock pulses are required to complete transfer of one row of information to the floating gate amplifier.

Floating Gate Amplifier – The charge packets from the horizontal register are sensed by a floating gate whose potential changes linearly with the quantity of signal charge and which drives a first MOS transistor. The output signal from the transistor in turn drives the gate of an output n-channel MOS transistor which produces the video output signal at terminal VIDEO$_{OUT}$.

DC CHARACTERISTICS: $T_A = 25°C$

SYMBOL	PARAMETER	RANGE			UNITS	CONDITIONS
		MIN	TYP	MAX		
V_{DD}	DC Supply Voltage		12	15	V	
V_{AB}	Anti-Blooming Bias Voltage	2.0	5.0	15	V	Note 1
V_{SF}	Source of Floating Gate Amplifier	5.0	8.0	10	V	Note 9
TP1, TP3 TP4, TP5	Test Points		15		V	
TP2, TP6, TP7, TP10, TP11	Test Points		0		V	
TP8, TP9	Test Points	No Connection				

CLOCK CHARACTERISTICS: $T_A = 25°C$

SYMBOL	PARAMETER	RANGE			UNITS	CONDITIONS
		MIN	TYP	MAX		
$V_{\phi PL}$	Photogate Clock LOW		0.0		V	$C_{\phi P} = 1250$ pF
$V_{\phi PH}$	Photogate Clock HIGH		5.0	10	V	$C_{\phi P} = 1250$ pF
$V_{\phi BEL}$	Bias Electrode of FGA Clock LOW		−5.0		V	
$V_{\phi BEH}$	Bias Electrode of FGA Clock HIGH		5.0	10	V	
$V_{\phi H1L},$ $V_{\phi H2L}$	Horizontal Analog Transport Register Clock LOW		−5.0		V	$C_{\phi H1} = C_{\phi H2} = 115$ pF
$V_{\phi H1H},$ $V_{\phi H2H}$	Horizontal Analog Transport Register Clock HIGH		5.0	10	V	$C_{\phi H1} = C_{\phi H2} = 115$ pF
$V_{\phi V1L},$ $V_{\phi V2L}$	Vertical Analog Transport Register Clock LOW		−5.0		V	$C_{\phi V1} = C_{\phi V2} = 1250$ pF
$V_{\phi V1H},$ $V_{\phi V2H}$	Vertical Analog Transport Register Clock HIGH		5.0	10	V	$C_{\phi V1} = C_{\phi V2} = 1250$ pF
$f_{\phi H1},$ $f_{\phi H2}$	Horizontal Analog Transport Register Clock Frequency	0.5	7.0	15	MHz	Note 2

AC CHARACTERISTICS: $T_A = 25°C$, $f_{\phi 1H} = f_{\phi 2H} = 7.0$ MHz, Light source is 2854°K Tungsten illumination with a Corning 1-75 IR filter.

SYMBOL	PARAMETER	RANGE			UNITS	CONDITIONS
		MIN	TYP	MAX		
DR	Dynamic Range	200	300			Note 3
SE	Saturation Exposure	0.1	0.2		$\mu J/cm^2$	Note 4
V_{SAT}	Saturation Output Voltage	100	200		mV	Note 5
R	Responsivity		1.0		$V/\mu Jcm^{-2}$	
HR	Horizontal Resolution		142		L/PH	Note 6
VR	Vertical Resolution		244		L/PH	Note 6
CB	Column Blooming		20		%	Note 7
S	Shading		5.0		%	Note 8
DS	Average Peak Dark Signal		0.6		mV	Note 10

NOTES:
1. Adjustment is required over the indicated range for optimum anti-blooming operation.
2. Clock rates shown are typical rates at which the device operates. Operation of the devices at lower or higher frequencies will not damage the device.
   Three factors contribute to the fundamental low frequency limit: a) Integration time. b) Dark current contributions from the photosites and associated dark current non-uniformities. c) Dark current contributions in the register which will result in an average dark signal at the output.
3. Measured by adjusting incident illumination to the signal saturation point, then attenuating the incident light with a neutral density filter of density N.D. = 2.3 (which corresponds to a reduction in incident light of 200 times). The resultant off chip video signal should be ≥ 0.5 mV$_{P-P}$.
4. 1 $\mu J/cm^2$ = 0.02 fcs at 2854°K
   1 fcs = 50 $\mu J/cm^2$ at 2854°K
5. Measured with a 100% contrast bar pattern as a test target. The saturation level is where the video peaks just start to flatten out as the incident illumination is increased.
6. L/PH = lines per picture height. Measured using a standard E.I.A. resolution chart as a target.
7. Measured with a point source of incident illumination level of 1000 times saturation light level and with a diameter of approximately 6 elements. Blooming is expressed as a percent of the width of the bloomed column to the width of the picture.
8. Measured with an incident light level which produces a video output equal to $V_{sat}/2$ (50% saturation). Shading is defined as the slow variation in photoresponse across a line or a field. It is expressed as a percentage variation relative to the V_{sat} level. Measurement does not include outermost rows or columns of sensors.
9. Adjustment is required in the range of 5.0 to 10.0 V for optimum operation.
10. This is the value of the dark current peaks averaged over all elements.

AMPLIFIER POWER DISSIPATION AND OUTPUT RESISTANCE VERSUS EXTERNAL SOURCE RESISTOR, R_S

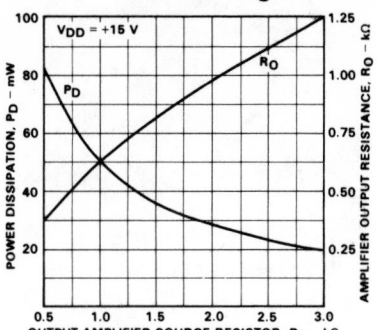

SATURATION OUTPUT VOLTAGE VERSUS DATA RATE AND OUTPUT SOURCE RESISTANCE

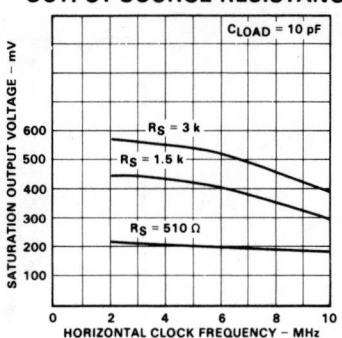

SPECTRAL RESPONSE

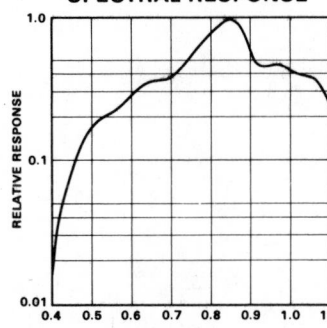

BLOOMING CHARACTERISTICS

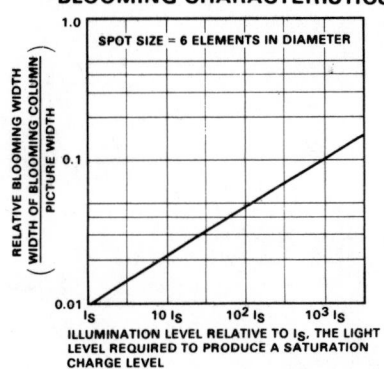

CONTRAST TRANSFER FUNCTION

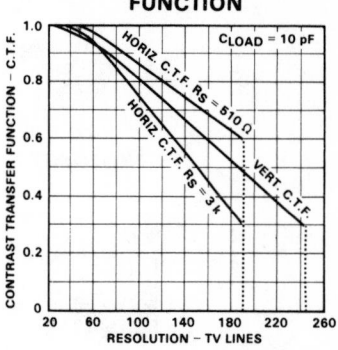

TIMING DIAGRAM DRIVE SIGNALS

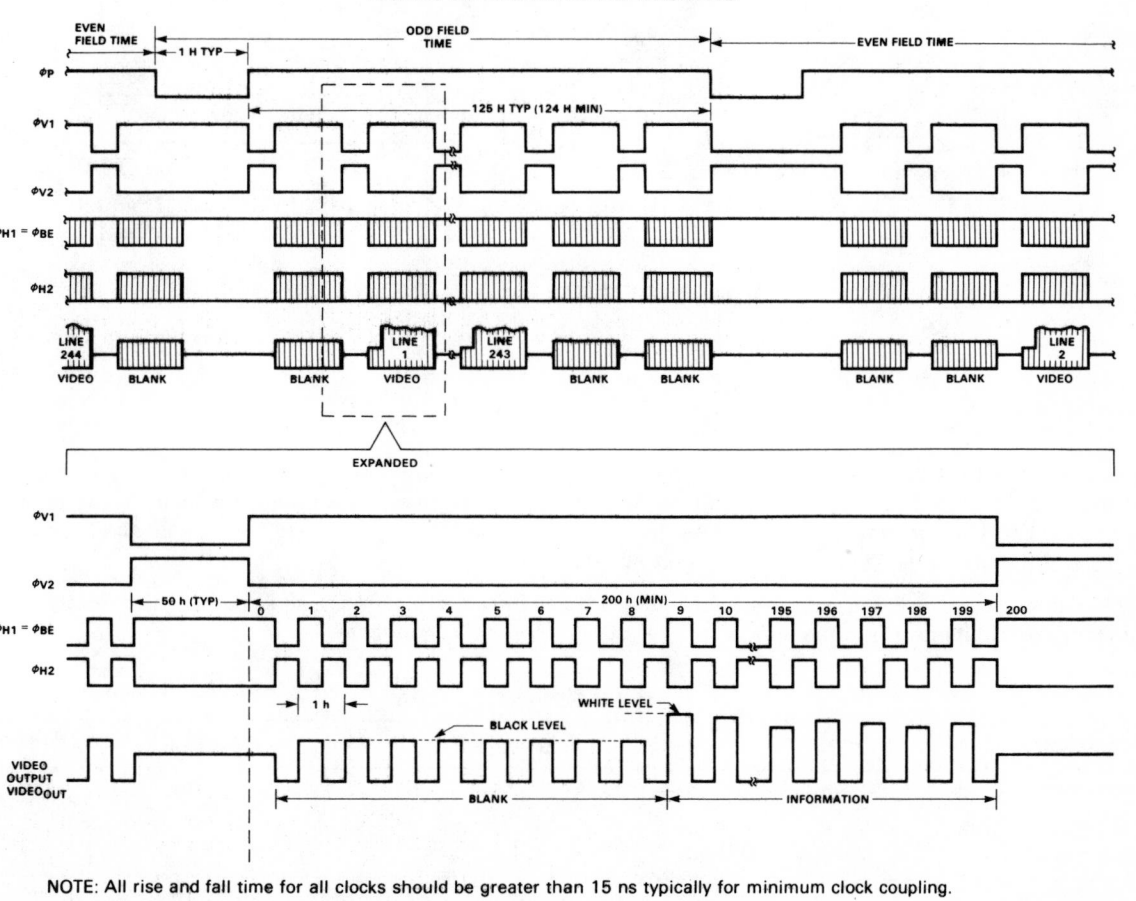

NOTE: All rise and fall time for all clocks should be greater than 15 ns typically for minimum clock coupling.

221

TEST LOAD CONFIGURATION

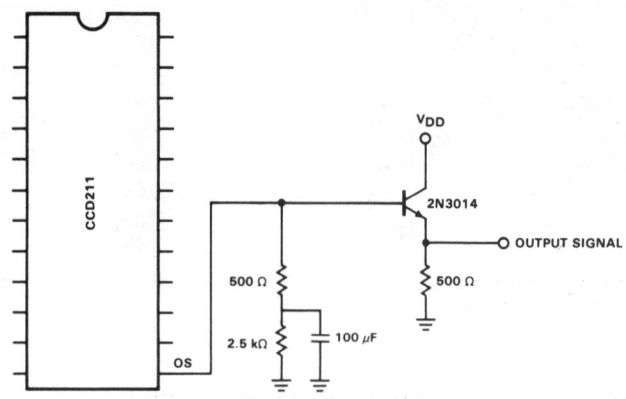

PHOTOSITE DIMENSIONS

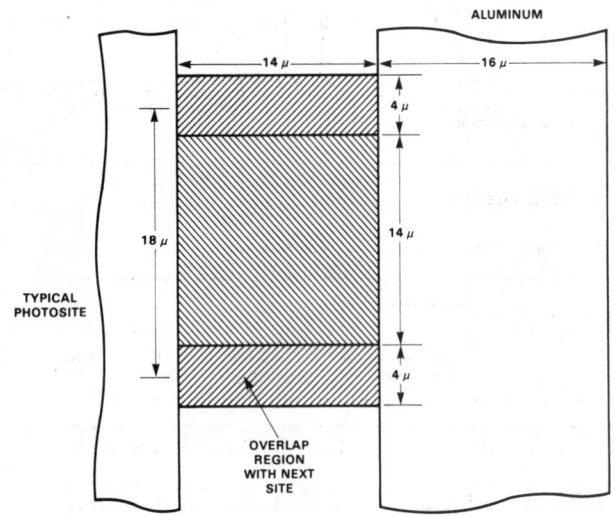

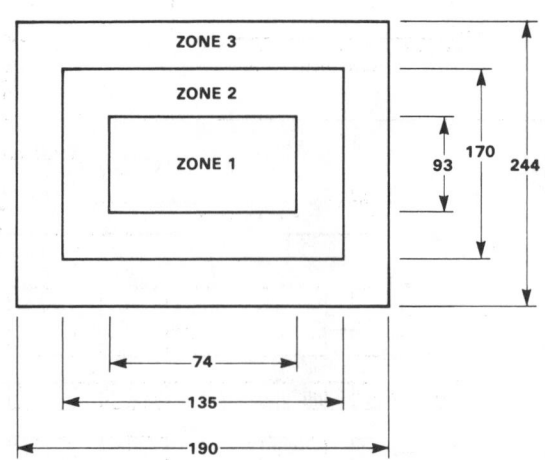

NOTE: All dimensions in number of elements.

CLASSIFICATIONS — CCD211s are classified in terms of maximum number of image blemishes allowed and their position on the image format. The array is divided into three zones (see Figure), since blemishes near the periphery of the array are usually less objectionable than those near the center. The area of the array has a 4 x 3 aspect ratio. Zone 1 encompasses 15% of the elements. Zone 2 is 35% and Zone 3 is 50%.

An image element is blemished if it shows a spurious output $\geqslant$10% of saturation. The output waveform of the array is analyzed under two conditions. (1) in the dark and (2) at 50% of the saturation level.

BLEMISH SPECIFICATION

CLASSES	ZONE 1	ZONE 2	ZONE 3	TOTAL
A TYPE	3	5	8	16
B TYPE	4	7	11	22
C TYPE	6	10	15	31

ORDER INFORMATION – The CCD211 as described in this data sheet can be ordered in its various classes as a standard device. The CCD211 chip provides the following:

1. A charge injection port where analog information can be fed to an input horizontal analog transport register in series form. This register is similar to the output register and is situated at the bottom of the block diagram on page 1. This information is then clocked in parallel through the vertical column registers to the output horizontal analog transport register which is then clocked in series and provides at the output of the floating gate amplifier a line by line reproduction of the electrical input information. The device is thus organized in a series-parallel-series (SPS) configuration. The device can be used in conjunction with an imaging array for frame to frame comparison applications. For further information regarding that device refer to the CCD361 data sheet.

2. In addition to the floating gate amplifier, a 12-stage distributed floating gate amplifier for very low noise, low light level applications where low irradiance imaging is required.

   Standard CCD211 devices are not tested to have either their input registers or their distributed floating gate amplifier (DFGA) operational. Devices that exhibit the above two features of the CCD211 die can be ordered on special request. Further information on the DFGA can be obtained from any Fairchild office.

Therefore the CCD211 can be ordered in the following distinct categories:

1. A standard CCD211 with the floating gate amplifier (FGA) and an optical window. The device is also selected for its various classes depending on blemish count as shown in the classifications table.

2. A specially selected CCD211 having both the FGA and DFGA amplifier operational with an optical glass window and an A class blemish count.

To order CCD211s in the various forms, order the device types outlined in the table below.

DESCRIPTION	DEVICE TYPE
STANDARD CCD211, FGA, CLASS A BLEMISH SPEC.	CD211ADC
STANDARD CCD211, FGA, CLASS B BLEMISH SPEC.	CD211BDC
STANDARD CCD211, FGA, CLASS C BLEMISH SPEC.	CD211CDC
SPECIAL CCD211, FGA, DFGA, CLASS BLEMISH SPEC.	SL62818

PACKAGE OUTLINE

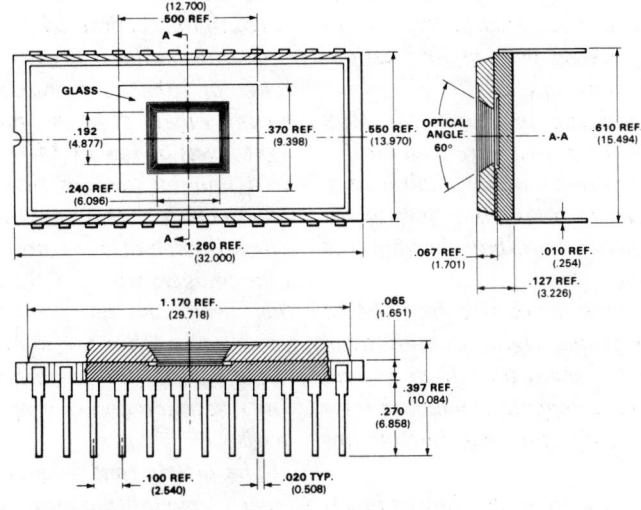

NOTES:
All dimensions in inches (bold) and millimeters (parentheses)
Header is black ceramic A1$_2$O$_3$
Lid is plastic with transparent glass window
Pins are gold plated kovar

The tradeoffs in monolithic image sensors: MOS vs CCD

With two kinds of solid-state imaging devices now available, designers have a choice in systems that will replace image tubes, especially in low-light-level applications

by Roger Melen, *Stanford Electronics Laboratories, Stanford, Calif.*

Designers of information-display and recording systems have long dreamed of the coming of solid-state imaging devices that are small and fast, operate at low power without high voltages, and work at wide dynamic ranges of ambient light. Those dreams are becoming a reality as the charge-coupled device comes to the marketplace to join the older MOS photodiode image sensor.

Both types of monolithic image sensors offer fundamental improvements over earlier imaging methods, especially for optical character recognition, facsimile systems, and video communications, where high-voltage devices often requiring high light levels are being used. But the capabilities of the two types of devices overlap, and designers are evaluating the strong points of each.

Preliminary experiences indicate that both the MOS diode array and the CCD imaging array are suitable for OCR and facsimile displays that require only small arrays, while the CCD display appears to be the only one of the two suitable for television applications, both at high and unusually low light levels.

MOS diode linear arrays ranging in size from 64 to 1,000 diode elements (photograph, right, from Reticon Corp.) have been available for some time from several manufacturers. These devices are capable of imaging rates higher than 5 megahertz, and they offer real-time facsimile-quality performance.

The MOS diode array also is useful for small-area imaging applications where resolution need not be of television quality. In these applications, the MOS diode array is an excellent replacement for low-resolution image tubes because of its capability of operating at low light levels with self-contained power supply, drive circuitry, and displays.

Complete camera systems with 50-by-50 diode arrays are already on the market and are being used for surveillance, OCR, and defect-detection systems.

The value of the CCD must be considered as lying mainly in its potential for supplying full video-quality imaging at both high and low light levels, a performance which is beyond the capability of present MOS diode technology and costs. Moving from concept to marketplace in less than three years, the CCD is already available in linear device with 500 elements (photograph, far right, from Fairchild), an incredible achievement when compared to the earliest, 64-element MOS image sensors.

CCD technology already offers products with a dynamic range of 1,000 to 1, making it possible to image objects having widely different intensities. The sensor can detect light levels as low as 15 microfootcandles. Full-scale CCD area imaging devices that are expected within the next year include video cameras with resolution of 250 lines, which is adequate for most data-communications systems. Line-imagers with 1,500 elements for page readers are on their way, and not far behind is the ultimate imaging goal—well within the immediate developmental capabilities of CCDs—full-video-quality cameras that have 550-line resolution and that operate at ordinary ambient-light levels.

The article that follows compares the two competitive types of monolithic image sensors from the standpoint of important performance criteria—dynamic range, sensitivity, noise, and image clarity. But it is emphasized that the comparison must be based on MOS arrays that have a product history of three to four years, while the CCDs are only now entering production. —*Laurence Altman*

☐ Monolithic image sensors provide a major new dimension in the fabrication of information system displays—be they video cameras, facsimile equipment, or process control instrumentation for optical character-recognition systems. Both MOS photodiode arrays, now three or four years old, and charge-coupled devices, which are now entering the marketplace, offer the user a new standard in small size, high speed, high reliability, and ease of use.

But because of subtle differences in structure and readout, each has its unique performance characteristics. The design engineer should clearly understand the operation and performance limitations of each device before committing it to an expensive, complex system design.

Superficially, both types of arrays appear to be similar in operation. Both are fabricated with basically the same integrated-circuit technology. Images formed on the face of semiconductors are scanned off in conventional shift-register fashion. However, their performance differs because of different methods of projecting the image on the chip and reading out the signals.

The MOS image sensor (Fig. 1a) is essentially a high-performance diode scanning circuit built with standard photodiode and either metal- or silicon-gate technology [*Electronics,* Nov. 8, 1971]. The scanning circuit is made up of MOS transistors that are embedded in the same monolithic structure containing the array of photodiodes. After an object is imaged onto the surface of the photodiode array, the MOS scanning circuit shifts the signals off the chip by accessing the diodes sequentially through an analog switch to a common bus line.

A simple CCD (Fig. 1b) is essentially an analog-signal shift register (a delay line) fabricated from a closely spaced array of MOS capacitors [*Electronics,* March 29, p.25]. It also is usually built with some sort of silicon-gate buried-layer technology. The input signal takes the form of minority carriers generated in the semiconductor beneath the capacitor plates by the absorption of incident light.

The signal charge, consisting of minority carriers, is stored in packets in the semiconductor beneath the capacitor plates. Since the signals are stored in packets, they appear at the output as sampled signals, with each sample representing a packet of charge.

In operation, the signal charge may be transferred from capacitor to capacitor throughout the array by application of a sequence of biasing pulses. The charge-transfer efficiency is typically greater than 99.9% because of the close spacing of the capacitors in the arrays. Recent devices have efficiencies as high as 99.999%.

Differences in devices

Despite the similarities in fabrication technologies, the performance of the MOS and CCD image sensors are different because they have different methods of imaging the light and different techniques of reading out sig-

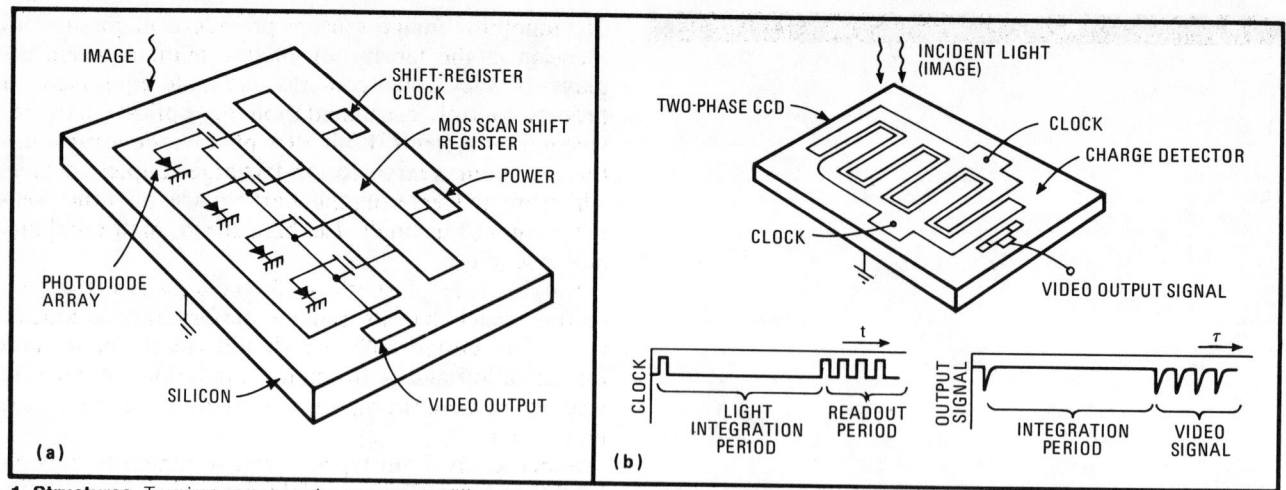

1. Structures. Two image-sensor types operate differently. MOS sensor (a) passes charge directly from the imaging photodiode into an MOS shift register, which then carries the charge to a detection circuit. CCD sensor (b) uses the imaging array itself as the transfer mechanism.

nal charges. In the CCD image sensor, the signal charge is collected by a field-induced junction beneath an MOS capacitor electrode, and readout is accomplished by multiple transfers of charge through the array of induced junctions to the output circuitry. But in the MOS image sensor, the charge is collected by a diffused junction in the photodiode, and readout is accomplished by a single charge transfer from the diffused junction to the video-out circuitry.

These rather subtle differences in structure and readout result in wide differences in performances at high and low light levels, in image clarity, and in device complexity. At low light levels, the minimum light that can be resolved by the image sensor depends on the efficiency with which the image sensor can collect the light incident on it, as well as noise introduced by the sensor and its associated circuitry.

The MOS sensor converts light to signals more efficiently than does the CCD, a property that results from the differences in the amount of light reflected from the imaging surface of each device and from the differences in the site at which the signal charge (generated by the incident light) is collected.

Illumination: front vs. back

Two common techniques (Fig. 2) are used to illuminate the semiconductor substrate in monolithic image sensors—front and back illumination. Although either technique could be used with either CCD or MOS image sensors, only back illumination is used for CCDs because most CCD structures have electrodes on the front that are opaque.

Unfortunately, back lighting introduces fabrication problems and performance limitations. The substrate die must be made very thin so that the light-generated carriers, which are generated within 4 micrometers of the semiconductor surface for visible light, may be efficiently collected and stored in the depletion layer beneath the capacitor electrodes on the front side.

About the thinnest substrate that can be fabricated has a thickness of about 25 micrometers. This means that device elements cannot be spaced less than 25 μm apart—thicker substrates would cause charges to spread from one electrode to another—a restriction that limits the potential resolution of back-illuminated CCDs. This limitation on element spacing is especially damaging for image sensors containing large numbers of elements because it means that a great deal of silicon must be used to accommodate the density.

Clearly, front illumination is desirable for simple structures to give good resolution. MOS image sensors, fortunately, have silicon oxide covering the semiconductor substrate. Not only is this oxide transparent, but it also acts as an optical coating that matches the optical impedance of the silicon to the impedance of air.

Some CCD sensors also have been built with polycrystalline electrodes that can be illuminated from the front, but these polycrystalline structures, unfortunately, pro-

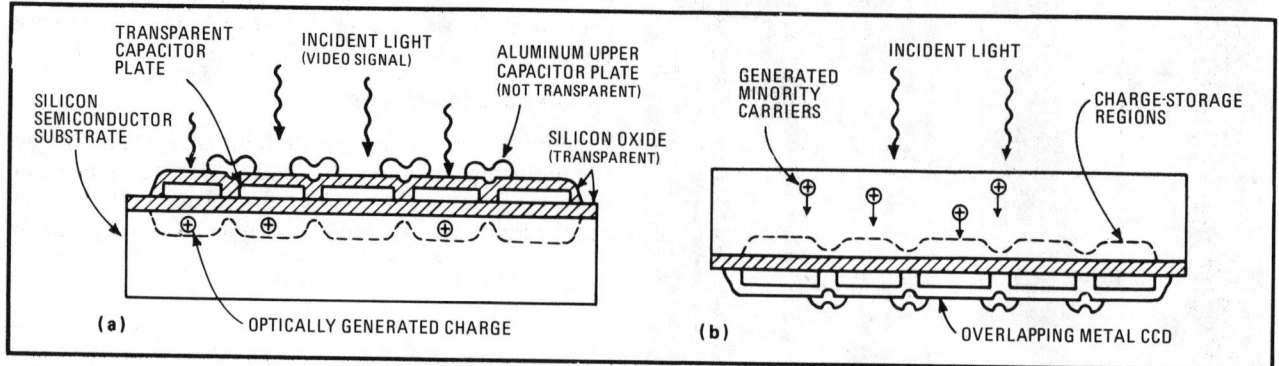

2. Backlighting. Monolithic image sensors can be imaged either on the back or the front of the substrate. Most CCDs, on the other hand, have front metalization that is opaque to light and therefore requires the imaging to be done from the back.

vide poor impedance matches with the oxide beneath, which causes reflection at the poly-oxide interface. These mismatches create interference patterns in the surface reflections, resulting in a decrease in the photocurrent output.

That villain, noise

But whether the array is illuminated from the front or the back, noise introduced into the video signal by the image sensors and associated circuitry is probably the greatest factor that limits operation at low light levels. The noise, which masks small photosignals in both types of arrays, comes from mismatches in parasitic capacitances and thermally generated carriers. Moreover, CCDs suffer noise from transfer losses.

In MOS image sensors, capacitor noise results from mismatches between parasitic gate-source and gate-drain MOS capacitance of transistors in the scanning circuit and photodiodes and video output port, with which these capacitances are in series. These MOS transistors are analog switches that address the individual photoelements in the array.

When these transistors are turned on or off, there is a corresponding voltage spike on the analog photosignal line being switched. Although these spikes may be reduced by filtering, because they occur at twice the maximum video frequency, they can not be eliminated completely.

The variation in the magnitude of these spikes throughout the MOS photoarray gives rise to fixed-pattern noise (FPN) in the video passband—noise that cannot be filtered. Fortunately, the variation in the noise is small compared to the absolute magnitude of the spikes. Indeed, with no incident illumination, a low-

Night vision

Monolithic image sensors may be operated at incident light levels below those found in the average office, a capability that has resulted in their being used in monitoring and surveillance applications. For this kind of work, exposure range is the important parameter for evaluating the sensor's ability to perform at low light levels.

The exposure range is the ratio of the maximum to minimum intensity that can be resolved by the image sensor. The exposure range is often expressed in f stops (factors of two in light intensity).

Exposure range may be calculated by:

$$ER = 0.301 \log_{10} I_{max}/I_{min}$$

where:

ER = exposure range in f stops
I_{max} = maximum resolved light intensity
I_{min} = minimum resolved light intensity

Maximum resolved light intensity can be limited by either the brightness range of the scene or the saturation level of the image sensor. At the low illumination levels being considered, the exposure range is often limited by the brightness range of the scene. Typical values of exposure range for MOS image sensors are six to 10 f stops, which translates into a dynamic range of 20 to 1. Experimental MOS image sensors and some recent CCD products have already been built with dynamic ranges as high as 1,000 to 1.

level noise image resulting from FPN may be observed at the output of the image sensor.

Spike noise, indicated in Fig. 3 as observed at the sensor output, is referenced to an equivalent noise voltage across the capacitance of the photosensing element in a

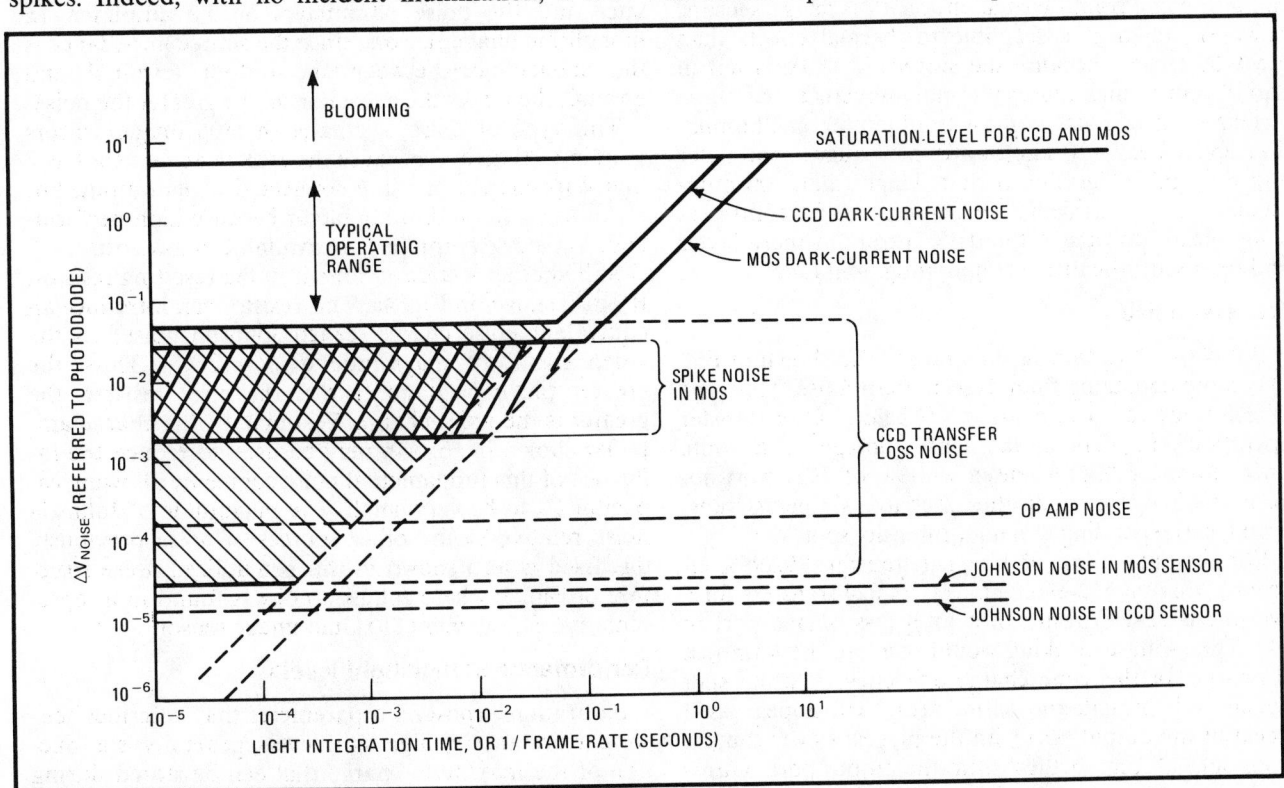

3. The limit. Noise is the limiting factor in the level of light that can be detected by a monolithic image sensor. In MOS imagers, spike noise in the range of 0.5×10^{-2} volt is low enough to allow use in poorly lit rooms. In CCDs, charge transfer noise can be as low as 10^{-5} V.

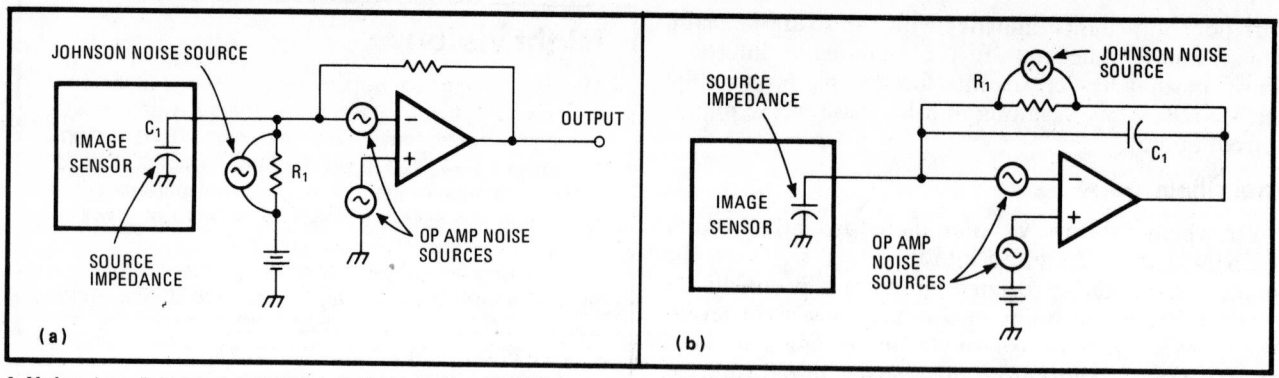

4. Noise. Low-light-level detection is limited by noise in the detection amplifier and reset resistor (a). Noise from the latter (R_1) is more prevalent in MOS images and dictates the use of low-noise amplifiers, such as the charge amplifier shown in (b).

representative 512-element device. Values of noise range from 10^{-3} to 0.5×10^{-2} volts, well within practical operating levels. The saturated output signal referred to the diode is typically 5 volts, resulting in dynamic ranges of 100 to 1 and more.

While CCDs are not affected by FPN from the spikes in switching transistors, they have fixed-pattern noise resulting from capacitance between clock lines and the output lines. Luckily, these noise pulses are all the same height and can be filtered out by low-pass filters, but the filters consume power and occupy space.

A better method of reducing this parasitically coupled noise is to fabricate video preamplifiers on the same image-sensor chips. The noise is thereby reduced because the magnitude of the parasitic coupling capacitance may be made smaller for amplifiers on the same chips than for off-the-chip amplifiers.

Fixed-pattern noise in both MOS sensors and CCDs can also come from thermal effects. CCD image sensors, however are more susceptible to thermal effects than are MOS sensors because the surface of CCDs is not in equilibrium, which causes thermal imbalance.

This form of noise is most troublesome at illumination levels below 10 microwatts per square centimeter and for light-integration periods longer than 100 milliseconds for typical devices because the noise comprises a significant portion of the dark current at these levels and represents the ultimate operating limitation.

Transfer noise

But with CCDs, transfer-loss noise (also shown in Fig. 3) is more damaging than fixed-pattern noise. This type of noise, the result of charges left behind after transfer operations, appears in the sensed image as a white smear to one side of a sensed white spot. It is most noticeable when large quantities of charge are being transferred, corresponding to a high-intensity spot.

For example, a loss of 10^{-5} per transfer (99.999% efficient) will in a 512-element array (1,024 transfers for a two-phase device) result in a total loss of one part in 100. Three-phase clocking would increase the total loss of charge for the same charge efficiency. A white spot transferred through the entire array will appear as a smear at the output port with the biggest smear coming from dots starting farthest from the output port. Transfer-loss noise also reduces a CCD's exposure range, and it basically decreases the contrast the sensor can detect.

One method of reducing transfer noise is to bury the

transfer channels about 1 micrometer beneath the surface of the substrate by ion implantation. Charges transferred in the buried channels are not subjected to transfer inefficiencies caused by charges trapped in surface states at the semiconductor-oxide interface. A 500-element buried-channel device just appearing on the market has a transfer efficiency of 99.999%.

Fighting noise

However, some noise sources cannot be readily overcome, such as thermally generated noise, which will always be present. This limits sensor performance at low light levels. All amplifiers and all resistors are subject to thermal noise; in an imaging system, the circuitry connected to the output of the image sensor (Fig. 4a) generates this noise.

In this example, the thermal noise signal appearing at the amplifier output is a function of the source impedance and the noise parameters of the amplifier. In monolithic image sensors, since the source impedance is the capacitance between the output terminal and ground, the larger the capacitance, the greater the noise.

This type of noise is greater in MOS image sensors than in CCD arrays, because the MOS image sensor has a high-capacitance bus line connected to its output, but this noise is not a limiting factor because high-performance low-noise amplifiers are available at low cost.

Still another source of noise is in the resetting resistor, R_1, also shown in Fig. 4a. This resistor can introduce an equivalent noise charge (called Johnson noise) on the video signal of magnitude $q_{noise} = KTC_1$. Thus, the greater the capacitance across the reset resistor, the greater is the noise charge. Fortunately, the charge amplifier shown in Fig. 4b may be used to reduce the influence of this fundamental noise source by allowing capacitor C_1 to be very small. The magnitude of Johnson noise relative to the other sources of noise previously discussed is also shown in this generalized noise structure of Fig. 3, where Johnson noise is found in a representative 512-element MOS line-image sensor.

Performance at high light levels

Saturation exposure, a parameter that describes sensor performance at high light levels, generally is a function of the maximum charge that can be stored during the light-integration period of the sensing elements in the photoarray. The light-integration period is the time used by the photoelements to collect charges represent-

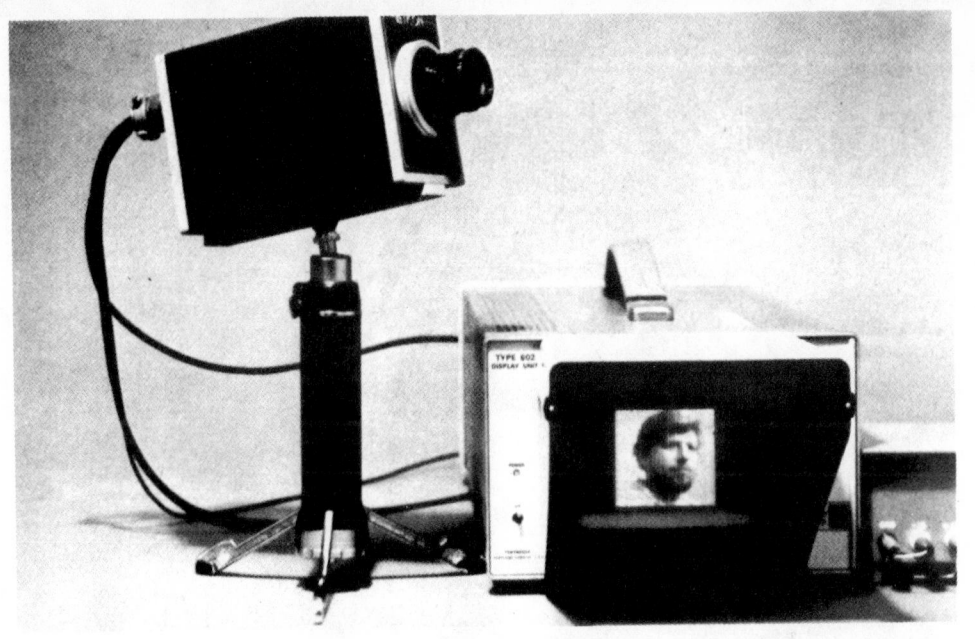

5. Image systems. Both MOS and CCD imaging arrays can be used in area imaging systems. CCD cameras potentially are capable of TV-quality resolution. MOS arrays, like this one from Reticon, are useful for lower-resolution systems.

ing the illuminated image. Typically, the light-integration period corresponds to the frame period.

For MOS image sensors, the maximum signal charge that can be stored depends on the bias applied to the photodiodes. For CCDs, it depends on the potential of the storage surface. Because the photoelements of CCDs and MOS image sensors have similar geometries and similar storage potentials, the saturation-light levels of both devices are similar.

The high-light-level capability of both sensors can be maximized by increasing the storage capacitance of the photoelements while masking the other regions from incident light. This type of structure, called a monolithic aperture, allows the light-handling capability of line-scanning arrays to be significantly increased, while keeping noise in the unexposed areas small. Area-sensing monolithic arrays do not benefit as much from this technique because of the loss of spatial resolution resulting from the large area required to achieve increased capacitance

MOS image sensors benefit most from this technique because a pn junction is the photoelement, and it can be read out quickly. In a CCD, on the other hand, an adjacent-capacitor photoarray technique would increase the size of the photoelements, and, in turn, the time required to transfer the signal charge from the adjacent photo capacitor to the analog CCD shift register.

CCDs, which have a low charge-transfer efficiency, are subject to blurring where the charges that are left behind during transfer between electrodes may appear later at the device's output terminal. Unfortunately, the transfer losses that cause blurring in CCDs increase not only with the number of transfers but also with lower light levels. However, the MOS imager does not blur because there is only one transfer. The signal flows only through a single analog switch before reaching the output.

Large-area imaging

There is nearly universal agreement that CCD image sensors, larger because of their smaller cell size, are more likely than MOS sensors to achieve television quality of 525 lines in a two-dimensional scanned-area image sensor. The highest-density MOS area image sensor

available has a cell size of 2 mils by 2 mils, whereas CCD area image sensors have already been built in the laboratory with cell sizes on the order of 1 mil by 1 mil—4 times denser. Indeed, a CCD area image sensor with half the resolution of that required for data-transmission systems, such as Picturephone, has already been built on today's LSI-size silicon dice. Industry observers are hopeful that a sensor with full TV-resolution can be built in the next couple of years by using larger 500-by-500-mil dice.

However this device won't have the simple structure that was first conceived for the CCD scanner. It will probably incorporate diffusions for low-noise charge-detection, blooming control, and high charge-transfer efficiency. The fabrication process will most likely include ion implantation, two layers of metal, special annealing steps, and multiple diffusions to obtain the necessary high performance. The elegance and simplicity of the original device may have to be sacrificed to attain the high level of technology required to mass-produce a competitive CCD device having television-quality resolution.

On the other hand, TV-quality systems can be constructed with existing 512-bit MOS line-image sensors by adding a rotating mirror to optically scan the images. However, the mechanical scanning mirror necessitates adding volume in the camera. But less silicon real estate, which is expensive, is required for the mirror-scanned line-image sensor than for a corresponding area image sensor, which results in a correspondingly lower component cost. But the 512-element MOS line scanner requires higher light levels because less time is available for integration of each scanned element than in area sensing devices.

These tradeoffs make it clear that the system designer must evaluate carefully the relative merits and disadvantages of both technologies. CCD image sensors, which are free of spike noise, are more likely to be built in high-density arrays, whereas MOS image sensors tend to be less susceptible to image degradation.

In any case, since both types of sensors are fabricated with silicon semiconductor technologies and offer similar performance, the interchange of the two types of device should be straightforward. ☐

Part IV
Memory

The field of CCD research which has progressed most dramatically in the past few years is that of digital memory. A few years ago, CCD memories were still on the drawing boards. Today, however, several different 16-kbit CCD IC's have been demonstrated, and two of these are commercially available. In addition, a 64-kbit IC has been reported, and it seems almost certain that 64-kbit IC's will be commercially available in 1976.

CCD's are inherently serial devices, so in a sense they functionally resemble disks except that they are volatile (i.e., they lose their information when they lose power). However, in the semiconductor industry which has had tremendous success with random access memories (RAM's), the natural question is: "Who wants CCD's?". To be sure, CCD's are slower than RAM's, but in a sense, they compete with RAM's in some applications because they have the advantage of higher density and hence lower cost per bit. The all-important question is: "How much cheaper than MOS RAM's must CCD's be in order to make up for their slower speed?". The answer to this question is extremely involved, but the consensus seems to be that CCD's having an access time of a few hundred microseconds will find widespread use if their cost is one-third to one-fourth that of MOS RAM's. Roughly speaking, the 64-kbit CCD must cost approximately the same as the 16-kbit RAM in order to be successful as a high-volume memory component.

Paper 1 discusses the use of CCD's as drum and disk equivalents. The conclusion of this paper is that, for serial storage in the range of 8 Mbit capacity, CCD's are superior to disks in terms of faster access time, smaller volume, lighter weight, lower power, and perhaps most important of all, improved reliability. Elementary considerations of CCD organization and chip design tradeoffs are discussed, and some early experimental results are presented.

In paper 2, Terman and Heller assess CCD memory technology as compared to MOS RAM technology. This comparison is useful because 1) the technology for manufacturing CCD memories is very similar to that of MOS RAM's, 2) CCD techniques can be used to improve RAM's, and 3) in computer memory hierarchies, serial CCD's can be used as virtual memories or fast auxiliary memories to replace RAM's if the CCD is cheap enough [1]–[3]. By combining some rather fundamental design considerations with data on existing memory IC's, Terman and Heller conclude that for some time to come, CCD's will have an area per bit which is one-half to one-fourth that of MOS RAM's. Based upon this, they conclude that general penetration of CCD's into memory hierarchies of large computers is still uncertain, but in the long term, advanced concepts such as electrode-per-bit operation and multilevel storage could result in an order of magnitude cost advantage for CCD's over MOS RAM's.

Papers 1 and 2 both discuss some of the options available for organizing CCD memory IC's, and the 16K IC's developed to date illustrate this wide diversity in design. Three of the 16K CCD's are covered in the next three papers. Paper 3 discusses the Intel 2413, paper 4 discusses the Fairchild CCD-460, and paper 5 discusses the 16K chip developed at Bell Northern Research. These three designs were selected because they illustrate different design options. Two other 16K designs are presented in paper 1 and in [4].

The Intel 2416 is a four-phase, surface-channel CCD made with n-channel Si-gate technology. It is organized into 64 tracks, each 256 bits long, with on-chip decode which uses a 6-bit address to select one of the 64 tracks. The device is driven with off-chip clock drivers at frequencies of up to 1 MHz, and the data in each track are phase multiplexed so that the maximum output data rate is 2 Mbit.

The clock line capacitance presented by the Intel 2416 is 500 pF on ϕ_1 and ϕ_3 and 700 pF on ϕ_2 and ϕ_4. In an effort to avoid the large capacitance inherent in this architecture, Fairchild utilized the line-addressable random access memory (LARAM) concept, as discussed in paper 4. This device requires a single master clock which is required to drive only 120 pF, and the CCD clock drivers are on-chip. Because of the LARAM architecture, the chip dissipates only 13 μW/bit at the maximum data rate of 5 Mbit. The Fairchild CCD-460 is organized into four 4096-bit blocks of memory operating in parallel. Each 4096-bit block consists of 32 128-bit tracks which are selected by on-chip, 5-bit decoding. The LARAM architecture, together with the buried-channel CCD structure, permits high clock frequency and short average access time (13 μs).

The Bell Northern Research 16K CCD represents a third totally different CCD architecture. It utilizes the series–parallel–series (SPS) structure together with the electrode/bit (E/B) configuration in what is called condensed SPS (CSPS). The advantage of the CSPS architecture is small cell size at the expense of longer access time and more complex on-chip clocking. The BNR 16K is organized as four blocks of 4096-bit serial memory in parallel. Each 4096-bit block consists of two phase-multiplexed CSPS blocks. At a 5-MHz clock frequency, each block operates at a 10-Mbit data rate. In spite

TABLE I
16 384-Bit CCD Comparison

	Area/Bit	Organization	Maximum Data Rate	Access Time	On-Chip Clock Drivers
Intel 2416	1.1 mil^2 (cell) 2.1 mil^2 (total)	1 block 6-bit decode 256 bits/track	2 Mbit	64 μs	No
Fairchild CCD 460	1.5 mil^2 (cell) 2.7 mil^2 (total)	4 blocks 5-bit decode 128 bits/track	5 Mbit	13 μs	Yes
Bell Northern Research	0.57 mil^2 (cell) 1.4 mil^2 (total)	4 blocks No decode 4096 bits/track	10 Mbit	205 μs	Yes

of the complex clocking required for the CSPS structure, the area is only 1.4 mil^2/bit and the on-chip power dissipation at 10 Mbit is only 20 μW/bit.

The three 16K memories discussed in papers 3, 4, and 5 and the accompanying spec sheets are compared in Table I. The area per bit of the memory cell and the total area per bit including periphery are compared. The maximum data rate in column 3 applies to each block in cases where multiple blocks operate in parallel. The access time is an average when the chip is operating at the maximum data rate indicated.

CCD memories are very important to military electronics, and the U.S. Department of Defense has hastened their development with research and development funding. In paper 6, Belt discusses the applications of CCD memories to radar signal processing, and compares the different 16K CCD memory IC's from the standpoint of U.S. Air Force electronic systems needs.

16K CCD IC's offer four times the number of bits per package as 4K MOS RAM's. However, 16K MOS RAM's are being developed for introduction in 1976, and in order to maintain the 4:1 ratio of bits per package, 64K CCD's must be developed. Such a chip has been reported by Mohsen et al. of Mnemonics, Inc. in paper 7. The Mnemonics CCD utilizes the offset-mask CCD electrode structure [5], [6], fabricated using an n-channel Si-gate process. The chip is large (217 × 240 mil^2), but densely packed (0.8 mil^2/bit overall). It is organized into 16 4096-bit tracks with on-chip 4-bit decode. The device can be clocked at up to 10 Mbit, giving a mean access time of around 200 μs. Six different off-chip clocks are required to drive the 4096-bit SPS tracks, but the SPS architecture affords considerable saving in power.

The major emphasis in CCD memory development has been the serial structures discussed thus far. However, it has been recognized for some time that charge-coupling principles can be applied advantageously to the design of MOS RAM's [7], [8]. Recently, a RAM concept has been developed which is analogous to the uniphase CCD and is sometimes called the "zero transistor RAM." This concept is presented in paper 8 and suggests that RAM's of the future will not be designed in terms of conventional circuit elements (capacitors, FET's, etc.), but will utilize concepts of electrons in potential wells which are generated by sophisticated ion-implantation profiles, concepts which are being developed through CCD research.

As mentioned earlier, one of the principle drawbacks to CCD's as drum replacements is their volatility. CCD's themselves are volatile, but they can be combined with MNOS or

other nonvolatile storage locations to achieve nonvolatility [9]–[11]. This concept has evolved to the nonvolatile charge-addressed memory (NOVCAM) presented in paper 9 in which charge-coupling principles are used to address, read, and write MNOS storage locations.

For CCD dynamic serial memories, the major emphasis of future development will be lower cost per bit, and the extent to which CCD memories are accepted depends upon their success in achieving this goal. On the other hand, low cost will be achieved only if production volume is high. The critical experiments are now being performed in the marketplace as users weigh the cost/performance tradeoffs between CCD's and competitive memory technologies such as MOS RAM's, magnetic bubbles, etc. These experiments will determine the ultimate success of CCD memories.

References

[1] Y. S. Lin and R. L. Matson, "Cost-performance evaluation of memory hierarchies," IEEE Trans. Magn., vol. MAG-8, pp. 390–392, Sept. 1972.

[2] A. V. Pohm, "Cost/performance perspectives of paging with electronic and electromechanical backing stores," Proc. IEEE, vol. 63, pp. 1123–1128, Aug. 1975.

[3] D. P. Bhandarkar, "Cost performance aspects of CCD fast auxiliary memory," in CCD '75 Proc., San Diego, CA, Oct. 1975, pp. 435–442.

[4] A. M. Mohsen, M. F. Tompsett, E. N. Fuls, and E. J. Zimany, Jr., "A 16-kbit block addressed charge-coupled memory device," IEEE J. Solid-State Circuits (Joint Special Issue with Transactions on Electron Devices on Charge-Transfer Devices), vol. SC-11, pp. 40–58, Feb. 1976.

[5] R. W. Bower, A. M. Mohsen, and T. A. Zimmerman, "The two-phase offset gate CCD," and "Performance characteristics of the offset gate charge-coupled device," IEEE Trans. Electron Devices, vol. ED-22, pp. 70–73, Feb. 1975.

[6] A. M. Mohsen and T. F. Retajczyk, Jr., "Fabrication and performance of offset-mask charge-coupled devices," IEEE J. Solid-State Circuits (Joint Special Issue with Transactions on Electron Devices on Charge-Transfer Devices), vol. SC-11, pp. 180–188, Feb. 1976.

[7] W. E. Engeler, J. J. Tiemann, and R. D. Baertsch, "Surface-charge RAM system," in 1972 IEEE ISSCC Dig. Tech. Papers, Philadelphia, PA, Feb. 1972, pp. 18–19, 209.

[8] ——"A surface-charge random-access memory system," IEEE J. Solid-State Circuits, vol. SC-7, pp. 330–335, Oct. 1972.

[9] Y. T. Chan, B. T. French, and R. A. Gudmundsen, "Charge-coupled memory device," Appl. Phys. Lett., vol. 22, p. 650, 1973.

[10] M. H. White, D. R. Lampe, and J. L. Fagan, "CCD and MNOS devices for programmable analog signal processing and digital nonvolatile memory," in Tech. Dig., Int. Electron Devices Meeting, Dec. 1973, p. 130.

[11] K. Goser and K. Knauer, "Nonvolatile CCD memory with MNOS storage capacitors," IEEE J. Solid-State Circuits, vol. SC-9, pp. 148–150, June 1974.

CCD's AS DRUM AND DISC EQUIVALENTS

J. M. Chambers, D. J. Sauer, RCA Van Nuys, CA and
W. F. Kosonocky, David Sarnoff Research Center, Princeton, N. J.

ABSTRACT

This paper discusses the application of Charge Coupled Device (CCD) technology to mass memories equivalent in performance to discs and drums. The characteristics of drums and discs are discussed and the reasons for needing a replacement for them are explored. Trade-off considerations related to CCD chip design at both the system and chip levels are presented. Developmental effort in CCD and related chips is reviewed and a drum-equivalent CCD memory system configuration is discussed.

ROTATING MEMORIES

Typical Characteristics of Rotating Memories

Rotating memories have existed since the early days of digital technology. They provide a means of extending the directly accessible memory capacity of a computer beyond that available from the high performance, high cost-per-bit, mainframe memory. Rotating memories have matured in parallel with all other facets of computer technology. The performance and physical characteristics of discs and drums cover a wide range of values and in many respects span several orders of magnitude. One commonly used tool to relate memory forms is a plot of capacity versus access time. Such a plot is shown in Figure 1. Current state-of-the-art rotating memories range in capacity from a few million, to billions of bits while access times vary from less than five milliseconds for fast, fixed head devices to over a hundred milliseconds for large moveable head disc files. As is evident from Figure 1, a large gap exists between the performance of electromechanical peripheral memories and their all-electronic counterparts.

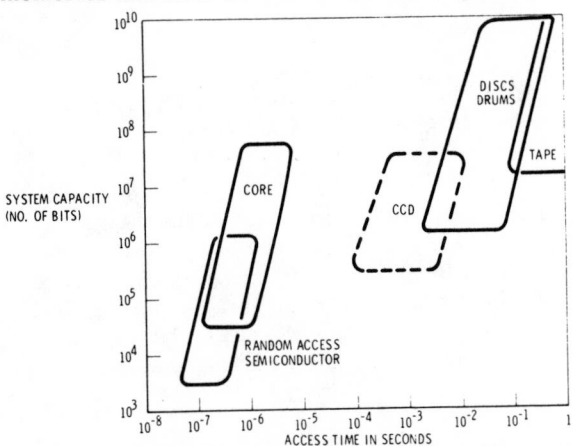

Fig. 1. Memory Hierarchies

Other important characteristics of rotating memories are summarized in Table I. The subject of this paper is to discuss the duplication of these system characteristics by using CCD chips specifically designed for that purpose.

Table 1. State-of-the-Art Rotating Memory Characteristics

Capacity	0.5 to >1000 megabits
Access Time	2.5 to >100 milliseconds
Data Rate	1.0 to 6.0 megabits/sec per channel
Power	50 watts and up
Size	0.5 cubic feet and up
Weight	20 pounds and up
Error Rate	1 in 10^{11}
Non-Volatile Memory	

Limitations of Rotating Memories

Before addressing CCD memory considerations let us first review the inherent shortcomings of rotating memories.

Reliability is probably the most significant limitation for rotating memories. A low Mean-Time-Between-Failure (MTBF) figure results from predictable bearing failure mechanisms and the need for relatively complex electronic circuitry for track selection and read and write functions. Planned periodic refurbishments provide a solution for the bearing failure rate problem but they are costly and troublesome.

Maintainability is another cumbersome aspect of rotating memories. Because of the unique characteristics of discs and drums, they require special skills, facilities and equipment to maintain. Further aggravating the situation is that an increasing number of small computing systems are situated in remote locations where specialized maintenance is less available and more expensive. Thus, reliability and maintainability considerations will become even more significant in the future.

Power consumption, and the corresponding cooling requirement, of disc and drum memories is disproportionately high. In addition to the overhead

Reprinted with permission from *1974 Western Electron. Show and Conv. Tech. Papers*, Session 6, vol. 18, Sept. 10–13, 1974, pp. 1–9.

power to turn the rotating medium, disc and drum memories often require pressure, vacuum or hydraulic systems to position head assemblies, as well as a multiplicity of dc voltages to drive the track selection and read/write circuitry. Although power consumption is not critical in many applications, it is in others and it always represents additional cost.

Size and weight of discs and drums may be critical in some applications but not in others. Airborne and transportable systems could benefit from smaller, lighter weight devices.

In many applications the throughput of a computing system is limited by the access time of its peripheral memory system. Rotating memory access times are limited by the physical constraints of the rotating medium and the aerodynamics of head flight. Improvements in access time beyond those achievable today are expected to be slight. This fundamental limitation frequently presents a severe constraint on the performance of systems which rely on rotating memories for bulk storage.

Discs and drums are normally considered to be low cost storage devices. However, for the capacity range up to 1.0 megabits the cost per bit of memory is excessive. The relatively fixed overhead cost of a motor to rotate the medium, a set of read/ write and track selection circuits, and the necessary power supplies, places the cost per bit of these small memories at an unacceptable level. Thus, rotating memories are not practical for this capacity range and other memory forms must be used.

CCD MEMORIES

Chip Organization

Various system organizations have been proposed for the construction of charge-coupled memories[1-5]. The choice of the system organization depends on the desired system performance and involves trade-offs concerning clock power, access time, chip overhead for peripheral circuits, frequency range, temperature range, and the number of CCD clock phases. Three arrangements which can be used to form the basic memory element are illustrated in Figure 2. As shown, the signal flow in System A follows a serpentine pattern and has signal refreshing stages at each corner. In this serpentine system, all bits traverse the same path through the loop at the same frequency. The number of bits between regenerate amplifiers is determined either by transfer efficiency or the lowest operating frequency desired in the standby or idle mode of operation. The number of bits between the data

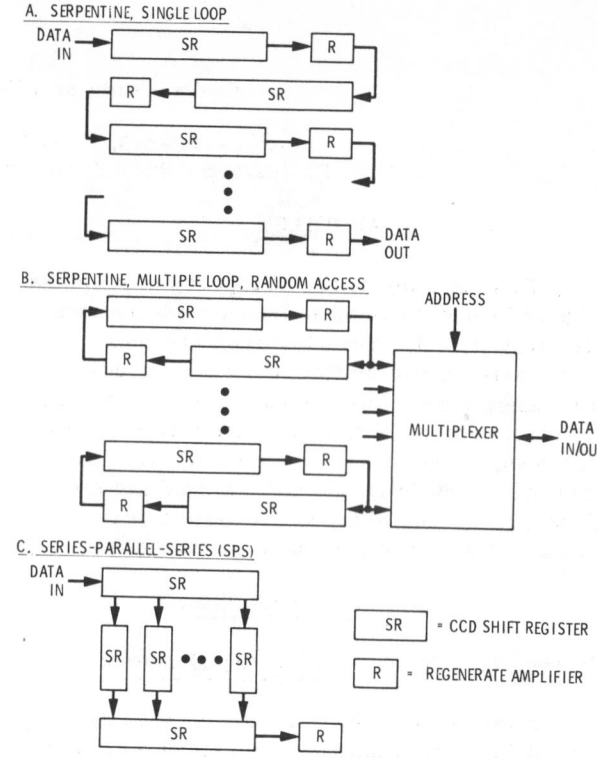

Figure 2. Arrangements for Basic Memory Elements

input and output determines the average access time. A slight modification to the serpentine structure results in System B which provides random access to a number of CCD memory loops. As shown, each loop consists of two CCD registers with a regenerate amplifier at each end. Address decoders are provided for each loop so that the maximum access time for an arbitrary random bit becomes

$$t_A = 2 \frac{N_r}{f_c}$$

where N_r is the number of bits between regenerate amplifiers and f_c is the clock frequency.

System C, as an alternative, is in the form of two serial registers and one large multi-channel parallel shift register. The data stream is introduced serially into the upper serial high speed register. The data is then transferred in parallel into the first stage of each vertical channel. All of the parallel vertical channels in the middle area are clocked together at a lower frequency. At the output the process is reversed, the lower high speed serial register is loaded in parallel and read out through a regeneration amplifier to the data output. In this series-parallel-series (SPS) system, all of the bits do not traverse the same path. If there are N_s bits across and N_p bits down, then each bit is transferred through $N_s + N_p$ stages and N_p of them are at a lower frequency. Thus a single regeneration amplifier can support a larger number of bits than is possible in the serpentine system.

Chip Design Trade-Offs

Power Requirements. The power consumed in a CCD memory (aside from regeneration and I/O stages) is essentially that required to charge the gate capacitances. This power is dissipated mostly in the clock drivers, not on the CCD chip itself. Each bit requires $CV^2 f$ watts where C is the total clock line capacitance per bit, V is the clock voltage, and f is the clock frequency. Thus, the total clock power necessary for a serpentine system is $CV^2 f_c N_b$ where N_b is the total number of bits and f_c is the data rate of the loop. In a large system all unselected loops could be operated at a low idle frequency to minimize power. The lowest practical idle frequency is limited by thermal charge generation during the integration time a particular bit spends between regeneration stages. Typically, the time between regeneration stages should not exceed 0.1 to 1 second at room temperature. Thus, the lowest idle frequency is related to the number of stages between regeneration (N_r) according to:

$f_{c, idle} \geq 10 N_r$. At higher temperatures, a factor of two increase in f_c is required for every 10°C. Frequent regeneration permits low idle frequency and hence low clock power but decreases the number of storage sites per chip. A basic design trade-off involving power and cost thus relates to the selection of the number of stages between regeneration.

The clock power required for the SPS memory loop is lower than the serpentine system at the same data rate because most of the gates in an SPS system operate at a reduced frequency. For an SPS system with N_s bits in the serial register and N_p bits in the parallel registers, the clock capacitance is $2N_s C$ and $N_p N_s C$ for the serial and parallel registers respectively. Then the serial clock power is

$$2N_s CV^2 f_c$$

and the parallel clock power is

$$N_p N_s CV^2 \frac{f_c}{N_s}.$$

Thus, the total clock power is

$$P_{SPS} = CV^2 f_c (2N_s + N_p) \qquad (1)$$

Since $N_b = N_p N_s$, this may be expressed as:

$$P_{SPS} = CV^2 f_c \left(2 \frac{N_b}{N_p} + N_p \right) \qquad (2)$$

It is interesting to note that for the SPS organization there is an optimum ratio N_s/N_p which minimizes the average clock power per bit. Considering the relation for P_{SPS}, define the function

$$g(N_p) = 2 \frac{N_b}{N_p} + N_p.$$

To minimize this function,

$$\frac{dg(N_p)}{dN_p} = 0 = 1 - \frac{2N_b}{N_p^2} = 1 - \frac{2N_s}{N_p} \qquad (3)$$

Therefore, $\dfrac{N_s}{N_p} = \dfrac{1}{2}$ results in a minimum

clock power for the SPS memory loop of:

$$P_{SPS}(min) = CV^2 f_c N_b \left(\dfrac{2}{N_s} \right) \qquad (4)$$

Comparing this to the clock power for the serpentine system, we see that the clock power for the SPS system is reduced by the factor $\dfrac{2}{N_s}$ over the serpentine structure for the same I/O data rate.

Access Time. One drawback of a CCD memory is the long access time associated with its serial nature. The average access to a random bit is determined by the number of bits in an addressable memory loop and the clock frequency.

$$t_{access} = \dfrac{N_{loop}}{2f_c} \qquad (5)$$

The trade-off between access time and number of addressable memory loops is the most significant trade-off decision involved in designing a CCD memory. Short access time ($\sim$ 20 μsec) can be obtained only at the expense of many I/O stages, significant on-chip address decoding and high power consumption because of high clock frequency. The serpentine structure would be necessary to achieve these low access times, and the chip area used for overhead circuitry would compromise the packing density advantage of a CCD memory.

At the other extreme, overhead circuitry could be minimized by utilizing only one I/O to a chip. Assuming 32, 768 bits per chip and a clock frequency of 5 MHz, the average access time would be 3.2 milliseconds. Thus, a rather wide latitude in access time is possible and no particular structure is optimum for all applications.

Clocking Circuitry. Another important trade-off between the serpenting and SPS organizations is the difference in clock drive complexity. In a large CCD memory system, the number of clocks per chip and the total number of clock drivers required are both important considerations. For example, using a two-phase CCD structure, the serpentine system requires a minimum of 2 clock lines per chip whereas the SPS structure requires a minimum of 5 clock lines (2 serial clocks, 2 parallel clocks, and 1 strobe for performing the serial-to-parallel and parallel to serial transfers). On the other hand, if $N_s = 32$ for the SPS structure, the total clocking power is only $\dfrac{2}{32} = \dfrac{1}{16}$ that of the serpentine approach. By driving a number of SPS chips in parallel the total number of clock drivers in the SPS system is only $\dfrac{1}{16}$ of those required in the serpentine system.

Volatility. A CCD memory loop is inherently a volatile form of memory because if the signal charge is not periodically refreshed, thermally generated charge will eventually fill up the potential wells destroying stored information. For certain applications, this presents a fundamental problem unless it is feasible to create a non-volatile memory at the system level.

One approach to achieving a non-volatile memory system using CCD's is to employ standby battery power to permit data retention capability, if necessary. In order for this to be practical, a major design goal must be reduce to power dissipation to an absolute minimum. This is particularly true of the idle power. Since the remainder of the computing system is not operational in the absense of prime power, the memory will not be accessed and all of the loops will be shifted at the idle frequency. It is also very important that the clock driver be highly efficient so that its bias power is commensurate with its output power at idle frequency.

In addition, the average system power for overhead circuitry, such as refresh stages and address decoders should be substantially less than the fundamental $CV^2 f_c$ power required to clock the CCD's at a low idle rate. Also, in the idle mode, techniques to de-power peripherial circuits not required for data retention should be used. It is estimated that in the future a large CCD memory system may be designed having an idle power of $\sim$ 0.1 μW/bit. For a CCD memory of 10 megabits, the total idle power would be 1 watt, permitting battery powered data retention for several days.

EXPERIMENTAL CCD MEMORY CHIPS

Two CCD memory chips are under development at RCA to demonstrate the feasibility of implementing an all-electronic drum memory equivalent. The technology employed in the design of these chips is the two-phase double-level gate structure developed at RCA[6,7]. Figure 3 shows a cross-section of a shift register employing this technology.

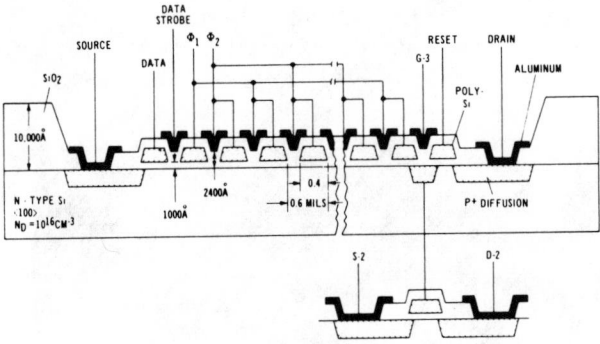

Fig. 3. Two-Phase Shift Register

Since the intent of this development has been to devise a memory equivalent to a drum with an access time of several milliseconds, relatively long loops using an SPS structure have been chosen thereby favoring minimization of power and overhead circuitry rather than fast access times.

In conjunction with the memory chips, a support chip employing Complementary MOS (COSMOS) technology has been designed. COSMOS technology was chosen in order that power consumption be kept at the absolute minimum level. In the system, support chips are paired with memory chips to form functional memory modules.

CCD Memory Test Chip

The CCD memory test chip was designed to (1) demonstrate operation of the individual SPS arrays and their regenerate circuits; (2) demonstrate operation of the on-chip peripheral circuitry used with the CCD memory elements; (3) and permit device characterization so that the COSMOS support chip and a large capacity CCD memory chip could be designed in the next iteration. Design and test features of the chip include:

1. An input translator that accepts COSMOS data input signals and converts them to levels compatible with the CCD input stage.

2. An output translator that accepts the CCD data output and converts it to a level compatible with COSMOS levels.

3. Four of the basic "large geometry" CCD SPS segments. The SPS segment is an array having 16-bit serial input and output registers and 64-bit parallel registers providing 1024 bits of capacity per SPS segment.

4. The regenerate stage which restores the output of the SPS segment to its full 1 or full 0 level for input to the next SPS segment.

5. A threshold variation normalization circuit. This circuit produces a voltage which is a function of the device thresholds on the chip. Its output is the bias for the CCD input source diffusion and is necessary to insure that an optimum charge is introduced into the CCD channel.

6. A reference generator circuit which produces a voltage level half way between the 1 and 0 output levels from a CCD register. This voltage is used in the regenerate circuit as a reference to which the output signal is compared before being restored to its full 1 or full 0 level.

7. A "standard geometry" 16 by 64, or 1024-bit SPS segment for evaluation of the effects of shrinking the width of the CCD channel in an attempt to increase packing density.

8. Test devices to characterize processing related chip characteristics.

The CCD memory test chip is shown packaged in Figure 4 and graphically in Figure 5, along with a location key which identifies the major functional elements on the chip. The four "large geometry" 1024-bit SPS cells are at the bottom of Figure 5 and the single "standard geometry" 1024-bit cell is at the top. The large geometry cell has a 1.2 mil channel width, 0.3 mil diffused channel stops, and a length per stage of 1.2 mils which results in a basic cell area of 1.8 mils^2 per bit. The standard geometry cell has a 0.9 mil channel width, 0.3 mil diffused channel stops, and a length per stage of 1.2 mils yielding a basic cell area of 1.44 mils^2 per bit.

Testing of these devices has revealed that all features of the chip operate as designed. The range of frequency of operation has encompassed serial data rates from 32 kHz to 2 MHz (corresponding to shift rates in the parallel section of 2 to 125 kHz). Performance of the standard geometry 1K SPS

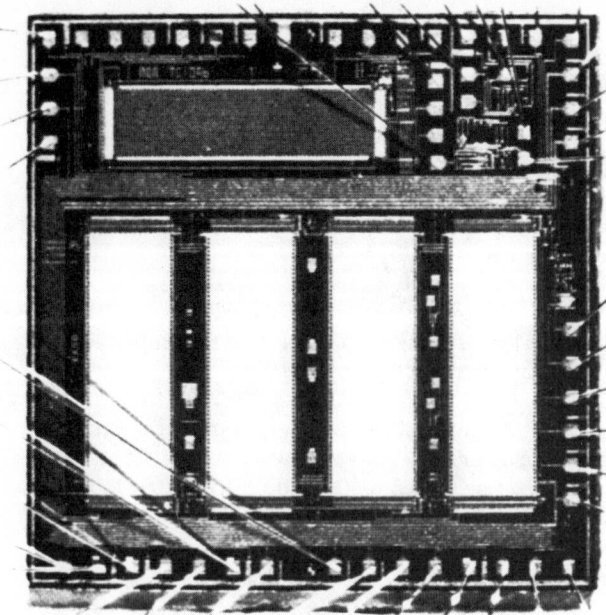

Fig. 4.   Packaged Test Chip

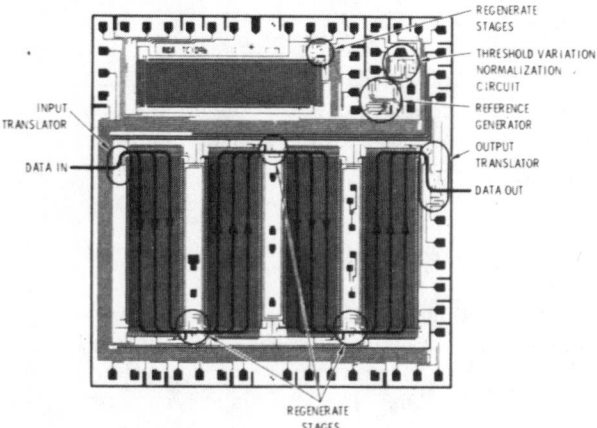

Fig. 5.   Test Chip Features

segment was compared to that of the large 1K segment and found to be essentially the same.   Slightly smaller charge levels are employed in the standard geometry segment so that output signal levels are smaller; however, signal-to-noise ratios are more than adequate.   As a result, a decision to employ the standard geometry for subsequent chips was made.   Transfer efficiencies measured on the input register of the first 1k SPS segment, a simple 16-bit shift register operated at 200 kHz, ranged from 0.990 when operating without fat zero to 0.9985 with fat zero.   Transfer efficiency in the parallel section was 0.9992 with fat zero.   Transfer efficiencies in

the standard geometry section were observed to be comparable to those of the large geometry section. The effects of temperature variations were examined on the chip.   The minimum frequency of operation can be expected to double for every $10^\circ C$ temperature rise. This was experimentally verified.   On most units, a minimum frequency of 2 kHz for the parallel shift rate was achievable at room temperature.

Figure 6 shows the detected analog output waveforms from two of the 1024 bit cells connected serially by means of a signal regeneration stage.   The operating frequency is 400 kHz and fat zero is used.

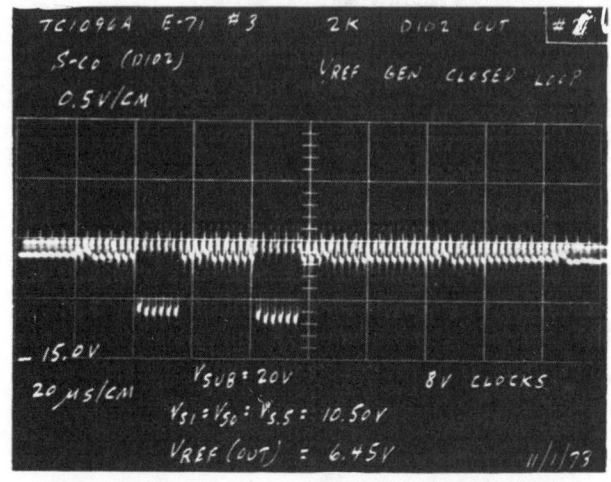

Fig. 6.   Analog Output Waveform

Figure 7 shows the analog signal (from a 1024 bit cell) applied to the output translator as well as the resultant NRZ COSMOS compatible output signal.   The operating frequency is 750 kHz and the output levels are 0 to 10V.

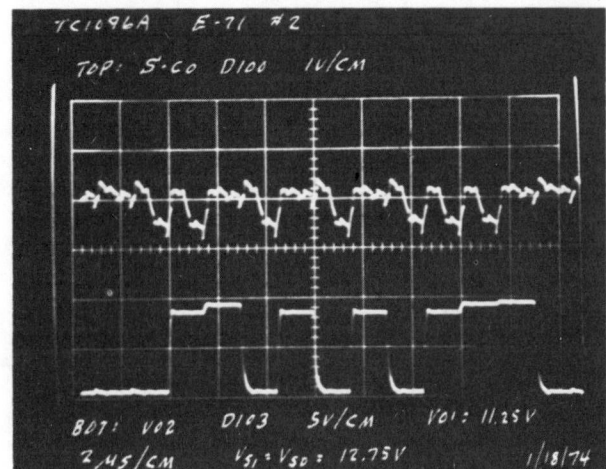

Fig. 7.   Output Translator Waveform

16K CCD Memory Chip

The 16k CCD memory chip was designed to be a useful memory device. It contains two 8192-bit CCD shift registers and its design was directly extrapolated from that of the test chip in most respects.

The 16k chip is shown packaged in Figure 8 and graphically in Figure 9, along with identifiers for the major functional elements on the chip. The 16k memory chip size is 224 × 240 mils. The 8192-bit Register A consists of an input translator, four SPS cells, and an output translator connected in series. The cell size is the same as that of the standard geometry cell on the test chip. Register B is identical to Register A except that it is reflected about the X-axis. Clock signals for both registers are driven in parallel; however, the data paths are separated by individual COSMOS input and output translators for each register.

A 2048-bit SPS structure was chosen as a result of testing performed on the Test Chip which indicated that the charge transfer efficiency in the serial register was high enough to use a 32-bit input register rather than the 16-bit register used in the earlier SPS cells. The 2k cell results in higher packing density and fewer regnerate stages than are

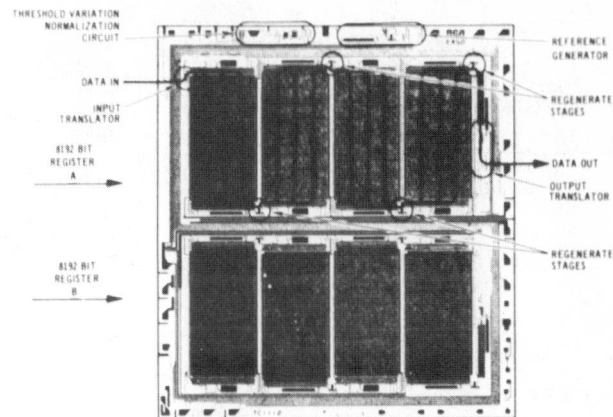

Fig. 9. 16K CCD Memory Chip Features

required using 1k cells. In addition, as was shown earlier, for a given cell size, the minimum CV^2f power per bit occurs when $N_s = 1/2\ N_p$, where N_s and N_p are the lengths of the serial and parallel registers, respectively.

COSMOS Support Chip

The COSMOS support chip is designed to provide the clock drive and logic necessary to convert the 16k memory chip into a two-loop recirculating memory. As discussed earlier, COSMOS technology was selected for this support chip in the interest of maintaining absolute minimum power. The functions included on the COSMOS support chip include clock drivers, clock pulse generation and shaping logic, address decode logic, recirculate gating, and data control logic. The chip also features automatic selection of fast or slow modes to switch between access and idle clock rates, depending on whether that particular chip has been addressed. The clock driving capability of the support chip is sufficient to drive 32k of memory making it useful for the next generation of devices. Figure 10 is a functional block diagram of the COSMOS ancillary chip. On the right side of the diagram is the interface with the CCD chip while on the left is the interface with the remainder of the system. Note that all signals to and from the CCD chip interface directly with the COSMOS chip and none with the remainder of the system. The clock signals are all derived from the system master clock.

Functional Memory Module

Figure 11 shows a COSMOS support chip and a 16k memory chip which together form a functional Memory Module. The Functional Memory Module is the basic building block for the memory system

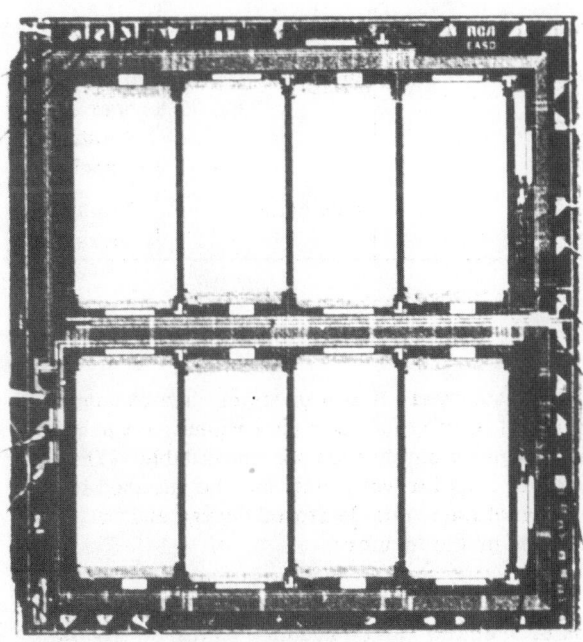

Figure 8. Packaged 16K CCD Memory Chip

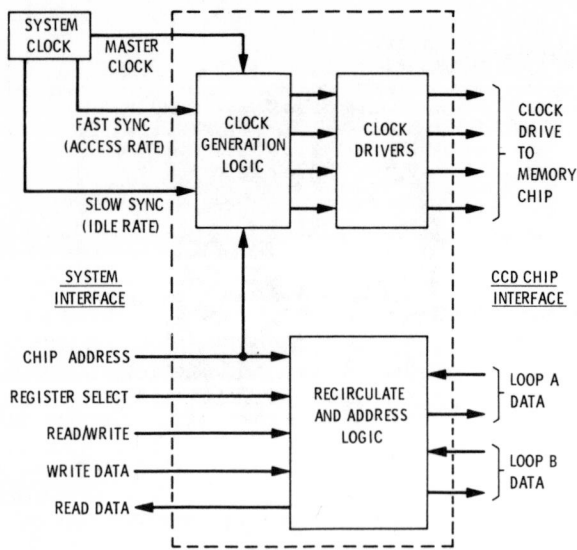

Fig. 10. COSMOS Support Chip

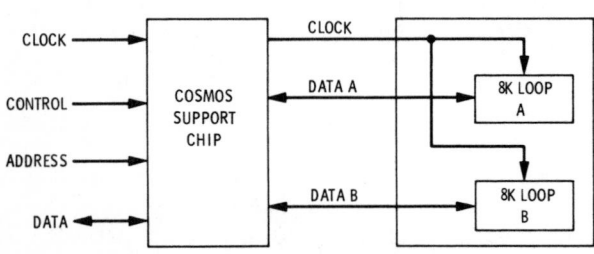

Fig. 11. Functional Memory Module

and in many respects may be equated to a track on a drum or disc. Data recirculates past the tap in the loop in much the same fashion that data on a rotating medium passes the recording head. Addressing of the data is accomplished via logic on the support chip rather than through a track selection matrix as is the case in most drum or disc memories.

TYPICAL SYSTEM CHARACTERISTICS

Using the devices described in the previous section an equivalent to a typical drum memory has been configured. The basic design ground-rule was to duplicate the characteristics of the drum insofar as possible. However, in some areas, the CCD equivalent naturally outperforms the drum. Figure 12 shows how the individual Functional Memory Modules emulate the drum. Table 2 compares the typical characteristics of a drum with those expected for an equivalent CCD memory.

As discussed earlier the inherent characteristics of CCD's provide volatile memory. The only practical solution to this problem, for those

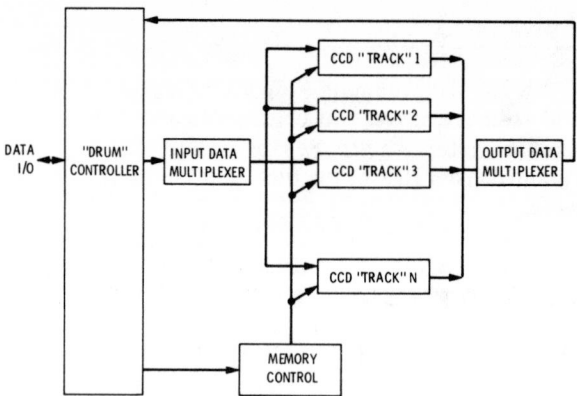

Fig. 12. CCD "Drum" Organization

Table 2. Drum and CCD Equivalent Characteristics

	Drum	CCD Equivalent
Tracks (Loops)	256	1024
Bits per track (net)	32,768	8192
Data Rate	2 MHz	2 MHz
Access Time (max) (avg)	20 msec, 10 msec	4 msec, 2 msec
Useable capacity	8,388,608	8,388,608
Volume	3 cubic feet	1/3 cubic feet
Weight	125 pounds	15 pounds
Power	300 watts	5 watts operating, 2 watts standby
MTBF	3500 hours	20,000 hours

applications where it is a problem, is a backup source of power to sustain data retention capability when primary power becomes unavailable. The length of time for which data may be retained is a function of the available stored energy and the power required by the memory.

The 2 watts of standby power for the 8.4 million bit memory can be supplied easily by a relatively small battery. For example, if a 20 hour data retention capability is needed it can be provided by the equivalent of 8 "D" cells. Rechargable nickel-cadmium cells, which are trickle charged when power is available, can automatically supply the required 2 watts if for any reason power is removed. This approach provides a memory that is non-volatile at the system level. As long as the system design provides sufficient energy storage to sustain data retention capability for all anticipated periods of power outage, the volatile memory characteristic of CCD's is not a problem.

FUTURE EVOLUTION OF CCD MEMORIES

The sole, overwhelming consideration which will determine the long range success or failure of CCD memories is their cost. Since their access time is generally poorer than that of other forms of all-electronic memory, they must offer the user a cost advantage to be accepted. Compared to MOS RAM's, which seem to offer the closest competition costwise, CCD memories will have to ultimately be no more than half and perhaps as low as one fifth the cost in order to be widely used. The potentially large number of bits per chip and the very low system overhead costs should make the one half factor readily achieveable and the one fifth factor a challenging goal.

The issue of whether CCD memories will evolve in the direction of absolute minimum cost or as a compromise between cost and performance is not yet clear. The strong possibility exists that these two divergent directions will both be pursued. The absolute minimum cost may be achieved through packing as many bits as possible on the CCD chip and by employing an uncomplicated clocking scheme to minimize system overhead. Typically, long loop, slow access time devices would result. If faster access devices are required, they can be implemented via shorter loops. However, this implies more gating (either on-chip or off-chip), more addressing capability, and perhaps a more complex clocking system. All of this represents additional power requirement and system cost. Shorter access time, however, may be appropriate for some applications.

The application of magnetic recording media to peripheral memories has provided inherent non-volatility. Many potential CCD memory users who are accustomed to magnetic media react negatively to the thought of an inherently volatile memory device. However, when their application and operating procedures are reviewed in detail, it is often determined that the non-volatile feature of their memory is not exploited and often totally unnecessary. Usually after exploring the situation in some detail, a potential user recognizes that non-volatile

memory is not required and that, when it is, interim power from a battery is an acceptable approach.

CCD memory devices will probably first be applied as direct replacement for rotating memories. Airborne military equipment requiring extreme reliability will be among the first to exploit their advantages. As costs are reduced, CCD's will find their way into commercial memory systems. A new family of memories filling the access time void in the sub-millisecond region is another likely early application. Another void cited earlier exists in the availability of a truly practical memory in the 0.1 to 1.0 megabit range. CCD's can fill this void and in so doing are likely to find application in microprocessor systems which require extended memory.

In summary, CCD's offer a direct answer to the problem of finding a cost-competive, all-electronic counterpart to rotating mass memories. The only apparent shortcoming of the technology is the volatile nature of the memory devices, but for most applications this does not appear to be a genuine problem.

REFERENCES

[1] Kosonocky, W. F., and J. E. Carnes, "Charge-Coupled Digital Circuits," Digest of Technical Papers IEEE Solid State Circuit Conference, p. 162, Feb. 19, 1971.

[2] Collins, D. R., J. B. Barton, D. D. Buss, A. R. Kmetz and J. E. Schroeder, "CCD Memory Options," 1973 IEEE Int'l. Solid-State Circuits Conference, Digest of Technical Papers, p. 136, Philadelphia, Pa.

[3] Wegener, H. A. R., "Appraisal of Charge Transfer Technologies for Peripheral Memory Applications," CCD Applications Conference Proceedings, p. 43, San Diego, California, September 1973.

[4] Agusta, B. and T. V. Harroun, "Conceptual Design of an Eight-Megabyte High Performance Charge-Coupled Storage Device," CCD Applications Conference Proceedings, p. 55, San Diego, California, Sept. 1973.

[5] Carnes, J. E., W. F. Kosonocky, J. M. Chambers, and D. J. Sauer, "Charge-Coupled Devices for Computer Memories," National Computer Conference, Chicago, IL, May 1974.

[6] Kosonocky, W. F. and J. E. Carnes, "Two-Phase Charge-Coupled Devices with Overlapping Polysilicon and Aluminum Gates," RCA Review, 34, No. 1, p. 164, March 1973.

[7] Kosonocky, W. F. and J. E. Carnes, "Design and Performance of Two-Phase Charge-Coupled Devices with Overlapping Polysilicon and Aluminum Gates," 1973 International Electron Devices Meeting Technical Digest, p. 123, Washington, D.C.

Overview of CCD Memory

LEWIS M. TERMAN, FELLOW, IEEE, AND LAWRENCE G. HELLER, MEMBER, IEEE

(Invited Paper)

Abstract—This paper summarizes the status and potential of charge-coupled device (CCD) memories. Cost-performance tradeoffs for serial memories are reviewed, and the CCD chip organizations for slow and fast access systems are discussed. Comparisons are made between CCD and MOS random access memory (RAM) chips on the basis of cell area, support circuits, cell operation, and technology.

T HE INDUSTRY is now seeing the widespread emergence of charge-coupled device (CCD) memory. A number of 16-kbit chips have been announced, others have been published, and a 64-kbit chip is reported in this issue [1]–[6]. It is interesting to note, as can be seen in Fig. 1, that this emergence of large CCD memory chips continues the yearly doubling of bits per chip that began with MOS random access memories (RAM's) in 1968 and which had come to a halt (or at least a pronounced slowdown) for RAM's after 1973.

CCD MEMORY SYSTEM PERFORMANCE REQUIREMENTS

CCD memory is serial, and access to a given bit of information requires a long latency time compared to RAM's, since the information must be shifted to an output port. In some applications, this is not a drawback, since the information is generated or handled in a serial manner. Also, the shift times of CCD's are generally shorter than RAM cycle times, resulting in possibly shorter transfer times where block transfer is important.

In general, however, the longer access times of CCD memory mean that it must be cheaper than RAM's to compete effec-

Manuscript received November 14, 1975.

The authors are with the IBM Thomas J. Watson Research Center, Yorktown Heights, NY 10598.

Reprinted from *IEEE Trans. Electron Devices*, vol. ED-23, pp. 72-78, Feb. 1976.

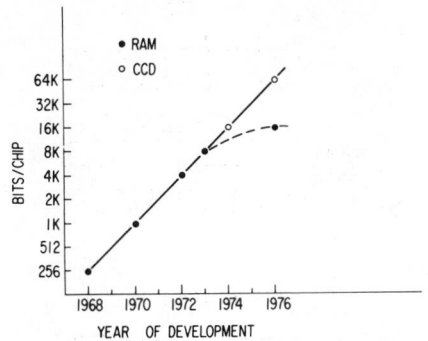

Fig. 1. Bits per chip versus year of development. Points are based on papers presented at the International Solid-State Circuits Conference, except for the 1974 CCD point, which is from WESCON.

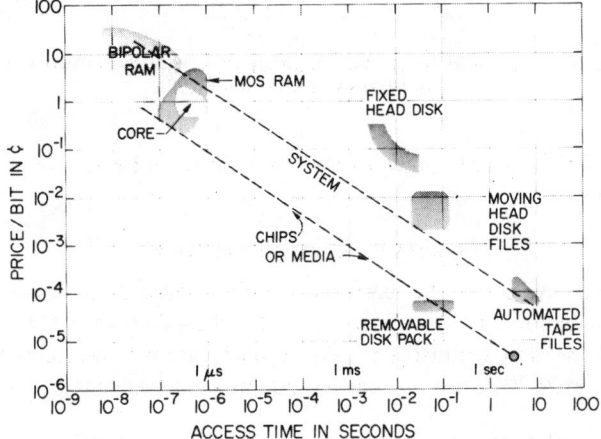

Fig. 2. Price per bit as a function of access time for memory technologies in mid-1975. The lower line is for chips and media, the upper for systems. (After Feth [12].)

tively. The fundamental question is how much cheaper. Clearly, in applications where the CCD performance is sufficient, either because the information is serial in nature, or because the system is slow enough, any cost advantage over RAM's will be adequate, and such applications will be numerous enough to give CCD's a toehold in the commercial market.

Of major interest is whether CCD's can find a place in the memory hierarchies of main frame computing systems. Fig. 2 is the familiar plot of price per bit as a function of access time, showing the well-known "access gap" [12]. The curve has a slope of around $-\frac{1}{2}$, implying that a two order of magnitude increase in access time must be accompanied by at least a one order of magnitude reduction in cost to stay on the curve. However, such an approach is too simplistic; the basic concern is the effect on performance of the memory hierarchy if a CCD store is used. Tradeoffs in memory hierarchies have been the subject of considerable study and speculation, and there is in fact no simple single answer, since the results depend strongly on a great number of factors, including the characteristics, size, and organization of the hierarchy, the job stream mix performed by the system, the processor and bus characteristics, computing system performance, etc.

The following studies are not definitive, but they do give an indication of where CCD memory may be able to fit into memory system hierarchies. In a rather idealized study, Lin

and Matson [13] calculated the cost per access for two- and three-level memory systems which were optimized for a specific job stream. The results indicate that a three-level system with a serial backing store tends to become advantageous in cost per access over a two-level system as the system gets larger and the memory bus wider; however, the overhead costs of the additional level of memory and wider bus were not included. Pohm [14] looked at the cost effectiveness of replacing a drum backing store with a higher speed electronic backing store such as a CCD. The analysis was done for a specific set of processor and main memory costs and backing store size, with three types of job streams, and two processor sizes. For the interactive job stream and the larger (2.8 MIPS) processor, the backing store can be cost effective if it has a cost advantage of 2:1 over main memory, and an access plus page transfer time around a half millisecond. For the other combinations, the backing store would have to be roughly from four to ten times cheaper than main memory for the same sort of performance range. Bhandarkar [15] analyzed the use of fast auxiliary memory, and concluded that a cost reduction of two to four can make such memories attractive for block read time ranging from 200 to 500 μs.

In fact, present computer systems are pretty well designed around the access gap. If the processor is kept busy, simply adding a backing store or replacing an existing one with a faster one will not have much effect on system performance. On the other hand, redesigning the total hierarchy may result in a better system; as discussed above it depends upon a large number of factors. One simply cannot be definitive as far as cost versus performance; optimization must be done for a specific set of parameters.

CCD MEMORY SYSTEM PERFORMANCE

The worst case access or latency time of a CCD memory depends upon the chip organization, and can be varied over rather wide limits, with some tradeoff in chip area. For a given number of bits per chip, longer shift registers result in longer latency times, but require less on-chip decoding and sense/regeneration circuits, while short shift registers reduce latency, but require more support circuits.

For large low cost systems the most common organization is the serial-parallel-serial (SPS) block [16], generally with around 4K cells per block. It has the same worst case latency as a serpentine or loop register of the same number of bits clocked at the same frequency as the serial sections, but has a number of advantages.

1) The number of transfers between input and output is greatly reduced, allowing the elimination of internal refresh amplifiers. Serpentine layouts generally require a refresh amplifier every 128 bits or so.

2) There is no need for clock line crossovers in the array.

3) Power is reduced due to the slow clocking in the parallel section.

4) The slow clocking of the parallel section lends itself to electrode-per-bit (E/B) operation, as discussed below.

Since a high density CCD cell is normally longer than it is wide, the series-parallel interface may result in nonminimum pitch between the paralled tracks. Interfacing each track with a half cell overcomes this [4]; odd-numbered tracks are loaded from the left-hand cell halves, a new string of bits is shifted in, and the even tracks are loaded from the right-hand cell halves. The number of transfers that a bit experiences in the serial registers is halved.

Shorter access time can be achieved with shorter shift registers in serpentine or loop organizations. A typical chip may consist of 256- or 128-bit registers, clocked in parallel, each with its own sense/refresh circuitry, and on-chip decoding to connect the selected register to the chip I/O.

An organization optimized for short access time is the line-addressable random access memory (LARAM) [2]. A LARAM block consists of a number of short shift registers which share common input and output buses. The registers are the single phase type, with each register having its clock line running along the register, rather than across the entire block. Thus, by decoding the clock lines, the registers may be activated individually while the others remain quiescent. Clock driver power and loading are reduced, and only a single set of I/O circuits is needed, which is switched to the active shift register. Since the register length is comparatively short, access time can be over an order of magnitude less than that obtainable with large blocks. A drawback is that appreciable time is lost to refresh, since all bits in a block must be refreshed by a single sense amplifier, and registers unaccessed in a retention period must interrupt the memory operation to be refreshed. In [2], 14 percent of real time is spent in refreshing at a 5-MHz clock rate for a 6-ms retention interval; this is much more than for RAM's, which generally run around 1 percent. In conventional CCD organizations, all registers on a chip are clocked in parallel and refreshed independently. No time is lost in refresh, and no controls for it are needed. Another problem arising from the lack of motion of charge in unaccessed cells is that dark current accumulates nonuniformly in the storage wells just as in a RAM, unlike the averaging effect which occurs in a CCD when the charge is kept moving.

Table I is a summary of data on a number of large CCD memory chips, along with some area data on RAM chips for comparison. The data in the table represent the state of the art, but it must be remembered that the CCD is a less mature technology than RAM's, although it has benefited considerably from RAM experiences in technology and circuit design. Two chips in the table—the 16-kbit RAM and the 64-kbit CCD—are more aggressive than the others, and are included as representative of the next generation of such chips. From the table, it can be seen that slow access SPS chips have smaller cell areas and area per bit at the chip level approaching a factor of 2 better than for the non-SPS chips. The effect of additional on-chip overhead for the fast access non-SPS chips is not as noticeable. In fact, the chip utilization (defined as the ratio of total cell area to total chip area) is best for the LARAM chip. However, this appears to be due to a large cell area, rather than efficient on-chip support circuitry. For the 16-kbit chips, the slow access SPS chips do have less total nonarray chip area than the fast access non-SPS chips, and the overall chip area per bit is better by around a factor

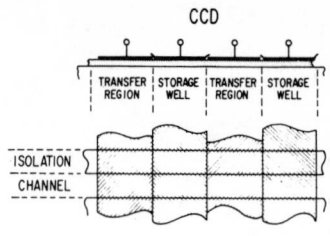

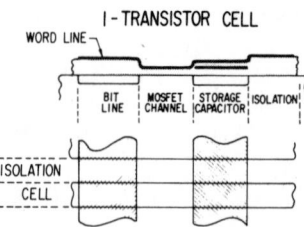

Fig. 3. Diagram of theoretical minimum area two-/four-phase CCD and one-transistor MOS RAM cells.

of 2. Thus, chip organization has a significant effect on chip density.

COMPARISON WITH MOS RAM's

To overcome the performance disadvantage in nonserial applications, CCD memory must be cheaper than RAM. In this section comparisons will be made against one-transistor cell MOS RAM's for actual and potential cost advantages.

A. Cell Area

Fundamentally, CCD and one-transistor cells have about the same ultimate limit in cell area. Consider Fig. 3. Both the two- or four-phase CCD cell and the one-transistor cell require four separate regions in the X direction; the CCD has two storage wells and two transfer regions, while the one-device cell has the bit line, channel region, storage plate, and gap to the adjacent cell. Both require two regions in the Y direction—the actual cell area and the gap to the adjacent cell. Given a hypothetical ideal technology, capable of producing the desired features with a linewidth l without misregistration, the area of both cells will be $8l$. This is an indication of the minimum cell area possible (a three-phase CCD cell would require $6l$). In practice the CCD cell becomes larger, although not greatly so. Effects such as practical processing and layout considerations, and the desirability of nonminimum cell area to increase charge carrying capability, as well as edge effects and transfer efficiency, must be considered. CCD shift registers with cells having an area near the minimum number of squares have been operated with good transfer efficiency and clock rate [17].

The actual minimum cell area possible depends upon the particular technology employed to fabricate the cell, and the tolerances. Four of the most common are shown in Fig. 4. On the basis of a technology capable of producing unit lines and spacings of width l and a misregistration tolerance of r, expressions for the minimum cell areas can be derived. Two expressions are given for the offset gate cell: i) the case for the minimum well and transfer regions which must be at least

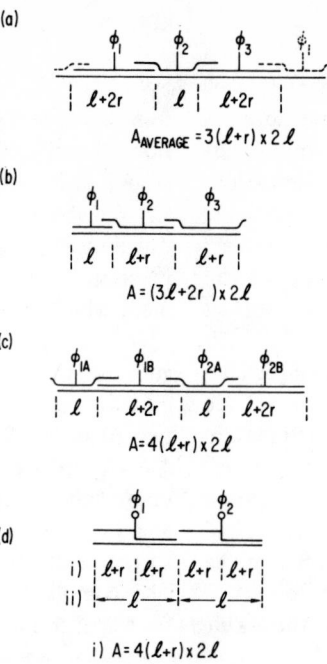

Fig. 4. Minimum cell area for four common CCD gate structures. (a) 3ϕ-2 gate levels. (b) 3ϕ-3 gate levels. (c) 2/4 ϕ-2 gate levels. (d) Offset gate. Area expression assumes processing capable of producing features with linewidth l and misregistration r, with minimum well and transfer regions equal to l [except for (d)-ii, see text], and a cell-to-cell pitch of 2l (into the paper).

TABLE I

CCD CHIPS

Ref. No.	Bits/Chip	Worst Case Access @ f (us @ MHz)	On-chip Power @ 1 MHz (uW/bit)	Chip Size (mils)	Area per Bit Chip/Cell (mils^2)	Chip Utilization	Technology
(1)	64r x 256b	128 @ 2	9	143x237	2.1/1.1	0.53	4∅ 2 level polysilicon gates
(2)	4x32rx128b (LARAM)	26 @ 5	10	220x200	2.7/1.5	0.57	1∅ 2 level polysilicon gates buried channel
(3)	9r x 1024b	512 @ 2	15	135x200	2.9/1.3	0.44	2∅ 2 level polysilicon gates buried channel
(4)	4x4k SPS	400 @ 10	1.5	136x169	1.4/0.57	0.40	2∅ 2 level polysilicon gates E/B operation
(5)	4x4k SPS	400 @ 10	10	124x350 (89x246)	2.7/1.2 (1.35/0.7)	0.45 (0.52)	3∅ 3 level 9 um polysilicon gates (same, with 7 um gates)*
(6)	16x4k SPS	400 @ 10	0.5	235x213	0.76/0.4	0.52	2∅ offset gate process 2 level polysilicon gates

RAM CHIPS

Ref. No.	Bits/Chip	Worst Case Access @ f (us @ MHz)	On-chip Power @ 1 MHz (uW/bit)	Chip Size (mils)	Area per Bit Chip/Cell (mils^2)	Chip Utilization	Technology
(7)	4k x 1			122x165	4.9/2.0	0.40	
(8)	4k x 1			105x178	4.6/1.5	0.32	
(9)	4k x 1			129x139	4.4/1.6	0.37	
(10)	4k x 1			159x181	7.0/2.0	0.29	
(11)	16k x 1			145x234	2.1/0.7	0.34	

*The 7 um gate technology is felt to be more representative of the state of the art and is used in the comparisons in the text.

l in width, and ii), the minimum cell area case where the cell area is determined by mask layout only. This layout does not exist for $r/l \geqslant \frac{1}{2}$ because with worst case misregistration the transition regions or storage wells vanish. As dimensions shrink, the i) case will be more representative. For $r/l = \frac{1}{2}$ the cell areas run about 8–12 squares. If the chips in Table I have been made in approximately 5-μm technology, the cell areas run 15–25 squares, indicating that CCD cells are not yet the minimum number of squares.

In the present state of the art the one-transistor cell is considerably larger than the minimum number of squares, due mainly to the sensitivity of the sense amplifiers, which require more than a minimum area storage capacitor to provide sufficient sense signal. Cells run 35–50 squares. Smaller ones and improved sense amplifiers have been proposed [18]–[20], and as sense amplifiers improve, the number of squares may be reduced.

From Table I, comparing the 16-kbit CCD's with the 4-kbit RAM's, the RAM cells are 1.4–2.0 mil^2, the SPS CCD's 0.7–0.9 mil^2, and the fast access CCD's 1.1–1.5 mil^2. For the more aggressive chips, the 16-kbit RAM cell is less than a factor of 2 larger than the 64-kbit CCD cell.

Neither the CCD nor RAM cells are minimum area at present, and as technology improves, both can be expected to shrink. It seems likely that the cell area ratio will remain around 1.5 to over 2:1 for some time.

B. Support Circuits

CCD on-chip support circuits have been expected to be simpler than that of RAM's, but they must carry out a number of functions, including decode, drive, sense, regeneration, gating, off-chip interface, on-chip timing, clock generation, etc. It is evident from the recent CCD papers that much effort has been focused on optimizing these circuits, and many circuit techniques used are adapted from RAM chips. MOS dynamic logic is generally used to reduce power dissipation. Sensing is done by the familiar single-ended conditional turn on of an inverter, or by balanced schemes reminiscent of the balanced latch sensing used in RAM's [4], [5]. Two papers report the use of on-chip feedback circuits to generate a calibrated reference voltage or charge which tracks the "1" and "0" levels, greatly reducing the effect of processing tolerances, and ensuring good margins for sensing [1], [3].

A measure of the support circuits needed is the chip utilization. Table I shows it to be larger for the CCD's than for RAM's, in a 5/4 to 3/2 ratio. It should be kept in mind, however, that the CCD chips are larger than the RAM chips, and some of the better chip utilization is due to the diminished effect of fixed overhead such as pads, kerf, TTL compatability, on-chip clock generation, etc, with larger chips.

C. Cell Operation

By appropriate operation, a basic CCD cell can be made to store more than a single bit, although at the cost of additional support circuit complexity. E/B operation [21] is one way. N bits are stored on $N+1$ wells, and the bit transfers within the $N+1$ wells occur sequentially. Compared to two-phase operation, density increases by $2N/(N+1)$, bit rate decreases

by $2/(N+1)$, $N+1$ clocks are required, and the number of transfers seen by a bit is reduced by $(N+1)/2N$. E/B operation is readily used in the parallel section of an SPS organized block, since the slow clocking requirement there overcomes any drawback in the slower bit rate. However, generation and distribution of additional clocks are required. This, in conjunction with the monotonically decreasing density improvement which results as N increases, normally limits N to about 3 [4], but this can be overcome if a shift register is used to generate the clocks. A "1" in a field of "0"s sequentially generates the clocks; a two-phase shift register acts as a simple two phase to N phase converter. N may be made large, and $N = 16$ has been reported [22]. For large N the bit density improvement approaches the asymptotic limit of 2; when the area of the shift register and associated circuits are considered, an actual density improvement of more like 1.5 is reasonable. The complexity of clock crossovers makes E/B unattractive for serpentine organization.

Another way of obtaining more than 1 bit per cell is by using the analog properties of the CCD to store more than 1 bit in a charge packet, or multilevel storage (MLS). Using 4 levels of charge, 2 bits can be stored, 8 levels for 3 bits, etc. For a given dynamic range, the second bit shrinks the sense windows by a factor of 3, and each additional bit reduces it by a further factor of approximately 2. A theoretical study has been made of the probability of false detection as the sense window gets smaller [16]; however, long before this becomes a factor, other effects become limiting. Sensing with small windows is much more sensitive to leakage and incomplete charge-transfer effects, and parameter tracking between input and output circuits must be very good. Actually, for the same dynamic range as the MOS RAM cell, an eight- or sixteen-level CCD cell will give about the same net sensing voltage window at the sense amplifier as a RAM cell. However, the sensing procedure and reference generation is much more complex. Also, the information charge is sensed by placing it on a capacitor (a MOSFET gate). The resulting voltage depends upon the capacitor area, which must be close to its nominal value or the sensing window will vanish. Further, since the CCD well is a minimum area capacitor, the sensing capacitor must also be close to minimum area, which is unfavorable for dimensional reproducibility. A self-calibration scheme, similar to the reference generation approaches described above, may be advantageous.

Another problem arises from the complexity and performance of the input and output circuitry. It is more complex than for conventional CCD's, and must be shared over more cells to achieve the same chip utilization. More cells implies longer latency at the same clock rate, but since circuit operation is more complex, especially for sensing, the circuits will probably be slower. For example, a 16-level/4-bit system would require 15 simultaneous or 4 sequential compares. An alternative would be some sort of pipeline sensing, operating at the clock speed.

Thus, there is a difficult density or cost versus latency tradeoff, especially, since as the number of levels increases, the proportional effect on density decreases, and a point of diminishing returns is rapidly reached. It does not seem unreasonable that a 2-bit/4-level system eventually could be designed. No work on MLS has been reported to date.

MLS is unattractive for RAM's for several reasons. First, the RAM cell is already designed for minimum sensible signal, and to further divide the sense signal is undesirable. Second, the generation of reference levels is much more complex then for present RAM's. Third, there are many fewer bits per sense amp in a RAM, and a more complex sense amplifier would not be a good tradeoff, and might have trouble matching the bit line pitch. Finally, there is a severe geometry reproducibility requirement, since the cell is written by impressing a voltage upon the bit line, while sensing depends upon the amount of charge stored in the cell, which is proportional to the storage plate area. The CCD's seem much more amenable to MLS.

D. Technology

The most common processes for present CCD memory chips use two or three levels of polysilicon for the gate structure. The offset gate process is used in the 64-kbit chip. Surface n-channel devices are used universally, except for [2] and [3], which are buried channel. CCD technology is similar to that for MOSFET devices, with at least one additional mask step for each additional gate conductor level. It has been claimed that CCD technology will have superior yield, but no data have been published to verify the contention. One basis for such claims is the absence of contacts in the array area. RAM one-device cells can be made without contacts [18], but this does remain a CCD advantage at present. Another advantage is the averaging of leakage which occurs in conventional CCD organizations due to the continual motion of the charge packets. RAM's and the LARAM organization are sensitive to worst case leakage. The time dependent suppression of dark current due to exposure of the midband states to previous charge packets [23] is important in reducing leakage current.

CONCLUSIONS

CCD memory is in the process of finding its place in the industry. Despite serial operation, it will find a place because it is lower cost than RAM's, and performance is adequate for many applications. Some impact in specific computer system memory hierarchies can be expected, but general penetration into the memory systems of large computers is not clear. From Fig. 1, it is evident that fixed head disk memory is a prime target.

There are two major trends in chip organization. One is for large low cost slow access systems, using SPS blocks with E/B operation. The other is towards short access, using relatively short shift registers in serpentine or loop organization to reduce latency. There may be as much as a factor of 2 difference in cost and an order of magnitude difference in latency time between the two approaches. MLS, probably with 2 bits per packet, is an eventual possibility for low cost systems.

Compared to MOS RAM's, CCD's require between two and four times less chip area per bit. The cost difference seems likely to hold in the reasonably near future. Eventually, full exploitation of minimum area cells, E/B operation, and MLS

could result in an order of magnitude cost advantage for CCD's over MOS RAM's.

A simple "back of the envelope" calculation can indicate the potential of CCD memory. Consider a chip fabricated with a cell area of 10 squares, 50 percent chip utilization, 1.5× density improvement from E/B operation, 2-bit MLS, and an area of 200 × 200 mils. If it can sell for the same price as present day 4K RAM's and processing improves to 0.1 to 0.14 mils linewidth capability, at the chip level the cost will be in the millicent per bit range; this is clearly sufficient for widespread usage.

ACKNOWLEDGMENT

The authors would like to thank M. F. Tompsett for many valuable discussions. The support of D. L. Critchlow and B. Agusta is appreciated.

REFERENCES

[1] S. Chou, "Design of a 16 384-bit serial charge-coupled memory device," this issue, pp. 78–86.
[2] K. C. Gunsagar, M. R. Guidry, and G. F. Amelio, "A CCD line addressable random-access memory (LARAM)," *IEEE J. Solid-State Circuits*, vol. SC-10, pp. 268–273, Oct. 1975.
[3] R. C. Varshney, M. R. Guidry, G. F. Amelio, and J. M. Early, "A byte organized NMOS/CCD memory with dynamic refresh logic," this issue, pp. 86–92.
[4] S. D. Rosenbaum, C. H. Chan, J. T. Caves, S. C. Poon, and R. W. Wallace, "A 16 384-bit high density CCD memory," this issue, pp. 101–108.
[5] A. M. Mohsen, M. F. Tompsett, E. N. Fuls, and E. J. Zimany, Jr., "A 16-kbit block addressed charge-coupled memory device," this issue, pp. 108–116.
[6] A. M. Mohsen, R. W. Bower, M. Wilder, and D. Erb, "A 64-kbit block addressed charge-coupled memory," this issue, pp. 117–126.
[7] R. C. Foss and R. Harland, "Peripheral circuits for one-transistor cell MOS RAM's," *IEEE J. Solid-State Circuits*, vol. SC-10, pp. 255–261, Oct. 1975.
[8] Intel 2107 (data obtained by measurement of chip).
[9] MOSTEK 4096 (data obtained by measurement of chip).
[10] Texas Instruments 4030. Data obtained by measurement of chip.
[11] J. Koo, N. Ahlquist, J. Breivogel, J. McCallum, W. Oldham, and A. Renninger, "A 16k dynamic RAM," to be presented at the IEEE Int. Solid-State Circuits Conf., Philadelphia, PA, Feb. 18–20, 1976.
[12] G. C. Feth, oral presentation at a panel session on "System implications of advancing storage technology," Nat. Comput. Conf., Anaheim, CA, May 19–22, 1975.
[13] Y. S. Lin and R. L. Mattson, "Cost-performance evaluation of memory hierarchies," *IEEE Trans. Magn.*, vol. MAG-8, pp. 390–392, Sept. 1972.
[14] A. V. Pohm, "Cost/performance perspectives of paging with electronic and electromechanical backing stores," *Proc. IEEE (Special Issue on Large Capacity Digital Storage Systems)*, vol. 63, pp. 1123–1128, Aug. 1975.
[15] D. P. Bhandarkar, "Cost performance aspects of CCD fast auxiliary memory," in *Proc. 1975 Int. Conf. Application Charge-Coupled Devices*, Oct. 29–31, 1975.
[16] C. H. Séquin and M. F. Tompsett, *Charge Transfer Devices*, suppl. 8 for *Advances in Electronics and Electron Physics*. New York: Academic, 1975, pp. 238–247.
[17] N. A. Patrin, "Performance of very high density charge coupled devices," *IBM J. Res. Develop.*, vol. 17, pp. 241–248, May 1973.
[18] K.-U. Stein and H. Friedrich, "A 1-mil² single-transistor memory cell in n silicon-gate technology," *IEEE J. Solid-State Circuits*, vol. SC-8, pp. 319–323, Oct. 1973.
[19] K. Hoffmann, "Continuously charge-coupled random access memory (C³ RAM)," to be presented at the IEEE Int. Solid-State Circuits Conf., Philadelphia, PA, Feb. 18–20, 1976.
[20] L. G. Heller, D. P. Spampinato, and Y. L. Yao, "High sensitivity charge transfer sense amplifier," in *ISSCC Dig. Tech. Papers*, Feb. 1975, pp. 112–113.

[21] D. R. Collins, J. B. Barton, D. D. Buss, A. R. Kmetz, and J. E. Schroeder, "CCD memory options," in *ISSCC Dig. Tech. Papers*, Feb. 1973, pp. 136–137.

[22] W. E. Tchon, B. R. Elmer, A. J. Denboer, S. Negishi, K. Hirabayashi, I. Nojima, and S. Kohyama, "4096-bit serial de- coded multiphase serial-parallel-serial CCD memory," this issue, pp. 93–101.

[23] D. G. Ong and R. F. Pierret, "Thermal carrier generation in charge-coupled devices," *IEEE Trans. Electron Devices*, vol. ED-22, pp. 593–602, Aug. 1975.

Design of a 16 384-Bit Serial Charge-Coupled Memory Device

SUNLIN CHOU, MEMBER, IEEE

Abstract—This paper describes a 16 384-bit serial charge-coupled memory device designed primarily for low cost and compatibility with existing high-volume manufacturing techniques. To obtain low access time, the device was organized as 64 recirculating shift registers each 256 bits long. Any one register can be selected at random for reading or writing, by means of a 6-bit address input.

The alternatives considered in choosing the charge-coupled device (CCD) structure and chip organization are discussed. Data regeneration circuits are described in detail. The device was fabricated on a silicon chip, with an area of 2.07 mil^2/bit (including all peripheral circuitry). It operates at data rates exceeding 2 MHz, and has a minimum average access time of under 100 μs.

I. INTRODUCTION

THE POSSIBILITY of high-density serial memories using charge-coupled devices (CCD's) has been suggested for several years [1]. These memories could offer access times considerably shorter than those available from rotating mechanical memories such as drums and disks, but would have low costs relative to the faster random-access memories. Potential applications would include small serial memory systems previously employing relatively high-cost shift registers, as well

Manuscript received September 12, 1975.
The author is with Intel Corporation, Santa Clara, CA.

as large systems in which the higher speed of CCD memories would make them more attractive than rotating mechanical memories.

CCD memories of different capacities have been discussed in the literature [2]–[6], but apparently none has so far been introduced for large scale commercial applications. In this paper we describe a 16 384-bit CCD (designated the 2416) which was designed primarily for high-volume production. The specific design objectives are listed in Table I. The indicated packing density is at least a factor of 2 higher than those achieved by commercially available 4096-bit random-access memories, but is entirely compatible with existing high-volume manufacturing technology. This should result in a relatively low manufacturing cost. The minimum average access time to any bit of 100 μs is more than an order of magnitude lower than that available from the fastest rotating mechanical memories.

II. CHOICE OF CCD STRUCTURE AND TECHNOLOGY

A variety of CCD structures and associated fabrication techniques have been suggested. Our choices were oriented towards low manufacturing cost and ease of fabrication on standard production lines, rather than unusually high density or performance. A list of alternatives is given below, with an explanation of the choice made in each case.

Reprinted from *IEEE Trans. Electron Devices*, vol. ED-23, pp. 78–86, Feb. 1976.

TABLE I
DESIGN OBJECTIVES

Area/bit including peripheral circuits $\simeq 2$ mil^2/bit (1300 μm^2/bit).
Operating junction temperature range = $0° - 80°$C.
Minimum average access time to any bit = 100 μs.
Maximum read or write rate > 2 Mbits/s.
Ratio of maximum to minimum operating frequency > 10.
On-chip power dissipation:

$\leqq 10$ μW/bit at maximum shift frequency or data rate;
$\leqq\ \ 2$ μW/bit in standby recirculating mode.

Voltage supplies (relative to $V_{SS} = 0$ V):

$V_{DD} = 12$ V nominal (for circuit power);
$V_{BB} = -5$ V nominal (for substrate bias).

Clock and input voltage levels (except data input):

Nominal low level = V_{SS};
Nominal high level = V_{DD}.

Data input voltage levels:

Nominal low level = V_{SS};
Nominal high level = 3.5 V.

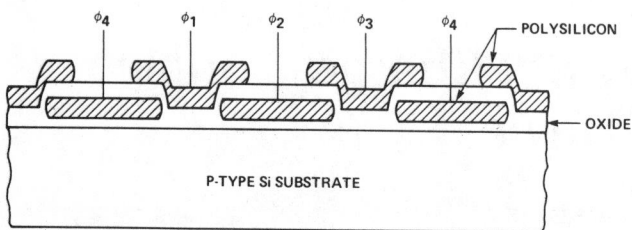

Fig. 1. Cross section of four-phase CCD structure.

A. Single- Versus Dual-Level CCD Electrode Structure

The single-level structure has the apparent advantage of easier fabrication, but may actually require very tight control on processing parameters or other special techniques in order to overcome limitations associated with having gaps [4], [7]. The dual-level structure, in which the entire channel is covered by electrodes, is inherently more tolerant to processing variations and permits greater flexibility in layout. For these reasons we selected the four-phase overlapping gate structure illustrated in Fig. 1. Both layers of electrodes are of polysilicon. The four-phase structure can be converted into a two-phase structure by suitable adjustments in oxide thickness or substrate doping concentrations. However, a reduction in charge-carrying capacity usually occurs [8].

B. p-Channel Versus n-Channel

With n-channel MOS technology already in high-volume production, it was natural to choose an n-channel approach for the higher speed and performance it offered in both the CCD array and supporting circuitry.

C. Buried Channel Versus Surface Channel

For memory applications, the advantage of buried-channel devices [9] is significant when high-frequency performance is required. Although the buried-channel device also has better charge-transfer efficiency at low frequencies, the surface-channel device gives very adequate results for charge-transfer

rates under 1-2 MHz, provided the data are refreshed after every 100-200 charge transfers. An advantage of surface-channel devices is their greater charge-carrying capacity, which could considerably simplify the sensing and refreshing operations. Since the memory product to be designed was not intended for operation at charge-transfer rates exceeding 2 MHz, we chose the surface-channel approach for its easier fabrication and greater charge carrying capacity.

III. CHIP ORGANIZATION

In order to achieve low access time, the memory array was organized as 64 independently accessible, recirculating shift-register loops each 256 bits long (see Fig. 2). By applying a 6-bit address, the user can select any of the loops for reading or writing.

Two different approaches may be taken with regard to data circulation in the loops. The first ("selective clocking") approach is to shift and recirculate only the selected loop. This means that only the CCD electrodes of one selected loop would be clocked at any moment, while the electrodes of other loops would be held at steady levels. In the second ("parallel clocking") approach all loops are clocked in parallel (i.e., in synchronism). A distinct advantage of the first approach is the relatively light loading on the CCD clock driver; however, the additional circuitry needed to steer the driver to the selected loop increases the complexity and cost of the memory device. A further penalty for the selective clocking approach is the need for external logic to control the refresh functions and synchronize the different loops. Also, in order to satisfy the refresh requirement, the minimum shift frequency for this approach is of necessity much higher than that for parallel clocking. This may narrow the frequency range of operation of the device and/or constrain both the CCD array and the sensing circuitry to operate at higher frequencies where their performance is intrinsically worse.

Despite the drawback of high clock loading, it was decided that the second parallel clocking approach would be more effective for most applications. Note that the use of a lower CCD clock frequency in the parallel clocking approach affects the access time but need *not* affect the data rate. A high data rate can be maintained even at low shifting rates by sequentially accessing several or all of the loops in the time interval *between* consecutive shift operations. In this mode of operation, the device enjoys the same advantage offered by the serial–parallel–serial (SPS) [3] type of CCD register—namely, high data rates with low CCD clock rates in the storage array. Of course, a single SPS structure of the same total capacity as the 2416 would not be able to offer comparably low access times unless the sensing circuitry could operate at greatly increased speeds.

The block diagram of Fig. 2 shows further details of the chip organization. Each recirculating loop has two refresh amplifiers, one on each side of the array. The refresh amplifiers on one side of the array interface with peripheral circuitry for reading and writing.

Briefly, the device operates as follows: Shifting and refreshing operations are controlled entirely by the four-phase CCD clocks ($\phi_1 \cdots \phi_4$). A simplified timing diagram for the four-

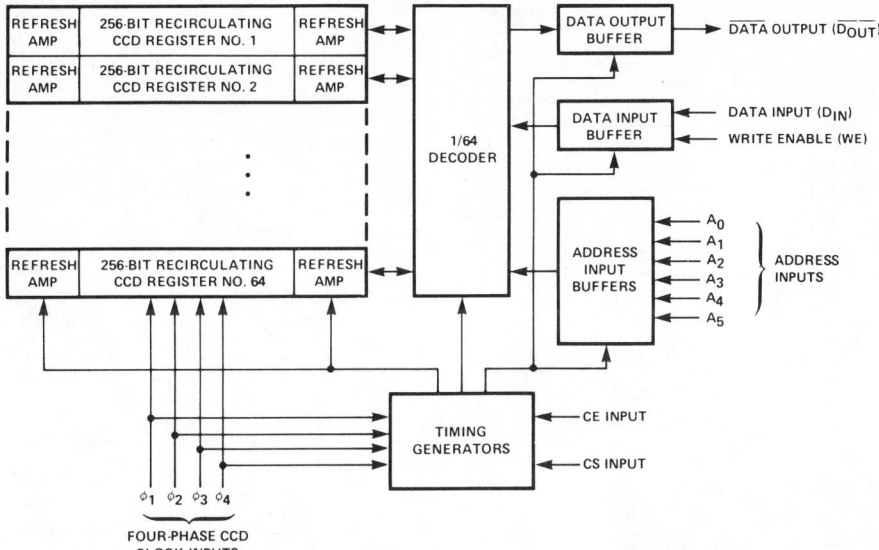

Fig. 2. Simplified block diagram of 16 384-bit CCD memory device.

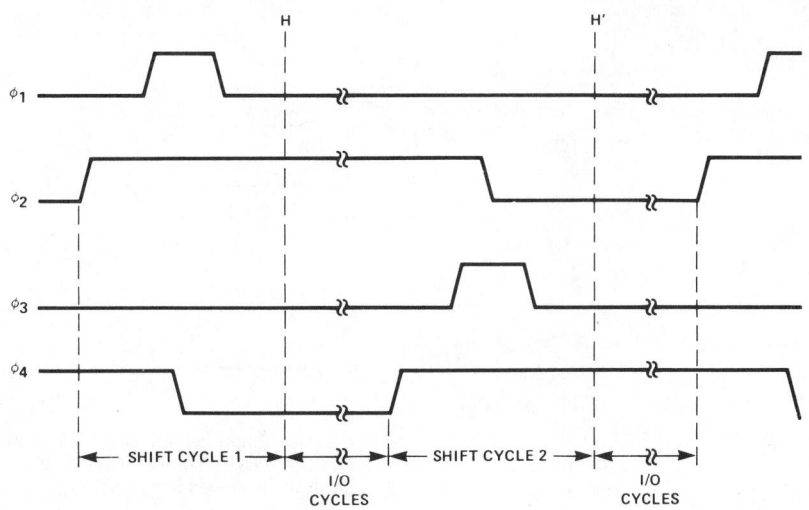

Fig. 3. Timing diagram for four-phase CCD clocks.

phase clocks is shown in Fig. 3. A low-to-high transition of ϕ_2 initiates a one-bit shift in each register loop. Shifting is completed after the ϕ_1 pulse occurs. Some additional time is then required for the refresh amplifiers to detect the data received and to write it into the next stage of the recirculating loop. These actions are complete at time H (Fig. 3), following which reading and writing operations may be performed at the option of the user. The user can (within limits of the refresh time requirement) do consecutive READ/WRITE cycles on one or more of the 64 address locations. These I/O cycles are controlled by the CE clock input; details of the timing will be discussed later. The second shift cycle is initiated by a low-to-high transition of ϕ_4, and I/O cycles may be executed again after time H'. During the second shift cycle, the contents of the 64 accessible refresh amplifiers (whether modified or not) are transferred forward into the respective registers, to be replaced by new data from the next bit location in each register. I/O functions are inhibited while a shift–refresh cycle is in progress. Note that shifting occurs at twice the frequency of

the four-phase clocks; this is the result of a multiplexing scheme to be described below. Consecutive shift cycles can also be executed without intervening I/O cycles; in this case the second shift cycle of Fig. 3 can be initiated at time H.

Only one input and one output line are provided on the 2416. This appears to be a desirable organization for systems using a relatively large number of devices, especially if an error correcting scheme is employed. Should the need arise, the organization could easily be changed to provide more inputs and outputs.

IV. DESIGN OF ARRAY AND REFRESH CIRCUITS

A. Layout Considerations

Perhaps the most crucial task in the layout of the CCD array is fitting the refresh amplifiers onto the ends of the CCD channels, without sacrificing the packing density of the array. The task is rendered more difficult by the fact that the sensing device must be situated extremely close to the output of the

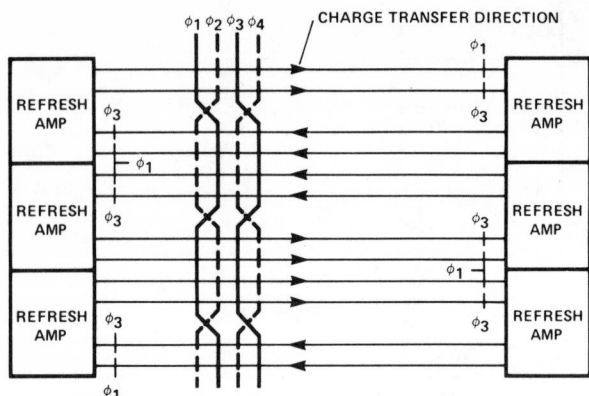

Fig. 4. Schematic illustration of array layout.

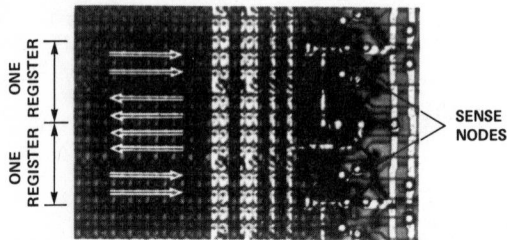

Fig. 5. Photomicrograph of a portion of CCD array (left side) and refresh amplifiers (right side). Arrows indicate direction of charge transfer in each channel.

CCD channel in order to minimize the sense node capacitance and to take full advantage of the available output signal.

In the 2416 a satisfactory layout was achieved by a multiplexing arrangement in which channels were paired to share a common refresh amplifier. Clocks ϕ_1 and ϕ_3 are used to gate the charge outputs of the channels in each multiplexed pair, so that the refresh amplifier alternately receives data from one channel and then the other. An analogous scheme is used to multiplex the inputs to a pair of channels. Fig. 4 shows further details of the layout scheme. Each recirculating loop consists of two pairs of channels. The pairs transfer charge in opposite directions. An alternative arrangement is to have all channels transfer charge in the same direction, and to recirculate data by means of a metal line returning the length of the channels. In addition to yield losses which could be expected from the multitude of long metal lines, this approach places much greater demands on the refresh amplifiers because of the high capacitance load on each metal line, and the noise coupled to each line from the underlying clock electrodes. With the presently adopted approach, an apparent penalty exists in the requirement to interweave clock electrodes to reverse the direction of charge transfer. However, using the layout scheme of Fig. 4, the criss-crossing of electrodes takes place only once every four CCD channels, and the slight area penalty that results is compensated for by a simplification in the design and layout of the refresh amplifiers.

Fig. 5 is a photomicrograph showing the actual layout of a portion of the CCD array as well as the refresh amplifiers. The average cell size in the array itself is 1.09 mil²/bit (702 μm²/bit). The channel width in the array is nominally 7 μm. Measurements on similar structures indicated that the transfer inef-

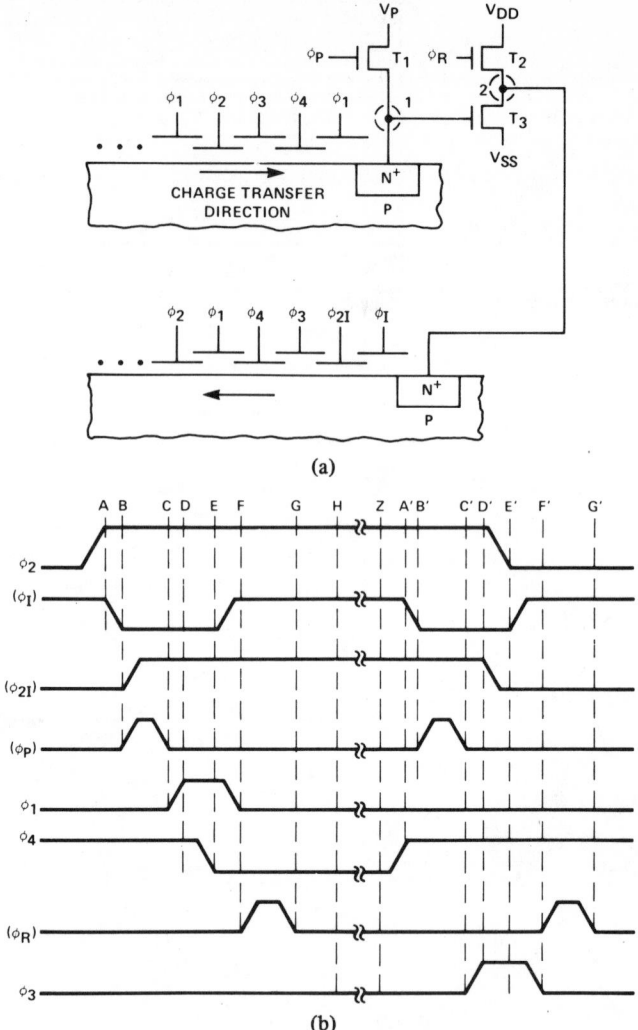

Fig. 6. (a) Simplified schematic of refresh amplifier and input–output stages of CCD channel. (b) Timing diagram. Symbols for internal signals are enclosed in parentheses.

ficiency was less than 10^{-3} per transfer with use of a "fat zero." Since the data in each channel are refreshed after every 128 charge transfers (i.e., 64 bits), the transfer losses between refresh operations is limited to about 10 percent.

B. Refresh Amplifier Design

A simplified schematic of the refresh amplifier and the associated timing diagram are shown in Fig. 6. The multiplexing arrangement has been omitted here since it does not contribute to the understanding of the operation of the refresh amplifier and would obscure the following discussion. Thus Fig. 6(a) is drawn as if only one CCD channel was connected to each refresh amplifier.

The refresh amplifier consists of a presetting transistor T_1 and a dynamic inverter stage comprising transistors T_2 and T_3. The sense node (node 1) is designed so as to have the minimum parasitic capacitance. (The total capacitance of the sense node is approximately 0.07 pF.) The output of the refresh amplifier (node 2) is used to regenerate the data under the input metering electrode ϕ_{2I}. The clock signals ϕ_I, ϕ_{2I}, ϕ_P, and ϕ_R are generated within the memory chip. All have high levels equal to the V_{DD} supply (nominally 12 V).

The detailed operation of the refresh circuitry may be understood with the aid of the timing diagram in Fig. 6(b). A shift-refresh cycle is initiated at time A following a low-to-high transition of ϕ_2. During the interval B–C node 1 is preset to the voltage V_P, which is typically 1.5 V above the threshold of T_3 and is regulated by another circuit to be described later. During the interval C–F, charge from the last stage (ϕ_4) of the top channel [Fig. 6(a)] is dumped onto node 1, causing its voltage to fall. If a "zero" (i.e., a relatively small charge, $\simeq 0.03$ pC) is received by node 1, device T_3 remains conducting at time F. On the other hand, a "one" (i.e., a relatively large charge, $\simeq 0.2$ pC) dumped onto node 1 would cause the voltage on that node to fall below the threshold of T_3 and turn this device "off." During the time F–G, device T_2 is momentarily turned "on" by the ϕ_R pulse, and the voltage on node 2 is raised. Following time G, the voltage on node 2 will either be reset to 0 V (V_{SS} potential) if T_3 is conducting, or remain high (at one threshold voltage below the V_{DD}) if T_3 is "off." The detection of the data is thus completed a short time later, at time H.

The regeneration of the data takes place under electrode ϕ_{2I} concurrently with the detection of the data. ϕ_{2I} is clocked high at time B, but does not receive any charge until ϕ_I goes high after time E. Between times E–F, node 2 is still at approximately 0 V, and thus the potential well under ϕ_{2I} is filled with charge to its maximum operating capacity. If node 2 is maintained at 0 V after time G, the potential well under ϕ_{2I} remains filled with charge, constituting a "one." However, if node 2 becomes high at time G, most of the charge under ϕ_{2I} is pulled back out. A small amount of charge, corresponding to the difference in minimum surface potentials under ϕ_I and ϕ_{2I}, remains and constitutes a "fat zero." The "fat zero" level is controlled by differences in oxide thickness under the ϕ_I and ϕ_{2I} electrodes, and is approximately 15 percent of the full "one" level. Note that each refresh amplifier inverts the data; because there are two refresh amplifiers in each recirculating loop, data consistency is maintained.

The regenerated charge under ϕ_{2I} is not transferred to the next storage site until after the subsequent refresh cycle is initiated at time A'. Immediately following time A', ϕ_I turns off so as to preserve the charge level under ϕ_{2I} while the refresh amplifier is reset. During the second refresh cycle A'–G', the data in the companion channel of each multiplexed pair [not shown in Fig. 6(a)] is detected and regenerated. The charge previously regenerated under ϕ_{2I} is transferred forward immediately following time D'. The timing of ϕ_I is arranged so that regeneration of data in the companion channel has no effect on the one under consideration.

C. Reference Voltage (V_P) Regulator

In order for the refresh amplifier to detect data reliably, the reference voltage V_P used to preset the sense node must be precisely controlled. Experiments on test structures indicated that V_P could not be safely varied by more than ±0.5 V from its optimum value. Furthermore, the optimum value was affected by process parameters as well as operating voltages. Be-

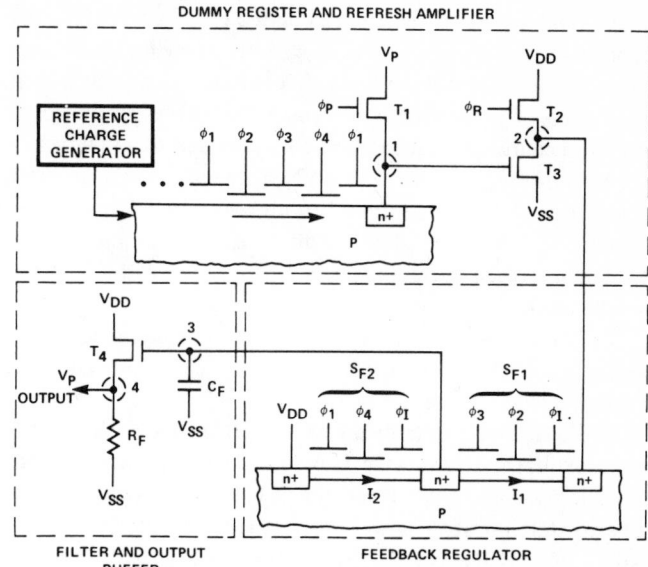

Fig. 7. Simplified schematic of V_p regulator.

cause of the stringent requirements on V_P, a feedback regulation scheme was used to generate it.

Fig. 7 gives a schematic diagram for the V_P regulator. Again, this schematic is highly simplified and includes only those basic elements required to understand the principle of operation. The circuit can be divided into three functional blocks as shown. The geometrical layout of the output section of the "dummy register and refresh amplifier" is identical to that used for the other CCD register loops in the array. The output (V_P) of the circuit at node 4 is fed to all the refresh amplifiers in the array, including the one attached to the end of the dummy register. The timing of signals is identical to that indicated in Fig. 6(b).

The principle of operation can be explained quite simply. The reference charge generator feeds the dummy register a charge level roughly midway between that of a "one" and a "fat zero." The circuit then adjusts V_P such that after node 1 (the sense node) receives this charge, the voltage on node 1 just equals the threshold voltage of T_3. When this condition is achieved, a "one" or "zero" charge level received on the sense node of an array register would cause the corresponding transistor T_3 to be "off" or "on," respectively.

To understand in greater detail how the circuit works, assume that V_P is initially too low. Device T_3 is turned off by the reference charge and node 2 is pulled to a high voltage level by ϕ_R. This signal on node 2 is used by the "feedback regulator" to change the voltage on node 3 and hence the output V_P level. The feedback regulator consists of two short CCD sections labeled S_{F1} and S_{F2}. These CCD sections transfer electrons to and from node 3, respectively, so as to decrease or increase the voltage on node 3. The equivalent directions of current flow are indicated in Fig. 7 with the notations I_1 and I_2. In the case under discussion where node 2 is high after the ϕ_R pulse, no charge is transferred by section S_{F1}, so the voltage of node 3 (and hence also V_P), rises incrementally with each CCD clock cycle due to the action of S_{F2}. The capacitance of the filter capacitor C_F is large compared to those of the ϕ_2 and ϕ_4 electrodes in S_{F1} and S_{F2}. This ensures the stability of the feedback scheme by only allowing the voltage on node 3 to change slowly. As V_P continues to rise with each clock cycle, T_3 will eventually remain "on" after receiving the reference charge, and node 2 will be discharged to a low potential after the ϕ_R pulse. This will cause section S_{F1} to start transferring electrons to node 3 to prevent its voltage from rising further. A steady-state condition is thus obtained where the time-averaged currents I_1 and I_2 are equal, with V_P at the desired level.

The output bleeder resistor R_F is used to maintain a low impedance output which is not affected by capacitive coupling transients in the positive direction.

The V_P regulator described here will generate the correct reference voltage over a wide range of process parameters and operating voltages. However, the V_P level does require some time to stabilize after power supplies and CCD clocks are first turned on. Hence no I/O functions should be performed on the memory device until the CCD clocks have executed at least several hundred cycles with power supplies at operating voltages. After this start-up period, the clocking requirements for refresh functions will keep the V_P level stable.

V. I/O AND PERIPHERAL CIRCUITRY

A. Timing for I/O Cycles

The timing during READ or WRITE operations is controlled by the user's CE and WE inputs. The user initiates a READ cycle by setting up the desired address inputs, raising CS high, and then pulsing CE from low to high. (I/O functions are inhibited if the CS input is low.) This results in the following actions within the chip (refer to Fig. 2). An internally generated CE pulse activates the address buffers, whose outputs act on the decoders so as to select one of the registers. After selection has taken place, the input stages of the address buffers are disabled so that transitions in address inputs are disregarded for the remainder of the cycle. The output of the selected register is gated onto the READ bus and activates the data output buffer. If a WRITE-ONLY operation is desired, a positive pulse is applied to the WE input shortly after CE becomes high. The user's data input is gated into the data input buffer during the time WE is high. By delaying the WE pulse until after the data output becomes valid, a READ-MODIFY-WRITE cycle can be performed. Both reading and writing are then accomplished during the same CE cycle. Further details on the internal timing during a WRITE operation will be given below.

The I/O cycle is terminated when the user returns CE to a low level. The internal $\overline{CE}$ clock resets the decoders, address buffers, data input and output buffers, and I/O timing circuits.

B. I/O Buffers

These include the address buffers and the data input and output buffers. As with all other circuits on the chip, dynamic circuit operation was used as much as possible in order to minimize power dissipation.

Table I defines the nominal input voltages required by the 2416. Except for the data input, all other inputs require nominal high levels of V_{DD}. Most of these inputs could have been designed to accept TTL input levels, with some sacrifice in speed and power dissipation.

An open-drain data output is employed. Outputs of several memory chips can be OR-tied and made to interface directly with TTL logic by using an external pull-up resistor.

C. Interface between CCD Array and I/O Circuits

Fig. 8(a) is a simplified schematic diagram showing how the CCD refresh amplifier on one side of the array interfaces with I/O circuitry. The refresh amplifier circuitry is identified and has the same configuration as that shown in Fig. 6(a), except that the source terminal of T_3 is not directly connected to V_{SS}.

When the register is not selected for reading or writing operations, the output S of the decoder [see Fig. 8(a)] is low. Transistors T_5, T_8, and T_9 are therefore turned "off," and the refresh amplifier is isolated from the peripheral I/O circuitry. Transistor T_4 is kept "on" by ϕ_P acting on T_7, and effectively connects the source terminal of T_3 to V_{SS}. In this mode, the refresh amplifier recirculates data exactly as described previously in Section IV-B.

To explain how reading and writing operations are performed, Fig. 8(b) gives the simplified timing for a READ-MODIFY-WRITE type of I/O cycle. Note that the timing diagram in Fig. 6(b) is still valid here during shift–refresh operations. The execution of I/O cycles is restricted to the time interval H–Z [Fig. 6(b)] between shift cycles. The notation for timing locations in Fig. 8(b) is consistent with that in Fig. 6(b).

Referring to Fig. 8(b), an I/O cycle commences at time H with a positive transition of CE. The internal voltages at this time have been preset such that the READ bus R is high, the data output is an open drain, and other lines are held low. We have assumed that the data output is connected to a pull-up resistor R_L so that the output is at the terminating voltage V_{CC}. At time H, the refreshed data from the CCD register are already valid at node 2, so that transistor T_6 is either "on" or "off." At time I the output of the selected decoder has made a positive transition (while the outputs of all unselected decoders remain low). This turns on T_5 and connects T_6 to the READ bus R which now is floating at a high voltage level. Depending on the output data of the register as reflected in the state of T_6, the READ bus R will either remain high or be discharged low. The signal on R is amplified by the data output buffer to give a valid $\overline{D_{out}}$ output at time J.

The internal voltages then remain steady until the WRITE operation is started at time K by a WE pulse input from the user. The ϕ_W generator responds with a positive transition on its output. For the selected register, the output of the AND gate $G1$ then becomes high, turning on T_8 and discharging node 4 to V_{SS}. This disables the refresh amplifier by disconnecting T_3, so that node 2 is free to respond to the data input which now appears on the WRITE bus W and is gated into node 2 of the selected register via T_9. Since T_5 is still "on," the written data may cause the READ bus R to discharge if it remained high during the READ operation; if so, this will result in a transition in $\overline{D_{out}}$. Hence $\overline{D_{out}}$ should not be considered valid after a WRITE command has been given.

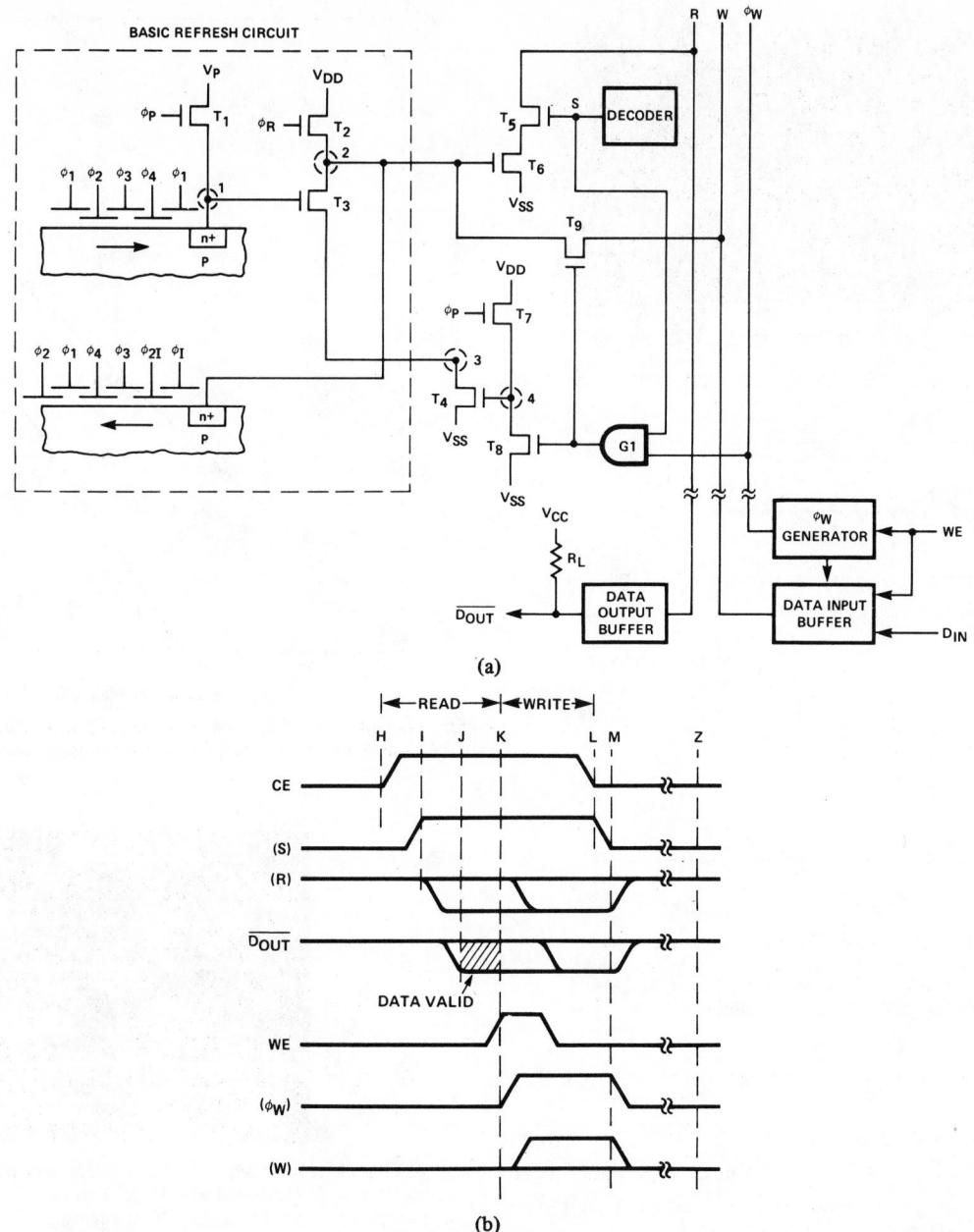

Fig. 8. (a) Simplified schematic of interface between refresh amplifier and I/O circuitry. (b) Timing diagram. Symbols for internal signals are enclosed in parentheses.

After the voltage on the WRITE bus has stabilized, the I/O cycle can be terminated by bringing CE low at time L. The output of the selected decoder is reset low first (at time M), so as to isolate node 2 from the WRITE bus W before the latter is also reset low. Note that once the refresh amplifier at a selected address location has been disabled, it remains so until the subsequent shift cycle, at which time the ϕ_P pulse turns T_4 "on" again.

From time M until the next shift cycle, additional I/O cycles can be executed. CE should be reset low for about 100 ns between consecutive cycles.

D. CCD Timing Generators

These circuits generate the internal clocks (ϕ_P, ϕ_R, ϕ_I, ϕ_{2I}, etc.) needed for controlling the refresh and recirculating oper-

ations. In order to minimize the dependence of internal voltages on process parameters, the high-level outputs of these generators are all held at the V_{DD} supply voltage by means of bootstrapped output stages. The operation of the CCD timing generators is controlled only by the four-phase CCD clock inputs.

VI. MEMORY CHIP PERFORMANCE

Fig. 9 is a photomicrograph of the 2416 CCD memory chip. The chip dimensions are 143 mils × 237 mils (=3.63 mm × 6.02 mm).

The input capacitances on the CCD clock phases are typically 350 pF each on ϕ_1 and ϕ_3, and 500 pF each on ϕ_2 and ϕ_4. In order to ease driver requirements, the chip was designed to accept transition times as long as 200 ns on the four-phase CCD

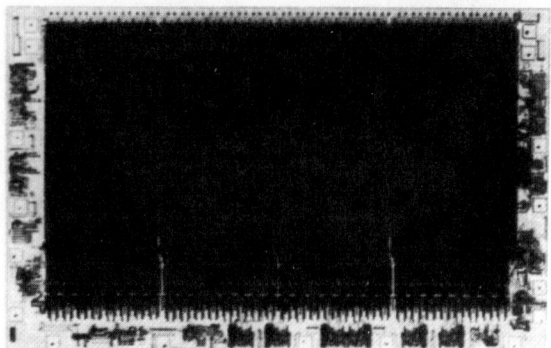

Fig. 9. Photomicrograph of 16 384-bit CCD memory chip (the 2416).

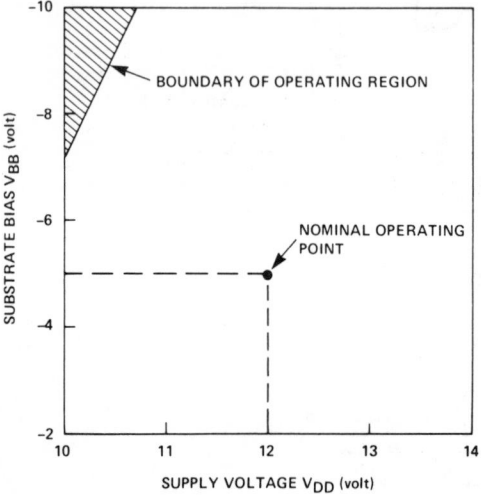

Fig. 10. Operating range of substrate bias V_{BB} versus supply voltage V_{DD} for a typical 2416 device at 80°C. Shift frequency was 100 kHz. All clocks were at nominal levels as defined in Table I. Cross-hatched area above is outside region of functionality.

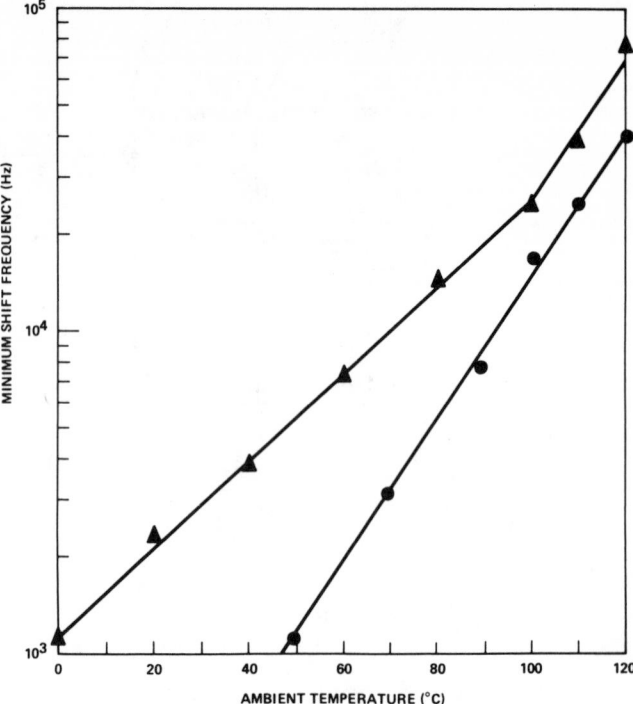

Fig. 11. Minimum shift frequency versus temperature for two different 2416 devices. Power supplies and clock levels were at nominal values (see Table I).

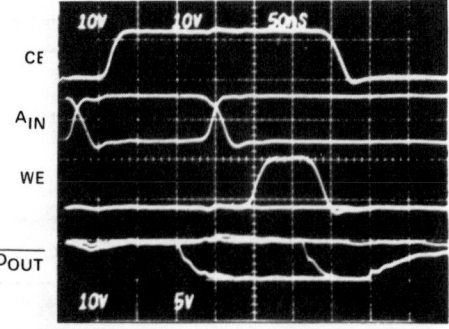

Fig. 12. Oscilloscope display of CE, address, and WE inputs, and $\overline{D_{out}}$ output for a 2416 device at 25°C. The $\overline{D_{out}}$ output pin was connected to +5 V dc through a 500 Ω resistor. The capacitive load on $\overline{D_{out}}$ was approximately 100 pF. All power supplies and clock levels were at nominal voltages. Vertical scale = 10 V/div for CE, address, and WE inputs; 5 V/div for $\overline{D_{out}}$. Horizontal scale = 50 ns/div.

clocks. Typically, transition times of about 50 ns were used in our tests.

Electrical evaluation of the memory device indicated that it met the design objectives listed in Table I. The data pattern used to exercise the device was chosen so that each CCD channel would store isolated "ones" embedded in a long string of "zeros," as well as the complement of this pattern. The device was operated successfully with multiple (up to 64) READ, WRITE, or READ-MODIFY-WRITE cycles to different addresses between consecutive shift cycles.

The device was able to operate over a reasonably wide range of supply voltages. Fig. 10 shows the typical operating range for the substrate bias V_{BB} as the supply voltage V_{DD} is varied ±2 V from its nominal value of 12 V.

The device performs comfortably at a shift frequency of 1.3 MHz, which corresponds to an average access (or latency) time of under 100 μs. The minimum shift frequency is limited by thermal carrier generation, which is strongly dependent on temperature and varies from device to device. Fig. 11 shows this temperature dependence for two devices.

Fig. 12 is a photograph of an oscilloscope display of the CE, address, and WE inputs, and the output signal $\overline{D_{out}}$, during READ-MODIFY-WRITE cycles. The signals from many different cycles are superimposed. Note that the address inputs are held stable only for about 120 ns following the leading edge of CE; address inputs are ignored by the chip for the remainder of the cycle. Between I/O cycles when CE is low, the $\overline{D_{out}}$ level rises to the terminating voltage on the external pull-up resistor (5 V in the case of Fig. 12). In Fig. 12, the output data are valid within 150 ns after the leading edge of CE. The previously mentioned high-to-low transition of $\overline{D_{out}}$ in response to the WE pulse can also be seen. The device has been operated with data rates as high as 4 Mbits/s with READ-ONLY or WRITE-ONLY cycles.

VII. Conclusion

The present work has demonstrated that a useful 16 384-bit CCD serial memory chip can be produced using existing manufacturing techniques. By designing conservatively and taking full advantage of the capabilities of the chosen charge-coupled structure, a low cost has been achieved.

ACKNOWLEDGMENT

The author wishes to acknowledge the contributions of M. Geilhufe in defining the chip organization. Thanks are due to P. Keshtbod, I. Lee, R. Simko, and C. Steele for their direct involvement in the design, fabrication, and electrical evaluation of the device.

REFERENCES

[1] W. F. Kosonocky and J. E. Carnes, "Charge-coupled digital circuits," *IEEE J. Solid-State Circuits (Special Issue on Semiconductor Memories and Digital Circuits)*, vol. SC-6, pp. 314–322, Oct. 1971.

[2] B. Augusta and T. V. Harroun, "Conceptual design of an eight megabyte high performance charge-coupled storage device," in *1972 Fall Joint Comput. Conf., AFIPS Conf. Proc.*, vol. 41, pt. 2. Montvale, NJ: AFIPS Press, 1972, pp. 1261–1268.

[3] D. R. Collins, J. B. Barton, D. C. Buss, A. R. Kmetz, and J. E. Schroeder, "CCD memory options," in *1973 IEEE Int. Solid-State Circuits Conf., Dig. Tech. Papers*, Feb. 1973, pp. 136–137, 210.

[4] N. A. Patrin, "Performance of very high density charge-coupled devices," *IBM J. Res. Develop.*, vol. 17, pp. 241–248, May 1973.

[5] A. Ibrahim and L. Sellars, "4096-bit charge coupled device serial array," presented at the 1973 IEEE Int. Electron Devices Conf., Paper 7.6.

[6] R. H. Krambeck, T. F. Retajczyk, Jr., D. J. Silversmith, and R. J. Strain, "A 4160 bit C4D serial memory," *IEEE J. Solid-State Circuits (Special Issue on Analog Circuits)*, vol. SC-9, pp. 436–443, Dec. 1974.

[7] C. N. Berglund and R. J. Strain, "Fabrication and performance considerations of charge-transfer dynamic shift registers," *Bell Syst. Tech. J.*, vol. 51, pp. 655–703, Mar. 1972.

[8] A. M. Mohsen, T. C. McGill, and C. A. Mead, "Charge transfer in overlapping gate charge-coupled devices," *IEEE J. Solid-State Circuits*, vol. SC-8, pp. 191–207, June 1973.

[9] R. H. Walden *et al.*, "The buried channel charge coupled device," *Bell Syst. Tech. J.*, vol. 51, pp. 1635–1640, Sept. 1972.

 **2416**

16,384 BIT CCD SERIAL MEMORY

- ▪ **Organization: 64 Recirculating Shift Registers of 256 Bits Each**

- ▪ **Avg. Latency Time Under 100 μs**
- ▪ **Max. Serial Data Transfer Rate —2 mega bits/sec.**
- ▪ **Address Registers Incorporated on Chip**
- ▪ **Standard Power Supplies— +12V, −5V**

- ▪ **Open Drain Output**
- ▪ **Combined Read/Write Cycles Allowed**
- ▪ **Compatible to Intel® 5244 CCD Driver**

The Intel® 2416 is a 16,384 bit CCD serial memory designed for low-cost memory applications requiring average latency times to under 100 μs. To achieve low latency time the memory was organized in the form of 64 independent recirculating shift registers of 256 bits each. Any one of the 64 shift registers can be accessed by applying an appropriate 6-bit address input.

The shift registers recirculate data automatically as long as the four-phase CCD clocks ($\phi_1 \ldots \phi_4$) are continuously applied and no write command is given. A one-bit shift is initiated in all 64 registers following a low-to-high transition of either ϕ_2 or ϕ_4. After the shift operation the contents of the 64 registers at the bit location involved are available for non-destructive reading, and/or for modification. I/O functions are accomplished in a manner similar to that of a 64-bit dynamic RAM. At the next shift cycle, the contents of the 64 accessible bits (whether modified or not) are transferred forward into the respective registers and the contents of the next bit of each register become accessible. No I/O function can be performed during the shift operation itself.

The Intel 2416 generates and uses an internal reference voltage which requires some time to stabilize after the power supplies and four phase clocks have been turned on. No I/O functions should be performed until the four-phase CCD clocks have executed at least 4000 shift cycles with power supplies at operating voltages. After this start-up period, no special action is needed to keep the internal reference voltage stable.

The 2416 is fabricated using Intel's advanced high voltage N-channel Silicon Gate MOS process.

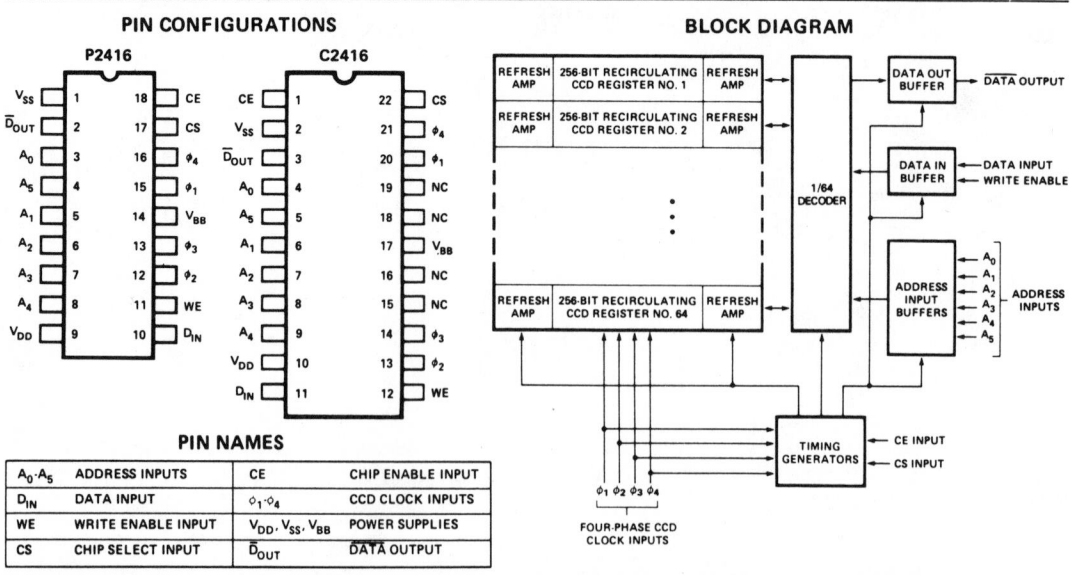

Reprinted with permission from *Intel Data Catalog*, pp. 4-15–5-2, 1976.

Absolute Maximum Ratings*

Temperature Under Bias	$-10°C$ to $80°C$
Storage Temperature	$-65°C$ to $+150°C$
All Input or Output Voltages with Respect to the most Negative Supply Voltage, V_{BB}	$+25V$ to $-0.3V$
Supply Voltages V_{DD} and V_{SS} with Respect to V_{BB}	$+20V$ to $-0.3V$
Power Dissipation	1.0W

*COMMENT:

Stresses above those listed under "Absolute Maximum Ratings" may cause permanent damage to the device. This is a stress rating only and functional operation of the device at these or any other conditions above those indicated in the operational sections of this specification is not implied. Exposure to absolute maximum rating conditions for extended periods may affect device reliability.

D.C. and Operating Characteristics

$T_A = 0°C$ to $70°C$, $V_{DD} = +12V \pm 5\%$, $V_{BB}[1] = -5V \pm 5\%$, $V_{SS} = 0V$, unless otherwise specified.

Symbol	Parameter	Min.	Typ.	Max.	Unit	Test Conditions
I_{LI}	Input Leakage Current		1	10	μA	$V_{IN} = 0V$
I_{LO}	Output Leakage Current		1	10	μA	$CE = 0V$, $V_{OUT} = 0V$
I_{OL}	Output Low Current	3			mA	$V_{OL} = .45V$
I_{OH}	Output High Current			10	μA	$V_{OH} = +5V$
I_{DDAV1}	Average V_{DD} Supply Current for Shift Cycles Only			Note 2	mA	
$I_{DDAV2}[3]$	Average V_{DD} Supply Current		15	25	mA	
I_{BB}	Average V_{BB} Supply Current		100	200	μA	
V_{IL}	Input Low Voltage, All Inputs Except $\phi_1 \ldots \phi_4$	-1.0		0.8	V	
V_{IH1}	Input High Voltage, All Inputs Except D_{IN} and $\phi_1 \ldots \phi_4$	$V_{DD}-1$		$V_{DD}+1$	V	
V_{IHD}	D_{IN} Input High Voltage	3.5		$V_{DD}+1$	V	
$V_{ILC}[4]$	$\phi_1 \ldots \phi_4$ Input Low Voltage dc	-2.0		0.6	V	
V_{ILCT}	$\phi_1 \ldots \phi_4$ Input Low Voltage w/Coupling	$-2.0[5]$		$1.2[6]$	V	
V_{IHC1}	ϕ_1 and ϕ_3 Input High Voltage dc	$V_{DD}-1$		$V_{DD}+2$	V	
V_{IHCT1}	ϕ_1 and ϕ_3 Input High Voltage w/Coupling	$V_{DD}-1.6[6]$		$V_{DD}+2[5]$	V	
V_{IHC2}	ϕ_2 and ϕ_4 Input High Voltage dc	$V_{DD}-0.6$		$V_{DD}+2$	V	
V_{IHCT2}	ϕ_2 and ϕ_4 Input High Voltage w/Coupling	$V_{DD}-1.2[6]$		$V_{DD}+2[5]$	V	
t_{PWT}	Cross Coupling Voltage Pulse Width			Note 7	ns	Pulse width measured at 0.8V and $V_{DD}-1.2V$ (ϕ_1 and ϕ_3) or $V_{DD}-0.8V$ (ϕ_2 and ϕ_4)

Notes: 1. The only requirement for the sequence of applying voltage to the device is that V_{DD} and V_{SS} should never be 0.3V more negative than V_{BB}.

2. For shift only mode $I_{DD} = 2.0mA + \dfrac{15mA}{t_{\phi}/2 \text{ (in } \mu s)}$

3. I_{DDAV2} is for combined shift and data I/O cycles.

4. The difference in the low level reference voltages between all four clock phases must not exceed 0.5 volts.

5. These voltage levels with coupling are within the specified dc range and are not, therefore, subject to t_{PWT} restrictions.

6. These voltage levels with coupling are outside specified dc ranges and must be restricted to t_{PWT} pulse widths.

7. The maximum clock cross coupled pulse width is the sum of the clock transition time (t_T) plus 20ns.

$\phi_1 \ldots \phi_4$ CROSS-COUPLING

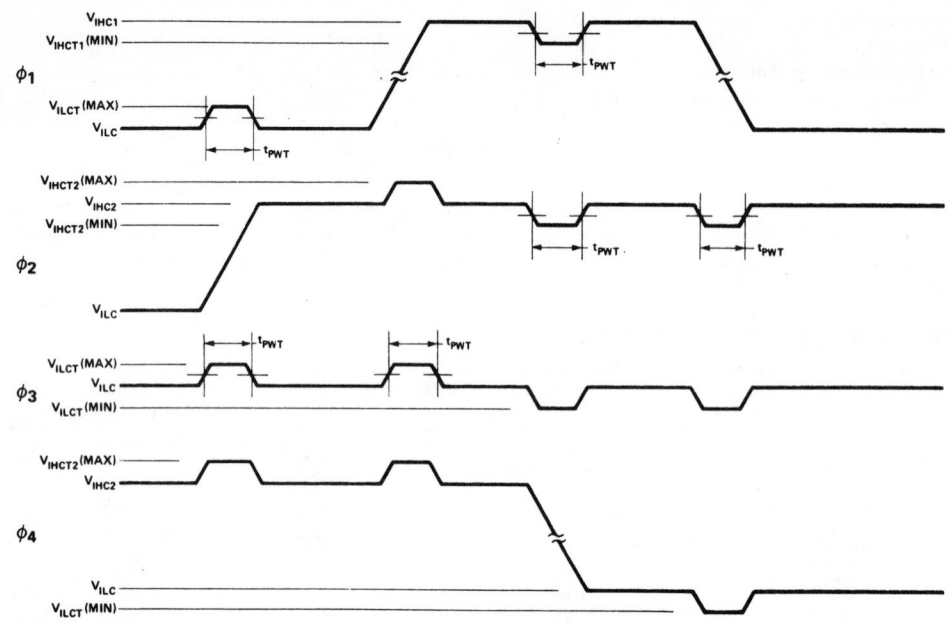

A.C. Characteristics $T_A = 0°C$ to $70°C$, $V_{DD} = 12V \pm 5\%$, $V_{BB} = -5V \pm 5\%$, $V_{SS} = 0V$, unless otherwise specified.

SHIFT ONLY CYCLES

Symbol	Parameter	Min.	Max.	Unit	Conditions
$t_{\phi/2}$	Half Clock Period for $\phi_1 \ldots \phi_4$	750[1]	10,000	ns	t_T = 40nsec
t_{PT}	ϕ_2 On to ϕ_1 On Time, ϕ_4 On to ϕ_3 On Time	200		ns	
t_{TD}	ϕ_1 to ϕ_4 Overlap, ϕ_3 to ϕ_2 Overlap	30		ns	
t_{DT}	ϕ_4 to ϕ_1 Hold Time, ϕ_2 to ϕ_3 Hold Time	40		ns	
t_{TP}	ϕ_1 Off to ϕ_4 On, ϕ_3 Off to ϕ_2 On	320		ns	
t_T	Transition Times for $\phi_1 \ldots \phi_4$	30	200	ns	

Note: 1. The 750ns Half Clock Period will be met for 30ns $\leq t_T \leq$ 40ns. Values of $t_T >$ 40ns lengthen $t_{\phi/2}$.

WAVEFORMS (Numbers in parentheses are for minimum cycle timing in ns)

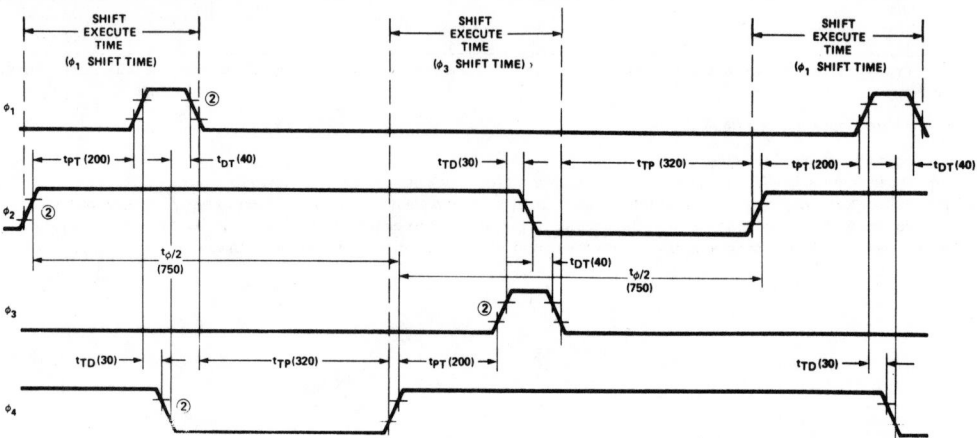

Note: 2. +2.0V and V_{DD}-2.0V are the reference low and high level respectively for measuring the timing of ϕ_1, ϕ_2, ϕ_3 and ϕ_4.

A.C. Characteristics

SHIFT—READ—READ—...—READ—SHIFT CYCLE

Symbol	Parameter	Min.	Max.	Unit	Conditions
t_{RCY}	READ Cycle Time	460		ns	
t_{PT}	ϕ_2 On to ϕ_1 On Time, ϕ_4 On to ϕ_3 On Time	200		ns	t_T = 40ns
t_{TD}	ϕ_1 to ϕ_4 Overlap, ϕ_3 to ϕ_2 Overlap	30		ns	t_{T1} = 20ns
t_{DT}	ϕ_4 to ϕ_1 Hold Time, ϕ_2 to ϕ_3 Hold Time	40		ns	
$t_{\phi/2}$	Half Clock Period for $\phi_1 \ldots \phi_4$		10,000	ns	
t_T	Transition Times for $\phi_1 \ldots \phi_4$	30	200	ns	
t_{T1}	Transition Times for Inputs Other Than $\phi_1 \ldots \phi_4$		100	ns	
t_{TC}	ϕ_1 or ϕ_3 Off to CE On	280		ns	
t_{SC}	CS to CE Set-Up Time	0		ns	
t_{AC}	Address to CD Set-Up Time	0		ns	
t_{AH}	Address Hold Time	240		ns	
t_{CS}	CE to CS Hold Time	0		ns	
t_{CC}	CE Off Time	140		ns	
t_{CP}	CE Off to ϕ_2 or ϕ_4 On	40		ns	
t_{CER}	CE On Time	280		ns	
t_{CF}	CE Off to Output High Impedance State	0		ns	
t_{CO}	CE to $\overline{D}_{OUT}$ Valid	250		ns	

WAVEFORMS[1] (Numbers in parentheses are for minimum cycle timing in ns)

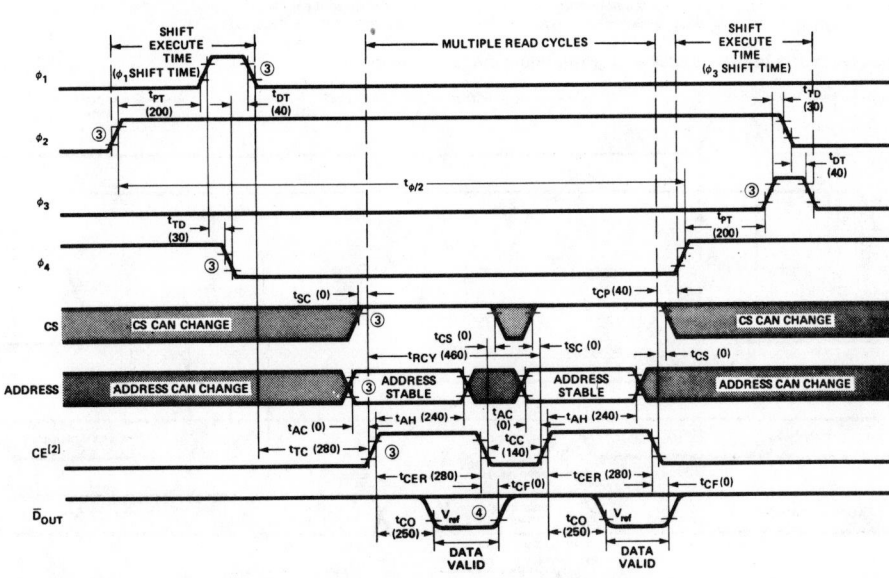

NOTES:
1. WE must be continuously low during the READ cycle.
2. When CE is off, the 2416 output level is determined by the external output termination.
3. +2.0V and V_{DD}-2.0V are the reference low and high level respectively for measuring the timing of $\phi_1 \ldots \phi_4$, CE, CS and addresses.
4. +0.8V is the reference level for measuring the timing of $\overline{D}_{OUT}$.

A.C. Characteristics
SHIFT—WRITE—WRITE—...—WRITE—SHIFT CYCLE

Symbol	Parameter	Min.	Max.	Unit	Conditions
t_{WCY}	WRITE Cycle Time	460		ns	
t_{PT}	ϕ_2 On to ϕ_1 On Time, ϕ_4 On to ϕ_3 On Time	200		ns	t_T = 40ns
t_{TD}	ϕ_1 to ϕ_4 Overlap, ϕ_3 to ϕ_2 Overlap	30		ns	t_{T1} = 20ns
t_{DT}	ϕ_4 to ϕ_1 Hold Time, ϕ_2 to ϕ_3 Hold Time	40		ns	
$t_{\phi/2}$	Half Clock Period for $\phi_1 \ldots \phi_4$		10,000	ns	
t_T	Transition Times for $\phi_1 \ldots \phi_4$	30	200	ns	
t_{T1}	Transition Times for Inputs Other Than $\phi_1 \ldots \phi_4$		100	ns	
t_{TC}	ϕ_1 or ϕ_3 Off to CE On	280		ns	
t_{SC}	CS to CE Set-Up Time	0		ns	
t_{AC}	Address to CE Set-Up Time	0		ns	
t_{AH}	Address Hold Time	240		ns	
t_{CS}	CE to CS Hold Time	0		ns	
t_{CC}	CE Off Time	140		ns	
t_{CP}	CE Off to ϕ_2 or ϕ_4 On	40		ns	
t_{CEW}	CE On Time	280[1]		ns	
t_{CW}	CE to WE Set-Up Time	100[1]		ns	
t_{DW}	D_{IN} to WE Set-Up	0		ns	
t_{WP}	WE Pulse Width	100[1]		ns	
t_{WC}	WE Off to CE Off	0[1]		ns	
t_{DH}	D_{IN} Hold Time	0		ns	

Note: 1. The minimum t_{CW}, t_{WP} and t_{WC} times with appropriate transitions do not necessarily add up to the minimum t_{CEW}. This allows the user flexibility in setting the WE Pulse Width edges without affecting either t_{CEW} or the WRITE Cycle Time, t_{WCY}.

WAVEFORMS (Numbers in parentheses are for minimum cycle timing in ns)

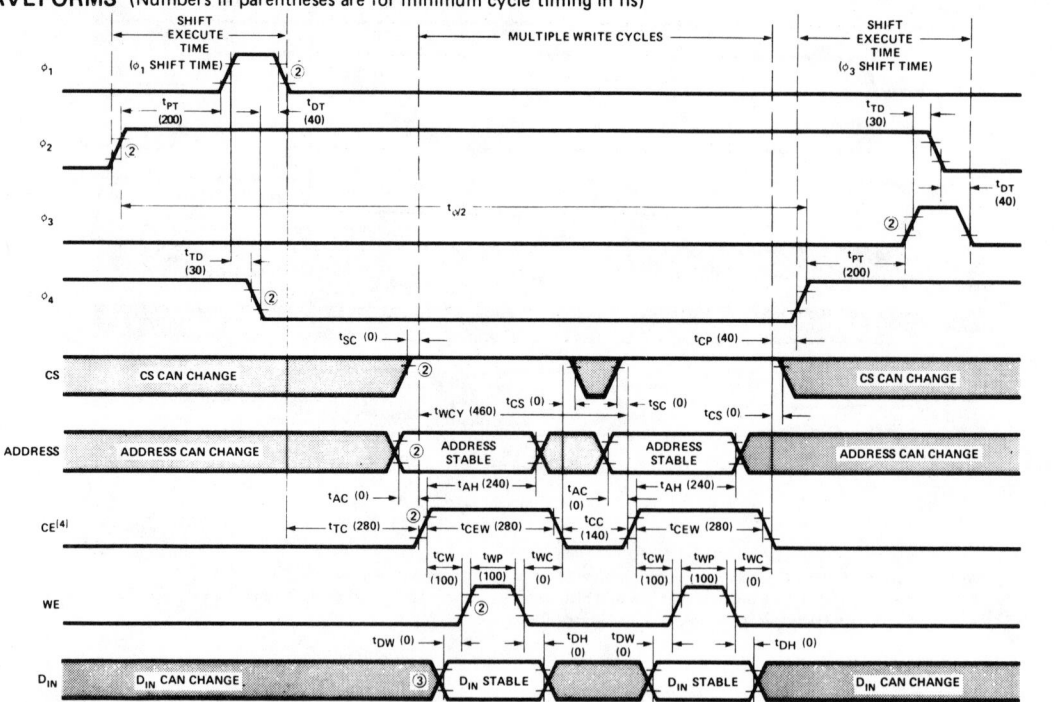

Notes: 2. +2.0V and V_{DD}–2.0V are the reference low and high level respectively for measuring the timing of $\phi_1 \ldots \phi_4$, CE, CS, WE, and addresses.
3. +1.5V and +3.0V are the reference low and high level respectively for measuring the timing of D_{IN}.

A.C. Characteristics SHIFT—RMW—RMW—...—RMW—SHIFT CYCLE

Symbol	Parameter	Min.	Max.	Unit	Conditions
t_{RWC}	READ-MODIFY-WRITE Cycle Time	620		ns	
t_{PT}	ϕ_2 On to ϕ_1 On Time, ϕ_4 On to ϕ_3 On Time	200		ns	$t_T = 40ns$
t_{TD}	ϕ_1 to ϕ_4 Overlap, ϕ_3 to ϕ_2 Overlap	30		ns	$t_{T1} = 20ns$
t_{DT}	ϕ_4 to ϕ_1 Hold Time, ϕ_2 to ϕ_3 Hold Time	40		ns	
$t_{\phi/2}$	Half Clock Period for $\phi_1 \ldots \phi_4$		10,000	ns	
t_T	Transition Times for $\phi_1 \ldots \phi_4$	30	200	ns	
t_{T1}	Transition Times for Inputs Other Than $\phi_1 \ldots \phi_4$		100	ns	
t_{TC}	ϕ_1 or ϕ_3 Off to CE On	280		ns	
t_{SC}	CS to CE Set-Up Time	0		ns	
t_{AC}	Address to CE Set-Up Time	0		ns	
t_{AH}	Address Hold Time	240		ns	
t_{CS}	CE to CS Hold Time	0		ns	
t_{CC}	CE Off Time	140		ns	
t_{CP}	CE Off to ϕ_2 or ϕ_4 On	40		ns	
t_{CRW}	CE On Time	440[1]		ns	
t_{CO}	CE On to $\overline{D}_{OUT}$ Valid	250		ns	
t_{DW}	D_{IN} to WE Set-Up Time	0		ns	
t_{WP}	WE Pulse Width	100[1]		ns	
t_{WC}	WE Off to CE Off	0		ns	
t_{DH}	D_{IN} Hold Time	0		ns	
t_{WD}	CE On to WE On	300[1]		ns	
t_{WF}	WE to $\overline{D}_{OUT}$ Undefined	0		ns	

Note: 1. The minimum t_{WD} and t_{WP} times with appropriate transitions do not necessarily add up to the minimum t_{CRW}. This allows the user flexibility in setting the WE Pulse Width edges without affecting either t_{CRW} or the READ-MODIFY-WRITE Cycle Time, t_{RWC}.

WAVEFORMS (Numbers in parentheses are for minimum cycle timing in ns)

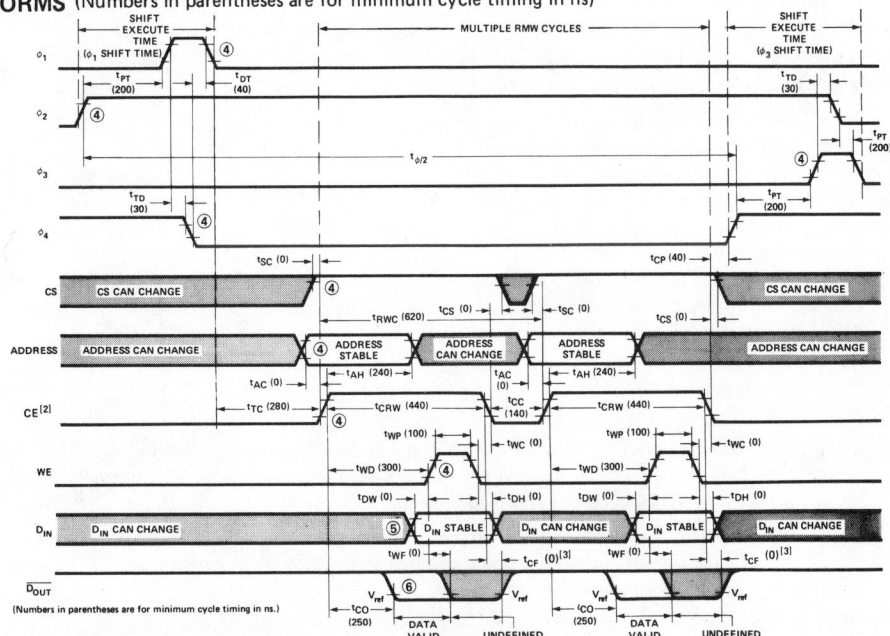

Notes: 2. When CE is off, the 2416 output level is determined by the external output termination.
3. The parameter t_{CF} is the same as in the Shift-Read-Shift Cycle on page 4.
4. +2.0V and V_{DD}–2.0V are the reference low and high level respectively for measuring the timing of $\phi_1 \ldots \phi_4$, CE, CS, WE, and addresses.
5. +1.5V and +3.0V are the reference low and high level respectively for measuring the timing of D_{IN}.
6. +0.8V is the reference level for measuring the timing of $\overline{D}_{OUT}$.

A.C. Characteristics

CAPACITANCE [1] $T_A = 25°C$

Symbol	Parameter	Typ.	Max.	Unit	Conditions
C_{IN}	Address, D_{IN}, CS, CE, WE Capacitance	4	6	pF	$V_{IN} = V_{SS}$
C_{OUT}	$\overline{D}_{OUT}$ Capacitance	3	5	pF	$V_{OUT} = V_{SS}$
$C_{\phi 1}$[1], $C_{\phi 3}$[2]	ϕ_1, ϕ_3 Input Capacitance	350	500	pF	$V_\phi = V_{SS}$
$C_{\phi 2}$[1], $C_{\phi 4}$[2]	ϕ_2, ϕ_4 Input Capacitance	480	700	pF	$V_\phi = V_{SS}$
$C_{\phi 1 - \phi 2}$	Clock ϕ_1 To Clock ϕ_2 Capacitance	120	175	pF	$V_\phi = V_{SS}$
$C_{\phi 1 - \phi 4}$	Clock ϕ_1 To Clock ϕ_4 Capacitance	150	200	pF	$V_\phi = V_{SS}$
$C_{\phi 3 - \phi 2}$	Clock ϕ_3 To Clock ϕ_2 Capacitance	150	200	pF	$V_\phi = V_{SS}$
$C_{\phi 3 - \phi 4}$	Clock ϕ_3 To Clock ϕ_4 Capacitance	120	175	pF	$V_\phi = V_{SS}$

Notes: 1. This parameter is periodically sampled and is not 100% tested.
2. The $C_{\phi 1}$ .... $C_{\phi 4}$ input clock capacitance includes the clock to clock capacitance. The equivalent input capacitance is given below.

Four-Phase Clock Inputs

The four-phase clock inputs are internally connected to long electrodes used for several thin-oxide gates, resulting in high capacitance to the substrate on the clock inputs. In addition, considerable cross-coupling between adjacent clock exists due to the overlapping structure of the electrodes. The figure to the right shows the circuit equivalent of the clock inputs, indicating maximum capacitance values.

The equivalent circuit suggests two opposed clock driver requirements:

1. Ability to drive high-capacitance loads quickly.
2. Ability to suppress cross-coupled current transients.

The first requirement could ordinarily be met rather easily, if it weren't for the fact that the cross-coupled current, I, is proportional to the rate of change of the voltage, i.e., $I = C\dfrac{dv}{dt}$.

For the quiescent driver to hold the coupled voltage to a minimum, the driver must have very low output impedance. However, when this driver becomes active the low output impedance increases the slope of the transitions which in turn increases coupling currents to the other drivers. This suggests that a driver have a controlled output transition time and a low output impedance characteristic in the quiescent state (high or low level). The Intel® 5244 meets these requirements.

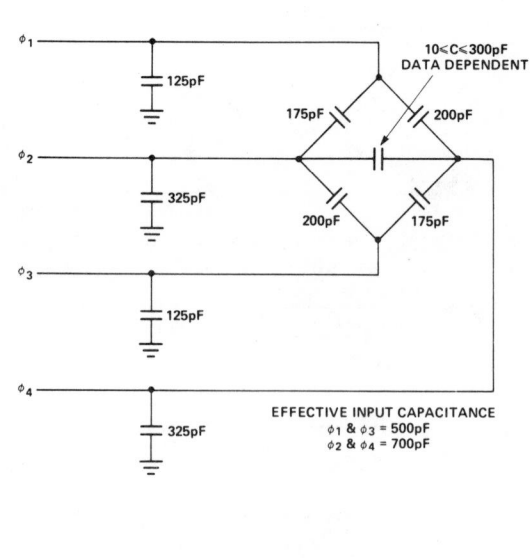

EFFECTIVE INPUT CAPACITANCE
ϕ_1 & ϕ_3 = 500pF
ϕ_2 & ϕ_4 = 700pF

5244 — CCD Clock Driver

The Intel® 5244 is a CMOS implemented fully TTL input compatible high voltage MOS driver, designed especially for the four phase clock inputs of the 2416. The device features very low DC power dissipation from a single +12V supply with output characteristics directly compatible with the 2416 clock input requirements.

The 5244 uses internal circuitry to control the cross-coupled voltage transients between the clock phases generated by the 2416. This internal circuitry limits the transition time to a specified range so that excessively fast transitions (<30ns) do not occur on the clock line. The entire operation is transparent to the user.

The 5244 is designed to drive four 2416s, but can drive fewer devices when loaded with additional capacitance to prevent a speedup in the transition times. Additional information on this and other aspects of the 5244 can be found on the 5244 data sheet.

Application Information

The Intel® 2416 is a charge coupled device (CCD) containing 16,384 bits of dynamic shift register storage available in a standard 18 pin plastic package. To minimize latency time (access time to any given bit in the device), the 2416 has been organized as 64 registers containing 256 bits each and, therefore, any bit can be accessed with a maximum of 255 shift operations. Since the minimum shift cycle requires 750 ns, the maximum latency time for the 2416 is less th _n 200μsec.

Access to the 64 recirculating registers is performed in a random access mode. A six bit address selects one of the 64 registers for read, write, or read/modify/write operations. These random access operations are performed between shift operations, and can be performed in any number or sequence as long as the basic shift frequency is maintained.

Because of substrate leakage currents the charge coupled storage mechanism is dynamic in nature. To satisfy the refresh requirements of the 2416, one shift operation must be performed every ten microseconds. A shift operation is completed on the falling edge of clock phase ϕ_1 or ϕ_3 and random access cycles may occur only between (1) the falling edge of ϕ_1 and the rising edge of ϕ_4 or (2) the falling edge of ϕ_3 and the rising edge of ϕ_2. This refresh requirement limits the number of random access cycles between successive shift operations to a maximum of 16.

Random access operations are performed in a manner which is very similar to any random access memory (RAM). All random access cycles are initiated with the rising edge and terminated with the falling edge of CE (Chip Enable). Read operations are performed when WE (Write Enable) remains low throughout a CE cycle. Data is strobed into the memory whenever WE is strobed high during a CE cycle as illustrated in the appropriate timing diagrams. CS (Chip Select) controls only the input and output circuits and is only effective when CE is high.

Typical Current Transients vs. Time

The oscilloscope photos in Figures 1 and 2 show typical I_{DD} current transients during shift and I/O cycles. The typical I_{BB} current during a shift cycle is shown in Figure 3.

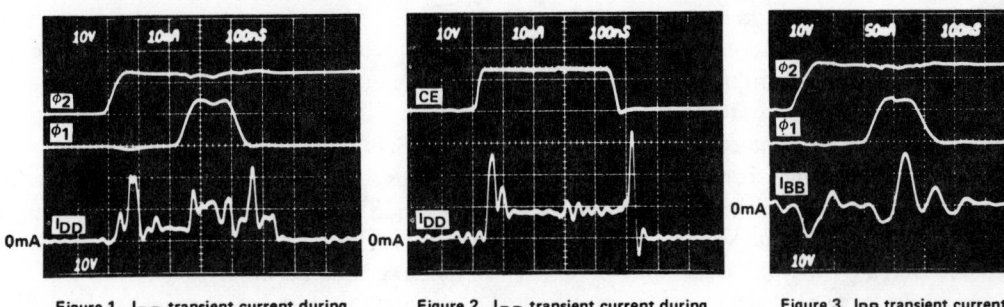

Figure 1. I_{DD} transient current during shift cycles.
I_{DD} scale: 10mA/div.

Figure 2. I_{DD} transient current during I/O cycles.
I_{DD} scale: 10mA/div.

Figure 3. I_{BB} transient current during a shift cycle.
I_{BB} scale: 50mA/div.

A CCD Line Addressable Random-Access Memory (LARAM)

KAMAL C. GUNSAGAR, MEMBER, IEEE, MARK R. GUIDRY, MEMBER, IEEE, AND GILBERT F. AMELIO, MEMBER, IEEE

Abstract—A novel approach to charge-coupled device (CCD) memory organization has been conceived and implemented in a 16 384-bit memory chip. It utilizes an isoplanar n-channel silicon gate MOS process in conjunction with self-aligned implanted barrier, buried channel CCD technology. The chip is organized in four parallel, identical sections of 32 independent lines with each line 128 bits long. The four sections are controlled in parallel. Any of the 32 lines (the same line in each of the four sections) can be randomly accessed; hence the name, line addressable random-access memory (LARAM). Each line can be brought to a halt at any of its 128 possible positions. Design features and test results of the memory are described.

I. INTRODUCTION

SINCE the emergence of the charge-coupled device (CCD) concept, CCD's have been the prime candidate for filling the gap between low-speed, low-cost disk memory and high-speed, relatively high-cost random-access memory (RAM) [1]. The objective of this paper is to describe the organization, design features, and performance results of a practical 16 384-bit CCD memory aimed at demonstrating CCD's commercial viability for filling this memory product gap.

In order to be truly cost-performance competitive, CCD memories should possess minimum chip area, access time, and clock-drive requirements. Currently available CCD memory devices [2], [3] have relatively long average access times to a desired bit and large capacitive loads on the clocks. The capacitive load limits the frequency of operation and makes the overhead drive requirements prohibitive in large systems. The line addressable random-access memory (LARAM) represents a new concept in memory that permits CCD's to overcome these limitations. Although access capability to a random bit is inherently precluded by the serial nature of the CCD storage medium, the LARAM organization does provide a pseudorandom access that combines short access time and high-serial data rates with low-power consumption and low-drive requirements.

The LARAM concept is presented in Section II. The overall chip organization of the 16 384-bit chip and its operation are discussed in Section III. Some of the circuit design features and performance results are outlined in Section IV.

II. THE LARAM CONCEPT

As the name implies, in the LARAM concept the memory is organized like a RAM except that the data address accesses

Manuscript received May 2, 1975; revised June 9, 1975. This work was supported by the U.S. Air Force Avionics Laboratory, Wright-Patterson AFB, Dayton, Ohio, under Contract F33615-73-C-1044.

The authors are with the Fairchild CCD Research and Development Laboratory, Palo Alto, Calif. 94304.

not one bit, but rather one CCD register (or line), as shown in Fig. 1. An m-bit (input) address properly decoded can address any one of the n-lines where $n = 2^m$. Selection of an address causes the clock waveform to be imposed on the selected line while all other lines are held stationary. Information data are "written in," "read out," or "refreshed" only in the selected line.

The important features of this organization are as follows.

1) Only one clock is necessary for the memory operation. This is the clock that has its waveform imposed on the addressed line. No additional clock is required to transfer charge in the CCD register since a single-phase clocking system is employed. Unidirectional charge transfer is obtained by employing an implanted barrier structure where one of the phases is operated by a dc voltage. The other phase swings both above and below the dc phase. This is elaborated upon in Section IV.

2) The capacitive drive requirements on the clocks are minimal since the external clock waveform that is imposed on the addressed lines is buffered by the line drivers.

3) The n-lines share a common input bus and a common output bus. This simplifies support circuitry on the chip. For example, a single charge sense amplifier serves the sense/refresh function for an entire 4K memory block in the 16K memory which is described in Section III.

4) Comparatively fast access times are available. The worst case access time is determined by the clock rate and the number of bits in each line. Thus, for a 128-bit line clocked at a 5 MHz rate, the longest latency time is 25.6 μs; the average latency time is half of this. The first bit access time for a newly selected line can be as short as 80 ns.

5) The n-lines are independent and can be brought to a halt at any bit position. This permits the address to be changed from one line to another without restoring the first line to its initial position.

6) Chip power dissipation is low because power is required to clock only one data line per section and also because a minimum of sense/refresh circuitry is needed.

III. 16 384-BIT LARAM CHIP ORGANIZATION AND DESCRIPTION

The LARAM concept has been implemented by a 16 384-bit chip using an isoplanar n-channel silicon-gate MOS process in conjunction with a self-aligned gapless buried channel CCD technology. The chip size is 220 mils × 200 mils. It is configured into four sections of 32 lines; each line is 128 bits long as shown by the block diagram in Fig. 2. Thus, each section is a 4096-bit memory block with its own 1 of 32 de-

Reprinted from *IEEE J. Solid-State Circuits*, vol. SC-10, pp. 268–272, Oct. 1975.

266

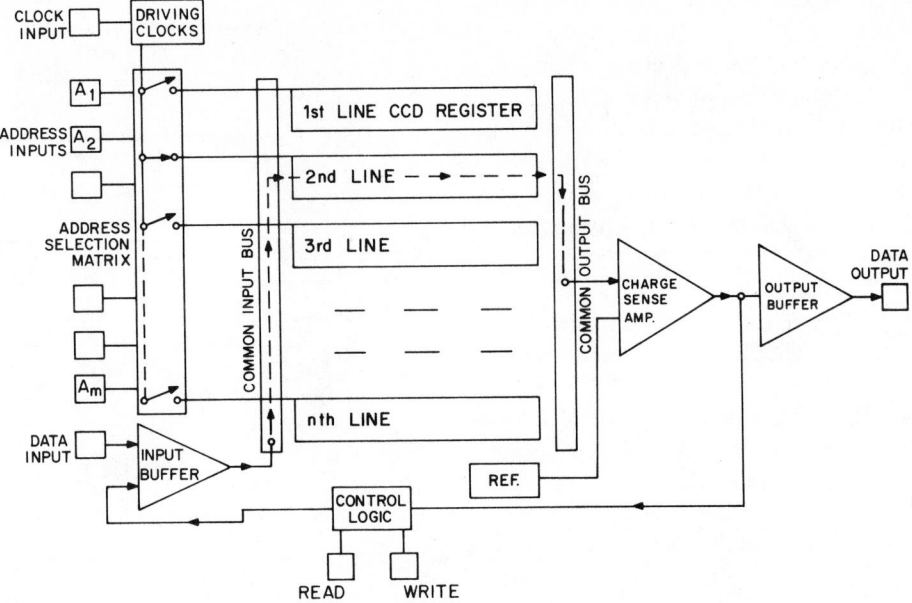

Fig. 1. Schematic illustrating the LARAM organization concept.

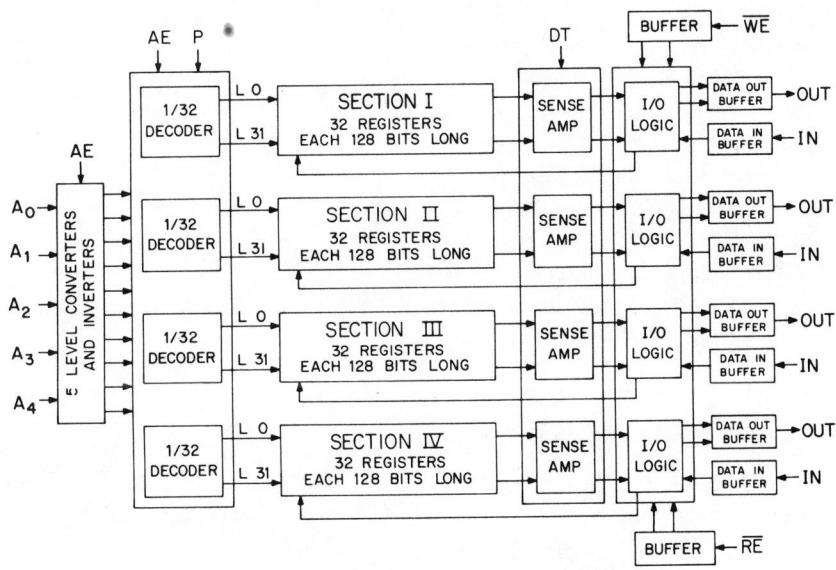

Fig. 2. Block diagram of the 16 384-bit LARAM chip.

coding matrix, recirculating loop and data input–output circuits. The external clocks P and data transfer (DT), the dc power supplies V_{DD}, V_{CC}, V_{BB} and V_{SS} (ground), the control signals address enable (AE), read enable ($\overline{RE}$), and write enable ($\overline{WE}$), and the addresses are common to all four sections.

A five-bit (A_0–A_4) address is decoded to impose the clock P waveform on one data line in each section. The remaining lines are not clocked. The addressing of a specific bit location is accomplished by the selection of the line address and by clocking that line until the desired bit is reached. Each clock pulse constitutes a one-bit transfer.

The operating modes of this chip are: read, write, read–

modify–write, refresh and halt. Since CCD memory utilizes the thermal nonequilibrium storage of charge, the memory must necessarily be refreshed. Address changing is accomplished through a gating signal called AE. AE is raised and lowered while the clock P is high. During AE high, address subcircuits are dynamically precharged. Lowering AE stores internally the last address information present during AE high and inhibits address changes that occur on the external pins from entering the circuit. Address pins are in a DON'T CARE state during AE low. It is to be noted that AE is not required every cycle; it is needed only when a new line is to be accessed, since address retention time is equal to or longer than the memory cell storage time. Power is thus saved in the addressing circuitry when information is required from successive bits in the same line.

In the read mode, data are advanced in the addressed lines at the rate of one bit per clock cycle. It appears at the output after a finite delay from the DT clock rise and it is also automatically recirculated to the input of the addressed lines.

At the output pin, data stay valid until they are cleared by the next bits of data or by a change of mode as controlled by the read enable logic signal ($\overline{RE}$).

In the write mode, the input data override the stored data which are erased to prevent it from being recirculated back to the beginning of the register. The output circuits are disabled during the write mode.

The read-modify-write mode requires an extended clock cycle which disables the recirculating loop when the output is being read out to permit the new input to be written into the same bit location.

In the halt mode, both clocks can be held stationary for a period determined by the storage time. The halt mode of operation can be used for two purposes. The first is to synchronize the DT to some demand rate other than the basic clock frequency. Data can be queued in some lines while other lines are being refreshed. The second purpose for the halt operation is to reduce power dissipation. From this standpoint, the preferred halt condition occurs when both the clocks P and DT are low.

The refresh mode occurs when the chip is not set to any of th other modes of operation. In this mode of operation, the I/O buffer circuits are disabled; when the clock P is applied, the data in the addressed lines are recirculated to the input of those lines. Since the storage device is dynamic, refreshing must be used at regular intervals. Each line is addressed and clocked through at least once in each refresh interval. However, by appropriate system design, the refresh can be made virtually transparent by interspersing refresh cycles between memory active cycles.

IV. CIRCUITS AND PERFORMANCE RESULTS

A photomicrograph of the 16 384-bit LARAM memory chip is shown in Fig. 3. Sections of the chip have been identified on the photograph to indicate their correspondence to the block diagram of Fig. 2.

All the circuits on the chip were simulated using an ISPICE program; a preliminary characterization of the circuit has indicated that the original design objectives have been achieved. All the inputs, except for the clocks P and DT, and the signal AE, are TTL compatible. Although P, DT, and AE are required to be MOS level signals (0-12 V swing) the capacitive load on each of these lines is nominally only 120, 20, and 40 pF.

The address level converters and inverters generate the true and the complement of the address inputs, and feed them into the decoder which is a dynamic NOR circuit as shown in Fig. 4. The output nodes of the decoder are unconditionally precharged during the AE high time. Before AE makes its transition from high to low, all the inputs to the decoder must be stable. When AE does make its transition from high to low, all the decoder outputs except the selected one, follow AE. Then,

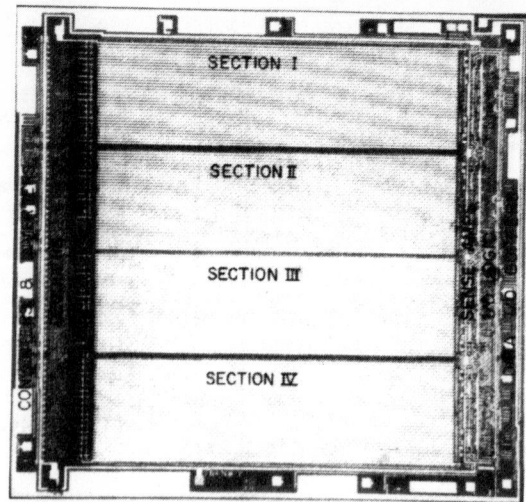

Fig. 3. Photomicrograph of the 16 384-bit LARAM chip.

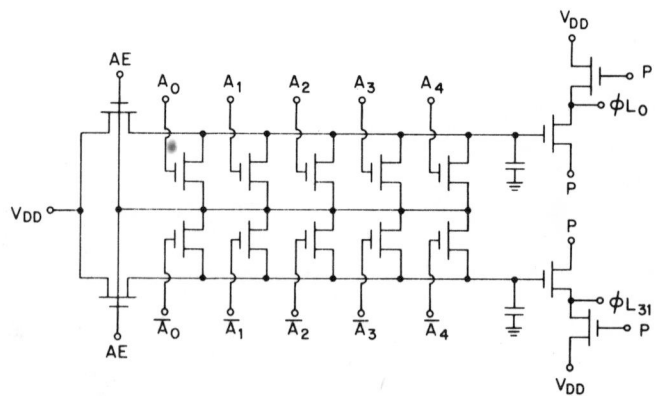

Fig. 4. Circuit schematic of the decoder and line driver circuit.

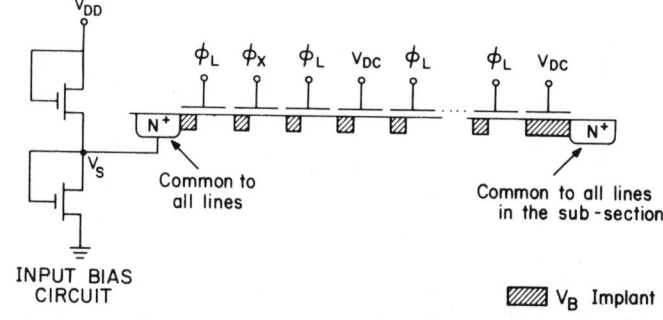

Fig. 5. Basic elements of a CCD line in the LARAM organization.

when P makes its transition from high to low, the addressed lines follow P, while the unaddressed lines stay in the high precharged state. It is evident from Fig. 4 that it is essential that P overlaps AE at both the rising and falling edges of AE.

The basic element of a line or memory register shown in Fig. 5 is a buried channel, implanted barrier CCD structure in which one phase (V_{dc}) is static while the other phase (ϕ_L) is driven by the line driver [4]. All the ϕ_L lines are precharged unconditionally during clock P high and stay in that state except for the addressed lines which follow the transitions of the clock signal P. A cross section of the memory register is shown in Fig. 6. Potential diagrams depicting charge transfer for ϕ_L low and ϕ_L high conditions illustrate the principle of the single-phase operation. The static electrode is held at a potential approximately midway between the potential extremes of the clocked electrode. One transition of ϕ_L from high to low and back to high moves the data by one bit in that line.

Fig. 5 also shows the input writing concept which enables the use of a common input bus. This bus, ϕ_X, is controlled by the read-write logic; when ϕ_X is high, it permits charge data to be written only in the addressed line. It is obvious that for this configuration to operate properly, the high level for ϕ_X be such that the maximum potential in its barrier region is lower than V_S, lest all the unaddressed lines write directly into ϕ_X. V_S is a dc voltage generated on chip and is a potential which maintains the first ϕ_L well in all the lines filled and precharged.

At the output end, several lines share a common output diffusion. Although, in principle, all the 32 lines of a 4K block can share a common output diffusion, the 4K blocks in the subject memory are configured differently. A 4K block is divided into two subsections, each comprised of 16 data lines and a reference line which shares a common output diffusion. When a line is addressed, in one subsection, the reference line in the other subsection is also automatically addressed. The two output diffusions are connected to a balanced flip-flop comparator similar to those used in most MOS RAM's. Thus, only one comparator is needed for each 4K section [5].

The flip-flop sets into its "1" or "0" state which depends on the subsection of the addressed line and on whether charge (QSAT $\simeq$ 100 fC) or no charge was delivered by the addressed line. The charge delivered by the reference line is always 1/2 QSAT ($Q_{ref} \simeq 50$ fC). The comparator drives a buffer circuit and the outputs of the buffer circuit are fed into the read-write-refresh control logic. The data output is a three-state circuit which permits the direct usage of a data bus.

The chip has been functionally tested. No pattern sensitivity at room temperature has been observed for operating frequencies ranging from 500 kHz to 5 MHz which were the design goals for the frequency extremes. Read access times (address valid to output valid) as low as 120 ns have been observed. Some of these characteristics are shown in the oscilloscope photographs of Fig. 7. Storage times in excess of 100 ms have been observed at room temperature. This extrapolates to 6 ms at an ambient temperature of 55°C. At this operating temperature, 14 percent of the real time will be spent refreshing at a 5 MHz clock rate.

V. APPLICATIONS

A qualitative comparison of the LARAM organization, as exemplified by the 16K LARAM, with other CCD memory organizations [6], [7] shows that in the critical areas of relatively fast access memories, LARAM is the preferred configuration. This choice is derived from the fast access times (τ read access $\simeq$ 120 ns, τ average latency $\simeq$ 12.8 μs, τ maximum latency $\simeq$ 25.6 μs), low clock capacitance (less than 120 pF) and the flexibility in data handling that is characteristic of

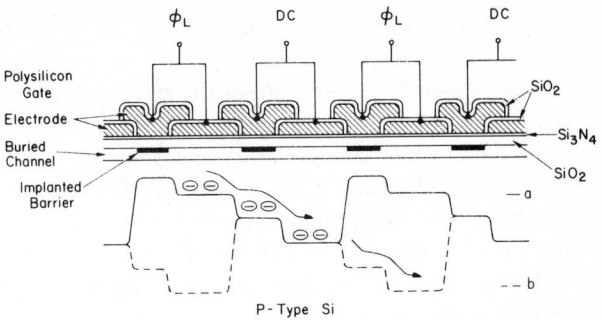

Fig. 6. Cross section of the memory cell with potential profiles illustrating charge transfer (a) when ϕ_L is low and (b) when ϕ_L is high.

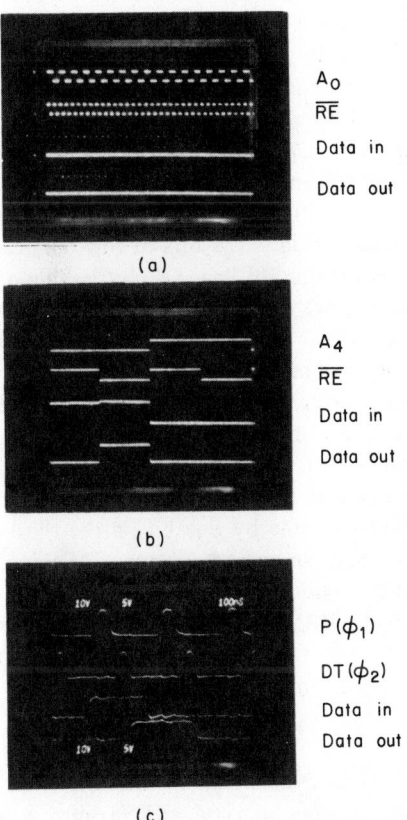

Fig. 7. CRO traces from a functioning 16 384-bit LARAM. (a) Fully functional 4K section, (b) one worst case pattern, and (c) clock timing showing 5 MHz capability.

this organization. In addition, the power dissipation is low (less than 200 mW) and the organization leads naturally to a high density chip layout.

Some of the functions which can be implemented with the 16 384-bit LARAM are as follows.

1) Quad 4K Bit Long Shift Register: The 16K LARAM device can be configured as four 4K bit long shift registers by addressing each of the 32 lines in sequence. This considerably simplifies addressing since address changes in this case are not random, but in sequence, and take place once every 128 clock cycles. Furthermore, because AE is required only when there is an address change, i.e., every 128 clock cycles, there is a considerable saving of power. The data stream is continuous even when an address change is made from one line to another. Another feature is that no special refresh interval is necessary.

2) Buffer Memory: It is well known that the provision of a fast buffer between archival storage such as disk and drum and the main memory can radically affect the overall system performance. The high data rate and reasonably fast access time offered by the LARAM organization make it a promising candidate for buffer memory or paging applications. For example, four LARAM chips can serve to provide access to 4K words with a word length of 16 bits.

3) Block-Oriented RAM (BORAM): Since a line in any section is independently addressable, the LARAM organization can be used to access 128-bit blocks of data. To perform this function, a given address is held for 128 clock cycles and four 128-bit blocks of data are written in or read out of the memory. Each LARAM section is comprised of 32 such blocks. Here again, since the address change occurs once every 128 cycles, address enable is required only once every 128 cycles.

4) Special Sort Operations: A matrix sorting application where data have been written serially, row by row, and needs to be read out column by column comes naturally to the LARAM configuration. 128 bits of row data can be written into the memory line by line, i.e., by leaving address stable for 128 clock cycles and then reading out column by column. To read, line address must be changed every clock cycle in sequence. One example where such an operation is required is in synthetic aperture radar memories [8] where 4K bits of range and azimuth information must be sorted out. The sorting operation is not limited to transposing row and columns. For example, a matrix of data can be skewed by shifting the data in the second line by one bit, in the third by two bits, and so forth. Such an operation is very difficult to accomplish with a more conventional memory architecture.

Because the LARAM organization is highly flexible, it should find application in fast access bulk storage, mini CPU, main memory, and disk enhancement.

VI. Conclusion

The LARAM CCD memory organization offers both a high memory bit packing density and reasonably short access times. The flexibility of random line access is achieved while the clock capacitance is kept to a minimum. A 16 384-bit LARAM chip has been designed using isoplanar n-channel silicon gate MOS and buried channel CCD technology. The CCD memory section employs charge writing and sensing circuit configurations that enhance access speed and signal-to-noise ratio. The CCD registers require only one clock for DT.

The circuit performance meets the design goals and confirms the applicability of the memory design to a variety of systems.

Acknowledgment

The authors wish to thank J. M. Early and H. C. Pao for contributions to the design of the device, A. Solomon for his review of the manuscript, J. Gaudagna for his assistance in circuit design and testing, J. Tran for his assistance in chip layout, S. Keller for his computer-aided design effort, and B. Choy and D. Means for fabricating the devices.

References

[1] R. H. Krambeck, T. F. Retajczyk, D. J. Silversmith, and R. J. Strain, "A 4160-bit C4D serial memory," *IEEE J. Solid-State Circuits (Special Issue on Analog Circuits)*, vol. SC-9, pp. 436–443, Dec. 1974.

[2] R. Davis, "CCD fills the memory gap," *Fairchild J. Semiconductor Progress*, vol. 3, pp. 20–21, Jan.–Feb. 1975.

[3] Intel Corp., "Intel charge coupled device 2416," preliminary specifications, 1975.

[4] R. D. Melon and J. D. Meindl, "One-phase CCD: A new approach to charge-coupled device clocking," *IEEE J. Solid-State Circuits (Special Issue on Solid-State Microwave Circuits)* (Corresp.), vol. SC-7, pp. 92–93, Feb. 1972.

[5] K. Gunsagar, J. Guadagna, M. Guidry, and G. Amelio, to be published.

[6] G. F. Amelio, "Charge coupled devices for memory applications," presented at the Nat. Comput. Conf., Anaheim, Calif., May 1975.

[7] G. F. Amelio, M. Guidry, K. Gunsagar, and H. Pao, "Application of CCD in memory systems," in *Proc. NAECON Conf.*, Dayton, Ohio, May 1974.

[8] N. Gutlove, G. Amelio, A. Green, and R. Wakeman, "Charge coupled device memories for synthetic aperture radar," in *Conf. Rec. NAECON*, 1973, p. 166.

CCD460/460A, CCD461/461A

PRELIMINARY* SPECIFICATION

16,384 BIT CCD DIGITAL MEMORY
DYNAMIC LINE ADDRESSABLE RANDOM ACCESS MEMORY, LARAM

ORGANIZATION: 4 BLOCKS OF 32 RECIRCULATING SHIFT REGISTERS OF 128 BITS EACH

DESCRIPTION

The Fairchild CCD460 family consists of 16,384 bit dynamic CCD memories designed for fast access cache, swapping store, mainframe, and other memory applications where LARAM performance features are required. The CCD461 and CCD461A are specified for memory applications requiring the guaranteed refresh period of 10 milliseconds at 55°C. CCD460 and CCD460A have a 2 millisecond refresh period at 55°C and are intended for less demanding serial memory applications. The standard devices operate to 2.5MHz and the faster "A" devices are specified to 4MHz.

The Line Addressable Random Access Memory organization provides a data rate of up to 16 megabits per second with an average random access time of 16 microseconds at an operating frequency of 4MHz, and with typically less than 200 milliwatts of power. It also provides very low clock drive capacitance loading. The five bit address selects one of 32 128-bit registers in each section and those registers deliver or receive data thru their input and output pins. Data is not inverted and is available 4 bits parallel. Recirculation is automatic in accessed registers.

Operation is straightforward and support circuitry kept simple by the TTL compatibility of data in, address lines and 3 state outputs. Ease of use is enhanced by the low drive capacitance loading on the 0-12V inputs to Address Enable (70 pF), Data Transfer clock, (15 pF), and Precharge (70 pF) pins. Readout is non-destructive and data out lines can be wired-OR for flexibility and ease of expansion. The device operates in four modes: read, write, read/modify/write, and low power standby recirculate. In the latter mode the recirculate/refresh period can be as long as 10 milliseconds with a power dissipation of only 50 mW.

The CCD460 features Isoplanar, NMOS, buried channel, and silicon gate structures for high density and reliable performance.

*This is a preliminary specification. Operating characteristics shown are based on the device design and measurements on prototype devices. Operating parameters and performance characteristics are subject to change. The final data sheet will specify guaranteed performance.

FEATURES

- FAIRCHILD <u>LARAM</u> PERFORMANCE

- High data rate - 16 megabits per second
- Fast average random access time - 16 μs
- Low power - 200 mW, @ 4 MHz, 50 mW standby recirculate @ 400 KHz
- Low clock capacitances - 70 pF and 15 pF
- Two clock operation - 0 to + 12V

- TTL compatible
- 3-State outputs
- Four mode operation: Read, Write, Read/Modify/Write, Recirculate
- Four inputs and four outputs
- Standard 22 lead DIP package
- Isoplanar®, NMOS, Buried Channel, Silicon Gate Structure

FUNCTIONAL DESCRIPTION

The "LARAM" CCD460 is a 16,384 bit dynamic memory. It contains four sections of 32 lines (each line of 128 bits length) as shown in the Block Diagram. Each section is a 4096 bit memory block which has its own 1 of 32 decoding matrix, recirculating loop and data input and data output circuits. The external clocks P (Precharge) and DT (Data Transfer), the d-c power supplies V_{DD}, V_{CC}, V_{BB}, and V_{SS} (Ground), the control signals AE (Address Enable) $\overline{RE}$ ($\overline{Read\ Enable}$) and $\overline{WE}$ ($\overline{Write\ Enable}$) and the Addresses ($A_0 - A_4$) are common to all four sections.

A five bit (A_0 thru A_4) address is decoded to impose the clock P on one line in each section. The remaining lines are not clocked. Addressing a specific location in each section is accomplished by selection of the intended line and by the clocking of that register until the desired location is reached. Each clock pulse advances the addressed register one bit.

The operating modes of the CCD460 are: read, write, read-modify-write, and standby recirculate. Since CCD memory utilizes non-equilibrium charge storage each register must be periodically refreshed. To refresh a register it must have been addressed and clocked a total of at least 128 bits during the refresh period.

272

Address changes are provided by a 0-12Volt gating signal called Address Enable (AE). AE is raised and lowered while the clock P is high. During AE high, address subcircuits are dynamically precharged. Lowering AE dynamically stores internally the last information present during the high state and inhibits address changes that occur on the external pins from entering the circuit. Address pins are in a "don't care" state during AE low. It is to be noted that AE is not required during every cycle --------- it is required only when a different line is to be accessed. This saves power in the addressing circuitry when data is required from the successive bits in the same line.

MODES OF OPERATION (REFER TO TIMING DIAGRAMS AND TRUTH TABLE FOR EACH MODE)

I READ MODE ($\overline{RE}$ is Low; $\overline{WE}$ is High)

In the read mode, data is advanced in the addressed line of each section at the rate of one bit per clock cycle. It appears at the output after a finite delay following DT clock rise and it is also automatically refreshed and recirculated to the input of the addressed lines. At the output pin data stays valid until it is cleared by the next bit of data or by a change of mode as controlled by the Read Enable ($\overline{RE}$) logic signal.

II WRITE MODE ($\overline{RE}$ is High, $\overline{WE}$ is Low)

In the write mode the input data overrides and erases the stored data before it can recirculate back to the register input. The output circuits are automatically disabled during the write mode.

III READ-MODIFY-WRITE MODE ($\overline{RE}$ is Low, goes High; $\overline{WE}$ is Low)

The read-modify-write mode requires an extended clock cycle which disables the recirculating loop when the output is being read out, thus enabling new input to be written into the same bit location.

IV REFRESH/SCAN MODE ($\overline{RE}$ is High, $\overline{WE}$ is High)

In the refresh mode, the input and output buffer circuits are disabled and the data in the addressed lines are automatically recirculated and refreshed. Since the device is dynamic each line must be addressed and clocked a total of 128 cycles each refresh interval. The refresh may be made virtually transparent by interposing refresh cycles between memory active cycles. This mode can be used for scanning a register to a bit location where read is to occur.

V HALT CONDITION

In the halt condition both clocks can be held stationary for a period determined by the device storage time. This condition can be used for two purposes. One would be to synchronize the data transfer to some demand rate other than the basic clock frequency. Another would be to reduce power dissipation, in which case the preferred halt condition is with the P clock high, and AE and DT low.

VI MEMORY CLEAR CONDITION ($\overline{RE}$ is LOW; $\overline{WE}$ is LOW)

In the Memory clear condition data inputs are ignored. Data stored in the addressed line is read out but not recirculated. The input to the addressed line is disabled. After a bit has been read once in this mode subsequent read operations will show a "0" in that bit location if addressed by A_0 Low or a "1" in that bit location if addressed by A_0 High.

VII LOW POWER OPERATION

Power dissipation can be minimized by reducing the operating frequency while increasing the P high time (t_p) to its maximum. AE width (t_{AE}) should be a minimum and AE should only occur when address changes are necessary.

VIII MEMORY START UP

Whenever the stored time of any bit in any register has exceeded the specified maximum refresh period that register must be purged before it can be used to store data. Purging is accomplished by clocking the register a minimum of 512 clock cycles. When the chip is first powered up the entire 16K memory must be purged by addressing each line and clocking it a minimum total of 512 cycles.

PIN NAMES

$DI_0 - DI_3$	DATA INPUT LINES	TTL level input, 5pF
$DO_0 - DO_3$	DATA OUTPUT LINES	TTL level output, 3 state, 7 pF
DT	DATA TRANSFER CLOCK	0-12V, 15 pF
P	PRECHARGE	0-12V, 70 pF
$\overline{RE}$	$\overline{READ\ ENABLE}$	TTL level, 5 pF
$\overline{WE}$	$\overline{WRITE\ ENABLE}$	TTL level, 5 pF
AE	ADDRESS ENABLE	0-12V, 70 pF
$A_0 - A_4$	ADDRESS LINES	TTL level, 5 pF
V_{BB}, V_{CC}, V_{DD}	POWER SUPPLIES	-5V, +5V, +12V
V_{SS}	GROUND	0 V

CONNECTION DIAGRAM DIP TOP (VIEW)

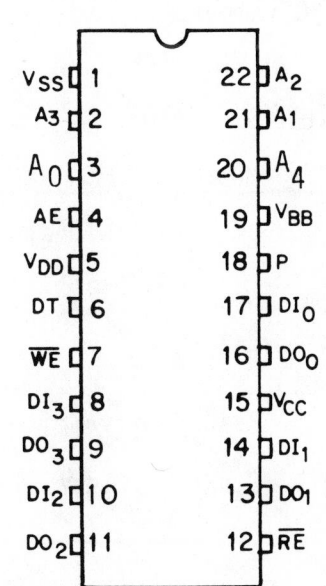

LOGIC SYMBOL

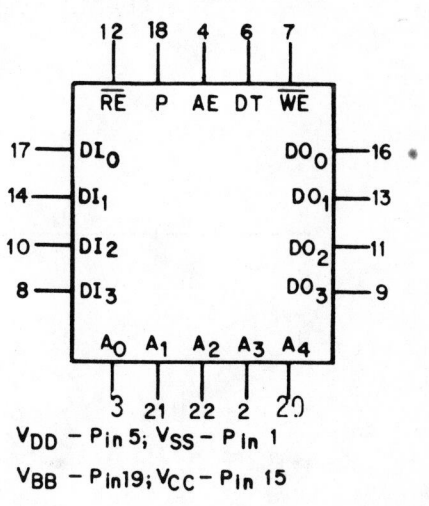

$V_{DD} - Pin\ 5;\ V_{SS} - Pin\ 1$
$V_{BB} - Pin\ 19;\ V_{CC} - Pin\ 15$

ABSOLUTE MAXIMUM RATINGS

(above which the useful life of the device may be impaired)

STORAGE TEMPERATURE	-65°C to +150°C
TEMPERATURE UNDER BIAS	-55°C to +70°C

LEADS WITH RESPECT TO V_{BB}:

V_{CC}, DO Pins	-.5V to +12.0V
V_{SS}	-.5V to + 7V
AE, $\overline{RE}$, $\overline{WE}$, DT, P, V_{DD}, A_0 - A_4, DI	-.5V to +20V

(Current limited to 10 mA)

MAXIMUM POWER DISSIPATION	1 Watt

ORDER INFORMATION

TO ORDER: CCD460 Specify CCD460DC ⎫ When D signifies ceramic DIP
           CCD461 Specify CCD461DC ⎬ and C signifies 0°C to 55°C
                                      ⎭ temperature range

           CCD460A Specify CCD460ADC ⎫ When A signifies 4.0 MHz
           CCD461A Specify CCD461ADC ⎬ Operation, D signifies ceramic
                                      ⎭ DIP, and C signifies 0°C to
                                        55°C temperature range.

TRUTH TABLE

$\overline{RE}$	$\overline{WE}$	DI	DO	MODE OR CONDITION
H	H	X	Z_H	Recirculate (Refresh)
H	L	DI	Z_H	Write
L	H	X	DO	Read
L→H	L	X→DI	DO→X	Read/Modify/Write
L	L	X	DO	Start up/Clear

DI = Data In   H = HIGH Voltage
DO = Data Out   L = LOW Voltage
Z_H = High Impedance State   X = Don't Care (HIGH or LOW)

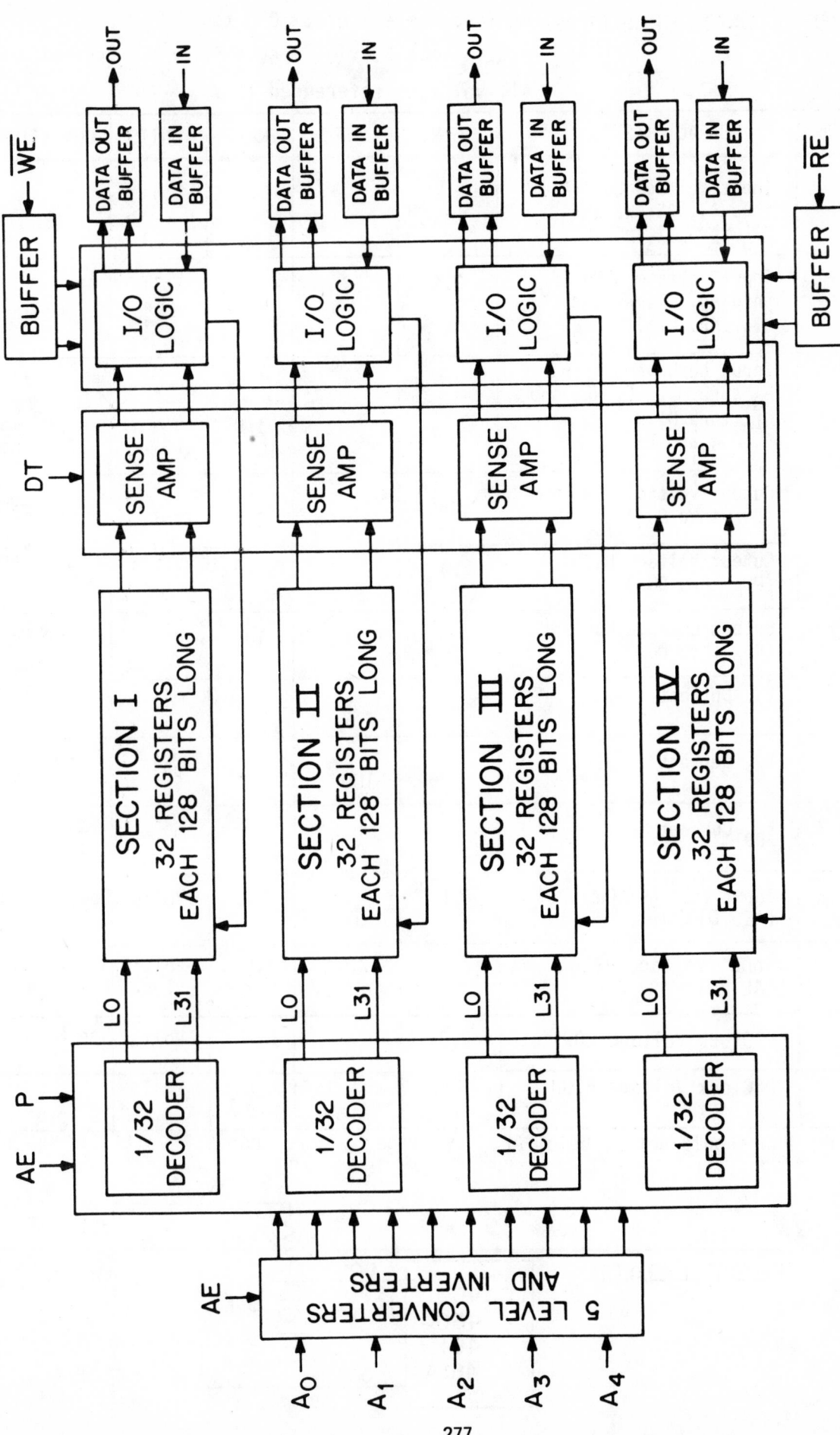

BLOCK DIAGRAM

DC CHARACTERISTICS

Ambient Temperature: 0°C to 55°C

V_{DD} = 12V $\pm$ 5% V_{CC} = 5V $\pm$ 5% V_{BB} = -5V $\pm$ 5%

All voltages referenced to V_{SS}

SYMBOL	CHARACTERISTIC	MIN.	TYP.	MAX.	UNITS	CONDITIONS
V_{IL}	Input Voltage LOW (A_0 thru A_4, RE, WE and DI_0 thru DI_3)			0.8	Volts	
V_{IH}	Input Voltage HIGH (A_0 thru A_4, RE, WE and DI_0 thru DI_3)	2.4			Volts	
I_{LI}	Input Leakage Current A_0 thru A_4, RE, WE, and DI_0 thru DI_3, P, DT, and AE			1.0 10.0	μA μA	5.25 max. 12.6V max.
V_{OL}	Output Voltage LOW (DO_0 thru DO_3)			0.4	Volts	I_{OL}= -2mA
V_{OH}	Output Voltage HIGH (DO_0 thru DO_3)	2.4			Volts	I_{OH}= +500μA
I_{LO}	Output Leakage Current (DO_0 thru DO_3)			10	μA	5.25V max.
I_{CC}	V_{CC} Current		4		mA	+5V
I_{BB}	V_{BB} Current		100		μA	-5V
I_{DD}	V_{DD} Current		14		mA	+12V
$V_{\emptyset L}$	Input Voltage LOW (AE, DT)	-0.5		0.8	Volts	See Note 1
$V_{\emptyset H}$	Input Voltage HIGH (AE, DT)	10.7	12.0	13.0	Volts	
V_{PL}	P Clock Voltage LOW	-0.5		0.4	Volts	See Note 1
V_{PH}	P Clock Voltage HIGH	11.4		14.0	Volts	

Note 1: V_{PL} should not be below AE_{low} voltage by more than 0.7 volts.

TEST LOAD

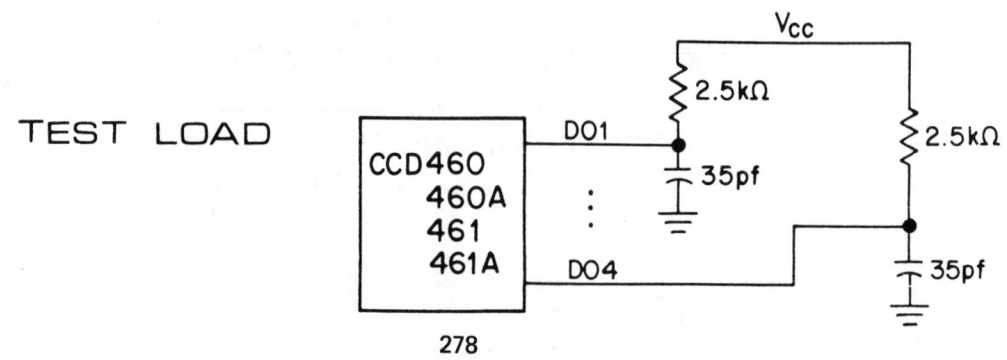

AC CHARACTERISTICS: V_{DD} = 12V $\pm$ 5%, V_{CC} = 5V $\pm$ 5%, V_{BB} = -5V $\pm$ 5%

T_A = 0° to +55°C   All voltages referenced to V_{SS}

AE, DT, P: t_{rise} = t_{fall} = 10 nsec. A_0 - A_4, DI_0 - DI_3: t_{rise} = t_{fall} = 5 nsec.

Reference levels for timing purposes: LOW level 0.8V, HIGH level 2.4V, HIGH level clocks 10.0V.

Symbol	Characteristics	460A, 461A			460, 461			UNITS	Notes
		Min.	Typ.	Max.	Min.	Typ.	Max.		

Read, Write, Refresh and Read/Modify/Write Cycles and Halt Condition

Symbol	Characteristics	Min.	Typ.	Max.	Min.	Typ.	Max.	UNITS	Notes
t_p	Precharge Width	100			160			ns	
t_{AE}	Address Enable Width	55			90			ns	
t_{PAE}	Precharge to Address Enable Delay	0			0			ns	
t_{AAE}	Address Capture Time	55			90			ns	
t_{AEA}	Address Enable to Address Change	10			10			ns	
t_{DS}	Decoder Stabilization Time	20			30			ns	
t_{SS}	Sense-node Stabilization Time	30			50			ns	
t_{WMC}	Write Signal Capture Time	50			80			ns	
t_{RMC}	Read Signal Capture Time	50			80			ns	
t_{PWC}	Precharge to Write Signal Change	10			10			ns	
t_{DTP}	Data Transfer to Precharge Delay	0			0			ns	
	Halt Condition @ 55°C: CCD460, 460A 2 ms CCD461, 461A 10 ms			Note 2			Note 2	ms	Note #2

Note 2: Maximum refresh period is a function of ambient temperature. Data is shown (in Figure 1) for the typical temperature dependence of maximum refresh period for the CCD460/A and for the CCD461/A.

FIGURE 1

TYPICAL TEMPERATURE DEPENDENCE OF
MAXIMUM REFRESH PERIOD

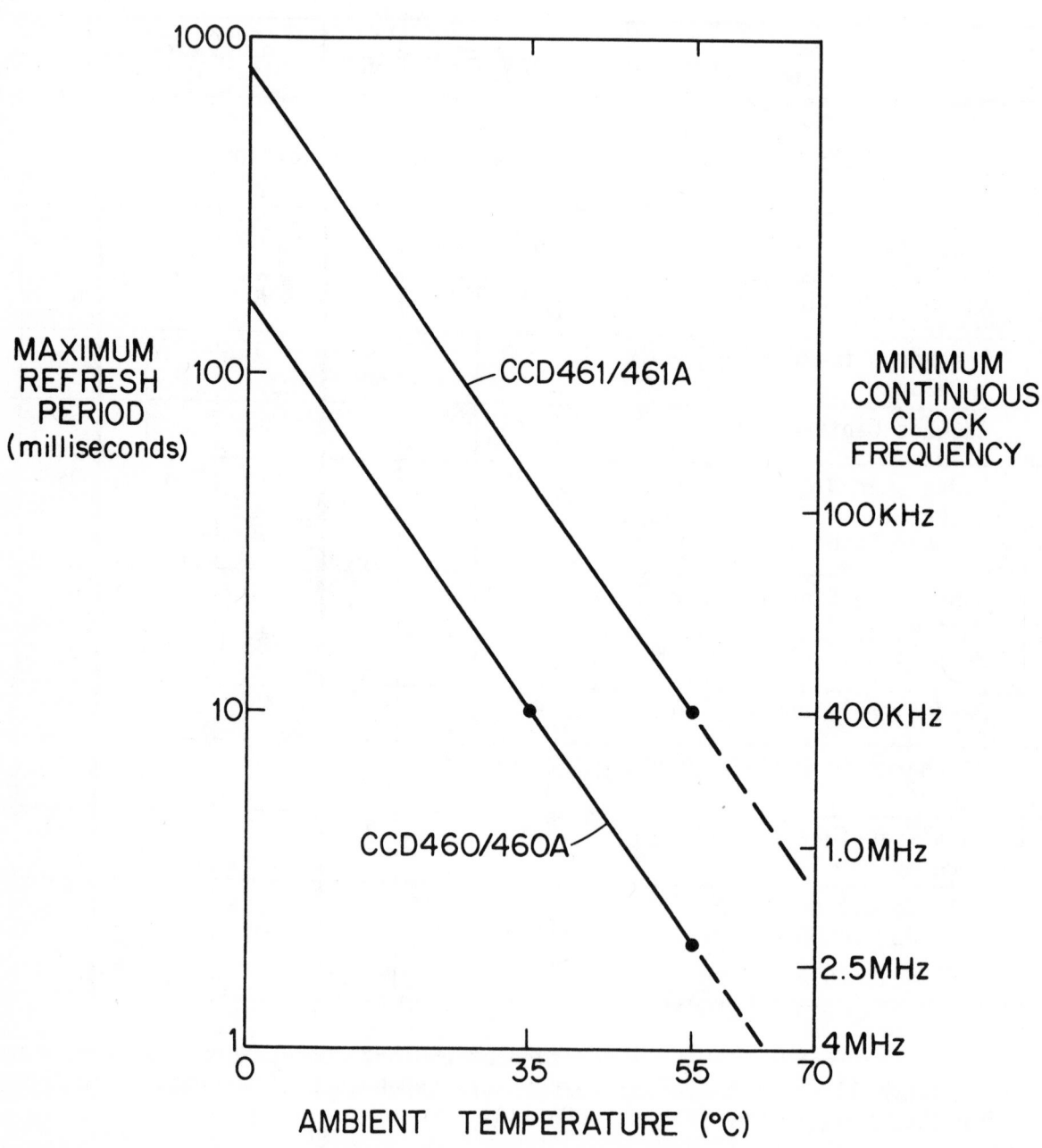

A.C. Characteristics - Continued

Symbol	Characteristics	460A, 461A			460, 461			Units	Notes
		Min	Typ.	Max.	Min.	Typ.	Max.		
Read Cycle									
t_{RCY}	Read Cycle Time	250			400			ns	
t_{DT}	Data Transfer Width	60			95			ns	
t_{TOD}	Data Transfer to Output Delay		80	100		130	160	ns	
t_{REZ_H}	P_{LOW} to Output Z_H			20			30	ns	
Write Cycle									
t_{WCY}	Write Cycle Time	250			400			ns	
t_{DT}	Data Transfer Width	60			95			ns	
t_{PRC}	Precharge to Read Signal Change	10			10			ns	
t_{DIC}	Data In Capture Time	40			65			ns	
t_{DTI}	Data Transfer to Input Change	10			10			ns	
Refresh Cycle									
t_{RECY}	Refresh Cycle Time	250			400			ns	
t_{DT}	Data Transfer Width	60			95			ns	
t_{PRC}	Precharge to Read Signal Change	10			10			ns	

A.C. Characteristics - Continued

Symbol	Characteristics	460A, 461A			460, 461			Units	Notes
		Min.	Typ.	Max.	Min.	Typ.	Max.		
Read Modify Write Cycle									
t_{RMCY}	Read modify write cycle time	385			595			ns	
t_{DT}	Data Transfer Width	215			345			ns	
t_{RMMC}	Read Modify Mode Capture Time	90			140			ns	
t_{DTRC}	Data Transfer to Read Signal Change	10			10			ns	
t_{IRE}	Input Valid Prior to Read Signal Change	30			50			ns	
t_{DTI}	Data Transfer to Input Change	10			10			ns	
t_{TOD}	Data Transfer to Output Delay		80	100		130	160	ns	
t_{DOV}	$\overline{RE}$ HIGH to Output not valid	20			30			ns	

Capacitance (CCD460, 460A, 461, 461A)   Min. Typ. Max.

Symbol	Characteristics	Min.	Typ.	Max.	Units				
C_P	Precharge Capacitance		70		pF				
C_{DT}	Data Transfer Capacitance		15		pF				
C_{AE}	Address Enable Capacitance		70		pF				
C_A	Address Input Capacitance		5		pF				
C_{RW}	Read/Write Input Capacitance		5		pF				
C_{DI}	Data In Input Capacitance		5		pF				
C_O	Output Capacitance		7		pF				

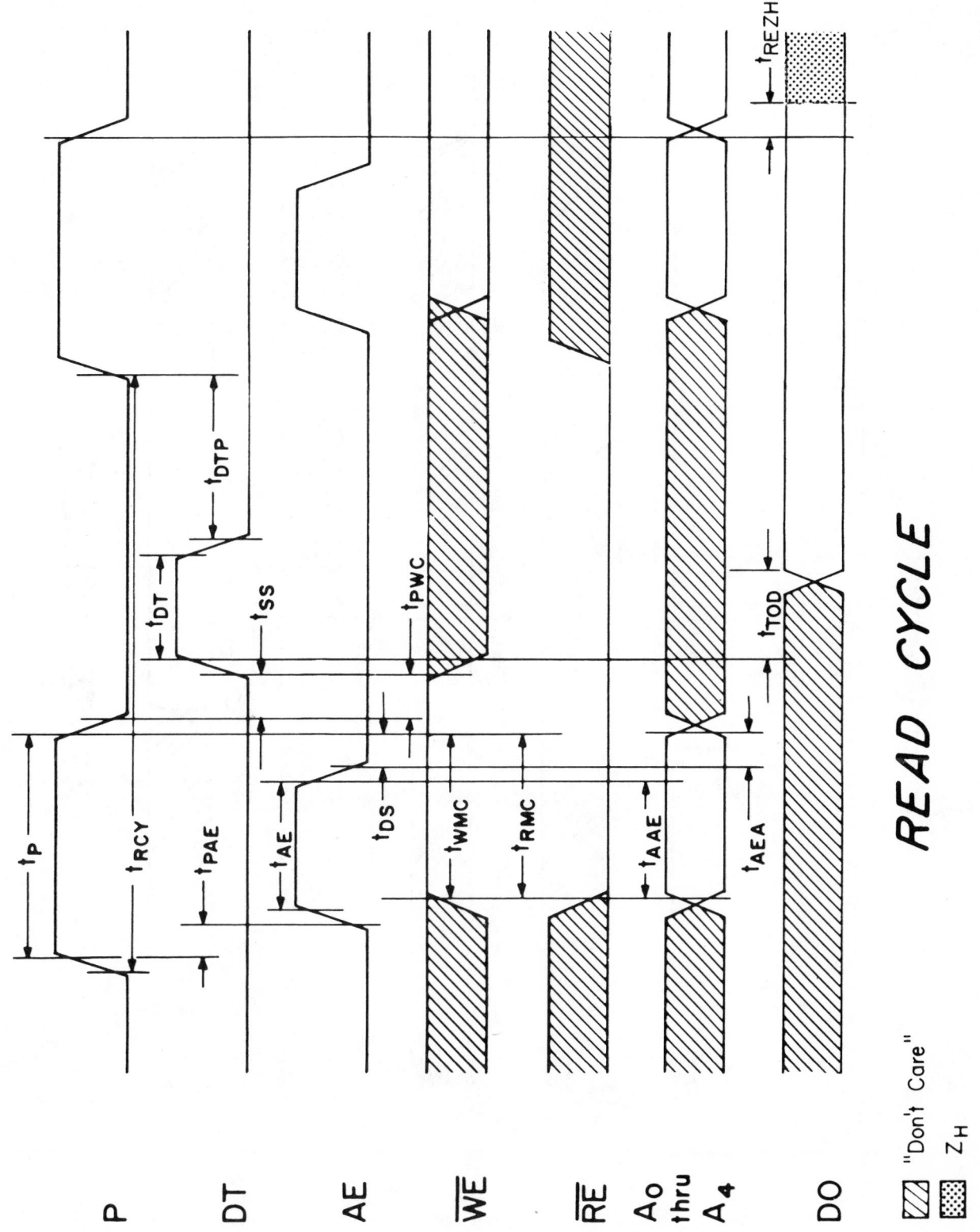

READ CYCLE

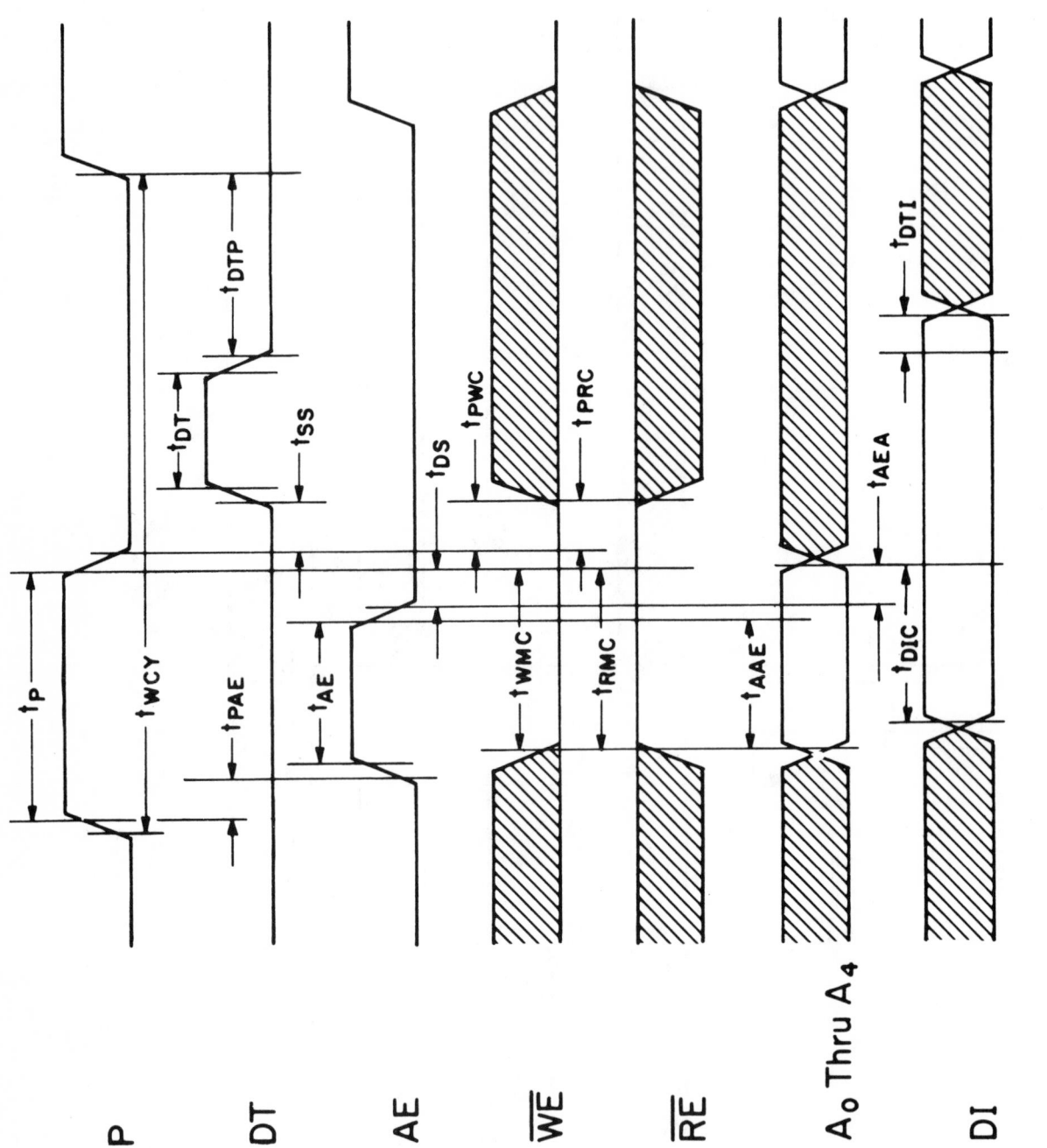

WRITE CYCLE

P

DT

AE

$\overline{\text{WE}}$

$\overline{\text{RE}}$

A_0 Thru A_4

DI

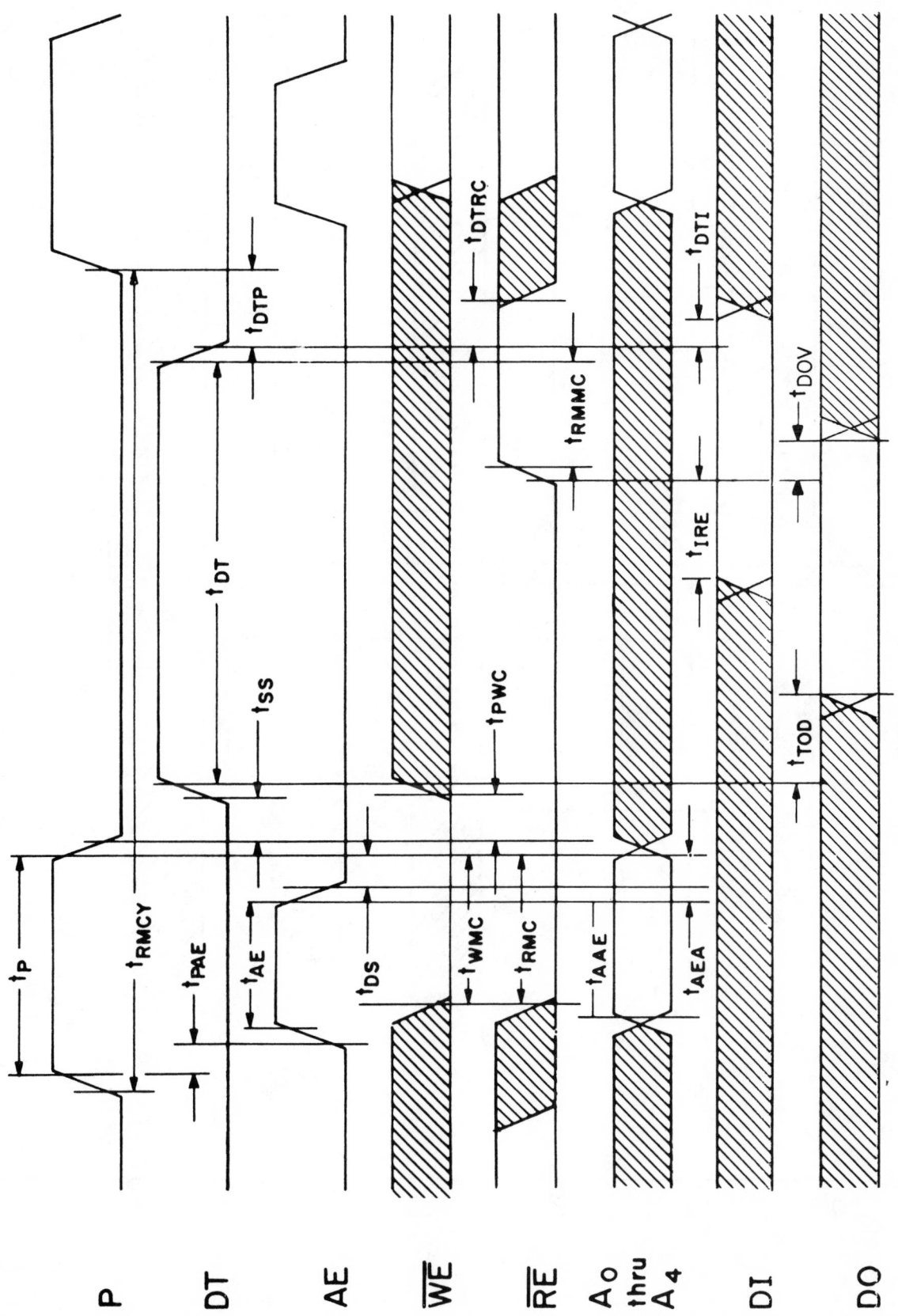

READ/MODIFY/WRITE CYCLE

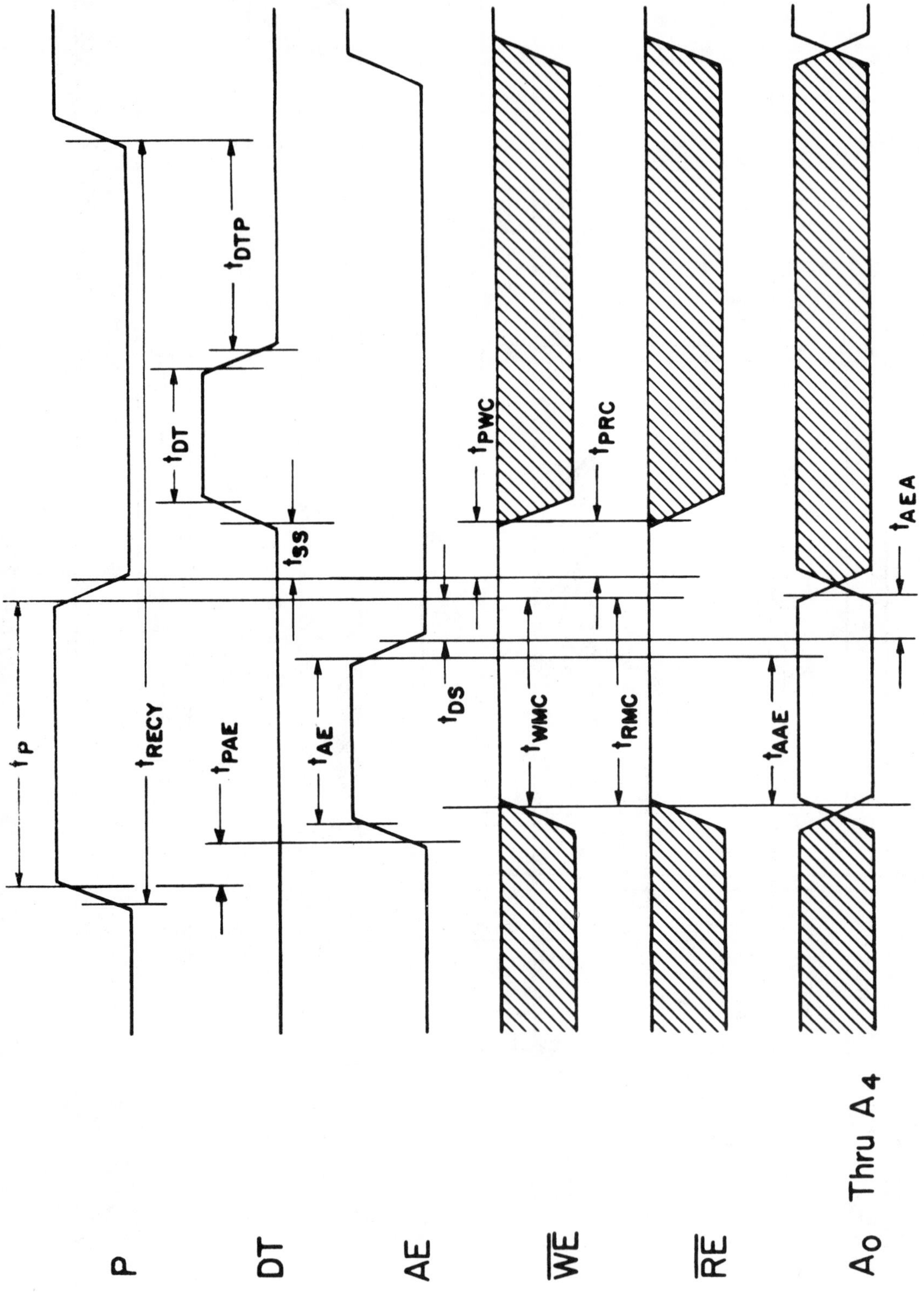

REFRESH CYCLE

t_{DTP}

t_{DT}

t_{SS}

t_{PWC}

t_{PRC}

t_{AEA}

t_P

t_{RECY}

t_{PAE}

t_{AE}

t_{DS}

t_{WMC}

t_{RMC}

t_{AAE}

P

DT

AE

$\overline{WE}$

$\overline{RE}$

A_0 Thru A_4

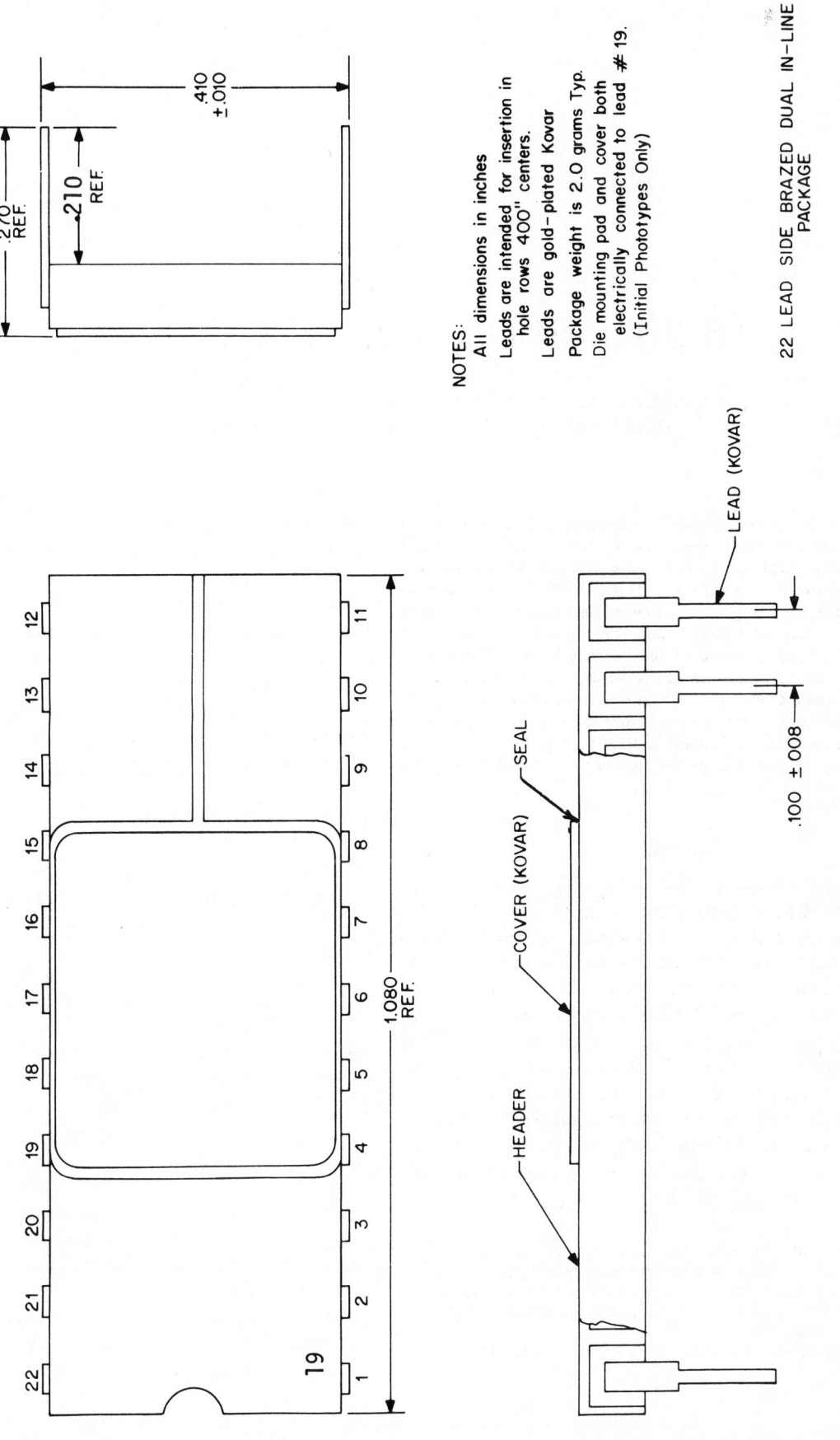

NOTES:

All dimensions in inches

Leads are intended for insertion in hole rows 400" centers.

Leads are gold-plated Kovar

Package weight is 2.0 grams Typ.

Die mounting pad and cover both electrically connected to lead # 19. (Initial Phototypes Only)

22 LEAD SIDE BRAZED DUAL IN-LINE PACKAGE

PACKAGE INFORMATION

A 16 384-Bit High-Density CCD Memory

STANLEY D. ROSENBAUM, CHONG HON CHAN, STUDENT MEMBER, IEEE,
J. TERRY CAVES, STEWART C. POON, AND ROBERT W. WALLACE

Abstract—A 16 384-bit charge-coupled device (CCD) memory has been developed for mass storage memory system application where moderate latency, high data rate and low system cost are required. The chip measures only 3.45×4.29 mm^2 (136×169 mil^2), fits a standard 16-pin package, and is organized as four separate shift registers of 4096 bits, each with its own data input and data output terminals. A two-level polysilicon gate n-channel process was used for device fabrication. A condensed serial-parallel-serial (CSPS) structure was found to provide the highest packing density. Only two external clocks are required driving capacitances of 60 pF each at one-half the data transfer rate. Operations at data rates of 100 kHz to 10 MHz have been demonstrated experimentally, the on-chip power dissipation at 10 MHz being less than 20 μW/bit.

I. INTRODUCTION

THE development of the 16-kbit CCD memory to be described here was aimed at filling the gap between the short access time of random-access memories (RAM's) and the longer access time of fixed-head disks with a cost-effective alternative. For the present device, a fourfold system cost advantage over RAM systems is projected. This, however, requires the acceptance of longer access times than those already achieved on CCD's, including block-addressable memories [1], [2] and line-addressable memories [3]. These designs provide access to relatively small blocks of serial data: 256, 256, and 128 bits, respectively, while differing in maximum data rates, operating power, and system overheads for clocking, data transfer and data refreshing. The new 16-kbit

Manuscript received October 7, 1975. This work was supported in part by the Defence Research Board of Canada under its Defence Industrial Research Program. This paper was presented at the International Conference on the Application of Charge-Coupled Devices, San Diego, CA, October 29-31, 1975.
The authors are with Bell-Northern Research Limited, Ottawa, Ont., Canada.

device described here achieves a higher packing density by using an improved form of serial-parallel-serial organization using relatively large arrays of 2048 bits each, two such arrays being paralleled for each block of 4096 bits. Despite this large block size, the worst case access time is only 410 μs, because the device is capable of operating at a data transfer rate of 10 MHz. However, this organization requires a larger number of clock waveforms, which therefore must be generated on-chip in order to minimize the number of external connections and to keep the system overheads low.

II. ORGANIZATION OF CCD ARRAYS FOR HIGH-DENSITY STORAGE

For any given set of geometrical layout rules, serial-parallel-serial (SPS) organizations offer high packing density because only one sense amplifier is required for a large array, such as the 2048-bit array chosen for the present design. This allows space for designing the sense amplifier to detect extremely small amounts of charge, and the CCD storage electrodes can be small, both for this reason and because the electrode dimensions are not limited by any need to pack the sense amplifier into the repeat spacing of every one or two parallel rows. The simplest form of SPS structure is shown in Fig. 1(a). Whenever the M-bit serial input register becomes filled with data, a serial-parallel transfer is made into the first stores of all the M parallel registers, and a parallel-serial transfer into the serial output register occurs whenever this register becomes empty. In the following discussion, it is assumed that these transfers can be made rapidly enough so that the serial data stream need not be interrupted. If there are N bits in every parallel register, and if the input and output dumps are made simultaneously (which is not essential), the total number of bits stored is $M(N+1)$. However, it is unnecessary to provide

Reprinted from *IEEE Trans. Electron Devices*, vol. ED-23, pp. 101-107, Feb. 1976.

288

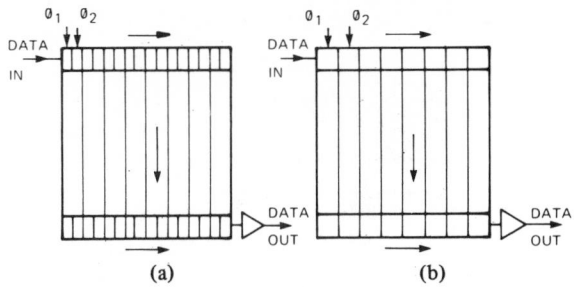

Fig. 1. Schematic diagram of the SPS structures for two-phase CCD's.
(a) Standard. (b) Interlaced.

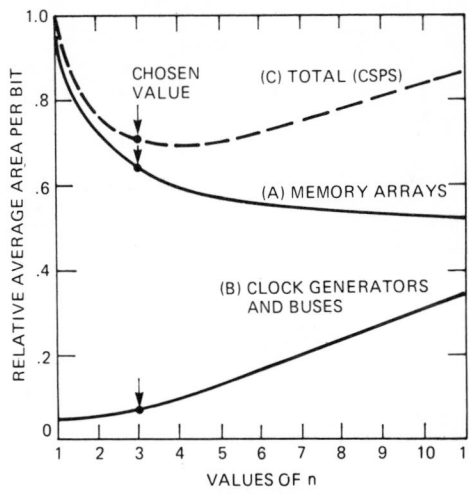

Fig. 2. Relative area per bit versus the number of adjacent bits in a
string n.

storage for as many as M bits in each of the serial registers; for example, a two-phase register contains two stores per bit, so that an $M/2$-bit register contains one store for every parallel register. An *interlaced* structure as shown in Fig. 1(b), having input and output serial registers of only $M/2$ bits each, permits a significantly smaller repeat spacing of the parallel registers, which otherwise would be limited by that of the serial registers. The first storage locations of all the parallel registers can then be loaded with data by two successive dumps from the input register, one dump being made from the ϕ_1 stores, and the second dump being made from the ϕ_2 stores after the input register has been refilled. A corresponding procedure can be followed for the output dumps.

After minimizing the repeat spacing of the parallel registers, the next step in achieving highest packing density (again for a given set of layout rules) is to minimize the repeat spacing along the parallel registers. A two-phase structure requires two storage locations per bit, but the density can be increased by almost a factor of two by using the "electrode-per-bit" (E/B) organization, in which a serial string of n bits can occupy adjacent storage locations [4]. The bits are transferred one at a time in sequence, making use of an extra, vacant location which therefore "travels" in the opposite direction to that of transfer. The relative area per bit is then represented by a factor $(n + 1)/n$, compared to a factor 2 for ordinary two-phase using the same layout geometries. The $n + 1$ storage electrodes require individually timed clock waveforms or "ripple clocks," each of which must have fast edges, because the individual transfers must take place one at a time, within a fraction $1/(n + 1)$ of the transfer period for the complete string. Each of the $n + 1$ transfer electrodes may be driven from the same clock as is applied to the following adjacent storage electrode, as is done in ordinary two-phase clocking. In practice, the charge-handling capability can be approximately doubled by using separate clocking of the transfer electrodes, making a total of $2(n + 1)$ clocks, because this enables the use of "full-bucket" charge storage and transfer, increasing the amount of charge reaching the sense amplifier, with resulting increase in noise margins.

Because of the need for a high net rate of data transfer, the ripple clock organization is best suited to a structure with parallel-multiplexed paths, where the parallel transfer rate is only a fraction $1/M$ of the incoming data rate. This has been called the "multiplexed electrode-per-bit" organization (ME/B) [4] and is particularly well-suited to being adapted to the SPS type of multiplexed structure. However, another problem is the difficulty of providing a sequence of $n + 1$ clocks on successive locations. It could be done by routing a single high-speed clock to each location in turn by using a decoder or ring-counter circuit, but the geometrical layout of this circuitry may actually require a larger repeat spacing than the CCD electrodes. This "pitch limitation" problem is comparable to that which limits the packing density in RAM's. The difficulty can be avoided by adapting the ripple clock principle to short data strings, requiring only a small number of clock waveforms which can be distributed along buses. The clock waveforms also become easier to generate, because more time is available for each individual transfer, and therefore fast edges are not required. Although the packing density factor $(n + 1)/n$ is less favorable when a small value of n is chosen, this disadvantage is more than outweighed by the advantage of being able to retain the same closely packed electrode structure as for ordinary two-phase clocking. As Fig. 2 illustrates, the area advantage obtained by increasing the value of n is relatively small for values of n greater than three, while the area occupied by clock drivers and distribution buses increases linearly with n. The total area reaches a minimum for quite small values of n, the actual position of the minimum depending in a complicated way on geometrical layout considerations. The dotted curve showing total area in Fig. 2 has been drawn by hindsight, using area parameters obtained on the 16 kbit chip, and extrapolated to other values of n.

The combination of an interlaced serial-parallel-serial organization with a ripple-clock scheme characterized by small values of n provides a high density of storage, and will be referred to here as "condensed serial-parallel-serial" or CSPS organization. The factor $n + 1$ will be referred to as the ripple spacing.

The choice of CSPS as the optimum organization for CCD memories is only justified if the array size is sufficiently large, otherwise the area occupied by the serial registers and sense amplifiers, and the power they consume, become disproportionately large. Fig. 3 illustrates the relationship between array size and average area per bit, for CSPS arrays having the same ripple spacing and multiplexing factors, cell dimensions, sense amplifier, and input and output buffers as used in the 16-kbit CCD. The averaging takes into account the areas oc-

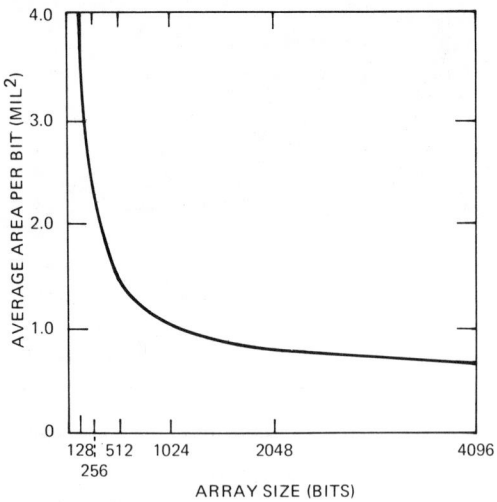

Fig. 3. Average area per bit versus the array size for CSPS structures with M = 32 and n = 3.

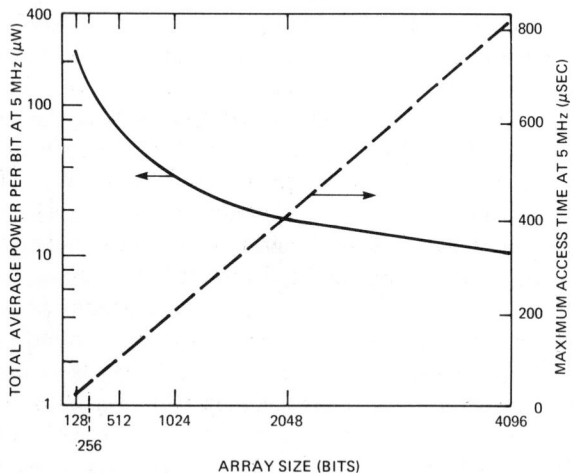

Fig. 4. Average power consumption and maximum access time as a function of array size for CSPS structures with M = 32 and n = 3.

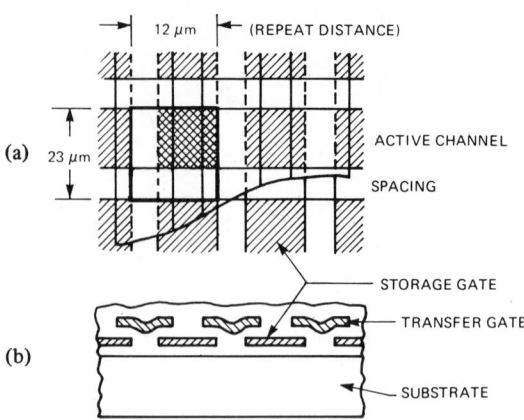

Fig. 5. Layout of the storage and transfer electrodes in the parallel array of the CSPS structure. (a) Plan view. (b) Cross section.

cupied by the above circuit features, but does not include the area occupied by the clock drivers, clock distribution buses, and other support circuitry, since these facilities can be shared by a number of arrays and therefore do not greatly influence the choice of size for each individual array. Fig. 4 illustrates the corresponding effect of array size on average power consumption per bit. Once again, the contribution made by the shared support circuitry is not included. It may be seen that the CPS organization is more favorable for larger arrays, containing at least 1 kbit each.

III. OPTIMIZATION OF MEMORY CHIP

Considerations such as those illustrated in Figs. 2, 3, and 4 led to the selection of a CSPS organization having the particular value n = 3 for the number of adjacent bits in each string, taking into account the area efficiency factor $(n + 1)/n$, and the need to avoid undue complexity and area in the clock generator circuitry and clock distribution buses. In the geometrical layout which was chosen, the area of each storage location is 278 μm^2 (0.43 mil^2), including the area of the associated transfer gate and isolation between adjacent parallel

rows, as illustrated in Fig. 5. Allowing for the factor $(n + 1)/n = 4/3$, the effective area occupied by each bit stored in the parallel part of the array is 368 μm^2 (0.57 mil^2). In order to increase the charge-handling capability of the array, and so to broaden the operating margins at the sense amplifier, the option previously discussed of clocking the storage and transfer electrodes with different waveforms was adopted. This required the generation of eight clocks, rather than four for the main parallel array. A multiplexing factor M = 32 was chosen, providing 32 parallel registers for each array. Since two arrays are paralleled to make up each block, the effective multiplexing factor is 64 relative to the external data stream, so that at a data rate of 10 MHz the parallel clock frequency is only 156 kHz. Providing two paralleled arrays is considerably more favorable than providing a single array with 64 parallel paths and the same overall block size, since the number of serial transfers experienced by each bit is halved, and the serial transfer frequency is also halved. An array size of 2048 bits was chosen, providing an average area per bit of 419 μm^2 (0.65 mil^2) for each CCD array, equivalent to 516 μm^2 (0.8 mil^2)/bit when the array support circuitry is included, as assumed in Fig. 3. Only a small advantage in area per bit and power per bit would have been achieved by choosing a larger array size than 2048 bits, while the access time would have lengthened in proportion. However, another reason for this choice was that it led to a quad-4-kbit configuration on the 16-kbit chip, which was convenient since a standard 16-pin package provides enough pins to permit individual input and output connections to each block, in addition to those required for the dc power rails, two external clocks, and two control inputs.

A third consideration was associated with leakage currents and the corresponding maximum refresh time of a CCD. Since present-day 4-kbit dynamic MOS RAM's are being manufactured to meet a specified refresh time of 2 ms at temperatures up to 70°C, the same considerations apply to CCD's, because the leakage currents have the same physical cause. A refresh time of 2 ms would indicate a minimum working data rate of 2 MHz, which is far enough removed from the maximum of 10 MHz to provide a flexible speed range. In practice a lower data rate than 2 MHz should be practicable, because the maximum storage time is determined

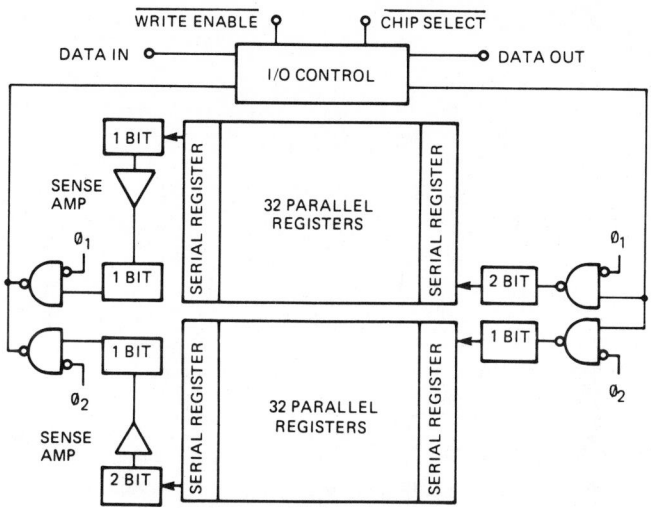

Fig. 6. Block diagram of one 4096 bit shift register on the 16-kbit chip.

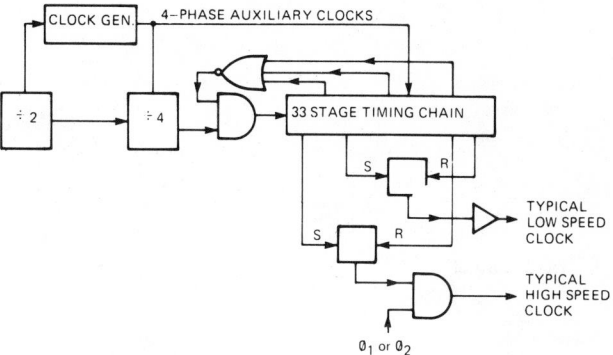

Fig. 7. Block schematic diagram of on-chip clock generation circuitry.

by the average leakage current of many storage locations, rather than being limited by the highest leakage location, as in the case of RAM's.

The use of a CSPS structure greatly reduces the undesirable effects of transfer inefficiency, compared to a conventional SPS structure storing the same number of bits. Whereas a conventional two-phase SPS array of 2-kbit size and having 32 parallel paths would subject each charge packet to approximately 66 high-speed transfers and approximately 126 low-speed transfers, the corresponding approximate figures for a CSPS array (assuming $n = 3$) would be only 34 high-speed transfers and 84 low-speed transfers. The benefits of paralleling two arrays could, of course be obtained in either case. The actual numbers of transfers for each charge packet in the 16-kbit chip are 37 high-speed transfers and 84 low-speed transfers, for the following reasons. In addition to the regular serial shifts and serial-parallel and parallel-serial shifts, there are several extra high-speed transfers. One extra transfer results from a preliminary storage electrode, used for defining the amount of charge launched. Another simplifies the paralleling of the two arrays for every block: an extra storage electrode is placed before one of the two CSPS arrays, while an extra storage electrode is placed after the other array. In this way, the two arrays accept data from the input buffer in opposite phases, while being clocked together by common clock waveforms. Another extra transfer is used in conveying charge from the CSPS array to the input node of the sense amplifier. Preliminary experiments showed that the 2-kbit CSPS array could transmit data at clock rates up to 4 MHz without use of a bias charge and with adequate operating margins. However, to achieve satisfactory margins at 5 MHz clock rate (10 MHz on the 16K chip) a bias charge (fat zero) of between 10 and 30 percent was found to be helpful.

Fig. 6 shows the organization of one 4-kbit block of memory, including the bit delays associated with the extra transfers. Each transfer causes a delay of one data bit, which is one half of the clock period. The extra bit delay which follows each sense amplifier is discussed in the next section. Despite the extra bit delays, and the complication resulting from the

two-phase interlacing of the serial-parallel transfers, the total number of bits stored in the block at any time has been set at 4096 by proper timing of parallel-serial transfers relative to the timing of the serial-parallel transfers.

IV. SUPPORT CIRCUITRY

A block schematic diagram of the on-chip clock generation circuitry is shown in Fig. 7. The external clocks ϕ_1, ϕ_2 drive a divide-two counter, which in turn generates a set of four-phase clocks for operating the internal circuitry. A divide-4 circuit and feedback logic provide an impulse at the parallel clock rate into a 33-stage timing chain, which in turn provides impulses at the proper times to each of the clock waveform generators. Because of the three feedback tappings on the timing chain, the divide-4 circuit actually functions as a divide-16 counter. Each clock generator can be represented as a set/reset flip-flop, which is set by an impulse from one point on the timing chain, and reset from another point. Those clock generators which operate on the main parallel arrays must drive high capacitances, but are not required to switch at high speed. The clock generators which are associated with transfers into and out of the serial registers drive only small capacitances, but must generate fast edges to synchronize with the serial transfer processes. Therefore, the MOS transistors employed in the output driver of each clock generator are all of comparable size, with channel breadth about 30 times the channel length. These provide more than adequate driving capability for the 16-kbit memory. The counters, timing chain, and clock generators have been designed to clear rapidly any illegal states which might be picked up as a result of external clock irregularities. The circuitry also recognizes a "start-up mode" whenever dc power is applied while both external clocks are low, which does not occur during normal operation. Following recognition of the start up mode, the internal clocks will commence with predictable phases relative to the first turn-on of either of the external clocks, and input data can be supplied within 43 or 45 clock cycles, depending on which external clock turns on first.

The data input buffer was designed to accept the incoming data at TTL levels and to supply the CCD input with the appropriate MOS levels. For a high-level input state, the buffer generates an intermediate voltage level on the CCD input gate to launch a "fat zero" charge of approximately 20–30 percent of a full bucket into the CSPS array. Conversely, for a low

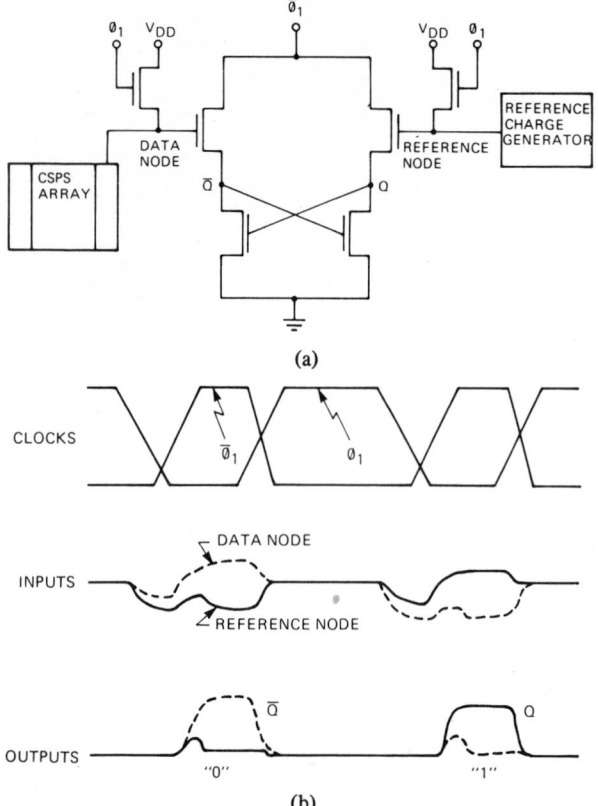

Fig. 8. Sense amplifier for the CSPS array. (a) Circuit schematic. (b) Clock and signal waveforms (from computer simulation).

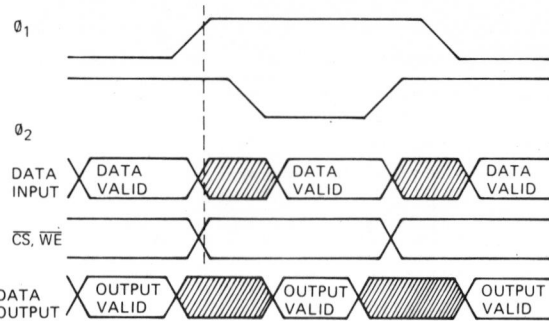

Fig. 9. Operating waveforms and timing diagram of the memory device.

level input state a high-level voltage is supplied to the CCD input and a full bucket of charge is launched. Locally generated $\bar{\phi}_1$ and $\bar{\phi}_2$ pulses were used at the CCD inputs to provide two-way data splitting.

The sense amplifier at the output of each 2-kbit array is a differential-input flip-flop, which differs, however, from the sense amplifiers commonly used in MOS RAM's [5], for the following reasons. In the RAM flip-flops, the refresh operation is achieved by positive feedback to the input nodes, one input being raised high, and the other being brought down almost to V_{SS} potential. To bring either input node to V_{SS} would cause errors in the present device, since the node could act as a source, allowing charge to run back into the last storage location of the CCD. This problem is inherent in two-phase CCD's [1]. To avoid this problem, a "load-steered" flip-flop was developed, as shown in Fig. 8, whose operation is as follows. The data node and reference node are precharged high during ϕ_1 clock high level, then ϕ_1 turns off, dumping the data charge packet and the reference charge packet onto their respective nodes. Almost simultaneously, internal clock $\bar{\phi}_1$ turns on, bootstrapping the input nodes to relatively high potentials. Depending on the state of the data, one load element passes more current than the other, and by positive feedback on the driving elements the flip-flop is driven into one of its two states. From computer simulation it was found that for a differential input of 0.5 V the realization time of the flip-flop connected to its load capacitance was approximately 30 ns, limited by the rise time of clock $\bar{\phi}_1$ [Fig. 8(b)]. To achieve such high speed it was necessary to

avoid loading the flip-flop with excessive capacitance. This was achieved by using dynamic MOS principles in designing the data output buffer. Functionally, this involved storing the output state of the flip-flop, and driving the output buffer during the following bit time. This also reduced the effective propagation delay between the clock transition and the appearance of valid data at the output terminal to the delay of the output buffer alone, so that the operating frequency was not limited by the output circuitry. The tristate output buffer was designed to drive at least one low power Schottky TTL gate. Individual 2-kbit CSPS arrays using this sense amplifier and output buffer were successfully operated at a speed of greater than 7 MHz, which would be equivalent to a data rate of 14 MHz on the 16K chip.

Two control inputs, CHIP SELECT ($\overline{CS}$) and WRITE ENABLE ($\overline{WE}$), and associated circuitry were provided, designed for operation up to 10-MHz data rate in three functional modes: RECIRCULATE, ($\overline{CS}$ high), READ and RECIRCULATE ($\overline{CS}$ low, $\overline{WE}$ high), READ and WRITE ($\overline{CS}$ low, $\overline{WE}$ low). In READ and WRITE mode, the 4-kbit blocks function as digital delay lines, which may be serially interconnected to form larger blocks with no penalty in maximum data rate. It was considered to be important that CHIP SELECT would operate from ordinary 5-V TTL signals, since these signals must be generated individually for all the memory devices which are OR tied to common data buses. WRITE ENABLE requires a full clock level, but this is not a serious disadvantage because this signal can be supplied to a group of devices from a common driver.

Fig. 9 shows the timing diagram of the operating waveforms of the 16-kbit memory. Only two high level clocks (ϕ_1, ϕ_2) and a high level $\overline{WE}$ control input are needed, all data inputs, data outputs and $\overline{CS}$ control input are TTL levels.

V. CHIP LAYOUT

The timing chain and clock generators which were common to all four memory blocks were distributed along the center axis of the chip, since this was found to minimize the area required by interconnections. The resulting layout is shown in Fig. 10. The four data input buffers were located along the central region also, so that the sense amplifiers, output buffers, and local clock generators could be situated along the two longer sides of the chip. The input/output control circuits were placed along the shorter sides, as were all the bonding pads for external connections, since this provided a rectangular

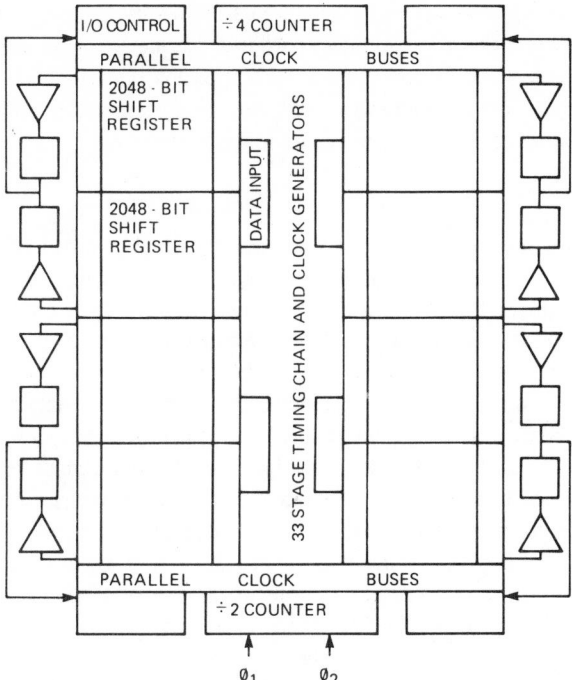

Fig. 10. Chip layout diagram of the 16 384-bit CCD memory.

Fig. 11. Photomicrograph of the 16-kbit chip.

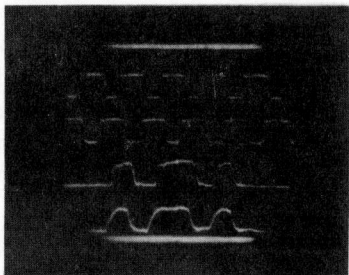

Fig. 12. Typical waveforms of the clocks and data. Top two traces: ϕ_1 and ϕ_2, 10 V/div. Bottom two traces: data input and data output, 5 V/div. Time scale: 200 ns/div.

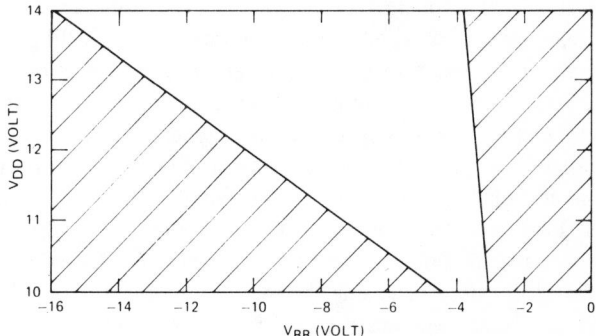

Fig. 13. Typical operating range of the supply voltages at a data rate of 10 MHz.

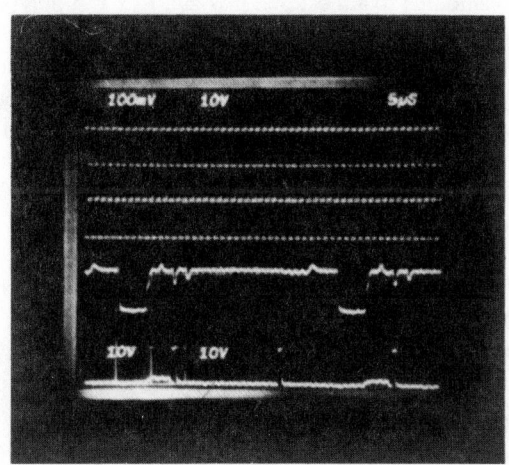

Fig. 14. Clock waveforms as observed on 16-kbit chip. Top two traces: external clocks ϕ_1 and ϕ_2. Bottom two traces: examples of internally-generated clocks: upper trace shows one of the four low speed clocks which drive the storage electrodes of the parallel array; lower trace shows the high speed serial-parallel strobe pulse. All traces 10 V/div, 5 μs/div.

chip format which was more economical in total area, and also more suited to being fitted into a standard 16-pin dual-in-line package. The overall dimensions of the chip are 3.45 mm (136 mils) by 4.29 mm (169 mils), the average area per bit of storage being 900 μm^2 (1.4 mil^2). Comparing this figure with the 516 μm^2 (0.8 mil^2)/bit occupied by the CSPS arrays and local support circuits alone, it can be seen that the common circuit overheads increase the area per bit by a considerable amount, although the area per bit is still very attractive compared to MOS 4-kbit RAM's which typically occupy a chip of the same size.

VI. EXPERIMENTAL VERIFICATION

The 16-kbit CCD memory was designed and fabricated using a two-level polysilicon gate n-channel process [6]. A photomicrograph of the finished device is shown in Fig. 11.

For most purposes of testing for data throughput, the input data were set up to fill the parallel arrays with diagonal patterns of isolated ONE's in a background of ZERO's, and then inverted to replace this by a pattern of isolated ZERO's in a background of ONE's. These patterns were generated by setting a ONE in a background of ZERO's, repeating the ONE every 66 bits for a total of 4096 bits, and then generating the complementary sequence for another 4096 bits. The output data were verified by comparison with a repeat of the input test data.

It was found that these patterns were effective for most purposes, and especially for measuring the operating margins of the device at high frequencies, because the effects of charge transfer inefficiency are especially pronounced for isolated ONE's or isolated ZERO's. An example of the input and output data waveforms at a data frequency of 10 MHz is shown in Fig. 12. Diagonal test patterns do not show up clearly on the monitoring oscilloscope, and therefore a nondiagonal test pattern was used to make this photograph. Fig. 13 shows an example of the operating range of dc and clock supply voltages, also at a data frequency of 10 MHz. Two examples of the internally-generated clock waveforms are shown in Fig. 14, together with the two external clocks ϕ_1, ϕ_2 for comparison, the external clock frequency being 5 MHz. One example shows one of the low-speed clocks applied to storage electrodes in the parallel array, running at the parallel transfer frequency of 156 kHz. The uneven amplitude of the clock high level is caused by capacitive coupling with adjacent electrodes, and is not a defect. The second example shows a strobe clock, used to enable the serial-parallel transfers. This clock produces two pulses at the parallel transfer frequency, one pulse having ϕ_1 timing, and the other having ϕ_2 timing, as required for the two-phase interlacing.

The principal contribution to operating power is the current drain from the positive dc supply, (V_{DD} to V_{SS}), and at 10 MHz this amounts to almost 300 mW at a +12 V supply, about 18 μW per bit. Half of this power is dissipated in the output buffers and sense amplifiers, for high speed sensing and to provide TTL compatible outputs. The other half of the power is dissipated equally among 1) the data input buffers with local clock inverters, 2) the $\overline{CS}$ inverters for I/O control, and 3) the on-chip clock generation circuits. It is evident that a fair proportion of the power is dissipated to achieve TTL compatibility. Notice that only a fraction (1/6) of the total power is needed to generate all the necessary waveforms to drive the CSPS arrays.

A smaller contribution of 60 mW is dissipated in the external clock drivers, resulting in a total operating power of approximately 22 μW per bit at 10 MHz. By monitoring the transient current drain for the dc power supply, it was verified that the major current drain occurs during ϕ_1 - ϕ_2 clock overlap, and overall drain is therefore reduced greatly by reducing the over-

lap to between 10 and 20 ns, which is readily achieved using ordinary circuit techniques. At low frequencies the overall drain is reduced, since the clock overlaps occupy a smaller fraction of the total cycle, and the remaining power drain is largely caused by the eight sense amplifiers.

VII. CONCLUSIONS

This work has demonstrated the practical realization of a 16 384-bit CCD memory on a silicon chip which is small enough for high yield, low cost manufacturing by present-day techniques, and suitable for applications requiring low overall system cost per bit of storage. The result was made possible by the introduction of a condensed serial-parallel-serial organization as a way to achieve optimum packing density, taking into account the practical problems of peripheral pitch limitations, and the need to generate the clock waveforms on-chip by circuitry of small total area. Although many improvements of detail can still be made to the design, to reduce chip size and power drain still further, the device as described should be a suitable candidate for filling the "access gap" in memory technologies. In addition, extension of the CSPS concept to memory chips of at least 64 kbits can be readily predicted, making use of new developments in processing technology which are already being applied to the development of other MOS and CCD structures.

ACKNOWLEDGMENT

The authors wish to thank D. R. Colton and C. R. Robinson for important contributions to the CSPS concept, and J. J. White for consultation regarding process tolerances and yields.

REFERENCES

[1] S. D. Rosenbaum and J. T. Caves, "8192-bit block addressable CCD memory," *IEEE J. Solid-State Circuits*, vol. SC-10, pp. 273–280, Oct. 1975.
[2] S. Chou, "Design of 16 384-bit serial charge-coupled memory device," this issue, pp. 78–86.
[3] K. C. Gunsagar, M. R. Guidry, and G. F. Amelio, "A CCD line addressable random-access memory (LARAM)," *IEEE J. Solid-State Circuits*, vol. SC-10, pp. 168–272, Oct. 1975.
[4] D. R. Collins, J. B. Barton, D. C. Buss, A. R. Kmetz, and J. E. Schroeder, "CCD memory options," in *1973 Tech. Dig. Papers, Int. Solid-State Circuits Conf.*, Feb. 1973, pp. 136–137.
[5] K. U. Stein, A. Sihling, and E. Doering, "Storage array and sense/refresh circuit for single-transistor memory cells," *IEEE J. Solid-State Circuits*, vol. SC-7, pp. 336–340, Oct. 1972.
[6] J. J. White, "Silicon gate device structures with two or more gate levels," Canadian Patent 941 072, Jan. 1974; U.S. Patent 3 897 282, 1975.

A CCD MEMORY FOR RADAR SIGNAL PROCESSING

R. A. Belt
Air Force Avionics Laboratory
Wright-Patterson AFB, Ohio 45433

ABSTRACT

The results of an Air Force Avionics Laboratory program to develop a CCD digital memory device are discussed. Chip design and performance characteristics are given and compared with other known CCD devices. The use of the 16K bit CCD memory for radar signal processing is discussed and a specific system example is shown. CCD memory cost is projected and compared with other leading semiconductor memory approaches.

INTRODUCTION

The Air Force has a significant number of applications for 10^6 to 10^8 bit memories which do not require non-volatile operation. Among these are synthetic aperture radar processors, scan converters, radar change detectors, high speed swapping discs in communications systems, and buffer memories for video sensor bandwidth compression. In 1972 the Air Force Avionics Laboratory initiated a program with Fairchild Camera and Instrument Co. (Ref 1) to develop a low-cost general purpose digital CCD memory chip that would offer significant cost and performance advantages for such systems. That program has now concluded with the demonstration of a 16K bit device and the partial development of a 32K bit device, both of which have potential application in military and commercial systems.

DISCUSSION OF CHIP DESIGNS

The Fairchild devices employ a novel memory organization to achieve fast access time along with low power dissipation. This organization is known as the Line-Addressable Random Access Memory (LARAM) approach. (Fig. 1) As implemented with a 1-1/2 phase CCD structure, this organization achieves bit densities comparable to either the serpentine or SPS types while reducing chip access time and clock load capacitance by an order of magnitude. The principle of operation involves clocking a string of bits into a selected register via a common input bus. The clocks are then halted and the information stored in a static fashion as in an MOS RAM type device. As long as the memory returns to this register before the leakage charge exceeds the unity threshold, any other register may be selected at random and undergo a READ, WRITE, or RESTORE operation. When the original register is again selected, the information is clocked out, sensed, and either recirculated or replaced with new data as appropriate. The key to obtaining a compact chip layout is the 1-1/2 phase CCD structure, which permits the single selectable clock line to run the entire length of each register.

16K BIT CHIP DESIGN

A block diagram of the 16K chip is shown in Fig. 2. The chip is organized as four parallel sections of 4K bits each. Each section has its own data input and output pin. A decoder for each section selects one of the thirty-two registers in that section, resulting in only four registers being clocked on the chip at any one time. The active power dissipation is thus reduced, allowing clock drivers on the chip itself, resulting in a simpler device to the memory user. On-chip control logic can switch the chip into either a WRITE mode,

Reprinted from *Proc. 1975 Naval Electron. Lab. Center Int. Conf. on the Application of Charge-Coupled Devices*, Oct. 29–31, 1975, pp. 413–421.

an NDRO READ mode, a R/M/W mode, or a data REFRESH mode. In the REFRESH mode the data is recirculated on-chip with the I/O circuits powered down.

Data rate is variable from 400 KHz to 5 MHz except in the R/M/W mode where the maximum data rate is 3 MHz. Data can be read successively into the same register or into adjacent registers with no loss of speed, although power dissipation will be higher in the latter mode due to the higher duty cycle of the address circuits. Data rate is continuous in all chip modes and all addressing patterns. Access time to a random register is 200 nanoseconds. The average access time is 12.8 microseconds, and the maximum access time to a random bit is 25.6 microseconds.

A picture of the chip is shown in Fig. 3. The chip is 201 by 219 mils, and is packaged in a standard 22 pin dual-in-line package. The chip dissipates 200 mW at a 5 MHz data rate, and less than 50 mW in the REFRESH mode. Only one 12V clock is required, which has a 200 pF capacitance load. All pins except the 12V clock are fully TTL compatible with no pull-up resistors required. The device has tristate outputs to facilitate paralleling the chips in a memory system.

32K BIT CHIP DESIGN

A block diagram of the 32K bit device is shown in Fig.4. The device is identical in design and operation to the 16K bit device, except that it has only one input pin and one output pin for the entire chip. This is achieved by means of an additional on-chip decoding circuit which selects one of the eight simultaneously clocked registers for routing to the output buffer. The other seven registers are automatically recirculated on-chip. A picture of the device is shown in Fig. 5. The entire chip is 240 X 240 mils, and is packaged in an 18 pin ceramic DIP with 0.4 mil pin spacing. Clock capacitance is only 200 pF for a single clock line.

CHIP CONSTRUCTION FEATURES

Both devices share the same CCD structure, sub-circuit designs, and fabrication sequence as the Fairchild 9K CCD memory already on the market (CCD450). The CCD registers are two-phase n-channel buried

channel structures with overlapping poly-silicon gates. A dual nitride-oxide gate dielectric is used, which places all gate electrodes on the same level of gate dielectric. Ion implantation through the first level polysilicon CCD electrodes generates the required asymmetry for two-phase operation. Dielectric isolation is used in the MOS support circuits to obtain higher packing density. The fabrication sequence requires nine photomasks, of which only two involve critical mask alignments.

Sub-circuits common to both devices include level shifters, decoders, line drivers, sense amplifiers, control logic, and I/O circuits. A dual differential sense amplifier is used as in the TI 4K RAM, with slight modification for sensing charge rather than current. Amplifier sensitivity is further increased by a novel capacitance decoupling circuit, which reduces the loading of the CCD outputs on the amplifier input node. The sense amplifier compares the charge levels in the selected register with a reference level in a dummy register located in each memory subsection. Since the reference level can be adjusted by geometric factors, threshold voltage tracking is achieved without the need for dedicated circuits. This feature, plus conservative design rules, permits wide operating margins and a higher device yield.

The present status of 16K chip development is characterized by the capability to produce small numbers (>100) of fully working prototypes at occasionally excellent yields (>30%). A large number of additional devices are fully functional except for one line or one section, indicating that further yield improvement is possible. Dark current is still a problem and, if not corrected, could restrict the temperature range to less than 50°C. Experiments with variations in process steps show promise for reducing the number of dark current spikes as well as the residual level of leakage charge, indicating that a temperature range of 0°C to 50°C may still be possible. Device modification and yield improvement is continuing under an AFML contract (Ref.2) as well as under company produce development funds. Sample quantities should become available to users by the early fall of 1975, with limited production quantities available approximately six months later. The device is expected to be competitive in a sizeable commercial market.

The development of the 32K bit device lags the 16K device by approximately six months. This is attributable to the greater complexity of the device, the tighter design rules, and larger chip size involved. It is possible that this device will be made available commercially also, although market volume maybe higher for the 16K device.

COMPETING 16K BIT DEVICES

Two other 16K bit CCD memory devices have been announced to date, the Intel 2416 device and the Bell Northern CC16M1 device. The Intel device is available commercially whereas the BNR device is basically a developmental prototype. However, both devices have characteristic features which can be contrasted with the 16K LARAM device.

INTEL 2416

A block diagram of the Intel CCD is shown in Fig. 6. The chip is organized into blocks of 256 bits which communicate with a single set of I/O pins through a 1 of 64 decoder, giving the chip a one bit swath. The decoder operates like a 64 bit RAM -which transfers one bit at a time from each of the 256 bit registers. On-chip control logic determines whether the chip is in a NDRO READ, WRITE, R/M/W, or standby SHIFT cycle. Cycle time for the RAM decoder is 460 ns. for a READ or WRITE cycle and 620 ns. for a R/M/W cycle. Random access time for the READ cycle is 250 ns. to the first bit of a block. The minimum shift cycle period is 750 ns. This gives a maximum chip access time of less than 200 microseconds to any random bit, and a maximum internal shift rate of 1.33 MHz in the search mode.

The decoder can be used to read consecutive bits out of the same 256 bit block or consecutive bits out of successive adjacent registers. The first alternative requires one RAM cycle and one SHIFT cycle for each bit transferred, limiting the maximum data rate to 800 KHz for the READ and WRITE modes or 730 KHz for the R/M/W mode. The second alternative permits a higher 2 MHz data rate for the READ and WRITE mode and a 1.33 MHz rate for the R/M/W cycle; however, in this case the data rate is not continuous because each time the RAM is completely loaded, a SHIFT operation must take place to transfer the data to the CCD registers. Minimum refresh considerations

require further that a SHIFT operation occur at least every 16 RAM cycles at 70°C.

A picture of the chip is shown in Fig. 7. It is 143 X 237 mils long and is packaged in a standard 18 pin plastic DIP. The device has four 12 to 14V clock inputs, two with a 500 pF load capacitance and two with a 700 pF load capacitance. Total power dissipation is 600 mW at 1 MHz, which decreases to 150 mW in the 125 KHz standby SHIFT mode. Chip I/O lines are TTL compatible, although pull-up resistors are required. The chip inverts data from input to output.

BELL NORTHERN CC16M1

The BNR device (Fig. 8) is organized as four parallel registers of 4K bits each, with separate data input and output pins for each 4K block. Each block consists of two parallel SPS registers of 2K bits each which operate in a push-pull fashion. The SPS registers have a 16 bit input register and 32 parallel 64 bit registers half of which are loaded on alternate clock phases. The data rate at each I/O pin is 1 to 10 MHz, which implies an average access time of 0.2 to 2.0 milliseconds for each 4K block. The chip dissipates only 325 mW at a 10 MHz data rate and only 85 mW at a 1 MHz rate.

The chip generates all its own internal clocks and timing from a single 2-phase clock input. Clock swing is 0-12V and capacitive loading is only 60 pF on each clock line. On-chip control logic switches the chip to either a WRITE, R/M/W, NDRO READ, or an on-chip RECIRCULATE mode. Data and control lines are fully TTL compatible, with no pull-up resistors required. Chip size is only 137 by 170 mils. The device is packaged in a standard 16 pin DIP.

COMPARISON OF 16K DEVICES

A comparison of the three 16K devices is shown in Table 1. The widely ranging characteristics reflect the fact that the devices are directed at different market objectives. The Intel device is directed primarily at the head-per-track disk and drum market, where its faster access time should yield significant system advantages. The Bell Northern device is directed mainly at the MOS shift register market, where it can be used in CRT refresh, terminal storage and communications buffering type applications. The Fairchild device is intended to be a general-purpose shift register type of device which will compete in both of the memory areas cited, but whose unique accessing capabilities might allow the exploitation of new memory system architectures.

Selection of the appropriate device depends strongly on the type of system contemplated. In a computer-type memory system, only a single bit, word, or block of information is accessed at any given time. The primary measure of memory performance is how fast the data can be made available to the CPU. Since only a small fraction of the system is active at any given time, standby power is the dominant feature. The larger the memory becomes, the more standby power must be emphasized.

For a serial buffer or signal processing type of system the primary measure of memory performance is the data transfer rate. Random accessability is not a major advantage since the data sequence is rarely re-ordered. Most importantly, the fraction of devices active at any time is usually quite high, making the active device power the dominant factor. This is especially true in serial signal processing systems, where all devices might be active at any given time.

Fig. 9 plots the total power dissipation versus serial data rate for some candidate serial devices. The standby power is seen to be quite similar for all the devices, reflecting the fact that leakage mechanisms are similar for these NMOS type devices. At clock frequencies near one megahertz, however, the power dissipation differs greatly, reflecting differences in chip organization and in the ratio of capacitive loading to DC loading per bit. Clearly, the power savings can be substantial for large serial data systems if the new CCD memories are used.

Keeping these comparisons in mind the BNR device would likely be preferred for shift-register memory applications due to its high data rate, low power dissipation, small package size, wide temperature range, ability to string registers, and potential for low chip cost (ie. small chip size and standardized process). The Fairchild device would rank a close second with its more

complicated overhead requirements, smaller temperature range, larger package size, and more complex fabrication process. The Intel device would rank a distant third, with a heavy penalty attached to its discontinuous data rate, single bit swath, data inversion, and increased overhead requirements. Cost considerations based on volume sales capability could influence the preference somewhat if low system cost is a primary objective.

A different order of ranking results if a disk-type memory system is considered. Here, the Intel device is most highly preferred for its fast access time, single bit swath, short time to refresh, small package size, higher temperature range, and lower cost potential. The Fairchild device compares favorably, but is penalized somewhat for its four bit swath, lower temperature range, more complicated fabrication process, and either a long or an unconventional refresh cycle. The BNR device ranks third for its slow access time and long time to refresh. All three devices can be improved somewhat for a disk-type memory application through simple design modifications. Such modifications might include the sharing of a common I/O pin, decoding on-chip for a single bit swath, and in general, simplifying the chip interface requirements.

APPLICATION OF CCD's TO RADAR SIGNAL

PROCESSING

Synthetic aperture radar systems are notorious for the digital memory capacities they require. A block diagram of a typical system is shown in Fig. 10. Essentially, the system collects high frequency radar returns at a low duty cycle, stretches them to achieve a lower bandwidth at a continuous duty cycle, filters the data to extract the imaging information, and presents the information to a display or recorder. Two memories are generally required, one in the processor and one with the display. With high resolution systems, these memories can reach capacities of 10^6 to 10^8 bits or more.

Figs. 11 and 12 show the detailed memory organization for a particular SAR processor known as Synthetic Aperture Precision Processor High Reliability (SAPPHIRE). This processor is currently being developed by Goodyear Aerospace Corporation under ASD/AFAL sponsorship. (Ref 3.) The small rectangles in Fig. 12 represent 1 Kbit blocks of data. Each block stores data from consecutive radar returns and presents it to the complex multipliers sixteen returns at a time. In the accessible memory, each block receives new data every sixteenth return in an interleaved pattern. The entire memory is 23.6 megabits in size and dissipates 2350 watts.

Fig. 13 shows how the Fairchild 16K CCD can be used to replace the Intel 2401 2K bit shift register in the accessible portion of the SAPPHIRE memory. Since the non-accessible portion is purely serial storage, direct CCD substitution is a straightforward matter. The entire memory is therefore replaceable with CCD storage with no re-organization required. It is expected that the CCD version will dissipate one tenth the power, occupy 1/3 to 1/4 the volume, and weigh 1/2 to 1/3 as much as the present MOS shift register version. In production, it should cost 1/3 to 1/4 as much as the MOS unit.

COST POTENTIAL OF CCD MEMORY

The cost of CCD memory is bounded by the cost of MOS RAM's to less than 0.1 cents per bit. It is expected that CCD's can achieve even lower costs because roughly four times as many bits can be placed on the same size chip using the same process and design rules.

The projected cost of CCD memory is shown in Fig. 14. This projection by an independent consulting firm agrees closely with estimates from potential CCD manufacturers and large volume memory users. Table II presents a second independent assessment of CCD memory market size. It is still too early to tell whether CCD's will fulfill these optimistic predictions, but the potential is definitely there. Actual system experience will provide the answer over the next two to three years.

The following individuals and companies are acknowledged for their discussions of the status of CCD technology and for furnishing detailed descriptions of device operation: Gordon Moore and Sun-lin Chow of Intel Corp.; Mark Guidry, Gil Amelio, Henry Pao, and Kamil Gunsagar of Fairchild; and Doug Colton, Bill Coderre, and Neil Waterhouse of Bell Northern. Thanks also go to J. Decaire for a critical reading of the manuscript and to Mrs. Penny Carpenter for typing the final version.

BIBLIOGRAPHY

1. Contract F33615-73-C-1044, entitled: "Memory for Radar", December 1972 to June 1975.

2. Contract F33615-74-C-5088, entitled "Manufacturing Method on Charge-Coupled Device (CCD) Memory Arrays", May 1974 to May 1976.

3. Contract F33657-74-C-0604, entitled "Synthetic Aperture Precision Processor High Reliability (SAPPHIRE), July 1974 to December 1975.

4. Quantum Science Corp., in Electronic Design 22, October 25, 1974, p. 43.

5. H.C. Wainright & Co., in Electronic News, February 5, 1975.

TABLE 1.   COMPARISON OF 16K CCD MEMORY CHIPS

CHARACTERISTICS	FAIRCHILD CD460	INTEL 2416	BNR CC16M1
Organization (Swath/select/block size)	4/random/128	1/random/256	4/serial/4K
Operating modes	R,W,R/M/W,RECIRC	R,W,R/M/W,SHIFT	R,W,R/M/W,RECIRC
On-Chip recirculate	YES	YES	YES
Off-Chip recirculate	R/M/W (latch Req'd)	NO	YES
Required power supplies	+125V, +5V, -5V	+12V, -5V	+12V. +5V, -5V
Clock voltages	12V	12V	12V
Clock phases	2	4	2
Clock capacitance	120pF, 15pF	2 @ 500pF 2 @ 700pF	2 @ 60pF
Clock rep rate	0.5 to 5.0 MHz	0.1 to 1.3 MHz	0.5 to 5.0 MHz
Data transfer rate (per pin)	0.5 to 5.0 MHz	0.2 to 2.0 MHz*	1.0 to 10 MHz
Worst case access time	25.6μs	192μs	0.4ms
Average access time	12.8μs	96μs	0.2ms
Time to refresh	0.8ms	49μs	0.4ms
No. refresh clock cycles	4096	64**	2048
Temperature range	0°C to 55°C	0°C to 70°C	0°C to 70°C
Chip power (max)	200mW @ 5 MHz	300mW @ 2 MHz	325mW @ 10 MHz
Standby power	50mW	24mW	85mW @ 1 MHz
Output structure	Tri-state TTL Compatible	Open drain Pull-up resistor req'd	Tri-state TTL Compatible
Chip size	219 X 201	237 X 144	137 X 170
Package size	22 pin ceramic	18 pin plastic	16 pin ceramic
Process	ISO-NMOS(BC)	NMOS(SC)	NMOS(SC)
Start-up clear req'ts	8192 cycles	4096 cycles	43 cycles
Unique features	--	inverts data	--

*   The data rate is interrupted every 16 cycles for a SHIFT operation.

** 128 cycles are required to restore the addresses to their initial state.

TABLE 2. MEMORY MARKET *

	1973		1978	
	Qnty. (Bill. Bits)	Value (Millions)	Qnty. (Bill. Bits)	Value (Millions)
Semiconductor				
Bipolar	16	$250	80	$ 500
MOS	45	208	280	565
SOS	--	2	15	65
CCD	--	--	60	45
TOTAL	61	460	435	1,175
Magnetic				
Core	62	280	50	135
Plated Wire	8	135	15	155
Buble	--	--	10	20
TOTAL	70	415	75	310
Other	5	95	5	65
TOTAL	136	$970	515	$1,550

_____

Note: Total market, including captive production

*Source: Ref. 5

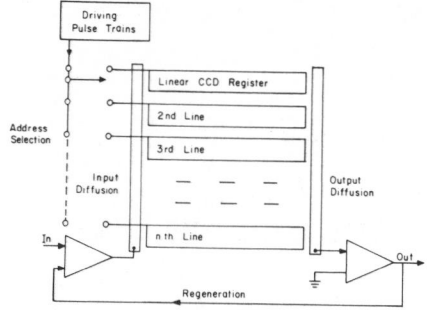

Fig.1. LARAM principle.

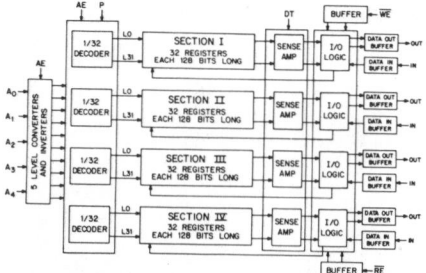

Fig.2. CCD460 block diagram.

Fig.3. CCD460 chip.

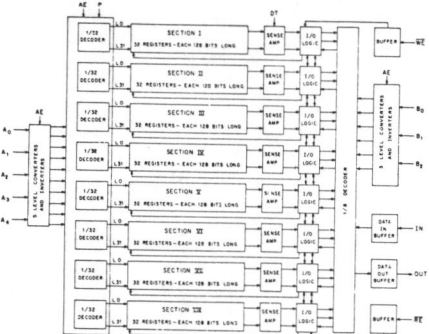

Fig.4. 32K block diagram.

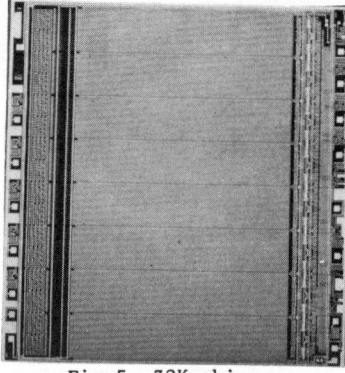

Fig.5. 32K chip.

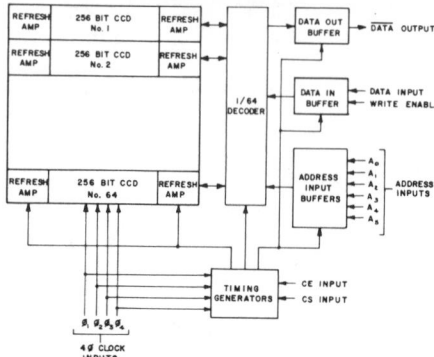

Fig.6. Intel 2416 block diagram.

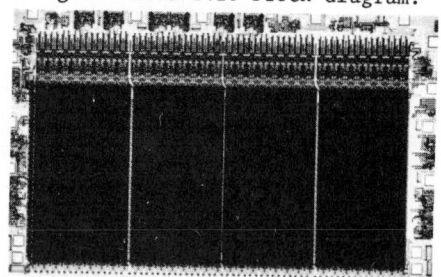

Fig.7. Intel 2416 chip.

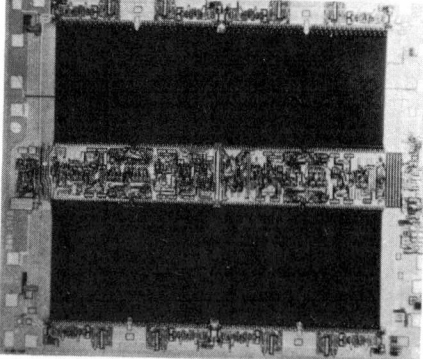

Fig.8. BNR CC16M1 chip.

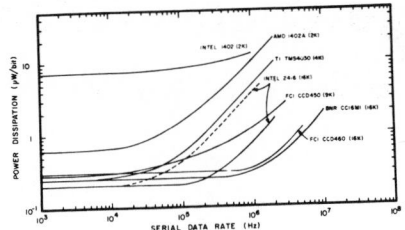

Fig.9. Active power vs. data rate.

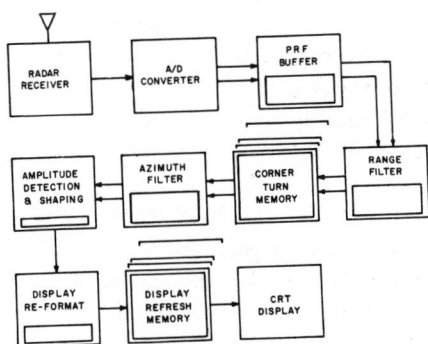

Fig.10. Typical SAR block diagram.

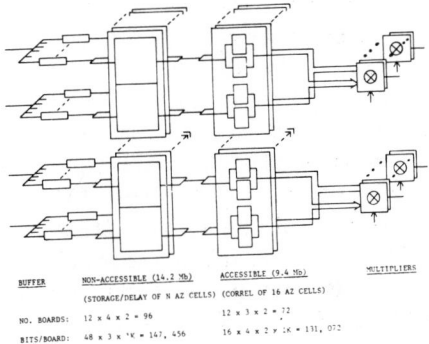

Fig.11. SAPPHIRE memory organization.

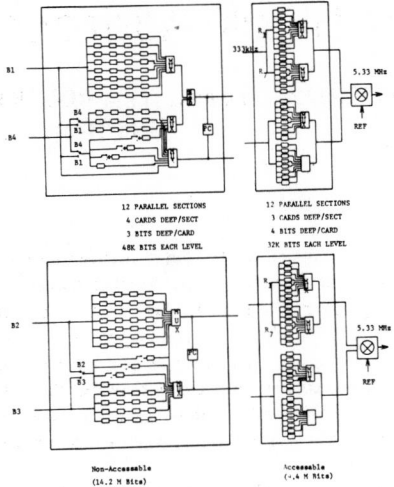

Fig.12. SAPPHIRE circuit boards.

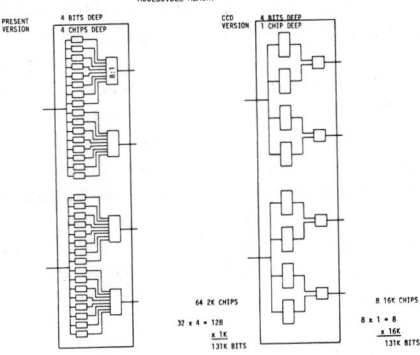

Fig.13. CCD vs. MOS version.

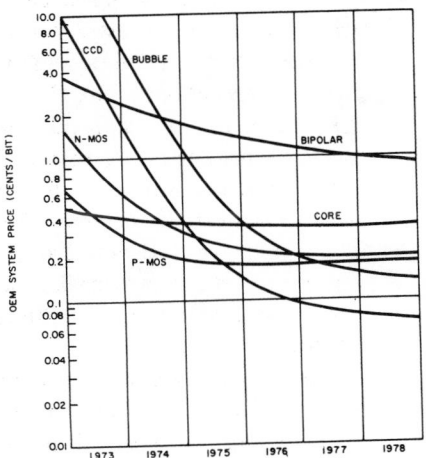

Fig.14. Projected CCD cost.

303

A 64-kbit Block Addressed Charge-Coupled Memory

AMR M. MOHSEN, MEMBER, IEEE, ROBERT W. BOWER, MEMBER, IEEE, E. MARSHALL WILDER, AND DARRELL M. ERB, MEMBER, IEEE

Abstract—This paper describes the design and performance of a 64-kbit (65 536 bits) block addressed charge-coupled serial memory. By using the offset-mask charge-coupled device (CCD) electrode structure to obtain a small cell size, and an adaptive system approach to utilize nonzero defect memory chips, the system cost per bit of charge-coupled serial memory can be reduced to provide a solid-state replacement of moving magnetic memories and to bridge the gap between high cost random access memories (RAM's) and slow access magnetic memories.

The memory chip is organized as 64K words by 1 bit in 16 blocks of 4 kbits. Each 4-kbit block is organized as a serial-parallel-serial (SPS) array. The chip is fully decoded with write/recirculate control and two-dimensional decoding to permit memory matrix organization with X-Y chip select control. All inputs and the ouput are TTL compatible. Operated at a data rate of 1 MHz, the mean access time is about 2 ms and the average power dissipation is 1 μW/bit. The maximum output data rate is 10 MHz, giving a mean access time of about 200 μs, and an average power dissipation of 10 μW/bit. The memory chip is fabricated using an n-channel polysilicon gate process. Using tolerant design rules (8-μm minimum feature size and ±2-μm alignment tolerance) the CCD cell size is 0.4 mil^2 and the total chip size is 218 × 235 mil^2. The chip is mounted in a 22-pin 400-mil wide ceramic dual in-line package.

I. INTRODUCTION

EARLY computer memories were made almost entirely with magnetic devices. Electrically addressed magnetic components such as cores naturally provided relatively high speed, high cost random access memory (RAM). Similarly, mechanically addressed magnetic devices provided relatively low speed, low cost, serial memories. A rather sizable gap resulted in memory access time and cost-per-bit between the electrically and mechanically addressable memories. Thus, this gap is a consequence of a mismatch in electrical and mechanical access time and cost in magnetic storage media. However, computer systems can take advantage of a continuous hierarchy of memory cost and access time.

High cost prevented the use of early solid-state devices for memory application except for latches and registers where high speed and direct interface with logic and control functions were required. As solid-state devices became more integrated and thus cheaper per function they began to compete with cores in high cost random access applications. As of this date, solid-state devices compete successfully with all classes of electrically addressable magnetic memories, but have not ventured into the domain of mechanically addressable memories or into the gap. In Fig. 1 this gap can be seen to span about three decades in access time and over a decade in cost [1]. No fundamental barrier in solid-state technology causes a gap such as the one which developed for magnetic devices. Therefore, it

should be expected that more highly integrated, lower cost solid-state devices will fill this gap.

Charge-coupled memories are presently being considered as contenders for the gap filling technology. Higher packing density and lower power consumption are the major reasons that charge-coupled memories should result in lower systems cost per bit than MOS RAM's. It is generally accepted that a charge-coupled device (CCD) must provide a cost advantage of four to six compared to an MOS RAM to emerge as an economically viable memory technology.

The relative manufacturing cost per bit of RAM and charge-coupled serial memory versus the number of bits per chip is plotted in Fig. 2. The random defects are assumed to follow a Poisson distribution [2] with a defect density of 12/cm^2. The wafer processing cost is assumed constant as is the cost of packaging, assembly, testing, and quality assurance. For small numbers of bits per chip the assembly and testing costs are dominant, while for large numbers of bits the cost of wafer processing is dominant because of yield loss.

The chip area per bit of a RAM is presently about 4.5–7 mil^2/bit while the overlapping gate four-phase CCD electrode structure with two levels of polysilicon is about 1.4–2.0 mil^2/bit [3], [4]. Fig. 2 indicates that the economical chip size for a RAM is 4K with these assumptions and 16K for a CCD. The ratio of RAM to CCD manufacturing cost from these curves is then only a factor of 2.8. This is considerably below the factor of 4–6 felt to be required for a CCD to emerge as a viable memory technology. The chip area per bit for the offset-gate CCD electrode structure [5]–[10] described in this paper is 0.8 mil^2/bit. The economical chip size in this case is 32 kbits. However, by using an adaptive system approach to utilize partially defective chips combined with the offset-gate CCD electrode structure the economical chip size increases to 64 kbits. The relative manufacturing cost of a CCD serial memory is then reduced by an order of magnitude compared with RAM. The parameters of these curves depend on defect density, feature size, and alignment and thus as progress is made in these areas all of the curves will tend to scale to lower cost per bit and greater chip sizes.

The memory chip to be described in this paper is a 64-kbit (65 536 bits) block addressed charge-coupled memory. This chip is designed to take advantage of the high packing density of the offset-gate electrode structure and the defect tolerant adaptive system approach. This combination will allow a CCD serial memory to fill a large portion of the cost and access time gap. The chip is organized as 64K words by 1 bit in 16 blocks of 4 kbits. In the following sections the organization, design, fabrication, and operation of the charge-coupled serial memory are described.

Manuscript received November 11, 1975.
The authors are with Mnemonics, Inc., Cupertino, CA 95014.

Reprinted from *IEEE Trans. Electron Devices*, vol. ED-23, pp. 117–126, Feb. 1976.

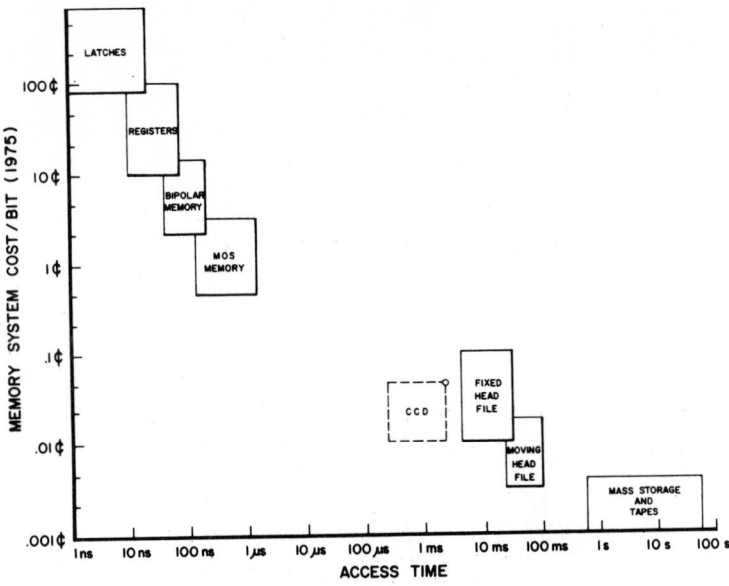

Fig. 1. Memory system cost per bit versus access time.

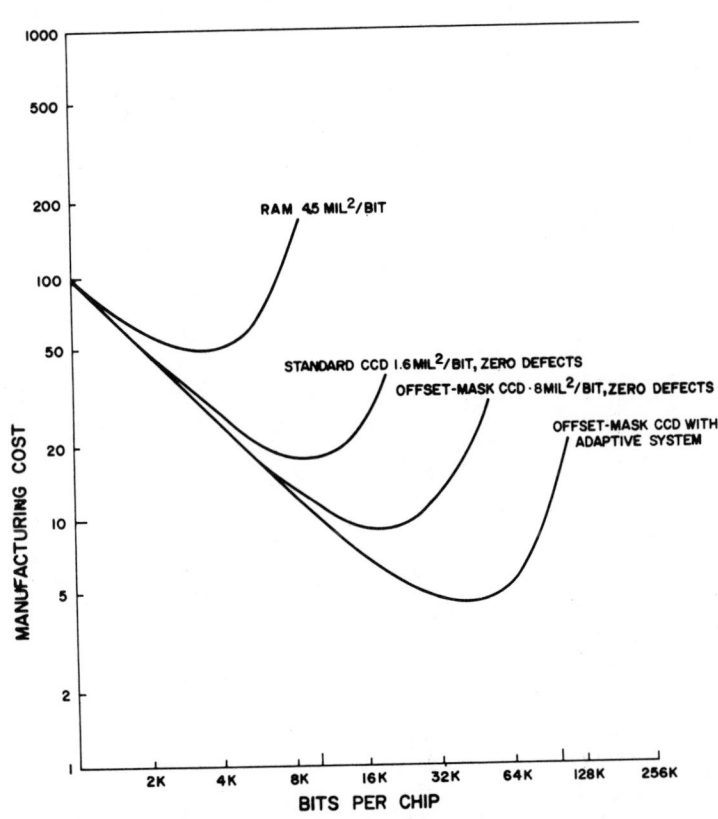

Fig. 2. Projected relative manufacturing cost per bit of serial CCD memories versus number of bits per chip.

II. MEMORY ORGANIZATION

A block diagram of the 64-kbit charge-coupled memory chip is shown in Fig. 3. The memory is organized as 64K words by 1 bit in 16 blocks of 4 kbits.

Each 4-kbit block is organized as a serial-parallel-serial (SPS) array [1] of 64 channels and 64 rows. Each block is designed in a modular form to include a balanced charge-detection circuit, a novel self-metered alignment-compensated charge-injection circuit, and an input/recirculate circuit (Fig. 4). The charge-detection circuit senses the signal charge and sends it to the input/recirculate circuit which either recirculates the old

data or writes new data into the memory. The charge-injection circuit meters the appropriate amount of charge and injects it into the input serial register. All peripheral circuits on the chip use four-phase dynamic circuits designed to operate up to 10 MHz.

Each CCD block in the memory chip can be independently accessed by applying the appropriate address signals. The four address buffers receive the four address bits A_0, A_1, A_2, and A_3 at TTL voltage levels and generate at MOS voltage levels the addresses and their complements. Four row address decoders and four column address decoders generate the appro-

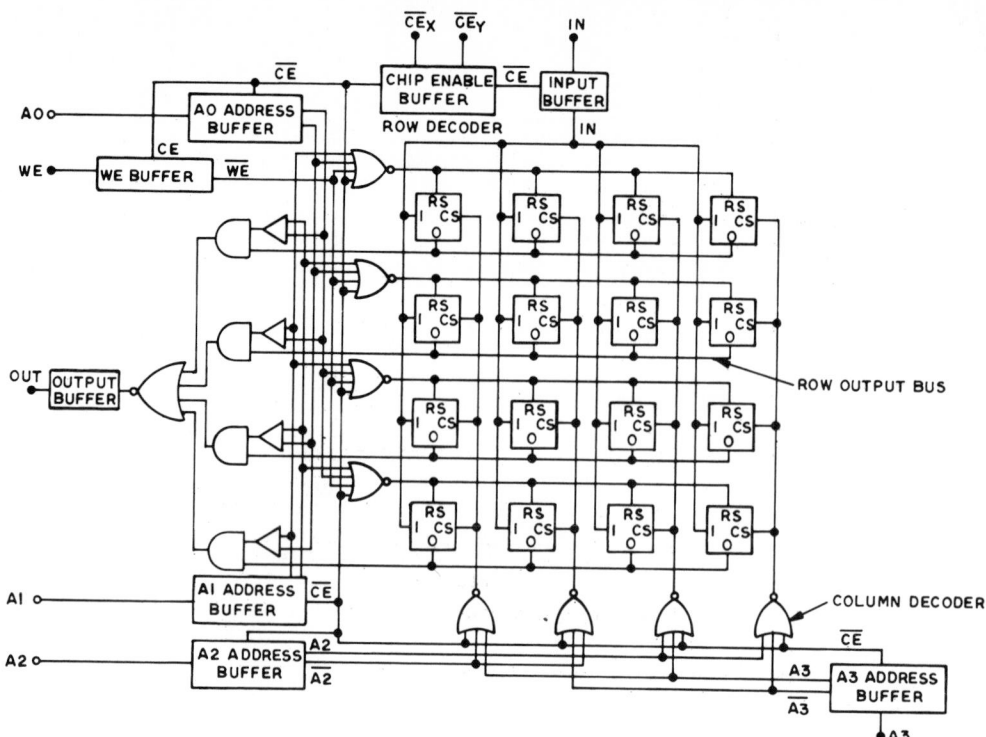

Fig. 3. Block diagram of the 64-kbit CCD memory chip.

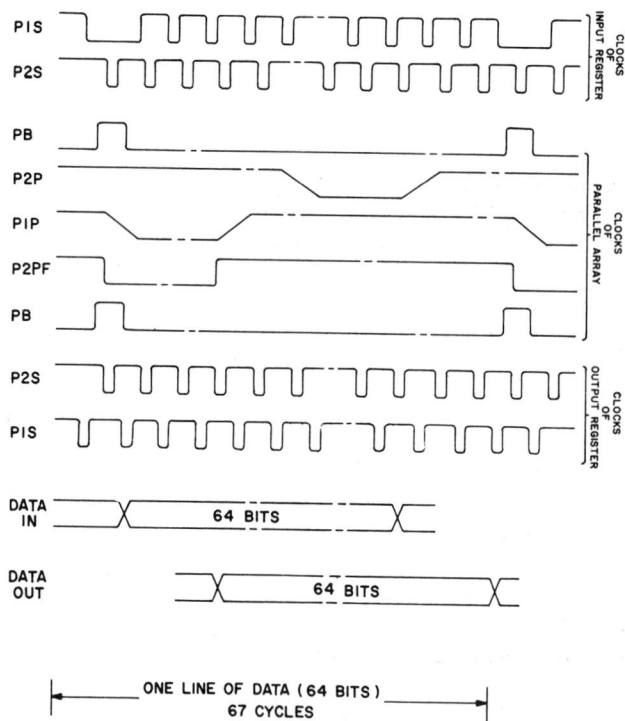

Fig. 4. Input–output and charge-transfer timing diagram of the CCD memory.

written into the memory. A chip-enable buffer receives two chip-select signals $\overline{CE}_x$ and $\overline{CE}_y$ at TTL voltage level and generates a complementary chip-enable signal ($\overline{CE}$). The ($\overline{CE}$) signal inhibits the input-data (IN), the write-enable complement ($\overline{WE}$), the row-select (RS), and the column-select signals (CS) when the chip is not selected. The column-select signals (CS) gate the outputs of the charge-detection circuits to the row output buses. A second set of row-select signals (RS) generated from the addresses route the appropriate row output to the chip output buffer.

The chip output data buffer receives the information from the CCD block selected by the row and column decoders and drives an open drain output transistor. This device is capable of sinking 3 mA at 0.4 V and 100°C and driving a 50-pF output load at frequencies as high as 10 MHz. When the memory chip is not selected, the output transistor is turned off, thus providing OR-TIE capabilities of the 64-kbit memory chips.

Two dc power supplies, VDD equal to +12 V and VBB equal to −7.5 V with ±10 percent tolerance, are required. A voltage regulator circuit on the chip derives a reference voltage used in the charge-injection circuit. This reference voltage tracks the threshold voltage on the chip and the power supplies and allows the CCD memory chip to function with wide operating margins. The on-chip peripheral circuits operate with four-phase clocks. The CCD arrays operate with a set of two-phase fast serial clocks, two other fast clocks which interface the serial registers with the parallel array, and two slow parallel clocks. The clock voltage swing is 0–12 V with ±10 percent tolerance.

III. INPUT AND OUTPUT TIMING DIAGRAM

The CCD arrays are operated with two fast overlapping serial clocks $P1S$ and $P2S$, two fast gating clocks PB and $P2PF$

priate row-select (RS) and column-select (CS) signals to select one of the 16 CCD blocks. An input-data buffer and a write-enable buffer receive the input-data and write-enable signals at TTL voltage levels and generate the (IN) and ($\overline{WE}$) signals. The (IN) signal is sent to the input/recirculate circuit of the 16 CCD blocks. The ($\overline{WE}$) signal is sent to the row decoders to inhibit the (RS) signals when no new input data are to be

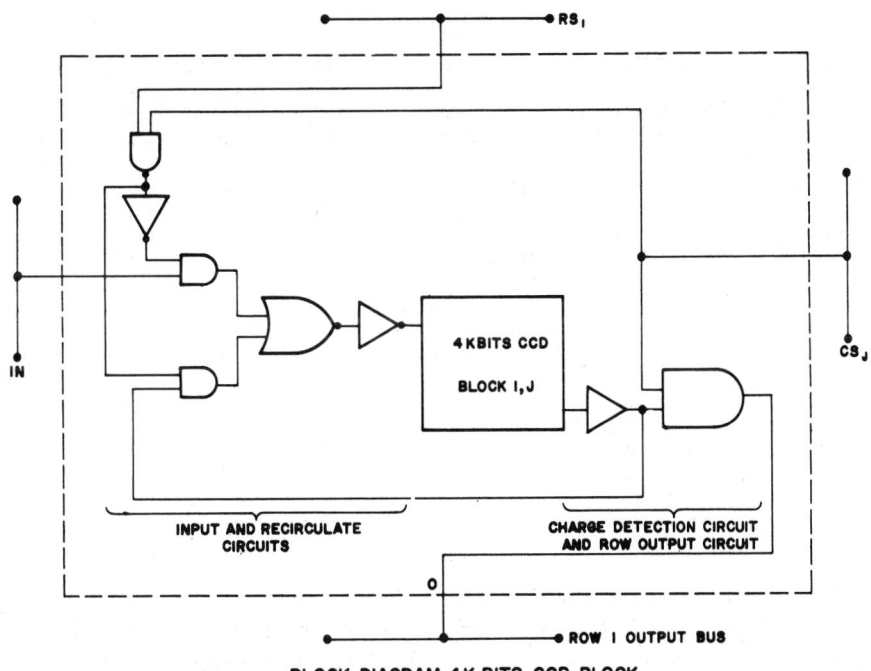

Fig. 5. Data flow in each 4-kbit block of the CCD memory.

which interface the input and output serial registers with the parallel array, and two parallel overlapping slow clocks $P1P$ and $P2P$. Timing diagrams of the transfer of charge from the input serial register to the parallel array, and from the parallel array to the output serial register are shown in Fig. 5. The transfer of charge between the electrodes of the input and output serial registers and the electrodes of the parallel array requires only one serial clock cycle. While the parallel clocks drive a large capacitance their drive requirements are reasonable since they are much slower than the serial clocks.

The input-data and control signals (IN, WE, A_0, A_1, A_2, A_3, $\overline{CE}_x$, and $\overline{CE}_y$) are sampled at the fast serial clock frequency rate. The output data rate is equal to the serial clock frequency. Since all peripheral circuits operate at the output data rate, one whole line of data can be read and recirculated, but any particular bit of data in this line can be overwritten. Information is written into and read out of the memory in a line format consisting of 64 bits of data separated by 3 cycle times. The three cycle dead time is required for the input and output data to progress sequentially through the peripheral circuits.

The mean access time T_A of the data stored in any of the CCD blocks is given by

$$T_A = \frac{2144 (\mu s)}{fc (\text{MHz})}$$

where fc is the clock frequency in MHz. So for a clock frequency range of 1–10 MHz, the mean access time varies from 2.144 ms to 214.4 μs.

IV. THE CCD CELL

The CCD array uses the two-phase offset-gate structure [5]–[10]. A pictorial cross-sectional view of this structure is shown in Fig. 6. Each gate element forms a transfer-storage

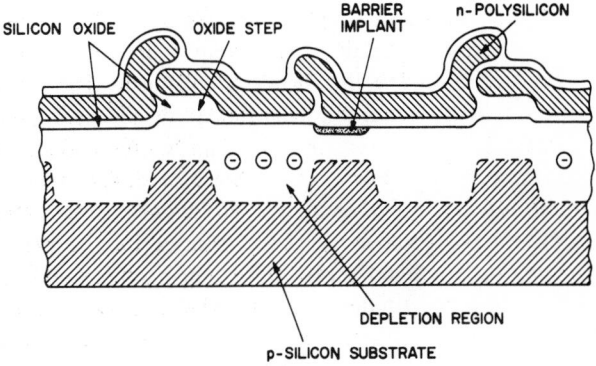

Fig. 6. Two-phase offset-mask electrode structure with polysilicon gates.

pair. The first gate level pair is formed by a step in the channel oxide aligned so that part of the gate is over the thin- and part over the thick-oxide regions. A barrier implant which self aligns to the thin-oxide region not covered by the first gate area forms the second gate transfer region, while the remaining portion of the second gate forms a storage region. The two-phase offset gate design was chosen because of its high packing density, small number of contacts, and its insensitivity to intralevel shorts.

The gate electrodes on each polysilicon level are connected to the same clock phase in the offset-gate structure. Therefore, only one essential contact is required per phase rather than at least one per transfer-storage pair in a standard overlapping gate two- or four-phase structure. Furthermore, intralevel shorts cause clock failure only if they occur between the serial register clocks and the parallel array clocks. While intralevel shorts on the first polysilicon may cause phase errors by blocking adjacent second-level transfer-storage pairs, similar shorts on second-level polysilicon cause no phase error. Two-phase

TABLE I

Electrode Type	Photolithographic Limitations	Physical Limitations	This Design (μm)	State of the Art (μm)	On the Horizon (μm)	Future (μm)
Two-phase	$4(F + R)$	$2(L_{tr} + L_{st})$	40	32	20	10
Two-phase offset gate	$2F$	$2(L_{tr} + L_{st} + 2R)$	16	14	10	6
Three-phase three level	$3F$	$3(L_{st} + R)$	24	18	12	6
L_{st} (μm)		minimum storage region	3	2	2	1
L_{tr} (μm)		minimum transfer region	1	1	1	1
R (μm)	alignment tolerance		2	2	1	0.5
F (μm)	minimum feature size		8	6	4	2

operation of the parallel array was chosen rather than the electrode per bit approach [11], [12] because full advantage of the insensitivity to intralevel shorts and small number of contacts is provided in this case.

The packing density of the two-phase offset-gate design is compared with other high density approaches in Table I [1]. The first column of this table represents the cell bit length based on photolithographic consideration where F is a minimum feature size and R the alignment tolerance. The second column indicates the cell bit length based on physical limitations where L_{tr} and L_{st} are the minimum transfer and storage length, respectively. For a given choice of design rules the minimum cell is the maximum of column one or two. Minimum cell size is compared in column three for the present design, the next column makes the comparison for more aggressive state-of-the-art design rules. The final two columns compare minimum cell size for future cases where technological advances are required. A case just on the horizon where projection and automatic alignment are used is considered in column five, while the last column considers a case where a combination of electron beams, automatic alignment, and plasma- and ion-beam process techniques [13] are used. This table indicates that both the two-phase offset gate and three-phase three-level designs have about half the bit cell length of the two-phase design.

Conservative design rules and ease of fabrication were emphasized in the layout and process procedure used in this memory chip.

The electrodes on the first-level polysilicon are designed to have 8-μm active length and 8-μm separation. The electrodes on the second-level polysilicon have an active length of 12 μm, thus overlapping the first level gates by 2 μm. The active charge-transfer channel width and separation are 8 μm. Thus, the cell size is 16 × 16 μm (0.4 mil²).

The charge-coupled memory is fabricated using an n-channel polysilicon gate process with self-aligned source and drain in the peripheral circuits. Surface-channel rather than bulk-channel devices are used because of their higher charge storage density. The substrate is ⟨100⟩ 6–8 Ω · cm p-type silicon. The second-level polysilicon is used for the active gates in the pe-

ripheral circuits. A low energy boron implant is used to define the barrier implant on the second-level polysilicon electrodes. The fabrication process of the charge-coupled memory requires two additional photolithographic masks compared to the standard n-channel MOS process. One mask is used for the offset-gate definition.

V. THE INPUT/RECIRCULATE AND CHARGE-DETECTION CIRCUITS

Full advantage of the two-phase offset-gate CCD can only be achieved by careful considerations of the input charge injection and output detection circuits. Maximum charge capacity for this electrode structure is a function of alignment variations and deviations in critical dimensions. Therefore, the injection circuit must either be designed to provide the minimum charge allowable under worst case misalignment and dimensional control, or provide a charge packet size which scales automatically to the smallest storage region encountered in the array. The latter approach was chosen in this design since improved operating margins are realized in all cases except where worst case misalignment and critical dimensions actually occur. The use of a high sensitivity balanced detection circuit combined with the self-sizing alignment-compensated injection circuit allows a minimum cell size to be chosen.

The block diagram of the input/recirculate and charge-injection circuit is shown in Fig. 7. The input/recirculate circuit sends a pulse to the charge-injection circuit to inject a fat zero charge for a logical ONE signal or a full charge well for a logical ZERO signal. The charge-injection circuit meters the appropriate amount of charge to be injected without overfilling the charge wells in the input and output serial registers and the parallel array. The charge packet is sized according to the CCD clock voltage amplitudes, the alignment of the offset masks relative to the gate electrodes, the linewidths of the gate electrodes, and the active CCD channels.

The recirculate/input circuit is implemented with four-phase dynamic MOS circuits. The write-enable signal (WE) is encoded into the row-select signal (RS). The row-select (RS) and column-select (CS) signals are combined in a dynamic NAND gate and then inverted. The output data from the

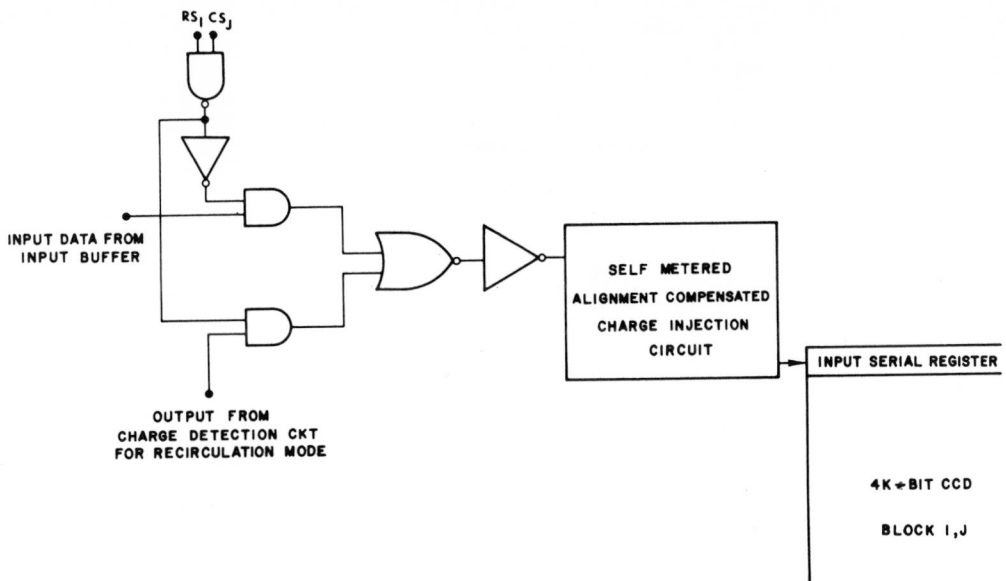

Fig. 7. Block diagram of the input/recirculate and charge-injection circuit.

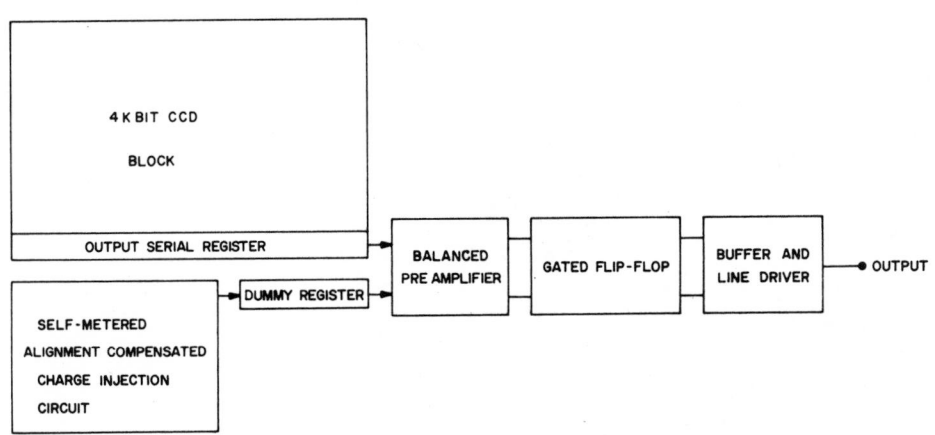

Fig. 8. Block diagram of the charge-detection circuit.

charge-detection circuit and the new data from the input buffer are then gated with these signals and then combined with a dynamic NOR gate. The resulting output is inverted and fed into the charge-injection circuit.

The block diagram of the charge-detection circuit is shown in Fig. 8. Balanced detection at the output of each CCD block is achieved by comparing the output signal charge with a balance charge generated from a dummy register [14]. The dummy register is one bit long and has an injection circuit similar to the injection circuit used at the input serial register. The size of the balance charge packet is equal to half the sum of the FAT ZERO and ONE charges. The output signal charge and the balance charge are detected by a low-input capacitance balanced preamplifier. The charge difference between the collected ONE or FAT ZERO charges and the balance charge is transmitted to the nodes of a gated flip-flop as a voltage difference. When the gated flip-flop is turned on, the small voltage unbalance latches up rapidly to a larger voltage difference due to the regenerative feedback in the flip-flop. The output of the gated flip-flop is sampled by a push–pull buffer that is capable of driving the required output capacitance. The use

of the separate balanced preamplifier, the gated flip-flop, and the output buffer improves the sensitivity and speed of the charge-detection circuit. Balanced regenerative charge detection has the advantages of high speed operation at low detected charge levels and is very insensitive to common-mode signals introduced by clock coupling and threshold voltage variations.

VI. THE INPUT- AND OUTPUT-DATA BUFFER

The input-data, write-enable, and address buffers in Fig. 9 sample the input signals at TTL voltage levels at the beginning of each clock cycle. Once the input signals are sampled and stored internally, the external signals may change during the rest of the clock cycle. The input buffers generate the input-data signal (IN), the write-enable complement signal ($\overline{WE}$), and the address signals and their complements (A and $\overline{A}$) at MOS voltage levels to drive the other peripheral circuits on the chip. The buffer circuits are implemented with four-phase dynamic low-power MOS circuits designed to operate at the maximum output data rate of 10 MHz. Thus, the block addresses and the write/recirculate mode can be changed at

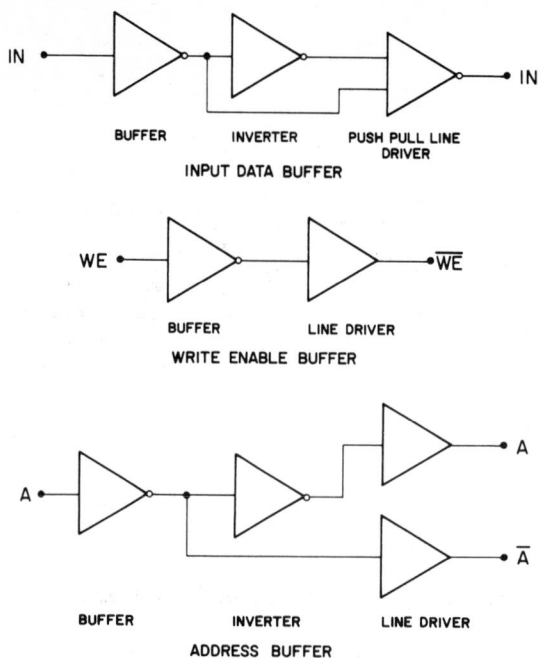

Fig. 9. Block diagram of the input-data, write-enable, and address buffers.

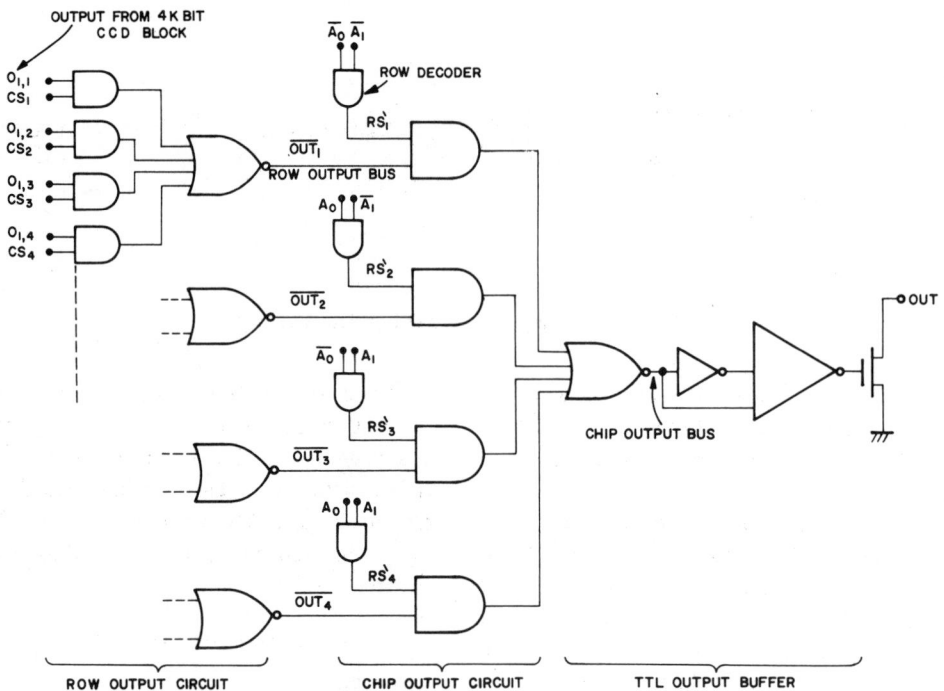

Fig. 10. Block diagram of the output circuits on the CCD memory chip.

the output data rate. The complement of the chip-enable signal ($\overline{CE}$) inhibits the outputs of the input buffers when the chip is not selected.

The input-data buffer consists of a buffer, a dynamic inverter, and a push–pull line driver. The write-enable buffer consists of a buffer and a source-follower driver. The address buffers consist of a buffer, a dynamic inverter, and bootstrapped source-follower drivers.

The block diagram of the output circuits on the CCD memory chip is shown in Fig. 10. The outputs of the 4-kbit CCD block on each row are gated by the column-select signals (CS) to a common row output bus. The output of the row output buses are in turn routed by the row-select signals ($R\overset{\vee}{S}$) to a common chip output bus. Thus, only the output of the CCD block selected by the row and column decoders is sent to the output buffer. The output buffer consists of a dynamic inverter and a

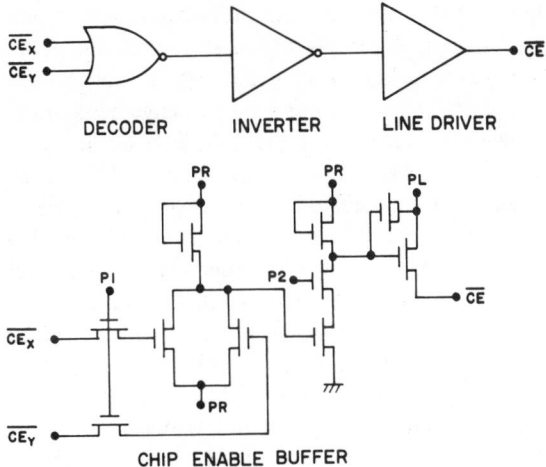

Fig. 11. Block and circuit diagram of the chip-enable buffer.

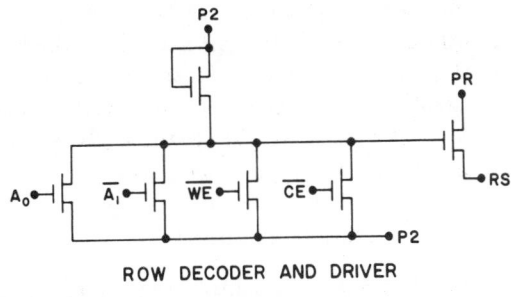

ROW DECODER AND DRIVER

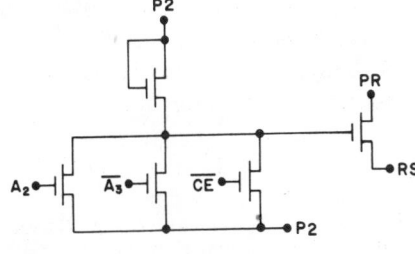

COLUMN DECODER AND DRIVER

Fig. 12. Circuit diagram of the row and column decoders.

Fig. 13. Photomicrograph of the 64-kbit CCD memory chip.

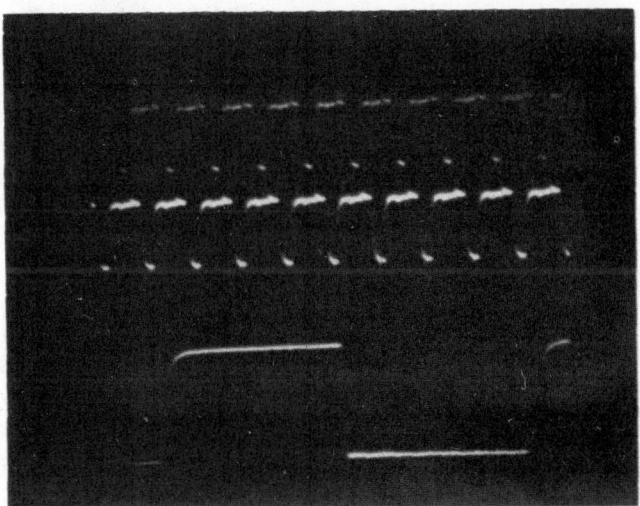

Fig. 14. Output waveform from the 64-kbit CCD memory chip at a clock frequency of about 1 MHz. The two top traces are the two fast serial clocks. The lower trace is the output waveform for a data pattern of four ONE's and four ZERO's.

push–pull driver which drives an open-drain output transistor capable of sinking 3 mA with 0.4 V drop at 100°C and driving a 50-pF load at 10 MHz. When the memory chip is not selected, the output transistor is turned off, thus providing OR-TIE capability for the memory chip. The output circuits are driven by four-phase clocks and the output data progress sequentially from the CCD block to the output buffer.

VII. THE CHIP-ENABLE BUFFER AND THE ADDRESS DECODERS

The chip-enable buffer samples two chip-enable signals $\overline{CE}_x$ and $\overline{CE}_y$ at TTL voltage levels at the beginning of each cycle. Once the chip-enable signals are sampled and stored internally, they may change during the rest of the clock cycle. The chip-enable buffer generates a chip-enable complement signal ($\overline{CE}$) at the MOS voltage level. When the chip is not selected the ($\overline{CE}$) signal inhibits the control and address signals on the chip to reduce power dissipation. Two chip-enable input signals ($\overline{CE}_x$) and ($\overline{CE}_y$) are used to simplify the use of the chip in a two-dimensional memory matrix organization. The block and

TABLE II
CCD CLOCK CAPACITANCE VALUES OF THE 64-KBIT CHARGE-COUPLED MEMORY CHIP

C_{P1S}, C_{P2S} each	60 pF
$C_{P1S-P2S}$	80 pF
C_{P1P}, C_{P2P} each	1000 pF
$C_{P1P-P2P}$	1200 pF
C_{PB}	100 pF
C_{P2PF}	60 pF

circuit diagrams of the chip-enable buffers are shown in Fig. 11. It consists of an input buffer-decoder, a dynamic inverter, and a bootstrapped source-follower driver. The row and column decoders (Fig. 12) are dynamic NOR decoders with source-follower drivers. The row decoders receive the addresses A_0, A_1, their complements, the write-enable complement signal ($\overline{WE}$), and the chip-enable complement ($\overline{CE}$), and generate the row-select signals (RS) to select one of the four rows on the chip. The column decoders receive the addresses A_2, A_3, their complements, and the chip-enable complement ($\overline{CE}$) and generate the column-select signals (CS) to select one of the four columns on the chip. When the chip is not selected the column- and row-select signals are pulled low.

VIII. DEVICE OPERATION

The 64-kbit CCD memory chip has been designed, fabricated, and operated. Two versions of the charge-coupled memory have been designed using 10-μm and 8-μm minimum feature size. With the 8-μm minimum feature size and ± 2-μm alignment tolerances, the total chip size is 218 $\times$ 235 mil^2 (0.8 mils^2/bit) and the chip is mounted in a 22-pin 400-mil wide ceramic dual in-line package. Internal signals are brought to separate pads so that internal margins can be monitored and evaluated during wafer testing. A photomicrograph of the 64-kbit CCD memory chip is shown in Fig. 13. The output waveform from the memory chip is shown in Fig. 14 with the CCD serial clocks. The data pattern consists of four ONE's and four ZERO's. The output data changes at the beginning of each clock cycle and remains valid until the next clock cycle.

Table II lists the calculated values of the capacitance loads of the parallel and serial CCD clocks. The calculated power dissipation at 1 MHz with 12-V clocks is 1 μW/bit including the reactive power in the clock drivers. Since dynamic circuits are used in all the on-chip peripheral circuits, the power dissipation scales with frequency. At the maximum output data rate of 10 MHz, the calculated power dissipation is about 10 μW/bit.

IX. DISCUSSION

A high-density slow access block addressed 64-kbit charge-coupled serial memory has been designed, fabricated, and operated. The memory chip is organized as 64K words by 1 bit in 16 blocks of 4 kbits.

The memory chip is fully decoded with write/recirculate control and two-dimensional decoding to permit a memory matrix organization with X–Y chip select control. All inputs and outputs are TTL compatible. Operated at a data rate of 1 MHz, the mean access time is 2 ms, and the average power dissipation is 1 μW/bit. The maximum output data rate is 10 MHz giving a mean access time of 200 μs and an average power dissipation of 10 μW/bit. High speed low power four-phase dynamic circuits are used for the on-chip peripheral circuits. The CCD arrays operate with two fast serial clocks, two fast gating clocks which interface the input and output serial registers with the parallel array, and two parallel slow clocks. The clocks amplitude is 12 V and two dc power supplies of +12 V and -7.5 are required.

The memory chip is fabricated using an n-channel polysilicon gate process. The CCD memory arrays use the two-phase off-

set gate electrode structure with polysilicon gates. Compared to other two-polysilicon level CCD structures, the offset-gate structure has a higher packing density, is more tolerant to intralevel shorts, and does not require a large number of small contact windows to connect the CCD gate electrodes to the clock bus lines. The two-phase operation of the parallel CCD array relaxes the demands of photolithography on the second polysilicon level. Using tolerant design rules (8-μm minimum feature size and ± 2-μm alignment tolerance) the CCD cell size is 0.4 mil^2 and the total chip size is 218 $\times$ 235 mils.

To take maximum advantage of the offset-gate two-phase design and to allow minimum cell size a self-metered alignment-compensated charge-injection circuit and a high sensitivity balanced detection circuit have been used.

In conclusion, the feasibility of a high density slow-access block addressed 64-kbit charge-coupled serial memory has been demonstrated. By using the offset-gate CCD electrode structure to obtain a small cell size, and the adaptive system approach to utilize nonzero defect memory chips, the system cost per bit of a charge-coupled serial memory can be reduced to provide a solid-state replacement for magnetic drums and fixed head disk memories, and to bridge the gap between high cost RAM's and slow access magnetic memories.

ACKNOWLEDGMENT

The authors acknowledge the contributions of J. P. Ray, D. Monroe, S. Colley, M. Hennessy, C. A. Mead, and R. McClure for system considerations and stimulating discussions; A. Louwerse and S. Bishop for the chip layout; and J. Dinjian for careful preparation of the manuscript.

REFERENCES

[1] C. H. Séquin and M. F. Tompsett, *Charge Transfer Devices*, suppl. 8 for *Advances in Electronics and Electron Physics*. New York: Academic, 1975.

[2] D. A. Hodges, "Chip yield and manufacturing costs," in *Semiconductor Memories*. New York: IEEE Press, 1972, p. 175.

[3] S. Chou, "Design of a 16 384-bit serial charge-coupled memory device," this issue, pp. 78–86.

[4] S. D. Rosenbaum, C. H. Chan, J. T. Caves, S. C. Poon, and R. W. Wallace, "A 16 384-bit high-density CCD memory," this issue, pp. 101–108.

[5] R. W. Bower, A. M. Mohsen, and T. A. Zimmerman, "A high density overlapping gate charge-coupled device array," in *Tech. Dig., Int. Electron Devices Meeting*, Dec. 1973, pp. 30–32.

[6] R. W. Bower, T. A. Zimmerman, and R. Huber, "A 4K bit high density charge-coupled memory," presented at the IEEE Int. Electron Devices Meeting, Washington, DC, Dec. 1974.

[7] A. M. Mohsen, T. F. Retajczyk, and C. H. Séquin "Offset-mask charge-coupled devices for memory applications," presented at the IEEE Int. Electron Devices Meeting, Washington, DC, Dec. 1974.

[8] R. W. Bower, T. A. Zimmerman, and A. M. Mohsen, "The two-phase offset gate CCD," *IEEE Trans. Electron Devices* (Corresp.), vol. ED-22, pp. 70–72, Feb. 1975.

[9] —, "Performance characteristics of the offset-gate charge-coupled devices," *IEEE Trans. Electron Devices* (Corresp.), vol. ED-22, pp. 72–73, Feb. 1975.

[10] A. M. Mohsen and T. F. Retajczyk, Jr., "Fabrication and performance of offset-mask charge-coupled devices," this issue, pp. 248–256.

[11] F. L. J. Sangster and K. Teer, "Bucket-brigade electronics–New possibilities for delay, time-axis conversion, and scanning," *IEEE J. Solid-State Circuits*, vol. SC-4, pp. 131–136, June 1969.

[12] D. R. Collins, J. B. Barton, D. D. Buss, A. R. Kmetz, and J. E. Shroeder, "CCD memory options," in *ISSCC Dig. Tech. Papers*, 1973, pp. 136–137.

[13] E. D. Wolf, L. O. Bauer, R. W. Bower, H. L. Gavin, and C. R. Buckey, "Electron beam and ion beam fabricated microwave switch," *IEEE Trans. Electron Devices*, vol. ED-17, pp. 446–449, June 1970.

[14] A. M. Mohsen, M. F. Tompsett, E. N. Fuls, and E. J. Zimany, Jr., "A 16-kbit block addressed charge-coupled memory device," this issue, pp. 108–116.

The Charge-Coupled RAM Cell Concept

AL F. TASCH, JR., ROBERT C. FRYE, AND HORNG-SEN FU, MEMBER, IEEE

Abstract—A new concept in MOS dynamic RAM cells is described and demonstrated. The charge-coupled RAM (CC RAM) cell combines the storage capacity and transfer gate of the one-transistor cell into a single gate. The resulting cell is simpler than the conventional one-transistor cell and possesses significant advantages in packing density and potentially higher yield. One of the variations of the CC RAM cell concept results in a cell whose operation is identical (voltage and timing) to that of the present one-transistor cell. In addition, the CC RAM cell fabrication is essentially the same as the present one-transistor cell process. The CC RAM is an attractive candidate for the next generation RAM's.

I. INTRODUCTION

SINCE the introduction of 1 kbit MOS dynamic random access memories (RAM's) in 1970, dynamic MOS RAM development has progressed rapidly to the point that semiconductor memories are preferred over core. The strong

Manuscript received October 17, 1975.
The authors are with Texas Instruments Inc., Dallas, TX 75222.

emphasis on reducing area per bit has resulted in the evolution of the RAM storage cell from six transistors to a single transistor, and the development of high yield, high density process technologies. This is evidenced by the introduction of the 4 kbit RAM in 1973 and the fact that the 16 kbit chip is actively being pursued for introduction in 1976. And there is speculation that a 64 kbit RAM will be introduced in 1979–1980 [1]. Moreover, an 8 kbit RAM on a 1×1.5 mm^2 $(40 \times 60$ mil$^2)$ chip has been built using E-beam technology [2]. This progression to greater size and density has and will continue to require new concepts and advances in the technology. This paper describes a new MOS dynamic RAM storage cell concept and its successful demonstration. The new cell is called the charge-coupled RAM (CC RAM) cell, and it is simpler than the widely used one-transistor cell [3] or the surface charge RAM [4]. It possesses important advantages in packing density and potentially higher yield, and one of the two structures which results from this concept has direct operational compatibility with the conventional one-

Reprinted from *IEEE Trans. Electron Devices*, vol. ED-23, pp. 126–131, Feb. 1976.

314

transistor cell. Thus, this cell could be a direct replacement for the present one-transistor cell.

The CC RAM cell concept is described and two different approaches using this concept are discussed. The evaluation results of the CC RAM storage cell are also given.

II. CHARGE-COUPLED RAM CELL WITH IMPLANTED TRANSFER REGION

In order to better understand the concept of the CC RAM cell, it is useful to begin by considering the expression for the surface potential ϕ_s at the SiO_2-Si interface as a function of the gate voltage, V_G, in dynamic operation.

$$\phi_s = V_0 + (V_G - V_{FB}) - \{V_0^2 + 2 V_0 (V_G - V_{FB})\}^{1/2} \quad (1)$$

where

V_{FB} = flatband voltage

$V_0 = \dfrac{q K_s N_A t_{ox}^2}{K_{ox}^2 \epsilon_0}$

K_s = dielectric constant of silicon
K_{ox} = dielectric constant of SiO_2
N_A = substrate doping
t_{ox} = gate oxide thickness.

For high resistivity material, ϕ_s is essentially linear with gate voltage because V_0 is very small, and the second term $(V_G - V_{FB})$ dominates in (1). For low resistivity material V_0 is larger, and ϕ_s varies less than linearly with V_G due to the square-root term in (1). Compare curves (a) and (b) in Fig. 1. The ϕ_s versus V_G behavior exhibited by curve (b) can also be realized by performing a boron (p-type) implant in the 15 $\Omega \cdot$ cm p-type material near the surface. That is, the implant is used to locally alter the substrate doping, and the result is curve (c) in Fig. 1. When comparing curves (a) and (c) note that for a given applied gate voltage ϕ_s is lower in curve (c). However, near zero gate voltage the ϕ_s values are almost equal in the two curves.

With the preceding discussion in mind, consider the CC RAM cell structure in Fig. 2. In this structure the storage gate and transfer gate have been combined into a single gate overlying two regions. The gate in the storage region overlies normal substrate material (i.e., no ion implants) and the ϕ_s versus V_G characteristic is shown by curve a in Fig. 1. In the transfer region a p-type ion implant has been added to the substrate near the surface so that the ϕ_s versus V_G behavior is given by curve (c) in Fig. 1. The operation of this storage cell can best be understood by examining the surface potential configurations in Fig. 2(b) for the READ, WRITE, and STORE modes. In order to write information into the storage cell the word line is turned on. Referring to Fig. 1 it can be seen that a potential well forms in the storage region while a barrier exists in the transfer region. If it is desired to write a ZERO into the cell, the bit line voltage is held higher than the surface potential in the transfer region so that no charge (electrons) is allowed to flow into the potential well beneath the storage region. A ONE is written into the cell by dropping the bit line voltage below the surface potential in the transfer region. In this case

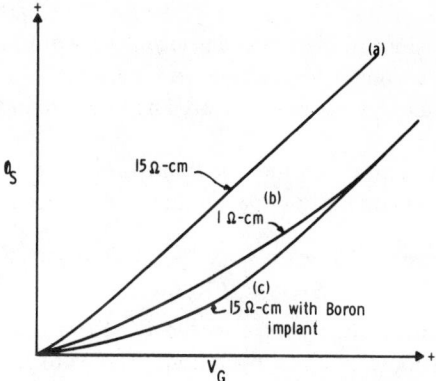

Fig. 1. Surface potential versus gate voltage. Curve (a) is for 15 $\Omega \cdot$ cm, p-type material, curve (b) is for 1 $\Omega \cdot$ cm p-type material, and curve (c) is for 15 $\Omega \cdot$ cm p-type material with a boron implant near the surface.

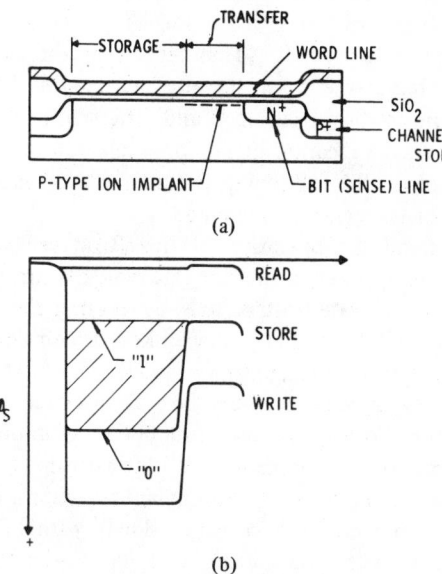

Fig. 2. CC RAM cell with implanted transfer region. (a) Cross section of cell structure. (b) Surface potential configuration for READ, WRITE, and STORE modes.

charge will flow from the bit line into the potential well to fill it. In this paper the following convention is used to designate a ONE or ZERO. If the potential well beneath the storage gate is filled with charge (electrons), a ONE is stored. If the well is empty, a ZERO is stored.

After writing information into the storage cell the word line voltage is dropped to an intermediate level for the STORE mode as shown by the surface potential in Fig. 2(b). This is necessary in order to isolate the cell from the bit line during write operations on other cells along the same bit line. When it is desired to read the contents of the storage cell, the voltage on the word line is dropped to or near ground, and the surface potential relaxes to the READ configuration shown in Fig. 2(b). As the surface potential drops, any charge contained in the potential well is pushed out and onto the bit line to be sensed.

Thus, in the CC RAM cell, the storage gate and transfer gate in the conventional one-transistor cell are combined into a single gate. By altering the ϕ_s versus V_G characteristics under a portion of the word line gate, the word line can be used simultaneously for storage and transfer. As a result, only two lines, word line and bit line, are required for the CC RAM cell.

This is in contrast to the one-transistor and surface charge cells which require a word line, bit line, and storage line.

III. Charge-Coupled RAM Cell with Implanted Storage Region

One potential disadvantage of the CC RAM cell shown in Fig. 2 is its more complex operation. The word line must be switched between three rather than two voltage levels, and the voltage levels required of the sense amplifier differ somewhat from those employed in the conventional one-transistor cell operation. There is another variation of the CC RAM cell concept which appears to be considerably more attractive because its operation (timing and voltages) is identical to that of the one-transistor cell. Thus, it offers the possibility of a direct replacement for the present one-transistor cell. This structure is shown in Fig. 3 and differs from the cell in Fig. 2 only in the location of the ion-implanted region. The storage region in the cell in Fig. 3 contains two ion implants while the transfer region is unaltered.

To understand the operation of this RAM cell note first that the surface potential versus gate voltage curve can be shifted along the gate voltage axis by altering the flatband voltage V_{FB}. [Refer to (1).] A well-known convenient way to shift the flatband voltage to a negative value is to implant n-type ions at or very near the SiO$_2$-Si interface [5], [6]. This approach is used in the structure in Fig. 3. In the storage region, p-type ions are implanted into the silicon at a depth of several thousand angstroms. This implant results in a ϕ_s versus V_G characteristic in which ϕ_s varies slowly with V_G in contrast to the stronger dependence of ϕ_s on V_G in the unimplanted transfer region. The p-type implant is then followed by a shallow n-type implant, which shifts the ϕ_s versus V_G curve to the left along the gate voltage axis. The resulting ϕ_s versus V_G characteristic for the storage region is given by curve (b) in Fig. 4. Also shown is the ϕ_s versus V_G curve [curve (a)] for the transfer region.

With the preceding discussion in mind refer to Fig. 3(b) to understand the operation of the RAM cell in Fig. 3(a). The operation to be described is identical (voltages and timing) to the operation of the conventional one-transistor cell. To write information into the storage cell the word line voltage is turned on. The surface potential configuration for the WRITE mode is shown in Fig. 3(b). If it is desired to write a ONE into the storage cell, the bit line voltage is lowered to or near ground. Charge (electrons) then flows into the potential well to fill it. If it is desired to write a ZERO, the bit line voltage is set high and no charge enters the storage region. The word line is then turned off to complete the WRITE operation and isolate the storage cell from the bit line. This is the STORE mode. Referring to Fig. 3 it can be seen that with the gate voltage off (STORE mode), a potential well exists for storing charge in the storage region. When the word line gate voltage is turned on, the surface potential in the transfer region is

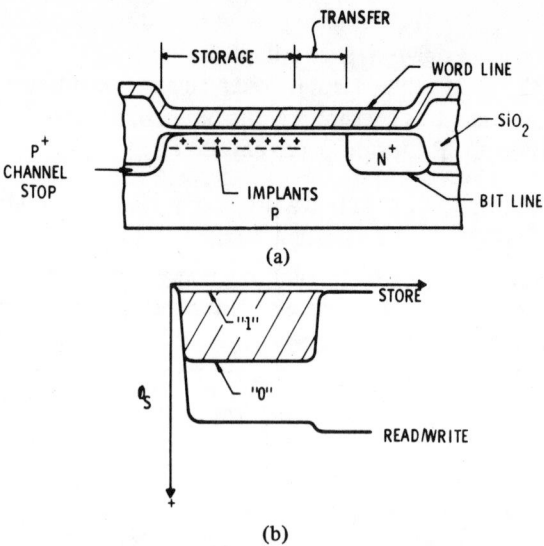

Fig. 3. CC RAM cell with implanted storage region. (a) Cross section of cell structure. (b) Surface potential configuration for READ, WRITE, and STORE modes. This cell is operationally compatible with the conventional one-transistor cell.

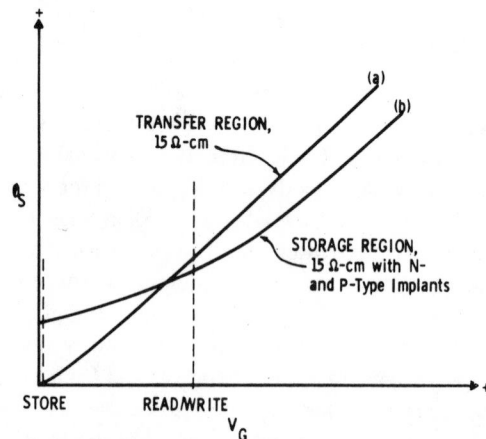

Fig. 4. Surface potential versus gate voltage. Curve (a) is for 15 $\Omega \cdot$ cm, p-type material and curve (b) is for the same material with combined n- and p-type implants near the surface.

more positive than that in the storage region so that charge can flow between the bit line and the storage region.

When it is desired to read the contents of the cell, the word line is turned on just as for the WRITE mode. Any charge present in the storage cell is then dumped onto the bit line and sensed.

IV. Experimental Verification of CC RAM Cell Concept

In order to verify the CC RAM cell concept, a test structure was designed containing four CC RAM cells of different sizes and a conventional one-transistor cell for comparison. A photograph of the five cells is shown in Fig. 5(a). From left to right, the first four cells are CC RAM cells with storage areas of 2.0, 1.5, 1.0, and 0.5 mil^2. The last cell is the conventional one-transistor cell consisting of a transfer gate (connected to the word line) and a storage gate with area of 1.0 mil^2. All cells share a common diffused sense line which is

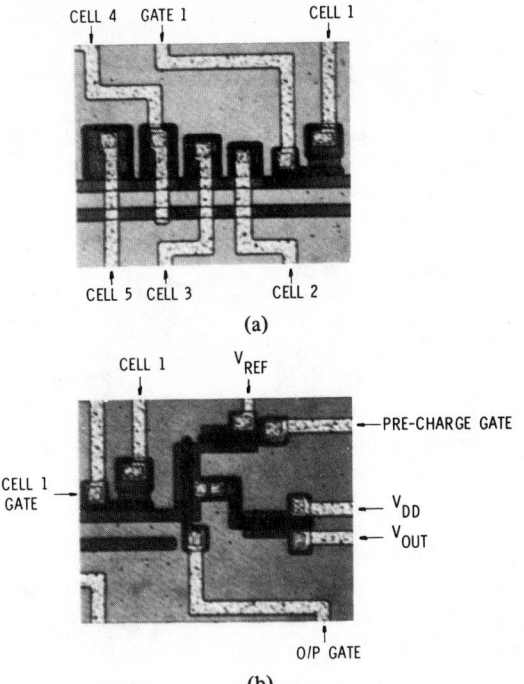

Fig. 5. Photographs of CC RAM cell test structure. (a) CC RAM cells (cells 2–5) and one-transistor cell (cell 1). (b) Output amplifier. The diffused bit line is 0.4 mils (10 μm) wide.

connected to the gate of a source-follower MOSFET to detect charge in the storage cell. An n-channel, self-aligned silicon gate process was used to fabricate the test structure on p-type, (100), 15 Ω · cm material. Both types of CC RAM structures were fabricated with a gate oxide thickness of 1000 Å. The p-type boron implant doses ranged from 1.0 to 1.5 × 10^{12} cm^{-2} at 60 keV through 1000 Å oxide for the CC RAM cell with the implanted transfer region (Fig. 2). Several variations of p- and n-type implant doses were employed in the fabrication of the CC RAM cell with the implanted storage region (Fig. 3). These doses both ranged from 1.5 to 3.0 × 10^{12} cm^{-2}.

Measurements were performed to verify the CC RAM concept and to determine parameters of the cell such as charge handling capacity (signal size) and the read, store, and write voltage levels required. Other tests were made in order to demonstrate that data can be written or read from one cell without interfering with data in other cells sharing the same bit line. The evaluation results for the CC RAM cell with the implanted transfer region are presented next followed by results of measurements on the CC RAM cell with the implanted storage region.

A circuit schematic of the CC RAM cell test structure is shown in Fig. 6. The following voltages were used for the word lines to operate the cell shown in Fig. 2.

$$V_{STORE} = 10 \text{ V}$$
$$V_{READ} = 0 \text{ V}$$
$$V_{WRITE} = 15 \text{ V}.$$

A dc bias voltage of 6 V was applied to the output gate and the substrate was held at ground potential. Data input was ac-

complished by switching the bit line to the input signal level through transistor Q1. The barrier voltage in the transfer region is about 4 V when the word line is in the WRITE mode (15 V). Thus, a ONE is written in if the bit line is forced to a potential less than 4 V during the WRITE pulse. If the bit line is held above 4 V during the WRITE command, no electrons can reach the storage area across the barrier, and a ZERO is written. Note that the operation of the cell implies that the cell is emptied of charge or read immediately before each WRITE operation.

A precharge output circuit was used as a sensitive voltage amplifier for the purpose of testing the CC RAM cell [Fig. 5(b)] [4], [7]. The precharge transistor Q2 is turned on immediately after each WRITE operation. The precharge restores the bit line to a potential which is a threshold voltage below the output gate voltage (approximately 6 V). The precharge pulse also leaves the output node at about 15 V. When a given word line is pulsed to the READ level (0 V) any signal charge in the cell is dumped onto the bit line which in turn falls below the preset level. This action turns on the transistor formed by the output gate and transfers all of the signal charge from the bit line to the preset output node [4], [7]. The output node is a very low capacitance node; thus the signal charge causes a large signal voltage swing. This output node voltage is monitored by the source follower Q3.

The sequence of pulses used to operate the cell for this particular test structure was as follows.

1) A word line is placed in the READ mode to remove any stored charge and then returned to STORE (10 V).

2) Input data are placed on the bit line by turning on the input gate.

3) The word line is pulsed into the WRITE mode (15 V) and returned to STORE.

4) The data input gate is turned off.

5) The bit line and the output node are precharged as described above.

6) The cycle is complete and the circuit is ready for the next READ or WRITE command.

With the sequence of steps described above and depending on the data input voltage, a ONE or a ZERO was written into a given cell, stored for a prescribed time, and then read by applying a READ pulse (0 V) to the storage cell of interest. If a ZERO (no charge) was written into the cell, no output signal was detected as would be expected. When a ONE (full well) was written into the cell, an output signal was detected whose amplitude depended on the particular cell addressed. For example, ONE's in cell 2 (0.5 mil^2 storage area) and cell 3 (1.0 mil^2 storage area) yielded output signal amplitudes of 3 and 5 V, respectively. This corresponds to a charge storage density of 1.6 × 10^{12} e/cm^2 for the above conditions. Changing the data level and the relative timing of the WRITE and READ pulses confirmed that one of the cells can be read and written without affecting the data in the other cell.

The circuit schematic shown in Fig. 6 also applies for the operation of the CC RAM cell with the implanted storage region except that now the word line switches between two levels (0 and 8–10 V) rather than three levels. A dc bias voltage of 8–10 V was applied to the output gate, and the

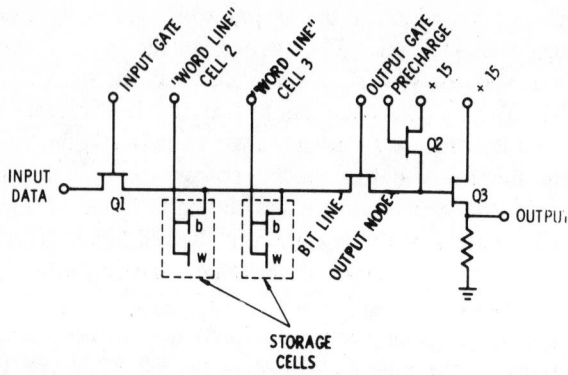

Fig. 6. Circuit schematic for CC RAM cell test structure. The gate area designated by *b* is the transfer region and the area designated by *w* is the storage region.

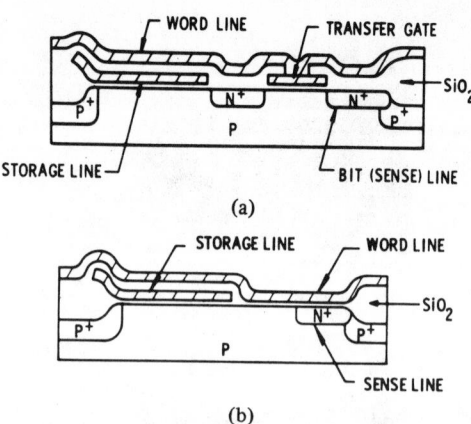

Fig. 7. Cross section of (a) one-transistor cell, and (b) surface charge cell.

substrate bias was −2 V. Data input was accomplished by switching the bit line to the input signal level through transistor $Q1$. A ONE is written in with the bit line low (near ground) and a ZERO is written in with the bit line high (∼10 V).

The precharge output circuit described earlier was used as the signal output amplifier. The sequence of pulses used to operate the CC RAM cell with the implanted storage area is given below.

1) With the word line off, the bit line, and output node are precharged as described earlier.

2) If it is desired to write in a ONE (full well), the input transistor $Q1$ is turned on and a low (near ground) signal level is applied to the bit line. If it is desired to write in a ZERO (empty well), the bit line is held at the high (10–12 V) level.

3) While $Q1$ is turned on the word line is pulsed to 8–10 V and then turned off. This allows charge (electrons) to enter from the bit line into the storage portion of the cell.

4) The data input gate $Q1$ is turned off.

5) The bit line and the output node are precharged as described above.

6) The cycle is complete and the circuit is ready for the next command.

With the above sequence of operations ONE's or ZERO's were successfully written, stored, and read. The measured storage charge densities ranged from 6 to 9×10^{11} cm^{-2} for the range of implant doses used in the fabrication of the test structure.

As mentioned earlier, the operation of the CC RAM cell in Fig. 3 is identical to the one-transistor cell operation. To illustrate this point the word line address was applied to the one-transistor cell which is located on the chip adjacent to the four CC RAM cells [refer to Fig. 5(a)]. The voltage on the storage capacitor of the one-transistor cell was set at 12 V while all other operating voltages remained unchanged. The one-transistor cell operated as expected with a signal amplitude of 7 V corresponding to a stored ONE. This allows a comparison to be made of the storage charge capacity between the one-transistor cell and the two CC RAM cells described in this paper. On a per unit area basis, the CC RAM cell with the im-

planted transfer region (Fig. 2) exhibited a charge capacity 70-percent of that measured for the one-transistor cell, while the CC RAM with the implanted storage region (Fig. 3) ranged from 28 to 41 percent of that for the one-transistor cell. It should be possible to increase these percentages with increases in the implant doses, but the practical upper limit has not yet been established.

V. DISCUSSION

When the CC RAM storage cell is compared with the one-transistor [3] and surface charge [4] storage cells shown in Fig. 7, a number of advantages become apparent. In the CC RAM cell the storage and transfer gates are combined into a single gate which eliminates the need for patterning two separate gates. Thus, the one-transistor and surface charge cells have three lines connecting each cell, while the CC RAM requires only two lines. This simplifies the cell structure significantly and allows a greater fraction of the cell area to be used for charge storage. In addition, the storage part of the cell contains no diffusions as does the one-transistor cell. As a result a higher packing density is achievable without resorting to tighter lithographic tolerances, and higher yield is probable.

Since the CC RAM cell contains only one gate, only one level of interconnect (besides diffusion) is required. An important consequence of this is that the contact to each cell can be eliminated. Although this feature is attractive, the density advantage gained by using two levels of interconnects will probably influence the circuit designer to use two levels. It should also be mentioned that the CC RAM cell fabrication process is very similar to that for the one-transistor cell. The major difference is the addition of a photolithographic step to selectively define the storage (or transfer) regions which receive ion implants.

The advantages of the CC RAM cell detailed above are not gained without expense. For comparable operating voltages there is a reduction in the charge storage capacity per unit area in the CC RAM cell compared to that of the one-transistor cell or the surface charge cell. The exact amount has not yet been established at this time for practical circuits. However, from the evaluation results presented in this paper and from computer calculations, the reduction is expected to be no more than 15–20 percent for the CC RAM cell with the implanted transfer region and around 30 percent for the CC RAM cell with the implanted storage region. In addition, the word line capacitance is increased due to the fact that the gate connected to the word line is used for storing charge in addition to gating charge.

VI. CONCLUSION

A new MOS dynamic RAM storage cell concept and its successful demonstration have been described. The CC RAM cell combines the storage capacitor and the transfer gate into a single gate which serves both functions simultaneously. The resulting cell is simpler than the conventional one-transistor cell or the surface charge cell. The CC RAM offers significant advantages in packing density and potential yield at the expense of some reduction in charge storage capacity per unit area. One variation of the CC RAM cell concept is identical in operation to the present one-transistor or surface charge cell. The charge-coupled RAM storage cell is an attractive candidate for the next generation high density RAM's.

ACKNOWLEDGMENT

The authors are grateful to the many people at Texas Instruments whose assistance made this work possible.

REFERENCES

[1] G. Kruschke, "Prices to fall, capacities to rise," *Digital Design*, pp. 32–36, June 1975.
[2] H. N. Yu *et al.*, "Fabrication of a miniature 8 K bit memory chip using a pattern exposure," presented at the 13th Symp. Electron Ion and Photon Beam Technology, Colorado Springs, CO, May 21–23, 1975.
[3] R. H. Dennard, "Field effect transistor memory," U.S. Patent 3 387 386, June 4, 1968.
[4] W. E. Engeler, J. J. Tiemann, and R. D. Baertsch, "A surface-charge random-access memory system," *IEEE J. Solid-State Circuits*, vol. SC-7, pp. 330–335, Oct. 1972.
[5] J. MacDougall, *Solid-State Technol.*, vol. 14, p. 46, 1971.
[6] R. M. Swanson and J. D. Meindl, "Ion-implanted complementary MOS transistors in low-voltage circuits," *IEEE J. Solid-State Circuits*, vol. SC-7, pp. 146–153, Apr. 1972.
[7] L. G. Heller *et al.*, "High-sensitivity charge transfer sense amplifier," in *ISSCC Dig. Tech. Papers*, 1975, p. 112.

A Nonvolatile Charge-Addressed Memory (NOVCAM) Cell

MARVIN H. WHITE, FELLOW, IEEE, DONALD R. LAMPE, MEMBER, IEEE, JOHN L. FAGAN, MEMBER, IEEE,
FRANCIS J. KUB, AND DOUGLAS A. BARTH, MEMBER, IEEE

Abstract—A nonvolatile charge-addressed memory (NOVCAM) cell is described in a 64-bit shift register configuration. The charge address is performed by a charge-coupled device (CCD) shift register and the information is stored in metal-nitride–oxide–silicon (MNOS) nonvolatile sites located in parallel with the CCD shift register. The tunneling electric field strength across the thin-oxide MNOS structure is controlled by the magnitude of the charge transferred from the CCD register. The write, erase, and read modes of operation are discussed with typical ±20 V 10 μs write/erase, and 2 V 2 μs read conditions. Readout is accomplished by parallel stabilized charge injection from a diffused p/n junction to minimize access time to the first bit.

I. INTRODUCTION

THERE has been recent interest in the metal–nitride–oxide–silicon (MNOS) structure for electrically alterable nonvolatile charge-storage memory arrays [1]-[3]. Historically, the MNOS structure was configured in the form of a transistor with source and drain terminals for "current accessing" the stored charge. The stored charge determined the threshold voltage of the transistor which in turn affected the magnitude of the current flow between source and drain terminals for a specified read voltage on the gate electrode. In this mode of operation the read-disturb effect must be minimized since a sizeable read voltage must be used to provide the necessary current to charge the external node capacitance of the detection circuitry. This is particularly true as the memory transistor is reduced in size and the (W/L) ratio is not sufficient to provide current drive. Thus, there is a tradeoff between read-disturb effect, packing density, speed, and array format. In addition, the MNOS memory transistor configuration mode of operation may have peak oxide electric field strengths $>10^7$ V/cm, $dV/dt > 150$ V/μs, and peak oxide tunneling densities >0.1 A/cm^2 which results in a deterioration of charge retention due to Si–SiO$_2$ interface-state creation [4]. These excessive conditions are caused by the almost instantaneous formation of an inversion layer beneath the gate electrode with the applied write voltage initially across the Si$_3$N$_4$/SiO$_2$ dielectric sandwich.

The MNOS structure may be charge addressed as an alterna-

tive method of cell operation. The oxide electric field strength and current density are controlled by the magnitude of signal charge beneath a deep depleted MNOS capacitor. The surface potential of the latter is collapsed and the oxide electric field strength increases with a concomitant increase in oxide tunneling current. The transfer of signal charge is accomplished with a charge-coupled device (CCD). The combination of charge transfer and MNOS principles for signal control and address, and nonvolatile storage has been discussed by several workers [5]-[7]. In the initial attempts to combine these structures the MNOS storage site was located inside a stepped dielectric, 2ϕ or 3ϕ CCD shift register [5], [12], as shown in Fig. 1(a). Certain problems were encountered such as inadequate charge handling capability which resulted in poor write characteristics, ineffective write/inhibit operation since high voltage clocks (for good transfer efficiency) caused spurious write operation, poor read operation due to the large access time to first bit and small detection window, and degraded memory retention caused by a small write window and read disturb effects. To alleviate these problems a compact, nonvolatile charge-addressed memory (NOVCAM) cell with parallel write and read-out injection was introduced [6] for block-oriented random-access memory (BORAM) applications. The NOVCAM cell is composed of a CCD shift register and a thin-oxide MNOS memory structure in parallel with the register to provide separate locations for signal address and storage [8], as shown in Fig. 1(b). This paper will describe the operation of the NOVCAM cell.

II. ANALYSIS OF A NOVCAM CAPACITOR

Fig. 2 illustrates a cross section of a thin-oxide MNOS capacitor under deep depletion. Continuity of electrostatic displacements require

$$K_o \xi_o - K_n \xi_n = Q_I/\epsilon_o \tag{1}$$

$$K_s \xi_s - K_o \xi_o = Q_{SS}/\epsilon_o \tag{2}$$

for the Si$_3$N$_4$/SiO$_2$ and SiO$_2$/Si interfaces, respectively, where K_n, K_o, K_s are the relative dielectric constants and ξ_n, ξ_o, ξ_s are the associated electric field strengths. Q_I and Q_{SS} represent the trapped charge and fixed charge densities at their respective interfaces. For simplicity, these charges are assumed to lie at their respective interfaces. Gauss' theorem for the silicon depletion region gives the semiconductor electric field at the SiO$_2$/Si interface,

Manuscript received March 28, 1975; revised June 20, 1975. This work was supported in part by the Department of the Army, U.S. Army Electronics Command (ECOM), Ft. Monmouth, N.J. This paper was presented in part at the 1974 International Electron Devices Meeting, Washington, D.C., December 1974.

The authors are with the Advanced Technology Laboratories, Westinghouse Electric Corporation, Baltimore, Md. 21090.

Reprinted from *IEEE J. Solid-State Circuits*, vol. SC-10, pp. 281–286, Oct. 1975.

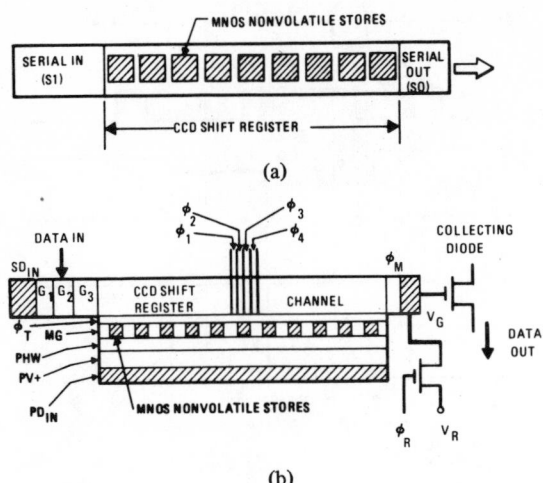

Fig. 1. Combination of CCD shift registers and MNOS nonvolatile stores. (a) MNOS nonvolatile stores within CCD channel. (b) MNOS nonvolatile stores adjacent to CCD channel.

$$\xi_s = \frac{q_s + qN_A x_d}{K_s \epsilon_o} \tag{3}$$

assuming a uniform distribution of fixed background charge N_A over a depletion width x_d and a signal charge q_s located at the SiO_2/Si interface. Poisson's equation for the bending of the semiconductor bands gives the surface potential ϕ_s,

$$\phi_s = \frac{eN_A x_d^2}{2K_s \epsilon_o}. \tag{4}$$

Kirchhoff's law for the summation of potentials yields

$$V_G = x_n \xi_n + x_o \xi_o + \phi_{MS} + \phi_S \tag{5}$$

where V_G is the applied voltage, ϕ_{MS} the metal-semiconductor work function difference, and x_n, x_o the nitride and oxide thickness, respectively.

Combining (1)-(5) yields an expression for the oxide electric field strength in the nonequilibrium condition $\phi_S > 2\phi_F$,

$$\xi_o = \frac{q_s + Q_B \left\{ 1 - [q_s/C_{\text{eff}}(V_G - V_{FB})] \right\}^{1/2}}{K_o \epsilon_o}$$

$$\phi_S = 2\phi_F \tag{6}$$

where

$$Q_B = [2eN_A K_s \epsilon_o (V_G - V_{FB})]^{1/2}$$

$$C_{\text{eff}} = \frac{K_o \epsilon_o}{x_o + (K_o/K_n)x_n}$$

$$V_{FB} = \phi_{MS} - \frac{Q_{SS}}{C_{\text{eff}}} - \frac{Q_I x_n}{K_n \epsilon_o}.$$

The equilibrium condition is defined at strong inversion $\phi_S = 2\phi_F$, and the signal charge fills the surface well with $q_s = C_{\text{eff}}(V_G - V_{FB})$, and

$$\xi_o \simeq \frac{C_{\text{eff}}(V_G - V_{FB})}{K_o \epsilon_o} \qquad \phi_S = 2\phi_F. \tag{7}$$

Fig. 3 illustrates the variation of oxide electric field strength

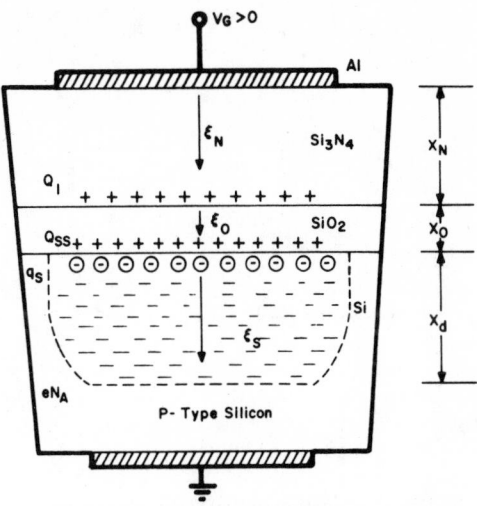

Fig. 2. Cross section of thin-oxide MNOS capacitor.

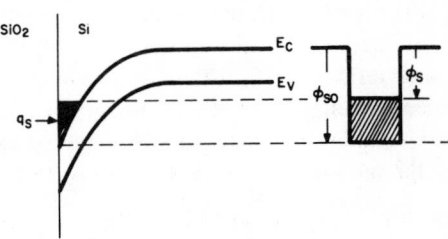

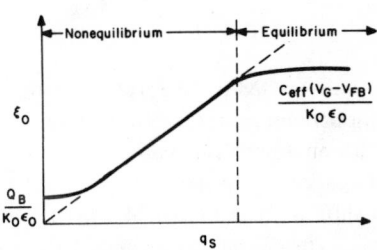

Fig. 3. Variation of oxide electric field strength ξ_o with signal charge q_s.

with injected signal charge q_s. The oxide electric field strength at low signal charge is typically $\sim 10^5$ V/cm and increases linearly with charge until saturation which is determined by the applied gate voltage V_G and the final V_{FB}. The latter is varying because Q_I is altered by the oxide tunneling current through the relationship

$$\frac{dQ_I}{dt} = J_n - J_o(\xi_o) \tag{8}$$

where J_n is relatively constant since $x_o \ll x_n$ in thin-oxide MNOS structures. Thus, the charge-controlled tunneling feature of the NOVCAM cell limits the peak fields and current densities in the thin-oxide MNOS structure. In the conventional MNOS memory transistor this is accomplished through careful control of the erase/write rise and fall times such that $|dV_G/dt| < 150$ V/μs and the use of a drain-source protected geometry [9].

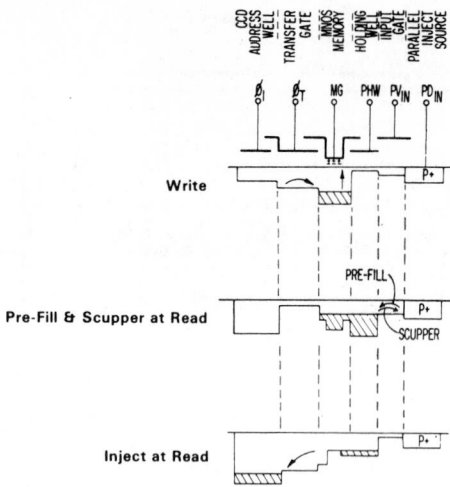

Fig. 4. Operation of the NOVCAM test cell.

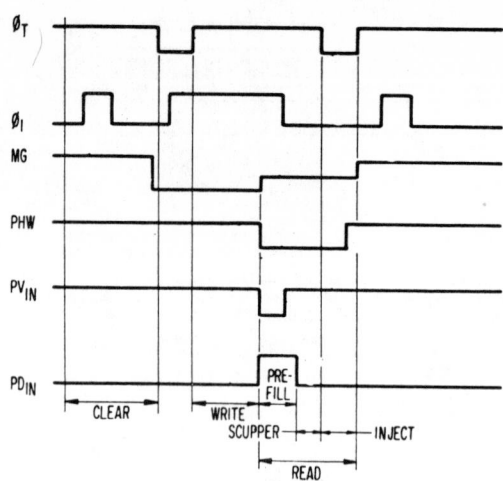

Fig. 5. Write/read parallel injection sequence.

III. OPERATION OF THE NOVCAM CELL

Fig. 4 illustrates the write and read operation of the NOV-CAM cell. The write operation employs a *surface potential amplification* method which consists of a MNOS storage/CCD well area ratio of 1:10 to allow sufficient collapse of the surface potential beneath the MNOS storage site. The ratio of CCD phase electrode capacitance to MNOS storage capacitance determines the surface potential amplification ratio.

$$\frac{\Delta\phi_S(\text{MNOS})}{\Delta\phi_S(\text{CCD})} = \frac{x_o(\text{MEM}) + (K_o/K_n)x_n}{x_o(\text{NONMEM}) + (K_o/K_n)x_n} \cdot \frac{A_{\text{CCD}}}{A_{\text{MNOS}}}$$

(9)

where "MEM" and "NONMEM" refer to memory and non-memory oxide thicknesses, respectively, and A is the associative area. With an equivalent oxide ratio of 1:5 the surface potential amplification ratio is 2:1. Charge is transferred from the CCD shift register to the MNOS structure under a transfer gate ϕ_T. The charge is transferred from the CCD to the MNOS site in push-clock fashion with ϕ_1 (CCD phase voltage) and ϕ_T collapsing in succession. Fig. 5 illustrates the timing diagram of the write/read sequence with the low clock voltage an attractive potential well for signal charge. Once the surface potential is collapsed, the oxide electric field strength increases and the tunneling of signal charge commences from the surface channel into deep traps located near the SiO_2/Si_3N_4 interface. Fig. 6 illustrates experimental data on the write operation of the NOVCAM cell. The data are obtained from an initial saturated erase or clear state.

The read operation is nondestructive (NDRO) and accomplished through the control action of the MNOS surface potential which gates the parallel charge injection from a p+ source diffusion into the CCD shift register. The read operation consists of pulsing the parallel diode (PD_{in}) and the subsequent "scuppering" performed together with the parallel input gate (PV_{in}). The read injection is a push-clock sequence with the successive collapse of the voltages on the parallel holding well (PHW) electrode, the memory gate (MG), and the transfer gate ϕ_T. Stabilized charge injection (surface

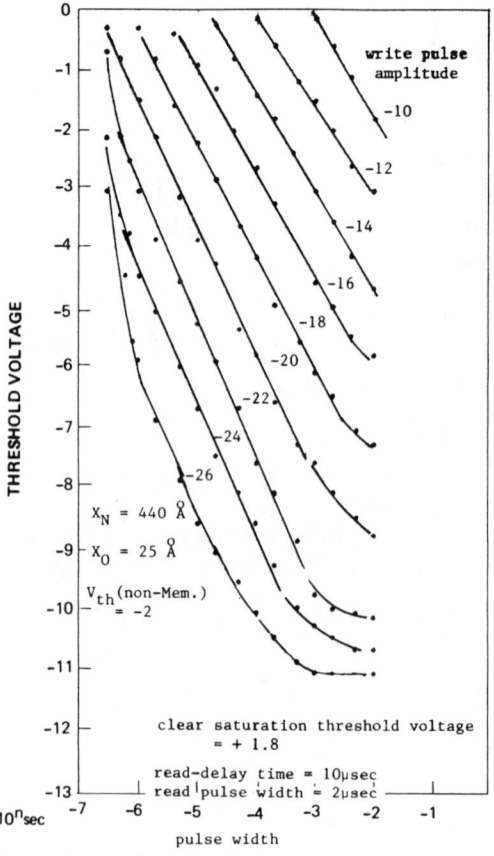

Fig. 6. Write operation of the NOVCAM cell.

potential equilibration) is accomplished with the clock waveforms of Fig. 5.

Fig. 7 illustrates a detailed profile of the surface potential in the read mode with the barrier height defined as

$$\phi_B = |V_{th}(\text{MEM "WRITE"}) - V_{th}(\text{NONMEM})| \quad (10)$$

where the measurement is performed at a fixed read-delay time t_{rd}, as shown in Fig. 7. The reduction in ϕ_B with increasing t_{rd} is due to the finite charge retention slope of the thin-oxide MNOS memory structure. The slope of the charge

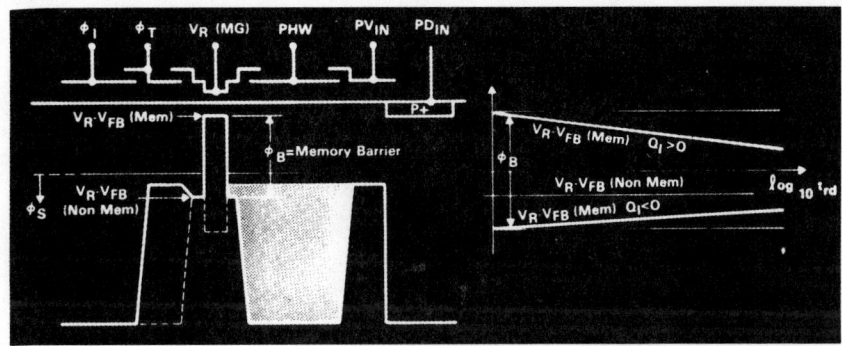

Fig. 7. Surface potential profile in the read mode with effective barrier height ϕ_B.

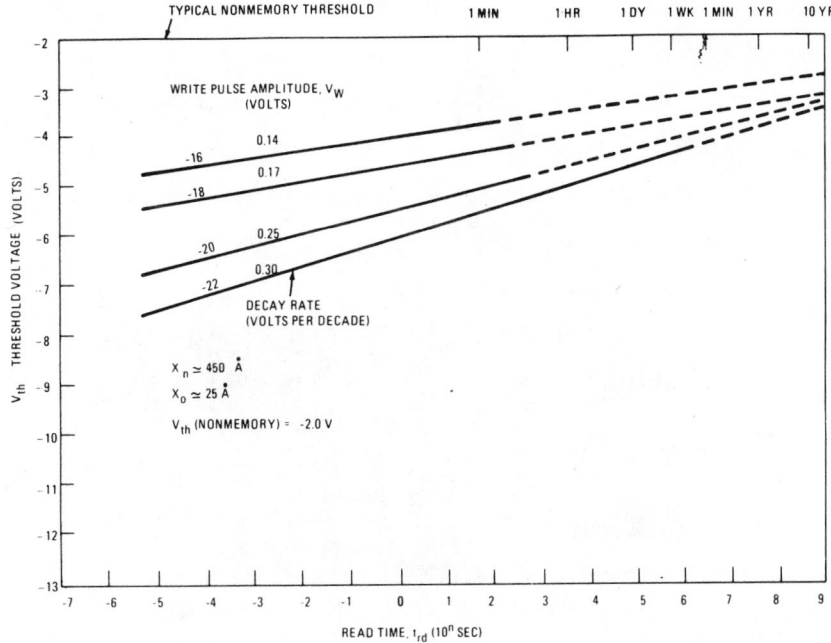

Fig. 8. NOVCAM threshold voltage retention of memory exposed to polysilicon thermal oxidation during fabrication.

retention curve is influenced by the initial charge stored in the deep traps, as shown in Fig. 8. The decay in the "clear" state threshold is typically a factor of 2 less than the "write" state threshold voltage [4]. Fig. 9 illustrates the test pulse waveform to exercise the NOVCAM cell.

IV. DESCRIPTION OF A 64-BIT NOVCAM CELL

The serial CCD operation is combined with the MNOS nonvolatile storage to achieve the NOVCAM cell. The insertion of signal charge in the CCD is performed with stabilized charge injection [10], [11], as shown in Fig. 10. The serial diode input (SD_{in}) is pulsed in combination with the serial input gate (G_3) to allow the data input (G_2) to be injected into the ϕ_1 electrode of the first bit. Fig. 10 illustrates a 4ϕ complementary pairs clock sequence which provides efficient signal transfer down the CCD shift register.

Also indicated is the output signal voltage V_G on the gate electrode of the on-chip gated integrator (i.e., a reset switch and MOS electrometer combination sometimes called a "precharge and float" circuit). The feedthroughs of the reset switch waveform ϕ_R and "mux" gate waveform ϕ_M (i.e., out-

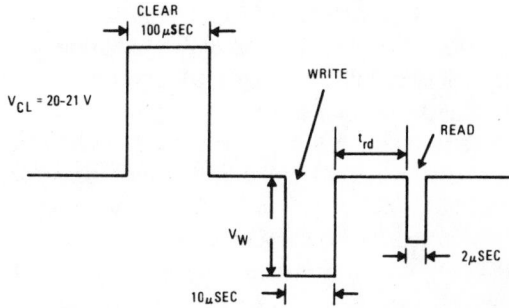

Fig. 9. Test pulse waveforms to exercise NOVCAM cell.

put gate) are illustrated in the figure. ϕ_M is pulsed, rather than dc operated, in order to allow a large fraction of the data bit time for "settling" in the output circuitry. The push-clock operation of ϕ_3 and ϕ_M going high (i.e., repulsive to signal charge) to force the signal charge onto the collecting diode is shown in the figure. For a MNOS structure of $x_n = 500$ Å, $x_o = 25$ Å typical write voltages are 20 V (10 μs), clear voltages are variable within a CCD line time, typically 20 V, with

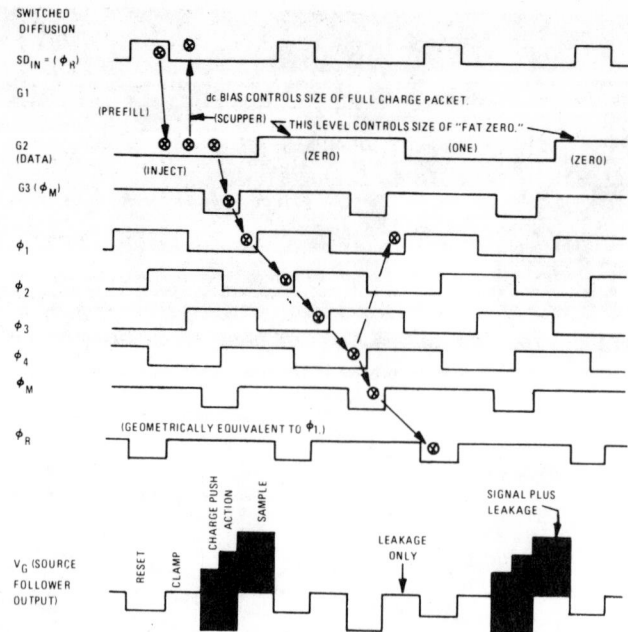

Fig. 10. NOVCAM shift register waveforms (see Fig. 1). Arrows denote direction of signal charge flow.

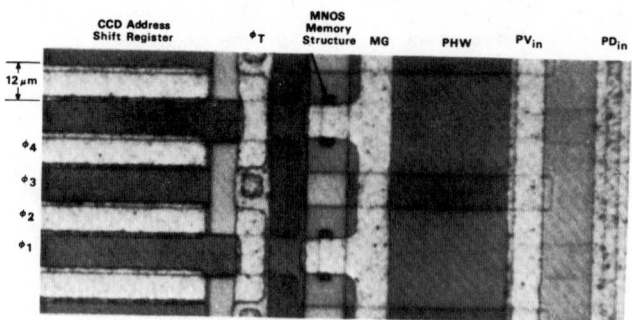

Fig. 11. Photomicrograph of the NOVCAM test cell.

typical read voltages of 2 V (2 μs). The CCD shift register operates with 20-V clocks and a 2-V substrate bias (e.g., V_{CL} = 20 V, V_{SS} = 22 V). Fig. 11 shows a photomicrograph of the NOVCAM test cell with the various electrodes labeled. The test cell is repeated 64 times in a 64-bit shift register. The fabrication sequence consists of 8 photographic mask steps including a low temperature phosphosilicate glass (PSG) for passivation. The basic CCD shift register is a dual dielectric SiO_2/Si_3N_4 system with coplanar, overlapping, aluminum/polysilicon electrodes to form a sealed surface-channel CCD, as shown in the cross section of Fig. 12. The electrode dimensions are L = 8-μm transfer gates and L = 12-μm storage gates with W = 50 μm to provide a transfer efficiency $\epsilon \leqslant 4 \times 10^{-5}$ at f_c = 200 kHz for a 10 percent "fat-zero" (FZ), and $\epsilon = 4 \times 10^{-4}$ for "no fat-zero" (NFZ), and a substrate bias of 2 V. The gate oxide is 1 kÅ (HCl grown), and the Si_3N_4 is x_n = 500 Å and deposited at 750°C in a N_2 atmosphere at a deposition rate of 125 Å/min. The final oxide thickness is estimated at x_o = 25 Å and is formed with a combination of an initial preclean procedure and *in situ* reaction. The polysilicon electrodes are 3.5 kÅ and deposited at 650°C with subsequent phosphorous doping at 950°C to provide a nominal 100 $\Omega/\square$ sheet resistance. The

polysilicon is oxidized thermally in steam at 1000°C to provide self-aligned technology. The Si_3N_4 does not oxidize appreciably which provides coplanarity for the polysilicon and aluminum electrodes.

Fig. 13 illustrates the memory operation of the NOVCAM test cell in a 64-bit shift register configuration with nonvolatile storage of a 4-zero/8-one data pattern. The upper waveform is the input data train, the middle waveform the unreconstructed analog output readout 4 μs after write, and the bottom waveform the analog readout 13 h after write. There is no noticeable degradation in the NOVCAM cell operation after 10^{10} cycles of erase/write and read, 10^{10} cycles of erase or write followed by a read, or a single erase/write followed by 10^{10} read cycles. The charge retention of the NOVCAM cell is shown in Fig. 8, and projected retention at room temperature is past 2 years and approaches 10 years barring an unforeseen change in the slope of the retention curve.

V. CONCLUSIONS

The NOVCAM cell has the following advantages: 1) the ability to address the MNOS memory structure without the need for cross-coupled "flip-flop" detection circuitry

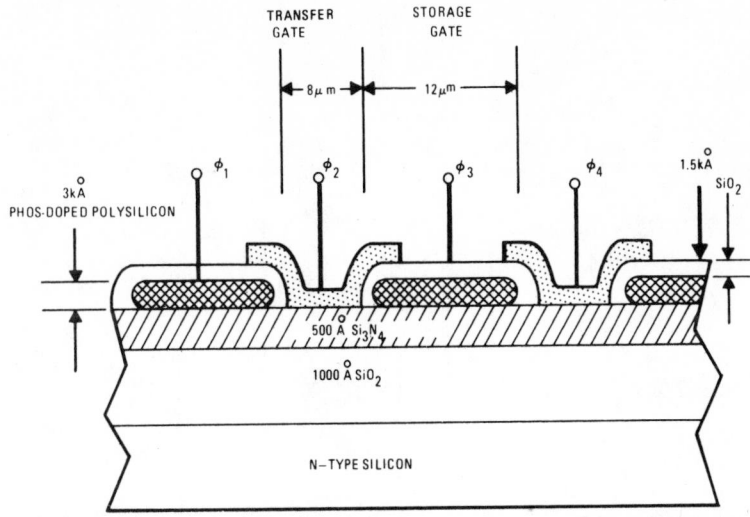

Fig. 12. Cross section of aluminum/polysilicon 4ϕ, coplanar electrode configuration.

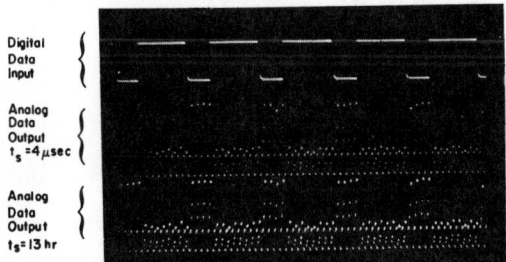

Fig. 13. Nonvolatile memory operation for various storage times t_s at $f_c = 125$ kHz.

placed on the "pitch" of the memory array, 2) the access time is limited by the charge-transfer time past several electrodes (e.g., nominally, 1 "bit") and not parasitic line capacitance, and 3) it has high density, nonvolatile semiconductor memories. The inherent advantage of charge addressing is a control of the oxide electric field strength which is important in achieving high endurance memory operation and minimizing read-disturb effects. This paper has demonstrated the NOVCAM operation in a 64-bit shift register with typical erase/write voltages of ±20 V for 10 μs, and read voltages of 2 V for 5 μs on a structure of $x_n = 500$ Å, $x_o = 25$ Å, where x_n and x_o are the nitride and oxide thickness, respectively. There is no noticeable degradation in the NOVCAM cell operation after 10^{10} cycles of erase or write followed by a single read, or a single erase/write followed by 10^{10} read cycles. Memory threshold voltage retention of the NOVCAM cell at room temperature is projected past 2 years and approaches 10 years. Although this paper discussed a digital NOVCAM readout the input signal may be analog or multilevel since the basic MNOS memory cell is an analog-storage device.

ACKNOWLEDGMENT

The authors wish to express their appreciation to R. M. McLouski for the diffusions and oxidations, P. R. Reid for the polysilicon depositions, J. Grossman for the chemical processing, C. J. Taylor for the photolithography, and D. S. Herman for the special aluminum/silicon evaporations. The authors would also like to thank W. S. Corak, Manager of the Solid-State Systems Technology Laboratory, and G. Strull, Manager of the Advanced Technology Laboratory, for their encouragement and support. This work was directed in part by W. H. Glendinning of the U.S. Army ECOM Laboratories, Ft. Monmouth, N.J. A special thanks is accorded to G. Mudd for the preparation of the manuscript.

REFERENCES

[1] J. E. Brewer, "MNOS secondary storage," presented at the Nat. Aerospace and Electronics Conf. (NAECON), Dayton, Ohio, May 13–15, 1974.
[2] L. G. Carlstedt and C. M. Svensson, "MNOS memory transistors in simple memory arrays," IEEE J. Solid-State Circuits, vol. SC-7, pp. 382–388, Oct. 1972.
[3] J. R. Cricchi, J. E. Brewer, D. W. Williams, F. C. Blaha, M. D. Fitzpatrick, D. R. Hadden, Jr., and D. Haratz, "Nonvolatile block-oriented RAM," in IEEE ISSCC Dig. Tech. Papers, vol. 17, Feb. 1974, p. 204.
[4] M. H. White and J. R. Cricchi, "Characterization of thin-oxide MNOS memory transistors," IEEE Trans. Electron Devices, vol. ED-19, pp. 1280–1288, Dec. 1972.
[5] Y. T. Chan, B. T. French, and R. A. Gudmundsen, "Charge-coupled memory device," Appl. Phys. Lett., vol. 22, p. 650, 1973.
[6] M. H. White, D. R. Lampe, and J. L. Fagan, "CCD and MNOS devices for programmable analog signal processing and digital nonvolatile memory," Tech. Dig., Int. Electron Devices Meeting, Dec. 1973, p. 130.
[7] K. Goser and K. Knauer, "Nonvolatile CCD memory with MNOS storage capacitors," IEEE J. Solid-State Circuits (Corresp.), vol. SC-9, pp. 148–150, June 1974.
[8] M. H. White, D. R. Lampe, J. L. Fagan, and D. A. Barth, "A nonvolatile charge-addressed memory (NOVCAM) cell," in Tech. Dig., Int. Electron Devices Meeting, 1974, p. 115.
[9] J. R. Cricchi, F. C. Blaha, and M. D. Fitzpatrick, "The drain-source protected MNOS memory device and memory endurance," in Tech. Dig., Int. Electron Devices Meeting, 1975, p. 126.
[10] D. R. Lampe, M. H. White, J. L. Fagan, and J. H. Mims, "An electrically-reprogrammable analog transversal filter," in IEEE ISSCC Dig. Tech. Papers, Feb. 1974, pp. 156–157.
[11] C. H. Séquin and A. M. Mohsen, "Linearity of electrical charge injection into charge-coupled devices," IEEE J. Solid-State Circuits, vol. SC-10, pp. 81–92, Apr. 1975.
[12] G. E. Smith, "Charge-coupled memory with storage sites," U.S. Patent 3 654 499, Apr. 4, 1972.

Part V
Analog Signal Processing

CCD's are inherently analog, and as such they are ideally suited to performing sampled data filtering functions in the analog domain. A sampled data filter is one which samples an analog signal $x(t)$ at a sample frequency f_s to produce analog samples x_n and then operates on these samples to produce the desired output. This type of filter is almost always implemented with digital hardware by digitizing the x_n, and for this reason, sampled data filters are usually called digital filters. Reference [1] gives an excellent overview of advanced digital filtering concepts.

CCD's however, offer the possibility of performing certain sampled data filtering functions in the analog domain, thereby eliminating analog-to-digital (A/D) conversion and simplifying the required electronics. In such applications, the CCD performance limitations, such as dynamic range, linearity, charge-transfer efficiency, leakage current, etc., limit the filter performance relative to a digital filter. However, for applications having modest performance requirements and sufficiently high volume that unit cost is a primary design goal, CCD's offer tremendous potential advantages.

In a sense, CCD's combine the best features of digital and analog techniques. One of the primary advantages of digital filtering is that everything is controlled by a master clock which permits a high degree of synchronization and stability. CCD's are also controlled by a master clock, and time delays are similarly insensitive to temperature and component drift.

CCD's are functionally similar to acoustic surface-wave devices (SWD's) with two important differences. 1) SWD's are not sampled-data devices. Their time delay is determined by the geometry of the device and the speed of Rayleigh waves. Time delay in CCD's may be electronically varied via the CCD clock frequency. 2) SWD's operate with high bandwidth (up to 1 GHz) and short time delays (up to 100 μs). CCD's, on the other hand, are limited in bandwidth (about 10-MHz maximum) but can achieve longer time delays (up to 1 s). CCD's and SWD's can be thought of as complementary technologies which perform similar functions in different regimes of time delay and bandwidth. CCD's can be operated at clock frequencies in excess of 100 MHz [2], but except for special applications [3], economics usually favors SWD's in this frequency range.

The applications for CCD's in analog signal processing are extremely diverse and not generally familiar to MOS IC designers. Their use is further complicated by the difficulty in configuring general-purpose components which can oe manufactured in high volume and purchased from a catalog. Most of the applications reported to date require custom IC's, and it is doubtful whether general-purpose analog signal pro-

cessing CCD's will be commercially available on a large scale in the near future.

In paper 1, White and Lampe give an introduction to CCD analog signal processing technology. They discuss some of the CCD's limitations in this application and present a set of functional CCD building blocks which can be combined to perform complex operations.

The CCD delay line is functionally the simplest CCD building block, and paper 2 discusses a delay line which can be used in a number of signal processing applications. The analysis of the performance of this device indicates the performance criteria that are important in analog signal processing. It also illustrates the potential for integrating non-CCD analog MOS components on the CCD IC.

In paper 3, Butler et al. present two applications of CCD delay lines to radar signal processing: 1) a delay canceller for rejecting stationary targets in a moving target indicator (MTI) radar, and 2) an analog memory for recording, delaying, and retransmitting signals in radar electronic counter measures (ECM).

A delay canceller is basically a multiplexed transversal filter which rejects dc and multiples of the radar pulse repetition frequency, but feedback is often employed to improve the frequency characteristic [4]. Butler et al. utilized BBD's, but delay cancellers have also been made using CCD's [4]. The delay canceller is a specific example of a general multiplexed recursive filter which can be used to realize more general frequency responses, such as the one briefly discussed in paper 1 [5].

The ability to control the clock frequency of CCD's is also important CCD delay line applications. In the ECM application, a false target is generated by varying the delay between read in and retransmission. In other applications, data are clocked into a CCD at one frequency and clocked out at a slower frequency [5], [6].

Paper 4 presents a time delay beam former for ultrasonic imaging in medical electronics. In this application, focusing is achieved by varying the clock frequency of the CCD "lens." This type of beam forming also has application in sonar systems, as discussed briefly in paper 1.

Integrated CCD transversal filters can be economically realized using the split electrode technique [7]. Paper 5 presents the design considerations of a practical CCD transversal filter and presents the achievable performance characteristics of a 63-stage low-pass filter. Such a filter has application to frequency division multiplexing and anti-aliasing in communication systems [8], [9].

Other types of filters which are potentially important to

communication systems include bandpass filters and matched filters. Paper 6 reviews the performance of CCD transversal filters for communication applications and presents a number of illustrative examples. The filters discussed in papers 5 and 6 require off-chip clock drivers and output amplifiers. However, these functions are amenable to integration, as demonstrated in [10].

Perhaps one of the most important potential applications of the CCD transversal filter is spectral analysis using the chirp z-transform (CZT) [11], [12]. The CZT is an algorithm for performing the discrete Fourier transform (DFT) in which the bulk of the computation is performed in a transversal filter. This makes it very attractive for CCD implementation [12]. Paper 7 discusses the design of a 500-point CCD CZT system, and paper 8 compares the CCD CZT with the digital fast Fourier transform (FFT). The CCD CZT has application to video bandwidth reduction [13], Doppler processing in MTI radar [14], speech processing, sonar spectral analysis, remote surveillance, and image enhancement.

Paper 7 also illustrates how the impulse response of a CCD transversal filter can be modified with a single photomask for diverse applications. In this respect, CCD transversal filters resemble factory programmable read-only memories (ROM's) which can be custom programmed in a short time at relatively low cost.

Since CCD's are inherently serial, most of the work to date has been done on serially organized devices. However, as with memory, charge-coupling principles can be utilized to advantage in other types of analog components. In paper 9, Tiemann, Engeler, and Baertsch present a surface-charge correlator which employees a parallel architecture in which signal charges are stored on MOS capacitors and addressed with digital electronics to perform the desired operation. Components which are functionally similar to the surface-charge correlator are presently commercially available [15]. The analog/digital hybrid approach to signal processing has advantages, as discussed in paper 9. CCD's are also being developed for purely digital signal processing [16].

Signal processing in advanced IR imaging systems is a specialized but very active area of CCD research and development. Paper 10 summarizes the diverse system concepts which have led to studies of CCD's at cryogenic temperature [17] and to CCD's on InSb [18], [19].

CCD's are not expected to make digital filters obsolete. However, for those signal processing applications which fulfill the twin requirements of modest performance and high volume, tremendous cost advantages can be gained with CCD's. In addition, more applications will certainly emerge as system designers become aware of CCD's and as new signal processing structures are conceived [20]. The magnitude of the ultimate impact of CCD's on analog signal processing is not clear at the present time. However, it seems certain that

CCD's will play an important role in analog signal processing electronics in the future.

REFERENCES

[1] L. R. Rabiner and B. Gold, *Theory and Application of Digital Signal Processing.* Englewood Cliffs, NJ: Prentice-Hall, 1975.

[2] L. J. M. Esser, M. G. Collet, and J. G. Van Santen, "The peristaltic charge-coupled device," in *Tech. Dig., 1973 Int. Electron Devices Meeting*, 1973, pp. 17–20.

[3] T. F. Linnenbrink, M. J. Monahan, and J. L. Rea, "A CCD-based transient data recorder," in *CCD '75 Proc.*, San Diego, CA, Oct. 1975, pp. 443–453.

[4] J. E. Bounden, R. Eames, and J. B. G. Roberts, "MTI filtering for radar with charge transfer devices," in *CCD '74 Proc.*, Edinburgh, Scotland, Sept. 1974, pp. 206–213.

[5] J. Mattern and D. R. Lampe, "A reprogrammable filter bank using charge-coupled device discrete analog-signal processing," *IEEE J. Solid-State Circuits (Joint Special Issue with Transactions on Electron Devices on Charge-Transfer Devices)*, vol. SC-11, pp. 88–93, Feb. 1976.

[6] W. Bailey, W. Eversole, J. Holmes, W. Arens, W. Hoover, J. McGehee, and R. Ridings, "CCD applications to synthetic aperture radar," in *CCD '75 Proc.*, San Diego, CA, Oct. 1975, pp. 301–308.

[7] D. D. Buss, D. R. Collins, W. H. Bailey, and C. R. Reeves, "Transversal filtering using charge transfer devices," *IEEE J. Solid-State Circuits*, vol. SC-8, pp. 138–146, Apr. 1973.

[8] R. D. Baertsch and J. J. Tiemann, "Applications of a CCD low-pass transversal filter," in *CCD '75 Proc.*, San Diego, CA, Oct. 1975, pp. 251–256.

[9] J. J. Tiemann and R. Sherrick, "Application of CCD's to single sideband generation and demodulation," in *Nat. Telecommun. Conf. Rec.*, vol. 1, New Orleans, LA, Dec. 1975, pp. 1, 12-1, 14.

[10] C. R. Hewes, "A self-contained 800-stage CCD transversal filter," in *CCD '75 Proc.*, San Diego, CA, Oct. 1975, pp. 309–318.

[11] L. R. Rabiner, R. W. Shafer, and C. M. Rader, "The chirp z-transform algorithm." *IEEE Trans. Audio Electroacoust.*, vol. AU-17, pp. 86–92, June 1969.

[12] H. J. Whitehouse, J. M. Speiser, and R. W. Means, "High speed serial access linear transform implementation," presented at the All Applications Digital Comput. Symp. Orlando, FL, Jan. 1973.

[13] R. W. Means, H. J. Whitehouse, and J. M. Speiser, "Television encoding using a hybrid discrete cosine transform and a differential pulse code modulator in real time," in *Nat. Telecommun. Conf. Rec.*, San Diego, CA, Dec. 1974, pp. 69–74.

[14] W. H. Bailey, D. D. Buss, L. R. Hite, and M. W. Whatley, "Radar video processing using the CCD chirp z-transform," in *CCD '75 Proc.*, San Diego, CA, Oct. 1975, pp. 283–290.

[15] R. R. Buss and G. P. Weckler, "Discrete time analog signal processing devices employing a parallel architecture," in *CCD '75 Proc.*, San Diego, CA, Oct. 1975, pp. 237–244.

[16] T. A. Zimmerman, "The digital approach to charge-coupled device signal processing, in *IEEE Advanced Solid-State Components for Signal Processing*, suppl. to *Proc. 1975 IEEE Int. Symp. on Circuits and Systems*, Boston, MA, Apr. 1975, pp. 69–82.

[17] K. Nummedal, J. C. Fraser, S. C. Su, R. Baron, and R. M. Finnila, "Extrinsic silicon monolithic focal plane array technology and applications," in *CCD '75 Proc.*, San Diego, CA, Oct. 1975, pp. 19–30.

[18] J. C. Kim, "InSb MIS technology and CID devices," in *CCD '75 Proc.*, San Diego, CA, Oct. 1975, pp. 1–17.

[19] R. D. Thom, R. E. Eck, J. D. Phillips, and J. B. Scorso, "InSb-CCD's and other MIS devices for infrared applications," in *CCD '75 Proc.*, San Diego, CA, Oct. 1973, pp. 31–41.

[20] P. Bosshart, "An integrated analog correlator using charge-coupled devices," in *1976 IEEE ISSCC Dig. Tech. Papers*, Philadelphia, PA, Feb. 1976.

Charge Coupled Device (CCD) Analog Signal Processing[*]

Marvin H. White   Donald R. Lampe

Westinghouse Defense & Electronic Systems Center, SDD,
Advanced Technology Laboratories, P.O.Box 1521, Balto.,Md. 21203

ABSTRACT

CCD basic building blocks provide a flexible approach to analog signal processing in systems. The Serial In/Serial Out (SI/SO) block provides time-base translation through electrically alterable time delay and recursive, programmable, filter-banks may be realized with the addition of PROM's or EAROM's (adaptive programming) to determine filter center frequency, bandwidth, and gain. The Parallel In/Serial Out (PI/SO) block may be used for time-division-multiplexing (TDM) of signals from a number of parallel channels into a serial data stream; and through the "delay and add" mode of operation, the PI/SO block permits sensor array beam forming and steering as well as convolution. The Serial In/Parallel Out (SI/PO) block provides variable tapped delay lines and transversal filters/correlators. Electrically reprogrammable analog weights combined with these building blocks offer adaptive filtering for communications. Combinations of the above linear or one-dimension blocks may be employed for Fourier transforming, filter banks and multiple correlators. Applications of CCD's are discussed for Radar, Sonar and Communication Systems.

1.0   Introduction

A simplification of the charge-coupled (CCD) analog shift register is shown in Figure 1. The timing is arranged so the switch toggling frequency is one half of the four-phase clock frequency. Thus, the input sampling switch, S_1, alternately samples data and zero reference. At the output, switch S_2 clamps during zero reference, samples during data, and holds when it is not actually sampling. The output holding capacitor, therefore, contains only the "time-stretched" data samples. In a shift register having N pairs of stages, there will be N signal samples and N zero reference samples, each of duration T/2.

Figure 2 illustrates some key waveforms which are applied to the CCD analog shift register. The waveforms $\emptyset_1$ through $\emptyset_4$ are the four-phase clocks whose function is to propagate charges down the line without dispersion. Waveform S_1 demonstrates the switch functions; data is sampled in the up

position and zero reference is sampled in the down position. V_0 demonstrates the appearance of the delay line output voltage with the alternate data and zero reference outputs confined to 3/8T. S_2 demonstrates the output processing functions: the data is sampled in the up position, the zero is clamped in the down position, and the data is held when it is not sampled (center). The interval from data sample to data sample is T and the total transport time is NT where N is the number of CCD analog delay line data stages. Thus, the signal delay for a serial in/serial out (SI/SO) CCD analog shift register is,

$$\tau = NT = N/f_c \qquad (1)$$

which illustrates the electrically alterable delay feature of the CCD delay line.

---

[*] Sponsored in part by Naval Research Labs., Washington, D.C.

Reprinted from *Proc. 1975 Naval Electron. Lab. Center Int. Conf. on the Application of Charge-Coupled Devices*, Oct. 29-31, 1975, pp. 189-197.

Since the Shannon Sampling Theorem requires the analog signal to be sampled at least twice during its period we may write,

$$f_s \leqq 1/2T = f_s(max) \qquad (2)$$

and the time-delay signal bandwidth product becomes,

$$\tau f_s \leqq N/2 \qquad (3)$$

The low frequency limit is set by the thermal leakage current which accumulates in each stage, and the upper frequency limit is determined by input injection limitations and transfer efficiency.

Figure 3 illustrates a cross-section of a four-phase electrode CCD with transfer and storage electrode dimensions. The CCD is fabricated with PMOS silicon-gate technology and the insulator is a dual dielectric comprised of silicon nitride (Si_3N_4) over thermal silicon dioxide (SiO_2). The electrodes are fabricated with polycrystalline silicon and aluminum, to give coplanar but overlapping electrodes. The overlapping electrode feature provides a "sealed" CCD surface and stable operation over temperature-bias excursions.

2.0 Signal Transport in a CCD

The CCD delay line[2,3] provides a unidirectional transfer of signal charge $q_s(x,t)$ from one storage cell to another adjacent cell. The signal charge is designated in cell x at time t, where x and t assume integer values; i.e., the unit of distance is the cell-to-cell separation X_o (center-to-center), and the unit of time is the stepping interval T (clock period).

The frequency response of the CCD delay line may be calculated from a discrete frequency expression for the signal charge:

$$q_s(x,t) = \epsilon q_s(x,t-1) + (1-\epsilon) q_s(x-1,t-1) \qquad (4)$$

where ϵ is the transfer inefficiency per stage delay with typical values of $\epsilon < 10^{-4}$ for f < 1 MHz. Since equation (4) is a discrete set of signal values in the time domain, we may transform the signal charge to the Z-domain:

$$Q_s(x,Z) = Z^{-1}[\epsilon Q_s(x,Z) + (1-\epsilon)Q_s(x-1,Z)] \qquad (5)$$

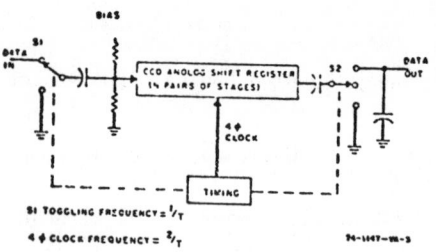

Figure 1. Charge Coupled (CCD) Analog Shift Register[1]

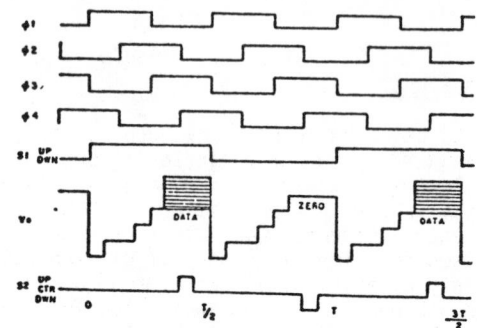

Figure 2. Basic CCD Four-Phase Clock Timing[1]

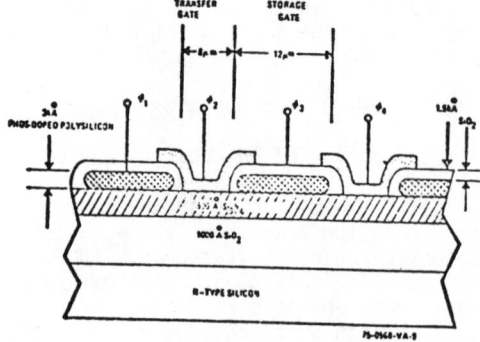

Figure 3. Cross-section of Four-Phase CCD Stage Delay

330

The transfer function of a N-stage CCD delay line becomes[4]

$$H(Z) = \frac{V_{out}(Z)}{V'_{in}(Z)} = \frac{g_m R_F C_{in}}{C_{out}} \left[\frac{1-\epsilon}{1-\epsilon Z^{-1}}\right]^N Z^{-N} \quad (6)$$

and the substitution of $Z = e^{j\omega t} = e^{j2\pi f_s/f_c}$ into equation (6) yields the amplitude and phase characteristics shown in Figures 4 and 5, respectively,[5] for various values of $N\epsilon$. The signal frequency, $f_s \leq 0.5 f_c$ as restricted by the Shannon Sampling Theorem, which states that a band-limited signal, f_s, may be reconstructed from samples taken at time intervals $T = 1/f_c$. The phase deviation $\Delta\emptyset(f_c)$ is the departure from linear phase shift. The insertion loss of the CCD delay line is less than 2 dB at the Nyquist limit ($f_s = 0.5 f_c$) if $N\epsilon < 0.10$ and the maximum phase deviation at $f_s = 0.25 f_c$ is less than $3°$.

2.1 Frequency Dispersion

This dispersion can be viewed in the frequency domain as a shift in the filter response frequency with the maximum shift at one-half the clock frequency.

$$f'_s = f_s + \frac{\epsilon f_c}{2\pi}\left[\mathrm{Sin}\frac{2\pi f_s}{f_c} - j(1 - \mathrm{Cos}\frac{2\pi f_s}{f_c})\right] \quad (7)$$

2.2 Amplitude Dispersion

The signal is "dispersed" or spread-out in time as a result of the finite transfer inefficiency ϵ and the number of cells, N. The dispersion of a single data sample of unit height (i.e., one propagating storage cell through the delay line) may be determined by the binominal expansion theorem:

$$[(1-\epsilon) + \epsilon]^N = \sum_{t=0}^{N} \binom{N}{t}(1-\epsilon)^{N-t}\epsilon^t \quad (8)$$

$$\doteq \sum_{t=0}^{N} q_s(N, t)$$

$$\binom{N}{t} = \frac{N!}{t!\,(N-t)!} \quad (9)$$

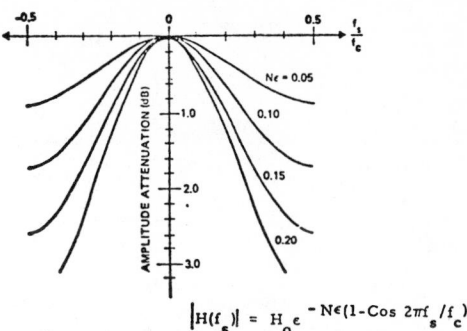

$$|H(f_s)| = H_o e^{-N\epsilon(1-\mathrm{Cos}\,2\pi f_s/f_c)}$$

Figure 4. Amplitude attenuation characteristics as a function of f_s/f_c for various values of $N\epsilon$.

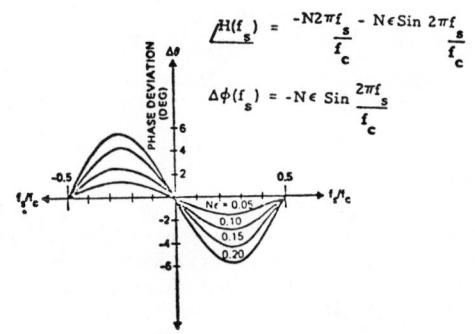

$$\underline{/H(f_s)} = \frac{-N2\pi f_s}{f_c} - N\epsilon \mathrm{Sin}\,2\pi\frac{f_s}{f_c}$$

$$\Delta\phi(f_s) = -N\epsilon \mathrm{Sin}\frac{2\pi f_s}{f_c}$$

Figure 5. Phase Deviation as a function of f_s/f_c for various values of $N\epsilon$.

Figure 6 illustrates the effect of $N\epsilon$ product[6] on the shape of a single data sample of unit height. The total area under the output waveforms is identical to the area under a single sample at $\epsilon = 0$. The peak of the signal lags by one clock period when $N \sim 1/\epsilon$ and dispersion is minimal for $N\epsilon \leq 0.10$.

3.0 CCD Basic Building Blocks for Discrete Analog Signal Processing (DASP)

Any signal processing system that involves the linear transformation of analog signals such as correlation, discrete Fourier transformation (DFT), filter banks, matched filters, multiplexing/demultiplexing, array scanning, orthogonal scan transformation, time base translation, etc., can be realized with combinations of CCD basic building blocks. In discrete analog signal processing (DASP),

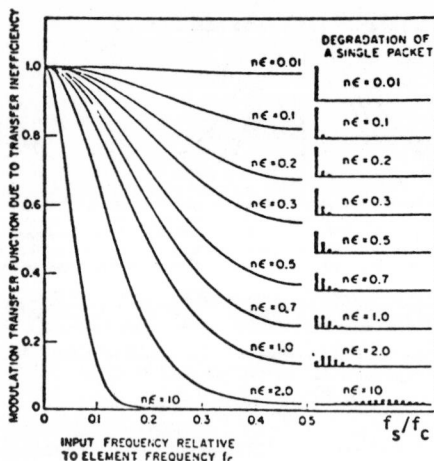

Figure 6. Degradation (Amplitude Dispersion)[6] of a Single Charge Packet as a function of Nε.

capabilities. Table 1 provides a partial listing of applications for these basic building blocks.

Table 1

Array Configuration	Basic Information Flow		
	Serial In; Serial Out (SI/SO)	Parallel In; Serial Out (PI/SO)	Serial In; Parallel Out (SI/PO) – Nondestructive Sensing Taps – Unweighted
Linear	Pure Delay; Time Base Interchange	Time Division Multiplexing; Array Scanning	Beam Focusing; Focus Scanning; Multiple-Beam Forming; **Beam Steering**
			Transversal Filtering; Correlation/Convolution; Adaptive Filtering; Sampled Data Smoothing or Interpolation (Scan Format **Converter**)
2-d Matrix (Area)	Bulk Serpentine Analog Storage*		**Corner Turn** (Orthogonal Scan Transformation)
			Discrete Fourier Transformers; Filter Banks; Multiple Cross Correlators

Bulk Serial-Parallel-Serial (SPS) Analog Storage*

*As for Video Refresh Memories

analog data samples are stored, transferred, and operated upon by analog means, whereas in conventional digital signal processing (DSP), digital or quantized samples are handled with binary logic. A major advantage of DSP is retained by DASP, namely the precise transport delay, particularly in relation to coherent signal processing. The dynamic range of an analog bit in DASP may be thought of as composed of 6-dB equivalent DSP digital bits. Thus, a typical example of 100 Stage (N = 100) CCD delay line with 60-dB dynamic range and transfer inefficiencies of $\varepsilon \sim 10^{-4}$ at 1-to-2 MHz clock rates will have an overall signal degradation of 1 percent (i.e., less than 0.1-dB insertion loss) without the need of A/D conversion.

One-dimensional basic building blocks[7] (linear arrays) may be classified according to the characteristic information flow patterns:

(1) Serial in/Serial out (SI/SO)
(2) Parallel in/Serial out (PI/SO)
(3) Serial in/parallel out (SI/PO)

These fundamental linear arrays may be combined to form area arrays (2-dimensional matrices) with increased signal processing

3.1 Serial In/Serial Out (SI/SO)

The SI/SO block is a simple CCD shift-register with the characteristics discussed in sections 1 and 2. In a linear array configuration the SI/SO block provides pure analog signal delay with the ability to provide time base translation. Typical dynamic range for present-day SI/SO blocks is 60-80dB with ±1 percent linearity and clock frequencies from 1KHz-1.0MHz for a 64 analog bit delay line. The clock requirement may vary from device to device with voltages varying from TTL to MOS compatible. In general, MOS-type voltage swings are needed to obtain dynamic range and frequency response. The capacitance loading for the drivers is typically $0.2pF/mil^2$ of active bit area; for bit areas of 1.5 mil^2 we have 0.3pF/analog bit. Thus, a 64 bit delay line will offer a loading of $\sim$20pF/driver. In general, CCD structures have not been built with interface/buffer circuits on the chip because of the advanced development nature of the work; however, CCD chips can be fabricated with MOS, CMOS, or bipolar interface circuits. In order to test SI/SO blocks without on-chip buffer circuits, a so-called "open collector" driver may be employed. This driver is relatively inexpensive and provides clock voltage swings of 30V up to 2 MHz clock frequencies. Clock shaping may be accomplished if desired by the use of a series resistor, which also protects the drivers in the pull-down transient. A CCD chip should have protective resistor/diode combinations, similar to MOS-type circuits, to limit the displacement current and prevent shorting of the input electrodes.

For analog signal processing, as discussed in the introduction, a desirable feature is the incorporation of an a-c zero reference between successive signal samples, particularly for PI/SO and SI/PO blocks. In addition, sample and hold techniques are required for analog signal reconstruction which attenuates the response with a $\dfrac{\sin \pi f_s/f_c}{\pi f_s/f_c}$ shape factor. The input to a CCD may also be filtered with a $\dfrac{\sin \pi/\Delta t f_c}{\pi/\Delta t f_c}$

roll-off where Δt is the sampling window aperture. The output after sample and hold requires filtering with a low-pass filter with ideal "brick-wall" cut-off at $f_c/2$, the Nyquist limit. Figure 7 illustrates an analog output swept frequency response of a CCD SI/SO block with sample/hold and a 7-pole Butterworth filter (-3dB @ 750 KHz) to filter the clock and limit the aliasing of frequencies higher than $f_c/2$. Thus, in a properly designed CCD Analog Delay Line the frequency response is limited by the sample/hold and low-pass filter characteristics.

A time multiplexed CCD filter bank[1] which used SI/SO blocks is shown in the block diagram of Figure 8. The Filter-Bank Characteristics are illustrated in Figure 9 for the case of uniform filter spacing. Storage and sequencing of the constants is accomplished digitally using programmable read only memories (PROM) or electrically alterable ROM's (EAROM's). Weighting of the analog signals by the filter constants

is accomplished by means of multiplying digital-to-analog converters. The serial output data is multiplexed onto N lines by the output sampler which stretches each sample to a width, $T_s = NT$. Timing circuitry provides the CCD clock waveforms, the PROM addresses, and the sampler address. Applications of this filter bank include Doppler spectrum processing in radar, sonar and communications systems with advantage of low total part count combined with variable filter parameters.

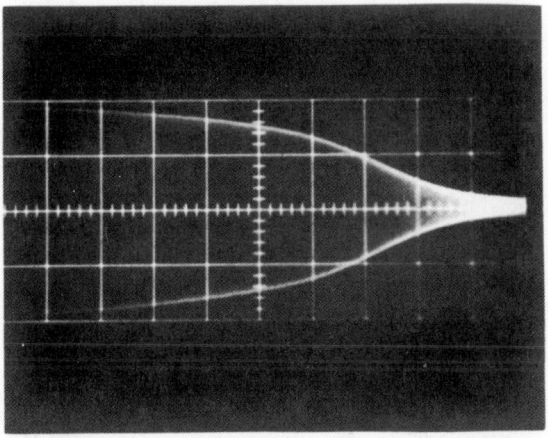

Figure 7. Frequency Response of Sampled Data CCD Analog Delay Line. 100 KHz/Div. Horizontal, f_c = 2.0 MHz. Transfer inefficiency ϵ (f_c = 2.0 MHz) = 2×10^{-3} for L = 12μm electrodes. Sample/Hold and Filter Responses are included in the overall response.

The main signal processing function in a moving target indicator (MTI) radar is the main beam clutter (MBC) canceller. A three pulse canceller using CCD's is illustrated in Figure 10. Each delay line contains a number of range cells (or bins) adequate for the required resolution and range coverage. Low pulse repetition frequencies (PRF's) with interpulse periods (IPP) of 0.5 to 5 milliseconds are use to provide unambiguous range detection. The delay in each CCD shift register for a given range cell is one IPP. In MTI radars with more than 500 to 1000 range cells, the CCD shift register may be arranged in the serial-parallel-serial (SPS) configuration to minimize the effects of charge transfer inefficiency. The dual sampling scheme of figures 1 and 2, then automatically

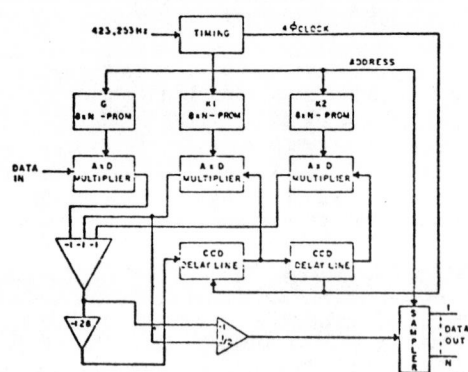

Figure 8. Block Diagram of N-channel CCD Filter Bank[1]

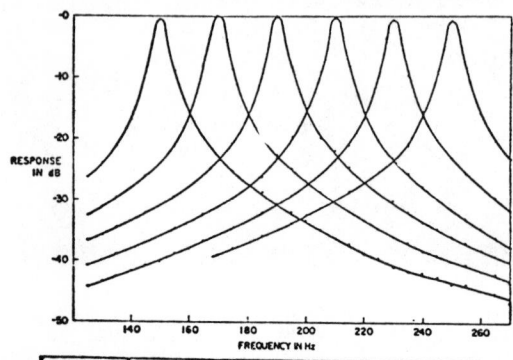

CENTER FREQUENCY IN Hz	BANDWIDTH IN Hz	C	R1	R2
150	5	1.00	0.75	-0.96
170	5	1.00	0.46	-0.96
190	5	1.00	0.15	-0.96
210	5	1.00	-0.15	-0.96
230	5	1.00	-0.46	-0.96
250	5	1.00	-0.75	-0.96

Figure 9. Uniform Filter Characteristics[1]

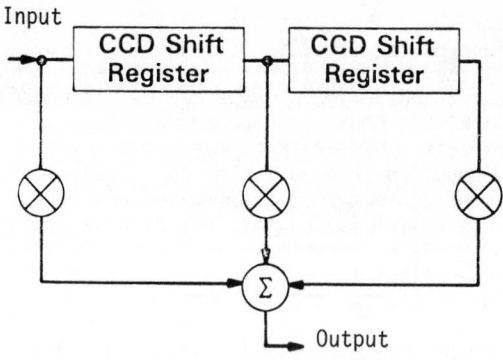

Figure 10. CCD Clutter Canceller for a Moving Target Indicator (MTI) Radar

eliminate most effects of leakage current nonuniformities since both "signal with reference" and "reference only" samples follow identical paths.

3.2 Parallel In/Serial Out (PI/SO)[8]

The PI/SO block may be used to time division multiplex a number of low data rate signal channels into a higher date rate output channel. The variations in electrical input may be minimized with the use of a stabilized charge injection circuit. An N-channel multiplexer converts N parallel input channels into a single-channel pulse amplitude modulated (PAM) signal. The input signals are synchronously sampled and the sampler information is entered into a unique spatial and temporal position in the CCD delay line. Applications include the multiplexing of many sensor input channels (e.g. electro-optical sensors, acoustical sensors) into a single video output channel. Figure 11 illustrates a photomicrograph of a PI/SO CCD chip with stabilized charge injection and parallel injection of alternate stage delays along the CCD delay line to minimize interchannel cross-talk and provide for the injection of an a-c reference for threshold voltage cancellation. Figure 12 illustrates the response of a PI/SO block with N = 20 and a simultaneous unit impulse at each parallel input. The injection of a reference and a signal and reference permits subsequent subtraction at the CCD output to remove input variations. Voltage variations, referenced to the input, not exceeding 100µV

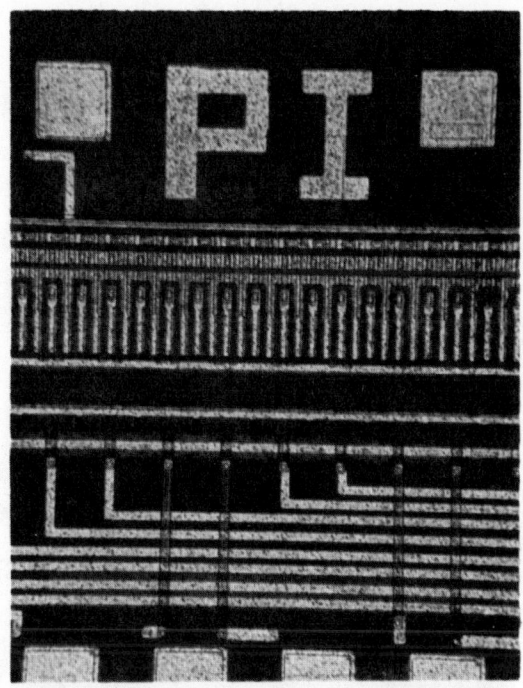

Figure 11. Photomicrograph of PI/SO CCD Basic Building Block.

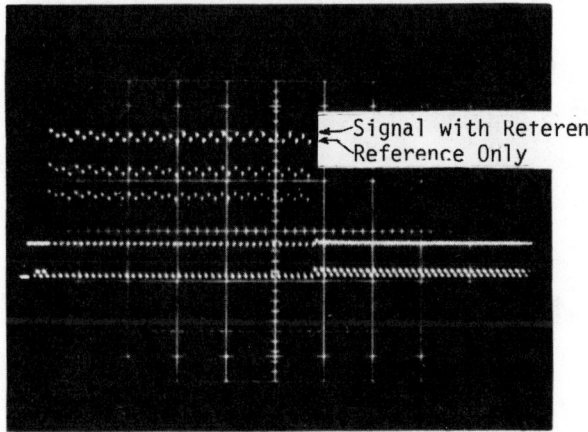

Figure 12. Impulse Response for Uniform Weighting Prior to Subtraction of Signal and Reference and Reference Inputs.

have been obtained to remove fixed pattern noise and place the limitation on noise with the charge injection uncertainty associated with the input capacitance.

The PI/SO block may also be operated in the time delay and add or integrate (TDI) mode to give such signal processing functions as sonar beam forming and steering or convolution. Formation of sonar or any acoustic beams using an array of transducers is illustrated in Figure 13.

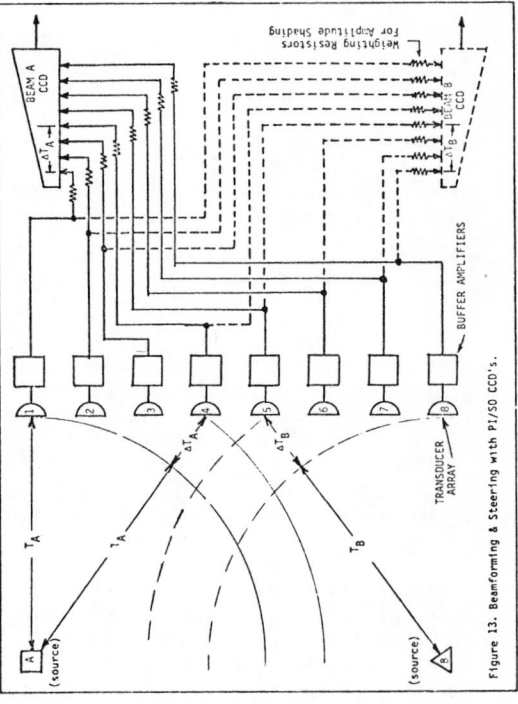

Figure 13. Beamforming & Steering with PI/SO CCD's.

Transducer "1" is first to receive the signal from source A which is suitably weighted and injected into the "Beam A" CCD, where it is delayed. The signal next arrives at transducer "2" and is weighted accordingly. When it is injected into "Beam A" CCD, the weighted signal from transducer "2" is added to the weighted signal from transducer "1". As the signal from source A arrives at each transducer in turn, it is weighted, injected into the "Beam A" CCD, and added to the previously accumulated signals that were injected and delayed from the closer transducers. The output charge integrated in any charge packet during transit through the "Beam A"

CCD may be written

$$Q_{out}(t) = \sum_{k=1}^{N} V_k (t-kT) \cdot W_k \cdot C_k \qquad (10)$$

where W_k = k^{TH} weighting coefficient, C_k = k^{TH} input capacitance, $V_k(t)$ = signal from k^{TH} transducer, T = CCD clock time. [Note: A common signal source applied to all weighting resistors, i.e., $V_k(t) \equiv V(t)$ for all k, gives a convolution of the signal V(t) with an impulse response function determined by the quantities $(W_k \cdot C_k)$.] Although the use of the PI/SO block in the TDI mode typically involves a progressively larger charge packet as the packet propagates from the initial input to the final output, such an approach has some advantages over the use of the SI/PO block for the same functions: ease of fabrication and yield, interface simplicity, and lower power dissipation.

3.3 Serial In/Parallel Out (SI/PO)

The SI/PO block features INDEPENDENT nondestructive, low-impedance voltage readouts of the analog signals at specified locations or taps corresponding to various delays through the CCD shift register. In general, the signal voltage at each tap may be multiplicatively weighted by conductance to give a current proportional to the PRODUCT of the signal voltage by the weighting conductance. Summation of the product currents provides such functions as transversal filtering, correlation, or sampled data smoothing/interpolation for line arrays. Two dimensional weighting matrices driven by the independent low-impedance taps of the SI/PO block can give discrete Fourier transformers, filter banks, or multiple cross correlators. Figure 14 illustrates a photomicrograph of a SI/PO block (N = 20 outputs) which uses a floating clock electrode sensor at alternate stage delays along the CCD delay line. This permits the use of a "reference-only" and "signal and reference" to compensate for nonuniformities in the SI/PO structure. Figure 15 shows the tapped output signal from the SI/PO block. Numerous taps with multiplicative analog weighting can be accomodated without signal amplitude degradation due to stray parasitic capacitance, by paralleling SI/PO blocks of feasible size due to the summation of product currents. Furthermore, the independent nondestructive

voltage taps of the SI/PO block can provide the analog voltage signals needed by multipliers to give CCD real-time analog correlation.[9] Use of programmable conductances such as the nonvolatile MNOS type or conventional MOS type permit such device applications as adaptive transversal line equalizer or programmable matched filter (or correlation detector) for secure voice/data communications systems

Figure 14. Photomicrograph of SI/PO CCD Basic Building Block

4.0 Conclusions

Many signal processing systems which involve the linear transformation of analog signals, such as matched filters or multiplexing/demultiplexing, can be realized with a finite number of CCD basic building blocks. To impact future electronic systems, the CCD basic building blocks should be flexible in the sense that systems design engineers can use them in a variety of applications. Thus independent unweighted taps keep the active device relatively simple yet require the use of external resistors and output buffer/reconstruction circuitry for trans-

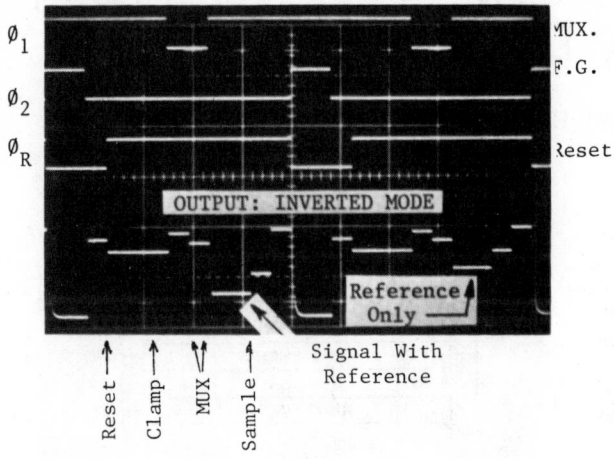

φ₁   MUX.

φ₂   F.G.

φ_R   Reset

OUTPUT: INVERTED MODE

Reference Only

Reset   Clamp   MUX   Sample

Signal With Reference

Figure 15. Floating Clock Electrode Sensor
for SI/PO Block

form operations. But the main advantage is
that a single device design may be used to
satisfy many applications requirements. In
systems where the external resistors are too
clumsy, but tap weight adjustment is desired,
electrically alterable tap weights are
appropriate. This is clearly the most
powerful method of tap weighting, which can
lead to real-time analog correlators and/or
adaptive filtering as well as tap error
compensation. Devices having electrically
alterable taps are substantially more complicated than fixed tap or unweighted devices
but can be used as universal filter/correl-
ator building blocks. Such universal blocks
can benefit from the economics of high volume
production and find diverse applications
ranging from one-of-a-kind R and D to production systems.

REFERENCES

1. John Mattern and Donald Lampe, "A Repro-
    grammable Filter Bank Using CCD Discrete
    Analog Signal Processing", 1975 IEEE
    International Solid-State Circuits Con-
    ference (ISSCC), Feb. 12-14, 1975,
    Philadelphia, Pa., pg. 148 Technical
    Papers Digest.

2. W. S. Boyle and G. E. Smith, "Charge
    Coupled Semiconductor Devices", B.S.T.J.,
    49, 587 (1970).

3. W. B. Joyce and W. J. Bertram, "Linear-
    ized Dispersion Related and Green's
    Function for Discrete Charge Transfer
    Devices with Incomplete Transfer",
    B.S.T.J., 50, 1741 1971).

4. D. D. Buss and W. H. Bailey, "Applica-
    tion of Charge Transfer Devices to
    Communication", 1973 CCD Applications
    Conference, Sept. 1973, San Deigo, CA,
    pg. 83 of Technical Papers Digest.

5. M. H. White and W. R. Webb, "Study of
    the Use of Charge-Coupled Devices in
    Analog Signal Processing Systems",
    Final Report, NRL Contract N00614-75-C-
    0069, May 1974.

6. C. H. Sequin, "Interlacing in Charge
    Coupled Imaging Devices", IEEE Trans.
    on Electron Devices, ED-20, 535 (1973).

7. D. R. Lampe, M. H. White, J. H. Mims
    and G. A. Gilmour, "CCD's for Discrete
    Analog Signal Processing (DASP)",
    INTERCON 74, March 1974, New York.

8. T. F. Cheek,Jr., et.al., "Design and
    Performance of Charge-Coupled Device
    Time-Division Analog Multiplexers",
    1973 CCD Applications Conference,
    Sept. 1973, San Diego, CA, p. 127 of
    Technical Papers Digest.

9. "Investigation of CCD Correlation Tech-
    niques", contract no. NAS 1-13674,
    NASA Langley Research Center.

A Dual Differential Charge-Coupled Analog Delay Device

DAVID A. SEALER, MEMBER, IEEE, AND MICHAEL F. TOMPSETT, SENIOR MEMBER, IEEE

Abstract—A dual differential charge-coupled analog device providing signal delays of 24 and 48 elements has been designed for sampled data analog signal processing applications. The aim of this design was to eliminate some of the disadvantages that have been associated with previous charge-coupled devices (CCD's). These include clock pickup, thermally generated dc offsets, and complex external control and amplification circuitry. The device has an input strobing circuit and an on-chip output amplifier. With a clock frequency of 8 kHz and a 400-mV rms input signal, the total harmonic distortion was below 0.2 percent and the signal-to-noise ratio was better than 70 dB with a 4-kHz bandwidth. The device gain was 6 dB and a gain variation of 0.2 dB was observed over a temperature range of 0 to 55°C.

I. INTRODUCTION

IN MANY analog signal processing applications [1] charge-coupled devices (CCD's) must compete against other available technologies. In particular, CCD's must show performance and cost effectiveness comparable to these technologies in order to gain widespread acceptance. All CCD's designed to date have a number of disadvantages for analog signal processing applications. These include performance limitations of excessive nonlinearity, clock pickup, noise, and thermally generated dc offsets, and operational requirements of external amplifiers and control circuitry. The nonavailability of devices without these disadvantages continues to be the dominant impediment to the use of CCD's by circuit designers. However, the inherent precision of charge transfer makes the potential analog performance of CCD's very attractive and their implementation with monolithic silicon integrated circuit technology promises extremely low cost. The device to be described was designed with these considerations in mind although not all the problem areas have been tackled yet. Applications of the device include analog delay and second-order recursive filtering [2] with time sharing [3], [4]. A preliminary discussion of the device was presented earlier [4].

II. DEVICE DESIGN

The organization is shown in Fig. 1. The device has two analog registers, one having 24 elements and the other having 48. With two registers, the device can be used for first- or second-order filter sections as well as for other delay applications. In time-shared applications, the capacity of the device depends on the number of isolation elements used between original elements in the charge transfer channels. If one isolation element is used, this device can provide time-shared delay for twelve inputs and for twelve functions.

Manuscript received October 11, 1975; revised October 18, 1975.
The authors are with Bell Laboratories, Murray Hill, NJ 07974.

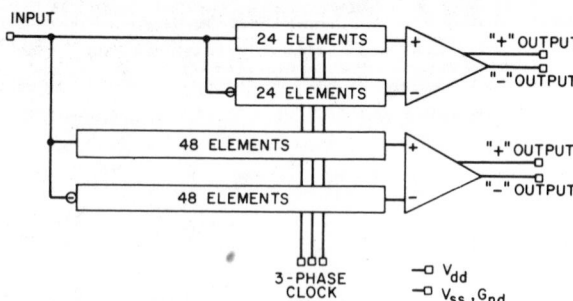

Fig. 1. Functional organization of differential charge-coupled dual delay device.

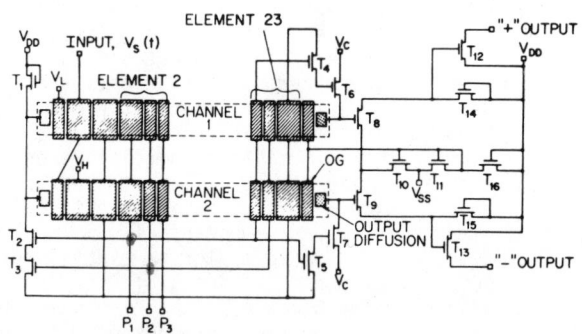

Fig. 2. Circuit diagram of 24-element differential charge-coupled delay device.

As shown in Fig. 1, the 24- and 48-element registers have common clock leads and a common input lead. Each register has a pair of differential channels. One of these channels has a normal input while the other has an inverting input. The pairs of channels drive on-chip differential amplifier stages. Balanced outputs from these channels permit the best performance of the differential amplifier to be used. A schematic circuit for the twenty-four element register is shown in Fig. 2. Details of each part of this register are described in the subsequent paragraphs.

The surface potential equilibration or charge preset method [5], [6] is used for injecting the signal into each charge transfer channel. This method has relatively low noise and nonlinearity and has minimal loading on the input signal source. In order to give a differential input without an inverter circuit, this method is used in two implementations as shown on the left of Fig. 2. For channel 1 the input signal, $V_S(t)$ is applied to the second electrode while the first electrode is held at the reference voltage V_L. This is the normal implementation and the charge injected into channel 1 is given by

$$Q_1 = C_{x1}(V_S(t) - V_L) \tag{1}$$

where C_{x1} is the effective storage capacitance of the second

Reprinted from *IEEE Trans. Electron Devices*, vol. ED-23, pp. 173–176, Feb. 1976.

gate of channel 1. For channel 2, the input signal $V_S(t)$ is applied to the first electrode while the second electrode is held at the reference voltage V_H. This is the inverting implementation and the charge injected into channel 2 is given by

$$Q_2 = C_{x2}(V_H - V_S(t)) \qquad (2)$$

where C_{x2} is the effective storage capacitance of the second gate of channel 2. If the input signal is biased so that

$$V_S(t) \longrightarrow V_S(t) + (V_L + V_H)/2, \qquad (3)$$

then the injected charge is given by

$$Q_1 = C_{x1}\left(V_S(t) - \frac{V_H - V_L}{2}\right) \qquad (4)$$

for channel 1 and by

$$Q_2 = C_{x2}\left(\frac{V_H - V_L}{2} - V_S(t)\right) \qquad (5)$$

for channel 2. The output signal is the difference of these two and is given by

$$Q_1 - Q_2 = 2C_x V_S(t), \qquad (6)$$

assuming that C_{x1} and C_{x2} have the same value C_x.

This method of injecting charge requires a strobe pulse on the input diode. As shown in Fig. 2, MOSFET $T1$ which is connected to V_{DD} pulls the input diodes to a high voltage. The gates of MOSFET's $T2$ and $T3$ are connected to clocks $P1$ and $P2$, respectively. During the overlap of $P1$ and $P2$, both $T2$ and $T3$ are turned on. The source of $T3$ is connected to clock $P3$ which is low when $P1$ and $P2$ are high. The result is that during the overlap of $P1$ and $P2$, the input diodes are pulled to a low potential flooding the input gate structure with charge. When $P1$ turns off, the diode potential is pulled high by $T1$ and the net injected charge packets Q_1 and Q_2 are stored under the second gates of the two channels. When $P3$ goes high, the net charge stored under the input gates transfers to the clocked electrodes.

The output circuitry is shown on the right in Fig. 2 and is clocked as follows. Clocks $P3$ and $P1$ are applied to MOSFET's $T4$ and $T5$. The $P1$ clock drives the gates of these MOSFET's and the $P3$ clock goes to their drains. When $P1$ goes high, $P3$ is already high and so the sources of $T4$ and $T5$ charge to $V_P - V_{th}$ where V_P is the clock voltage and V_{th} is the MOS threshold. This voltage turns on the gates of the reset MOSFET's $T6$ and $T7$ and the output diffusions are reset to V_C. During this overlap interval, the charge packets are stored under the wide $P3$ electrodes, being blocked by the output gates from transferring to the output diffusion. When $P3$ goes low, MOSFET's $T6$ and $T7$ turn off isolating the output diffusions, and the signal charges under the wide $P3$ electrodes are transferred under the output gates to the output diffusions. The width of these $P3$ electrodes was increased so that MOSFET's $T6$ and $T7$ will turn off before the transfer of signal charge occurs.

The signal voltages at the output diffusions drive the gates of a differential amplifier stage. Integrated MOS differential amplifier design has been discussed by Fry [7]. The design in Fig. 2 is a single-stage differential amplifier and uses seven

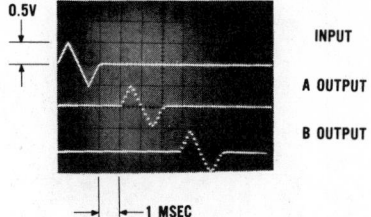

Fig. 3. Response of circuit to input ramp.

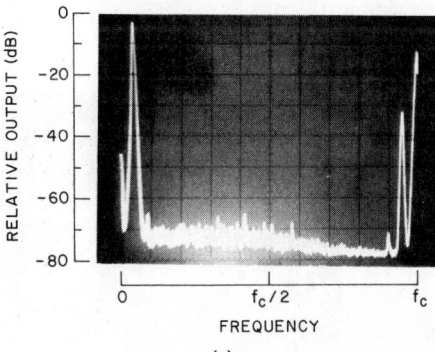

(a)

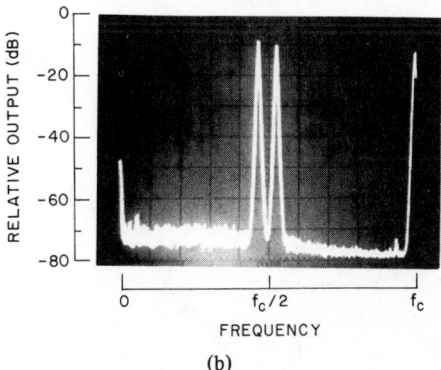

(b)

Fig. 4. Spectral distribution of output for 30-mV rms input. (a) $f_S = 0.04f_c$. (b) $f_S = 0.46f_c$.

MOSFET's. $T14$ and $T15$ are the load MOSFET's and $T8$ and $T9$ are driven by the signal voltage. A current source for the differential pair is provided by $T10$ operating in saturation. MOSFET's $T11$ and $T16$ are used as a voltage divider to get the gate voltage for the current source and for the output gates. The balanced output is obtained from $T12$ and $T13$ used as source followers. Although these outputs can be used single sided, best linearity is obtained when these two outputs are combined off-chip in a differential amplifier.

The devices were fabricated with an n-channel MOS self-aligned technology using three levels of polysilicon for the transfer gates. This process uses a selective oxidation technique to achieve the thick-field oxide and channel stopping diffusion with a single mask step[8].

III. DEVICE PERFORMANCE

To study the performance of this device a clock circuit was built using complementary MOS (CMOS) IC's. The reference voltages $V_{DD}/2$ and $V_{DD}/5$ for V_H and V_L, respectively, are provided by on-chip circuits. The output diffusion supply V_C was tied to V_{DD}. Normal operation was for $V_{DD} = 15$ V and with the substrate and V_{SS} connected to ground. The

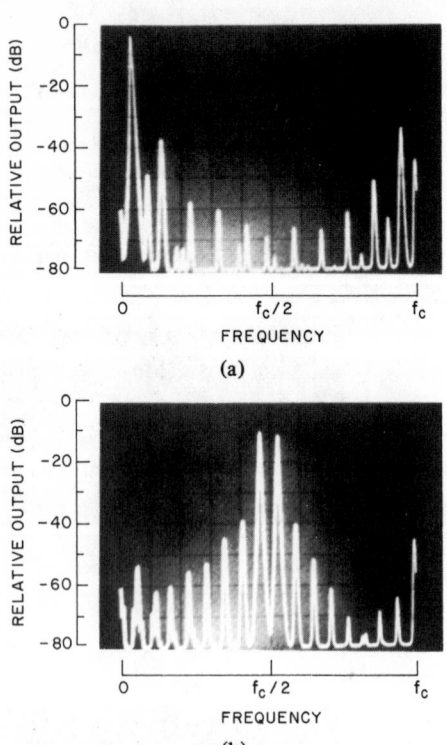

Fig. 5. Spectral distribution of output for 1-V rms input. (a)f_{in} = 0.04f_c. (b)f_{in} = 0.46f_c.

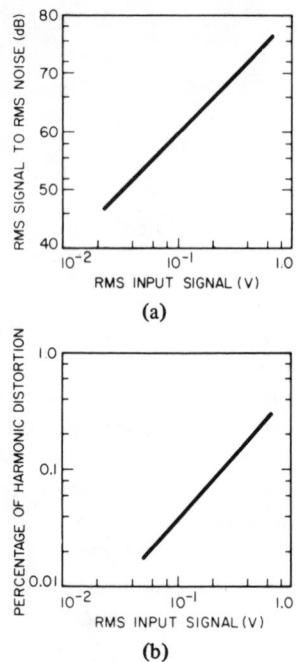

Fig. 6. Noise and harmonic distortion for the 24-element register operated at f_c = 8 kHz with a 500-Hz sine wave input and a 4-kHz bandwidth limit on the output.

balanced outputs were combined in a differential amplifier. Typical results for a 1-V peak-to-peak ramp are shown in Fig. 3. The clock frequency was 8 kHz. The A output has a delay of 3 ms and the B output has a delay of 6 ms.

The spectral distribution of the output is shown in Figs. 4 and 5 for two input levels, 30 mV rms and 1 V rms. The clock frequency was 20 kHz and the distributions are shown up to

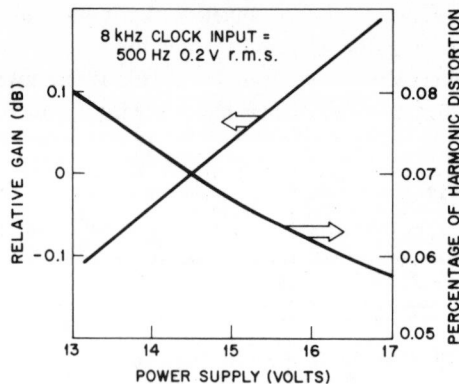

Fig. 7. Plot of gain and distortion as a function of supply voltage.

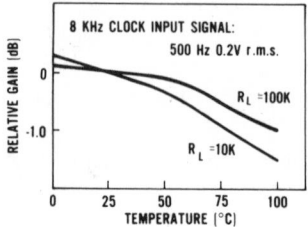

Fig. 8. Plot of overall gain as a function of temperature.

the clock frequency, i.e., twice the Nyquist sampling limit. Fig. 4 shows virtually no distortion of the input waveform except for slight second harmonic. In Fig. 5, the device is clearly being overdriven and the main distortion is at the odd harmonics.

Noise, harmonic distortion, power supply sensitivity, and temperature sensitivity have been studied using a distortion measurement system (Sound Technology Model 1700A). This unit consists of a very low distortion sine-wave variable oscillator, and ac voltmeter and tracking notch filter. To study the noise spectrum and to measure harmonic distortion at low levels, a wave analyzer (Hewlett-Packard Model 1900A) was used. For all measurements, a low-pass filter (Krohn–Hite Model 3202), set to one half the clock frequency of 8 kHz, was used to suppress clock noise and sidebands produced by the sampling process.

Fig. 6 shows the noise and harmonic distortion for the 24-element analog delay register for a 500-Hz input. The upper curve displays the ratio of the rms output with no input signal to the total rms output with an input signal. While the essential data on this curve could be stated as a single number, the curve is useful in studying the tradeoff between signal-to-noise ratio and harmonic distortion. The lower plot shows the percentage of total harmonic distortion as a function of input level. The main distortion component is the second harmonic. The second harmonic distortion is believed to be generated at the input of one of the differential channels [6].

Results of power supply sensitivity measurements are shown in Fig. 7. The supply voltage was varied between 13 and 17 V. The relative gain varied by 0.3 dB over the 4-V supply range. Harmonic distortion for a 0.2-V rms input signal decreased from 0.08 to 0.06 percent over the same voltage range.

The sensitivity of gain to operating temperature is shown in Fig. 8. The temperature was varied from 0 to 90°C. Measurements of gain were made using the 100-kΩ load resistors as shown in the schematic of Fig. 3, and using 10-kΩ resistors. Over a 0–55°C temperature range, the gain varied by 0.2 dB for the 100-kΩ load resistors and by 0.5 dB for the 10-kΩ load resistors.

IV. Conclusions

Complex operation has been a drawback of CCD's in signal processing applications. This design used on-chip circuitry for input diode strobing and to reset the output diffusions, thus eliminating two external pulse circuits. Differential channels were used to minimize noise and second harmonic distortion. The overall temperature dependence of gain, including an on-chip differential amplifier stage, was 0.2 dB for 0 to 55°C. For an input signal of 400 mV rms, the total harmonic distortion was 0.2 percent and the signal-to-noise ratio was 70 dB. Further optimization of the design should improve these values even further. The device shows that ease of use can be obtained in analog CCD's without loss in performance. This process of simplifying the use of analog CCD's by integration was not completed on this chip. However, further design efforts to provide on-chip prefiltering of the signal to avoid aliasing, more linear amplification, sample-and-hold circuits, and on-chip drivers is proceeding and will lead to completely self-contained analog CCD's.

References

[1] C. H. Séquin and M. F. Tompsett, *Charge Transfer Devices (Supplement 8 to Advances in Electronics and Electron Physics)*. New York: Academic, 1975.

[2] D. A. Smith, W. J. Butler, and C. M. Puckette, "Programmable bandpass filter and tone generator using bucket-brigade delay lines," *IEEE Trans. Commun.*, vol. COM-22, pp. 921–925, July 1974.

[3] J. Mattern and D. R. Lampe, "A reprogrammable filter bank using CCD discrete analog signal processing," in *Dig. Tech. Papers, Int. Solid-State Circuits Conf.*, Feb. 1975, p. 148.

[4] D. A. Sealer and M. F. Tompsett, "A dual-differential analog charge-coupled device for time-shared recursive filters," in *Dig. Tech. Papers, Int. Solid-State Circuits Conf.*, Feb. 1975, p. 152.

[5] M. F. Tompsett, "Surface potential equilibration method of setting charge in charge-coupled devices," *IEEE Trans. Electron Devices*, vol. ED-22, pp. 305–309, June 1975.

[6] C. H. Séquin and A. M. Mohsen, "Linearity of electrical charge injection into charge-coupled devices," *IEEE J. Solid-State Circuits*, vol. SC-10, pp. 81–92, Apr. 1975.

[7] P. W. Fry, "A MOST integrated differential amplifier," *IEEE J. Solid-State Circuits* (Corresp.), vol. SC-4, pp. 166–168, June 1969.

[8] J. T. Clemens, R. H. Daklan, and J. J. Nolen, in *Tech. Dig., Int. Electron Devices Meeting*, Dec. 1975.

Charge-Transfer Analog Memories for Radar and ECM Systems

WALTER J. BUTLER, MEMBER, IEEE, WILLIAM E. ENGELER, SENIOR MEMBER, IEEE, HOWARD S. GOLDBERG, MEMBER, IEEE, CHARLES M. PUCKETTE, IV, MEMBER, IEEE, AND HELMUT LOBENSTEIN, MEMBER, IEEE

Abstract—Charge-transfer devices (CTD's) offer new opportunities in analog signal processing since they provide analog memory in integrated circuit form. In this paper the successful application of serial charge-transfer structures to radar and to electronic countermeasure systems is described. The design and operation of a three-pulse, 50 dB clutter-suppression charge-transfer radar moving target indicator (MTI) and the implementation of a charge-transfer memory for radar electronic countermeasures (ECM) are discussed, and experimental results are presented.

I. INTRODUCTION

DIGITAL signal processing with its inherent stability has rapidly become an integral part of new radar systems with the advent of medium-scale integrated (MSI) and large-scale integrated (LSI) technology. One major disadvantage of high-performance digital systems, however, has been the expense and complexity of analog-to-digital (A/D) con-

Manuscript received July 11, 1975; revised September 26, 1975.
This research was supported in part by AFAL Contract F 33615-72-C-2203, Dr. R. Belt, Technical Monitor, and RADC Contract F 30602-73-C-0152, W. L. Simkins and D. Budzinski, Project Engineers. A preliminary version of this paper was presented at the 1975 IEEE International Symposium on Circuits and Systems, Boston, MA, April 1975.
W. J. Butler, W. E. Engeler, H. S. Goldberg, and C. M. Puckette, IV, are with the General Electric Company Corporate Research and Development Center, Schenectady, NY 12301.
H. Lobenstein is with the General Electric Company Aircraft Equipment Division, Utica, NY 13503.

version whenever a large dynamic range (eight or more bits) is required in conjunction with high bandwidth (5 MHz or more). An ideal solution to this problem is to use charge-transfer devices (CTD's) as low-cost analog sampled data processors, thereby eliminating the need for A/D conversion entirely, or in the case of a CTD preprocessor, at least reducing the cost of subsequent A/D conversion to the more economical realm of low-bit, high-speed technology.

CTD's are analog, sampled-data memories. The input analog signal is Nyquist sampled as in digital systems. The samples are not digitized, however; rather, they are stored and processed as discrete analog packets of charge. Hence, CTD's utilize a continuous range of amplitude, and are inherently large dynamic range devices. The bandwidth of a serial CTD, in addition to being Nyquist limited to one-half of the input sampling rate, may be further limited by the overall charge-transfer efficiency of the device. In applications that require many samples to be stored, therefore, the use of parallel or series–parallel architectures may be necessary. Other considerations, such as the linearity of the input–output relationship, or the effect of thermally generated dark current, or the pattern noise that can result from the nonuniformity of such dark current also have a bearing on the design of each particular application.

In this paper these aspects of CTD operation are considered in relation to the design of a CTD radar moving target indicator (MTI) and the implementation of a radar electronic

Reprinted from *IEEE Trans. Electron Devices*, vol. ED-23, pp. 161–168, Feb. 1976.

342

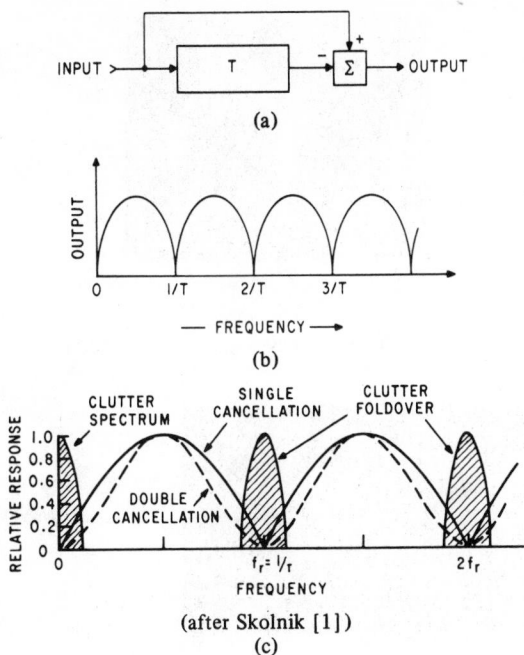

Fig. 1. Block diagram and frequency response of MTI filter.

TABLE I
MTI PARAMETERS

Type:	three-pulse vector canceler
Cancellation:	50 dB
Dynamic Range:	>50 dB
System Waveform:	multiple burst, 250 μs PRI
Pulsewidth:	1 μs
IF Bandwidth:	1 MHz
Input Format:	in-phase (I) and quadrature (Q) baseband video

countermeasures (ECM) memory. In both cases the system requirements are translated into a set of device specifications, and the significance of each such specification is established. Experimental data from several different CTD's are reported and some recent system results are presented.

II. CHARGE-TRANSFER MOVING TARGET INDICATOR

In many radar systems an MTI is used to enhance the signal-to-clutter ratio of the received signal by reducing signals due to sea or ground clutter. There are two basic MTI systems that may be used for this purpose, viz., a delay line canceller whose operation is based on the difference in Doppler shift of near-stationary clutter returns and moving targets, and a range-gated MTI that examines segments of the total range with Doppler filter banks to discriminate against stationary target returns and obtain target velocity signatures. In this section the implementation of a delay line canceler using CTD's is described and experimental data are presented.

The basic form of a delay line canceller is shown in Fig. 1(a). Functionally, the delay line canceller subtracts the radar return of one pulse repetition interval (PRI) from the return obtained during the previous PRI. The delay period T is made equal to the reciprocal of the radar pulse repetition frequency (PRF), which results in a frequency response magnitude characteristic such as is shown in Fig. 1(b). Spectral components of radar returns that occur around dc and at harmonics of the PRF are suppressed, thereby attenuating the clutter energy which exhibits a low Doppler spectral spread around the PRF lines (corresponding to returns from fixed or stationary targets), while passing the higher Doppler energy from moving targets [1].

The characteristics of the delay line will determine the extent to which the suppression can be maintained over the bandwidth of the radar signal and, therefore, the total clutter power reduction that can be achieved. For example, the fre-

quency response function of a single delay line canceller [such as is shown in Fig. 1(a)] may not have as broad a clutter rejection notch as desired. The notch can be widened by placing two cancelers in series. This configuration processes three pulses simultaneously, and is, therefore, known as a three-pulse canceler. The improved response is that of a squared sine wave [1] as is shown in Fig. 1(c).

A major problem in the use of conventional analog delay lines has been the matching of delay elements to the system PRF, particularly in the case of three-pulse cancellers where the first delay line usually controls the PRF, the second is slaved to the first, and complex temperature-controlled ovens surround the delay media which is normally quite temperature sensitive. Insofar as a CTD can be operated as an analog delay line, the delay period of which is accurately controlled by a digital clock, it appears to be well suited to the implementation of an MTI filter, and preliminary experimental circuits demonstrated notch depths of up to 40 dB in the output frequency response characteristic [2]. In order to further demonstrate feasibility, a charge-transfer MTI was designed as a direct replacement of an acoustic delay line canceller in a burst radar system. The technical performance parameters of the canceler are listed in Table I.

The required bandwidth for the I and Q channels was 500 kHz. In practice, the input signal is usually sampled somewhat above the Nyquist rate, and in this instance, a sampling frequency of $f_s = 1.6$ MHz was selected. For a serial CTD of N stages, the total delay T is determined by the clock period and is given by

$$T = \frac{N}{f_c} \tag{1}$$

where the clock frequency $f_c = f_s$. For $f_c = 1.6$ MHz, the number of stages required for the memory is 400.

For this application it was proposed to use p-channel MOS bucket-brigade devices with charge-transfer efficiency of approximately 99.9 percent per transfer, which therefore prohibited the use of a single serial device. Rather, a multiplex of eight channels was used as shown in Fig. 2.

For a multiplexed structure such as shown in Fig. 2, there are two methods of clocking the CTD's. For example, each device can be clocked at 1.6 MHz rate for a portion of the PRI. After enough samples have been taken to fill the first CTD, its clock is turned off, and the clock to the next CTD is turned on. This process is continued until the entire memory is full, with each CTD being clocked for a period equal to the PRI divided by M, where M is the number of parallel CTD's. During the next PRI the sequence is repeated with

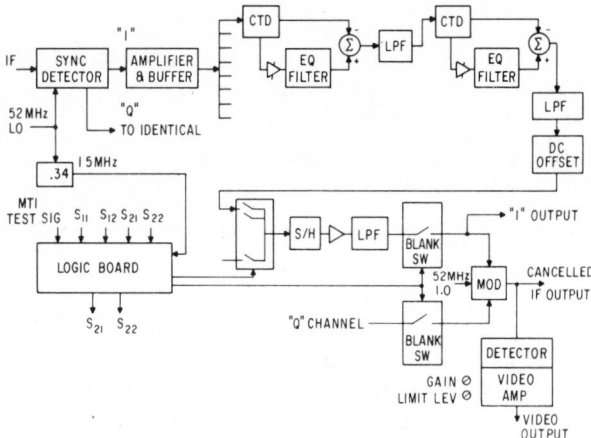

Fig. 2. Charge-transfer MTI block diagram.

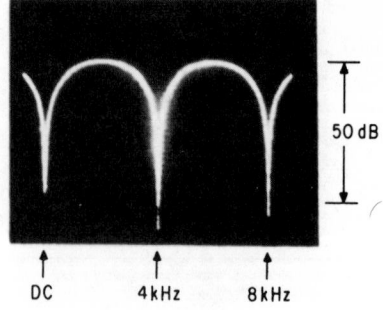

Fig. 3. Frequency-response of MTI filter.

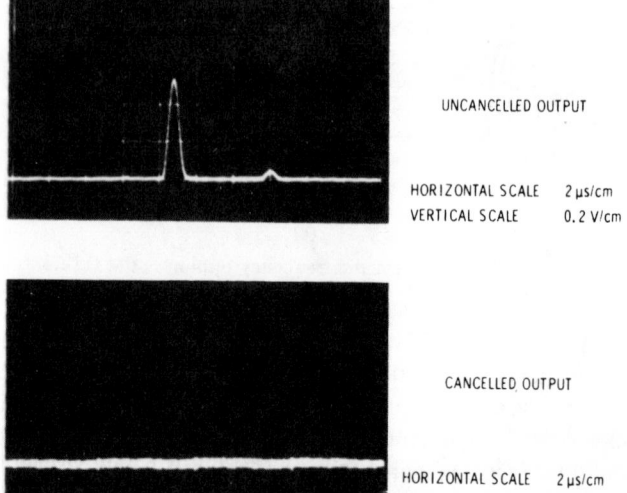

Fig. 4. MTI performance, multiplexed system.

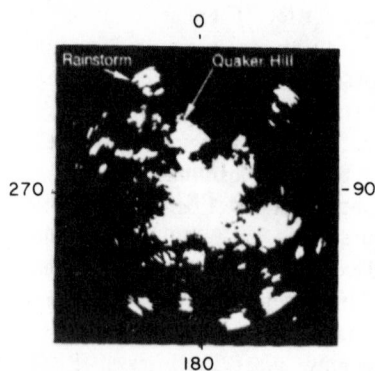

Fig. 5. Clutter map with MTI bypassed. Range: 14 mi. Quaker Hill at 335° to 355° and 5 to 8 mi approximately 60 dB above noise. Rainstorm from 310° to 335° at 10–12 mi.

the signal samples automatically being multiplexed at the output. A disadvantage of this scheme is that when the clocks are turned off to each CTD, the nonuniformity of the dark current that flows during that off period will superimpose a pattern noise on the stored signal. In addition, any variation in gain or offset from channel to channel will reflect itself as a spurious output at frequency f_s/M.

A second technique, and the one which was actually used in the present system, is to operate in an overlapping or skewed-clock mode [2]. In this mode the clocks to each CTD are run continuously at $f_c = f_s/M = 200$ kHz, but between adjacent CTD's the clocks are staggered, so that only one CTD is sampling the input at a time, and the actual sampling rate is thus maintained at the high rate (1.6 MHz). The skewed-clock mode of operation has the advantage of providing maximum time for charge transfer to occur. In addition, continuous clock operation serves to average the dark current over all samples in each CTD; however, variations in average dark current from device to device will still produce a spurious response at f_c.

The effect of charge-transfer inefficiency on the performance of the MTI filter is two-fold. From Fig. 2, it is clear that since the MTI subtracts a direct signal from a delayed signal, that the signal paths should have nearly identical characteristics for good suppression to occur. In order to achieve this, the direct signal was first passed through a short length of CTD, and subsequently passed through a simple *RC* equalization filter so that the frequency response of both paths was well matched. In this way, notch depths of 50 dB or more were obtained over the full Nyquist bandwidth of $f_c/2$ (=100 kHz). In Fig. 3, the filter frequency response in the vicinity of $f = 1/T$ is shown.

A second problem relating to charge-transfer inefficiency results from the use of a multiplexed structure. At the output of each CTD, the fraction of the kth-signal charge packet that has been misplaced due to charge-transfer inefficiency will not appear in the reconstructed train of signal samples until the $(k + M)$th sample. In other words, the effect of charge transfer inefficiency is to generate a "ghost" signal that is displaced in time by M samples. In the case of clutter returns, the extraneous signal bears a precise relationship to the clutter

patch and is thus subject to cancellation in the second stage of the canceler.

III. MTI EXPERIMENTAL RESULTS

Individual double-delay channels achieved a 58 dB suppression of simulated stationary targets when compared to an optimum moving target. The multiplexed system performance level decreased to approximately 51 dB (as shown in Fig. 4) due to multiplexing noise and residual unbalances between channels. Figs. 5 and 6 demonstrate actual system perfor-

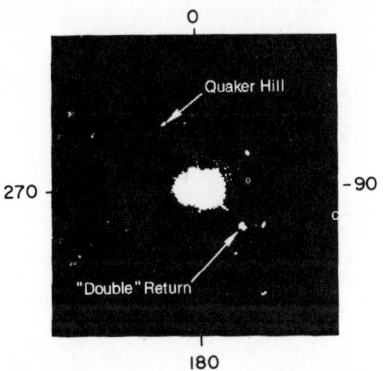

Fig. 6. Clutter map with MTI activated. Note that Quaker Hill is barely visible. Double return at 132° and 6 mi due to combination of large target and transfer inefficiency.

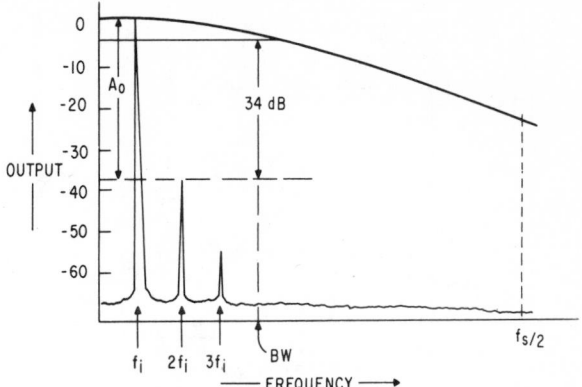

Fig. 7. ECM bandwidth definition.

TABLE II
ECM MEMORY SIGNAL CHARACTERISTICS

Signal	Bandwidth (MHz)	Length (μs)
1	4.5	20
2	6.0	12.8
3	1.5	60
4	0.078	200
5	1.0	6

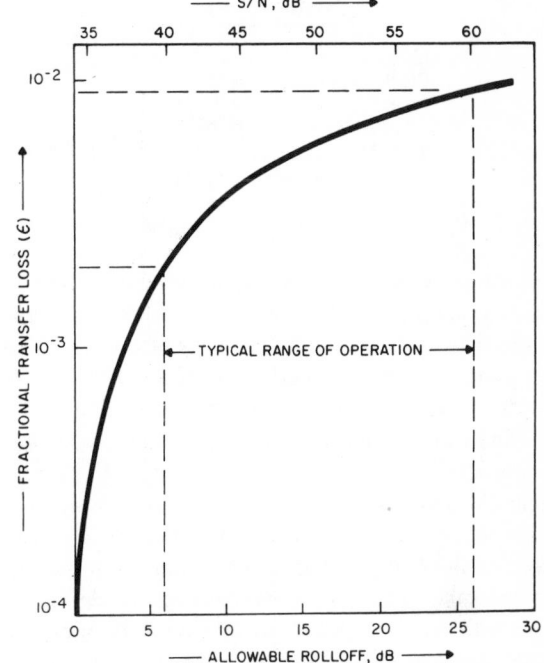

Fig. 8. Transfer efficiency requirements for an 80-stage device.

mance. The hill in Fig. 5 (PPI display without MTI) is approximately 60 dB above noise. This hill is barely visible in Fig. 6 (PPI display with MTI activated). Although a little time elapsed between the change over to the MTI mode, several targets (at 60°, 120°, etc.) previously inundated in clutter can be easily correlated. Note that several discrete returns due to the site location, plus a rainstorm, have been eliminated.

IV. CHARGE-TRANSFER MEMORY FOR ECM

The objective of the radar ECM memory system is to generate synthetic or false targets. This is achieved by acquiring the radar pulse of interest (linear FM chirp), storing it for a maximum of one or more PRI's, and recalling and retransmitting it on command at any time within that storage period. Some typical input signal characteristics are listed in Table II.

Preliminary design objectives were to achieve a 3-ms storage time over an operating temperature range of -55°C to +75°C. All spurious signal components due to harmonic or intermodulation distortion were to be at least 34 dB below the desired signal level. Having stored any one of the signals listed in Table II for up to 3 ms, and subsequently processing that signal with a pulse compression filter, a final system performance objective was to incur no more than 5 dB degradation in the main-to-sidelobe ratio of the compressed pulse.

In order to accomodate the reasonably large time delay bandwidth products of signals 1, 2 and 3 in a serial CTD, it was clear that (at clock frequencies of 6 MHz or greater) a higher level of charge transfer efficiency than that which could be achieved from p-channel bucket-brigade technology (such as was used in the MTI application) would be required. In the ECM memory, the bandwidth of the system is not defined in terms of the -3 dB point as is the case in a conventional analog

circuit. Rather, as is illustrated in Fig. 7, it is defined as the band of frequencies within which the signal level is at all points at least 34 dB above any noise or spurious response. The importance of achieving good overall linearity in addition to high charge-transfer efficiency from the CTD is thus clear. In Fig. 7, for example, were the second harmonic distortion suppressed by an additional 6 dB, the ECM bandwidth of the device would approximately be doubled. Using Tompsett's expression [3] for the frequency response of a CTD,

$$A(f) = \exp\left\{-n\epsilon\left(1 - \cos\frac{2\pi f}{f_0}\right)\right\} \qquad (2)$$

where n is the number of transfers and ϵ is the fractional loss per transfer, the tradeoff between linearity and required level of charge-transfer efficiency (for a device which is assumed to be 80 stages long) is plotted in Fig. 8. From these data, it is seen that with linearity at about 40 dB (corresponding to 1.0 percent harmonic distortion, a level that is well within the

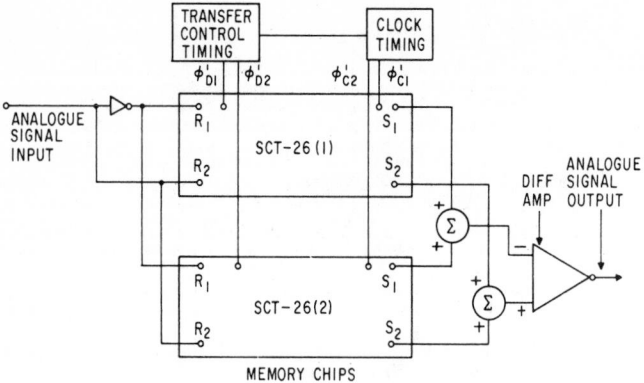

Fig. 9. Push–pull differential configuration.

TABLE III
ECM MEMORY DEVICE SPECIFICATIONS

Length:	64–80 stages
Clock Frequency:	6 MHz
Linearity:	>40 dB
Storage Time:	>3 ms
Temperature range:	–55 to +75°C
Transfer efficiency:	>99.8 percent

state of the art for CTD's), the required level of charge-transfer efficiency is approximately 99.8 percent.

In order to reduce the clock frequency at which that level of charge transfer efficiency had to be achieved, an "in-phase" and "quadrature" approach was again used, so that in fact, the signal bandwidths required from each I and Q CTD memory unit were one half the values listed in Table II. A push–pull differential configuration of CTD's (such as is shown in Fig. 9) was used to implement each I and each Q memory unit since it has been established [4] that such an arrangement makes good use of available device signal bandwidth, reduces clock feedthrough and hence relaxes post-memory filtering, and in addition provides automatic compensation for dc offset caused by uniform buildup of dark current during signal storage. The device specifications for each of the CTD's shown in Fig. 9 are listed in Table III.

The CTD's used to implement the ECM memory were p-channel surface charge transistor (SCT) devices, which have demonstrated performance levels well in excess of those listed in Table II. In Fig. 10, for example, the measured charge transfer inefficiency of a 64-stage device is shown to be approximately 10^{-4} at low frequencies, with a value of 5×10^{-4} at $f_c = 6$ MHz, a factor of 4 better than that required if linearity is 40 dB or greater. In Fig. 11 harmonic distortion is plotted as a function of input bias level for several values of input signal level. It is shown that up to 60 dB of linearity can be achieved for small input signal levels (0.5 V pp). Typical storage characteristics are shown in Fig. 12, where the upper or lower photographs show performance at $T = 25°C$ and $T = 75°C$, respectively. In each case, the lower trace shows the input signal which consists of eight cycles of a sine wave. The upper trace shows the output sampled data from the memory. The first 64 samples to be read out (covering three cycles of the sine wave followed by a dc segment) represent data which

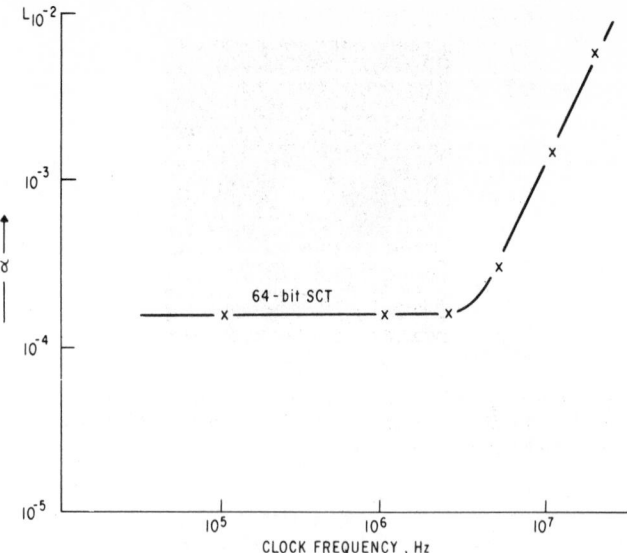

Fig. 10. Charge-transfer inefficiency of SCT device.

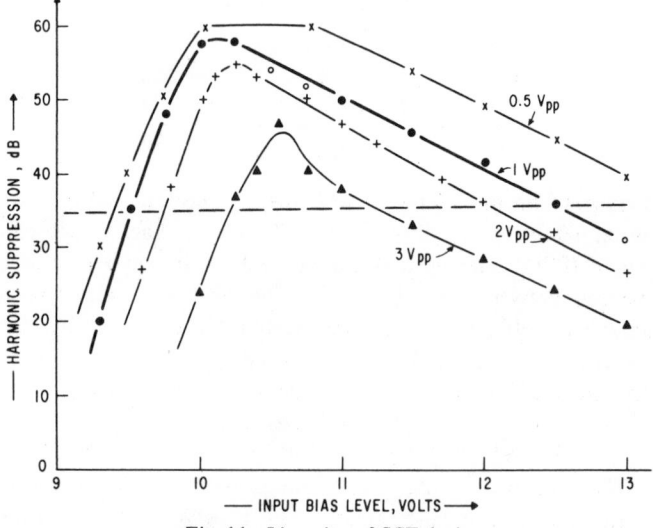

Fig. 11. Linearity of SCT device.

have been stored in the memory for the 3-ms storage time during which time the clocks have been interrupted. The remaining samples (corresponding to the 4 cycles of the sine wave on the right-hand side of the trace) represent data that have been sampled, and clocked straight through the memory without interruption. Thus, the effect of turning off the clocks for 3 ms can be observed by comparing the three cycles on the left side of the trace to the 4 cycles on the right side of the trace. At $T = 25°C$ very little degradation is visible. At $T = 75°C$ the pattern noise that results from the nonuniformity of the dark current is noticeable, but is still relatively small.

The effectiveness of push-pull operation is shown in Fig. 13. The output spectrum in the case of a single-channel device operating with a 5 MHz clock is shown in Fig. 13(a). This photograph is a multiple exposure record of the output line spectra that occur when an input sinusoid signal is stepped in 500 kHz steps from dc to 2.5 MHz. For a single-channel device, the bandwidth of the input signal must be limited to one-half the clock frequency (2.5 MHz) in order to prevent

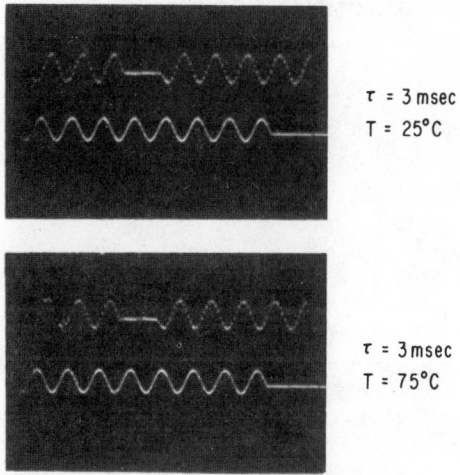

Fig. 12. Storage characteristics of SCT device.

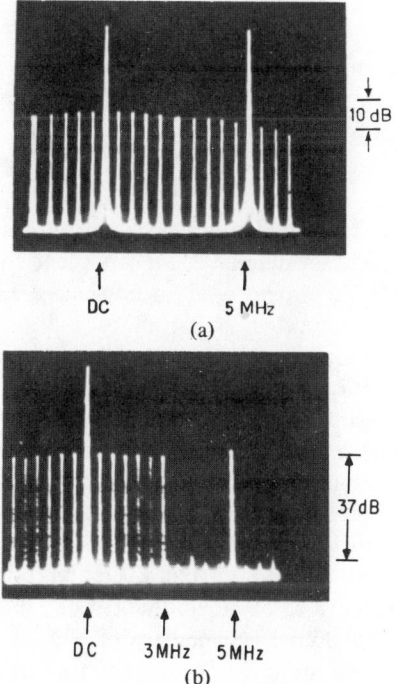

Fig. 13. Output frequency spectra. (a) Single-channel device. (b) Push–pull pair of devices.

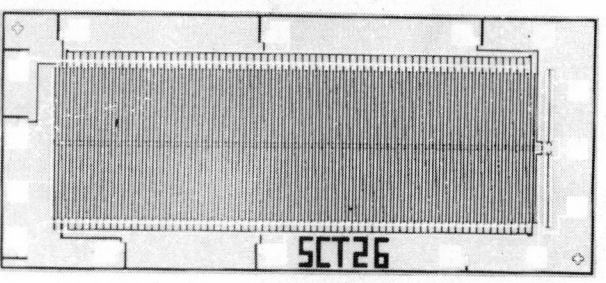

Fig. 14. Dual-channel 68-stage SCT device.

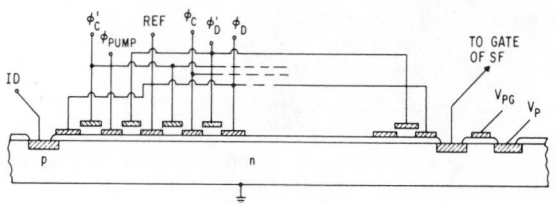

Fig. 15. Electrode structure of CTD.

aliasing (i.e., to prevent overlapping of the baseband component and the lower sideband of the clock). In push-pull operation, however, the clock waveforms to one channel are 180° out of phase to those in the second channel, so that on summing the outputs of the two channels, the output spectral components at the clock frequency and its sidebands are suppressed. In Fig. 13(b), for example, is shown the output spectrum of a push–pull pair of devices operating with a clock frequency of 5 MHz, and with the input signal stepped in 500 kHz steps from dc to 3.0 MHz. The clock component and its sidebands are suppressed to the point that in the overlap region (2–3 MHz), the peak-signal-to-peak-noise ratio is approximately 37 dB. Thus the input signal bandwidth to a push–pull pair of devices can exceed $f_c/2$, or equivalently, the push-pull configuration can be operated with a clock

frequency of less than twice the input signal bandwidth (which becomes increasingly important in high-frequency systems where transfer efficiency degrades rapidly with clock frequency).

In order to take advantage of the dc restoration capability and the inherent temperature insensitivity of a differential device [5], a new 68-stage dual-channel SCT device was designed for this application. A photomicrograph of the chip is shown in Fig. 14 and its electrode structure is shown in Fig. 15. Each channel of this device is serviced by an identical set of upper and lower electrodes with the exception of the third lower electrode, which provides the necessary analog input control of the charge transferred along each shift register, and is, therefore, brought out to a separate pad. Similarly, separate output stages are provided for each channel. The input scheme used for this device is the so-called "fill-and-spill" linear method similar to that proposed by Tompsett [6], and is further described in a companion paper in this issue [7]. The output of the device is obtained by measuring the charge dumped into an isolated output diffusion. In normal mode of operation the output diffusion is precharged to a negative potential which is not so negative that all of the charge is removed from below the final ϕ_D electrode, but is sufficiently negative so that adequate linear storage of a full packet of charge is possible on the output node. This mode of operation of the output circuit differs from that required for maximum sensitivity, and will introduce additional noise on the node because of the uncertainty of charging [8]. In this application, however, it is preferred because of its linearity and insensitivity to clock noise voltage.

For the ECM memory two of these dual-channel chips are connected in a push–pull differential configuration as shown in Fig. 9. Experimental results from a single dual-channel device are shown in Fig. 16, where separate channel and differentially summed outputs are shown, and in Fig. 17 where the output from a push–pull differential quad is shown.

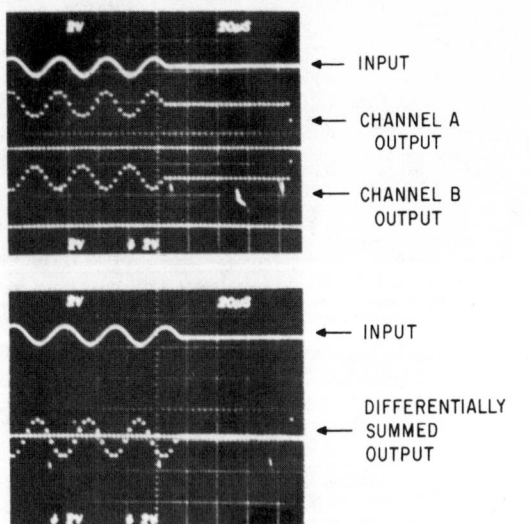

Fig. 16. Separate and differentially summed outputs of ECM memory device.

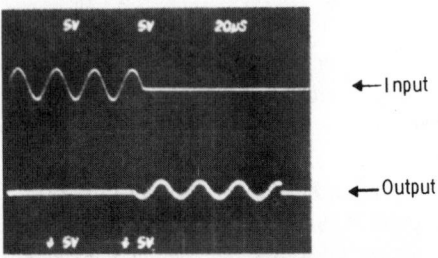

Fig. 17. Output of push–pull differential quad.

CCD COMPLEX SIGNAL STORAGE
TIME - BANDWIDTH · 90 (4.5 MHz, 20 μs)
HORIZONTAL · 5 μs/cm

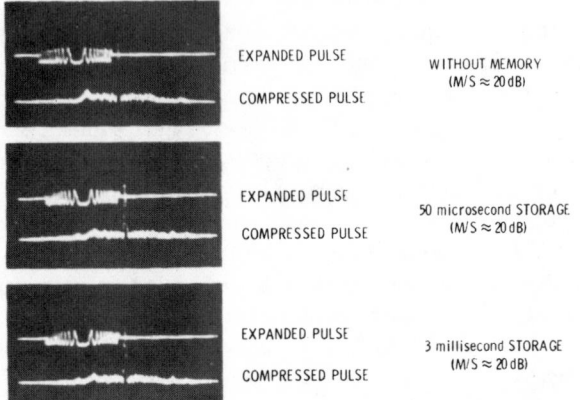

Fig. 18. ECM memory test results; no degradation of compressed pulse is evident.

System experimental results are shown in Figs. 18 and 19. In Fig. 18 the compressed pulse with and without the memory for the 4.5 MHz, 20 μs signal is shown. For these data, the pulse compression system was capable of providing a compressed pulse with 20 dB main/sidelobe ratio. No further degradation in this ratio occurred when pulse was stored in the memory for 3 ms. In Fig. 19 is shown similar experimental results for the 6 MHz, 12.8 μs signal. In this case, the system

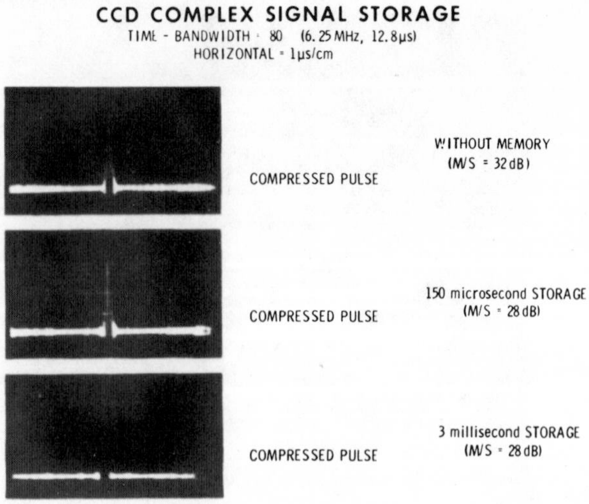

CCD COMPLEX SIGNAL STORAGE
TIME - BANDWIDTH · 80 (6.25 MHz, 12.8 μs)
HORIZONTAL · 1 μs/cm

Fig. 19. ECM memory test results; 4 dB degradation in M/S ratio occurs after 3 ms storage.

was capable of providing a main/sidelobe ratio of 32 dB without the memory. By storing the signal for 3 ms the main/sidelobe ratio was reduced to 28 dB.

V. Conclusion

In this paper, two successful applications of charge-transfer technology have been described. It is evident from the test results that CTD's can be used to implement an MTI with excellent clutter cancellation capability and that improvements in charge-transfer efficiency will reduce the level of multiplexing necessary for a given system. For the ECM memory it appears that state-of-the-art devices are presently capable of providing the required levels of performance in the areas of charge-transfer efficiency, linearity, and storage time. For performance over and above that reported here, it will become increasingly important to take full advantage of such techniques as differential push–pull operation, of utilizing the matched characteristics of monolithic devices, of nondestructive readout techniques, and perhaps of built-in redundancy for coherent noise cancellation so that truly optimum performance can be achieved.

Acknowledgment

The authors wish to acknowledge valuable contributions by Dr. R. D. Baertsch in the design of the SCT-26 peripheral and clock driver circuitry, and D. Ludington in the design and operation of the MTI system.

References

[1] M. I. Skolnik, *Introduction to Radar Systems.* New York: McGraw-Hill, 1962, pp. 123–133.
[2] W. J. Butler, C. M. Puckette, and M. B. Barron, "Implementation of a moving target indicator by bucket-brigade circuits," *Electron. Lett.*, vol. 9, pp. 543–544, Nov. 1972.
[3] M. F. Tompsett, "Charge transfer devices," *J. Vac. Sci. Technol.*, vol. 9, pp. 1166–1181, July–Aug. 1972.
[4] W. J. Butler, M. B. Barron, and C. M. Puckette, "Practical considerations for analog operation of bucket-brigade circuits," *IEEE J. Solid-State Circuits* (*Special Issue on Charge-Transfer Device Applications*), vol. SC-8, pp. 157–168, Apr. 1973.

[5] W. J. Butler, C. M. Puckette, and D. A. Smith, "Differential mode of operation for bucket-brigade circuits," *Electron. Lett.*, vol. 9, pp. 106–108, Mar. 1973.

[6] M. F. Tompsett, "Using CCD's for analog delay," in *Proc. CCD Applications Conf.*, San Diego, CA, 1973, p. 147.

[7] R. D. Baertsch, W. E. Engeler, H. S. Goldberg, C. M. Puckette, IV, and J. J. Tiemann, "The design and operation of practical charge-transfer transversal filters," this issue, pp. 65–74.

[8] S. P. Emmons and D. D. Buss, presented at the 1973 Device Research Conf., Boulder, CO.

CCD DYNAMICALLY FOCUSSED LENSES FOR ULTRASONIC IMAGING SYSTEMS

R.D. Melen   J.D. Shott   J.T. Walker   J.D. Meindl

Stanford University   Stanford University   Stanford University   Stanford University

ABSTRACT. Charge coupled device (CCD) delay lines offer many useful features when incorporated into multi-piezoelectric-element electronically focussed successors to the single element ultrasonic imaging systems currently in clinical use on humans and in non-destructive testing use on materials.

Two such electronically focussed systems are presented: (1) a microprocessor controlled delay line system and (2) a single chip C3D lens system.

INTRODUCTION

Ultrasonic imaging systems are currently in use: (1) on humans in a variety of medical imaging applications and (2) on materials in non-destructive testing applications. Single element piezoelectric transducers are used in these commercial instruments to obtain cross section scans using mechanical scanners and storage CRTs. Charge coupled device (CCD) delay lines offer many useful features when incorporated in multi-piezoelectric-element electronically focussed successor to these single transducer systems. Briefly stated the CCD acts as an electronically adjustable delay line performing the required delay-sum operation on the 1-5 MHz ultrasonic signals. This CCD capability makes economically feasible an ultrasonic imaging system with the following features:
1. High resolution
2. Dynamic focussing
3. Adjustable field of view

CCD ULTRASONIC IMAGING

The basic acoustic imaging problem being considered is shown in Fig. 1 along with a simple signal processing architecture. In operation a burst of ultrasound is transmitted from the "array" of piezoelectric transducers. Reflections from the "target" return to the array, arriving at the elements with a spherical time delay distribution associated with the target range (the farther away the target the smaller the time delay between elements in the spherical distribution). The basic signal processing task to be performed by the delay lines is to equalize the total propagation time from the target through the medium, piezoelectric transducers and individual channel electronics to the common signal output summing node. It is accomplished by spacing the input taps of the CCD delay line in a quadratic arrangement and dynamically sweeping the clock frequency to control the curvature of the quadratic delays to complement the curvature of the spherical wavefronts from targets T_1 through T_2 and beyond. It permits significant improvements in resolution at ranges less than ten times the aperture of the system.

One may consider a system as shown in Fig. 2 which utilizes a linear delay distribution (in cascade with the quadratic distribution) to steer the "focussed" beam providing two-dimensional display information. This is a simple steering technique which allows deflection of the formed beam in one direction from the perpendicular. Typical systems may be designed for thirty to forty-five degrees deflection. In operation the linear array of delay lines "electronically rotates" the piezoelectric array thereby providing the quadratic delay line with focus task similar to the "on-axis" focussing just discussed. One minor difference in focussing is that the apparent range for focussing is larger than the range to the midline of the array due to the "elec-

Reprinted from *Proc. 1975 Naval Electron. Lab. Center Int. Conf. on the Application of Charge-Coupled Devices*, Oct. 29–31, 1975, pp. 165–171.

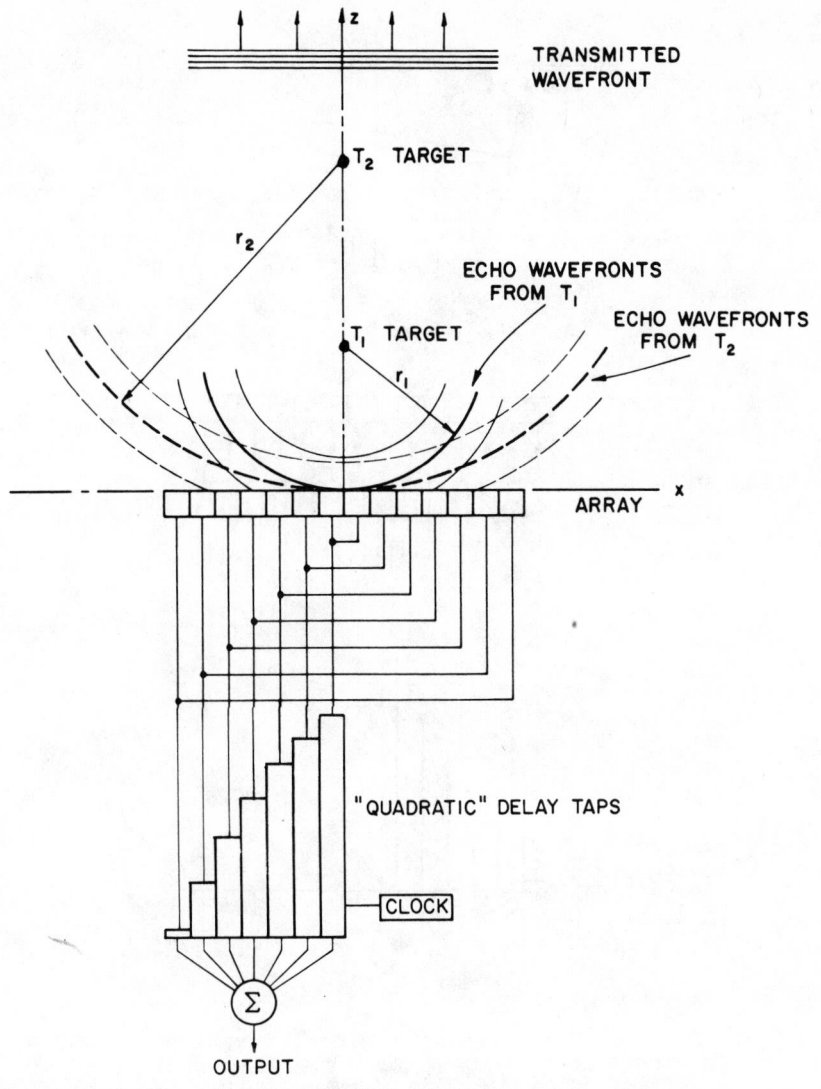

Fig. 1. Dynamic Focussing with Quadratically Tapped
         Variable Delay Line.

351

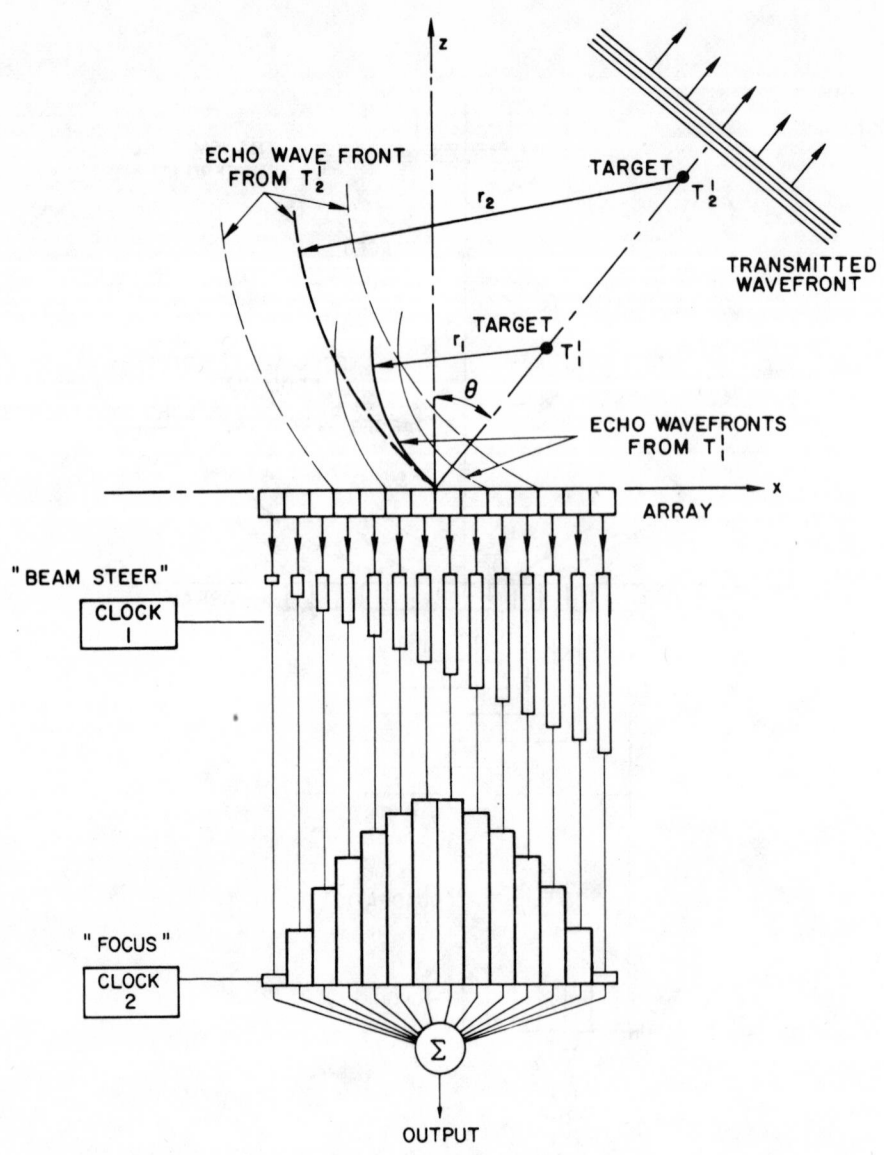

Fig. 2. Dynamic Focussing and Beam Steering with
Variable Delay Lines.

tronic rotation" being about the point of zero (beam steer) time delay (i.e. the far end of the piezoelectric array from the target).

In actual practice two problems exist with implementing this system shown in Fig. 2. First it is desirable to have as large a field of view as possible, thus the formed beam should be capable of being swept in either direction from the perpendicular. Secondly the clock frequency variations should be kept to a minimum to reduce the variation of CCD parameters with clock rate.

The two systems concepts described in the following sections achieve this set of goals through two quite different approaches. The first to be discussed achieves each channel delay function using a single serial-in-serial-out (SISO) CCD delay line. Thirty-one channels (one for each transducer element) is required for this system. The clock frequency applied to each CCD channel is independently adjustable. The frequency applied to a specific channel is a multiplicative product of two functions: the beam steering frequency function and the focus frequency function. Each CCD clock is applied uniformly along the delay line but is time-varying in frequency. The second approach utilizes a single specially-designed C3D lens integrated circuit for the entire system time delay function. This single chip approach has the potential of realizing a complete focussed ultrasound probe in a hand held package.

A MICROPROCESSOR-CONTROLLED IMAGING SYSTEM

The delay lines of an ultrasonic imaging system thus may perform two forms of time delay equalization: (1) parabolic (focus) and (2) linear (beam steer). An experimental ultrasonic imaging system has been built using thirty-one 200 element serial-in-serial-out CCD delay lines to perform the quadratic and linear delay functions of a single piezoelectric element in one device [6]. As shown in Fig. 3 the delay time of each of the thirty-one channels is determined by the digital rate multiplier assigned to each channel.

A voltage controlled oscillator (VCO) and a sweep generator are used to dynamically control the time delay for focussing, and an 8080 microprocessor is used to control a rate multiplier for beam steering. This system is capable of ± 30 degrees deflection and

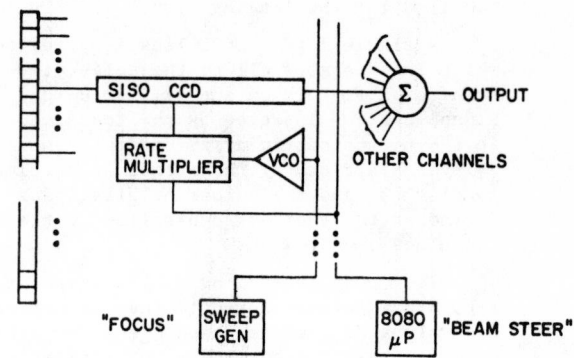

Fig. 3. Microprocessor Controlled System.

dynamic focussing of echoes from ranges of 2 to 40 cm when using a 31 element 3.1 cm linear array (element dimensions are .1 x 1 cm). This system uses a multiplicative control algorithm for the CCD time delays which provides the separation of focus and beam steer functions into tasks which are amenable to microprocessor control and simple sweep generation circuitry.

Unfortunately this system requires significant volume (1 cu.yd.) due to the large amount of electronics associated with each of the thirty-one channels. Over seven hundred integrated circuits and six thousand discrete components are used. There is little likelihood that with present technology this system will be configured in a hand-held probe.

THE C3D LENS ON A CHIP

The cascade charge coupled device (C3D) is a recent invention of the Stanford Integrated Circuits Laboratory [4] that has the potential of realizing a compact focus unit. This device is basically an array of charge coupled delay lines interconnected on a single integrated circuit substrate and having multiple sections of delay each clocked independently at a different frequency giving flexibility in design which can be utilized to realize sophisticated signal processing functions. Multiple input taps (first demonstrated in the Stanford Razorback CCD [5]) combined with multiple sections experiencing different clock rates yield devices capable of performing the high speed spatial Fourier transform required for the imaging task.

A simple form of C3D lens is a device which incorporates all of the delay electronics in Fig. 1b on one silicon chip. Signal charges injected by the transducers in the piezoelectric array undergo time delays determined by the frequency of clock 1, clock 2, and the number of bits transferred in the sections controlled by these two clocks.

The acoustic imaging system shown in Fig. 2 may be considered to have an optical equivalent of a wedge shaped lens for linear delay closely spaced with a spherically shaped lens for quadratic delay. The "thickness" of both lenses may be electronically varied to steer and focus the lens system on different target points.

Thus the C3D lens realizing this delay function may be considered as having two lens elements in one group on a single silicon chip. There is significant advantage in optics as well as in C3D lenses to have more than just two lens elements in an imaging system. The single parabolically shaped lens in Fig. 2, for instance, actually can be realized to advantage using multiple elements to correct for imaging and electronic aberrations. Fig. 4 shows photomicrographs of two element and three element

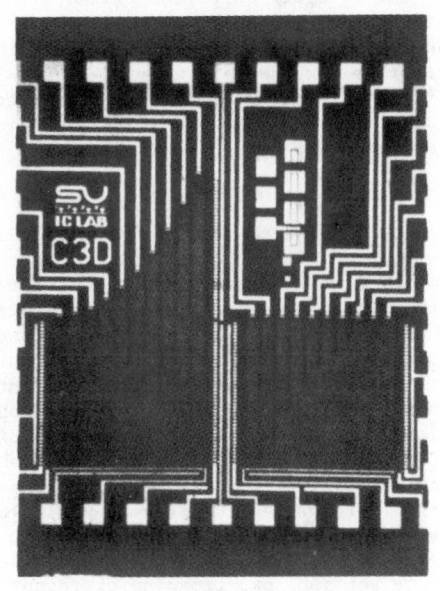

Fig. 4.   C3D Prototype Devices.

C3D lens versions designed and fabricated at Stanford to perform the parabolic lens function shown in Fig. 2.

The device on the right of this figure is the two element parabolic lens. This two element device uses the first element to <u>increase</u> the curvature of the arriving wavefront by a large constant amount (large in comparison to the curvature of the typical wavefronts being imaged). The second element in this two element lens focusses the highly curved wavefronts, i.e. <u>decreases</u> the curvature to a linear wavefront. The total curvature function focussed by the second element is a large constant curvature added to the curvature of the acoustic waves impinging upon the piezoelectric array. Thus the variations in the second element "thickness" (clock frequency) with changes in focussed range are small in comparison to the total second element thickness. This two element C3D lens requires significantly less clock frequency variation with focus range variation than single element equivalents. For an imaging system designed at Stanford this two element lens requires the 2 MHz variation of a 5 MHz clock while a single equivalent lens would require a 45 MHz variation (45 MHz variation is impractical for present day C3D lenses).

The three element lens shown on the left side of Fig. 4 can be used in several modes. The simplest is to simulate a two element lens by clocking the first two elements at an identical, time invariant frequency, and the third element at a frequency in keeping with the focus requirements. Since this two element lens has a first element with twice the number of sample storage positions as the previously described lens, this allows the clock frequency of this element to be twice that of the previously described two element lens. This design technique of allowing choice of clock frequencies provides a method for designing difference frequencies (cross modulation products) of the two clocks to be outside the passband of the ultrasonic channels used.

The goal of this C3D development program is to realize a device as shown in Fig. 5 which incorporates both the focus and beam steer functions on a single chip. As can be seen in this figure two wedge shaped "lens elements" are used to realize this beam steer function. Thus if clocks 1 and 2 are equal in frequency, there is no delay differ-

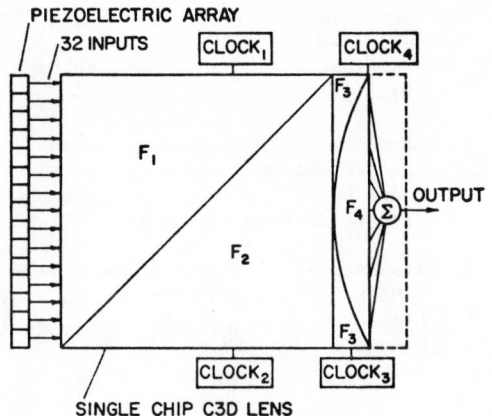

PIEZOELECTRIC ARRAY
32 INPUTS

CLOCK₁

CLOCK₄

F₁

F₃

F₄ Σ → OUTPUT

F₂

F₃

CLOCK₂

CLOCK₃

SINGLE CHIP C3D LENS

Fig. 5.   C3D Lens with Beam Steer
and Focus Elements.

ence between channels introduced by the two
wedge-shaped delay functions. Beam steering
off axis may then be accomplished by clocking
the two lens elements at different clock
frequencies (the larger the difference frequency the larger the formed beam deflection
off axis).

Thus the C3D shown in Fig. 4 is comprised
of isolated channels of delay (32 shown in
the figure) experiencing different zones
(F_1, F_2, F_3 and F_4) of signal propagation
velocity determined by the clock frequency
of the respective zone. Different channels
experience different quantities of delay
determined by the number of elements that
channel has in each of the zones and the
clock frequency of those zones.

There are three classes of focus aberrations associated with the C3D lens which
must be considered in any practical system
design. These are
    1. Dynamic focus aberrations
    2. Focus approximation aberrations
    3. Electronic distortion aberrations

Dynamic focus aberrations result from
"moving the lens" before all elements in the
array have received a given wavefront. In
the C3D lens design shown in Fig. 5 it is
assumed that the lens will focus only at one
depth during operation. In practice, the
focus range is scanned in synchronism with
the returning echoes. Initially after a
transmit burst the lens is focussed close
to the array to focus nearby targets, then is
dynamically scanned in synchronism with the
returning echoes from target at greater

depths. Unfortunately the lens must be moved
before the entire wavefront from a given
depth arrives to every element in the array.
The outer elements in the array are therefore focussed farther from the array than
they should to be exactly focussed on the
arriving wavefront. This focus error is
called dynamic focus aberration. In the
microprocessor controlled system shown in
Fig. 3 there is no dynamic focus error for
on-axis imaging. This system has the design
flexibility to focus the outer elements at
different depths than the inner elements
due to the separate focus oscillator on each
channel. In the C3D lens, however, extra
lens elements (clock zones) may be added to
connect for this form of abberation.

These are aberrations due to the focus
approximation used in the device design shown
in Fig. 5. This device uses parabolic delay
equalization for a spherical wavefront.
While a C3D lens could easily be designed to
exactly equalize (except for truncation
error) the delay for a specific spherical
wavefront, it could not be scanned by varying
the clock rate to equalize other spherical
wavefronts. Thus the parabolic approximation which is fairly accurate at ranges
greater than the aperature of the array is
used to acomplish a generally useful focus
function in a single device. As in the case
of dynamic focus aberration, additional lens
elements may be used to advantage in reducing
this form of aberration.

Electronic aberrations encompass all
of the imaging distortions due to the nonideal parameters of the C3D lens. Two significant sources of electronic aberrations
in the C3D lens are (1) charge transfer
efficiency (2) cross modulation distortion.
The efficiency of charge transfer between
storage electrodes determines the total
number of storage elements permissible
between device input and output. Large
charge transfer losses result in a reduced
delay line bandwidth. Typically transfer
loss considerations limit the total number
of storage elements in a channel to five
hundred and the maximum clock frequency to
fifteen megahertz. Intermodulation distortion results from the interaction of the
four clock frequencies shown in Fig. 5 with
nonlinearities at the C3D inputs, output and
lens element interfaces. While the input and
output cross modulation distortions may be
eliminated in principal by shielding these
points from the clocks, the lens element
interface cross modulation distortion is

primarily a function of the design of the lens interface and the respective clock frequencies of the lens elements on each side of the interface. This interface cross modulation distortion is the result of the asynchronous partitioning of charge packets which arrive at an interface with a propagation velocity different from the packets leaving the other side. In this case the packets are regrouped resulting in a difference frequency being generated. From a spectral point of view one may consider the signal packet as being resampled at each interface by a special type of sampler. If the resampling is not performed at frequencies large enough to encompass twice the highest frequency sampled, some aliasing will occur.

Unfortunately the spectrum to be sampled is very rich in harmonics so insuring twice that the largest frequency component be sampled is impractical. Fortunately some averaging (low pass filtering) occurs at the interface reducing the magnitudes of some of the aliased components. In a practical system design using C3D lenses the spectrum and magnitude of these difference frequencies relative to the desired signal passband should be considered. A detailed analysis of electronic distortion aberrations is beyond the scope of this paper.

SUMMARY

Two electronically focussed ultrasound imaging systems have been presented. Both systems offer significant improvement in resolution compared with single element systems for ranges from one to ten times the system aperature. The microprocessor controlled system offers fewer sources of imaging aberrations while the single chip C3D lens offers high performance in an extremely compact and potentially economical system configuration.

REFERENCES

1. M.G. Maginness, J.D. Plummer and J.D. Meindl, "An Acoustic Image Sensor Using a Transmit-Receive Array," Acoustical Holography, Vol. 5, ed. P.S. Green, Plenum, New York, 1974, pp. 619-631.

2. A.L. Susal, J.D. Plummer, M.G. Maginness and J.D. Meindl, "Ultrasonic Imaging in Cardiology," San Diego Biomedical Symposium, Feb 6-8, 1974, San Diego, CA.

3. J.D. Meindl, J.D. Plummer and M.G. Maginness, "A Monolithic Ultrasonic Image Sensor," Proc. 26 ACEMB, Vol. 15, 1973, p. 258.

4. J. Shott, R. Melen and J. Meindl, "The Cascade Charge Coupled Device for Compound Delay," Patent Disclosure, Stanford University, April 11, 1975.

5. J. Shott, R. Melen, "The Razorback CCD: A High-Performance Parallel Input Delay Line Architecture," 1975 Int. Solid State Circuits Conf. 13.5, 151, 227, Philadelphia, Feb 1975, pp. 150-51.

6. A.L. Susal, J.D. Plummer, M.G. Maginness and J.D. Meindl, "Ultrasonic Imaging in Cardiology," San Diego Biomedical Symposium Conference Digest, Feb 1974.

7. J.D. Plummer, J.D. Meindl and M.G. Maginness, "A New Ultrasonic Imaging System for Real Time Cardiac Imaging," ISSCC Conference Digest, Feb 1974.

8. M.G. Maginness, J.D. Plummer, J.D. Meindl, A.L. Susal, "A Cardiac Dynamics Visualization System," Proc. IEEE Ultrasonics Symp., Nov 1973.

9. R.D. Melen, "Solid-State Image Sensors for Biomedical Applications," Seventh Asilomar Conference on Circuits Digest, Systems and Computers, Nov 27-29, 1973.

10. R.D. Melen and J. Roschen, "A Parallel Input, High Speed Analog Delay Line," Naval Electronics Laboratory Command CCD Applications Conference Digest, San Diego, CA, Sept 1973.

11. J.D. Plummer, J.D. Meindl and M.G. Maginness, "An Acoustic-Image Sensor Using a Transmit-Receive Array," Fifth International Symposium on Acoustical Holography and Imaging Conference Digest, July 1973.

12. M.G. Maginness, J.D. Plummer, J.D. Meindl and A.L. Susal, "Real Time Imaging Using a Transmit-Receive Transducer Array," Proc. 4th Int. Symp. on Acoustic Holography, Plenum, New York, 1973.

This work was supported by the Department of Health, Education, and Welfare under Grant No. PO1 GM 17940-5.

The Design and Operation of Practical Charge-Transfer Transversal Filters

RICHARD D. BAERTSCH, MEMBER, IEEE, WILLIAM E. ENGELER, SENIOR MEMBER, IEEE, HOWARD S. GOLDBERG, MEMBER, IEEE, CHARLES M. PUCKETTE, IV, MEMBER, IEEE, AND JEROME J. TIEMANN, SENIOR MEMBER, IEEE

Abstract—Some of the design considerations for charge-transfer split-electrode transversal filters are discussed. Clock frequency, filter length, and chip area are important design parameters. The relationship of these parameters to filter performance and accuracy is described. Both random and tap weight quantization errors are considered, and the optimum filter length is related to tap weight error.

A parallel charge-transfer channel, which balances both capacitance and background charge, and a coupling diffusion between split electrodes greatly improves accuracy. A one-phase clock is used to simplify the readout circuitry. Two off-chip readout circuits are described, and the performance of two low-pass filters using these readout circuits is given. Signal to noise ratios of 90 dB/kHz and an overall linearity of 60 dB have been achieved with this readout circuitry.

I. INTRODUCTION

CHARGE-TRANSFER devices (CTD's) offer a number of very attractive advantages for signal processing [1] Applications in frequency filtering and time-domain matched filtering appear highly suited to one of these new devices—the split-electrode transversal filter [2], [3]. Very powerful design techniques exist for designing the tap weights for finite impulse response (FIR) transversal filters [4], and the split-capacitor CTD structure can implement these with high accuracy and low cost.

Manuscript received August 4, 1975; revised September 15, 1975.
The authors are with the Corporate Research and Development Center, General Electric Company, Schenectady, NY 12301.

Work is now being done at many locations to develop practical devices and circuits using charge-transfer transversal filters [5]–[9]. It is the purpose of this paper to discuss the properties of these devices in the context of some selected system applications and to present some specific examples where practical solutions could be obtained. The next section of the paper describes some of the design considerations that apply to the CTD transversal filter structure in the context of low-pass response. Section III is devoted to a discussion of tap weight tolerance aspects. In Section IV, clocking and output circuitry are considered. Some experimental data are presented in Section V, and Section VI summarizes the key points discussed in the paper.

II. PRACTICAL DESIGN CONSIDERATIONS

There are a number of tradeoffs to be made when designing a filter chip. To illustrate these tradeoffs, consider the design of low-pass filters and the problem of selecting the number of taps and clock frequency. Rabiner *et al.* [10] give an approximate empirical expression for the relationship between the parameters of optimum finite impulse response low-pass filters:

$$N = \frac{-10 \log_{10}(\delta_1 \delta_2) - 15}{14 (F_s - F_p)/F_c} + 1 \qquad (1)$$

where N is the number of taps, F_p, F_s, and F_c are the passband edge, stopband edge, and clock (sampling) frequencies, respec-

Reprinted from *IEEE Trans. Electron Devices*, vol. ED-23, pp. 133–141, Feb. 1976.

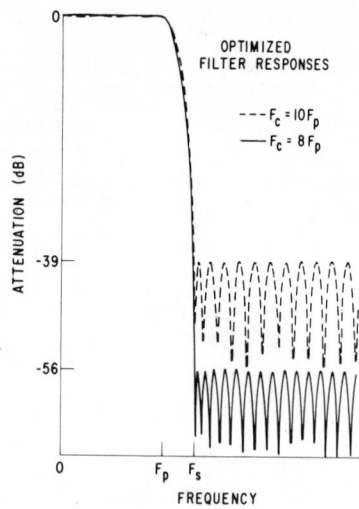

Fig. 1. Calculated filter response for two optimized filters. Passband and transition band specifications are identical, but ratio of clock frequencies is 5/4.

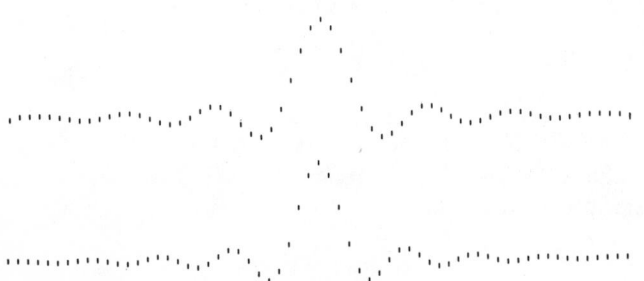

Fig. 2. Calculated impulse response (tap weights) of two low-pass filters of Fig. 4. Upper response corresponds to $F_c = 10\,F_p$; lower to $F_c = 8\,F_p$.

tively, and δ_1 and δ_2 are the amplitudes of the passband and stopband ripples, respectively. Typically, system requirements will fix F_s and F_p. Thus N and F_c must be appropriately chosen in order to obtain the desired levels of passband ripple and stopband attenuation. Some of the practical considerations involved in selecting a value for each of these parameters and for one related to these, namely the area of the CTD chip, are discussed below.

A. Choice of Clock Frequency

Consider the frequency response characteristics of the two designs shown in Fig. 1. The passband and transition band characteristics of these two low-pass filter designs are nearly identical, but the stop band attenuations differ by about 17 dB. They are both optimal designs [11] and they both have the same number of taps, but the sampling rates (clock frequencies) required to produce the same passband frequencies differ in the ratio of 5/4. Thus when the band edges are matched in frequency, the actual time duration of the impulse responses of these two filters differ as shown in Fig. 2, and the one with the longer duration (lowest clock frequency) has the better performance. Thus generally speaking, one should use the lowest possible clock frequency consistent with alias response considerations. The choice of clock frequency is usually determined by the permissible

complexity of the anti-alias analog filter which, in some form or other, must precede the sampled data CTD filter. This analog filter is necessary to prevent aliasing of signals near the clock frequency into the passband of the CTD filter. If one uses a sharp cutoff prefilter, one can use a lower clock frequency for the CTD filter and, therefore, obtain a longer time duration for the impulse response and have better CTD filter performance. However, this entails either a great number of components or tighter tolerances (or both), for the prefilter and, therefore, implies greater expense.

As a practical guideline, one can normally assume that if a 3-pole prefilter is used (requiring 1 op amp or resonator), the usable frequency region extends to about $\frac{1}{6}$ of the clock frequency. If a 5-pole filter is employed (requires 2 op amps or resonators), the usable fraction increases to about $\frac{1}{3}$.

B. Number of Taps

One might hope to increase the duration of the impulse response and hence improve performance by increasing the length of the filter (number of taps), but this approach also has limitations.

In general, the tap weights of frequency selective filters tend to zero at both ends of the filter. For example, an idealized low-pass transversal filter has tap weights given by

$$W_n = \frac{\sin 2\pi n F_p/F_c}{2\pi n F_p/F_c}, \quad -\infty < n < \infty.$$

In practice, two aspects of the device fabrication process limit the smallest tap weight which can be realized and hence the length of the filter. First, since computerized pattern generators are used in the photolithographic mask-making process, tap weights must be rounded off to integral multiples of a minimum coordinate spacing. Thus for any tap weight less than $\frac{1}{2}$ this minimum will be rounded to zero. Second, device process variations, such as nonuniform etching etc., will result in random tap weight errors and will put an effective lower limit on tap weight size. The effects of tap weight quantization and random errors are considered in more detail in Section III.

In addition to these fabrication limitations, insertion loss impacts the acceptable length of a filter. As one adds more taps to the general frequency selective filter, one is merely adding more and more small values at the ends. Thus the summed signal charge gets negligibly larger while the total electrode capacity continues to increase, resulting in a net decrease of signal voltage and a loss in signal-to-noise ratio.

C. Chip Area

Both types of tap weight error discussed above must be reduced to a level consistent with the desired filter length. In principle, this can be accomplished by making the chip suitably wide. However, increasing the chip area will reduce device yield and, therefore, increase the cost, and, since the capacitance of the electrodes is proportional to width, driver requirements also become more difficult.

There is another problem associated with the width of the transfer channel which has not yet been mentioned. If the

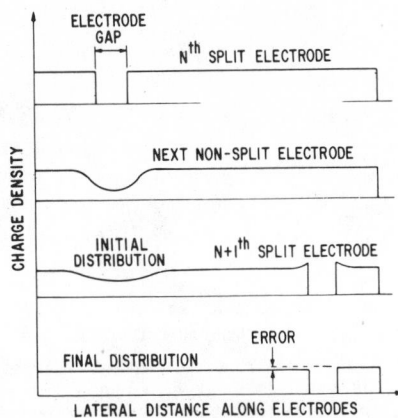

Fig. 3. Charge density is plotted as function of lateral distance along electrode for structure in which no charge is stored under electrode split. Error associated with this structure is illustrated.

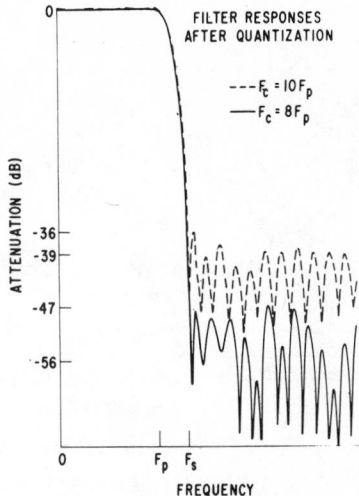

Fig. 4. Calculated filter response after quantization to one part in 600 of two filters of Fig.1.

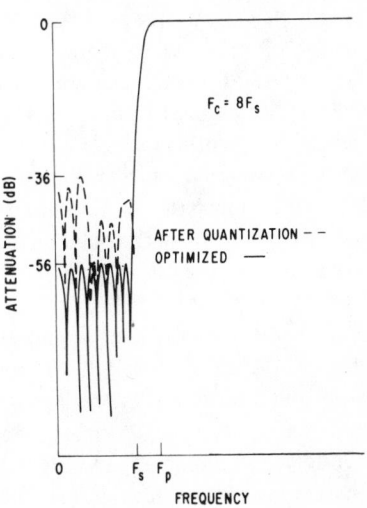

Fig. 5. Calculated response of high-pass analog of low-pass filter of Figs. 1 and 4 with $F_c = 8 F_p$. Both quantized and nonquantized responses are shown.

channel is wide, the time constant for equilibration of surface potential across the width of a particular reservoir can become quite long. If the gap cut in a split electrode produces a non-uniformity in the charge density and if this density fluctuation does not have time to equilibrate while the charge packet is within the adjacent nonsplit electrode, the charge will not be split into the correct proportions when it is transferred to the succeeding stage. Transversal filter designs which use a channel stop or thick oxide at the electrode split will be limited by this error. This error is particularly severe when the impulse response varies rapidly from stage to stage or when high frequency operation is contemplated.

Consider the charge distribution across the width of the signal channel within the Nth stage as shown in Fig. 3. As soon as the charge is transferred to the adjacent nonsplit stage, it starts to equilibrate the hole in the charge density caused by the gap cut, but unless this density fluctuation has a chance to fully equilibrate before the packet is transferred to the next stage, an error will result as shown. This error appears at much lower frequencies than those for which transfer losses become important, because the channel width is much larger than the length of a stage in the transfer direction. The time constant for lateral equilibration [12] is given by

$$T = W^2 / \mu (V_{app} - V_t) \qquad (2)$$

where μ is the mobility, V_{app} is the electrode voltage, V_t is the threshold voltage, and W is the longest distance over which lateral equilibration occurs.

If W is 10 mils, the lateral equilibration time constant is about 3×10^{-7} s for p channel and 1×10^{-7} s for n channel. One solution to this problem is to eliminate the hole in the charge density profile completely by bridging the gap cut with a coupling diffusion. This diffused region will assure that the surface potential is continuous across the gap, and when the charge packet is transferred to the adjacent nonsplit electrode, a uniform charge density will be transferred.

III. TAP WEIGHT TOLERANCE STUDIES

As was noted in the preceding section, tap weight quantization and device processing variables limit the precision of tap weights. These two factors are discussed in more detail in this section.

Tap weight quantization, in general, presents the designer with an interesting challenge. Typically one designs a filter with a given set of parameters, scales the tap weight vector so that the maximum tap weight equals the largest value permitted by the details of the mask construction, and then rounds off all remaining tap weights to the nearest integral multiple of the minimum coordinate spacing mentioned above. When this process is completed, the resulting response may or may not meet the stopband specifications even though the ideal design far exceeded them. It is usually possible to gain a few decibels improvement in the stopband characteristics of the quantized tap weight response by making minor variations in the design parameters, e.g., cutoff frequency, and picking the one that has the best quantized response characteristics.

The "dynamic range" of a set of tap weights varies quite differently with different types of filters and, therefore, places varying requirements on the tap weight accuracy

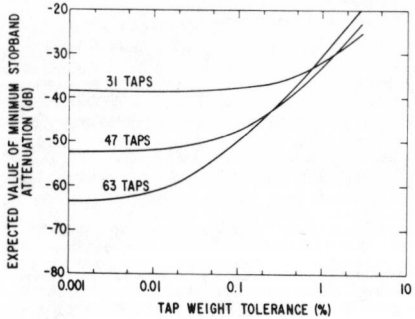

Fig. 6. Monte Carlo calculation of expected value of minimum stopband attenuation versus tap weight tolerance for three optimized low-pass filter designs. See text for details of calculation.

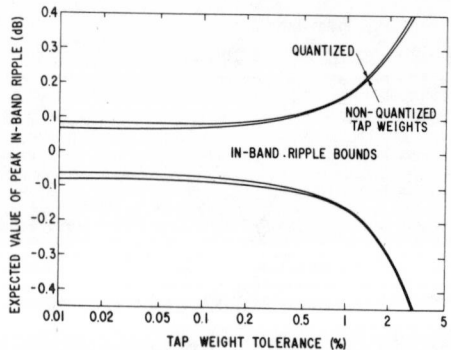

Fig. 7. Monte Carlo calculation of expected value of peak in-band ripple versus tap weight tolerance for 63-tap low-pass filter of Fig. 6. In one case, tap weights were quantized to 1 part in 600.

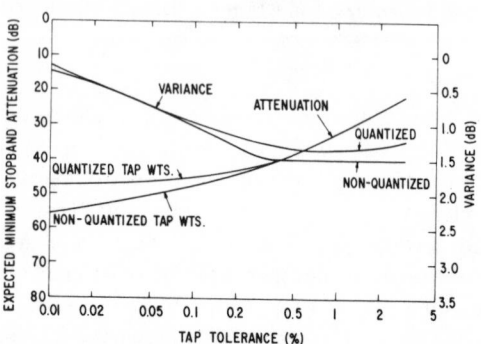

Fig. 8. Monte Carlo calculation of expected value and variance of minimum stopband attenuation versus tap weight tolerance for 63-tap low-pass filter of Fig. 6. Quantization was to 1 part in 600.

needed to realize them. A high-pass filter, for example, can be modeled as a low-pass filter in parallel with a unity gain path with the outputs of the two parallel paths being subtracted to yield the high-pass filter output. As a result, the tap weights of a high-pass filter may be viewed as a single large central tap weight superimposed on a low-pass filter tap weight vector of opposite sign and equal total weight. Since the large tap weight sets the maximum scale and thereby the absolute value of the quantization step for a given resolution, fewer levels of quantization are available for the tap weights of the high-pass function than would be if the scaling was controlled by the low-pass function alone. Fig. 4 shows the effect of quantization to 1 part in 600 of the two optimum low-pass filters shown in Fig. 1. Fig. 5 shows the effect of quantization on the high-pass "analog" to the low-pass filter with $F_c = 8F_p$. While the stopband attenuation of the low-pass filter is degraded from 56 to 47 dB, the same quantization reduces the stopband rejection of the high-pass filter to 36 dB.

The effects of random tap weight errors have been analyzed using Monte Carlo [13] techniques, and Fig. 6 shows a typical result. To generate the data of this figure, a family of low-pass filters was designed using the Parks–McClellan algorithm [11] to achieve optimum equiripple response characteristics. Filter lengths of 31, 47, and 63 taps were selected with the normalized passband cutoff set equal to 0.125. The stopband cutoff was set equal to 0.175 and the ratio of the passband to stopband ripple was set at 5. An ensemble of frequency responses was then calculated for each filter using tap weights that were calculated by adding an independent random error to each of the optimum values. The equation for the random error is

$$\Delta_n = T * \max \{W_n\} * RN$$

where

$$\max \{W_n\} = \text{maximum tap weight}$$

$$T \qquad = \text{tolerance}$$

$$RN \qquad = \text{random number uniformly distributed between } +1 \text{ and } -1.$$

The response ensemble was scanned in the stopband to find the minimum stopband attentuation; these values were then averaged and the data thus derived were plotted. Similar data have been generated for other designs. The general conclusion is that longer filters require increased tap weight accuracy,

i.e., smaller random errors, if they are to perform better than shorter filters.

The combined effects of quantization and random errors are illustrated in Figs. 7 and 8. Fig. 7 shows how the average value of the peak positive and negative passband ripples varies as a function of the random tap weight error tolerance, with and without tap weight quantization. Note that tap weight quantization to 600 levels has very little effect on the in-band characteristics. In Fig. 8, the average amplitude of the minimum stopband ripple, expressed as decibels of attenuation, is plotted together with its variance. Here the effect of quantization is more apparent. This is not surprising because the attenuation is due to a delicate vector cancellation between all of the taps. These results indicate that random tap weight errors in the range of 0.1 to 0.3 percent can be tolerated for this 63-tap design.

IV. CLOCKING AND SIGNAL RECOVERY

A. General

A wide variety of techniques have been suggested for clocking, charge insertion, and signal recovery in CTD's, and their advantages and disadvantages have been discussed elsewhere [14]–[16]. The primary considerations, apart from complexity, are signal to noise ratio and linearity. Recovery of the output signal from transversal filters raises some new problems beyond those encountered in other CTD's, such as simple delay lines. First, the desired output signal voltage is smaller. It is derived by subtracting the output signals from the two split-electrode

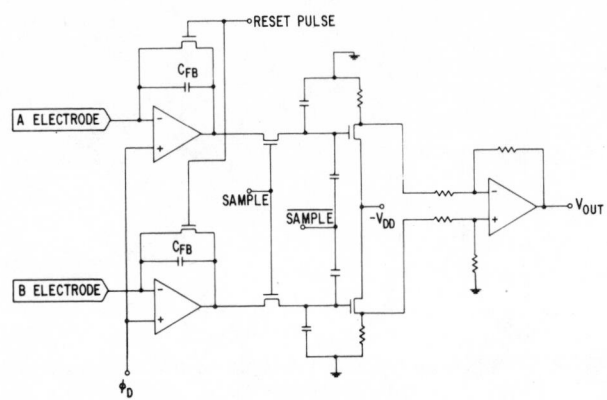

Fig. 9. Circuit which keeps split-electrode segments clamped at fixed voltage.

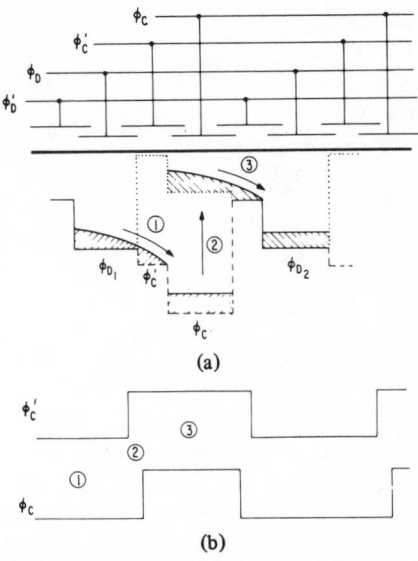

Fig. 10. Surface potentials and clock waveforms for p-channel overlapping gate structure using one-phase clock. Electrodes labeled ϕ_D and ϕ_D' are connected to dc voltages.

portions, and each of these is usually much larger than the difference. Thus even though the sense electrode output voltages are large, the dynamic range available for the actual output signal is usually reduced. Signal-to-noise and dynamic range considerations are, therefore, more important in transversal filters than in simple delay lines. Second, an uncorrectable error is introduced if the depletion capacitances of the reservoirs depend on the electrode output voltages. This error arises because the depletion capacitance is a function of both the charge in the individual packet and the potential of the overlying sense electrode. Although the dependence on the signal charge itself can be overcome by compensation at the input, the change due to the electrode potential depends on the total charge in all of the reservoirs and, therefore, cannot be compensated. This error will be called "crosstalk." One solution to this problem is to clamp the electrode potentials with operational amplifiers as shown in Fig. 9. This strategy forces the electrode voltages to remain fixed and eliminates the "crosstalk" between the charge samples, but it imposes severe requirements on the amplifiers. First, since uncorrelated amplifier noise is introduced before subtraction takes place, the amplifier must have low noise, and second, even if the sense electrodes are not clocked, the amplifier must handle the dynamic range of the full electrode signal rather than that of the difference signal. In practice, these problems are worse than the "crosstalk" problem, and we have found it preferable to permit the electrode potentials to change. In this case, a single differential op amp can be used, and the common mode signal can be eliminated with the common mode rejection of the amplifier. Two circuits using this technique will be described later.

A problem common to all data recovery schemes is the presence of large clock transients, and any useful scheme must have some method of suppressing them. This suppression can be greatly aided if a differential output is used and the transients can be made to balance. In the transversal filter, a differential output is required in any case, but the transients induced on the A and B portions of the split electrodes will not be equal unless the tap weights sum exactly to zero. Since a filter cannot have a response down to dc unless the tap weight sum differs from zero, balanced transient pick-up will not be obtained in the general case. To solve this problem, a

parallel charge-transfer channel has been incorporated in the experimental structures whose area is exactly equal to the difference in the total areas of the split electrode portions. Since all overlap capacitances are the same in the parallel channel as in the signal channel, transients from all sources are balanced. No signal charge is sent down this channel, but only a background charge which is equal in density to the background charge of the main channel. The background charge transferred under the two halves of the split electrodes is thus identical, and, when the difference is taken, the output signal is independent of background charge. Thus the filter will respond to a change in dc signal but not to a change in dc background. The parallel channel is *necessary* for any filter which requires a frequency response which is flat to dc and in which the output dc devel is to be independent of background charge.

B. One-Phase Clock Scheme

A schematic cross section of the charge transfer structure is shown together with a surface potential plot at two different times in the cycle in Fig. 10(a), and the corresponding clock voltage waveforms are shown in Fig. 10(b). The split electrodes, on which the output signal appears, are labeled ϕ_D, and these are interleaved by the ϕ_C electrodes which are not split. The primed electrodes are transfer gates, whose function is simply to prevent the flow of charge in the reverse direction. The voltage waveforms on the transfer gates are essentially identical to those for the corresponding reservoirs, but their potentials are always lower (closer to ground). Although the circuit operates correctly with the same timing on both clock waveforms, the actual waveforms shown have been used for many of the experiments. Three points in time within one cycle are shown on the waveforms, and the corresponding potentials are also indicated. Potentials at time 1 are shown dashed, while those at time 3 are shown dotted.

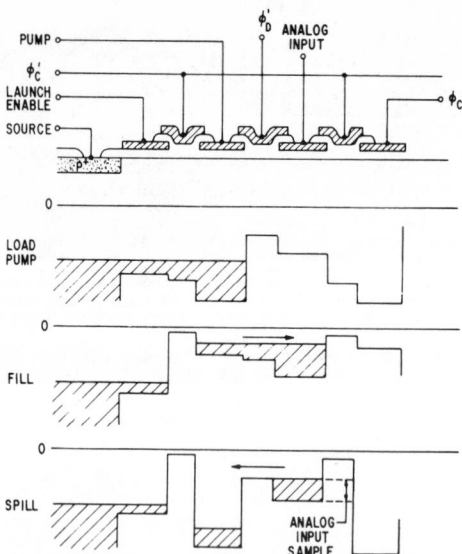

Fig. 11. Surface potentials at three different times illustrating "fill and spill" input method. Charge launched is proportional to difference between ϕ'_D and analog input.

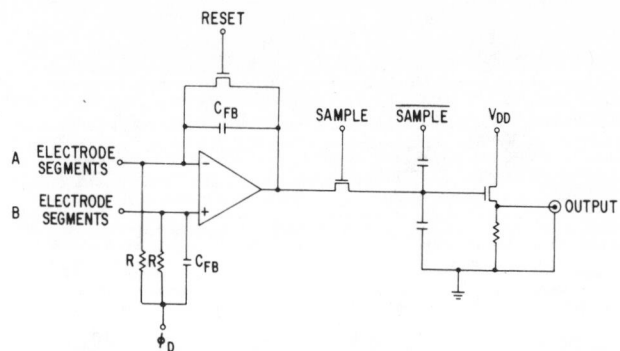

Fig. 12. Charge-sensitive amplifier circuit using resistors to reset voltage on electrodes.

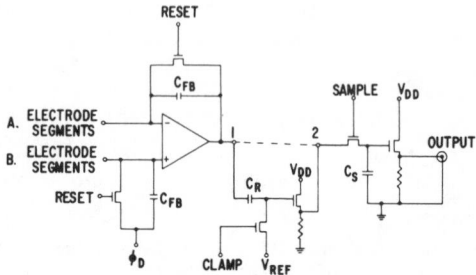

Fig. 13. Precharge and float circuit using charge-sensitive amplifier.

Note that the split electrodes ϕ_D are held at relatively constant potential throughout the cycle, and that charge transfer is accomplished by driving the intervening ϕ_C electrodes alternately higher and lower. There were two reasons for choosing this one-phase clocking method. First, the signal recovery is easier, since the signal does not ride on top of a high-amplitude clock, and second, the number of drivers required is reduced.

C. Fill and Spill Input

A linearized input scheme [14], [15] is shown schematically in Fig. 11. Here an excess quantity of charge is loaded under an electrode-called pump when the ϕ'_C electrode voltage drops. This excess charge is transferred under the analog input electrode when ϕ'_C and the pump electrode move toward ground. A quantity of charge proportional to the difference between the voltage applied to the analog input and the ϕ'_D electrode remains under the analog input electrode after the pump electrode voltage has dropped and the excess charge spilled.

Note that this input scheme requires that only a dc voltage be applied to the input diffusion, so that the possibility of charge injection is minimized. The charge launched is linear with the input analog voltage when a dc voltage is applied to the ϕ'_D electrode. It is invariant to the threshold to the extent that the upper and lower electrode thresholds track, which will depend on the details of the particular fabrication process employed.

D. Output Circuitry

The signal recovery circuit is shown schematically in Fig. 12. It is basically a current integrator or "charge-sensitive" amplifier followed by a sample and hold. In operation, the reset FET is first closed, thereby discharging the integration capacitor C_{FB} and then opened again. After the reset transistor is opened, the split electrodes are essentially floating, so that when the signal charge packets are moved under them, their potentials

change. The operational amplifier tries to keep both input terminals at the same potential by driving charge through the feedback capacitor. It can be seen from the symmetry of the input terminals that the charge supplied must be equal to the difference between the charges induced on the split-electrode portions. The output voltage is, therefore, proportional to this difference. The electrodes are reset to the "constant" potential ϕ_D by the resistors labeled R. The function of these resistors is to reset the split electrodes to the voltage ϕ_D by the end of the clock cycle. The resistor should be chosen such that RC_T is approximately $\frac{1}{10}$ the clock period, where C_T is the total capacitance to ground at either of the amplifier inputs including that of the transversal filter. This circuit operates best with an amplifier with sufficient slew rate to integrate the charge difference before the split electrode voltages have decayed appreciably. In addition, the resistors and on-chip and off-chip capacitance must be accurately matched so that the RC time constant of both halves is the same. Under these conditions the circuit shown in Fig. 12 will give accurate results. It should be pointed out that the output voltage of the charge sensitive amplifier will correctly reflect the difference in charge transferred to the two halves of the split electrode *even after* the voltages on the split electrodes have decayed to their initial value. Because the split electrodes return to the same voltage each clock cycle, this circuit does not have the crosstalk error described previously.

An alternate circuit is shown in Fig. 13. In this case, the two halves of the split electrode are reset to the ϕ_D voltage by turning on two FET's. Just prior to transfer of charge, both FET's are turned off and the difference in charge transferred is integrated on the feedback capacitor. This circuit, which will be referred to as precharge and float, suffers from the

TABLE I
TRADEOFF BETWEEN LINEARITY AND RMS-SIGNAL-TO-RMS-NOISE RATIO
FOR THE PRECHARGE AND FLOAT CIRCUIT OF FIG. 13

	Signal-to-Noise Ratio	
Linearity	Single Sampling	Double Sampling
40 dB	62 dB	68 dB
50 dB	59 dB	65 dB
60 dB	54 dB	60 dB

crosstalk error mentioned previously since the voltage which the two split halves finally reach depends on the total charge in the filter. It should be noted that an operational amplifier with a relatively slow slew rate can be used in this circuit.

An essential feature of both circuits described above is that feedback around the operational amplifier maintains the two portions of the electrodes at the same potential. If this were not done, charge that was initially transferred into one of the two portions of the split electrodes could transfer through the coupling diffusion to the other portion.

The precharge and float circuit of Fig. 13 can be operated with points 1 and 2 connected as shown by the dotted line, or by using an FET to clamp one end of the capacitor C_R to a reference voltage while the electrodes are being reset. If the clamp FET is turned off after the reset FET but before the charge has transferred, the noise associated with resetting the electrodes will be eliminated. This technique is referred to as correlated double sampling [17]. Both circuits, implemented with external components, have been used to obtain the results which will be described in the following sections.

E. Linearity and Noise

Measurements of the overall linearity and signal to noise ratio obtained with the CTD filter have been made. The linearity is obtained by measuring the harmonic content at the output when a sinusoidal input voltage is applied. The linearity is observed to depend on the signal level. This implies that a measurement of signal-to-noise ratio is meaningful only if the linearity of the output signal is specified.

A linearity of 50 dB is obtained with an input signal corresponding to $\frac{1}{2}$ the maximum input signal for the precharge and float circuit (see Fig. 13), while a linearity of 57 dB is obtained for the circuit of Fig. 12 under the same condition. As discussed above, the improved linearity in the latter case is obtained because the output electrodes return to the same voltage every clock period after the charge transfer. However, the resistors are an additional noise source and the signal-to-noise ratio of this circuit is reduced. Note that the linearity of either circuit can be improved by loading the electrodes with additional capacitance, but this also decreases the signal-to-noise ratio.

The tradeoff between linearity and signal-to-noise ratio is illustrated in Table I. The data in Table I show the ratio of the rms signal to rms noise integrated over a 50-kHz bandwidth for three different signal levels corresponding to an output linearity of 40, 50 and 60 dB for the precharge and float circuit of Fig. 13 with correlated double sampling and without

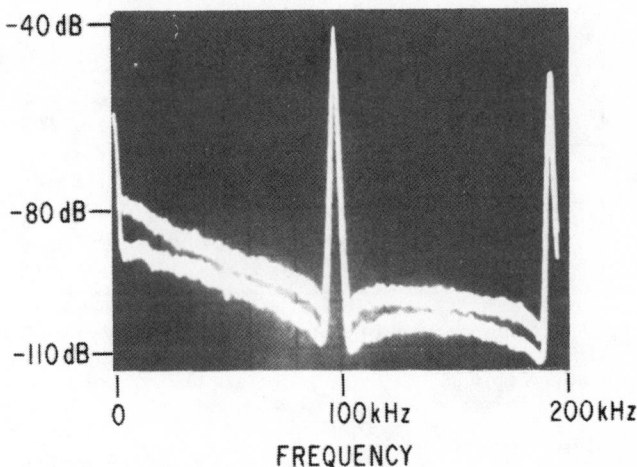

Fig. 14. Noise spectra at output of CTD low-pass filter with (lower trace) and without (upper trace) correlated double sampling for 100-kHz clock frequency. Analyzer bandwidth was 1 kHz.

(i.e., "single" sampling). The signal and noise at the sampled and held output were measured with a wide-band rms voltmeter after filtering with an *L-C* low-pass filter with a 50-kHz cutoff. For these measurements, the CTD low-pass transversal filter cutoff and clock frequencies were 10 and 80 kHz, respectively. While the results given in Table I refer specifically to the precharge and float circuit, virtually identical signal to noise ratios are obtained for the circuit of Fig. 12 when 10-kΩ resistors are used to reset the electrodes. Thus although the noise levels of the two circuits differ, the signal-to-noise ratio obtainable for a specified linearity is roughly the same because the circuit with the higher noise level has a higher degree of linearity. Correlated double sampling will eliminate reset noise and reduce $1/f$ noise associated with the amplifier. This technique reduced the integrated noise over a 50-kHz bandwidth by approximately 6 dB as shown in Table I.

The noise spectra obtained at the sample and hold output with (lower trace) and without (upper trace) correlated double sampling are shown in Fig. 14. The analyzer bandwidth was 1 kHz. Since 0 dB corresponds to an output signal with a linearity of 40 dB, the rms signal-to-noise ratio is observed to be about 90 dB/kHz in the passband of the CTD low-pass filter. A noise power of -90 dB/kHz is equivalent to a noise voltage at the amplifier output of 1 μV/$\sqrt{\text{Hz}}$. The two peaks observed in Fig. 14 are the fundamental and second harmonic of the clock. Although a sample-and-hold circuit was used to obtain these spectra, there is some clock feedthrough due to the finite sampling time.

V. EXPERIMENTAL RESULTS

The two low-pass filters whose computed frequency responses are shown in Fig. 4 have been fabricated. They are both 63-tap p-channel overlapping gate structures fabricated on 4 $\Omega \cdot$cm material. The filter parameters are: filter *A*: $F_p/F_c = 0.1$, $F_s/F_c = 0.131$, $\delta_1/\delta_2 = 1$, and filter *B*: $F_p/F_c = 0.1238$, $F_s/F_c = 0.165$, $\delta_1/\delta_2 = 5$. The optimum taps were quantized to 600 levels. Since the coordinate quantization of the available artwork generator was 0.00005 in, this quantization results in a channel width of 0.030 in. The total elec-

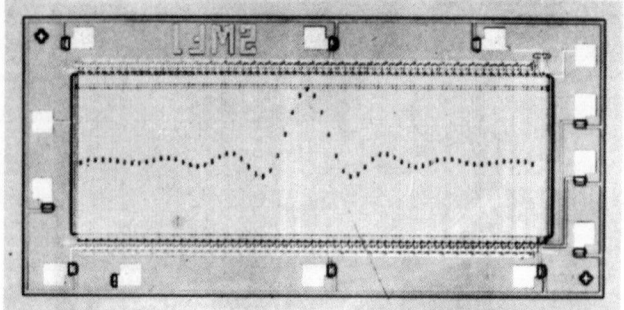

Fig. 15. Photomicrograph of 63-tap low-pass filter which will be referred to as filter A. Parallel channel is shown near top of charge-transfer channel.

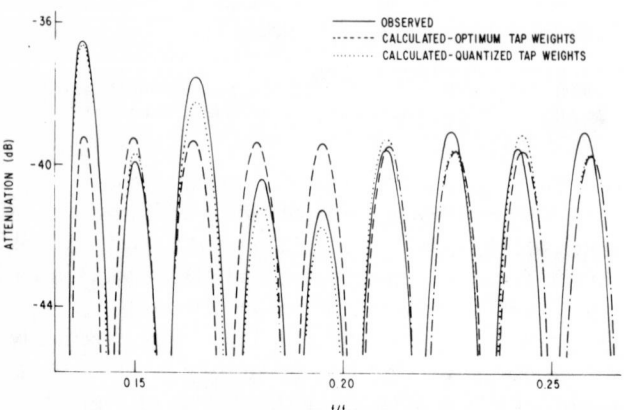

Fig. 16. Comparison of calculated and observed stopband attenuation of filter A.

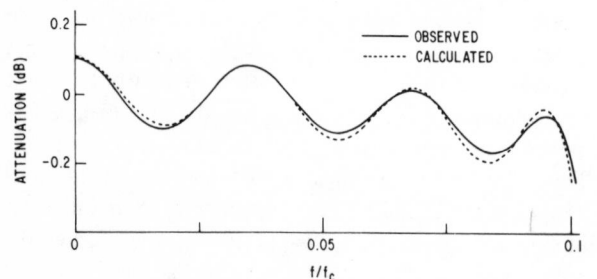

Fig. 17. Comparison of the calculated and observed passband response of selected sample of filter A.

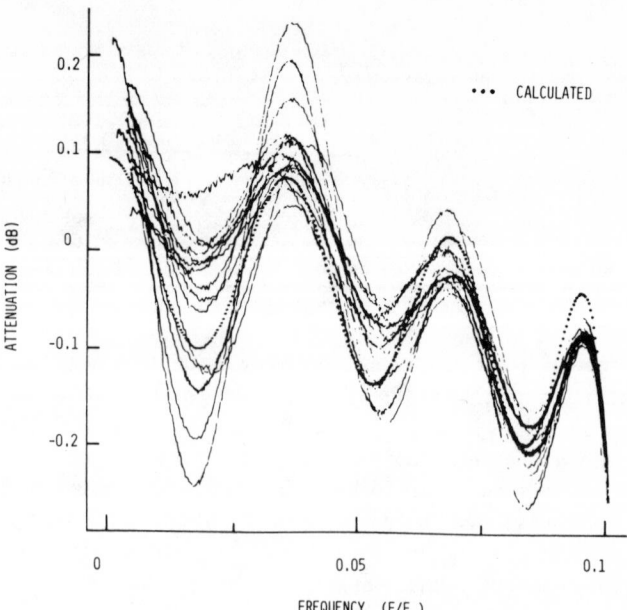

Fig. 18. Comparison of calculated and observed passband response for several different samples of filter A.

trode capacitances are of the order of 100 pF per phase, which is within the acceptable range for simple drivers. A photomicrograph of filter A is shown in Fig. 15. Note that the parallel channel which was discussed in Section IV may be seen immediately above the upper halves of the split electrodes.

Figs. 16–18 show details of the correspondence between the computed and observed frequency responses in the stopband and passband regions of filter A. The performance in the stopband of a typical device is shown in Fig. 16. Here, both the theoretical optimum response as well as that of the rounded off taps are shown. In this region, the relative importance of errors is greatly magnified, since the entire response is at −40 dB. From the fact that the experimental curve lies closer to the rounded off curve than to the opti-

mum, it can be seen that the round-off error of 1/600 of full scale is larger than the random tap weight error for this device.

The expected and the observed responses within the passband are shown in Fig. 17 for another device. The agreement, which was within 0.02 dB, also indicates a tap weight accuracy of about 0.1 percent. The slight droop in both response curves was due to the sample and hold circuit used at the output which was included in the calculated response. Unfortunately, the passband response of Fig. 17 was not typical. Fig. 18 shows an ensemble of passband response curves superimposed on the calculated response. The deviations are seen to be of order of 0.1 dB. Referring to Fig. 7 which shows passband ripple dependence on tap weight tolerance, it may be seen that a random error of about 1 percent would be required to account for this level of deviation. This amount of error is inconsistent with the stopband results. Therefore, this excess passband deviation cannot be caused by random error but could be some form of correlated error. While random errors contribute to the frequency response inaccuracies as $\sqrt{N}$, correlated errors can be expected to contribute as N, i.e., correlated errors will be about 8 times more serious for a 63-tap filter.

The simplest forms of correlated error are mask translation and rotation. Even relatively large mask rotations will result in negligible changes in the passband response of low-pass filters. On the other hand, mask translations will cause significant perturbations of the passband response [7]. However, a calculation of the effects of a simple mask translation does not agree with the responses shown in Fig. 18. The excess passband deviation may be the result of a more complex correlated tap weight error, or it may be unrelated to tap weight error but rather may relate to errors in the method of charge detection.

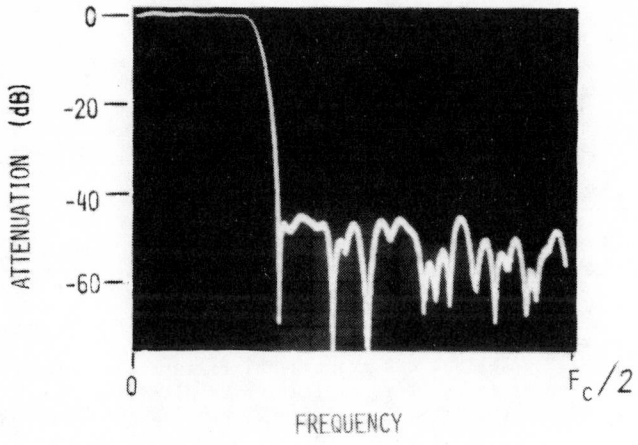

LOW PASS FILTER RESPONSE

$$F_p = 1/8 \ F_c$$

Fig. 19. Performance of filter *B*. Stopband attenuation appears to be limited by random tap weight errors.

Preliminary measurements of the frequency response of filters of type *B* have been taken, and a sample is shown in Fig. 19. Typical minimum stopband attenuation appears to fall in the range −42 to −45 dB. Reference to Fig. 8 indicates that these devices also have a tap weight tolerance in the range of 0.1 to 0.3 percent which is then consistent with the measurements taken on the type *A* filters.

VI. Summary

This paper discusses some of the considerations involved in designing CTD split electrode transversal filters. Much of the discussion on tap weight error and all of the experimental data relate specifically to low-pass filters. However, most of the observations made apply to a broader range of frequency selective filters.

Generally, the "best" transversal filter performance for a fixed number of taps is obtained when the filter frequency response occupies a large fraction of the Nyquist bandwidth. However, this increases the requirements on any anti-aliasing filters.

If the clock frequency is fixed, the performance is theoretically improved by increasing the filter length (i.e., number of taps). In practice, however, the filter length is limited by tap weight inaccuracy (charge-transfer inefficiency also becomes a problem when *N* gets large). For the low-pass filter described above, 63 taps give adequate performance for a tap weight accuracy of 0.1 percent, and a further increase in length would not improve performance.

The major sources of tap weight inaccuracy are random processing variations and tap weight quantization. Increasing the chip width will minimize these tap weight inaccuracies, but this tactic is limited by yield and peripheral circuit con-

siderations. The results above indicate that a random tap weight accuracy of 0.1 percent can be achieved with moderate chip area. When, for example, 50 dB out-of-band rejection in a low-pass filter is desired, this level of accuracy is required.

Accurate signal recovery circuitry is required for good transversal filter performance. A one-phase clock simplifies the output circuitry since the split electrodes can be operated at a dc voltage. A parallel channel which balances capacitance and background charge and a coupling diffusion between the split electrodes greatly improve accuracy. The experimental results attest to the accuracy that has been achieved with these techniques. Signal to noise ratios of 90 dB/kHz and an overall linearity of 60 dB have been achieved with this signal recovery circuitry.

References

[1] R. D. Baertsch, W. E. Engeler, H. S. Goldberg, C. M. Puckette, and J. J. Tiemann, "Two classes of charge-transfer devices for signal processing," in *CCD 74 Int. Conf. Proc.*, Sept. 1974, pp. 229–236.
[2] F. L. J. Sangster, "The bucket brigade delay line, a shift register for analog signals," *Philips Tech. Rev.*, vol. 31, pp. 97–110, 1970.
[3] D. R. Collins, W. H. Bailey, W. M. Gosney, and D. D. Buss, "Charge-coupled-device analogue matched filters," *Electron. Lett.*, vol. 8, no. 13, pp. 328–329, 1972.
[4] L. R. Rabiner and B. Gold, *Theory and Application of Digital Signal Processing*. Englewood Cliffs, NJ: Prentice-Hall, 1975.
[5] D. D. Buss, D. R. Collins, W. H. Bailey and C. R. Reeves, "Transversal filtering using charge-transfer devices," *IEEE J. Solid-State Circuits*, vol. SC-8, pp. 134–146, Apr. 1973.
[6] D. A. Sealer, C. H. Séquin, A. M. Mohsen, and M. F. Tompsett, "Design and characterization of charge-coupled devices for analog signal processing," in *1975 Int. Conf. Communications Rec.*, vol. 1, June 1975, pp. 2–10.
[7] H. S. Goldberg *et al.*, "Design and performance of CTD split-electrode filter structures," in *1975 Int. Conf. Communications Rec.*, vol. 1, June 1975, pp. 2–10.
[8] A. Ibrahim, L. Sellars, T. Foxall, and W. Steenaart, "CCD's for transversal filter applications," in *Tech. Dig., Int. Electron Devices Meeting*, Dec. 1974, pp. 240–243.
[9] J. A. Sekula, P. R. Prince, and C. S. Wang, "Nonrecursive matched filters using charge-coupled devices," in *Tech. Dig., Int. Electron Devices Meeting*, Dec. 1974, pp. 244–247.
[10] L. R. Rabiner, J. F. Kaiser, O. Herrmann, and M. T. Dolan, "Some comparisons between FIR and IIR digital filters," *Bell Syst. Tech. J.*, vol. 53, pp. 305–331, Feb. 1974.
[11] J. H. McClellan, T. W. Parks, and L. R. Rabiner, "A computer program for designing optimum FIR linear phase digital filters," *IEEE Trans. Audio Electroacoust.*, vol. AU-21, pp. 506–526, Dec. 1973.
[12] W. E. Engeler, J. J. Tiemann, and R. D. Baertsch, "Surface-charge transport in a multielement charge-transfer structure," *J. Suppl. Phys.*, vol. 43, pp. 2277–2285, May 1972.
[13] C. M. Puckette, W. J. Butler, and D. A. Smith, "Bucket brigade transversal filters," *IEEE Trans. Commun. Technol.*, vol. COM-22, pp. 926–934, July 1974.
[14] M. F. Tompsett, "Surface potential equilibration method of setting charge in charge-coupled devices," *IEEE Trans. Electron Devices*, vol. ED-22, pp. 305–309, June 1975.
[15] S. P. Emmons and D. D. Buss, "Noise measurements on the floating diffusion input for charge-coupled devices," *J. Appl. Phys.*, vol. 45, pp. 5305–5306, 1974.
[16] C. H. Séquin and A. M. Mohsen, "Linearity of electrical charge injection into charge-coupled devices," *IEEE J. Solid-State Circuits*, vol. SC-10, pp. 81–92, Apr. 1975.
[17] M. H. White, D. R. Lampe, F. C. Blaha, and I. A. Mack, "Characterization of surface channel CCD image arrays at low-light levels," *IEEE J. Solid-State Circuits*, vol. SC-9, pp. 1–13, Feb. 1974.

COMMUNICATION APPLICATIONS OF CCD TRANSVERSAL FILTERS

D. D. Buss, R. W. Brodersen, C. R. Hewes
and A. F. Tasch, Jr.

I. Introduction

Charge-coupled devices (CCD's) are analog sampled-data delay lines and as such they are ideally suited to a broad range of filtering functions in communication systems which are presently performed digitally. CCD's have performance limitations compared to digital filters but for applications which fall within their performance ranges, CCD's offer advantages of lower cost, smaller size, lighter weight, lower power, and improved reliability over digital filters . In a sense CCD filters combine the best features of analog and digital implementations: time delays are precisely controled by a master clock but digitizing is eliminated.

The CCD transversal filter is a fundamental building block which is particularly cost effective in terms of its simplicity and versatility. In Section II the design, operation and performance limitations of CCD transversal filters will be described and some examples will be given.

Section III discusses several communcation related applications of CCD transversal filters. Matched filtering in MODEM's is an obvious and potentially important application of simple transversal filters. However, transversal filters can be integrated with other MOS components to achieve more complex functions such as the chirp z-transform (CZT) which has potential applications to video bandwidth reduction and speech processing. Adaptive or programmable transversal filters can also be realized using fixed weighting coefficient filters.

II. CCD Transversal Filters

The block diagram of a CCD transversal filter is given in Figure 1. It consists of M delay stages D together with circuitry for performing the weighted summation of node voltages V_k. Each delay stage consists of p transfer electrodes in a p-phase CCD. The filter is sampled and clocked at clock frequency f_c, and the z-transform characteristic H(z) is given by

$$H(z) = \sum_{k=1}^{M} h_k z^{-k} \qquad (1)$$

where h_k, k=1, M are the weighting coefficients.

Design

The CCD transversal filter is an important component primarily because the weighting and summing circuitry is quite simple. Using the split electrode weighting technique,[1] a CCD delay line can be made into a transversal filter by splitting one of the electrodes in each delay stage and putting a differential current integrator (DCI) in the clock line instead of the conventional output amplifier. The split electrode technique was first developed for use with bucket-brigade devices (BBD's)[2] and has been widely used for both CCD and BBD filters.[3,4,5,6,7,8] The design techniques are further described in References 1 and 8.

Limitations

CCD transversal filters have a number of important performance limitations each of which will be discussed.

Maximum clock rate. In spite of the fact that CCD's have been operated at clock frequencies in excess of 100 MHz,[9] practical CCD transversal filters are limited to 20 MHz or less. This limitation results from the non-CCD electronics (clock drivers, summing and weighting circuitry etc.) but is no less of a limitation because of that.

Minimum clock rate. The slowest that a CCD can be clocked depends upon thermal leakage. An M-stage CCD clocked at frequency f_c is required to hold charge for a total delay time $T_d = M/f_c$. The CCD gradually loses its ability to hold signals, however, due to thermal leakage, and the potential wells are completely filled within the storage time t_s which in good devices is $t_s \approx 50$ sec at room temperature. The minimum clock frequency is therefore determined by the requirement that $T_d \ll t_s$. Delay times of up to $T_d \approx 1$ sec are possible at room temperature. However, leakage current increases by a factor of two for each 8°C increase in temperature, and the minimum clock frequency increases correspondingly.

Charge transfer loss. Transfer of charge from one electrode t_c, the next is not perfect and each time charge transfers, a fraction α is left behind. For a p-phase CCD, the loss per delay stage is defined to be $\varepsilon \equiv p\alpha$ and the effect of charge transfer loss is to modify the ideal filter characteristic given in equation 1 to

$$H^\varepsilon(z) = \sum_{k=1}^{M} h_k \left(\frac{1 - \varepsilon}{1 - \varepsilon z^{-1}} \right) z^{-k} \ . \qquad (2)$$

The amount of loss which can be tolerated depends upon the filtering application, but for many applications which have been considered in detail $M\varepsilon \leq .1$ is acceptable. The loss parameter ε is typically on the order of $\varepsilon \approx 10^{-4}$, and thus, filters having up to $M = 10^3$ stages are possible without resorting to the complication of multiplexing.

Linearity. Design techniques exist for inputting electronic charge q_n into a CCD with good linearity between q_n and input voltage.[10] However, the split electrode output circuitry does not sense q_n but rather $q_s = q_n + q_b$ where q_b is the depletion layer bulk charge. Therefore, for filters using the split electrode output, the input should be designed so that the total charge q_s is proportional to the input signal. A further discussion of filter linearity is found in Reference 8.

Noise. Noise in CCD transversal filters has two sources (1) noise within the CCD itself and (2) noise associated with the DCI. The CCD is inherently a low noise device[11] especially if buried channel CCD filters are used. In addition, the weighted summation performed within the filter gives processing gain against CCD noise in much the same way as it does against input noise. On the other hand, DCI noise is added after the weighted summation and for this reason it usually dominates filter noise.

Reprinted from *IEEE Nat. Telecommun. Conf. Rec.*, vol. 1, Dec. 1-3, 1975, pp. 1.1-1.5.

The principal sources of DCI noise are (1) kTC noise[11] on the split electrode clock lines and (2) voltage amplifier noise. Filters tested to date have been limited by voltage amplifier noise. However, by improving the amplifiers, 80 dB dynamic range should be achievable.

Weighting Coefficient Error. Error in the weighting coefficients is an important limitation of CCD transversal filters for spectral analysis applications because it limites the out-of-band rejection or sidelobe suppression which can be achieved. Errors result from; stepping quantization error in photomask generation, photomask misalignment, non-uniform etching of the split electrodes, non-uniform oxide thickness, and non-uniform substrate resistivity. All of these errors can be reduced by making wide CCD channels at the expense of larger size and higher clock line capacitance. Filters made to date with 5 mil. channel width[8] have weighting coefficient error on the order of .5%, and filters having 30 mil. channel width have error between .1% and .2%.[12]

Examples

The frequency response of a 500-stage CCD bandpass filter is shown in Figure 2. It has a passband at $f_o = f_c/4$ and a 3dB bandwidth of $\Delta f/f_c = .011$. Hamming weighting was used, and for an ideal filter the highest sidelobe would be −42.8 dB. However, weighting coefficient error degrades the highest measured sidelobe to −41 dB.[8]

Figure 3 shows the impluse response and correlation response of a 100-stage CCD filter matched to a pseudorandom pn sequence code. The highest time sidelobe is 16 dB below the correlation peak.

III. Applications

Matched Filtering

CCD's provide a cost effective implementation for a number of matched filtering applications which fall within the performance limits described in Section II. Since the weighting coefficients in a CCD transversal filter can be arbitrary, there exists a great deal of flexibility in the selection of the signaling waveform.

Figure 4 shows the impulse responses and correlation responses of two 200-stage BBD filters developed for a spread-spectrum MODEM.[13] The weighting coefficients are samples of the chirp or linear FM functions. The charge transfer loss of BBD's is somewhat greater than that of CCD's, and the devices shown in Figure 4 have $\varepsilon \approx 2 \times 10^{-3}$ which attenuates the impluse response visibly. The mismatch between the filter and an incoming chirp signal results in a calculated sensativity loss of .1 dB and the overall MODEM sensitivity was measured to be within .5dB of the theoretical limit for noncoherent FSK.[4]

Figure 5 shows how the filters of Figure 4 are used in a MODEM. Both the COS and SIN filters are on the same IC, and the entire filtering function is performed on two identical MOS IC's which could be cheaply produced in volume. CCD filters are also attractive for wireline MODEM's where the volume is high and the cost is required to be low.

Chirp z-Transform

The chirp z-transform (CZT) is an algorithm for performing the discrete fourier transform (DFT) in which the bulk of the computation is performed is a chirp transversal filter and for this reason, it is particularly attractive for CCD implementation.[14,15] A CZT system for measuring the power density spectrum is shown in Figure 6. The COS and SIN filters are chirp transversal filters like those shown in Figure 4. This system has been demonstrated using 500-stage CCD filters[8,16] with the results shown in Figure 7. The CCD CZT is discussed further in References 8 and 17.

Video Bandwidth Reduction

One of the potentially important applications of the CCD CZT is transform encoding for video bandwidth reduction.[18] DFT and discrete cosine transform (DCT) on "typical" video images have resulted in variance compaction approaching that of optimum non-adaptive transforms,[18] and both the DFT and DCT can be cost effectively implemented with the CCD CZT. The CCD CZT becomes particularly attractive in remote sensing applications such as RPV's where small size, light weight, and low power are essential.

Speech Processing

Spectral analysis is one of the most important functions in speech processing,[19] and speech processing requirements in terms of sample rate (10kHz), delay time (40ms) and dynamic range (40 dB) are well within the CCD capabilities.

The simplest speech processing systems decompose speech into its spectral components as in the early channel vocoder systems. Channel vocoders perform well only at high bit rates[20] and are therefore not useful for bandwidth reduction. However this approach is useful in word recognition systems.

In an effort to reduce speech bandwidth, several powerful techniques have recently been developed.[19] Among the most successful has been homomorphic deconvolution[21] which is particularly well suited for CCD implementation. In Figure 8 a block diagram of such a system is shown.[19] Since only the magnitude of transform 1 is required, the postmultiply can be eliminated, and in addition, the postmultiply of the inverse transform 2 cancels the premultiply of transform 3. Therefore the only chirp multiplications remaining are the premultiplies of transforms 1 and 2 which can be implemented at the input to the CCD.[22] After transform 2, the pitch of the voiced speech is detected and then windowed out so that only the smoothed speech spectrum remains at the output. This spectra can be used to extract formant data and if this information is efficiently encoded data rates as low as 1000 bits/s are realizable.[19] Homomorphic deconvolution is costly to implement digitally in real time because of the three sequential transforms which are required. However, the CCD CZT holds the promise of truly low cost implementations of this type of processing.

Adaptive Filtering

Up to now, this paper has dealt with fixed weighting coefficient filters because of their ultimate simplicity. However, for some communication applications, adaptive or electronically programmable filters are required.

Filters can readily be constructed for which the weighting coefficients are binary programmable[23] (+1, −1 or +1, 0 for example) but achieving analog programmable filters is considerably more difficult.[24] The optimum CCD solution to time domain programmable convolution appears to be the binary/analog scheme shown in Figure 9, where a bank of binary programmable filters are programmed with binary representations of the weighting coefficients. This scheme is well suited to adaptive equalization where the weighting coefficients are calculated and stored digitally, but it sacrifices much of the inherent CCD simplicity when compared with the fixed weighting coefficient filter, and it may not be cost competitive with an all digital approach. In addition, other CCD related binary/analog schemes exist[25,26] which pose strong competition for the system of Figure 9.

An alternative approach to programmable convolution is to use the CZT to go to the frequency domain where convolution becomes multiplication. This is illustrated in Figure 10 which shows that the Fourier transform $H(f)$ of the impulse response must be supplied to the filter.

IV. Conclusion

It has been the purpose of this paper to demonstrate the computational power of state-of-the-art CCD transversal filters in communication applications. This type of CCD filter provides the ideal compromise between flexibility and circuit simplicity by retaining the ultimate cost, size, weight, power and reliability advantages of analog CCD's.

Requirements do exist for adaptive filters, but the price of this increased flexibility is a significant increase in CCD circuit complexity. For some applications, the requirement for electronic variability justifies the added circuit complexity, but the cost advantage (if any) over conventional digital implementation is decreased. However, for fixed weighting coefficient filtering applications, the mask programmable approach provides the cost effective solution. As with read-only memories only a single photomask is required to change the code of a filter, and this photomask is generated under computer control. The cost is very low and becomes negligible when amortized over a few thousand devices.

Because of its extreme simplicity, the CCD transversal filter is expected to be one of the key building blocks for CCD analog signal processing in the future, and because of their extremely low cost, new applications for transversal filters will evolve. A case in point is the CZT. This algorithm is important because most of the computation is performed in a transversal filter. Many other applications have been identified which capitalize on the compactness of CCD transversal filters, and it is anticipated that this component will have significant impact on design of signal processing systems of the future.

Acknowledgement

This paper has reviewed CCD development performed under the sponsorship of Naval Electronics Command (Dr. David F. Barbe), Rome Air Development Center (Charles N. Meyer), Army Electronics Command (Robert H. Sproat and Ted J. Lukaszek), Naval Undersea Center (Dr. Robert W. Means) and National Aeronautics and Space Administration (Harry F. Benz).

References

1. D. D. Buss, D. R. Collins, W. H. Bailey, and C. R. Reeves, "Transversal Filtering Using Charge Transfer Devices," IEEE J. Solid-State Circuits, SC-8, pp 134-146, April 1973.

2. F. L. J. Sangster, "The Bucket-Brigade Delay Line, A Shift Register for Analogue Signals," Philips Tech. Review, 31, pp 97-110, 1970.

3. R. D. Baertsch, W. E. Engeler, H. S. Goldberg, C. M. Puckette, and J. J. Tiemann, "Two Classes of Charge Transfer Devices for Signal Processing," Proc. International Conf. Technology and Applications of CCDs, Edinburgh, Sept. 1974, pp 229-236.

4. D. D. Buss and W. H. Bailey, "Application of Charge Transfer Devices to Communication," Proc. CCD Applications Conference, San Diego, Sept. 1973, pp 83-93.

5. A. Ibrahim and L. Sellars, "CCDs for Transversal Filter Applications," IEEE International Electron Devices Meeting, Technical Digest, Washington, Dec. 1974, pp 240-243.

6. J. A. Sekula, P. R. Prince and C. S. Wang, "Non-Recursive Matched Filters Using Charge-Coupled Devices," IEEE International Electron Devices Meeting, Technical Digest, Washington, Dec. 1974, pp 244-247.

7. M. F. Tompsett, A. M. Mohsen, D. A. Sealer, and C. H. Séquin, "Design and Characterization of CCD's for Analog Signal Processing," IEEE Advanced Solid-State Components for Signal Processing, IEEE International Symposium on Circuits and Systems, Newton, Mass., pp 83-89, April 1975.

8. R. W. Brodersen, C. R. Hewes and D. D. Buss, "Spectral Filtering and Fourier Analysis Using CCD's," IEEE Advanced Solid-State Components for Signal Processing, IEEE International Symposium on Circuits and Systems, Newton, Mass., pp 43-68, April 1975.

9. L. J. M. Esser, M. G. Collet, and J. G. VanSanten, "The Peristaltic Charge-Coupled Device," 1973 International Electron Devices Meeting, Technical Digest, pp 17-20.

10. C. H. Séquin and A. M. Mohsen, " Linearity of Electrical Charge Injection into Charge-Coupled Devices," IEDM Technical Digest, Washington, D.C., p 229, 1974.

11. J. E. Carnes and W. F. Kosonocky, "Noise Sources in Charge-Coupled Devices," RCA Review 33, p 327, June 1972.

12. H. S. Goldberg, R. D. Baertsch, W. J. Butler, W. E. Engeler, O. Mueller, C. M. Puckett and J. J. Tiemann "Design and Performance of CTD Split-Electrode Filter Structure," International Conference on Communications, Conf. Record, San Francisco, Vol I, pp 2/10-2/14, June 1975.

13. D. D. Buss, W. H. Bailey and L. R. Hite, "Spread Spectrum Communication Using Charge Transfer Devices," Proc. of 1973 Symposium on Spread Spectrum Communications, San Diego, Calif., pp 83-92, March 1973.

14. L. R. Rabiner, R. W. Schafer, and C. M. Rader, "The Chirp Z-Transform Algorithm," IEEE Trans. on Audio and Electroacoustics, AU-17, pp 86-92, June 1969.

15. H. J. Whitehouse, J. M. Speiser and R. W. Means, "High Speed Serial Access Linear Transform Implementations," presented at the All Applications Digital Computer Symposium, Orlando, Fla., Jan. 1973, NUC TN 1026.

16. R. W. Brodersen, H. S. Fu, R. C. Frye and D. D. Buss, "A 500-point Fourier Transform Using Charge-Coupled Devices," 1975 IEEE International Solid-State Circuits Conference, Digest of Tech. Papers, Phila., Feb. 1975, pp 144-145.

17. H. J. Whitehouse, R. W. Means, and J. M. Speiser, "Signal Processing Architectures Using Transversal Filter Technology," IEEE Advanced Solid-State Components for Signal Processing, IEEE International Symposium on Circuits and Systems, Newton, Mass. April 1975, pp 5-29.

18. R. W. Means, H. J. Whitehouse and J. M. Speiser, "Television Encoding Using a Hybrid Discrete Cosine Transform and a Differential Pulse Code Modulator in Real Time," 1974 National Telecommunications Conference Record, San Diego, Dec. 1974, pp 69-74.

19. R. W. Schafer and L. R. Rabiner, "Digital Representations of Speech Signals," Proc. of the IEEE 63, pp 662-677, April 1975.

20. R. W. Schafer and L. R. Rabiner, "Design of digital filter banks for speech analysis," BSTJ, 50, pp 3097-3115, Dec. 1971.

21. A. V. Oppenheim and R. W. Schafer, "Homomorphic Analysis of Speech," IEEE Trans. Audio. Electro-acoustic, Vol AU-16, pp 221-226, June 1968.

22. Private Communications, D.A. Hodges, UC at Berkeley

23. C. S. Hartmann, L. T. Claiborne, D. D. Buss and E. J. Staples, "Programmable Transversal Filters Using Surface Wave Devices, Charge Transfer Devices, and Conventional Digital Approaches," Proc. International Specialist Seminar on "Component Performance and Systems Applications of Acoustic Surface Wave Devices", Aviemore, Scotland, Sept. 1973, pp 102-114.

24. M. H. White, D. R. Lampe, and J. L. Fagan, "CCD and MNOS Devices for Programmable Analog Signal Processing and Digital Nonvolatile Memory," 1973 Int. Elect. Devices Meeting, Technical Digest, Washington, pp 130-133, Dec. 1973.

25. J. J. Tiemann, W. E. Engeler, and R. D. Baertsch, "A Surface Charge Correlator," IEEE J. Solid-State Circuits, SC-9, pp 403-410, Dec. 1974.

26. G. P. Weckler, "The Serial Analog Processor," 1975 IEEE International Solid-State Circuits Conference, Digest of Technical Papers, Philadelphia, Feb. 1975, pp 142-143, 226.

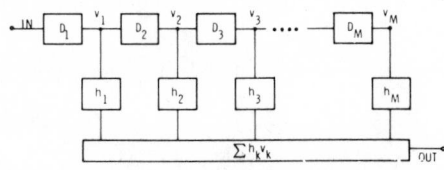

Figure 1. Block diagram of a transversal filter having M delay stages D_k.

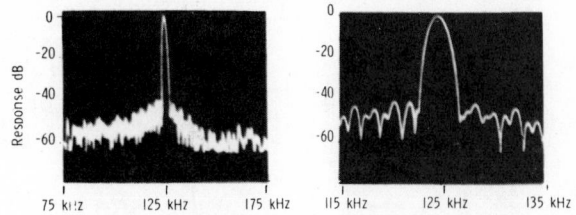

Figure 2. Measured frequency response of a 500-stage CCD bandpass filter having Hamming weighting.

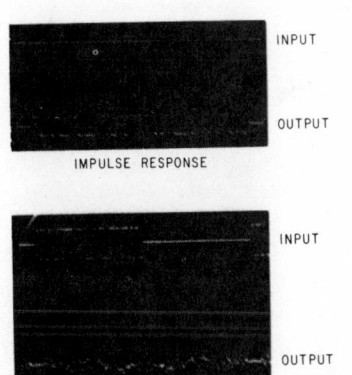

Figure 3. Impulse response (top) and correlation response (botton) of a 100-stage CCD filter matched to a pseudorandom pn sequence code. Note the correlation peak in the output waveform.

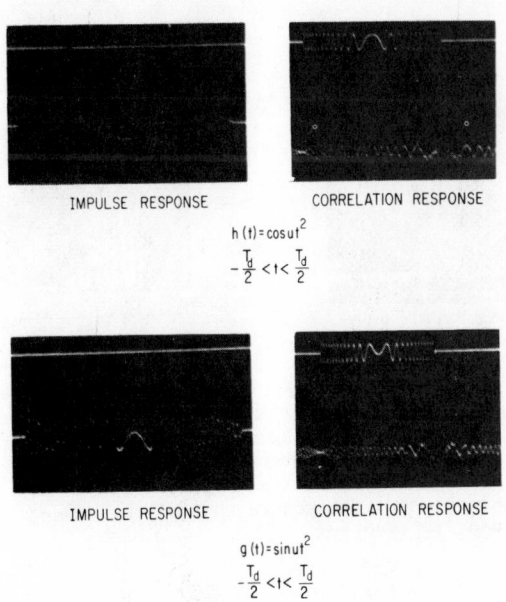

Figure 4. Impulse responses and correlation responses of two 200-stage BBD filters developed for a spread spectrum MODEM.

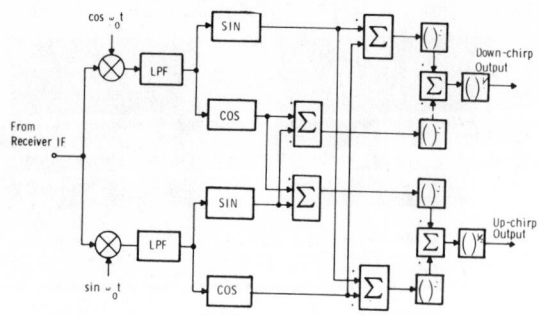

Figure 5. Block diagram showing how the filters of Figure 4 are used to demodulate binary chirp waveforms.

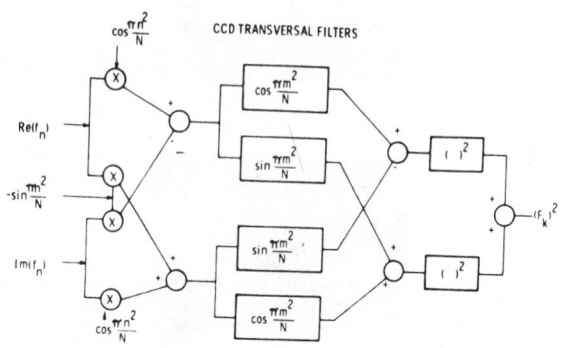

Figure 6. Block diagram of the CCD chirp z-transform for taking the power density spectrum.

Input: Frequency sweeps
50 kHz to 100 kHz in
$T_d = .11$ msec.

CZT Output: Line spectrum
having period
$T_d^{-1} = 9$ kHz

Square root of the
power density spectrum
displayed using an HP
spectrum analyzer

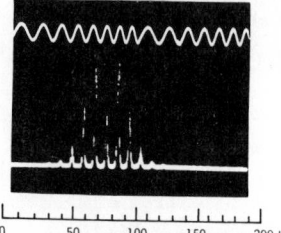

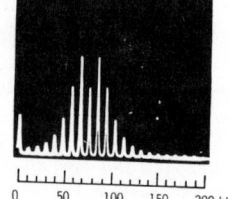

Figure 7. A comparison of the 500-point CZT power density spectrum (upper photograph) with the output of a conventional spectrum analyzer (lower photograph) which shows the square root of the power density spectrum.

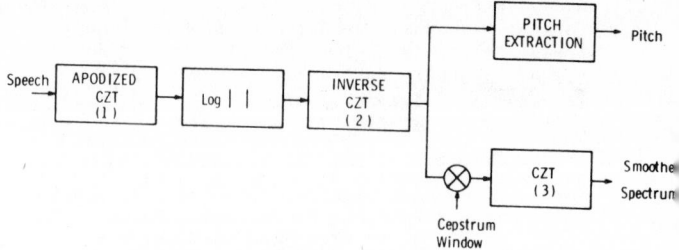

Figure 8. Block diagram of a system to perform homomorphic deconvolution of speech for bandwidth reduction (Reference 21). This is one type of speech processing which is particularly amenable to implementation with CCD's.

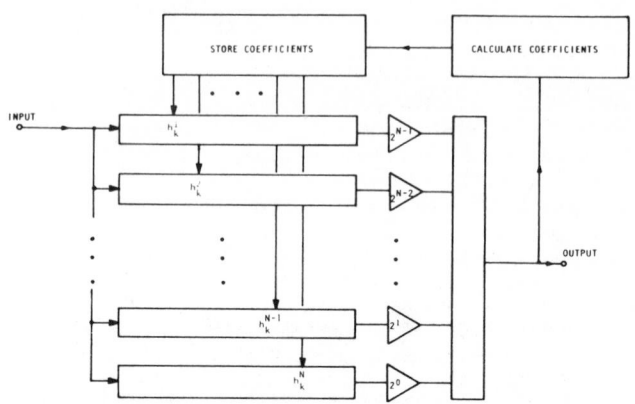

Figure 9. Adaptive channel equalization using the binary/analog filtering scheme. The input signal is processed in analog form whereas the coefficients are handled digitally. The mth binary digits of all the M weighting coefficients, h_k^m, k = 1, M, are supplied to the mth filter from memory.

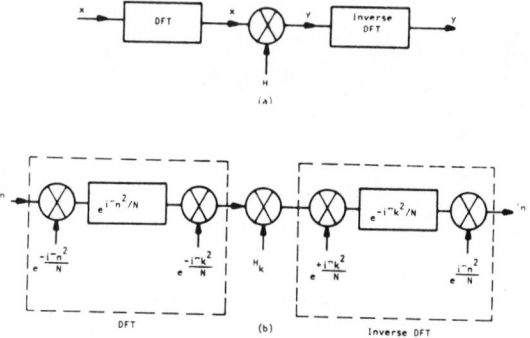

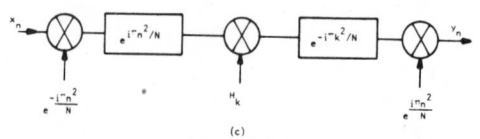

Figure 10. Block diagram of a CZT programmable correlator. The post-multiply operation of the DFT cancels the premultiply of the inverse DFT resulting in the simplified block diagram of (c).

A 500-Stage CCD Transversal Filter for Spectral Analysis

ROBERT W. BRODERSEN, C. ROBERT HEWES, MEMBER, IEEE, AND DENNIS D. BUSS

Abstract—The operation and design of 500-stage charge-coupled device (CCD) transversal filters are described. The filters have been mask programmed for implementing two spectral analysis techniques: 1) bandpass filtering and 2) Fourier analysis using the chirp z transform (CZT) algorithm. The bandpass filter has a measured fractional 3 dB bandwidth of 0.0108 and 38 dB sidelobe rejection. The dynamic range is 75 dB with less than 45 dB total harmonic distortion. A sliding transform is defined which is useful for calculations of the power spectral density and is shown to be particularly advantageous in a CCD–CZT implementation. Using the sliding transform, a 500-point spectrum is calculated using CCD's with resolutions which can be varied from 1-200 Hz. The dynamic range of the power spectral output was measured to be 80 dB. A discussion is given of the performance limitation of a general CCD filter due to the inherent characteristics of the device. The results are evaluated for the 500-stage devices described above and indicate that sample rates from 25 Hz to 10 MHz are possible with a dynamic range approaching 100 dB while retaining high linearity.

I. INTRODUCTION

CHARGE-COUPLED device (CCD) transversal filters are very efficient for implementing sampled-data filtering functions because they operate in the analog domain and perform a large number of multiplications simultaneously in real time. Fig. 1 illustrates what is meant by a transversal filter. The delay stages D represent one clock period of CCD delay. The signal is nondestructively sampled at each stage, multiplied by weighting coefficients h_n, $n = 0, N - 1$, and the products are summed. The technology for implementing such filters is developing rapidly [1]–[6], and it appears inevitable that CCD transversal filters will become increasingly important for sampled-data filtering.

This paper describes a versatile CCD IC which contains two 500-stage transversal filters [7]–[9]. By recoding the filter metallization pattern, a variety of CCD filter characteristics have been implemented. This paper discusses two quite different spectral analysis techniques which utilize the same device: 1) bandpass filtering and 2) Fourier analysis.

Section II discusses the design and fabrication of the CCD IC, the input–output circuitry required for its operation, and its inherent performance limitations. Section III discusses the design and operation of the CCD IC as a transversal bandpass filter, and Section IV discusses its operation as a chirp filter in a chirp z transform (CZT) system for performing Fourier analysis.

Manuscript received August 7, 1975; revised October 18, 1975.
R. W. Brodersen was with Texas Instruments Inc., Dallas, TX 75222. He is now with the Department of Electrical Engineering and Computer Science, University of California, Berkeley, CA 94720.
C. R. Hewes and D. D. Buss are with Texas Instruments Inc., Dallas, TX 75222.

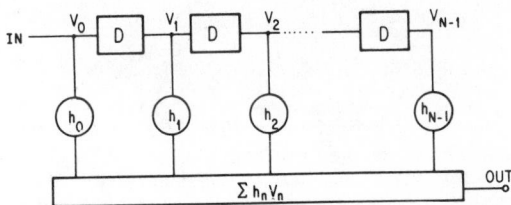

Fig. 1. Block diagram of a transversal filter showing delay stages D and weighting coefficients h_n for $n = 0, N - 1$.

II. DESIGN, OPERATION, AND LIMITATIONS

The 500-stage filter was designed for optimum flexibility and has been coded with three separate photomasks to generate six different filter impulse responses. This section presents the design, operation, and inherent performance limitations of the 500-stage filters with general weighting coefficients. Numerical evaluations are made which are appropriate to the bandpass and chirp filters discussed in Sections III and IV.

A. CCD Design and Fabrication

The IC layout is illustrated by the photograph of the second level metal photomask in Fig. 2. Each 500-stage filter is composed of four 125-stage segments connected by three corners. The photomasks were designed so that either the polysilicon aluminum [10] or the anodized aluminum [11] 4-phase double level process could be used. The bandpass filters discussed in Section III are anodized aluminum, and the chirp filters discussed in Section IV are polysilicon aluminum.

The filter weighting coefficients were implemented using the split electrode technique [1], and the pattern of the gaps in the ϕ_4 electrodes is clearly visible in Fig. 2. The impulse responses for the filters shown are chirp (linear FM) waveforms without windowing (see Section IV).

The devices were made on 50 $\Omega \cdot$ cm p-type Si. Surface channel was chosen because 1) adequate charge transfer efficiency could be obtained with surface channel and 2) surface channel gives higher linearity than buried channel when the split electrode weighting technique is used (see Section II-C). The electrode length is 0.3 mils, and the electrode width is 5 mils.

One of the novel design aspects of this filter is the corner which permits the folding of the two 500-stage filters into an IC of modest dimensions (130×200 mil^2). The corner consists of a diffusion followed by a bias gate as shown at the top of Fig. 3. The surface potential corresponding to a time interval just before transfer through the corner is shown in Fig. 3(a). At this time the signal charge is being stored under the phase 1 (ϕ_1) electrode, and the corner diffusion is at a potential set by the bias gate (the diffusion has charged up to this level by the previous transfer of charge through the corner). In Fig. 3(b),

Reprinted from *IEEE Trans. Electron Devices*, vol. ED-23, pp. 143–152, Feb. 1976.

371

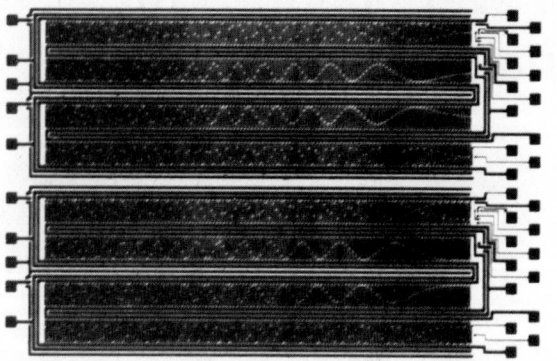

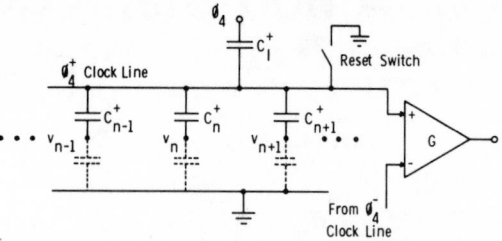

Fig. 4. Circuit schematic of the DCI showing the capacitance of the split electrodes C_n^+ in series with the integrating capacitance C_I^+. The depletion layer capacitances are shown with dashed lines.

Fig. 2. Metal mask for the CCD filter IC. Each of the two 500-stage filters consists of four 125-stage sections connected by three corners. The weighting coefficients which are coded into the split electrodes correspond to chirp waveforms.

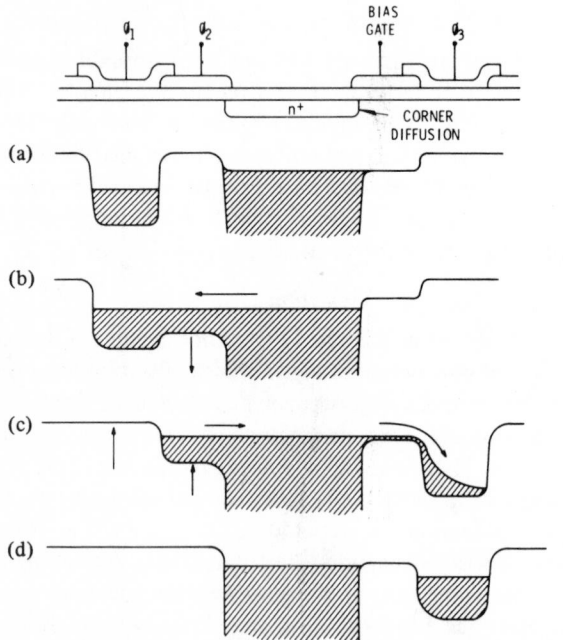

Fig. 3. Cross section of the CCD corner with potential diagrams showing how charge transfers through the corner (see text).

the ϕ_2 clock has turned on, and since the potential is set by the bias gate to be less than the potential resulting from the largest signal, the charge will transfer backwards from the corner diffusion to the ϕ_1 and ϕ_2 gates. In Fig. 3(c), the ϕ_1 clock has turned off and the ϕ_2 clock is collapsing. Charge continues to flow off the diffusion in order to maintain the diffusion potential below the level set by the bias gate. At this point transfer through the corner is complete, and the diffusion is again at the bias level necessary to transfer the next charge packet.

The corner is capable of very fast operation, because the main part of the transfer is by majority carriers in the n⁺ diffusion. The limiting process is the transfer of charge over the bias gate. This electrode can be considered as the gate of a MOSFET which has the corner diffusion as its source and the

receiving ϕ_3 well as an effective drain. Using the standard MOSFET equation [12]

$$I_D = \frac{\beta}{2}(V_{GS} - V_T)^2, \tag{1}$$

it is possible to determine the time it takes for the charge to be transferred from the capacitance of the corner diffusion C_{CD} into the receiving ϕ_3 well. The condition which must be met so that there is a negligible transfer loss due to the corner is

$$T_t q_s \gg \frac{C_{CD}^2}{(\beta/2)}, \tag{2}$$

in which T_t is the time available for transfer and is approximately the time between the turnoff of ϕ_2 and the turnoff of ϕ_3, and q_s is the signal charge. Using $\beta \approx 2 \times 10^{-4}$ A/V², $C_{CD} \approx 0.4$ pF, and q_s greater than 10 percent of a full well, then (2) is satisfied for frequencies up to 10 MHz. On good devices, corner loss cannot be detected experimentally over the frequency range of the tests (500 Hz–5 MHz).

B. Transversal Filter Operation

The use of the split electrode weighting technique places certain constraints on the input and output circuits if good linearity and high dynamic range are to be obtained. The differential current integrator (DCI) which is used for the output of a split electrode filter [1] will be discussed and a discussion of the CCD input technique which is required for overall linearity will follow.

The DCI consists of two capacitive divider networks formed by the split electrodes and integrating capacitors, together with a differential voltage amplifier and reset switches shown in Fig. 4. The N node voltages v_n, where $0 \leqslant n \leqslant N - 1$, represent the surface potentials under the split electrodes at a time in the clock cycle when the signal charge packets reside under the split electrode. The capacitors $C_n^{\pm}$ are formed by the split clock electrodes. They are related to the weighting coefficients h_n and the oxide capacitance of one CCD electrode C_e by

$$C_n^{\pm} = C_e \left(\frac{1 \pm h_n}{2}\right) \tag{3}$$

where $-1 < h_n < +1$, and the + and - signs are associated with the two clock lines supplying the clock signals to the split electrode phase.

The DCI operates in the following way. The ϕ_4 clock pulse is applied to the clock lines of the split phase, ϕ_4^+ and ϕ_4^-, through the series integrating capacitors C_I^+ and C_I^-. The dc level of these clock lines is maintained by a reset switch (see Fig. 4). Superimposed on the clock pulse waveform, there exists a differential signal voltage $v^\pm$ which is due to the node voltages v_n. For filters in which $\sum_n h_n \approx 0$, $v^\pm$ can be expressed as

$$v^\pm = \left(\frac{1}{RN}\right) \sum_{n=0}^{N-1} \left(\frac{1 \pm h_n}{2}\right) v_n \qquad (4)$$

where R is the capacitance ratio

$$R = \frac{C_I^\pm + \frac{N}{2} C_e + C_s}{N C_e} \qquad (5)$$

and C_s is a stray capacitance primarily due to interelectrode capacitance. The signal voltages $v^\pm$ are sampled and held (to eliminate the clocking pulses) and differenced in a differential amplifier with gain G. The node voltages v_n are related to the sampled input signal $v_{\text{in}}(kT_c)$ by

$$v_n(kT_c) = v_{\text{in}}(kT_c - nT_c), \qquad (6)$$

and the output voltage of the differential amplifier is given by

$$v_o(kT_c) = \left(\frac{G}{RN}\right) \sum_{n=0}^{N-1} h_n v_{\text{in}}(kT_c - nT_c). \qquad (7)$$

This illustrates that the transversal filter output is in the form of a sampled-data convolution of the input signal with the filter coefficients h_n [1].

For practical operation, the value of $C_I^\pm$ is selected large enough to obtain sufficient clocking voltage on the $\phi_4^\pm$ lines, but not so large that there is excessive attenuation of the signal voltages $v^\pm$. For the 500-stage filters $C_s \approx N/2\, C_e$ and C_I was selected to be $2NC_e$, resulting in $R = 3$.

In analyzing the DCI it is sometimes convenient to describe the CCD filter as having an insertion loss which is defined as the ratio of the peak amplitude of $(v^+ - v^-)$ to the peak amplitude of the input signal. Insertion loss depends upon the filter impulse response and upon the capacitance ratio R. For the bandpass filter with Hamming windowing [14] (see Section III) the insertion loss is 21 dB, while the unwindowed chirp filter (see Section IV) has 18 dB insertion loss.

The linearity of the overall filter depends on a linear relationship between the input signal voltage and the surface potential associated with each charge packet. This requirement contrasts with most other CCD applications in which a linear relationship is desired between the input voltage and the free electron charge q_f in the CCD [13]. The surface potential under the CCD electrodes is proportional to the total charge $q_t = q_f + q_b$ where q_b is the nonlinear bulk charge arising from the nonlinear depletion capacitance shown in Fig. 4. Although the potential equilibration or "fill and spill" input [13] has the advantages of low noise and threshold insensitivity, it sacrifices the inherent linearity of the split electrode CCD filter. To achieve linearity in a split-electrode filter the input signal is directly applied to an n^+ diode diffusion at the input to the CCD, and an input gate is used to sample the potential during the time when the first transfer electrode in the CCD is on.

C. Performance Limitations

CCD transversal filters of the type discussed above have certain inherent performance limitations due to 1) nonlinearity, 2) noise, 3) weighting coefficient accuracy, 4) sample rate limitations, and 5) charge transfer efficiency. Although ultimate performance has not been experimentally achieved in all cases, the limitations are presented here for a general 500-stage filter, and numerical evaluations are given for a Hamming windowed bandpass filter and an unwindowed chirp filter.

Nonlinear Effects: The operation of the filter as described above effectively eliminates output signal nonlinearities due to the dependence of the depletion layer capacitance on signal size. However, one potential source of nonlinearity is the variation of effective clock voltage with signal amplitude which results from variation of the signal voltage appearing across the integrating capacitors. This variation results in small changes in the depletion layer charge which is nonlinear. However, because of the small size of the maximum signal on the integration capacitors (<0.5 V peak to peak) and the use of high resistivity substrates, this source of nonlinearity gives a total harmonic distortion of less than −70 dB.

This distortion level is significantly lower than the measured levels (see Section III-B). The nonlinearities which actually limit performance are difficult to model but are believed to be clock transients during the input sampling operation, edge effects in the CCD channel, and a nonlinear partitioning of charge under the input sampling gate as it is turned off [13].

As mentioned in Section II-A, split electrode filters implemented with buried channel CCD's are inherently nonlinear. This is because the charge storage area and the depth of the charge depend strongly on the signal size [15]. Therefore, surface channel devices were selected for this work in spite of the fact that buried channel CCD's can operate at higher clock rates and have better transfer efficiency.

CCD Filter Noise: CCD transversal filters have the capability of achieving very wide dynamic range because CCD's are very low noise devices [16]. Also, this noise is effectively reduced even further by the processing gain of the filter [17]. However, because of the insertion loss discussed in Section II-B, the differential amplifier and associated output circuitry in the DCI must have very low noise if the ultimate filter dynamic range is to be achieved, and in fact the filters to be discussed in Sections III and IV were limited by this DCI noise. In this section, however, only those noise sources inherent to the 500-stage CCD filter are discussed in order to determine the limiting achievable dynamic range.

The CCD filter itself (prior to the sample and hold circuits and differential amplifier) has five noise sources: 1) input noise, 2) dark current noise, 3) fast interface state trapping noise, 4) corner noise, and 5) reset noise on the clock lines. The general expression for each of these noise contributions will be obtained for an arbitrary CCD filter, and then evaluated for the case of the bandpass and chirp filters.

Filter noise can be related to the noise voltages associated with the charge packets in the CCD by a modification of (7). The rms output noise voltage ν_o is given by

$$\nu_o = \frac{G}{R} \frac{1}{N^{1/2}} \left[\frac{1}{N} \sum_{n=0}^{N-1} \sum_{m=0}^{N-1} h_n h_m \langle \nu_n \nu_m \rangle \right]^{1/2} \quad (8)$$

where $\langle \nu_n \nu_m \rangle$ is the covariance of the noise voltages or surface potentials associated with the charge packets in the CCD channel. From (8) it can be seen that in general the output noise voltage is a function of both the filter weighting coefficients and the statistics of the various noise sources.

The input noise is limited by the uncertainty in setting the surface potential on the input electrode of capacitance C_e. Assuming this is characterized by kTC noise, the covariance is [16]

$$\langle \nu_n \nu_{n+j} \rangle = \left(\frac{kT}{C_e} \right) \delta_{j,o} \quad (9)$$

where $\delta_{i,j}$ is the Kronecker delta function which is unity for $i = j$ and zero for $i \neq j$. Inserting (9) into (8) gives

$$(\nu_o)_{\text{input}} = \left(\frac{G}{RN^{1/2}} \right) \left[\frac{1}{N} \sum_{n=0}^{N-1} h_n^2 \right]^{1/2} \left(\frac{kT}{C_e} \right)^{1/2} \quad (10)$$

The summation in the square brackets is approximately independent of N, and the $N^{-1/2}$ dependence of $(\nu_o)_{\text{input}}$ results from the processing gain of the filter. In practice the measured noise associated with the diode input technique exceeds the kTC limit by a factor of 2 to 3 due to partitioning of charge under the input gate [18].

The effects of thermal shot noise can be important if the device is to be operated at high temperatures or low frequency. The covariance function for dark current noise is [16]

$$\langle \nu_n \nu_{n+j} \rangle = \left(\frac{q J_D A_e p}{C_e^2 f_c} \right) n \delta_{j,o} \quad (11)$$

where J_D is the average dark current, p is the number of clock phases, and A_e is the CCD electrode area. This yields an rms output noise voltage of

$$(\nu_o)_{J_D} = \left(\frac{G}{R} \right) \left[\frac{1}{N^2} \sum_{n=0}^{N-1} n h_n^2 \right]^{1/2} \left(\frac{q J_D A_e p}{C_e^2 f_c} \right)^{1/2} \quad (12)$$

Trapping noise due to fast surface states is correlated between adjacent charge packets [19] and gives rise to a covariance function of the form [20]

$$\langle \nu_n \nu_{n+j} \rangle = v_{fs}^2 n (2\delta_{j,o} - \delta_{j,-1} - \delta_{j,+1}) \quad (13)$$

where [16]

$$v_{fs}^2 = \frac{q A_e N_{ss} kTp \ln 2}{C_e^2} \quad (14)$$

and N_{ss} is the density of fast interface states $(\text{eV}^{-1} \cdot \text{cm}^{-2})$. The output noise voltage due to fast interface states is

$$(\nu_o)_{N_{ss}} = \left(\frac{G}{R} \right) v_{fs} \left[\frac{1}{N^2} \sum_{n=0}^{N-1} \right.$$

TABLE I
CALCULATED NOISE LEVELS AT $f_c = 100$ kHz

Noise Source	Bandpass Filter	Chirp Filter
Input	7.8 μV	13 μV
Dark Current	1.2 μV	2 μV
Surface States	46 μV	74 μV
Corner	15 μV	25 μV
Reset	42 μV	42 μV
Total	65 μV	90 μV
V_{out} max (peak to peak)	5 V	7 V
Dynamic Range	98 dB	101 dB

$$\left. \cdot n(2h_n^2 - h_n h_{n-1} - h_n h_{n+1}) \right]^{1/2} \quad (15)$$

The noise in the three corners (see Section II-A) results from the uncertainty in the potential of the corner diffusion capacitance C_{CD} which is present at the end of the transfer through the corner. The corner resembles a bucket-brigade stage, and the resulting noise is correlated like bucket-brigade transfer noise [21]. The noise voltage covariance resulting from a corner located at $n = n_c$ is

$$\langle \nu_n \nu_{n+j} \rangle = \begin{cases} 0 & n < n_c \\ \dfrac{kT}{C_e^2} C_{CD}(2\delta_{j,o} - \delta_{j,-1} - \delta_{j,+1}) & n > n_c. \end{cases} \quad (16)$$

This yields an rms output noise voltage of

$$(\nu_o)_{\text{corner}} = \left(\frac{G}{RN^{1/2}} \right) \left[\frac{1}{N} \sum_{n_c} \sum_{n=n_c}^{N-1} \right.$$
$$\left. \cdot (2h_n^2 - h_n h_{n-1} - h_n h_{n+1}) \right]^{1/2} \cdot \left(\frac{kT}{C_e^2} C_{CD} \right)^{1/2} \quad (17)$$

in which n_c indexes over the three values 125, 250, and 375.

Finally the noise associated with resetting the clock line capacitance RNC_e of the split electrode phase by the reset switch is given by

$$(\nu_o)_{\text{reset}} = G \left(\frac{2kT}{RNC_e} \right)^{1/2} \quad (18)$$

The five noise expressions 10, 12, 15, 17, and 18 were evaluated for the 500-stage filters using the following parameters: $A_e = 9.7 \times 10^{-6}$ cm^2, $G = 10$, $R = 3$, $C_e = 0.3$ pF, $C_{CD} = 0.4$ pF, $N_{ss} = 5 \times 10^9$ cm$^{-2} \cdot$ eV^{-1}, and $J_D = 2$ nA $\cdot$ cm^{-2}. The bracketed averages of the noise expressions were evaluated for the bandpass and chirp filter weighting coefficients. Typical values of these averages range from 0.1 to 0.5. The calculated rms output noise voltages from each of the five noise sources are summarized in Table I.

It will be shown in Section III-B that acceptable linearity can be obtained for input signals of up to 5 V peak to peak. Therefore, the maximum achievable dynamic range (defined as maximum peak-to-peak output signal to rms output noise) for the inherent noise sources is 98 dB for the bandpass filter and

101 dB for the chirp filter. This difference is due to the different insertion losses of the two filters (see Section II).

Weighting Coefficient Accuracy: In a split electrode filter the weighting coefficients are coded into the metal mask as gaps in the gate electrode structure, and if higher accuracy is required, channel stop diffusions can be placed under the gaps, thereby making the coefficients insensitive to small photomask misalignments [22]. The predominant weighting coefficient error then becomes the location of these gaps on the photomasks. When computer generation is used these gaps are located at quantized intervals δ. Therefore, the fractional quantization step δ/W, is dependent on the channel width W.

This tap weight quantization is directly analogous to the quantization of filter coefficients in a digital filter [14]. In order to calculate this error it is convenient to express the input signal as an amplitude V_{in} times a sequence g_n which is normalized to unity like the weighting coefficients h_n. The rms error at the filter output is

$$\Delta_{rms} = \frac{G}{R} \frac{\delta/W}{\sqrt{3}\sqrt{N}} \left[\frac{1}{N}\sum_n g_n^2\right]^{1/.} V_{in} \qquad (19)$$

and the peak output signal level is

$$V_o = \frac{G}{R}\left[\frac{1}{N}\sum_n (h_n g_n)\right] V_{in}. \qquad (20)$$

This gives an rms error-to-peak signal

$$\frac{\Delta_{rms}}{V_o} = \frac{\delta/W}{\sqrt{3}} \frac{1}{\sqrt{N}} \frac{\left[\frac{1}{N}\sum g_n^2\right]^{1/2}}{\left[\frac{1}{N}\sum h_n g_n\right]} \qquad (21)$$

which decreases like $N^{-1/2}$ as a result of the lack of correlation between weighting coefficient errors.

For the 500-stage filters, the parameters of (21) have the value $\delta = 0.05$ mil and $W = 5.0$ mil. The term $1/N\sum_n h_n g_n$ is approximately $\frac{1}{4}$ for the windowed bandpass filter and $\frac{1}{2}$ for the unwindowed chirp filter, and in both cases, $1/N\sum_n g_n^2 \approx \frac{1}{2}$. This yields an rms error-to-peak signal of -63 and -69 dB, respectively, for these two filters and corresponds to 7-bit accuracy in an equivalent digital filter. If higher accuracy is required, δ can be made smaller and W can be made wider. Using $\delta = 0.01$ mil [22] and $W = 30$ mil [3], $11\frac{1}{2}$ bit equivalent accuracy can be obtained.

Sample Rate: The operation of transversal filters is limited to sample rates less than about 10 MHz because of the difficulty of providing clock drivers and analog circuitry associated with the filter output as well as the rapid rise in transfer inefficiency which occurs in surface channel CCD filters at data rates above 10 MHz. Bulk channel devices which have higher frequency response (>100 MHz) cannot be used for filters where good linearity is required for reasons which were discussed in Section II-C. In addition at frequencies above 10 MHz other technologies, such as surface wave devices are more suitable.

The lowest sample rate, f_c^{min}, which can be attained for an N-stage filter is limited by the thermally generated dark current J_D. The dark current acts to reduce the dynamic range

since it adds to the signal charge, and the size of the signal must be reduced in order to avoid over filling of the CCD potential wells. If a maximum loss of dynamic range of 6 dB is allowed, then the sample rate must be greater than

$$f_c^{min} = \frac{4J_D N}{q Q_{max}} \qquad (22)$$

where Q_{max} is the maximum signal charge which is determined by linearity considerations. Typical values obtained for the 500-stage filters were $J_D = 2.0$ nA·cm^{-2} and $Q_{max} \approx 1 \times 10^{12}$ cm^{-2} which yields a minimum frequency of 25 Hz.

Charge Transfer Efficiency: The effect of charge transfer efficiency in a transversal filter has the effect of modifying the filter weighting coefficients from their desired values. The resultant effective tap weights h_n' are related to the original weights h_n by [2]

$$h_n' = \sum_{j=0}^{n-1} h_{n-j} \binom{n-1}{j} (p\epsilon)^j (1-p\epsilon)^{n-j} \qquad (23)$$

where ϵ is the loss per transfer.

Using the expression, the modified transfer function of a filter can be calculated. It is apparent that the effect of transfer efficiency strongly depends on the particular filter response under consideration so that separate discussions will be made of its effect on the bandpass and Fourier analysis applications.

III. BANDPASS FILTERING

There exists a vast literature on transversal or finite impulse response (FIR) digital filters for spectral filtering [14], and the basic digital filtering concepts can be applied directly to CCD's. CCD transversal filters offer the advantages of digital filters without the necessity of digitizing the electrical signal. A 500-stage narrow bandpass transversal filter has been designed and tested to illustrate the use of CCD's in this application. The selection of the filter weighting coefficients is discussed, and the measured performance is presented.

A. Narrow Bandpass Weighting Coefficients

The 500-stage bandpass filter presented here was designed using windowing to suppress sidelobes [14]. The weighting coefficients are of the form

$$h_n = w_n \sin(n\pi/2) \qquad 0 \leq n \leq N-1 \qquad (24)$$

which gives a passband center frequency $f_o = f_c/4$. The w_n are Hamming window coefficients which are of the form

$$w_n = 0.54 + 0.46 \cos[2\pi(n - N/2 + \tfrac{1}{2})/N]. \qquad (25)$$

Hamming windowing was selected because it provides a very narrow bandwidth with good sidelobe rejection. The calculated 3 dB bandwidth is 1.04 percent of f_o and the highest sidelobe response is -42.8 dB.

B. Filter Performance

Frequency Response: The measured bandpass characteristic of this filter is shown in Fig. 5. The filter was operating at a clock frequency of 200 kHz with the center frequency at 50 kHz. The measured bandwidth is 540 Hz or 1.08 percent of

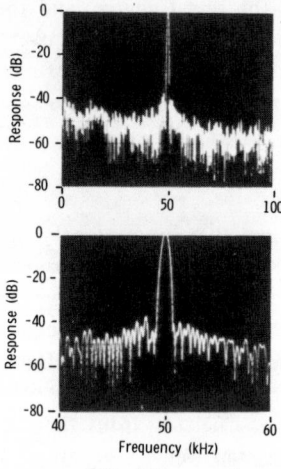

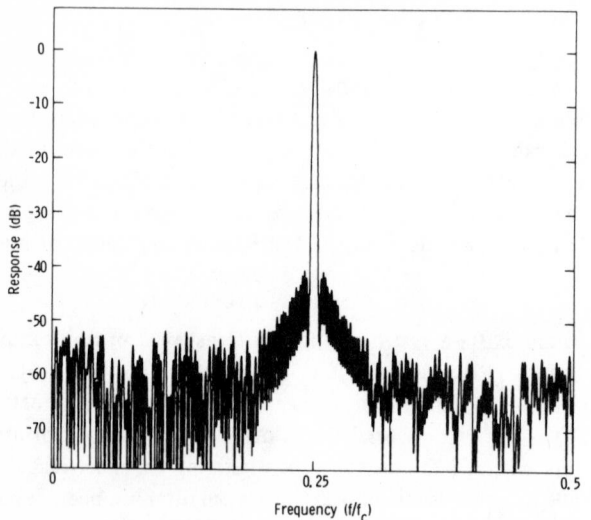

Fig. 5. Frequency response of the 500-stage Hamming windowed band-pass filter at a clock frequency of 200 kHz.

Fig. 6. Results of a computer simulation of the 500-stage bandpass filter, which includes the effects of weighting coefficient quantization and charge transfer inefficiency.

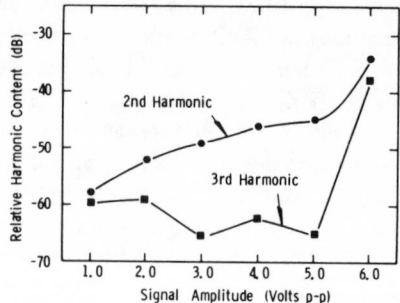

Fig. 7. Harmonic distortion appearing at the filter output as a function of input signal amplitude for a clock rate of f_c = 200 kHz.

the center frequency compared to 1.04 percent for the ideal Hamming weighted 500-stage filter. The highest sidelobe response is -38 dB in contrast to -42.8 for the ideal case. As expected, the response characteristics shown scale with frequency over the range of experimental testing from f_c = 20 kHz to f_c = 2 MHz.

As described in Section II-C, deviations from the ideal response are expected due to the weighting coefficient quantization error and transfer inefficiency. In order to determine the expected response, the filter was modeled with a computer simulation that included weighting coefficient quantization ($\delta/w = 0.01$) and the measured charge transfer inefficiency ($\epsilon = 3 \times 10^{-4}$), and is plotted in Fig. 6. The simulation has a 3 dB bandwidth of 1.08 percent and its highest sidelobe response is -40 dB. The uneven sidelobe structure is due to the weighting coefficient quantization which results in an rms error -63 dB below the passband. Charge transfer efficiency has no significant effect on the bandwidth but it contributes a small frequency dependent attenuation which increases with frequency. At $f = 0$ there is no attenuation, at $f = f_c/4$ the

attenuation increases to -2 dB, and at $f_c/2$ the attenuation is -4 dB.

There is exact agreement between the simulation and the measured bandwidth, and general agreement with the experimental sidelobe response. However, the average sidelobe rejection predicted by the computer simulation is a few decibels lower than the measured response and comparison of Figs. 5 and 6 reveals that the detailed structure is not identical. The major source of these discrepancies is direct coupling of the input signal into the output circuit at a level of -65 dB relative to the passband response.

Linearity: The linearity of the filter was determined by measuring, at f_c = 200 kHz, the harmonics of an input signal generated by the device. The procedure was to first insert an input signal at the passband frequency $f_c/4$ to determine the gain for the passband frequency. Then the input frequency was changed to $f_c/8$ or to $f_c/12$ such that the second or third harmonic of the input signal would lie in the passband of the filter. The observed response to the second and third harmonics relative to the response to the fundamental is plotted in Fig. 7. The total harmonic distortion is less than -45 dB for signal amplitude less than 5 V peak to peak. The observed nonlinearity is believed to be limited by clock transients during the input sampling operation.

Noise: While clocking the CCD at 200 kHz the noise was measured for a 1 kHz bandwidth over the frequency range $0 < f < 100$ kHz. Integrated over a 100 kHz bandwidth and referred to the filter input the rms noise voltage is 900 μV. The measured dynamic range defined as the maximum peak-to-peak signal to the rms noise level is 75 dB, which is approximately 25 dB under the inherent limitations of the device presented in Section II-C. The measured noise was dominated by the sample and hold circuits and the differential amplifier in the output circuit.

IV. FOURIER ANALYSIS USING THE CZT

The power spectrum of a waveform can be calculated using the CZT algorithm [24]. The CZT gets its name from the fact that it is implemented by premultiplying the signal with a chirp (linear FM) waveform which is followed by filtering with chirp filters. To obtain the power spectrum, the outputs of the filters are squared and summed (if the actual Fourier coefficients are required then the outputs are postmultiplied by another chirp waveform). When implemented digitally, the CZT has no particular advantages over the other fast Fourier transform algorithms. However, the CZT lends itself naturally to implementation with CCD transversal filters [25].

A. Conventional CZT

Using the definition of the discrete Fourier transform (DFT)

$$F_k = \sum_{n=0}^{N-1} f_n e^{-\frac{2\pi i n k}{N}} \tag{26}$$

and the substitution [24]

$$2nk = n^2 + k^2 - (n-k)^2 \tag{27}$$

the conventional CZT algorithm for power spectrum can be obtained [26]:

$$|F_k|^2 = \left| \sum_{n=0}^{N-1} (f_n e^{-i\pi n^2/N}) \, e^{i\pi(k-n)^2/N} \right|^2 . \tag{28}$$

This equation has been factored to emphasize the premultiplication and convolution operations which make up the CZT algorithm. Equation (28) can be implemented with filters which have $2N-1$ stages with chirp impulse responses which linearly vary in frequency from $-f_c$ to $+f_c$. The input data are premultiplied by a chirp having frequency from $-f_c/2$ to $f_c/2$ with a duration of NT_c (half of the total filter delay), and the input is then blanked during the next N clock periods, (i.e., multiplied by zero). During the first $N-1$ clock periods the filter is loaded with the premultiplied data, and during the succeeding N clock periods the convolution and squaring operations sequentially yield the magnitude squared of the Fourier coefficients. This approach has the disadvantage of requiring filters of length $2N$ for an N-point transform and is also inefficient in the use of the CCD's since only half of each CCD filter contains useful information at any point in time.

B. Sliding Transform

A modification of the DFT which in a CCD implementation only requires N-stage filters and operates continuously (no blanking of the input data) is called the sliding transform and is defined by

$$F_k^s = \sum_{n=k}^{k+N-1} f_n e^{-2\pi i n k/N} . \tag{29}$$

The difference between this transform and the DFT (26) is that the input data are shifted by one sample each time a spectral component is calculated. For a general waveform, this procedure destroys the phase information. However, since the power spectrum does not contain any phase information the sliding transform yields a power spectral density which is as useful as that which is obtained using the DFT.

To calculate the power spectrum using the sliding CZT, the substitution of (27) is used with (29) to obtain the following expression:

$$|F_k^s|^2 = \left| \sum_{n=0}^{N-1} (f_{k-n+N} \, e^{-i\pi(k-n+N/2)^2/N}) \, e^{i\pi(n-N/2)^2/N} \right|^2 \tag{30}$$

(for convenience in (30) the filter weighting coefficients and premultiplying waveforms are shifted by $N/2$). From (30) it can be seen that the premultiplying waveform $P_{n'}$ multiplying the input data $f_{n'}$ is a chirp from $-f_c/2$ to $f_c/2$ given by

$$P_{n'} = e^{-i\pi(n'-N/2)^2/N} \tag{31}$$

The premultiplying chirp has the property $P_{n'+N} = P_{n'}$ so that blanking of the input data is not required. The filter itself has N stages with weighting coefficients

$$h_n = e^{i\pi(n-N/2)^2/N} \qquad 0 \leqslant n \leqslant N-1 . \tag{32}$$

The subject of windowing discussed in Section III-A applies to Fourier analysis as well. To incorporate windowing into the sliding CZT it is only necessary to multiply h_n of (32) by the desired weighting function [e.g., w_n of (25)].

Fig. 8 gives a pictorial comparison between the conventional CZT and the sliding CZT for the simple case of a 3-point transform. With the conventional CZT all three Fourier coefficients F_0, F_1, F_2 are calculated using the first three time samples (f_1, f_2, f_3). These coefficients are being calculated by the filter during the next three clock periods, so that time samples f_4-f_6 must be blanked. Then the cycle repeats as shown in Fig. 8. Using the sliding CZT, F_0^s is calculated on the sample record (f_1, f_2, f_3) as before, but F_1^s is calculated on the sample record (f_2, f_3, f_4), F_2^s on the record (f_3, f_4, f_5) and the next F_0^s computation is made on the sample record (f_4, f_5, f_6). The sample record is continually updated by replacing the oldest sample with a new one. The above description shows that N Fourier coefficients are obtained for N time samples (100 percent duty cycle).

C. Sliding CZT Implementation and Results

A block diagram of a system to implement the complex arithmetic of (30) is shown in Fig. 9. This system was implemented using the 500-stage filters discussed in Section II-A. The weighting coefficients for the cos and sin chirp filters are

$$h_n^{\cos} = w_n \cos [\pi(n-N/2)^2/N] \tag{33}$$

$$n = 0, N-1$$

$$h_n^{\sin} = w_n \sin [\pi(n-N/2)^2/N] . \tag{34}$$

The weighting coefficients used for the experimental results to be presented were unwindowed $w_n = 1$ although Hamming weighted chirp filters (25) were also fabricated and tested.

The experimental implementation of Fig. 9 utilized two CCD IC's each containing two 500-stage filters. The premultiply chirps were stored with 8-bit precision in ROM's. Multiplication was performed in discrete multiplying digital to analog converters (MDAC). The DCI's required for the filter were implemented in discrete form as discussed in Section II-B, and the squaring operations were performed in analog multipliers.

Performance of the CCD CZT: The CCD chirp filters were processed at a later time from those discussed in Section III, and significantly better CCD performance was obtained. The devices showed an average transfer inefficiency of less than 1×10^{-4} including the effects of the corners.

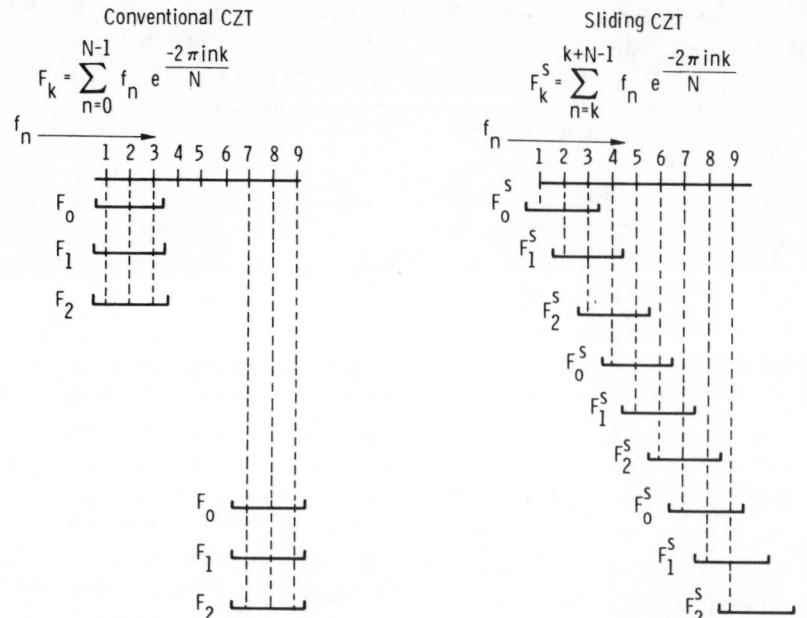

Fig. 8. Comparison between the operation of the conventional CZT and the sliding CZT.

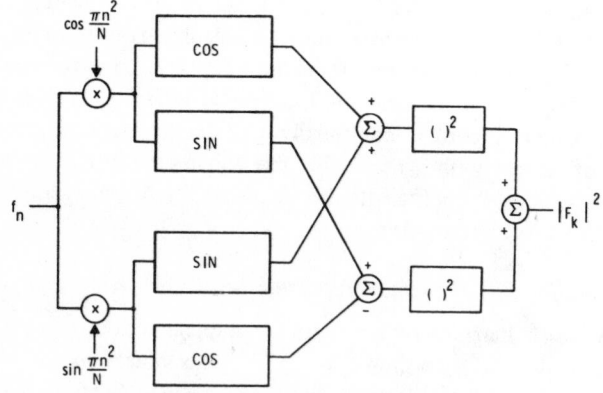

Fig. 9. Block diagram of the complex arithmetic of the CZT algorithm. The CCD transversal filters are designated by COS and SIN.

The CCD filters themselves had similar performance in terms of dynamic range, noise and linearity to that observed with the bandpass filters described in the Section III-B. This performance was sufficiently good that it did not limit the system operation. For example, the system noise level was limited by the squaring devices at the output of the filters, which had an equivalent dynamic range at their input of only 40 dB (which corresponds to 80 dB at their output).

Power spectra were obtained experimentally with rates ranging from 500 Hz to 100 kHz by simply varying the master clock frequency. The range of resolution corresponding to these sample rates was from 1–200 Hz. Equivalent performance was obtained at all frequencies.

In Fig. 10 the output of a sliding 500-point CZT is shown for three different sine wave inputs. The clock rate is 20 kHz so that a complete spectrum is obtained every 25 ms with a resolution of 40 Hz. The time scale is adjusted so that each horizontal division corresponds to 500 Hz. The peak (which consists of one sample) is observed to be delayed as the input frequency increases from 200 Hz to 4500 Hz.

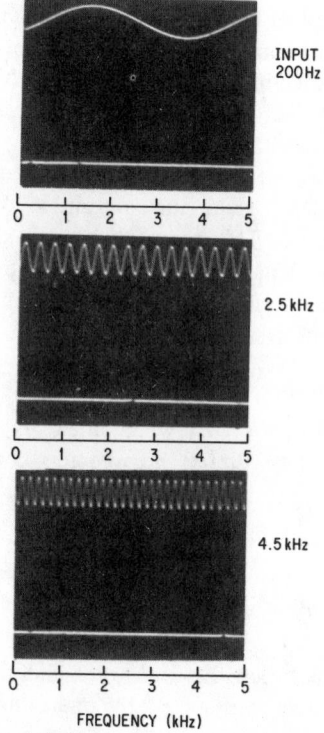

Fig. 10. Power density spectra for three sinewaves obtained using the 500-point sliding CCD CZT at a sample rate of 20 kHz.

In Fig. 11 a more complicated spectrum is shown. The input signal is a 200 Hz square wave which is shown in Fig. 11(a). The sliding transform spectrum is shown in Fig. 11(b) for a sample rate of 25 kHz. The time scale is adjusted so that each division corresponds to 200 Hz which includes four Fourier coefficients. The fundamental of the square wave (labeled f_o) is the largest peak and the 3rd through 9th odd harmonics are seen as peaks which decrease in amplitude as $1/n^2$ (where n is the harmonic index) as expected for the power spectrum

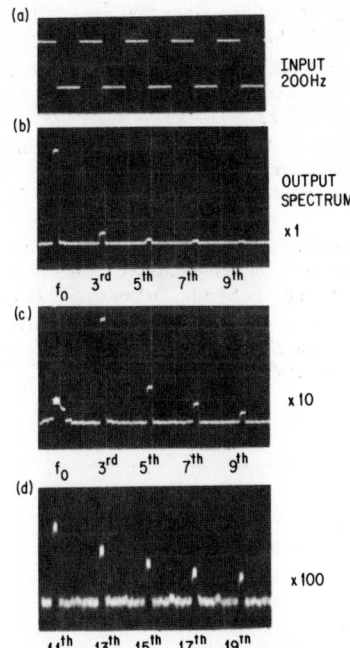

Fig. 11. 500-point power spectrum of a 200 Hz square wave for a sample rate of 25 kHz: (a) input waveform, (b) power spectrum output showing the first nine harmonics, (c) the output amplified by 10 which shows a small trailing pulse due to transfer inefficiency which is pointed out by the arrow, and (d) the 11th through 19th harmonics amplified by 100.

of a square wave. In Fig. 11(c), the amplitude scale has been increased by a factor of 10 and a small trailing pulse is observed (pointed out by an arrow) in the Fourier coefficient following the fundamental f_o. This trailing pulse is due to transfer inefficiency and is approximately equal to $(NP\epsilon/2)^2$ times the amplitude of the previous coefficient. This dispersion, which is similar to that obtained for an analog delay line, has the effect in the CCD CZT of slightly degrading the output resolution. In Fig. 11(d), the amplitude has been increased by a factor of 100 over the scale in Fig. 11(b) and the time scale has been shifted so that the 11th through the 19th odd harmonics can be observed. The noise from the squaring devices can be seen which limits the dynamic range of this particular implementation. However, it is possible to observe up to 49 harmonics before the noise level obscures the output.

V. Conclusions

CCD transversal filters have broad applicability in analog signal processing systems and a basic filter design can be mask programmed to perform a variety of diverse functions in much the same way as ROM's are programmed. The flexibility of CCD transversal filters is illustrated by the filter IC discussed here. Once the basic filter structure has been designed, it can be reprogrammed by a change in the metallization mask which determines the split electrode weighting coefficients. The metallization mask can be generated under computer control at very low cost.

Most of the performance limitations of the 500-stage filters are well understood, and it has been found that in some respects ultimate performance is very good. The ultimate perfor-

mance has not been experimentally achieved in all cases, but measured results are adequate for many applications, and the key areas for further improvement have been identified.

Work reported here has concentrated on the CCD itself, and peripheral circuitry has been designed with discrete components. Future work will be directed toward integrating the peripheral circuitry [22] (clock drivers, multipliers, DCI, etc.) on the CCD IC in order to make CCD signal processing compact and simple from the user's standpoint.

Acknowledgment

The authors would like to acknowledge the cooperative efforts of L. R. Hite, in designing the CCD, A. F. Tasch, Jr. and H. S. Fu in processing the devices, and F. G. Wall in designing the test electronics.

References

[1] D. D. Buss, D. R. Collins, W. H. Bailey, and C. R. Reeves, "Transversal filtering using charge-transfer devices," *IEEE J. Solid-State Circuits (Special Issue on Charge-Transfer Device Applications)*, vol. SC-8, pp. 138–146, Apr. 1973.

[2] D. D. Buss and W. H. Bailey, "Application of charge transfer devices to communication," in *Proc. CCD Applications Conf.*, San Diego, CA, Sept. 1973, pp. 83–93.

[3] R. D. Baertsch, W. E. Engeler, H. S. Goldberg, C. M. Puckette, and J. J. Tiemann, "Two classes of charge transfer devices for signal processing," in *Proc. Int. Conf. Technology and Applications of CCDs*, Edinburgh, Scotland, Sept. 1974, pp. 229–236.

[4] A. Ibrahim and L. Sellars, "CCDs for transversal filter applications," in *IEEE Int. Electron Devices Meeting, Tech. Dig.*, Washington, DC, Dec. 1974, pp. 240–243.

[5] J. A. Sekula, P. R. Prince, and C. S. Wang, "Non-recursive matched filters using charge-coupled devices," in *IEEE Int. Electron Devices Meeting, Tech. Dig.*, Washington, DC, Dec. 1974, pp. 244–247.

[6] M. F. Tompsett, A. M. Mohsen, D. A. Sealer, and C. H. Séquin, "Design and characterization of CCD's for analog signal processing," in *IEEE Advanced Solid-State Components for Signal Processing*, IEEE Int. Symp. Circuits and Systems, Newton, MA, Apr. 1975, pp. 83–89.

[7] R. W. Brodersen, H. S. Fu, R. C. Frye, and D. D. Buss, "A 500-point Fourier transform using charge-coupled devices," in *1975 IEEE Int. Solid-State Circuits Conf., Dig. Tech. Papers*, Philadelphia, PA, Feb. 1975, pp. 144–145.

[8] C. R. Hewes, R. W. Brodersen, and D. D. Buss, "Frequency filtering using charge-coupled devices," in *Proc. 29th Annu. Frequency Control Symp.*, Atlantic City, NJ, May 1975.

[9] R. W. Brodersen, C. R. Hewes, and D. D. Buss, "Spectral filtering and Fourier analysis using CCD's," in *IEEE Advanced Solid-State Components for Signal Processing*, IEEE Int. Symp. Circuits and Systems, Newton, MA, Apr. 1975, pp. 43–68.

[10] W. F. Kosonocky and J. E. Carnes, "2ϕ CCD's with overlapping poly-Si and Al gates," *RCA Rev.*, vol. 34, pp. 164–202, 1973.

[11] D. R. Collins, S. R. Shortes, W. R. McMahon, R. C. Bracken, and T. C. Penn, "Charge-coupled devices fabricated using Al-anodized Al-Al double level metalization," *J. Electrochem. Soc.*, vol. 120, pp. 521–526, Apr. 1973.

[12] For a definition of the terms in (1), see R. A. Crawford, *MOSFET in Circuit Design*, Texas Instruments Electronics Series. New York: McGraw-Hill, 1967.

[13] C. H. Sequin and A. M. Mohsen, "Linearity of electrical charge injection into charge-coupled devices," *IEEE J. Solid-State Circuits*, vol. SC-10, pp. 81–92, Apr. 1975.

[14] For a discussion of the digital filtering concepts used in this paper, see L. R. Rabiner and B. Gold, *Theory and Application of Digital Signal Processing*. Englewood Cliffs, NJ: Prentice-Hall, 1975.

[15] R. W. Brodersen and S. P. Emmons, "Noise in buried channel charge-coupled devices," this issue, pp. 215–223.

[16] J. E. Carnes and W. F. Kosonocky, "Noise sources in charge-coupled devices," *RCA Rev.*, vol. 33, pp. 327–343, June 1972.

[17] S. P. Emmons and D. D. Buss, "The performance of charge-coupled devices in signal processing at low signal levels," in *Proc. CCD Applications Conf.*, San Diego, CA, Sept. 1973, pp. 189–205.

[18] ——, "Noise measurements on the floating diffusion input for charge-coupled devices," *J. Appl. Phys.*, vol. 45, pp. 5303–5306, Dec. 1974.

[19] K. K. Thornber and M. F. Tompsett, "Spectral density of noise generated in charge transfer devices," *IEEE Trans. Electron Devices* (Corresp.), vol. ED-20, p. 456, Apr. 1973.

[20] D. D. Buss, W. H. Bailey, and D. R. Collins, "Analysis and applications of analog CCD circuits," in *Proc. 1973 Int. Symp. Circuit Theory*, Toronto, Ont., Canada, Apr. 1973, pp. 3–7.

[21] D. D. Buss, W. H. Bailey, and W. L. Eversole, "Noise in bucket-brigade devices," *IEEE Trans. Electron Devices*, vol. ED-22, pp. 977–981, Nov. 1975.

[22] C. R. Hewes, "A self-contained 800 stage CCD transversal filter," in *Proc. CCD '75*, San Diego, CA, Oct. 1975, pp. 309–318.

[23] L. J. M. Esser, M. G. Collet, and J. G. Van Santen, "The peristaltic charge-coupled device," in *1973 Int. Electron Devices Meeting, Tech. Dig.*, pp. 17–22.

[24] L. R. Rabiner, R. W. Schafer, and C. M. Rader, "The chirp z-transform algorithm," *IEEE Trans. Audio Electroacoust. (Special Issue on Fast Fourier Transform)*, vol. AU-17, pp. 86–92, June 1969.

[25] H. J. Whitehouse, J. M. Speiser, and R. W. Means, "High speed serial access linear implementations," presented at All Applications Digital Computer Symp., Orlando, FL, Jan. 1973.

[26] R. W. Means, D. D. Buss, and H. J. Whitehouse, "Real time discrete Fourier transforms using charge transfer devices," in *Proc. CCD Applications Conf.*, San Diego, CA, Sept. 1973, pp. 95–101.

COMPARISON BETWEEN THE CCD CZT AND THE DIGITAL FFT

D. D. Buss, R. L. Veenkant, R. W. Brodersen, and C. R. Hewes

Texas Instruments Incorporated
Dallas, Texas 75222

ABSTRACT. The CCD analog transversal filter is a tremendously cost-effective component in terms of its simplicity compared to equivalent digital hardware. In view of this, the chirp z-transform (CZT) algorithm for performing spectral analysis is ideally suited to CCD implementation because, in this algorithm, the bulk of the computation is performed in a transversal filter. The CCD CZT has some performance limitations relative to the digital fast Fourier transform (FFT), and for this reason, it is not applicable to all military signal processing systems. However, for those applications which fall within the CCD performance capabilities, the CZT offers significant potential cost saving over the digital FFT. The performance of the CCD CZT is evaluated and compared with the digital FFT. A discussion is given of selected military applications where the CCD CZT can be used to advantage.

I. INTRODUCTION

Charge-coupled device (CCD) analog transversal filters can be used to perform the discrete Fourier transform (DFT) on electrical signals using an algorithm called the chirp z-transform (CZT).[1,2] The CCD CZT has been demonstrated,[3,4] and it has been proposed as a replacement for the digital fast fourier transform (FFT), in a number of military signal processing applications.[5] It is the purpose of this paper to compare the CCD CZT with the digital FFT.

In a general way, the comparison can be summarized as follows. The CCD CZT has performance limitations when compared with the digital FFT, and it is somewhat less flexible. However, it has a tremendous projected cost advantage when manufactured in high volume, and has additional advantages in smaller size, lighter weight, lower power and improved reliability. Although the CCD CZT has modest performance, significant cost advantages can be realized for those applications which fall within its performance capabilities. It is the goal of this paper to quantitatively compare the CCD CZT with the digital FFT and to identify a few major military applications in which the CCD CZT is certain to be important.

Section II reviews conventional digital spectrum analysis techniques; the digital FFT and the deltec spectrum analyzer. Section III discusses the CCD CZT and a related prime transform. Section IV compares the CCD CZT with the digital FFT in error, resolution and implementation, and Section V discusses selected military applications of modest required performance for which the CCD CZT has clear cost advantages over digital implementation.

II. REVIEW OF DIGITAL SPECTRAL ANALYSIS METHODS

The FFT is currently the most widely used technique for digital spectral analysis. It is important because it requires only $N/2 \log_2 N$ complex multiply operations to perform an N-point DFT, as compared with N^2 complex multiply operations required to directly implement the DFT formula

$$F_k = \sum_{n=0}^{N-1} f_n e^{-i2\pi nk/N} \qquad (1)$$

The deltec spectrum analyzer historically preceeded the FFT, and has been largely replaced except in certain sonar spectrum analysis applications.

Reprinted from *Proc. 1975 Naval Electron. Lab. Center Int. Conf. on the Application of Charge-Coupled Devices*, Oct. 29-31, 1975, pp. 267-281.

FFT

The FFT algorithm is discussed from an elementary point of view in References 6, 7 and 8. In this section, it will be reviewed for the purposes of establishing nomenclature for discussion in later sections.

FFT algorithms can be classed as decimation in time or decimation in frequency. The former is discussed here, but the latter is quite similar. The most common FFT algorithms are radix 2, meaning that the entire FFT is performed by sequential operations involving only pairs of elements. The fundamental operation in a radix 2 FFT is the so-called butterfly which takes two complex inputs A and B and combines them to give X and Y through the operation

$$X = A + W_N^k B$$
$$Y = A - W_N^k B$$
(2)

where $W_N \equiv \exp[-i2\pi/N]$, and the W_N^k are the so-called twiddle factors. The butterfly structure is indicated schematically in Figure 1a, using the the notation of Reference 8.

An 8-point, radix 2, decimation-in-time, FFT algorithm is illustrated in Figure 1b. The data are first reordered by bit reversal,[8] and the first set of butterflies essentially performs the 2-point DFT on the reordered input data by pairs. The second set of butterflies combines the 2-point DFT's using twiddle factors to achieve two 4-point DFT's on the even and odd numbered input data. The final set of butterflies combines the 4-point DFT's using twiddle factors to achieve the final 8-point DFT.

Several important facts about radix 2 FFT's are apparent from the above discussion: (1) N must be an integral power of two. (2) There are $\log_2 N$ stages each requiring N/2 butterfly operations for a total of N/2 $\log_2 N$ butterflies required, and (3) Each Fourier coefficient is processed through $\log_2 N$ butterfly operations so that quantization errors are cumulative.

In many applications, the FFT is used to obtain a transform which will later be inverted to obtain the original signal. In such applications, the phase of the transform must be maintained. However, in many other cases it is desired to obtain a measure of the spectral energy density of a quasi-periodic waveform or a quasi- stationary random process. For these applications, the phase of the transform is not important. In addition, the true DFT is not performed, but the input data are apodized or windowed by an appropriate windowing function w_n to suppress frequency sidelobes in the spectrum.[6,7,8] The desired result is

$$\left| F_k \right|^2 = \left| \sum_{n=0}^{N-1} f_n w_n e^{-i2\pi nk/N} \right|^2$$
(3)

For even moderate values of N the computational saving of the FFT algorithm is extremely significant, and for common spectrum analysis applications the savings is easily a factor of 10 to 100. The FFT is flexible enough and powerful enough to be considered a general purpose spectrum analysis tool.

DELTIC SPECTRUM ANALYZER

The deltic spectrum analyzer has been implemented in analog, digital and hybrid technologies. It is essentially a straightforward implementation of equation (1) in which the time series f_n is stored in a circulating delay line, usually in digital form. A complete circulation of the time sequence is performed in the time between each newly acquired input sample. During this circulation, the f_n are operated on either digitally or analog to form a spectral estimate at a selected frequency. To obtain a complete spectrum at N frequencies requires N^2 operations. However, N is not restricted to be a power of 2, and the frequencies are not restricted to be of uniform spacing as in the DFT. Often, one-bit quantization is used to represent the f_n. This minimizes hardware requirements but results in significant detection loss and small signal suppression effects.[9]

III. CCD SPECTRAL ANALYSIS TECHNIQUES

From the standpoint of minimizing the number of digital operations required to perform the DFT, the FFT algorithm is optimal. However, in determining the optimum algorithm for implementation with analog CCD's, a whole new set of ground rules exists. It is no longer important to minimize multiplications, because CCD transversal filters can be built which perform a large number of multiplications simultaneously in real time.[10,11,12,13,14,15] Consequently, for

CCD implementation, algorithms should be selected in which the bulk of the computation is performed by a transversal filter. Two such algorithms have been identified for cost-effective CCD implementation; the CZT[1,2] and the prime transform.[16,17] Both algorithms are discussed in this section.

CCD CZT

The DFT can be performed using the chirp z-transform (CZT) algorithm.[2] The CZT gets its name from the fact that it can be implemented by (1) premultiplying the time signal with a chirp (linear FM) waveform, (2) filtering in a chirp convolution filter, and (3) postmultiplying with a chirp waveform. When implemented digitally, the CZT has no clear cut advantages over the conventional FFT algorithm.[2] However, the CZT lends itself naturally to implementation with CCD transversal filters.[1]

Starting with the definition of the DFT given in equation (1), and using the substitution

$$2nk = n^2 + k^2 - (n - k)^2 \qquad (4)$$

the following equation results:

$$F_k = e^{-i\pi k^2/N} \cdot$$

$$\sum_{n=0}^{N-1} \left(f_n e^{-i\pi n^2/N} \right) e^{i\pi(k - n)^2/N} \qquad (5)$$

This equation has been factored to emphasize the three operations which make up the CZT algorithm. It is illustrated in Figure 2.

To implement the conventional N-point CZT, the CCD filters are chirp filters of length 2N-1 which chirp from $-f_c$ to $+f_c$, and the premultiply waveform has a time duration N/f_c and chirps from zero to $-f_c$. A physical interpretation in terms of correlation of the input chirp with the filter is given in Figure 3. When the input signal has zero frequency, the product with the premultiply chirp results in an input waveform to the filter which chirps from 0 to $-f_c$. The samples corresponding to frequencies near $f = 0$ are clocked into the filter first, and those near $f = -f_c$ are clocked in last. This sequence of samples results in a correlation peak at $t = t_o$, when the product wave-

form has been clocked into the first half of the filter. When the input frequency is $f_1 \neq 0$, the product with the premultiply chirp results in an input to the filter which chirps from f_1 to $-f_c + f_1$. The input waveform (V_{in} x chirp) in Figure 3 corresponds to an input signal at a frequency f_1 at time $t = t_o$. This waveform is shifted to the right as t increases resulting in a correlation peak at t_1. The shift in time relative to the dc correlation peak is

$$t_1 - t_o = \frac{N}{f_c^2} f_1 \qquad (6)$$

In this way, the time axis of the output is calibrated in frequency. The postmultiply is needed to obtain the proper phase of the DFT coefficients and can be omitted when phase is not required.

Several undesirable features of this implementation become apparent from the above description. The output must be blanked during the loading of the chirp into the filter, and the input must be set to zero during the calculation of the coefficients. Also undesirable is the inefficient use of the CCDs since only half of the CCD filter has useful information at any point in time.

For DFT applications, such as video bandwidth reduction,[18] in which the transform is to be inverted to regain the original signal, the CZT is performed in the way described above. However, when the spectral density is required, the CZT can be simplified greatly. Using the substitution of equation (4) in equation (3) gives

$$|F_k|^2 = \left| \sum_{n=0}^{N-1} f_n w_n e^{-i\pi n^2/N} \cdot \right.$$
$$\left. e^{i\pi(k-n)^2/N} \right|^2 \qquad (7)$$

In this case, simplification of the CZT algorithm results from two observations: (1) The postmultiply operation can be eliminated and (2) The sliding CZT can be used.

The sliding DFT is defined in this paper to be

$$F_k^s = \sum_{n=k}^{k+N-1} f_n e^{-i2\pi nk/N} \qquad (8)$$

and it gives a windowed power density spectrum

$$\left| F_k^s \right|^2 = \left| \sum_{n=0}^{N-1} f_{n+k} w_n e^{-i2\pi nk/N} \right|^2 \qquad (9)$$

$$= \left| \sum_{n=0}^{N-1} f_{n+k} w_n e^{-i\pi n^2/N} \right.$$

$$\left. e^{i\pi(k-n)^2/N} \right|^2 \qquad (10)$$

Comparison of equations (9) and (10) with equations (3) and (7) indicates that the sliding CZT differs from the conventional CZT in that the sliding CZT indexes the data each time a spectral component is calculated. For a periodic waveform, indexing results in a phase factor which does not affect the result, and for a stationary random signal, the time record is different for each spectral component but stationarity insures that the result is unaffected. For these two classes of signal the sliding CZT gives the same result as the conventional CZT.

Figure 4 gives a pictorial comparison between the conventional CZT and the sliding CZT for the simple case of a 3-point transform. With the conventional CZT, all three Fourier coefficients F_0, F_1, F_2 are calculated using the first three time samples f_1, f_2, f_3. These coefficients are being calculated by the filter during the next three clock periods, so that time samples $f_4 - f_6$ must be blanked. Then the cycle repeats as shown in Figure 4a. Using the sliding CZT, F_0^s is calculated on the sample record f_1, f_2, f_3 as before, but F_1^s is calculated on the sample record f_2, f_3, f_4, F_2^s on the record f_3, f_4, f_5, and the next F_0^s computation is made on the sample record f_4, f_5, f_6. The sample record is continually updated by replacing the oldest sample with a new one. The above description shows that N Fourier coefficients are obtained for N time samples (100% duty cycle).

The advantages of the sliding CZT are (1) For an N-point transform, N-stage filters are required which chirp through a bandwidth f_c ($-f_c/2$ to $+f_c/2$ for example). (2) No blanking is required. The filters operate with 100% duty cycle; i.e., one spectral component out for each time sample in. (3) Windowing can be achieved by weighting the chirp impulse response of the filter with the desired window function. (4) The degradation due to imperfect charge transfer efficiency is less for the sliding CZT than for the conventional CZT.

The block diagram for obtaining the spectral density using the sliding CZT is shown in Figure 5. The rectangles represent CCD filters having impulse responses $w_n \cos \pi n^2/N$ and $w_n \sin \pi n^2/N$, $-N/2 < n < N/2-1$. This system has been implemented using 500-stage CCD filters. The window function is coded into the metal photomask and systems have been demonstrated both without windowing ($w_n = 1$) and with Hamming windowing.[5] Spectra obtained using Hamming windowing are shown in Figure 6.

CCD PRIME TRANSFORM

Another algorithm exists for computing the DFT which is also suitable to CCD implementation in the analog domain because the bulk of the computation is performed in a transversal filter.[16,17] The prime transform algorithm is implemented in 3 steps as indicated in Figure 7; (1) permutation of the input data, (2) Transversal filtering and (3) permutation of the output Fourier coefficients. (for details see Reference 17) The advantages of the prime transform over the CZT are (1) the multipliers are replaced by permuting memories. (This may or may not be an advantage depending upon the speed and dynamic range required) and (2) For a real input, only two filters are required instead of the four shown in Figure 5. The disadvantages are (1) the zero order Fourier coefficient (dc term) must be computed separately (2) Imperfect charge transfer efficiency does not result in a simple degradation in resolution as it does in the CZT and (3) The sliding DFT cannot be implemented with the prime transform.

The CCD prime transform has not yet been tested but preliminary estimates suggest that it may be important for applications in which (1) the phase of the DFT is required thus ruling out the sliding CZT and (2) high speed and high dynamic range make on-chip multipliers difficult to implement.

IV. PERFORMANCE COMPARISONS

In this section some important comparisons are made between the CCD CZT and the digital FFT. Perhaps the most important point of comparison is accuracy. CCD's, being analog, limit the accuracy of the CCD CZT. The major CCD limitations are (1) Charge transfer efficiency (CTE) (2) thermal noise (3) accuracy of the filter weighting coefficients (4) accuracy of the pre and post multipliers and (5) linearity of the CCD filters. With proper operation of the CCD filters it is expected that CCD operation can be made sufficiently linear that errors due to nonlinearities are not important. Each of the first four limitations is discussed in this section in the context of digital FFT comparison.

CTE affects the resolution of the DFT and is discussed separately. Thermal noise, weighting coefficient accuracy and multiplier accuracy are best discussed on terms of rms error in the transform. These sources of error are evaluated and related to the number of bits required to achieve equivalent error in an FFT. Finally, a state-of-the-art custom I^2L FFT is discussed and compared with the CCD CZT.

CTE

The effect of imperfect CTE is somewhat different for the conventional CZT than for the sliding CZT. In this section the sliding transform will be discussed.

The CZT is of course a sampled-data system, and the DFT is only defined for integral values of k. It is useful, however, to treat k as a continuously varying envelope which determines the spectral value at each integral k. Assuming the input is a complex sinusoid at frequency f, then equation (3) gives

$$\left|F_k\right|^2 = \left|\sum_{n=0}^{N-1} w_n e^{-i2\pi n\left(k - \frac{Nf}{f_c}\right)/N}\right|^2 \quad (11)$$

We can treat F as a function of continuously varying k by writing

$$\left|F(k)\right|^2 = G\left(k - \frac{Nf}{f_c}\right) \quad (12)$$

where the envelope function G(k) is the transform of the window function

$$G(k) = \left|\sum_{n=0}^{N-1} w_n e^{-i2\pi nk/N}\right|^2 \quad (13)$$

The frequency resolution, sidelobe level etc. are all determined by G(k), and the effect of imperfect CTE is to broaden and shift the peak of G(k).

Equation (11) can be rewritten as

$$\left|F_k\right|^2 = \left|\sum_{n=0}^{N-1} e^{-i\pi(n+k-N\ f/f_c)^2/N}\ h_n\right|^2 \quad (14)$$

where h_n is the ideal impulse response of a windowed chirp filter

$$h_n = w_n e^{i\pi n^2/N} \qquad n=0, N-1 \quad (15)$$

Imperfect CTE modifies the weighting coefficients to h_n' and changes the envelope G'(k) to

$$G'(k) = \left|\sum_{n=0}^{N-1} e^{-i\pi(n-k)^2/N}\ h_n'\right|^2 \quad (16)$$

Physically G(k) represents the correlation of a chirp waveform in a windowed chirp filter and equation (16) shows how the correlation response degrades as the filter decorrelates due to imperfect CTE. The decorrelating effect can be crudely estimated by scaling the clock frequency down by a factor on the order of $1 - \varepsilon$, where ε is the fractional loss per stage.[11] This has the effect of decreasing the df/dt of the chirp filters by a similar factor with the result that the correlation peak broadens due to mismatch and shifts by an amount $\Delta t \sim N\varepsilon/2f_c$. This approximate behavior is confirmed by the calculations of Figure 8 which show the response of the 500-point sliding CZT with Hamming windowing to an input sinusoid having frequency $f = 3.3\ f_c/N$.

Conclusions regarding the effect of imperfect CTE on the sliding CZT are summarized below. (1) The resolution degradation is on the order of $N\varepsilon$ times the ideal resolution f_c/N. For $N\varepsilon \sim .1$, the degradation is negligible. (2) The degradation is the same for all frequencies whereas in the conventional CZT, the degradation is three times worse for the high frequencies than for the low frequencies, (3) The degradation can be eliminated by modifying the filter coefficients to compensate for imperfect CTE or by modifying the premultiply chirp.

ERROR ANALYSIS

The accuracy which can be achieved using the CCD CZT is perhaps the most important performance criterion. In this section, the error sources are identified and evaluated in terms of bits in an equivalent digital FFT.

FFT Accurancy. Finite word length effects in a digital FFT fall into three categories.[8,19,20] (1) Quantization of the data at the input A/D converter. (It is assumed that no external gain control is used with either the FFT or the CCD CZT). (2) Errors due to the finite word lengths used to represent the twiddle factors.[19] (3) Truncation and roundoff effects generated within the butterflies.[20] In treating truncation effects a block floating point truncation algorithm is assumed in which all words are shifted right each time overflow occurs in any butterfly.

If the A/D converter has an accuracy of b_1 bits plus a sign bit, the quantization step size is 2^{-b_1} normalized to unity, and an rms error can be defined by

$$\Delta_Q = \left[\frac{1}{N} \sum_{n=0}^{N-1} \left(f_n - f_n^Q \right)^2 \right]^{\frac{1}{2}} \qquad (17)$$

$$= 2^{-b_1} / \sqrt{12} \qquad (18)$$

This noise does not scale with signal size. It represents a noise level below which signals cannot be processed. It dominates at low signal levels but other errors which scale with signal size dominate FFT error at large signal size.

If the twiddle factors are quantized to b_2 bits plus sign, the resulting rms error is given by[19]

$$\Delta_T = \frac{\log_2 N}{6} \ 2^{-b_2} \ \sigma_F \qquad (19)$$

where σ_F is the rms level of the output signal and is related to the rms input signal σ_f through $\sigma_F = \sqrt{N} \ \sigma_f$.

The most important source of error in a digital FFT is usually overflow and roundoff of data words during butterfly computation. If the data words are carried with b_3 bits plus sign, an upper bound on error in a block floating point machine can be determined assuming overflow occurs at every stage. The result is[20]

$$\Delta_B = .3 \ \sqrt{8} \ N^{\frac{1}{2}} \ 2^{-b_3} \qquad (20)$$

If the twiddle factors are quantized to the same accuracy as the data words ($b_2 = b_3$) Δ_B dominates FFT error. Although equation (20) does not contain the input signal size explicitly. Δ_B does scale in a general way with signal because for smaller signals, overflow does not occur at every stage. The dependence of Δ_B on the length of the transform indicates that higher accuracy (large b_3) is required for longer transforms.

CCD CZT Accuracy. The sources of error in a CCD CZT are (1) thermal noise, (2) quantization of the pre and post multiply chirp waveforms, (3) weighting coefficient error in the CCD transversal filters, and (4) CTE. When the criterion of rms error to rms signal is applied, imperfect CTE generates large errors, because the errors add coherently. Because of this fact, however, CTE effects can be treated as a resolution degradation as discussed above and not as "random" error.

Thermal noise is analogous to input quantization in a digital FFT because it generates an error which is independent of signal size. Thermal noise in the CCD CZT is dominated by noise in the output amplifier of the filter.[5] Assuming the rms noise referred to the input is 60 dB below the maximum peak signal, the equivalent quantization accuracy is $b_1 = 8$ bits plus sign. If 80 dB can be achieved[5] this will correspond to $b_1 = 11\frac{1}{2}$ bits plus sign. At higher signal levels thermal noise, like input quantization noise in a digital FFT, is dominated by signal dependent errors.

In implementing the CCD CZT, the pre and post multiply chirp waveforms can be stored digitally in a ROM and multiplication can be performed in multiplying D/A converters.[21] If the waveforms are stored with an accuracy of b_4 bits plus sign, the errors are analogous to twiddle factor quantization in a digital FFT. The calculated rms output levels Δ_M for both the pre and post multipliers is

$$\Delta_M = \frac{2^{-b_4}}{\sqrt{12}} \ \sigma_F \qquad (21)$$

In Figure 9 the rms error to rms signal (Δ_M/σ_F) is plotted as b_4. The solid line is calculated using equation (21). The points are obtained from computer simulation of a 32-point CCD CZT. The input data were normally distributed random numbers. However, similar results are obtained using sine waves.

Weighting coefficient error arises from a number of sources, but let us assume as a model, that the placement of the gap in the split electrodes is quantized in steps of δ during photomask fabrication. δ is typically 10 μin and the channel width W is typically 5 mil giving $\delta/W = .002$ this is equivalent to quantizing the weighting coefficients to 8 bits plus sign and is again analogous to twiddle factor quantization. The error which results is given by

$$\Delta_W = \frac{1}{\sqrt{6}} \; \frac{\delta}{W} \; \sigma_F \qquad (22)$$

A plot of this expression together with computer simulated results are given in Figure 10.

Comparison. The major source of FFT error increases like $\sqrt{N}$ (equation 20) whereas the major sources of CCD CZT error are independent of transform length.

Figure 11 compares the digital FFT with the CCD CZT using as the criterion the ratio of rms error to rms signal for large signals. The results are plotted as a function of the transform length N. A word length of 13 bits plus sign was assumed for the data and twiddle factors and on this case Δ_B dominates FFT error. For the CCD CZT, the multiplying chirp waveforms are quantized to 7 bits plus sign and $\delta/W = .002$.

CUSTOM FFT

In evaluating the potential of the CCD CZT, it is important to compare it, not with currently existing digital implementations of the FFT, but with projected state-of-the-art implementations which are under development. A potential digital competitor for the analog CCD spectrum analyzer is a custom FFT, and a low-power LSI FFT system using all I^2L technology is presented here for comparison.

A few introductory statements concerning the state of I^2L and the design philosophy of the custom FFT presented here are appropriate. I^2L is a very low-power, high density technology currently in early stages of product development. Current devices are running at speeds up to 2 MHz, but it is expected that speeds will improve significantly as the technology matures. The hypothetical FFT design presented here is custom in the sense that the architecture is tailored to perform the algorithm relatively efficiently and has matched the memory and computation speeds. None of the chips have actually been built, but are believed to be within the state-of-the-art. Flexibility normally expected in an FFT is provided, and with additional firmware, other vector oriented algorithms could be implemented. Therefore, this design can be considered as a general purpose digital signal processing module.

Table 1 lists the specifications of a hypothetical FFT processor.

Table 1.   FFT Specifications

Algorithm	Inplace
Cycle Clock	2 MHz
Butterfly Time	4 μsec
Transform Speed*	N/2 log (N+1) x Butterfly Time
Transform Lengths	N=1,2,3,..,512 complex points
Arithmetic	16 bit block floating point with rounding
Coefficient Word Length	8 bits
Data Word Length	16 bits

*Include unscrambling.

The processor is designed from three basic I^2L chips: a single control chip, a 4 bit arithmetic unit slice, and a 1024 x 4 bit RAM slice. The control chip includes the coefficient memory, the FFT microporgram memory, and the control and address generation circuitry. The arithmetic slice is a complex arithmetic unit (CAU) tailored to perform a parallel FFT (radix-2) Butterfly operation. Although the processor can be configured in multiplies of these 4 bit slices, a 16 bit configuration, sketched in Figure 12, has been chosen as a prototype model. Itemized power estimate of this processor with a 500 nanosecond cycle time is 300 mW dominated by the memory.

The I^2L hypothetical FFT processor described above has overwhelming advantages over the CCD CZT in (1) flexibility and (2) accuracy in a sense that additional slices can be added. However, when flexibility and high accuracy are not required the CCD CZT compares favorably in two important aspects. (1) Number of packages: The 512-point sliding CZT illustrated in Figure 5 can be implemented with 3 IC's; 2 CCD filter IC's and one ROM containing the premultiply chirp. The I^2L digital processor requires 9 IC's (see Figure 12) and in addition, requires A/D conversion. Even when state-of-the-art custom digital hardware is postulated, the CCD CZT maintains a clear cut advantage in package count and hence cost. (2) Speed: The I^2L FFT operating at it's maximum clock rate (2 MHz) generates spectral points at a 50 kHz rate. In general digital FFT hardware is limited to this kind of speed, and increased speed through parallel processing can be achieved only by a proportional increase in hardware. The CCD CZT, on the other hand operates in real time at speeds up to 5 MHz.

CCD's have also been proposed to perform the digital functions required to implement the FFT,[22,23] and it has been claimed that, compared to competing devices from other technologies, "charge-coupled devices appear to enjoy a ten to one advantage in device density and greater than a ten to one advantage is speed-power product".[23] Detailed comparisons with competing technologies need to be made to substantiate this claim.

V. APPLICATIONS OF THE CCD CZT

For a given spectral analysis application to be considered as a candidate for CCD CZT implementation it must satisfy two criteria: (1) It must be of modest performance which lies within the CCD performance limitations and (2) It must be required in sufficiently high volume that low cost is a dominant design specification. These two criteria rule out a large class of applications. However, there have been identified, several applications of great military importance which do satisfy both of the above criteria. These will be discussed in this section.

VIDEO BANDWIDTH REDUCTION

Transform encoding of video images for the purpose of bandwidth reduction is a potentially important application of the CCD CZT. A hybrid transform system has been developed[18] which performs a discrete cosine transform (DCT) in one dimension and differential pulse code modulation (DPCM) in the other dimension.

DFT and DCT on "typical" video images have resulted in variance compaction approaching that of optimum transforms,[18] and both the DFT and DCT can be cost effectively implemented with the CCD CZT. The CCD CZT becomes particularly attractive in remote sensing applications such as RPV's where small size, light weight, and low power are essential in addition to low cost.

SPEECH PROCESSING

Spectral analysis is one of the most important functions in speech processing,[24] and speech processing requirements in terms of sample rate (10 kHz), delay time (40 ms) and dynamic range (40 dB) are well within the CCD capabilities outlined above.

The simplest speech processing systems decompose speech into its spectral components as in the early channel vocoder systems. Channel vocoders perform well only at high bit rates[25] and therefore do not achieve optimal bandwidth reduction, but they are useful in word recognition systems.

There exist several algorithms which do achieve adequate bandwidth reduction and of these, one, is particularly well suited to CCD implementation. It is called homomorphic deconvolution[26] and operates upon the principle of the deconvolution of speech into pitch and to vocal tract resonances. In Figure 13 a block diagram of such a system is shown.[24] Since only the magnitude of transform 1 is required, the postmultiply can be eliminated, and in addition, the postmultiply of the inverse transform 2 cancels the premultiply of transform 3. Therefore the only chirp multiplications remaining are the premultiplies of transforms 1 and 2 which can be implemented at the input to the CCD.[27] After transform 2, the pitch of the voiced speech is detected and then windowed out so that only the smoothed speech spectrum remains at the output. This spectra can be used to extract formant data and if this information is efficiently encoded data rates as low as 1000 bits/s are realizable.[24] Homomorphic deconvolution is costly to implement digitally in real time because of the three sequential transforms which are required. However, the CCD CZT holds the promise of truly low cost implementations of this type of processing.

DOPPLER PROCESSING IN MTI RADAR

MTI (moving target indicator) radar operates upon the principle of detecting moving targets of small cross section in the presence of stationary background having much larger cross section. The doppler shift of the radar return is determined, and from this, the target velocity parallel to the radar line of sight, can be determined.

Radar returns are quasi-periodic, and the sliding CZT can be used. Typical transform lengths are 10 to 100 and typical pulse repetition frequency (PRF) is 1 kHz to 100 kHz. A doppler processor may be required to process thousands of DFT's on parallel, (one for each range gate) so reducing the cost of the DFT has a large cost impact on the overall system.

A doppler processor IC has been developed [28,29] which performs ten 17-point CCD CZT's of the type shown in Figure 5. The IC contains all the integrated amplifiers and squaring circuitry required to obtain the power density spectrum in each range bin. A doppler processor for thousands of range bins can be implemented by cascading 10-bin IC's at a projected cost of approximately one-third that of an all digital processor designed using state-of-the-art digital hardware.[28]

OTHER APPLICATIONS

Other potential applications of the CCD CZT include processing for FLIR images, sonobuoy signal processing, and remote surveillance.

The CCD CZT is not expected to make the digital FFT obsolete in military sistems. However, for those spectral analysis applications which fulfill the twin requirements of modest performance and high volume, tremendous cost advantages can be gained using the CCD CZT. More applications will certainly emerge, but in the meantime,

the potential cost impact in the application areas already identified guarantees the importance of the CCD CZT in military electronic signal processing systems of the future.

ACKNOWLEDGEMENT

Development of the CCD CZT at Texas Instruments has been sponsored by the Army Electronics Command (Robert H. Sproat, contract monitor) and by the Defense Advanced Research Projects Agency/Naval Undersea Center (Dr. Robert W. Means, contract monitor).

REFERENCES

1. H. J. Whitehouse, J. M. Speiser and R. W. Means, "High Speed Serial Access Linear Transform Implementations," presented at the All Applications Digital Computer Symposium, Orlando, Florida, 23-25 January 1973. NUC TN 1026.

2. L. R. Rabiner, R. W. Schafer, and C. M. Rader, "The Chirp Z-Transform Algorithm," IEEE Trans. on Audio and Electroacoustics, AU-17, pp. 86-92, June 1969.

3. R. W. Means, D. D. Buss, and H. J. Whitehouse, "Real Time Discrete Fourier Transforms Using Charge Transfer Devices," Proc. CCD Applications Conference, San Diego, pp. 127-139, September 1973.

4. R. W. Brodersen, H. S. Fu, R. C. Frye, and D. D. Buss, "A 500-point Fourier Transform Using Charge-Coupled Devices," 1975 IEEE International Solid-State Circuits Conference, Digest of Technical Papers, Phila, February 1975, pp 144-145.

5. R. W. Brodersen, C. R. Hewes and D. D. Buss, "Spectral Filtering and Fourier Analysis Using CCD's," IEEE Advanced Solid-State Components for Signal Processing, IEEE International Symposium on Circuits and Systems, Newton, Mass., pp 43-68, April 1975.

6. G. D. Bergland, "A Guided Tour of the Fast Fourier Transform," IEEE Spectrum, Vol. 6, pp. 41-52, July 1969.

7. W. M. Gentleman, and G. Sande, "Fast Fourier Transform - for Fun and Profit," 1966 Fall Joint Computer Conf. AFIPS Proc., Vol. 29, Washington, D.C.:

Spartan, 1966, pp. 563-578.

8. L. R. Rabiner and B. Gold, Theory and Application of Digital Signal Processing, Prentice-Hall, 1975, Ch.6.

9. P. R. Hariharan, and R. J. Scott,"Effects of Ideal Clipping on Signal-to-Noise Ratios - A Comparison with Linear Systems," 84th Meeting of Acoustical Society of Am, 30 Nov. 1972, Paper Y.6, The Magnavox Co, Fort Wayne, Indiana, 46804.

10. D. D. Buss, D. R. Collins, W. H. Bailey, and C. R. Reeves, "Transversal Filtering Using Charge Transfer Devices," IEEE J. Solid-State Circuits, SC-8, pp 134-146, April 1973.

11. D. D. Buss and W. H. Bailey, "Application of Charge Transfer Devices to Communication," Proc. CCD Applications Conference, San Diego, September 1973, pp 83-93.

12. R. D. Baertsch, W. E. Engeler, H. S. Goldberg, C. M. Puckette, and J. J. Tiemann, "Two Classes of Charge Transfer Devices for Signal Processing," Proc. International Conf. Technology and Applications of CCDs, Edinburgh, Sept. 1974, pp 229-236.

13. A. Ibrahim and L. Sellars, "CCDs for Transversal Filter Applications," IEEE International Electron Devices Meeting, Technical Digest, Washington, Dec. 1974, pp 240-243.

14. J. A. Sekula, P. R. Prince and C. S. Wang, "Non-Recursive Matched Filters Using Charge-Coupled Devices," IEEE International Electron Devices Meeting, Technical Digest, Washington, Dec. 1974, pp 244-247.

15. M. F. Tompsett, A. M. Mohsen, D. A. Sealer, and C. H. Sequin, "Design and Characterization of CCD's for Analog Signal Processing," IEEE Advanced Solid-State Components for Signal Processing, IEEE International Symposium on Circuits and Systems, Newton, Mass., pp 83-89, April 1975.

16. H. J. Whitehouse, R. W. Means, and J. M. Speiser, "Signal Processing Architectures Using Transversal Filter Technology," IEEE Advanced Solid-State Components for Signal Processing, IEEE International Symposium on Circuits and Systems, Newton, Mass., April 1975, pp 5-29.

17. C. M. Rader, "Discrete Fourier Transforms When the Number of Data Samples is Prime," Proc. IEEE 56, pp 1107-1108, June 1968.

18. R. W. Means, H. J. Whitehouse and J. M. Speiser, "Television Encoding Using a Hybrid Discrete Cosine Transform and a Differential Pulse Code Modulator in Real Time," 1974 National Telecommunications Conference Record, SAn Deigo, Dec. 1974, pp 69-74.

19. C. J. Weinstein, "Roundoff Noise in Floating Point Fast Fourier Transform Computation," IEEE Trans. Audio Electroacoustic, AU-17, pp 209-215, September 1969.

20. P. D. Welch, "A Fixed-Point Fast Fourier Transform Error Analysis," IEEE Trans. Audio Electroacoustic, AU-17 pp 151-157, June 1969.

21. R. W. Means, NUC Private Communications.

22. C. S. Miller and T. A. Zimmerman, "The Application of Charge-Coupled Devices to Digital Signal Processing," International Conference on Communications Conf. Record Vol.I, San Francisco, June 1975, pp 2/20-2/24.

23. T. A. Zimmerman, "The Digital Approach to Charge-Coupled Device Signal Processing," IEEE Advanced Solid-State Components for Signal Processing, IEEE International Symposium on Circuits and Systems, Newton, Mass., April 1975, pp 69-82.

24. R. W. Schafer and L. R. Rabiner, "Digital Representations of Speech Signals," Proc. of the IEEE 63, pp 662-677, April 1975.

25. R. W. Schafer and L. R. Rabiner, "Design of Digital Filter Banks for Speech Analysis," BSTJ, 50, pp 3097-3115, Dec. 1971.

26. A. V. Oppenheim and R. W. Schafer, "Homomorphic Analysis of Speech," IEEE Trans. Audio. Electroacoustic, Vol AU-16, pp 221-226, June 1968.

27. D. A. Hodges, U.C. Berkeley, Private Communication.

28. D. D. Buss, W. H. Bailey, M. M. Whatley and R. W. Brodersen, "Application of Charge-Coupled Devices to Radar Signal Processing," 74 NEREM Record, Boston, Nev. 1974, pp 83-98.

29. W. H. Bailey, D. D. Buss, L. R. Hite and M.M. Whatley, "Radar Video Processing Using the CCD Chirp z-Transform," This Conference.

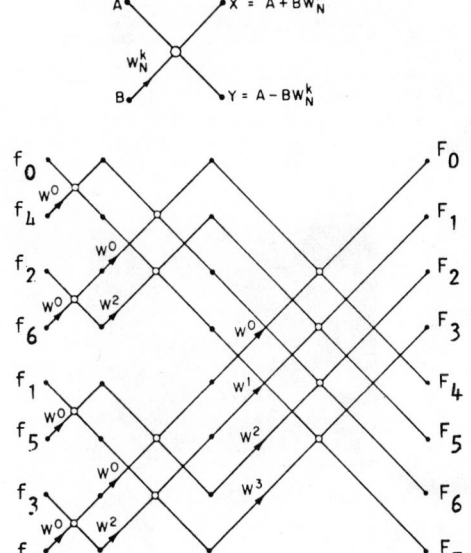

Fig. 1 a) Schematic of the butterfly. (see Eq. (2)) b) Schematic of an 8-point, radix 2, decimation in time FFT (Ref. 8)

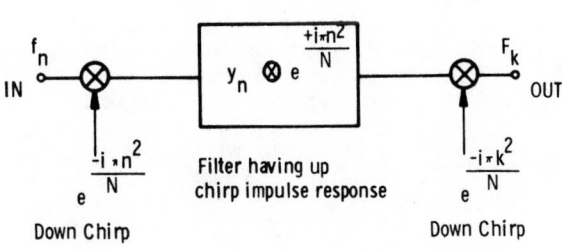

Fig. 2 Schematic of the CZT algorithm.

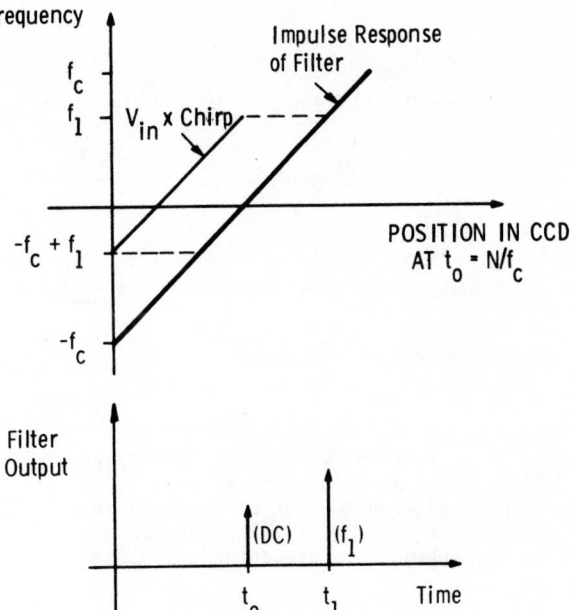

Fig. 3 Interpretation of the CZT in terms of chirp input waveforms in chirp filters.

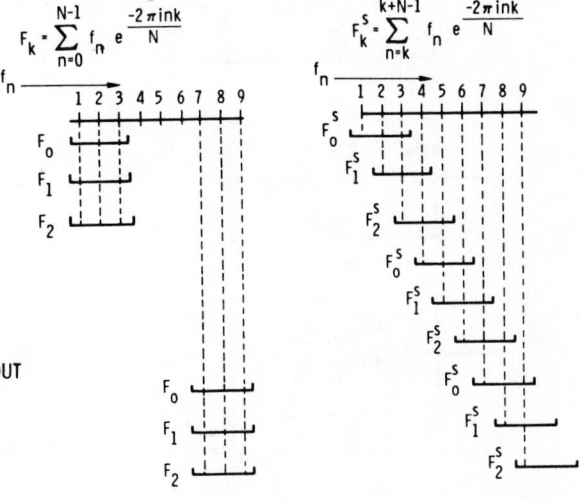

Fig. 4 Comparison between the conventional CZT and the sliding CZT for the case of a 3-point transform.

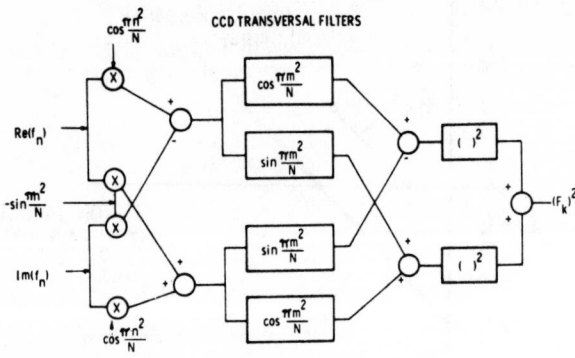

CCD TRANSVERSAL FILTERS

Fig. 5 An implementation of the CZT
algorithm using real components.
The power density spectrum $|F_k|^2$
is computed for an input which
has both real (in-phase) and
imaginary (quadrature) components.

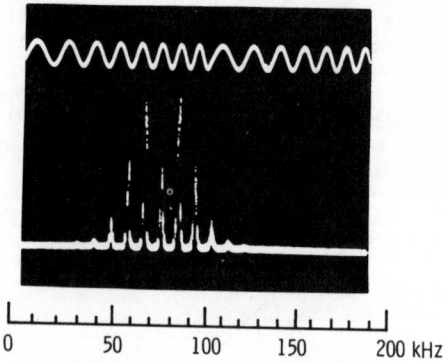

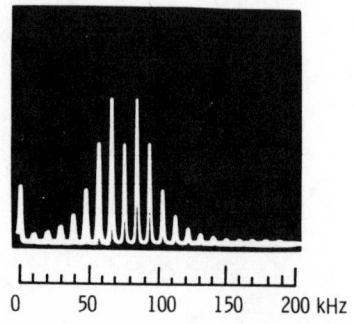

Fig. 6 A comparison of the 500-point
sliding CZT power density spectrum (upper photograph) with
the output of a conventional
spectrum analyzer (lower photograph which shows the square
root of the power density
spectrum).

The input signal shown at the
top of the upper photograph
chirps from 50 kHz to 100 kHz
with a repetition period of
T_d = 110 μsec. This results
in a line spectrum having
period T_d^{-1} = 9 kHz.

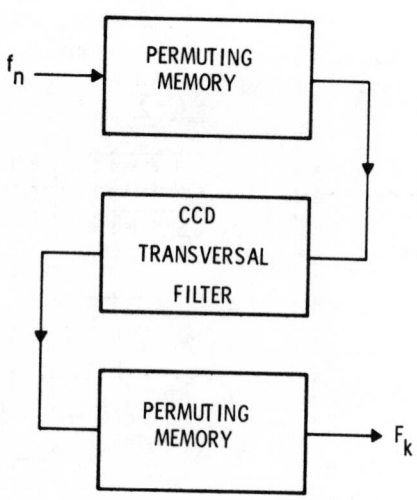

Fig. 7 Block diagram of the prime
transform implemented with
CCDs.

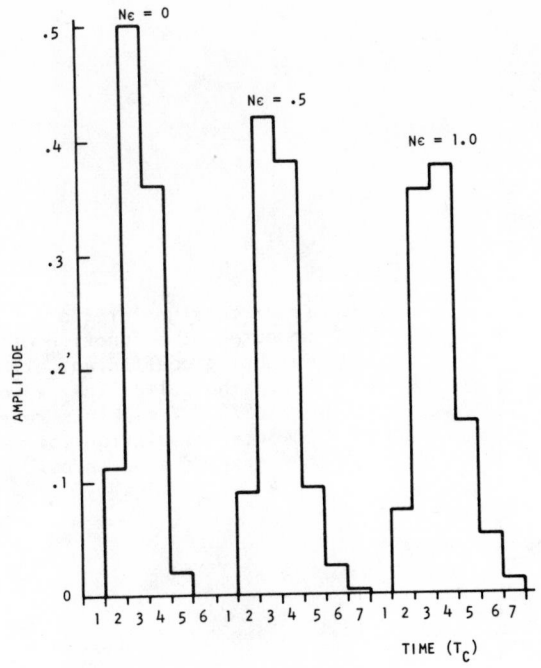

Fig. 8 The calculated response of the 500-point sliding CZT with Hamming weighting to a complex input sinusoid of frequency $f_{in} = \frac{3.3}{N} f_c$. The response is calculated for different values of CTE.

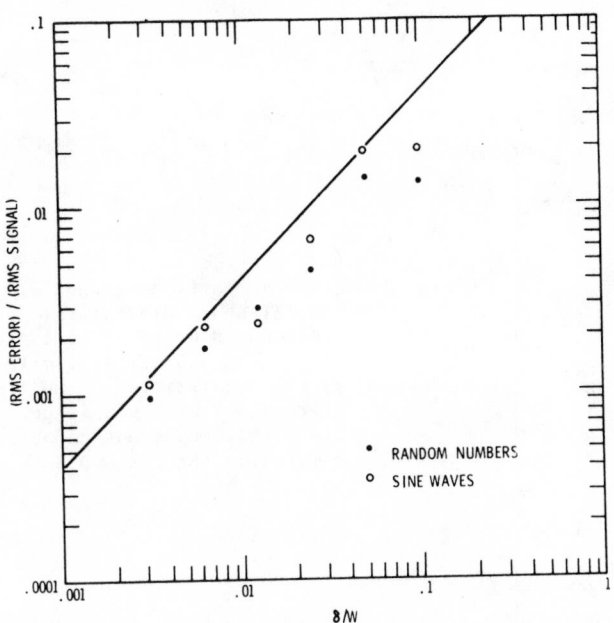

Fig. 9 Error to signal ratio Δ_M computed as a function of the number of bits b_4 used to quantize the premultiply and postmultiply chirp waveforms. The line represents Eq. (21). The points represent computer simulation in which just the premultiply or postmultiply is quantized.

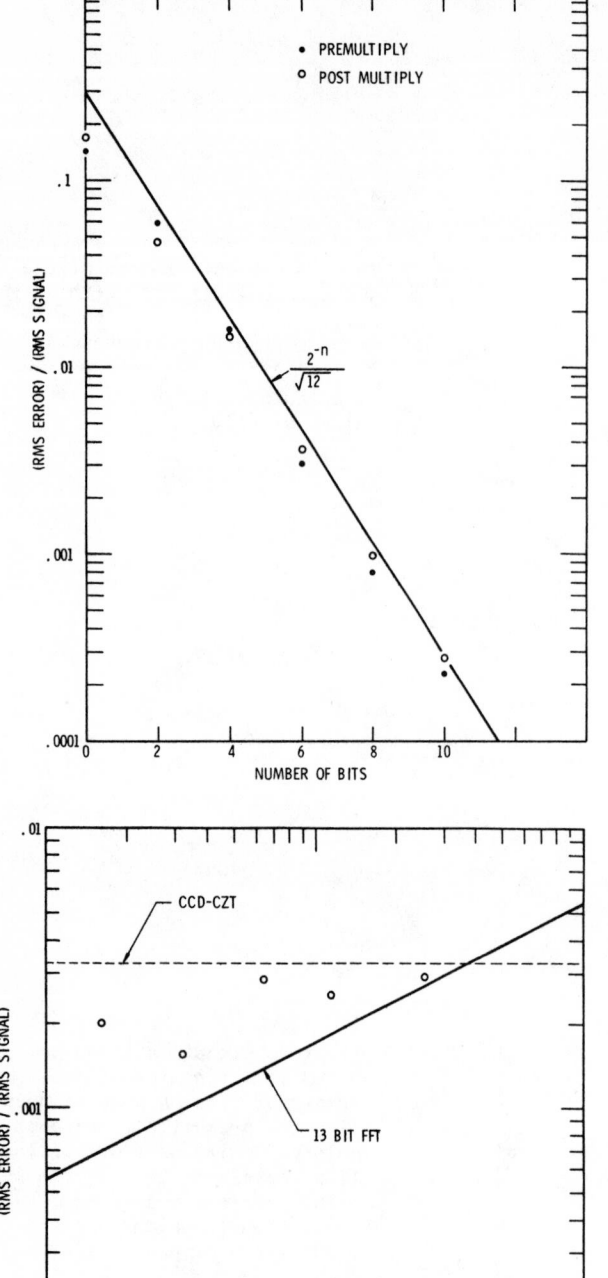

Fig. 10 Error to signal ratio Δ_W computed as a function of weighting coefficient error δ/W. The solid line represents Eq. (22). The points represent computer simulation for random numbers and sine waves.

Fig. 11 Error comparison between a CCD CZT and a digital FFT implemented using 13 bits plus sign. The CCD CZT is limited by the quantization of multiply chirps to 7 bits plus sign. The points represent computer simulation for the CCD CZT.

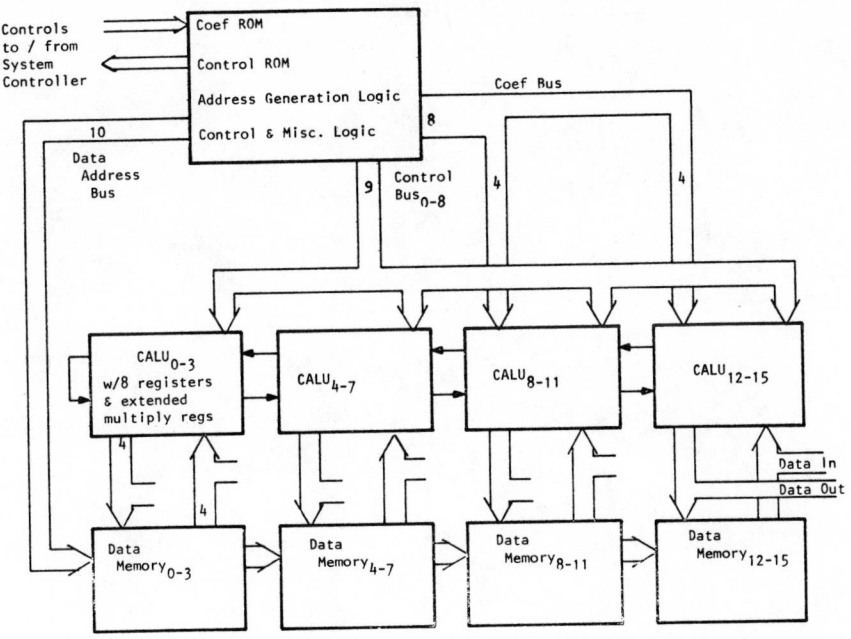

Fig. 12 A custom 16 bit I²L FFT requiring
9 IC's.

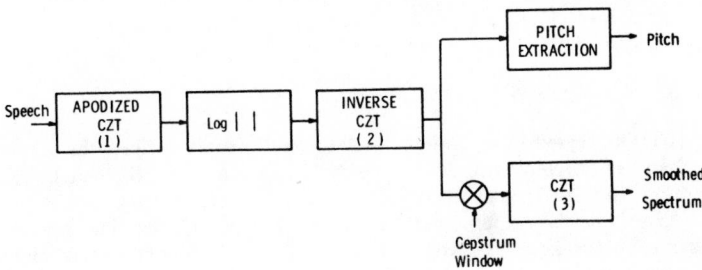

Fig. 13 Block diagram of a system to
perform homomorphic deconvolution of speech for bandwidth
reduction (Ref. 24). This is
one type of speech processing
which is particularly amenable
to implementation with CCDs.

A Surface-Charge Correlator

JEROME J. TIEMANN, SENIOR MEMBER, IEEE, WILLIAM E. ENGELER, MEMBER, IEEE, AND RICHARD D. BAERTSCH, MEMBER, IEEE

Abstract—Charge-transfer devices (CTD's) represent a major advance for analog signal processing. Transversal filters with fixed tap weights are naturally suited to these devices, and they can perform any linear shift-invariant process directly on the analog input signal. A cross correlator performs the same function as a transversal filter, but the tap weights of a correlator are externally controllable. Thus a correlator can be programmed to perform any linear process.

Surface-charge transfer techniques have been used to implement a cross-correlator module in which the tap weights are restricted to values of plus and minus one. This compromise permits most of the advantages of CTD's to be retained, and the disadvantage of fixed tap weights can be overcome by using a weighted binary code for the tap weight function and separate binary correlators for each binary digit.

A new architecture (parallel transfer) was used which eliminates the cumulative effect of charge-transfer inefficiency, thereby permitting longer impulse responses to be implemented than is practical with the conventional serial transfer approach.

Manuscript received May 1, 1974; revised July 28, 1974. This paper is an expanded version of papers presented at the International Solid-State Circuits Conference, Philadelphia, Pa., February 1974 and IEEE INTERCON, New York, N. Y., March 1974.

The authors are with the Corporate Research and Development Division, General Electric Company, Schenectady, N. Y. 12301.

The experimental device, which contains 32 stages, was implemented in p-channel MOS technology, and was designed to operate at 4 MHz. Test results are presented showing that a tap weight accuracy of order 1/2 percent can be achieved on a single chip. Presuming that the modular approach employed permits selection of well-matched chips with leakage at least as low as was obtained in the experimental units, it would be possible to implement an impulse response covering 40 000 samples with tap weight accuracy of perhaps 1 percent.

I. INTRODUCTION

A TRANSVERSAL filter can be implemented with a serial transfer charge-transfer device (CTD) in which charge packets representing sequential samples of the input signal are passed from stage to stage along a linear path. During each clock cycle, the amount of charge in each packet is nondestructively measured, and this measurement is used to define the output from the "tap" at that point. Note that the measurement in no way removes any charge from the packet, and does not degrade the signal nor cause any reflections. To form a transversal filter, the tap outputs must then be appropriately weighted and summed.

The simplest way to assign tap weights is to divide

Reprinted from *IEEE J. Solid-State Circuits*, vol. SC-9, pp. 403–409, Dec. 1974.

396

the clock electrode associated with each stage into two portions and sense the charge transferred under each portion separately. Summation of the charge for all of the selected portions can be performed by connecting them to a common drive circuit and monitoring the drive current required to charge or discharge them. Thus the multiplicative tap weights are determined by the location of cuts in the electrodes, and the addition is performed by strapping the corresponding portions of the electrodes together [1], [2].

A cross correlator performs the same function as a transversal filter, but it must accept the tap weights as the (sampled data) values of a second input signal. Since such a device can implement all possible transversal filters, a cross correlator is a very powerful general-purpose signal processing element.

Note that the method for obtaining tap weights discussed above is not suitable for a cross correlator because the tap weights cannot be changed once they are built into the electrode structure; in a correlator, the tap weights must be electrically determined in some manner. Although summation can be performed in any surface-charge device by strapping electrodes together, there is no convenient way of obtaining an electrically variable multiplicative fraction of the surface charge in a storage reservoir.[1] Thus there does not appear to be an ideal way to implement a cross correlator with surface-charge technology alone, and some compromise must be considered. One solution to this problem has been proposed which employs post multiplication of the output of each tap of a serial transfer transversal filter by the conductance of an electrically alterable MNOS transistor [3]. In this paper, an additional hybrid approach will be presented in which one set of signal samples is represented by a continuously variable packet of surface charge, while the other signal is restricted to two fixed values. Summation over samples is performed as described above by employing common drive circuitry for all cells, and the device automatically samples the analog input signal. Thus most of the efficiency of charge-transfer techniques is retained [4], [5].

Although limitation to only two tap weight values may seem to be a serious problem, it can be overcome by using additional binary correlators. Suppose, for example, that the analog signal is fed in parallel to a number of correlators, and that the first is assigned weights of ± 1, the second weights of $\pm 1/2$, and so on in descending powers of 2. These tap weights could be set either by combining the outputs of identical correlators with a resistor ladder, or a family of devices having reservoir areas that correspond to the desired weights could be used. A combination of these two methods could also be employed.

<hr />

[1] Charge transfer can be partially blocked by controlling the potential of the region between the reservoirs, but this does not result in the transfer of a multiplicative fraction of the initial charge.

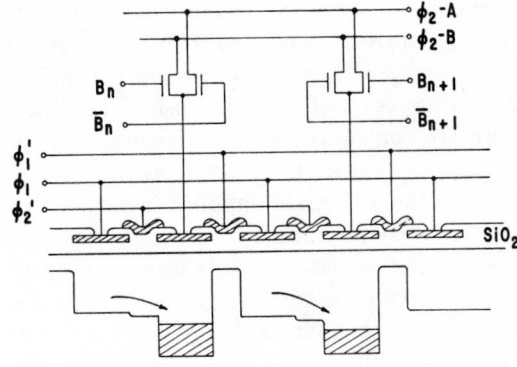

Fig. 1. Schematic of a method for obtaining binary tap weights using a serial charge transport structure. The surface potentials during charge transfer are shown.

If the outputs of these separate correlators are summed together, then every signal sample can be assigned an arbitrary weight corresponding to the binary number whose digits are sent to each of the correlators. Usually four to eight binary correlators are required to achieve a tap weight resolution appropriate for typical applications.

There are at least two approaches for obtaining binary tap weights. In one approach, the electrodes associated with a particular cell of a serial transfer structure are connected to either of two drive lines by series switches [6] (such as MOSFET's), only one of which is turned on for each cell at one time (see Fig. 1). In this way, the displacement current due to the transfer of the signal charge in each packet is made to appear as a load on either of two output lines. The total load on each drive line can be separately measured, and each of the two outputs can be assigned one of the two fixed tap weights by external means.

This approach is feasible for correlations of up to a few hundred samples, but for longer sequences, inaccuracy due to charge-transfer inefficiency becomes important. The main contribution of this paper is the presentation of an alternative architecture in which transfer losses do not cause inaccuracy, and which permits a long correlator to be partitioned into modules of convenient size without degrading the accuracy of the signal samples.

II. A New Charge-Transfer Architecture

In the new approach, the charge packets are not clocked along from cell to cell as they are in serial transfer structures. Instead, each packet is gated into a separate cell where it remains until it is finally replaced by a new sample. Nondestructive readout is accomplished without moving the charge by "sloshing" it back and forth within the confines of the cell, and sensing the corresponding displacement current induced in the overlying clock electrodes.

The key idea for implementing multiplicative tap weights in this approach is to provide two separate output regions at opposite ends of every cell, and to control the *direction* of charge transfer in each cell so that the displacement current associated with the charge transfer appears alternatively on either of the two output electrodes. The *direction* of charge transfer in each cell must be externally controlled, and this means that the basic charge-transfer element must derive its directional asymmetry from the externally applied potentials and not from any intrinsic asymmetry. Although there are many device structures which have this feature, a particularly simple one in which a transfer gate between each reservoir controls the transfer of charge is the surface-charge transistor [7]. Since the charge packets representing each signal sample move back and forth only within a single cell, transfer losses are not cumulative, and it is impossible for charge representing one sample to become mixed with charge representing any other sample.

Since the charge samples do not move from tap to tap in this approach, it is necessary that the tap weights move from cell to cell in order to obtain the shift operation of a transversal filter. The tap weights in this structure depend on the direction of charge transfer, which is a binary function of the potentials on the transfer gates. Thus digital shift registers can be used to propagate the tap weight signals from cell to cell to effect the desired shift operation.

III. OPERATION OF THE CORRELATOR

The basic structure, which is shown pictorially in Fig. 2 and schematically in Fig. 3, has an analog signal bus adjacent to a series of charge-storage regions which are represented by the heavily outlined rectangular areas. Scan gates overlap the barriers that separate these cells from the diffused analog bus, and these gates are turned on and off one by one by the scan shift register. As each gate is turned on, charge either enters or leaves until the potential in the corresponding reservoir equilibrates with that of the analog input.

After the scan is complete, the amounts of charge in the charge-transfer cells will represent the sampled data values of the input signal. The action taking place during a single clock period is further explained in Fig. 4. Here the surface potentials on a cross section through the center of one storage cell are shown. The diffused analog input bus is shown at the left of each diagram, and the potentials of the three regions within each cell are shown under the corresponding electrodes in the schematic representation at the top of the figure. The barrier regions between the reservoirs are controlled by the transfer gates, and these potentials are represented by the heights of the "walls" between the reservoir regions. The functions being performed at various times in the clock period are described in the figure.

There are three electrodes serving all cells in the system which drive the charge back and forth within each cell. During the first part of each readout cycle, the charge is "recollected" to the center of the cell by opening both transfer gates while the center electrode (electrode

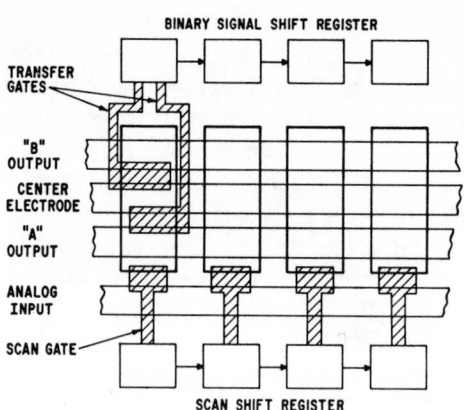

Fig. 2. Plan view of the surface-charge transistor (SCT) portion of the correlator.

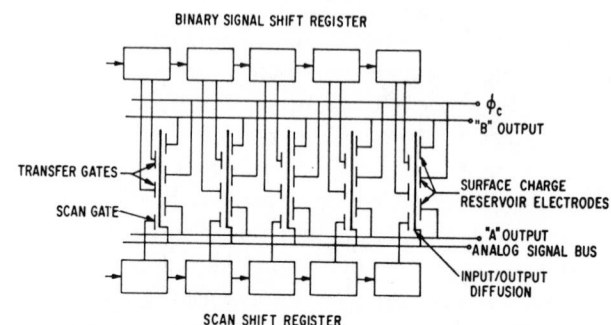

Fig. 3. Schematic representation of the same portion of the correlator as in Fig. 2.

C) is charged to an attractive potential. Next, the transfer gates are closed and the electrode potentials are changed so that it is now energetically favorable for the charge to transfer to either of the two output regions. After charging, the two output electrodes (electrodes A and B) are floated free, and their potentials are measured by sample-and-hold circuits. One or the other of the transfer gates is then opened, depending on the control signal from the binary shift register, causing transfer to one or the other of the output regions. This charge transfer is accompanied by a displacement current which causes the potentials of the two floating electrodes to drop toward ground by amounts that are approximately proportional to the total charge transferred. Since the A and B electrodes serve all of the cells, the final potentials differ from their initial values by an amount proportional to the sum of all charge transferred to them, regardless of which cells contained the charge. Similarly, if more than one chip is involved, summation over chips is accomplished simply by connecting the corresponding electrodes of each chip together. The output signals from the A and B electrodes are determined by substracting their potentials after transfer from their original values with a differential amplifier. Since the electrodes are floating during the entire operation, any noise charge that was trapped on them during the float operation cancels out and does not affect the output. Standard

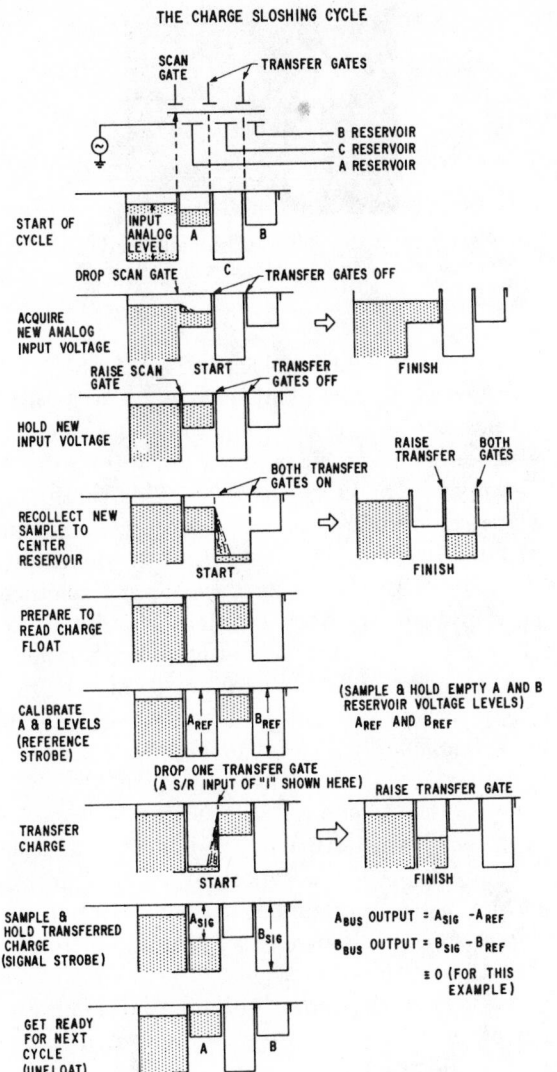

Fig. 4. The surface potentials under the various electrodes at different times in the cycle.

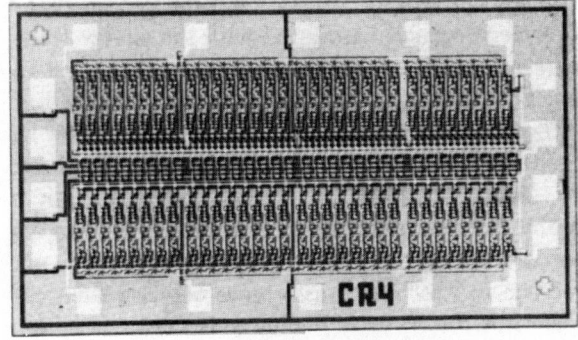

Fig. 5. Photomicrograph of a 32-bit ion-implanted silicon gate correlator chip.

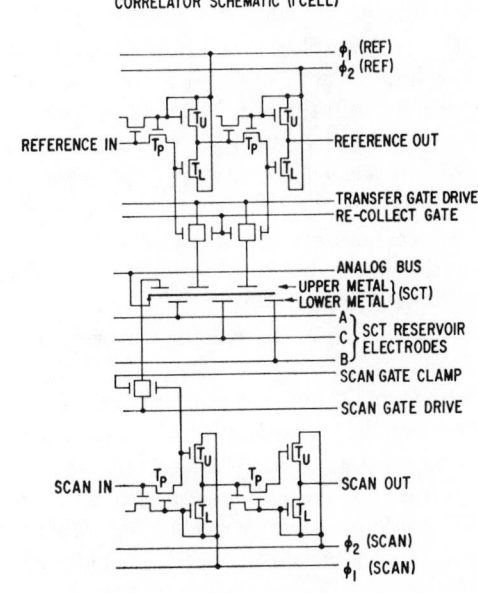

Fig. 6. Complete schematic of a single cell of the correlator. This cell is repeated 32 times in the chip.

MOS level translators can drive the structure since no electrode requires more than two voltage levels: the gates are either blocking or open, and the reservoirs are either receiving or sourcing charge.

IV. EXPERIMENTAL STRUCTURE

The chip contains 32 stages and measures 0.066-by-0.115 in. A photomicrograph is shown in Fig. 5. The binary signal and scan shift registers can be seen at the top and bottom of the chip, respectively, and the charge-transfer cells are at the center. The layout of the charge-transfer cells is as shown in Fig. 2, with three of the four horizontal runs corresponding to the three reservoir electrodes. The fourth run serves as the diffusion stop for the analog input bus, and was not used in the experiments presented here. This electrode can be used for linearized charge insertion techniques [8], [9]. Separate pads have been provided for the clock lines of the two shift registers, so it is possible to load the two signals independently of each other, but the chip is normally operated with both shift register clock lines tied together. The complete schematic diagram of a single element of the experimental device is shown in Fig. 6. The shift registers utilize a ratio-less MOS dynamic design [10] incorporating bootstrap MOSVAC capacitors [11] to provide the necessary high-output voltage. One feature of the design is the absence of a ground bus. Instead of a separate ground line, the clock lines themselves are used. Since one phase line is at ground potential at the time the other phase is active, the same line can be used both as the clock voltage line for one phase and the ground of the opposite phase. The switch transistors are all of minimum dimensions, and the MOSVAC capacitors, which are shown as FET's with floating drains, have about three times the capacitance of the storage nodes.

The polarity of these bootstrap capacitors is opposite to the usual case, and this feature deserves comment. Here it is desired to enhance the potential of the internal nodes only when their potentials are above threshold, but

399

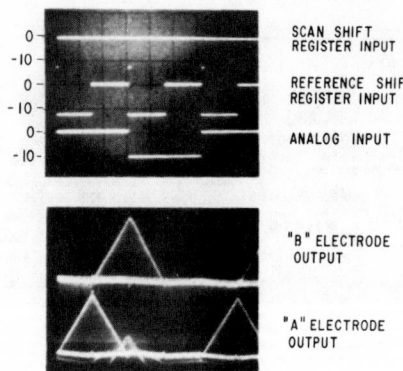

Fig. 7. Input and output waveforms from the correlator. In this case, the sample-and-hold circuit was not used and the voltages on the *A* and *B* electrodes are shown. Time scale is 50 μs/cm.

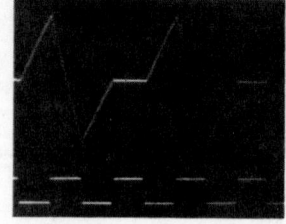

Fig. 8. Waveform at the output of the differential amplifier after the signal in Fig. 7 has been sampled and held. Time scale is 80 μs/cm.

not to enhance them when they are below threshold. That is, the voltage of a "one" (high negative voltage) should be enhanced, while that of a "zero" should be left alone. By connecting the *gate* electrode of the MOSVAC to the node, the capacitance of the bootstrap is large when its action is needed, and it is small when its action would be harmful.

Note that the shift registers do not drive the transfer gates directly, but merely connect the selected gates to drive lines which actually provide the control potentials. This was done so that the recollect operation could be performed independently of the state of the shift register. By providing an alternative switch which is turned on during the recollect operation, the state of the shift register becomes immaterial. This permits a shorter cycle time by permitting the shift operation to commence as soon as the readout operation is complete. This produced an additional benefit which was not realized at first, namely, the capacitive coupling between the drive line and the storage nodes of the shift register provides an additional boost to the output voltage. By allowing the "right" amount of overlap between the drive lines and the storage nodes, it was possible to overcome the threshold drop of the pass transistors, and to deliver an output voltage to the transfer gates equal to the magnitude of the voltage change of the shift register clock waveforms.

V. EXPERIMENTAL RESULTS

Although exhaustive testing of the experimental devices has not been completed, available preliminary results are very encouraging. In one test, the analog signal was held at a voltage corresponding to a "full" well for 64 pulses and then at a potential corresponding to an "empty" well for a like period of time. For each condition, 32 binary "ones" (causes readout on the *A* side) were inserted into the binary channel, followed by 32 binary "zeros" (causes readout on the *B* side). The outputs on the *A* and *B* electrodes are shown in Fig. 7, along with the input signals. At first, the *A* output increases linearly from zero to a maximum level, reflecting the fact that each clock pulse has one more full reservoir of charge than the preceding one. After 32 pulses, all

of the reservoirs contain charge and all the binary stages contain "ones." At this point, binary "zeros" start being loaded into the shift register, and as each one enters, the signal from one reservoir is taken away from the *A* side, and it appears on the *B* electrode. This continues until the signal on the *B* electrode reaches a maximum and the *A* signal reaches zero. At this time, the analog voltage is set to the level for an empty well, and another scan bit is inserted in the scan register. This bit causes the cells to empty one at a time as they propagate along the shift register until all 32 stages are empty. During this time, binary "ones" were being inserted into the binary channel, but since all reservoirs opposite these "ones" were emptied by the action of the scan bit, no output appears on the *A* side. At the end of this period, all reservoirs are transferring to the *A* side, but no charge is present. During the last portion of the trace, binary "zeros" are being inserted, but since no charge is present, no output appears either on the *A* or *B* electrodes.

When these two output signals are sampled and held and are fed to the output differential amplifier, the (much cleaner looking) waveform shown in Fig. 8 results. This double polarity signal can be recognized as the correlation of two square waves whose periods are in the ratio of 2 to 1.

In another mode of operation, the scan bit appeared only once during every 4096 cycles. That is, the charge packets remain in the reservoirs for 4096 clock periods before they are replaced by new samples. In the experiment, which is shown in Fig. 9, the analog input levels were again chosen so as to alternately fill and empty the wells, but the time axis of the scope trace has been compressed so that all 8192 readout cycles are shown on a single trace. In this case, the triangular patterns caused by the alternation of the binary reference appear blurred together, and the levels appearing as points in Fig. 8 now appear as horizontal lines.

These lines are actually made up of 256 separate replottings of the output signal corresponding to a specific number of full buckets of charge on each output electrode. Note that the lines diverge slightly as time goes on, and that the "empty" readings also have a slightly triangular envelope. This linear increase in amplitude is due to the accumulation of thermally generated leakage current. For the data shown here, the leakage would amount to a 1 percent error after approximately 10 000

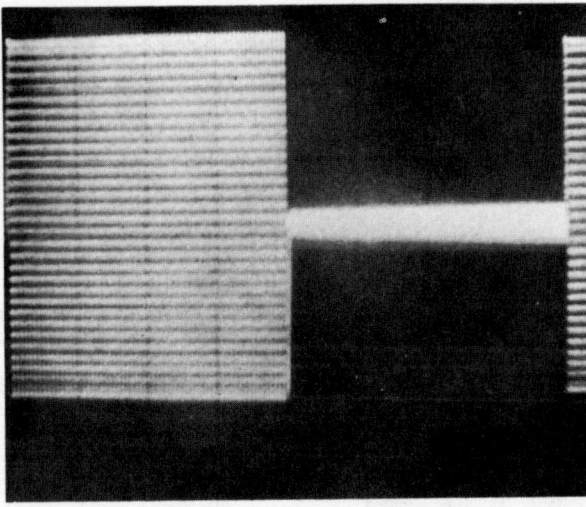

Fig. 9. Scope trace of the output of the differential amplifier. In this case, the charge packets remain in the reservoirs for 4096 clock periods before they are replaced by new samples. The buildup shown at the right is due to dark current. Time scale is 2 ms/cm.

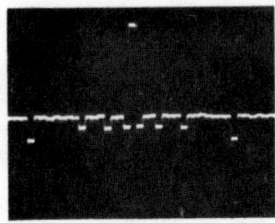

Fig. 10. Scope trace of the output of the differential amplifier. The input to the binary channel and analog channel is a 32-bit p-n sequence.

transfers. In other experiments on a similar device, tests involving up to 1 million transfers verified that there is no cumulative charge-transfer loss in this structure [12].

As a further test, a p-n sequence of length 32 with optimum autocorrelation properties [13] was fed to both the analog and the binary channels. The output, shown in Fig. 10, has one value each of +32 and −8, six values of −4, and 24 values of zero. If all the tap weights were exactly equal, all of the zero values would have fallen on the same line. As can be seen from the figure, however, there is a slight spread of these values. The measured value of this spread for the trace shown in the figure corresponds to an rms tap weight error of ±1/2 percent. It is believed that the cause of this error is the variation in the areas of the individual charge-storage regions. If chips from several different wafers were combined into a single system, one would expect additional variability due to differences in oxide thickness, etc. How much additional error would be introduced from this source obviously depends on the amount of device selection that can be done, and it is a question beyond the scope of this paper (see below).

V. Sources of Error

A number of different sources of inaccuracy are expected in the device described above. These include the following.

1) Input and output nonlinearities due to the voltage variation of the depletion capacitances of the charge-storage reservoirs. These nonlinearities, together with an additional nonlinearity inherent in the charge-transfer operation itself, may cause a small cross-modulation effect. A more complete analysis and discussion of these effects has been presented elsewhere [12].

2) Variations in the geometric areas of the individual charge-transfer cells will appear as a corresponding modulation of the tap weights. Since the charge in each packet is determined by the equilibration of the surface potential of the cell with the input voltage, a larger cell will take in more charge than a smaller one. This error source is essentially the same as that encountered in determining the resistance ratio of two diffused resistances on the same chip, and by analogy with that experience, variation of the order of 1 percent can be expected. This tap weight error is a constant property of each individual system, and in many instances, it can be taken into account.

3) The overall level of thermally generated leakage current limits the storage time of the analog signal samples [14]. This effect is common to all CTD devices, and it varies strongly with temperature. The implications of nonuniform leakage are more serious in the structure described here than in serial structures because each sample spends all of its time in a specific cell, whereas in a serial structure, each sample spends equal time in every cell. Thus unequal leakage current shows up as a temperature-dependent pattern noise. Experience with

self-scanned solid-state imagers has shown that thermal leakage appears as a smoothly varying component of relatively low value punctuated by isolated regions of much larger magnitude. Thus if a modular approach is used in which the individual chips are smaller in dimension than the average distance between areas of high leakage, the defects can be eliminated at the chip test level without seriously compromising yield. This problem is therefore not as serious as it may appear at first glance. In any case, the error introduced by leakage can be kept to an acceptable level by appropriately limiting the total processing delay.

If the total processing delay is shorter than a few percent of the storage time,[2] it appears that none of the above-mentioned sources of error or nonlinearity will exceed 1 percent. And it seems unlikely that geometrical errors much below this will be achieved in the near future. It therefore appears that an overall accuracy in the range of 1 percent is a reasonable expectation.

VI. DYNAMIC RANGE

For the reasons discussed above, the usable dynamic range at the input appears limited to the vicinity of 40 dB. The output dynamic range may be much greater than this, however, because the errors from individual tap weights will tend to cancel as the number of reservoirs in the system increases. In this case, the input errors can be expected to cancel out, and the dynamic range will be limited by the characteristics of the output circuit. If the limiting factor is the noise of the output amplifier, the dynamic range is determined by the ratio of the noise voltage of this amplifier to the maximum signal voltage. For a signal bandwidth of a few megahertz, input noise levels of 10–100 μV are achievable, whereas the output signal itself will be of the order of 10 V. This implies that the output dynamic range will be limited to less than about 100 dB. On the present device, an output dynamic range of approximately 66 dB was measured, but no particular pains were taken to achieve low noise in this test. In any case this would be expected to improve as larger systems are formed.[3]

[2] Storage times of a few seconds are commonly achieved at present.

[3] Note that the maximum output signal is independent of the number of reservoirs. This is because additional output charge is accompanied by additional output capacitance in exactly the

ACKNOWLEDGMENT

The authors are grateful to H. J. Whitehouse of USNUC for inspiring this work and partially supporting it, to D. M. Brown for contributions relating to device fabrication, and to L. Petrucco and J. Richotte for technical assistance.

REFERENCES

[1] F. L. J. Sangster, "The bucket brigade delay line, A shift register for analog signals," *Philips Tech. Rev.*, vol. 31, pp. 97–110, 1970.
[2] D. R. Buss, D. R. Collins, W. H. Bailey, and C. R. Reeves, "Transversal filtering using charge-transfer devices," *IEEE J. Solid-State Circuits*, vol. SC-8, pp. 138–146, Apr. 1973.
[3] D. R. Lampe, M. H. White, J. H. Mims, and J. L. Fagan, "An electrically programmable LSI tranversal filter for discrete analog signal processing," in *Proc. CCD Appl. Conf.*, Sept. 1973, pp. 111–125.
[4] J. J. Tiemann, R. D. Baertsch, and W. E. Engeler, "A surface-charge correlator," in *Dig. Tech. Papers, Int. Solid-State Circuits Conf.*, Feb. 1974.
[5] J. J. Tiemann, W. E. Engeler, and R. D. Baertsch, "A programmable filter using charge transfer device," presented at IEEE INTERCON, New York, N. Y., Mar. 1974.
[6] T. A. Zimmerman and R. W. Bower, "The use of CCD correlators in a spread spectrum communications example," in *Proc. CCD Appl. Conf.*, Sept. 1973, pp. 141–146.
[7] W. E. Engeler, J. J. Tiemann, and R. D. Baertsch, "The surface-charge transistor," *IEEE Trans. Electron Devices*, vol. ED-18, pp. 1125–1136, Dec. 1971.
[8] S. P. Emmons and D. D. Buss, "The performance of charge-coupled devices in signal processing at low signal levels," in *Proc. CCD Appl. Conf.*, Sept. 1973, pp. 189–205.
[9] M. F. Tompsett, "Using CCD's for analog delay," in *Proc. CCD Appl. Conf.*, Sept. 1973, pp. 147–150.
[10] R. F. Herlein and A. V. Thompson, "An integrated associative memory element," in *Dig. Tech. Papers, Int. Solid-State Circuits Conf.*, Feb. 1969, pp. 42-3–42-19.
[11] R. E. Joynson, J. L. Mundy, J. F. Burgess, and C. Neugebauer, "Eliminating threshold losses in MOS circuits by bootstrapping using varactor coupling," *IEEE J. Solid-State Circuits*, vol. SC-7, pp. 217–224, June 1972.
[12] J. J. Tiemann, W. E. Engeler, R. D. Baertsch, and D. M. Brown, "Intracell charge-transfer structures for signal processing," *IEEE Trans. Electron Devices*, vol. ED-21, pp. 300–308, May 1974.
[13] R. C. Titsworth, "Correlation properties of cyclic sequences," Jet Propulsion Lab., Pasadena, Calif., Tech. Rep. 32-388, July 1963.
[14] D. K. Schroder and J. Guldberg, "Interpretation of surface and bulk effects using the pulsed MIS capacitor," *Solid-State Electron.*, vol. 14, pp. 1285–1297, 1971.

same proportion. Thus the maximum output signal remains comparable to the clock voltage amplitude, while the noise and error levels get smaller.

Application of Charge-Coupled Devices to Infrared Detection and Imaging

ANDREW J. STECKL, MEMBER, IEEE, RICHARD D. NELSON, MEMBER, IEEE, BARRY T. FRENCH,
RICHARD A. GUDMUNDSEN, SENIOR MEMBER, IEEE, AND DANIEL SCHECHTER

Abstract—A review of infrared sensitive charge-coupled devices (IRCCD) is presented. Operational requirements of typical IRCCD applications are briefly introduced. IRCCD devices are divided into two major categories: a) Monolithic devices, which essentially extend the original CCD concept into the IR. Monolithic IRCCD's discussed include inversion-mode devices (with narrow bandgap semiconductor substrate), accumulation-mode devices (extrinsic wide bandgap semiconductor substrate), and Schottky-barrier devices (internal photoemission). b) Hybrid devices, in which the functions of detection and signal processing are performed in separate but integratable components by an array of IR detectors and a silicon CCD shift register unit. Hybrid IRCCD's discussed include both direct injection devices (in conjunction with photovoltaic IR detectors) and indirect injection devices (in conjunction with pyroelectric and photoconductive devices).

I. INTRODUCTION

SINCE the invention of charge-coupled devices (CCD's) by Boyle and Smith [1] in 1970, the remarkable versatility of this class of devices has led to their application in a diverse number of fields. These applications include visible-light imagers (e.g., TV cameras, optical character reader), digital and semianalog memories, communications and signal processing.

In this paper we explore a relatively new area of applications for CCD's and concepts: infrared detection and imaging. A variety of infrared CCD's (IRCCD's) which have been explored or proposed fall into two main categories: monolithic and hybrid devices.

The monolithic IRCCD concept generally uses the standard CCD structure with the substrate consisting of a narrow bandgap or an extrinsic semiconductor sensitive to IR radiation [2], [3]. Other monolithic IRCCD devices being investigated combine a Schottky-barrier internal photoemission sensing array in conjunction with a CCD readout structure [4]. It is important to note that the extrinsic semiconductor substrate IRCCD is the only one operating in the accumulation mode (AMCCD) and using majority carrier transport, while the other monolithic IRCCD's are depletion-mode devices.

By contrast to the monolithic IRCCD, the hybrid versions consist of the coupling of any one of various types of IR photodetectors to a silicon CCD shift-register unit. In the hybrid structure, the functions of detection and signal processing are performed in distinct but integratable components. The role of the silicon CCD in this case is that of a signal processor performing appropriate functions, such as multiplexing, amplification, correlation, delay-and-add, etc. The category of hybrid IRCCD's can be further subdivided into two subclasses: direct and indirect injection devices. In the former,

Manuscript received August 12, 1974; revised August 15, 1974. Portions of this paper were presented at the International Conference on the Technology and Applications of Charge-Coupled Devices, Edinburgh, Scotland, September 25-27, 1974.

A. J. Steckl, R. D. Nelson, B. T. French, and R. A. Gudmundsen are with Electronics Research Division, Rockwell International, Anaheim, Calif. 92803.

D. Schechter is with the Department of Physics–Astronomy, California State University, Long Beach, Calif. 90840.

Reprinted from *Proc. IEEE*, vol. 63, pp. 67-74, Jan. 1975.

403

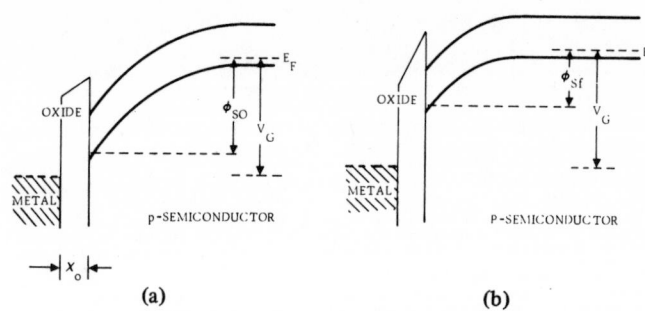

Fig. 1. Energy band diagram for an MIS structure. (a) Initial conditions. (b) Steady state.

photogenerated charge is directly injected from the detector into the CCD [5], while in the latter, a buffer stage exists in between the two components. Infrared detectors used in hybrid IRCCD devices include: photovoltaic and photoconductive quantum detectors and pyroelectric detectors.

II. OPERATIONAL CONSIDERATIONS

The number of IR applications has grown tremendously over the years in all fields: military, industrial, medical, scientific, etc. (cf. Hudson [6]). Existing and envisioned applications for IRCCD's fall within all of these fields. From reconnaissance to surveillance to nondestructive testing and inspection to earth resource surveys, the introduction of a self-scanned solid-state IRCCD can result in an IR system with lower cost, lower package size and weight, increased lifetime and sensitivity. In this section, therefore, an attempt is made to discuss some of the common operational requirements and limitations of IRCCD's.

Due to CO_2 and water molecule absorption, IR devices generally operate in one or more of the atmospheric transmission windows present at 2.0–2.5 μm, 3.5–4.2 μm (sometimes extended to 5.5 μm), and 8–14 μm (cf. Kruse et al. [7]). Here a major operating difference between IR and visible CCD's is the level of background radiation present. Assuming that the earth and most objects have an average temperature of 300 K, the spectral content of the radiation emitted is given by Planck's law [7] for a blackbody of the same temperature. In the visible region of the spectrum, the emitted background radiation is negligible but increases rapidly in each successive IR window from $Q_B = 10^{12}$ photons/s · cm^2 to 10^{17} photons/s · cm^2. The large Q_B results in increasingly shorter *background* charge-up or saturation time with increasing wavelength of operation, going from about 1 s for the 2.0–2.5-μm window to about 10 μs for the 8–14-μm window. In addition, internally generated charge is simultaneously integrated, further reducing the saturation time (see Section III). Thus one limitation of practical IRCCD devices is the length of the exposure time.

Another important consideration, especially for IR thermal imaging applications, is the contrast ratio between the photon flux generated by an incremental change in background temperature and the total photon flux. For 0.1 K change (typical for IR imaging) of the average 300 K background, the contrast ratio is on the order of 1.0 percent in the 2.0–2.5-μm window and less than 0.1 percent in the 8–14-μm window. The presence of these low-contrast ratios points to another practical limitation: the uniformity of IR material properties. Nonuniformity in the responsivity of the IR detectors of the same order or higher than the scene contrast ratio seriously degrades the minimum resolvable temperature and results in severe fixed pattern noise.

Finally, it should be pointed out that most IR detectors operate at reduced temperatures, pyroelectric detectors being a notable exception. Therefore, the presence of either a passive or an active cooling system is an operation requirement for IRCCD operation.

III. MONOLITHIC CCD'S

The category of monolithic IRCCD devices includes both inversion-mode devices and accumulation-mode devices, and they will be discussed in turn below. The distinguishing feature of these monolithic devices is that both the sensor and the transfer section are built up on the same semiconducting sub-

strate. Indeed, with the exception of the Schottky-barrier IRCCD, all the other monolithic IRCCD's use the same metal–insulator–semiconductor (MIS) structure.

A. Inversion-Mode Devices

The operation and structure of these devices is essentially similar to that of the Si CCD and, therefore, is based on the generation of an inversion region at the insulator–semiconductor interface where photogenerated minority carriers are collected. The bandgap of the semiconductor substrate determines the absorption peak and, therefore, the useful atmospheric IR window. These narrow-bandgap semiconductors are found among the binary and ternary III–V, II–VI, and IV–VI compounds [7], [8]. In this section calculations for MIS parameters essential for CCD operation are presented. These include the surface potential, thermal generation current, background current, charge transfer efficiency, etc. These parameters are necessary for the calculation of the storage time, the critical factor in establishing the feasibility of a practical IRCCD. The following representative semiconductors have been used in the calculations: InAs, InSb, $Hg_{1-x}Cd_xTe$. Similar calculations have been performed by Tao et al. for PbTe and $Pb_{1-x}Sn_xTe$ [9].

The energy band diagram for an MIS structure with a gate voltage of V_G is shown in Fig. 1. The surface potential ϕ_s for the depletion region approximation of Poisson's equation [10] is related to the gate voltage by

$$V_a = V_G - V_{FB} = \frac{x_0 Q_{inv}}{\epsilon_0 \kappa_0} + \beta \sqrt{\phi_s} + \phi_s \quad (1)$$

where

$$\beta = \frac{x_0}{\kappa_0} \frac{\sqrt{2N\kappa_s}}{\epsilon_0}$$

V_{FB} is the flat-band voltage, x_0 and κ_0 are the oxide thickness and relative dielectric constant, N and κ_s are the semiconductor doping and dielectric constant. At $t = 0$ [Fig. 1(a)] with no minority carrier charge present at the interface ($Q_{inv} = 0$), the surface potential is given by

$$\phi_s(t=0) = \phi_{s0} = \frac{\beta^2}{4}\left(\sqrt{1 + \frac{4V_a}{\beta^2}} - 1\right)^2 \quad (2)$$

and the depletion region has maximum width. At steady-state [Fig. 1(b)], charge Q_{inv} has accumulated at the interface reducing the surface potential [11]

$$\phi_s(t\to\infty) = \phi_{sf} = \frac{2kT}{q}\ln\left[\frac{N}{n_i}\right] \quad (3)$$

as well as the depletion width. The rate of change of the charge in the inversion layer is given by the summation of all current generation terms

$$\frac{dQ_{inv}}{dt} = J_G + J_S + J_D + J_B \tag{4}$$

$$J_G = \frac{qn_i x_d(t)}{2\tau} \qquad \text{depletion region generation} \tag{5}$$

$$J_S = \frac{qn_i S_0}{2} \qquad \text{interface state generation} \tag{6}$$

$$J_D = \frac{qn_i^2 L_D}{N\tau} \qquad \text{bulk generation} \tag{7}$$

$$J_B = \eta q Q_B \qquad \text{background generation} \tag{8}$$

where

n_i	intrinsic carrier concentration
τ	minority carrier lifetime
$x_d(t)$	depletion depth
S_0	surface recombination velocity
L_D	minority carrier diffusion length
η	quantum efficiency
Q_B	background photon flux.

Since the last three terms (6)–(8) are constant as a function of time, they can be combined in one term

$$\frac{dQ_{inv}}{dt} = J_G + J_X. \tag{9}$$

By introducing (9) in (1), the charge-up time for the MIS structure is obtained:

$$\tau_c = \frac{2}{\alpha}(\phi_{s0}^{1/2} - \phi_{sf}^{1/2}) + \frac{2}{\alpha^2}\left(\frac{\alpha\beta}{2} - \gamma\right) \ln\left(\frac{\gamma + \alpha\phi_{s0}^{1/2}}{\gamma + \alpha\phi_{sf}^{1/2}}\right) \tag{10}$$

where

$$\alpha = \frac{x_0}{\epsilon_0\kappa_0}\frac{qn_i}{2\tau}\sqrt{\frac{2\kappa_s\epsilon_0}{qN}}, \qquad \gamma = \frac{x_0 J_x}{\epsilon_0\kappa_0}.$$

Let us now turn to the other major parameter of all CCD's, the charge transfer efficiency. As pointed out by Boyle and Smith [1] there are three mechanisms involved in the charge transfer process: a) self-induced drift, b) thermal diffusion, and c) drift due to fringing fields. The transport of charge packet Q is described by the continuity equation [12]

$$\frac{\partial Q}{\partial t} = \frac{\partial}{\partial x}\left(D_{eff}\frac{\partial Q}{\partial x}\right). \tag{11}$$

D_{eff} is an effective diffusion constant which incorporates both thermal diffusion and drift

$$D_{eff} = \mu\Delta\phi\left(\delta + \frac{Q}{Q_s}\right) \tag{12}$$

$$\Delta\phi = |\phi_{sf} - \phi_{s0}| \tag{13}$$

$$\delta = \frac{D/\mu}{\Delta\phi} \tag{14}$$

where $\Delta\phi$ is the difference between the surface potentials in equilibrium and in deep depletion, respectively, and Q_s is the

equilibrium value of the charge packet. Initially, the contribution of the self-induced drift current dominates the transfer of initial transfer charge Q_{IT}

$$\frac{\partial Q_{IT}}{\partial t} = \mu\frac{\Delta\phi}{2Q_s}\frac{\partial^2 [Q_{IT}^2]}{\partial x^2}. \tag{15}$$

Based on their numerical results, Carnes *et al.* [13] point out that the spatial and temporal dependence of the charge as it decays under the influence of the self-induced field is independent. One can, therefore, use a separation of variables technique to solve (15):

$$Q_{IT} = \frac{Q_s}{1 + t/t_0} \tag{16}$$

where

$$t_0 = \frac{4L^2}{\pi^2\mu\Delta\phi}, \qquad L = \text{gate width.} \tag{17}$$

During the final stage of the charge transfer process, thermal diffusion and fringing field current are the dominant contributions. Daimon *et al.* [14] show that these processes result in an exponential decay of the form

$$Q_{FT} = C_1 e^{-t/\tau_f} \tag{18}$$

where τ_f is the final stage transfer time constant and Q_{FT} is the charge during the final transfer process:

$$\frac{1}{\tau_f} \approx \frac{4}{\tau_D} + \frac{\mu^2 E_m^2}{4D} \tag{19}$$

$$\tau_D = \frac{4L^2}{\pi^2 D}. \tag{20}$$

The fringing field process appears in the first term of (19) through the decrease of a factor of four of the relative importance of the thermal diffusion time constant τ_D. The second term of (19) is entirely due to the fringing field and involves the minimum value of that field [15]:

$$E_m = \frac{2\pi}{3}\frac{\kappa_s}{\kappa_0}\frac{x_0(V_G - V_T)}{L^2}\left[\frac{5x_d/L}{(5x_d/L) + 1}\right]^4. \tag{21}$$

The constant in (18) can now be obtained by matching the amplitudes and slopes of the expressions given by (16) and (18) at time $t = t_m$:

$$Q_{IT} = Q_{FT} \tag{22a}$$

$$\frac{dQ_{IT}}{dt} = \frac{dQ_{FT}}{dt}. \tag{22b}$$

Solving (22a) and (22b), t_m is found to be simply

$$t_m = \tau_f - t_0. \tag{23}$$

As a convenient yardstick, the time required for the transfer of 99.99 percent of the charge was defined at t_4:

$$\frac{Q_{FT}}{Q_s} = 10^{-4} = \frac{C_1 e^{-t_4/\tau_4}}{Q_s} \tag{24}$$

$$t_4 = \tau_f \ln\left[\frac{10^4 e^{t_m/\tau_f}}{1 + t_m/t_0}\right] \tag{25}$$

TABLE I
MATERIAL PROPERTIES, CHARGE-UP, AND TRANSFER TIMES FOR
SELECTED METAL–INSULATOR–IR SEMICONDUCTORS

Semiconductor		InAs		InSb		$Hg_{0.8}Cd_{0.2}Te$	
Substrate Doping		n		p	n	n	
Substrate Temperature	°K	77	150	77		77	
E_g	eV	0.41	0.39	0.23		0.09	
λ_c	μm	3.0	3.2	5.4		14	
Spectral Band $(\lambda_1 - \lambda_2)$	μm	2.0-2.5		3.5-4.2		8-14	
Q_B (a) (300°K, $\lambda_1 - \lambda_2$)	Photons/cm²-s	1.8×10^{11}		1.6×10^{14}		5.9×10^{16}	
$\dfrac{\Delta Q_B}{Q_B}$ (b)		7×10^{-3}		3×10^{-3}		5×10^{-4}	
Dielectric Constant (c)	κ_S	14.5		17		18	
Minority Carrier Lifetime τ	s	7×10^{-8}(d)	8×10^{-9}(d)	2×10^{-10}(e)	5×10^{-8}(e)	10^{-7}(c)	
Intrinsic Carrier Conc. n_i	cm⁻³	6.8×10^{2}(f)	2.5×10^{10}(f)	6.4×10^{8}(f)	6.4×10^{8}(f)	2×10^{13}(c)	
Impurity Conc. N	cm⁻³	6×10^{16}	6×10^{16}	6×10^{15}	6×10^{15}	10^{15}	
Inversion Layer Mobility μ_{eff}	cm²/V-s	350(g)	280(g)	8×10^{4}(e)	3×10^{3}(e)	600(h)	
						SiO_2	TiO_2
Initial Surface Potential ϕ_{so} (i)	V	0.8(j)	0.8(j)	2.5(j)	2.5(j)	3.7(j)	4.9(k)
Final Surface Potential ϕ_{sf}	V	0.42	0.38	0.21	0.21	0.05	0.05
Max. Inversion Region Charge Density Q_{inv}	C/cm²	5.5×10^{-8}	6.3×10^{-8}	1.4×10^{-7}	1.4×10^{-7}	1.6×10^{-7}	3.3×10^{-6}
Thermal Charge-Up Time τ_{th} (l)	s	2.7×10^{6}	1.8×10^{-2}	1×10^{-2}	2.1	5.2×10^{-5}	9×10^{4}
Background Charge-Up Time τ_B (m)	s	1.7	1.9	5.9×10^{-3}	5.9×10^{-3}	1.8×10^{-5}	3.6×10^{-4}
Combined Charge-Up Time τ_C	s	1.7	1.8×10^{-2}	3.6×10^{-3}	5.8×10^{-3}	1.3×10^{-5}	2.5×10^{-4}
Charge Transfer Time t_4 (n)	s	3.2×10^{-7}	2.2×10^{-7}	1.1×10^{-9}	2.9×10^{-8}	1.3×10^{-7}	1.3×10^{-7}

(a) FOV = 30°
(b) $\Delta T = 0.1$°K
(c) For comprehensive list of materials parameters and references see Moss et al [34]
(d) See [16]
(e) See data and references of [17]
(f) Calculated
(g) Estimated from μ_{bulk} values of [19] as discussed in [11]
(h) See [18]
(i) $V_G \cdot V_{FB} = 5v$
(j) SiO_2, 10^3 Å, $\kappa_o = 3.9$
(k) TiO_2, 10^3 Å, $\kappa_o = 75$
(l) Nominal Surface Recombination Velocity $S_o = 200$ cm/s
(m) $\eta = 1$
(n) $L = 10$ μm, $\epsilon_Q = 10^{-4}$

In Table I, charge-up and transfer times for representative MIS structures are presented along with the pertinent material parameters. To present a balanced picture, three IR semiconductors each with a long wavelength cutoff response falling in or near one of the three atmospheric windows were chosen: InAs ($E_g \simeq 0.4$ eV, $\lambda_c \simeq 3.0$ μm), InSb ($E_g \simeq 0.23$ eV, $\lambda_c \simeq 5.4$ μm), and $Hg_{0.8}Cd_{0.2}Te$ ($E_g \simeq 0.09$ eV, $\lambda_c \simeq 14$ μm). SiO_2 is the insulator in all structures except for $Hg_{0.8}Cd_{0.2}Te$, where calculations for TiO_2 have also been included. The substrate is taken to be n-type in all cases, except for InSb where both n- and p-types are considered. The device is considered to be at 77 K in all cases, except InAs where $T = 150$ K is also included. The field of view is taken to be 30°. Under these circumstances, InAs has the longest charge-up time of almost 2 s at 77 K and 20 ms at 150 K. The InSb *thermal* charge-up time increases two orders or magnitude in going from p-type ($\tau_{th} \sim 10$ ms) to n-type ($\tau_{th} \sim 2$ s). However, since the background charge-up time is the dominant term in both cases, the combined charge-up time changes only by 60 percent, from 3.6 to 5.8 ms. Finally, for the $Hg_{0.8}Cd_{0.2}Te$ case, the combined charge-up time is 13 μs when SiO_2 is the insulator and 250 μs when TiO_2 is the insulator. The charge transfer time required for 99.99-percent transfer efficiency, t_4, has been calculated assuming a 10-μm gate width. For InAs, t_4 is equal to

320 ns at 77 K and 220 ns at 150 K. As expected, p-type InSb has the shortest t_4 of only 1.1 ns due to its high minority carrier mobility, while n-type InSb has a t_4 of 29 ns. The influence of the insulator on the value of t_4 is shown to be minimal in the case of HgCdTe, where a t_4 of 130 ns has been calculated.

Both the charge-up and charge transfer times are dependent not only on material properties but on the device structure as well. Therefore, the numbers presented in the foregoing and in Table I do not necessarily indicate ultimate performance but rather some typical values to be expected.

B. Accumulation-Mode Devices

Accumulation-mode operation of an MIS structure results when the proper polarity gate voltage induces the same majority carrier type at the insulator–semiconductor interface as in the bulk of the semiconductor. In this manner, an extrinsic (rather than a narrow-bandgap intrinsic) semiconductor substrate can be used for monolithic IRCCD structures [2], [3]. The IR absorption peak is determined by the ionization energy of the impurity levels. For example, the extrinsic peak response gold-doped germanium (Ge:Au) occurs at ~8 μm [20] and Si:Ga peak response occurs at about 17 μm [21]. One can thus take advantage of the already developed silicon CCD tech-

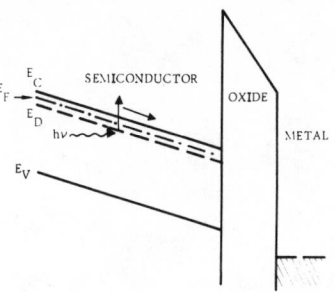

Fig. 2. Energy band diagram for an MIS structure biased into accumulation.

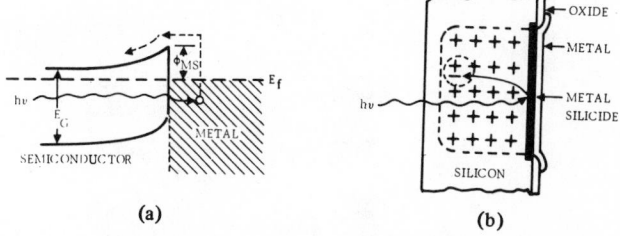

Fig. 3. Internal photoemission in a Schottky-barrier detector [4]. (a) Energy band diagram. (b) Device operation.

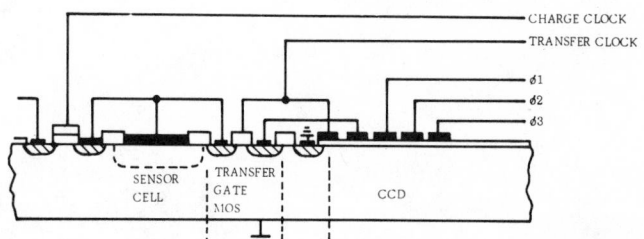

Fig. 4. Schottky-barrier IRCCD proposed by Sheperd and Yang [4].

nology and extend it to the IR region. However, to minimize the thermal generation current, temperatures below impurity freezeout are generally required. Fig. 2 shows the energy band bending in an accumulation-mode MIS structure at a sub-freezeout temperature. The bands have a linear spatial behavior through the semiconductor substrate due to the absence of space charge. The large spatial penetration of the electric field gives rise to considerable fringing fields [15] which enhance the charge transfer efficiency [22], [23]. Furthermore, the electric field at an accumulated semiconductor surface is much smaller than at an inverted surface because the surface potential appears across the whole thickness of the substrate, rather than only a thin depletion layer. As a consequence, the width of the accumulation layer charge packet is as much as an order of magnitude greater than an inversion layer packet. Accumulation-mode devices built and investigated by our laboratory [3] have a four-phase two-layer overlapping-gate structure. A 64-cell n-channel device was built on a 10^{15}-cm^{-3} phosphorous-silicon substrate. The observed transfer efficiency was 0.99 per gate at a chip frequency of 25 kHz and a temperature of 4.2 K. Calculations [3] for this device indicate a charge packet depth of 300 Å. Having the charge stored at this distance from the interface reduces the importance of surface trapping states and, hence, further aids transfer efficiency.

C. Schottky-Barrier Photoemissive Devices

Another approach to solving the nonuniformity limitations of monolithic IRCCD devices, as proposed by Sheperd and Yang [4], is based on internal photoemission from metal–semiconductor Schottky-barrier arrays on a silicon substrate. The photoemission process is shown in Fig. 3(a): photons of energy $h\nu < E_g$ are absorbed in the metal resulting in the excitation of a hot carrier. Carriers with energy larger than the contact barrier and with sufficient momentum in that direction are emitted into the semiconductor. The depletion region set up by the application of a reverse-bias voltage is diminished in size by the emitted carriers neutralizing the immobile charges at its edge [Fig. 3(b)]. The photoemitted current is, therefore, integrated by a negative charge accumulation method. The readout of the Schottky-barrier detector into the CCD takes place via an MOS transfer gate which converts the initial majority carrier signal into the minority carrier transport required by the Si CCD (Fig. 4).

The Schottky-barrier IRCCD has several main features: a) the use of silicon and its well-known properties as the basis for the entire device; b) the use of photodetection in a metal film, resulting in highly uniform responsivity across a typical array (approximately 1 percent [4]), as well as excellent reproducibility from array to array (approximately 1 percent [24]); c) the photoemissive yield is still relatively low, $\gamma \approx$

1 percent at $\lambda = 3.1$ μm and $T = 77$ K for Au-p-Si contacts [25].

D. Other Monolithic Devices

Other monolithic IRCCD's receiving attention include: a) Germanium-substrate CCD [26]. By substituting Ge for Si, one extends the intrinsic CCD response in the near infrared out to about 1.8 μm. A 10-cell Ge CCD reported by Schroder [26] has been operated over the temperature range of 100 to 265 K at clock frequencies of 10 to 300 kHz. Transfer efficiencies of 95 percent without background charge and 97.5 percent with 20-percent charge have been measured. b) Charge injection devices [27], [28] use injection into the bulk for readout rather than the sequential transfer principle of CCD, and, therefore, fall in a different category of devices. Infrared applications of CID's include the work of Kim [28] with InSb.

IV. Hybrid IRCCD's

Since the hybrid IRCCD involves the coupling of two already fairly well developed technologies, it has received more attention [5], [29]–[31] to date than the monolithic concept. As mentioned in Section I, hybrid IRCCD's fall into two categories: direct and indirect injection devices.

A. Direct Injection Devices

In the direct injection IRCCD, the photogenerated charge is directly introduced into the CCD shift register [5]. Since this is in effect a dc coupled system, only those detectors with very small dc currents (e.g., photovoltaic, extrinsic detectors) can be fed into the CCD, due to the latter's limited charge handling capacity. The basic direct injection concept is illustrated in Fig. 5 for an n-on-p IR photodiode and an n-channel CCD. The IR photodiode is connected in parallel to a silicon coupling diode (SCD) diffused into the same silicon substrate on which the CCD is fabricated. The first MOS gate (input) of the CCD is used to reverse-bias the IR diode and the silicon coupling diode; the TRANSFER gate is used to introduce the photocharge into the CCD through a field induced n-channel. When the $\phi1$ gate is activated with a positive pulse, the po-

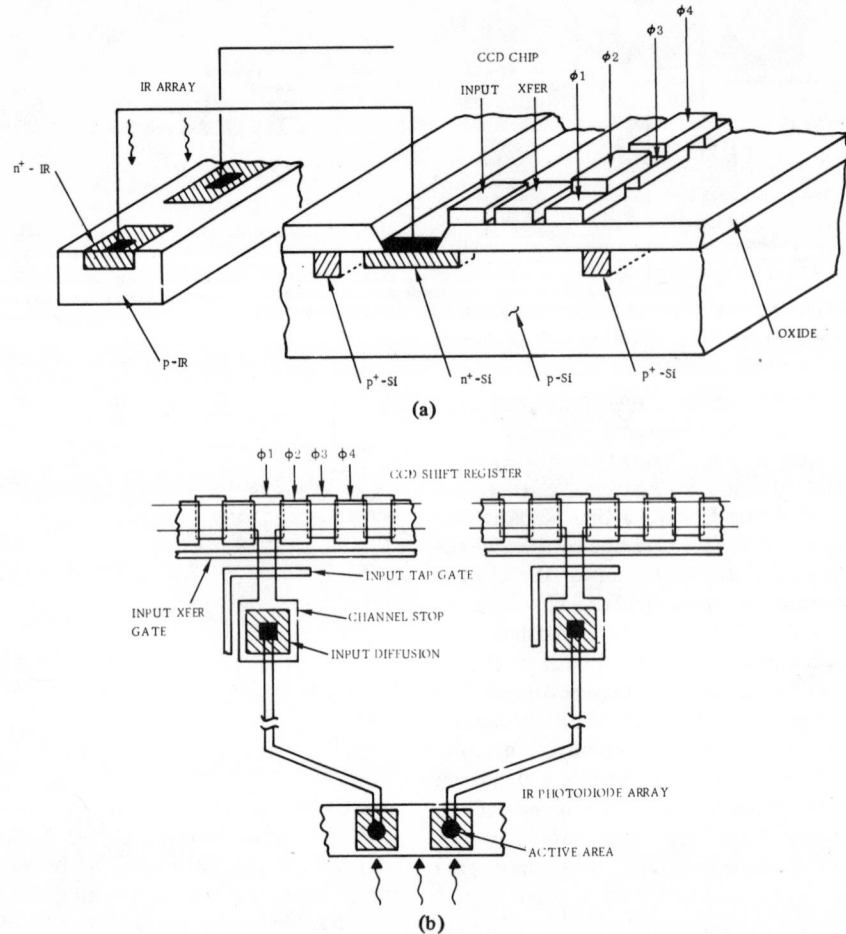

Fig. 5. Direct injection hybrid IRCCD. (a) Coupling concept. (b) Array layout.

tential well under the gate acts as a sink for the diodes and the photocurrent plus the saturation currents of the diodes will flow into the well for the duration of the $\phi 1$ pulse. When the $\phi 2$ gate is turned on and the $\phi 1$ gate is turned off, the current flow stops and the charge present is transferred on through the device.

Direct injection devices being investigated in the Electronics Research Division laboratory [32] include an InSb diode array coupled to a 100-bit p-channel CCD shift register through eight input taps. Both the detector array and the CCD have been operated at an 80 K temperature. Direct injection has been observed for CCD chip frequencies up to and higher than 1 MHz.

A potential application of the direct injection hybrid IRCCD is in the serially scanned TV-compatible Forward Looking Infrared (FLIR) system (cf. Milton [33]). The serially scanned FLIR uses a linear array of detectors raster-scanned across the scene, the scan direction being parallel to the array. Since each detector in turn scans the entire field of view, the dwell time per resolution element is approximately 150 ns. The amount of charge generated during this dwell time in even an 8–14-μm IR detector by a 300 K blackbody is well within the charge handling capacity of a typical CCD shift register. By processing the outputs of the detectors through a delay-and-add operation, the components of the signals corresponding to the same resolution element are summed linearly. Because the noise contribution of each detector is independent and thus uncorrelated, the total noise is obtained by an rms summation. For an array of m detectors, the delay-and-add operation could, therefore, result in a maximum improvement of $\sqrt{m}$ (in the detector noise limited case) in the signal-to-noise ratio (SNR) of the entire array over that of an individual detector. The upper limit of m and, therefore, of the maximum SNR improvement achievable is set by the total detector array signal versus the CCD charge capacity. The operation of the entire array appears at the CCD output essentially as one detector. This results in no fixed pattern noise, lowered detector response uniformity requirements, as well as built-in redundancy. These features together with the potential $\sqrt{m}$ improvement in SNR help circumvent the problems posed by the high background and low-contrast ratio present in thermal imaging systems at IR wavelengths.

An SNR analysis for a direct injection hybrid IRCCD operated in the delay-and-add mode was developed by Steckl and Koehler [5]. For a typical application consisting of an array of nine 12-μm (Hg, Cd) Te photodiodes receiving an incident flux of 3×10^{17} photons/cm$^2 \cdot$ s (T_B = 300 K, FOV = 90°), they calculate an SNR of 500, a dynamic range of 300, and a noise equivalent change in temperature of 0.1 K.

B. Indirect Injection Devices

Indirect injection hybrid IRCCD's use a buffer stage between the photodetection stage and the CCD shift register. The buffer stage can vary from one on-chip MOSFET to a number of off-chip amplifiers. It is the simplest version that is more appealing in its simplicity and ease of integration. Two indirect injection devices, one using pyroelectric detectors and the other photoconductive (PC) thin films, are discussed here.

1) Indirect Injection–Photoconductive Films: A technique which can be utilized with any photoconducting film which can be deposited on silicon dioxide without inducing fast states (or interface traps) at the silicon–silicon dioxide interface is described. The photoconductive film is used to directly influence the inversion of the channel of the input MOSFET. The degree of inversion, in turn, controls the rate of charge flow into the CCD. The photoconductive gate device is shown in Fig. 6. The advantage of this method is that the photoconductive film can be deposited by a variety of techniques after all the silicon MOS CCD process steps have been performed.

Operation of the device is described by the following sequence of steps.

1) A voltage is applied to the R_s bias line. This voltage has to be large enough to invert the oxide–semiconductor interface underneath the photosensitive elements.

2) A voltage is applied to the R_s control gate to provide an appropriate resistance in series with the photoconductive gate.

3) The PC gate bias is set just above the threshold value of voltage for this gate. This results in the transfer of maximum charge in the absence of optical input signal, thus providing blooming protection for the device.

4) Incident photons cause the resistance of the PC films to decrease accordingly. This causes a drop in the voltage across each PC film inversely proportional to the local density of incident photons.

5) The charge passing into the CCD is now a function of the illumination level falling upon the PC film. The response of the device to optical signal transients is governed by the ability of the biasing resistor to discharge the PC gate capacitance.

This device scheme, while hybrid and indirect, provides a straightforward and relatively easily fabricated method of IRCCD imaging.

2) Indirect Injection–Pyroelectric Detectors: Pyroelectric materials owe their photodetection property to a temperature-dependent polarization (for a review of pyroelectric detectors, see Putley [35]). In the steady state, the polarization is masked by surface charges and, therefore, the incident radiation must be chopped. This results in ac coupled detection and thus background subtraction. Additional features of pyroelectric detectors are inexpensive detector materials and room-temperature operation. However, the sensitivity of present pyroelectric detectors is considerably lower than that of quantum detectors.

Detection of the thermally induced electric polarization of the pyroelectric is accomplished by making the pyroelectric the dielectric of a capacitor. Two methods of introducing the pyroelectric signal into a CCD are 1) connecting the capacitor to an on-chip MOSFET or 2) fabricating a pyroelectric film between the MOSFET channel and the gate metal (viz., in series with the gate). Both methods are essentially the same, the latter requiring more sophisticated processing, but resulting in a more compact structure, as illustrated in Fig. 7. The device uses the voltage developed by the pyroelectric detector to modulate the "potential well" depth in an MOS structure and, in so doing, affects a charge transfer modulation in the charge carriers generated thermally and flowing to an adjacent deeper

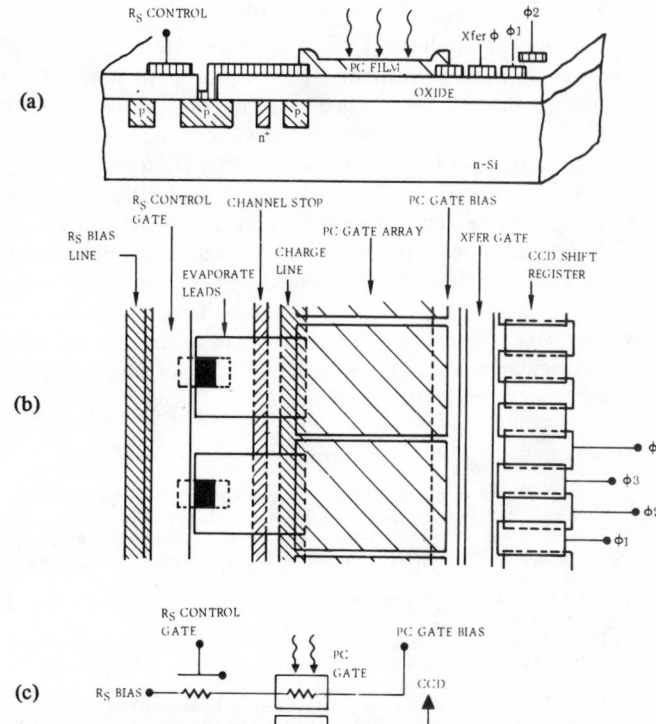

Fig. 6. Indirect injection: photoconductive thin-film hybrid IRCCD. (a) Device cross section. (b) Two-dimensional layout. (c) Equivalent circuit.

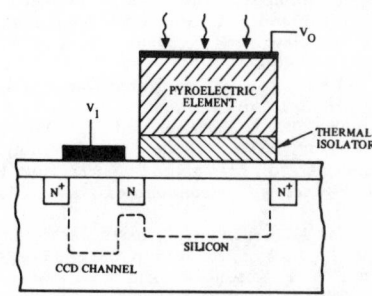

Fig. 7. Indirect injection: pyroelectric hybrid IRCCD.

but smaller area potential well in a CCD multiplexer which "passes by" the modulated well. The voltage V_0 is constant and just large enough to make the well several kT in depth. Charge (generated mainly by the surface states) keeps the well essentially full. The "drain" is over the N potential barrier and into the CCD channel. The voltages are adjusted until the modulation due to the scene (there is a shutter which modulates the incident radiation) is not "clipped" by the dark current flow and the dark current is equal to that required for essentially 100-percent modulation of the scene.

A first-order noise analysis of the device indicates [36] that the dominant noise contribution is the Johnson noise associated with the detector and input circuit. For a typical application, an IRCCD operating at 20 frames/s and using triglycine sulphate cells of 10^{-5}-cm² area placed on an oxide of 600-Å thickness should result in a minimum resolvable temperature of about 0.3 K in the 8–12-μm window. It should be noted that in order to achieve this predicted performance in practical devices, a high degree of thermal isolation is required in order to prevent thermal loading by the substrate as well as inter-element crosstalk.

V. CONCLUSIONS

A review of applications of CCD's for IR detection and imaging has been presented. It is obvious even at this fairly early stage that a large number of avenues to the development of IRCCD's exist. Indeed, no one approach seems to be versatile enough in the broad sense of the word to be developed as a building block for a wide range of applications. Consequently, the research and development of IRCCD's in the immediate future will in all probability continue along many paths, each closely tied to a specific end-use function. In this context, operating parameters and requirements (sensitivity, cooling, size, present and future costs, to name only a few) of the particular system in which the IRCCD is to be integrated become the key selection factors.

REFERENCES

[1] W. S. Boyle and G. E. Smith, "Charge coupled semiconductor devices," *Bell Syst. Tech. J.*, vol. 49, pp. 587–593, Apr. 1970.

[2] R. J. Keyes, J. O. Dimmock, and R. A. Cohen, "Long wave infrared CCD's," M.I.T. Lincoln Lab., Lexington, Mass., unpublished note.

[3] R. D. Nelson, "Accumulation mode charge coupled device," to be published in *Appl. Phys. Lett.*

[4] F. D. Sheperd and A. C. Yang, "Silicon Schottky retinas for infrared imaging," in *Tech. Dig.—Int. Electron Device Meet.*, pp. 310–313, Dec. 1973.

[5] A. J. Steckl and T. Koehler, "Theoretical analysis of directly coupled 8–12 μm hybrid IRCCD serial scanning," in *Proc. CCD Applications Conf.*, pp. 247–258, Sept. 1973.

[6] R. D. Hudson, *Infrared System Engineering*. New York: Wiley, 1969.

[7] P. W. Kruse, L. D. McGlauchlin, and R. B. McQuistan, *Elements of Infrared Technology*. New York: Wiley, 1962.

[8] R. K. Willardson and A. C. Beer, Eds., *Semiconductors and Semimetals*, vol. 5. New York: Academic Press, 1970.

[9] T. F. Tao, J. R. Ellis, L. Kost, and A. Doshier, "Feasibility study of PbTe and $Pb_{0.76}Sn_{0.24}Te$ infrared charge coupled imager," in *Proc. CCD Applications Conf.*, pp. 259–268, Sept. 1973.

[10] G. F. Amelio, W. J. Bertram, and M. F. Tompsett, "Charge coupled imaging devices: Design considerations," *IEEE Trans. Electron Devices*, vol. ED-18, pp. 986–992, Nov. 1971.

[11] S. M. Sze, *Physics of Semiconductor Devices*. New York: Wiley, 1969.

[12] C. K. Kim and M. Lenzlinger, "Charge transfer in charge-coupled devices," *J. Appl. Phys.*, vol. 42, pp. 3586–3594, Aug. 1974.

[13] J. E. Carnes, W. F. Kosonocky, and E. G. Ramberg, "Free charge transfer in charge coupled devices," *IEEE Trans. Electron Devices*, vol. ED-19, pp. 798–808, June 1972.

[14] Y. Daimon, A. M. Mohsen, and T. C. McGill, "Final stage of the charge transfer process in charge coupled devices," *IEEE Trans.*

Electron Devices, vol. ED-21, pp. 266–272, Apr. 1974.

[15] J. E. Carnes, W. F. Kosonocky, and E. G. Ramberg, "Drift aiding fringing fields in charge coupled devices," *IEEE J. Solid-State Circuits*, vol. SC-6, pp. 322–326, Oct. 1971.

[16] M. P. Mikhailova, D. N. Nasledov, and S. V. Slobodchikov, "Temperature dependence of carrier lifetimes in InAs," *Sov. Phys.—Solid State*, vol. 5, pp. 1685–1689, Feb. 1964.

[17] C. Hilsum and A. C. Rose-Innes, *Semiconducting III-V Compounds*. New York: Pergamon, 1961.

[18] A. F. Tasch, R. A. Chapman, and B. H. Breazale, "Field-effect measurements on the HgCdTe surface," *J. Appl. Phys.*, vol. 41, pp. 4202–4204, Sept. 1970.

[19] J. R. Dixon, "Anomalous electrical properties of p-type indium arsenide," *J. Appl. Phys.*, vol. 30, pp. 1413–1416, 1959.

[20] H. Levinstein, "Extrinsic detectors," *Appl. Opt.*, vol. 4, pp. 639–647, June 1965.

[21] R. A. Soref, "Extrinsic IR photoconductivity of Si doped with B, Al, Ga, P, As or Sb," *J. Appl. Phys.*, vol. 38, pp. 5201–5209, Dec. 1967.

[22] C. N. Berglund and K. K. Thornber, "Incomplete transfer in charge-transfer devices," *IEEE J. Solid-State Circuits*, vol. SC-8, pp. 108–116, Apr. 1973.

[23] L. G. Heller and H. S. Lee, "Digital signal transfer in charge-transfer devices," *IEEE J. Solid-State Circuits*, vol. SC-8, pp. 116–125, Apr. 1973.

[24] F. D. Sheperd, Air Force Cambridge Res. Lab., Solid State Sciences Lab., personal commun.

[25] J. Cohen, J. Vilms, and R. Archer, "Investigation of Schottky barriers for optical detection and cathodic emission," Air Force Cambridge Res. Lab. Rep. AFCRL-68-0651 and AFCRL-69-0287.

[26] D. K. Schroder, "A two-phase germanium CCD," presented at the Device Research Conf., Santa Barbara, Calif., June 1974.

[27] G. J. Michon and H. K. Burke, "Charge injection imaging," in *Dig. IEEE Int. Solid State Circuits Conf.*, pp. 138–139, Feb. 1973.

[28] J. C. Kim, "InSb MIS structures for infrared imaging devices," in *Tech. Dig.—Int. Electron Device Meet.*, pp. 419–422, Dec. 1973.

[29] D. M. Erb and K. Nummedal, "Buried channel charge coupled devices for infrared applications," in *Proc. CCD Applications Conf.*, pp. 157–167, Sept. 1973.

[30] D. E. French, M. Y. Pines, F. J. Renda, P. S. Chia, J. S. Balon, and A. H. Lockwood, "Result of PbSnTe hetero-structure detectors using charge coupled devices," unpublished, Hughes Aircraft Co., Culver City, Calif.

[31] J. C. Fraser, D. H. Alexander, R. M. Finnila, and S. C. Su, "Monolithic extrinsic silicon detectors with charge coupled device readout," unpublished, Hughes Aircraft Co., Culver City, Calif.

[32] A. J. Steckl, to be published (Internal Letter 74-520-011-100, Rockwell Int., June 1974).

[33] A. F. Milton, "Infrared detectors for serial scanning," unpublished, Naval Res. Lab., Washington, D.C.

[34] T. S. Moss, G. J. Burrel, and B. Ellis, *Semiconductor Optoelectronics*. New York: Wiley, 1973.

[35] E. H. Putley, "The pyroelectric detector," in *Semiconductors and Semimetals*, R. K. Willardson and A. C. Beer, Eds. New York: Academic Press, 1970, pp. 259–285.

[36] R. A. Gudmundsen, to be published (Internal Letter 73-520-67, Rockwell Int., Aug. 1973).

Author Index

A

Amelio, G. F., 15, 160, 266

B

Baertsch, R. D., 357, 396
Barbe, D. F., 130
Barth, D. A., 320
Belt, R. A., 295
Bertram, W. J., Jr., 160
Blaha, F. C., 167
Bower, R. W. 304
Boyle, W. S., 8
Brodersen, R. W., 70, 366, 371, 381
Burke, H. K., 186
Buss, D. D., 70, 366, 371, 381
Butler, W. J., 342

C

Carnes, J. E., 48, 87
Caves, J. T., 288
Chambers, J. M., 233
Chan, C. H., 288
Chou, S., 249
Cobbold, R. S. C., 104

E

El-Sissi, H., 104
Engeler, W. E., 342, 357, 396
Erb, D. M., 304
Esser, L. J. M., 127

F

Fagan, J. L., 320
French, B. T., 403
Frye, R. C., 314
Fu, H.-S., 314
Fuls, E. N., 179

G

Goldberg, H. S., 342, 357
Gudmundsen, R. A., 403
Guidry, M. R., 266
Gunsagar, K. D., 266

H

Heller, L. G., 77, 242
Hewes, C. R., 366, 371, 381

K

Kosonocky, W. F., 2, 28, 48, 87, 233
Kub, F. J., 320

L

Lampe, D. R., 167, 320, 327
Lee, H.-S., 77
Lobenstein, H., 342

M

Mack, I. A., 167
Meindl, J. D., 350
Melen, R. D., 224, 350
Michon, G. J., 186
Mohsen, A. M., 115, 304

N

Nelson, R. D., 403

P

Poon, S. C., 288
Puckette, C. M., IV, 342, 357

R

Ramberg, E. G., 48
Rodgers, R. L., III, 196
Rosenbaum, S. D., 288

S

Sangster, F. L. J., 23
Sauer, D. J., 2, 233
Schechter, D., 403
Sealer, D. A., 338
Séquin, C. H., 179
Shott, J. D., 350
Smith, G. E., 8, 15
Steckl, A. J., 403

T

Tasch, A. F., Jr., 70, 314, 366
Terman, L. M., 242
Tiemann, J. J., 357, 396
Tompsett, M. F., 15, 59, 115, 160, 179, 338

V

Veenkant, R. L., 381

W

Walker, J. T., 350
Wallace, R. W., 288
Wen, D. D., 207
White, M. H., 167, 320, 327
Wilder, E. M., 304

Z

Zimany, E. J., Jr., 179

Subject Index

Editors' Biographies

Roger D. Melen (S'70–M'73) received the B.S.E.E. degree from Chico State College, Chico, CA, in 1968, and the M.S.E.E. and Ph.D. degrees from Stanford University, Palo Alto, CA, in 1969 and 1973, respectively.

He spent the summer of 1968 designing circuits for nuclear monitoring instrumentation. Since 1972 he has been a member of the staff of the Stanford University Integrated Circuits Laboratory, working on the development of MOS integrated circuits and charge-coupled devices for medical and prosthetic applications. Since 1974 he has served as Associate Director of the Stanford University Integrated Circuits Laboratory.

Dr. Melen has served as a consultant to industry on CCD's applied to IR imagers, optical imagers, VTR time base-correctors, sonar systems, digital memories, and transversal filters.

Dennis D. Buss received the S.B., S.M., and Ph.D. degrees from the Massachusetts Institute of Technology, Cambridge, in 1963, 1965, and 1968, respectively.

His early research dealt with the lattice dynamics and magnetooptical properties of narrow-gap semiconductors. He spent a year on the M.I.T. electrical engineering faculty before joining the Central Research Laboratories of Texas Instruments, Inc., Dallas, in 1969. At TI he has been involved with the development of charge-coupled devices for analog signal processing and digital memory applications. In 1974 he was a visiting Associate Professor at M.I.T. and he is now managing a CCD R & D branch in TI's Central Research Laboratories.

Dr. Buss was co-recipient of the Outstanding Paper Award of the IEEE TRANSACTIONS ON ELECTRON DEVICES and he serves on the Program Committee for the IEEE International Solid-State Circuits Conference and the IEEE International Electron Devices Meeting.